ENGINEERING CHEMISTRY

AF644693

By :

Dr. Sunita Rattan

Addnl. Director
Amity Institute of Applied Sciences
Amity University Uttar Pradesh, Noida

S.K. KATARIA & SONS®

Publisher of Engineering Books

4885/109, Prakash Mahal, Dr. Subhash Bhargav Lane,
Daryaganj, New Delhi-110002 (India)
Tel.: +91-11-43551243, Mobile: +91-9871775858, 7678425209
e-mail: katariabook@gmail.com, katariabooks@yahoo.com
Website: www.skkatariaandsons.com
(e-books, subscriptions and bulk buying also available on our website)

Published by:

S.K. Kataria & Sons®
4885/109, Prakash Mahal, Dr. Subhash Bhargav Lane,
Daryaganj, New Delhi-110002 (India)
Tel.: +91-11-43551243, Mobile: +91-9871775858, 7678425209
e-mail: katariabook@gmail.com, katariabooks@yahoo.com
Website: www.skkatariaandsons.com

First Edition: 2008, Third Edition: 2012, Reprint: 2025

***Price :* ₹ 595/-**

Laser Type Setting by:
SARA ASSIGNMENTS, Patparganj, Delhi-110092.

Printed at:
A.S. OFFSET PRINTERS, Delhi-110051.

PREFACE

"I'd rather be a could-be if I cannot be an are;
because a could-be is a maybe who is reaching for a star.
I'd rather be a has-been than a might-have-been, by far'
for a might have-been has never been, but a has been was once an are."

— Milton Berle

It gives me immense pleasure to present revised textbook on Engineering Chemistry which covers the chemistry curriculum prescribed for the first year B.Tech course of Mahamaya Technical University in particular and other Indian Universities in general.

It has been a wonderful and fascinating experience for me to write a book which gives the students a clear, comprehensive and concise information to understand the various aspects of chemistry as applied to Engineering. The book has emerged from the lectures delivered during the past few years which have given me enough opportunity to understand the basic requirement of the students in assimilating the concepts of Chemistry.

I express my profound gratitude to the faculty and students of various engineering colleges for their *enthusiastic* response to the previous editions and the books written for other Technical Universities. Their tremendous appreciation has boosted my confidence and the book for students of Mahamaya Technical University continues with the same spirit: a presentation of the concepts in a clear, interesting and student friendly manner.

The book is written in a lucid style to enable the students for self study with minimal guidance. In order to make the concepts as clear as possible even to an average student, no effort has been spared to present the textbook in a systematic manner using simple and clear language.

This edition has been completely rewritten, revised, and expanded. The textbook is organized into six units. A comparatively concise presentation of various topics in each chapter has been made so as to instill in students, a confident grasp of the subject. Basic principles have been explained with the help of suitable examples and diagrams. A proper balance between theoretical and applied aspects have been maintained. Solved examples in the chapter will help in reinforcing the theoretical concepts. Wherever possible, commercial utilization of various principles have been discussed. Questions given at the end of each chapter ensure an adequate revision of the subject and help to build the confidence of the students.

Every day life applications of chemistry that should be of interest to the students appear through out in the text. These are 'screened' in order to set them apart from the body of the text.

The success of my endeavour will be decided by the teachers and the students for whom the book has been written. Suggestions and criticism are welcome as there is always a scope of improvement.

While every effort has been made to eliminate errors in this book, inadvertently some will have crept through. I would appreciate notification by readers of any errors or misprints.

I hope that readers will find the text interesting and informative at the same time.

Sunita Rattan

ACKNOWLEDGEMENTS

"At times our own light goes out and is rekindled by a spark from another person. Each of us has cause to think with deep gratitude of those who have lighted the flame within us."

— Albert Schweitzer

I would like to extend my most sincere appreciation to the individuals who have contributed in their best possible ways for the completion of this project.

My special thanks are due to my revered teacher, Professor B.N. Misra, Emeritus Fellow, Department of Chemistry, H.P. University, Shimla. The legacy of scholarly and scientific inquiry and the temperament which I have inherited from him, will remain with me as invaluable treasure to be invested in everything, I do and seek in life.

I will be failing in my duties if I do not acknowledge Dr (Mrs) I. K. Mehta (Himachal Pradesh University) whose affectionate advice and moral support have left a deep impression on my life.

It gives me great pleasure to place on record my profound gratitude and indebtness to our esteemed president Dr. Ashok Kumar Chauhan, Founder President RBEF, whose inspiring words, stimulating encouragement and uniflinching support have been a source of great strength in accomplishing this assignment.

I express my sincere gratitude to Mr. Atul Chauhan,Chancellor ,Amity University, Uttar Pradesh for his continued encouragement and support .

My heartfelt thanks are due to Dr. (Mrs.) Balvinder Shukla , Pro-Vice Chancellor, Amity University,Uttar Pradesh for boosting my confidence in all such type of ventures. I always admire the training given by her, which has always helped me to achieve my desired goals.

I am deeply indebted to Prof. A. L Verma, Director, Science & Technology, AUUP for his sincere support and inspiration to complete the present piece of work.

The text presented has been reviewed by the colleagues at other colleges and universities. I profoundly thank the reviewers for their valuable comments which have been used to polish the book in its present form.

I am extremely grateful to my colleagues at Amity University for their suggestions and criticisms which have been invaluable in bringing this book in its present form.

I also record my sense of appreciation to my students whose inquisitive questions during my classes provided valuable input into this effort.

Words elude me in expressing my sincere thanks to Praneesh who offered every possible help and supported the long hours of preparation that went into this effort. I feel greatly indebted to my kids, Tanishq and Tanvi without whose patience and sacrifice it was rather impossible to bring out the book in its present form.

My immeasurable gratitude and thanks are due to my family members who have been a driving force behind all my achievements. Their affection and benign blessings will ever remain to be cherished.

Finally, I am thankful to Mr. S.K. Kataria Publications and his entire staff for their efforts without which the book could not have seen the light of the day.

SUNITA RATTAN

CONTENTS

UNIT 1

Atoms to Molecules to Materials for Engineers

1

CHEMICAL BONDING-I

"What we observe is not nature itself, but nature exposed to our method of questioning".

— Werner Heisenberg

CHAPTER OUTLINE

1.1 RECAPITULATION

At ordinary temperatures atoms are rarely capable of free existence. Most of the substances exist as molecules like H_2, O_2, P_4, S_8, NH_3, CH_4, etc. But then a number of questions arise –

- Why and how do atoms combine?
- What kind of force holds these atoms together in a molecule?
- How these atoms are arranged in a molecule?
- Why do some atoms combine, while certain others do not. Two chlorine atoms combine to form Cl_2 molecule whereas two helium atoms do not combine to form He_2.

The study of Chemical Bonding attempts to answer all the above questions. **Chemical bond is defined as the force which binds the atoms together within a molecule**. Atoms combine to form molecules which are more stable and have lower energy than the individual atoms. A number of attempts were made to explain the formation of chemical bonds. Some of them are briefly explained below.

1. Electronic Theory of Valency: Kossel and Lewis (1916) put forward the first satisfactory explaination of the formation of chemical bonds. As this explaination was based on the electronic concept of atom, so it is referred to as "Electronic Theory of Valency."

The study of spectra of inert gases of zero group revealed that they are quite stable and do not combine chemically with any other element or with each other. Their electronic configurations are as follows:

Table 1.1

Element	*Atomic Number*	*Electronic Configuration*
Helium	2	2
Neon	10	2, (8)
Argon	18	2, 8, (8)
Krypton	36	2, 8, 18, (8)
Xenon	54	2, 8, 18, 18, (8)
Radon	84	2, 8, 18, 32, 18, (8)

The electronic configuration of these gases indicate that each of *s* and *p* orbitals of the outermost energy shell is completely filled with 2 and 6 electrons respectively i.e. a total of 8 electrons. Hence it was suggested that these constitute a structure of maximum stability (s^2p^6) and minimum energy. But in case of all other elements, which show chemical reactivity, the outermost orbitals are found to be incomplete. Therefore, all other elements, other than noble gases, tend to acquire stable gas configuration, by losing, gaining or sharing of electrons with other elements. This led to the conclusion that atoms of different elements participate in bond formation in order to complete their respective octets (i.e 8 electrons in the outer most shell) or duplet, (i.e 2 electrons in the outer most shell) in case of H, Li and Be, to attain stable inert gas configuration. This is the fundamental cause of chemical combination.

Thus in light of electronic theory of valency, it follws that, "*Chemical bonding may be defined as the combination of two or more atoms through rearrangement of electrons in their outer shells, either by sharing of electrons amongst themselves or by the transfer of electrons from one atom to another, so that all the atoms attain the stable configuration of nearest inert gas.*"

Limitations of Electronic Theory: Though octet rule has explained the formation of chemical bond in large number of compounds, still it has the following limitations:

(i) The formation of electron deficient compounds such as $AlCl_3$ or BF_3 where the central atom Al or B do not have their octet completed, cannot be explained on the basis of this theory.

(ii) Similarly the formation of compounds such as SF_6, PCl_5, where central metal atom has more than 8 electrons cannot be explained.

(iii) Noble gases having eight electrons in their valence shell are not expected to form any compounds on the basis of this theory, but compounds like XeF_2, XeF_4, XeF_6 have actually been prepared.

The formation of chemical bond can thus better be explained on the basis of lowering of energy.

2. Lowering of Energy: In the formation of a chemical bond, there is invariably a lowering of potential energy of the system. Energy considerations show that systems associated with less energy are more stable than the systems possessing more energy. Thus atoms combine with each other, to lower their energy and hence gain extra stability. This will be discussed in more detail a little later in this chapter.

1.2 TYPES OF CHEMICAL BOND

As discussed above, atoms combine together in order to complete their respective octets so as to acquire stable inert gas configuration. This can happen in two ways.

1. By complete transference of one or more electrons from one atom to another. This type of bond is called **Ionic or Electrovalent bond.**
2. By sharing of electrons: Sharing can take place in two ways:
   (a) When the shared electrons are contributed equally by the two combining atoms. This type of bond is called **covalent bond**.
   (b) When the shared electrons are contributed by one atom but shared by both, the bond so formed is known as **co-ordinate bond or dative bond.**

Considering the electronegativities of the elements, the following ideal types of bonds are formed.

Electropositive Element + Electronegative Element — **Ionic Bond**
Electronegative Element + Electronegative Element — **Covalent Bond**
Electro positive Element + Electropositive Element — **Metallic Bond**

The above mentioned bonds are extreme forms. In most of the cases the bond type is some where between these extreme forms.

1.3 ELECTROVALENT OR IONIC BOND

When a bond is formed by complete transference of electrons from one atom to another so as to complete their outermost shell by acquiring 8 electrons (or 2 electrons in case of hydrogen, lithium etc), the bond formed is called ionic or covalent bond. The atom that loses one or more electrons acquires positive charge and the ion so formed is called a **cation**. The atom that gains one or more electrons acquires negative charge and the ion so formed is called an **anion**. An ionic bond is the result of electrostatic attraction between positively charged cations and negatively charged anions.

Examples:

The most striking example of electrovalent bonds is the combination of sodium and chlorine to form sodium chloride.

Electronic configuration of sodium atom = $1s^2, 2s^2, 3s^1$.
Electronic configuration of chlorine atom= $1s^2\ 2s^2, 2p^6, 3s^2, 3p^5$.

By losing one electron from its outermost shell, sodium acquires the inert gas configuration of neon and changes into cation.

$$\underset{2,\,8,\,1}{Na} \longrightarrow \underset{2,\,8}{Na^+} + e$$

On the other hand, chlorine atom accepts one electron, released by sodium to complete its octet by attaining stable gas configuration of argon. In this process, chlorine is converted into chloride ion.

$$\underset{2,\,8,\,7}{Cl + e^-} \longrightarrow \underset{2,\,8,\,8}{Cl^-}$$

Now we have two oppositely charged ions. As soon as they approach each other, they are held together by strong electrostatic forces of attraction. Thus, formation of sodium chloride takes place-

$$Na^+ + Cl \longrightarrow NaCl$$

These steps can be represented by Lewis structures as follows:

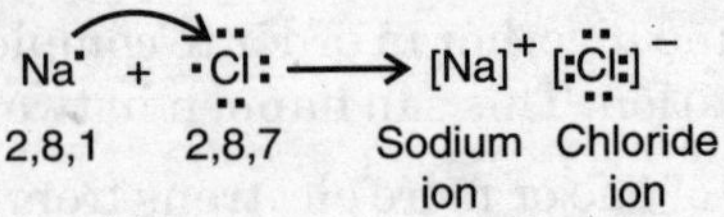

The number of electrons lost or gained during the formation of an electrovalent linkage is termed as the electrovalency of the element. For example, in the above case the electrovalency of sodium as well as chlorine is one. In other words, valency is equal to the charge on the ion.

It is worthwhile to mention here that an ionic compound consists of an aggregate of cations and anions thus having a number of ionic bonds.

X-ray analysis have confirmed that the crystals of NaCl contain charged ions in the form of Na^+ Cl^-. Such solids which consists of ions only are known as **Ionic Solids**. In a similar manner we can explain the formation of ionic bonds in a large number of compounds eg LiF, KCl, CaF_2, $MgCl_2$, MgO, $MgBr_2$, CaS etc. Some of which are depicted below.

1.4 COVALENT BOND

Lewis Theory: Lewis (1916) proposed a new concept of bonding among non-metallic elements. When two atoms taking part in a chemical combination are short of electrons than the nearest inert gas configuration, they can share their electrons in order to complete their octets. *The bond formed between the two atoms by mutual sharing of electrons between them, so as to complete their octets is called covalent bond. The number of electrons contributed by each atom is known as covalency.*

The electron pair shared between two atoms for the formation of a covalent bond is called the bonding pair of electrons.

Examples:

Formation of Hydrogen molecule: Hydrogen atom has one electron in its valence shell. In the formation of hydrogen molecule, both the atoms contribute one electron each for mutual sharing and thus acquiring the configuration of helium gas.

$$H\bullet \; \bullet H \longrightarrow H:H \text{ or } H\text{–}H$$

Formation of Chlorine molecule: A chlorine atom has 7 electrons in its valence shell. Each chlorine atom is short of one electron to complete its octet, so by sharing one electron with another chlorine atom both atoms attain noble gas configuration.

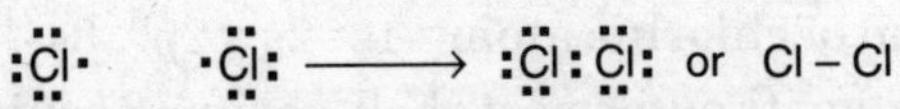

Formation of NH_3: Nitrogen atom has five outer electrons and in NH_3, it shares three of these forming three bonds and thus attaining an octet. Hydrogen has only one electron and by sharing it attains a stable configuration of helium.

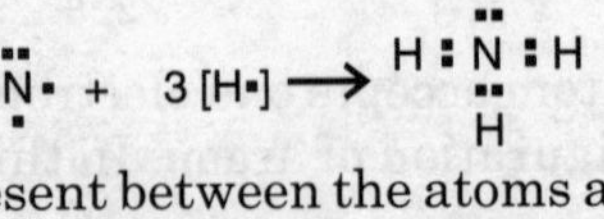

The shared pairs of electrons present between the atoms are called **bond pairs** because they are responsible for bonding between the atoms. On the other hand, the valence electrons not involved in bonding (i.e sharing) are shown as such and are called non-bonding pairs or **lone pairs**. Thus NH_3 has three bond pairs and one lone pair.

1.4.1 Valence Bond Theory

To overcome the shortcomings of Lewis concept of the formation of covalent bond Linus Pauling and J.C Slater proposed Valence Bond Theory. Infact this theory was first proposed by W. Heitler and F. London in 1927 and later developed by Pauling and Slater. The various points of the Valence Bond Theory are as follows:

1. An atom must possess one or more unpaired electrons to enter into a chemical combination.
2. Atoms with unpaired electrons tend to combine with other atoms which also have unpaired electrons. In other words a covalent bond is formed by the overlapping of two atomic orbitals belonging to two different atoms, having unpaired electrons. The electrons in the orbitals should have opposite spins.
3. The new orbital which is formed due to overlapping of atomic orbitals is more stable.
4. The strength of the bond is determined by the extent of overlapping of two half-filled orbitals. More is the overlapping, more is the decrease in energy and stronger is the bond formed.
5. The atoms which unite to form a molecule completely retain their identities in the resulting molecule.
6. The shape of the molecule is determined primarily by the directions in which orbitals point.
7. The number of bonds formed by an atom is equal to the number of unpaired electron it has.

The above points can be illustrated by considering a few examples–

Formation of hydrogen molecule, s–s overlap: Hydrogen has 1 electron in its valence shell i.e. 1s′. H_2 molecule is formed when two s-orbitals of two hydrogen atoms overlap along their axis. The electrons shared between the two atoms are located in the region of space between the nuclei of the atoms. Thus 1*s* orbital of the two atoms gets fully occupied. This bond is called *s-s* bond as it is formed by the overlapping of two *s* orbitals. (Fig. 1.1)

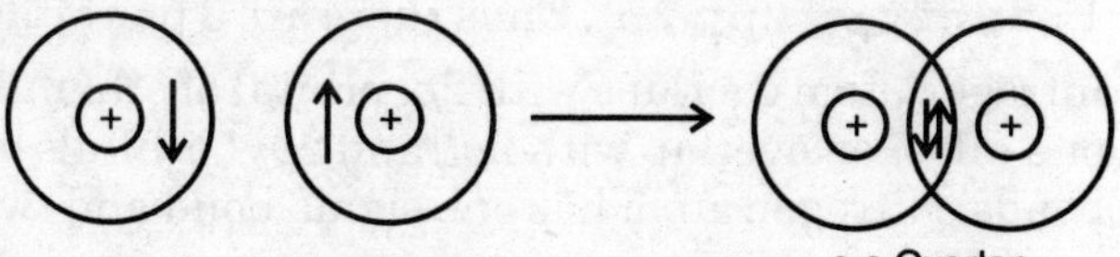

Fig. 1.1 s–s overlap

Formation of flourine molecule, p-p overlap: Electronic configuration of flourine = $1s,^2\ 2s,^2,\ 2p_x^2, 2p_y^2, 2p_z^1$ Thus its $2p_z$ orbital having unpaired electron overlaps with the $2p_z$ orbital of other flourine atom having electron with opposite spin. As the overlap is between two p-orbitals, this is called p-p overlapping. (Fig. 1.2)

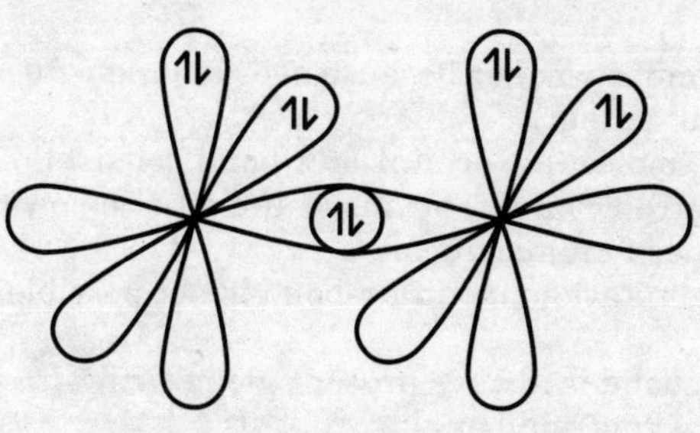

Fig. 1.2 p–p overlap along internuclear axis.

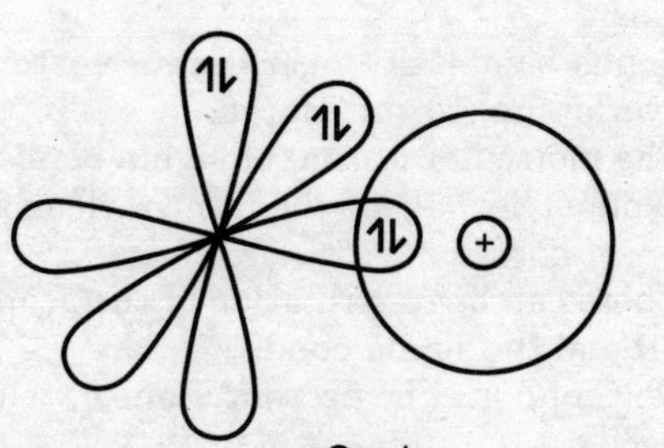

Fig. 1.3 s–p overlap

Formation of hydrogen flouride molecule s-p overlap: In the formation of hydrogen flouride, half filled $2p_z$ orbital of flourine overlaps with half filled 1*s* orbital of hydrogen atom resulting in s-p overlap. (Fig. 1.3)

Sigma and Pi Bonds

As is seen in the above examples the overlap is along the internuclear axis. When a bond is formed by the overlapping of the orbitals along the internuclear axis, it is known as **Sigma Bond**. On the other hand if the bond is formed by the sidewise overlapping or lateral overlapping of the atomic orbitals the bond is termed as **pi (π) Bond**. Formation of pi bond only takes place if sigma bond is already present i.e. it cannot be present alone. Formation of pi-bond is illustrated in the following example.

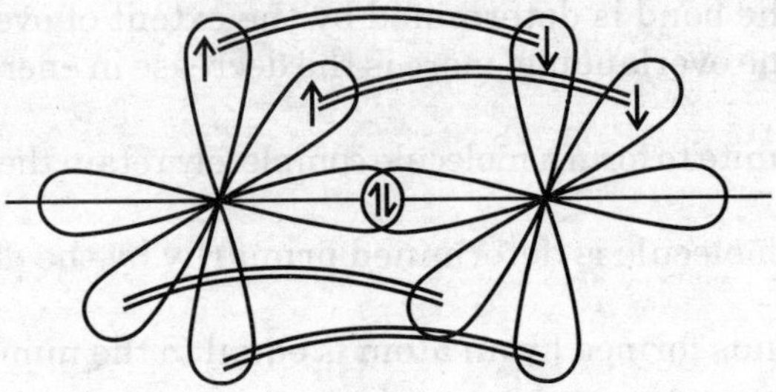

π bond

Fig. 1.4 Formation of N_2 molecule.

Formation of nitrogen molecule: Electronic configuration of nitrogen is $1s,^2\ 2s,^2\ 2p_x^1, 2p_y^1, 2p_z^1$. Thus there are 3 p-orbitals having unpaired electrons. $2p_z$ orbital of one nitrogen atom overlaps with $2p_z$ orbital of other nitrogen atom forming a σ bond. $2p_x^1$ and $2py^1$ of one nitrogen overlap with $2p_x^1$ and $2py^1$ orbitals of other nitrogen atom laterally to form two pi-bonds. Nitrogen atom has one sigma bond and two π bonds i.e a triple bond.

Table 1.2: Comparison of Sigma and Pi Bonds

Sigma Bond	*Pi Bond*
1. This bond is formed by the overlap of atomic orbitals along internuclear axis.	This type of bond is formed by the lateral (sideways) overlap of atomic orbitals.
2. This is formed by the overlapping of s-s, s-p, or p-p atomic orbitals along internuclear axis.	This is formed by the lateral overlapping of p-p atomic orbitals.
3. Sigma bond is stronger because extent of overlapping is quite large.	π-bond is weaker because overlapping is to a small extent.
4. The molecular orbital of sigma bond is symmetrical about the internuclear axis.	The molecular orbital of π bond consists of two electron clouds, one above and other below the plane of atomic nuclei.
5. There can be free rotation of the atoms around the sigma bond.	Free rotation about a π-bond is not possible.
6. The bond may be present alone or with π bond.	The bond is always present along with the sigma bond only.

1.4.2 Polar and Non Polar Covalent Bond

A covalent bond may be non-polar or polar depending upon the electronegativity of the bonded atoms. (cf : electronegativity is the tendency of an atom in a molecule to attract the shared pair of electrons towards itself.)

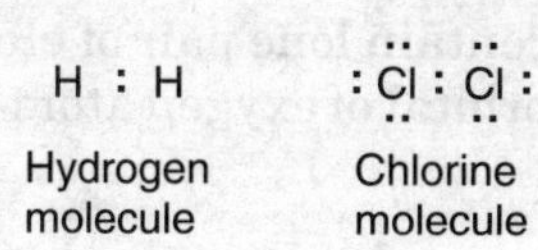

Non-Polar Covalent Bond: If the bond is formed between two identical atoms, the shared pair will be attracted equally by the two atoms and hence the bond formed will be non-polar covalent bond. For example-H_2, Cl_2 etc. In non-polar bond the electron cloud is completely symmetrical and there is no charge separation at all.

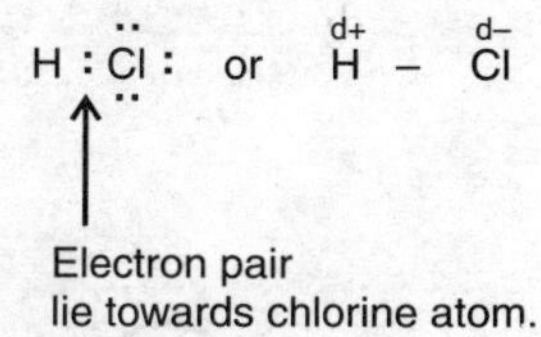

Polar Covalent Bond: If the bond is formed between two dissimilar atoms, the shared pair will be attracted more towards the atom having greater electronegativity. The distribution of electrons gets distorted as the electron cloud is displaced towards more electronegative atom. This leads to the development of partial negative charge near the atom which is more electronegative and partial positive charge near the other atom which is less electronegative. Such molecules having two oppositely charged poles are called polar molecules and the bond is said to be polar covalent bond. Examples-HCl, BrCl etc. The nonpolar molecules are called dipoles.

Partial Ionic Character of Polar Covalent Bond: The polar covalent bond is said to possess partial ionic character. The ionic character of the covalent bond depends upon the electronegativity difference between the two atoms. Several empirical equations have been proposed to calculate the percent ionic character of the covalent bond from the electronegativities of the bonded atoms. It has been proposed by Pauling that if the difference in electronegativities of the bonded atoms is 1.9 then the bond is 50% covalent and 50% ionic in character. If it is less than 1.9, then bond is predominantly covalent and if it is more than 1.9 then bond is predominantly ionic. He established the following approximate relationship-

Electronegativity Difference	0.2	0.6	0.8	1.2	1.9	2.2	2.8	3.2
Percent Ionic Character	1	10	15	30	50	70	86	92

1.4.3 Coordinate Bond or Dative Bond

It is a type of covalent bond formed by the mutual sharing of electrons between the two atoms but the shared pair of electron is contributed by one of the two atoms only. Coordinate or dative bond is formed between two atoms, one of which has a complete octet and possesses a pair of valency electrons while the other is short of a pair of electrons. The atom which contributes electron pair for sharing is called the **donor** while the other atom which accepts it for sharing is called **acceptor**. A coordinate bond is represented by an arrow head ⟶. Example-NH_4^+ ion.

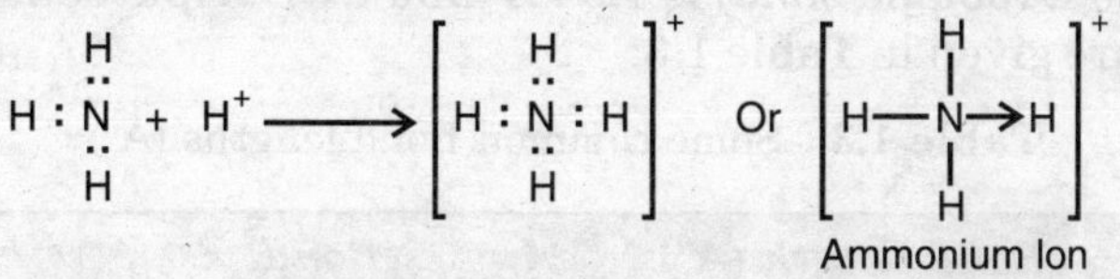

Ammonium Ion

Coordinate bond once formed is indistinguishable from a covalent bond.

Orbital Concept of Formation of Dative Bond: According to this concept a coordinate bond is formed when an orbital containing a lone pair of electrons overlaps with an empty orbital present in the other atom. This can be illustrated by considering the formation of hydronium ion, H_3O^+.

Formation of H_3O^+: The central atom oxygen in water has four orbitals $2s^2$, $2p_x^2$, $2p_y^1$, $2p_z^1$. Two of these are used up in formation of covalent bonds with hydrogen, while the other two contain lone pair of electrons. The H^+ ion has empty 's' orbital which overlaps with one of the p orbital of exygen atom containing lone pair forming hydronium ion, H_3O^+.

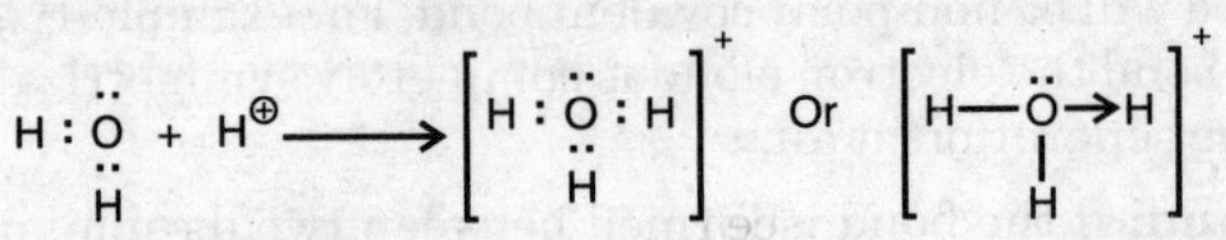

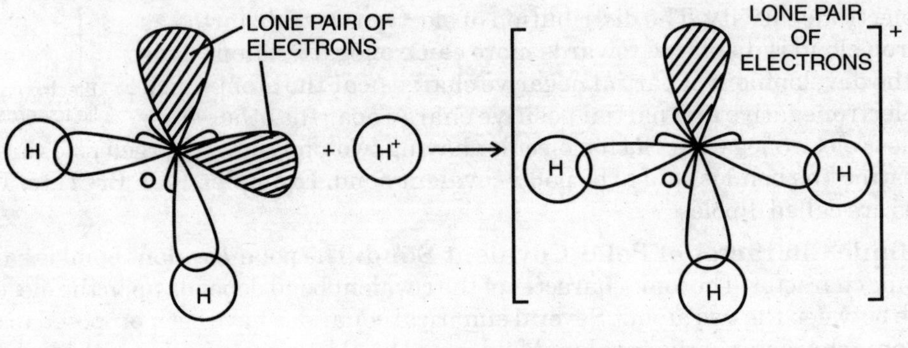

Fig. 1.5 Formation of hydronium ion

1.4.4 Covalent Bond Characteristics

The important characteristics of a covalent bond are:

1. Bond length
2. Bond Angle
3. Bond Energy

1. Bond Length: Bond length may be defined as the average equilibrium distance between the centres of the nuclei of the two bonded atoms. Since bond length is very small it is usually measured in Angstrom units ($1A^0 = 10^{-8}$cm) During the formation of diatomic molecule, let us say H_2, we have already seen that a stage is reached which corresponds to minimum energy. At this stage the atoms get bonded together. The distance between the nuclei of hydrogen atoms is called the bond length of H-H bond.

Bond length can be measured by certain physical methods such as X-ray diffraction in case of crystals.

The bond length of any multiple bond is shorter than that of single bond eg $C-C$ single bond length is 1.54 A° and $C=C$ (double bond) is 1.34 A° and $C \equiv C$ triple bond is 1.20 A°. Some of the common bond lengths are given in Table 1.3:

Table 1.3 Some common Bond lengths (A°)

Bond	*Bond length (A°)*	*Bond*	*Bond length A°*
H-H	0.74	C – C	1.54
Cl-Cl	1.98	C = C	1.34
Br - Br	2.28	C ≡ C	1.29
I-I	2.67	H-F	0.92
C-H	1.00	H-Cl	1.27
O-H	0.96	H-Br	1.41
C=O	1.20	H-I	1.61

2. Bond Angle: Covalent bond is formed by the overlap of atomic orbitals. The direction of overlap gives the direction of bond. Bond angle is defined as the internal angle between the bonded orbitals of the central atom in the molecule. For example the bond angles in case of H_2O, NH_3 and CH_4 are 104.5, 107° 109° 28′ respectively.

3. Bond Energy: The bond energy may be defined as the amount of energy required to break one mole of the bonds of that type in a substance in gaseous state. This energy is the measure of the strength of the bond. More is the energy required to break the bond, stronger is the bond. The value of bond energy for a particular bond is same only when present in similar molecules. For example, bond energy of C-H in CH_4 is not the same as in $CHCl_3$. Hence always average bond energy is considered.

Bond energy of some of the common bonds are given in Table 1.4 :

Table 1.4 Bond Energies

Bond	*Bond Energy kJ mol⁻¹*	*Bond*	*Bond Energy kJ mol⁻¹*
H–H	433	H–F	563
C–C	347	H–Cl	432
C = C	619.0	H–Br	366
C ≡ C	836.8	H–I	292

1.5 VALENCE SHELL ELECTRON PAIR REPULSION THEORY (VSEPR)

In 1940 Sidgwick and Powell reviewed the structures of various molecules known at that time. According to them, the shape of a molecule can be predicted from the number of electron pairs in the outer or valence shells of the central metal atom. The electron pairs arrange themselves around the central metal atom in such a way that there is minimum repulsion between them so that the molecule has minimum energy and maximum stability. Since there can be only one orientation of orbitals corresponding to minimum energy a molecule has invariably a definite shape i.e. a definite geometry.

In Sidgwick and Powell theory all the electron pairs whether bond pairs or lone pairs were considered equivalent. However this theory was further developed by Gillespie and Nyholm in 1957 to predict and explain molecular shapes and bond angles more exactly. This theory was called Valence Shell Electron Pair Repulsion (VSEPR) Theory. The basic points of this theory are as follows:

1. The central metal atom is bonded to the other atoms through electron pairs by single covalent bonds.
2. The electron pairs around the central metal atom are so arranged in space so as to minimize the coulombic repulsions between them, leading to stable arrangement.
3. The geometry of the molecule is determined by the total number of electrons pairs, bond pairs as well as lone pairs, surrounding the central metal atom.
4. When the central atom in a molecule is surrounded by bond pairs only, the molecule will have a regular shape or geometry as given in the following table.

Table 1.5: Molecular shapes predicted by VSEPR theory

No. of electron pairs	*Shape*	*Bond Angle*	*Example*
2	Linear	180°	Be F_2
3	Plane triangular	120°	BF_3
4	Tetrahedral	109° 28'	CH_4
5	Trigonal bipyramid	120° and 90°	PCl_5
6	Octahedral	90°	SF_6
7	Pentagonal biyramid	72° and 90°	IF_7

5. When the central atom in a molecule is surrounded by bond pair as well as lone pair the geometry is distorted or irregular. This is because a lone pair of electrons takes up more space around the central atom than a bond pair as lone pair is attracted to one nucleus whereas bond pair is under the influence of two nuclei. Consequently, the presence of lone pairs will alter the bond angle to a small extent which does not pertain to any regular geometry.
6. It is found that the repulsion between the lone pair electrons is greater than the repulsion between bond pair electrons. The order of decreasing repulsion being
   Lone pair – Lone pair > Lone pair – bond pair > Bond pair-bond pair.
7. Electronegativity difference between the central atom and the other atoms also influence the magnitude of repulsion.

Application of VSEPR Theory: In the light of above mentioned points, the geometry of various molecules can be predicted. Let us discuss a few examples–

1. **Shape of BeF_2 molecules:** In BeF_2, the central atom Be has two valence electrons. It shares three electrons with two flourine atoms i.e. the outer shell of Be has two electron pairs (bond pairs). According to VSEPR theory, for minimum energy and maximum stability, these bond pairs arrange themselves at an angle of 180° and molecule in such case will be linear.

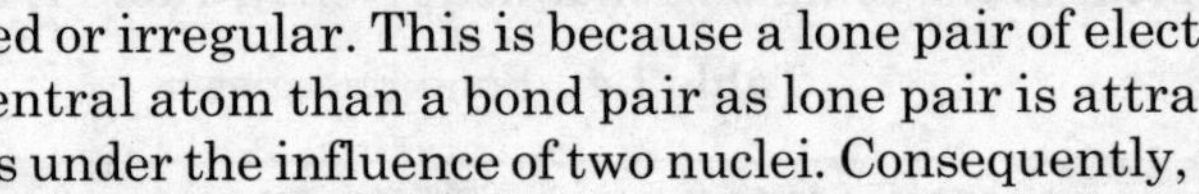

2. **Shape of BF_3:** Boron, the central metal atom has 3 electrons in the valence shell which are used to form covalent bonds with three flourine atoms Hence the central atom has three bond pairs which are directed towards the three corners of a equilateral triangle.

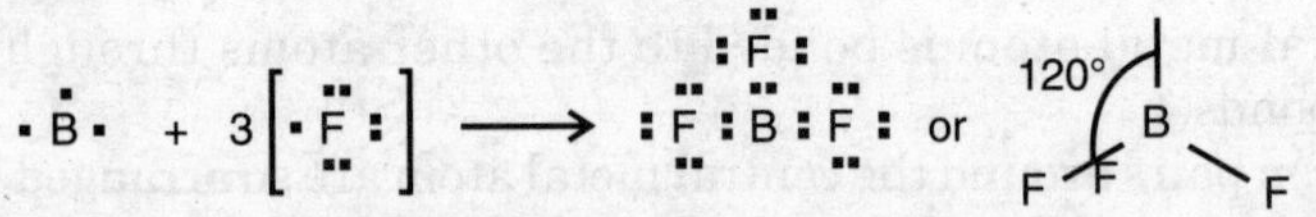

3. **Shape of CH_4:** Central atom carbon has four valence electrons which it shares with four hydrogen atoms. Hence 4 bond pairs are present around central atom giving tetrahedral shape to the molecule with a bond angle of 109°, 28′.

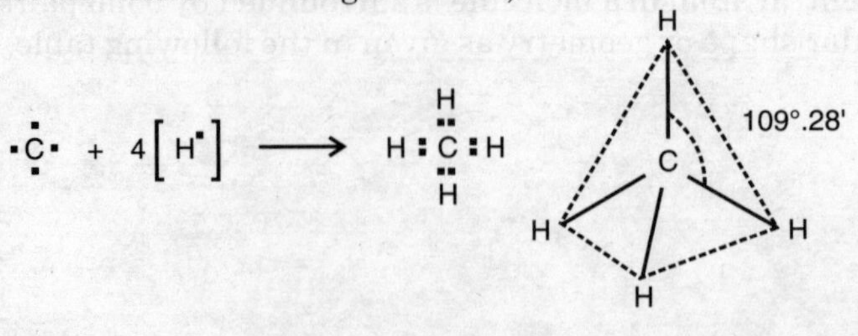

4. **Shape of PCl_5:** Phosphorus has five valence electrons which it shares with five chlorine atoms to form five bond pairs. Hence the geometry of the molecule is triagonal bipyramidal with bond angles of 120° and 90°.

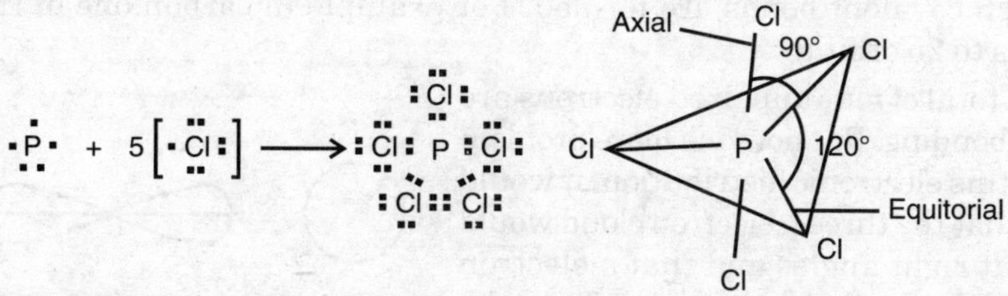

5. **Shape of SF_6:** Sulphur has six valence electrons and forms six bonds with flourine atoms. The geometry of the molecule is octahedral, with a bond angle of 90°.

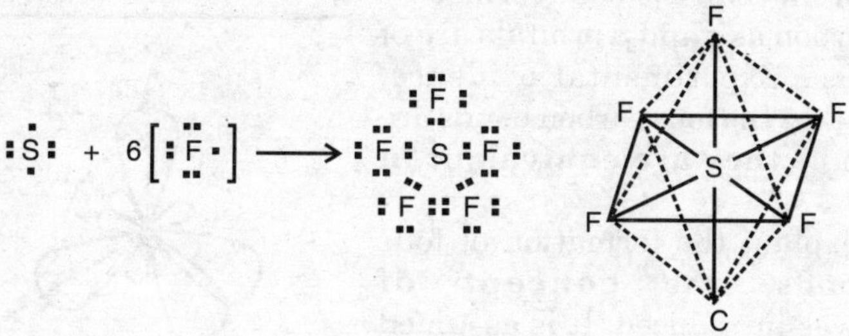

1.6 HYBRIDIZATION

In the theories of covalent bonding, discussed in the previous sections, it is discussed that half filled atomic orbitals undergo bond formation. The number of half filled atomic orbitals of the element is referred to as covalency of the element. However this concept could not explain the formation of beryllium, boron and carbon compounds. The electronic configuration of these elements can be represented schematically as-

	Z	*1s*	*2s*	*2p*
Beryllium	4	↑↓	↑↓	
Boron	5	↑↓	↑↓	↑ | |
Carbon	6	↑↓	↑↓	↑ | ↑ |

Thus, on the basis of electronic configuration of these elements and our knowledge of orbital overlap theory (VBT) we can say that beryllium should behave as noble gas since it does not have any unpaired electrons, whereas boron and carbon have covalency of one and two respectively. But this is not the case. Beryllium forms compounds like BeF_2, BeH_2 etc. in which it is bivalent. Similarly boron forms a number of compounds like BCl_3, BF_3 etc in which it is trivalent whereas carbon forms a large number of compounds in which it is tetravalent.

	2s	*2p*
Carbon (Ground State)	↑↓	↑ | ↑ |
Carbon (Excited State)	↑	↑ | ↑ | ↑

To explain this anomaly it was assumed that some of the paired electrons become unpaired before the atoms participate in bonding. These unpaired electrons then become available for the formation of covalent bonds. The energy required for excitation will be available from heat of reaction when covalent bonds are formed. For example in carbon one of the electron from 2s orbital jumps to $2p$ orbital-

Hence, a total of four unpaired electrons are available for bonding. But now one more problem arises. With this electronic distribution, it would be expected that the three p electron cloud would be oriented at right angles and that s-electron cloud will be spherically symmetrical. Thus the bonds formed by the overlap of p-orbital of carbon atom will be different than the bond formed by the s orbital of carbon as s and p orbitals are of different energies. Experimental evidence, however, showed that the four carbon bonds are of equal strength i.e **they are equivalent in all respects**.

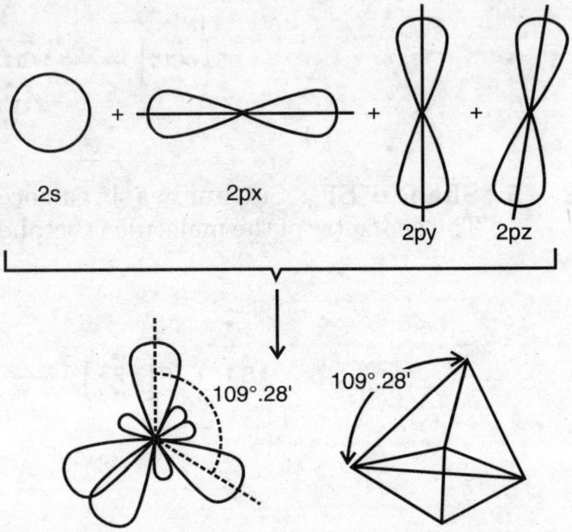

Fig. 1.6

In order to explain the formation of four equivalent bonds, the **concept of hybridization** was introduced. It is assumed that one s and $3p$ orbitals of carbon get mixed up and form four equivalent hybrids.

The phenomenon of mixing up of atomic orbitals of similar energies, so that a redistribution of energy takes place between them, resulting in the formation of new orbitals of equal energies and identical shapes is known ashybridization. The new orbitals thus formed are known as hybrid orbitals. The bonds formed by these orbitals are stronger than the bonds formed by pure s, p or d orbitals.

The four new bonds formed in the case of carbon are directed towards the four corners of a regular tetrahedron. Diagrammatically it can be represented as shown in Fig. 1.6.

The concept of hybridization is hypothetical but is extremely helpful in predicting the structure of various compounds. For understanding the phenomenon of hybridization, following points should be considered.

1. Hybridization involves the orbitals of similar energies belonging to the same atom or ion.
2. Number of hybrid orbitals produced is equal to the number of orbitals that participate in hybridization process.
3. Hybridization not only involves the half filled orbitals, even completely filled orbitals with slightly different energies can also participate in hybridization.
4. It is not necessary that all the half filled orbitals must hybridize.
5. Hybridization never takes place in isolated atom but it occurs only at the time of bond formation.
6. Type of hybridization indicates the shape of the molecule.

1.6.1 Types of Hybridization

1. sp hybridization: The combination of an s and a p-orbital leads to two hybrid orbitals known as sp orbitals. sp hybridization is well examplified by BeF_2 molecule. The electronic configuration of Be is $1s^2$, $2s^2$. It is assumed that one of the electron from $2s$ orbital jumps to $2p$ orbital in excited state of beryllium.

	1s	2s	2p
Be (Ground State)	↑↓	↑↓	
Be (Excited State)	↑↓	↑	↑ |
Hybridization	↑↓	↑	↑ |
			sp hybridization

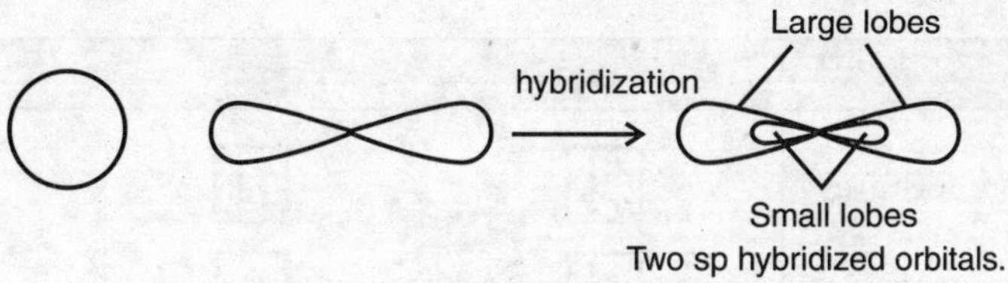

Fig. 1.7 Formation of sp orbitals

2*s* and 2*p* orbitals of beryllium mix to form two *sp* hybridized orbitals having equal energies. The size of the *sp* hybrid orbital has one lobe of larger size and other a smaller one. The broader shape of one of the lobes enables the hybrid orbital to bring about overlapping more effectively. This results in a stronger bond. These hybrid *sp* orbitals overlap with the *2p* orbitals from flourine atoms to form two bonds. The molecule is linear with a bond angle = 180°. Fig. 1.7 represents the pictorial representation of the formation of BeF_2 molecule.

2. sp² hybridization: Boron atom has electronic configuration as $1s^2, 2s^2, 2p_x^1$. It is assumed that 1 electron from 2*s* orbital of boron jumps to 2*p* orbital. Thus boron in the excited state has three unpaired electrons. The three orbitals i.e. 1s and two p orbitals hybridize to give three sp^2 hybridized orbitals which are coplanar and are directed at angles 120° to each other. sp^2 hybridized orbitals are able to form stronger bonds than *s* or *p* orbitals because of their larger size, they overlap more effectively.

	1s	2s	
Boron atom (Ground state)	↑↓	↑↓	↑ | |
Boron atom (Excited state)	↑↓	↑	↑ | ↑ |
Hybridization	↑↓	↑	↑ | ↑ |
			sp^2 hybridization

The three sp^2 hybridized orbitals overlap with *p* orbitals from three flourine atoms to form BF_3 molecule.

Other examples of sp^2 hybridization are SO_2, CO_3^{2-} etc.

3. sp³ hybridization: The electronic configuration of carbon atom is $1s^2, 2s^2, 2p_x^1, 2p_y^1$. One of the electrons from 2*s* orbital jumps to $2p_z$ orbital thus forming four orbitals having unpaired electrons. These four orbitals, one *s* and three *p* orbitals hybridize to form four sp^3 hybridized orbitals of equal energy.

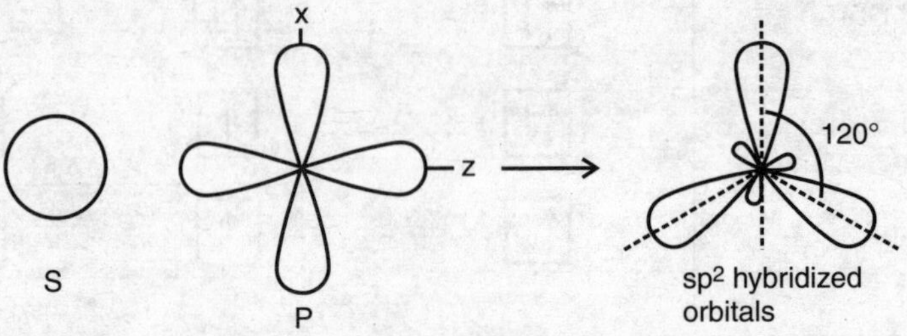

Fig. 1.8

	1*s*	2*s*	2*p*
C atom (Ground State)	↑↓	↑↓	↑ ↑ □
C atom (Excited State)	↑↓	↑	↑ ↑ ↑
Hybridization	↑↓	↑	↑ ↑ ↑
		sp^3 hybridization	

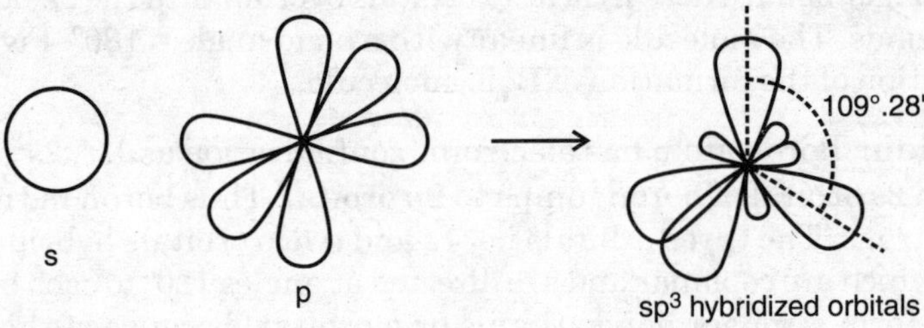

Fig. 1.9

Four sp^3 hybrid orbitals overlap with four 1*s* orbitals of hydrogen atoms to form four bonds in CH_4. The bond angle = 109°28′.

4. sp³ d hybridization: The combination of one *s*, three *p* and one *d* orbital to form five hybrid orbitals of equal energy is known as sp^3d hybridization. Example PCl_5 molecule.

	3*s*	3*p*	
Phosphorus (Ground State)	↑↓	↑ ↑ ↑	
Phosphorus (Excited State)	↑	↑ ↑ ↑	↑ □ □ □ □
Hybridization	↑	↑ ↑ ↑	↑ □
	sp^3d		

The shape of the molecule is trigonal bipyramid, the equitorial and axial angles being 120° and 90° respectively. (Fig. 1.10)

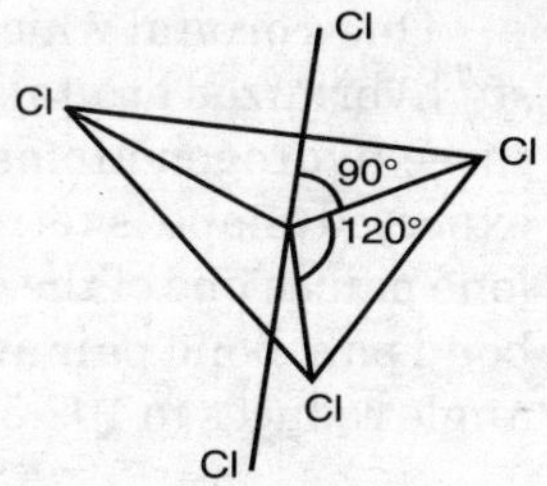

Fig. 1.10

5. sp^3d^2 hybridization: The combination of one *s*, three *p* and two *d* orbitals to form six hybrid orbitals of equal energy is known as sp^3d^2 hybridization. An example, is the structure of SF_6.

	3s	3p	3d
S (Ground State)	↑↓	↑↓ ↑ ↑	
S (Excited State)	↑	↑ ↑ ↑	↑ ↑
Hybridization	↑	↑ ↑ ↑	↑ ↑

sp^3d^2

The molecule has octahedral shape with bond angle of 90°.

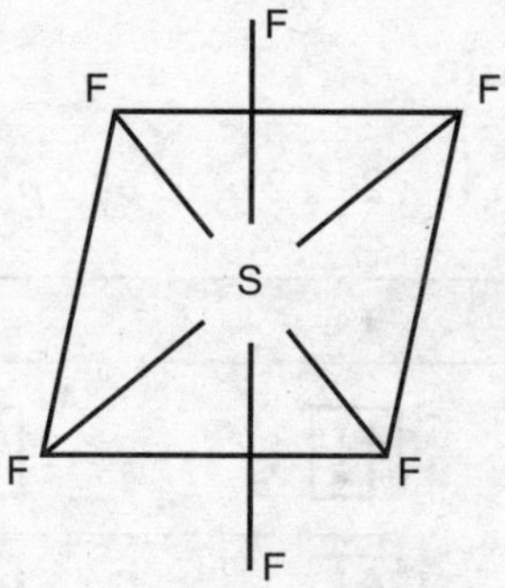

Fig. 1.11

So far we have discussed the molecules in which orbitals having unpaired electrons are involved in hybridization. But hybridization not only involves the half filled orbitals, *even completely filled orbitals with slightly different energies can participate in hybridization.* This can be illustrated by considering the following examples-

Hybridization in NH_3 Molecule

Nitrogen has electronic configuration $1s^2, 2s^2, 2p_x^1, 2p_y^1\ 2p_z^1$

	1s	*2s*	*2p*
Nitrogen	↑↓	↑↓	↑ ↑ ↑
Hybridization	↑↓	↑↓	↑ ↑ ↑
		sp^3 hybridization	

One *s* orbital which is completely filled and three half filled *p* orbitals hybridize to form four sp^3 hybridized orbitals. Out of four hybridized orbitals, three orbitals are used for bonding with three hydrogen atoms, while fourth orbital contains a lone pair. Due to sp^3 hybridization the expected shape is tetrahedral, but actual shape is pyramidal. This is because of the presence of lone pair in one of the orbitals. The force of repulsion between lone pair-bond pair is greater than bond pair-bond pair and hence the shape of the molecule is distorted from tetrahedral and bond angle reduces to 109.5° to 107°.

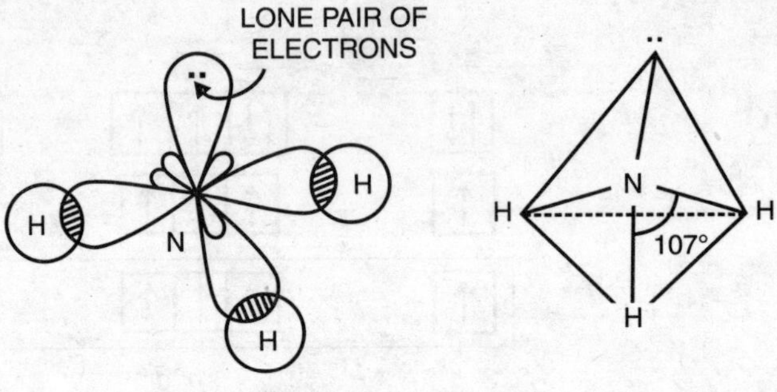

Fig. 1.12 Pyramidal structure of NH_3.

Hybridization in H_2O Molecule

Oxygen has electronic configuration $1s^2, 2s^2, 2p_x^2, 2p_y^1\ 2p_z^1$

	1*s*	2*s*	2*p*
Oxygen atom	↑↓	↑↓	↑↓ ↑ ↑
Hybridization	↑↓	↑↓	↑↓ ↑ ↑
			sp^3 hybridization

One *s* and three *p* orbitals hybridize to form four sp^3 hybridized orbitals. Out of these four hybridized orbitals, two contain lone pairs while the other two are utilized for bond formation with two hydrogen atom. The expected structure is tetrahedral but due to the presence of two lone pairs the geometry is V shaped. Presence of two lone pairs distort the bond angle which reduces to 104.5° from 109.5°.

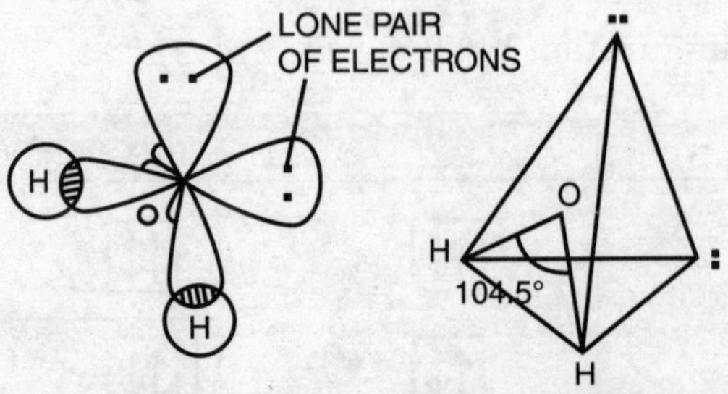

Fig. 1.13 Shape of water molecule

Hybridization in SF_4 molecule: The outer electronic configuration of central atom sulphur is $3s^2$, $3px^2$, $3py^1$, $3pz^1$.

	3s	3p	3d
S (Ground State)	↑↓	↑↓ ↑ ↑	□ □ □ □ □
S (Excited State)	↑↓	↑ ↑ ↑	↑ □ □ □ □
Hybridization	↑↓	↑ ↑ ↑	↑ □ □ □ □
		sp^3d	

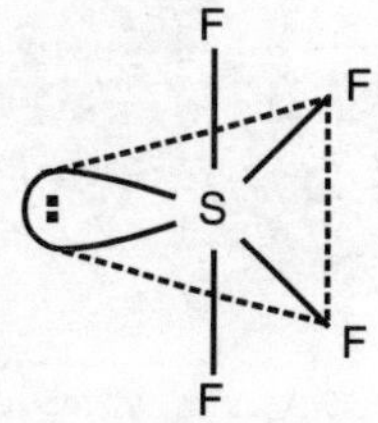

Fig. 1.14

One of the electron from 3*p* orbital jumps to empty 3*d* orbital, thus making available four unpaired orbitals to overlap with four flourine atoms. But prior to bond formation, one *s*, 3*p* and one *d* orbitals hybridize to form five sp^3d hybridized orbitals. Out of these five, one accommodates the lone pair and the other four form bonds with flourine atoms. The shape of the molecule is trigonal bipyramidal. The expected bond angles are 90°, 120° and 180°. But due to the presence of lone pair in one of the hybridized orbital, the angle reduces to 89°, 118° and 177° instead of 90°, 120° and 180° respectively. The lone pair occupies equatorial position.

Hybridization in $SnCl_2$ molecule: The outer electronic configuration of central atom tin is $5s^2$, $p_x{}^1$, $p_y{}^1$.

Sn (Ground State)	↑↓	↑ ↑ □
Hybridization	↑↓	↑ ↑ □

Thus sp^2 hybridizatin takes place. Out of the three hybrid orbitals one accomodates the lone pair and the other two form bonds with chlorine atoms.The shape of the molecule is trigonal planar.The expected angle is 120° but due to the presence of lone pair the angle is reduced and $SnCl_2$ has a V shaped geometry. (Fig. 1.15)

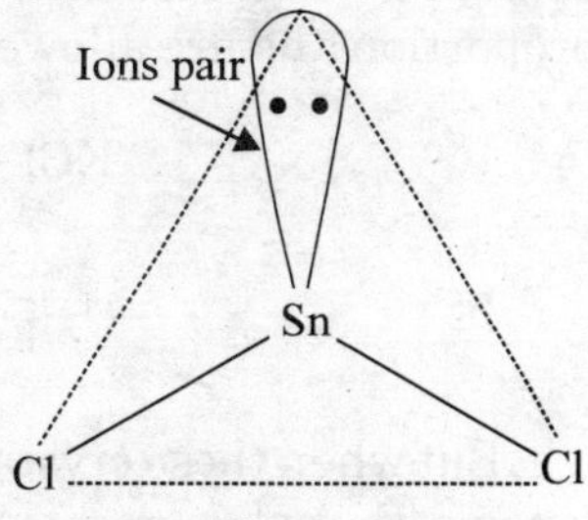

Fig. 1.15

Hybridization in CO_3^{2-} ion: The outer electronic configuration of carbon atom in ground state is $2s^2, 2p_x^1, 2p_y^1$ and in excited state it is $2s^1, 2p_x^1, 2p_y^1, 2p_z^1$. Out of three oxygen atoms of CO_3^{2-} ion, two carry one additional electron each, so that two of the oxygen atoms contain only one half filled *p* orbital. The third oxygen atom has 2 unpaired electrons. Carbon atom undergoes sp^2 hybridization. The three hybridized orbitals form σ bonds with three oxygen atoms as shown, whereas unhybridized p orbital of carbon atom overlaps laterally with *p* orbital of one of the oxygen atom to form π bond. Due to sp^2 hybridization the geometry of the ion is trigonal planar with bond angle of 120°.

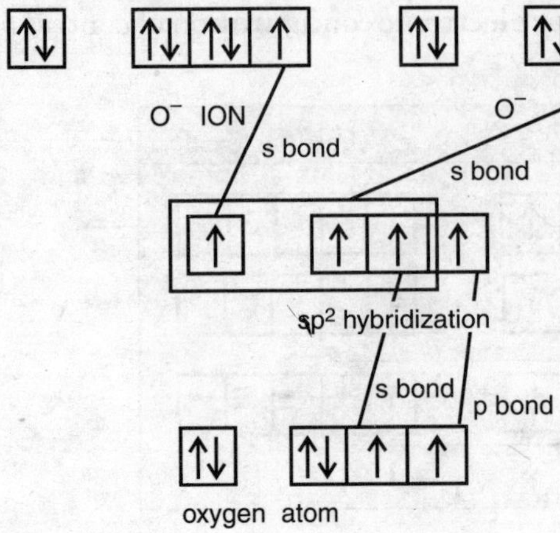

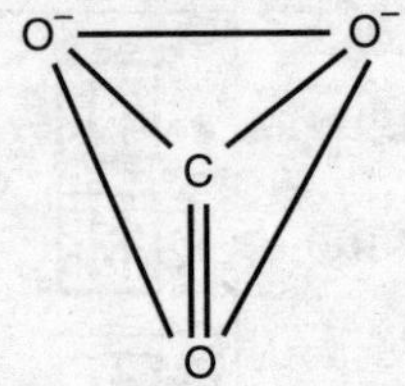

Fig. 1.16

Similarly hybridization of other molecules can be explained.

1.7 COORDINATION COMPOUNDS

The term coordination compound is widely applicable to the compounds in which the central metal atom is linked to a number of ions or neutral molecules by coordinate bonds.

When solution containing two or more salts in stoichiometric proportions are allowed to evaporate, crystals of new compounds separate out. These compounds fall under two catogories.

(a) Double Salts

(b) Coordination Compounds.

Double Salts: These are the molecular compounds which exist only in crystal lattices. When they are dissolved in water or any other solvent they break down into their constituent ions. Their physical and chemical properties are essentially the same as their individual components. For example, when solutions of two salts viz. KCl and $MgCl_2$ are mixed together in stoichiometric proportion and are allowed to evaporate, crystals of Carnellite, a double salt, are obtained

$$KCl + MgCl_2 + 6H_2O \longrightarrow \underset{\textit{Carnellite}}{KCl.MgCl_2.6H_2O}$$

$$KCl.MgCl_2.6H_2O \longrightarrow K^+_{(aq)} + Mg^{2+}_{(aq)} + 3C\ell^-_{(aq)}$$

But when these crystals are dissolved in water, the double salt break down into its constituent ions i.e. the solution shows the properties of K^+, Mg^{2+} and Cl^- ions. Thus double salts have their identity only in the crystalline state. Mohr's salt, potash alum, etc are some more examples of double salts.

$$FeSO_4(NH_4)_2\,SO_4.6H_2O \text{ (Mohr's Salt)}$$

$$K_2SO_4.Al_2(SO_4)_3.24H_2O \text{ (Potash Alum)}$$

Coordination Compounds: Coordination compounds are those molecular compounds which *retain their identity* even when dissolved in water or any other solvent. The properties of coordination or complex compound is entirely different from those of its components. For example, when a solution of ferrous cyanide is mixed with potassium cyanide, potassium ferrocyanide is formed. When this complex compound is dissolved in water it **does not** give the tests of its constituent ions Fe^{2+} and CN^- ions but gives the test of ferrocyanide ion $[Fe(CN)_6]^{4-}$, though a small dissociation does take place.

$$Fe(CN)_2 + 4KCN \longrightarrow K_4[Fe(CN)_6]$$

$$K_4[Fe(CN)_6] \rightleftharpoons 4K^+ + \underset{\text{Complex ion.}}{[Fe(CN)_6]^{4-}}$$

Thus coordination compounds do not lose their identity even when dissolved in water. Similarly tetrammine cupric sulphate when dissolved in water does not give the test of Cu^{2+} or NH_3 but instead gives $[Cu(NH_3)_4]^{2+}$ ions.

$$[Cu(NH_3)_4]SO_4 \rightleftharpoons [Cu(NH_3)_4]^{2+} + SO_4^{2-}$$

Terms Used in Coordination Compounds

The terms most frequently used in the study of coordination compounds are-

Central Ion: Central ion is the cation which accepts the electron pairs from the ligands to form coordinate bonds, e.g. in the complex $[Cu(NH_3)_4]SO_4$, Cu^{2+} is the central ion.

Ligands: The anions or neutral molecules which are attached to the central ion through coordinate bonds are called ligands, e.g. in the above complex $[Cu(NH_3)_4]SO_4$, NH_3 molecules act as ligands. The common ligands are NH_3, H_2O, CN^- etc. The ligands should have atleast a lone pair of electron, so that it may act as donor for the formation of covalent bond to the central metal atom. The ligands may have one or more than one point of attachment and hence are named accordingly.

Ligand	*Point of Attachment*	*Example*
Monodentate	1	H_2O, NH_3, CO
Bidentate	2	$C_2O_4^{2-}$, $NH_2(CH_2)_2NH_2$
Polydentate	2 or more than 2	EDTA

Further, ligands can be neutral, (H_2O, NH_3, ethylene diamine) or anionic ($C_2O_4^{2-}$, F^-, Cl^-, Br^-, CN^-, OH^-)

Complex ion: A complex ion is an electrically charged species which consists of central metal ion surrounded by a number of ligands. When dissolved in water the complex ion does not dissociate but retains its identity. For example $K[Ag(CN)_2]$ contains complex ion $[Ag(CN)_2]^+$ which does not dissociate into Ag^+ and CN^- ions in solution but behaves as a separate entity i.e. $[Ag(CN)_2]^+$ ion.

Coordination Number: Coordination number is the number of unidentate ligands attached to the central metal atom or simply it is the number of coordinate bonds formed with the central metal atom. In the complex ions $[Ag(CN)_2]^-$, $[Cu(NH_3)_4]^{2+}$ $[Cr(H_2O)_6]^{3+}$ the coordination number of Ag, Cu and Cr are 1, 2 and 3 respectively. If on the other hand, bidentate ligands are attached to the central metal atom, then the coordination number becomes double the number of bidentate ligands, because each ligand forms two coordinate bonds. For e.g. in the coordinate compound $[Fe(C_2O_4)_3]^{2-}$, the coordination number of Fe is 6 because each $C_2O_4^{2-}$ forms two coordinate bonds.

Oxidation Number: It is the charge which the central atom has or appears to have when attached to the ligands to form coordinate compound. For example in $[Co(NH_3)_6]Cl_3$, cobalt has oxidation number of +3, because charge on the complex ion $[Co(NH_3)_6]^{3+}$ is +3. Now as NH_3 is neutral so cobalt carries a oxidation number of +3.

Coordination Sphere: It consists of central metal atom and the surrounding ligands. It is always enclosed in square brackets. In [Pt $(NH_3)_4$] Cl_2, [Pt $(NH_3)_4]^{2+}$ constitutes the coordination sphere.

Ionization Sphere: The ions outside the coordination sphere constitute what is known as ionization sphere. In the example [Pt $(NH_3)_4$] Cl_2, Cl_2 constitute the ionization sphere.

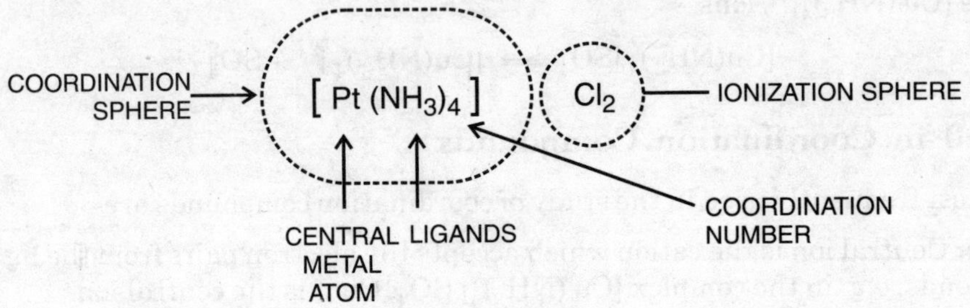

Fig. 1.17

1.7.1 Valence Bond (VB) Theory of Coordination compounds

This theory was extended to the coordination compounds by Pauling in 1931. The following are the main postulates of the theory.

(i) The metal-ligand bond arises by the donation of pairs of electrons by ligands to the metal atom/ion.

(ii) In order to accommodate these electrons, the metal ion must possess requisite number of vacant orbitals of equal energy. These orbitals are obtained by hybridization of the orbitals of the metal atom to give a set of hybrid orbitals of equal energy.

(iii) Sometimes, the unpaired $(n-1)d$ electrons pair up prior to hybridization thus making some $(n-1)d$ orbitals vacant. The central metal atom then makes available the number of empty orbital equal to its coordination number for the formation of coordinated bonds with suitable ligand orbitals.

(iv) With the approach of the ligands, metal-ligand bonds are then formed by the overlap of the orbitals with those of the ligands, i.e., by donation of electron pairs by the ligands to the empty hybridized orbital. Consequently, these bonds are of equal strength and directional in nature.

(v) Octahedral, square planar and tetrahedral complexes are formed as a result of d^2sp^3 (or sp^3d^2), dsp^2 and sp^3 hybridization respectively of the central metal atom or the ion.

A few examples are given below:

(A) Octahedral Complexes (Coordination Number: 6)

(1) Formation of $[Cr(NH_3)_6]^{3+}$. The oxidation state of chromium in $[Cr(NH_3)_6]^{3+}$ ion is +3. The electronic configuration of Cr is $[Ar]3d^54s^1$. Hence, we have

Cr^{3+} ion provides six empty orbitals to accommodate six pairs of electrons from six molecules of ammonia. The resulting complex $[Cr(NH_3)_6]^{3+}$ involves d^2sp^3 hybridization and is thus octahedral. The presence of these unpaired electrons in the complex explain its paramagnetic character.

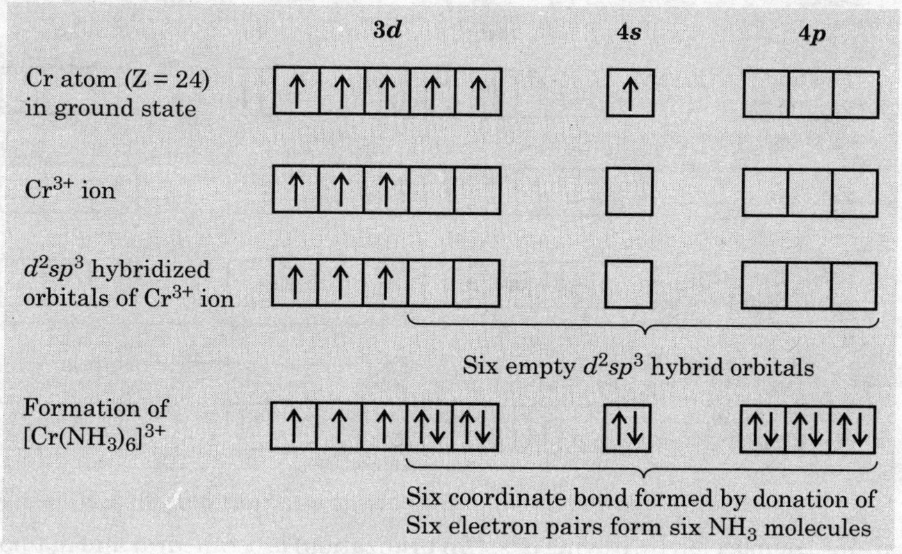

(2) Formation of ferricyanide ion, $[Fe(CN)_6]^3$: Oxidation state of Fe in this complex ion is +3. Hence, we have

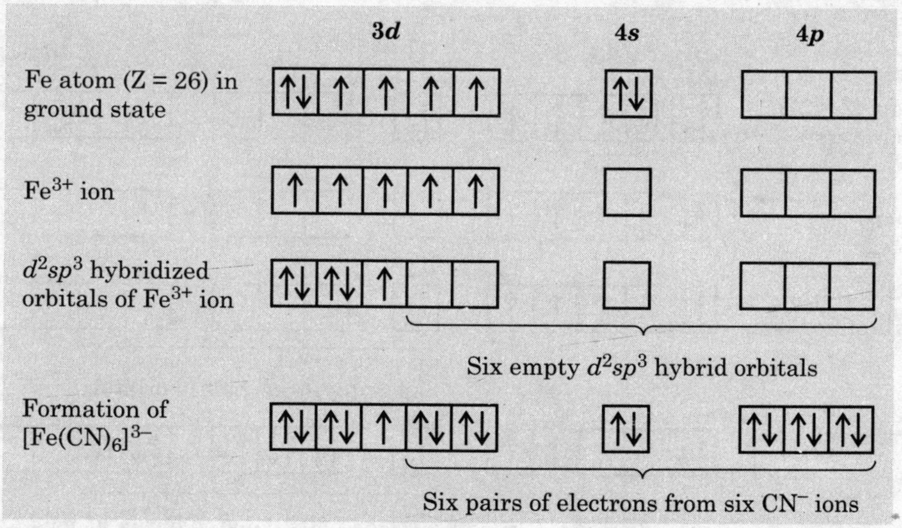

The resulting complex involves d^2sp^3 hybridization and is thus octahedral complex. Further, due to one unpaired electron, it should be slightly paramagnetic which is actually so.

(3) Formation of ferrocyanide ion, $[Fe(CN)_6]^{4-}$. Oxidation state of Fe here is +2. Hence, we have

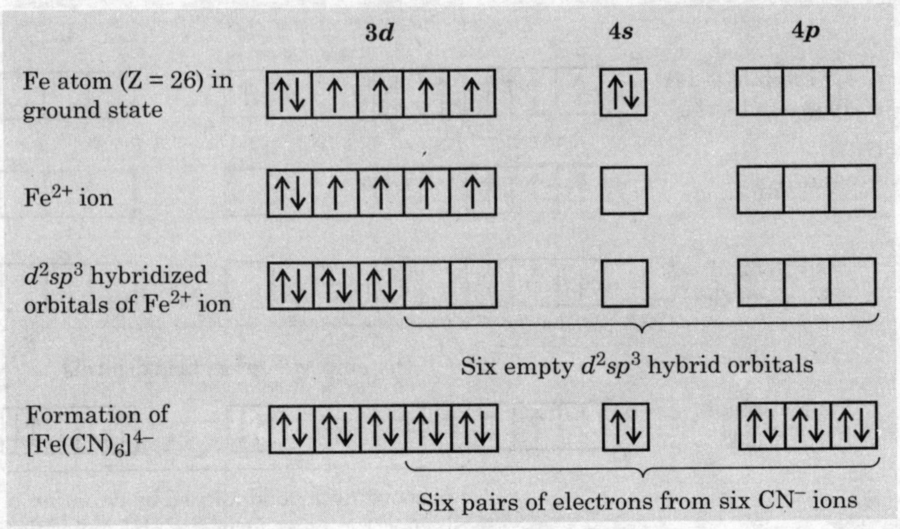

The resulting complex ion (involving d^2sp^3 hybridization) is octahedral and diamagnetic as it does not contain any unpaired electron.

(4) Formation of $[CoF_6]^{3-}$ ion. Oxidation state of cobalt in this complex is +3. Hence, we have

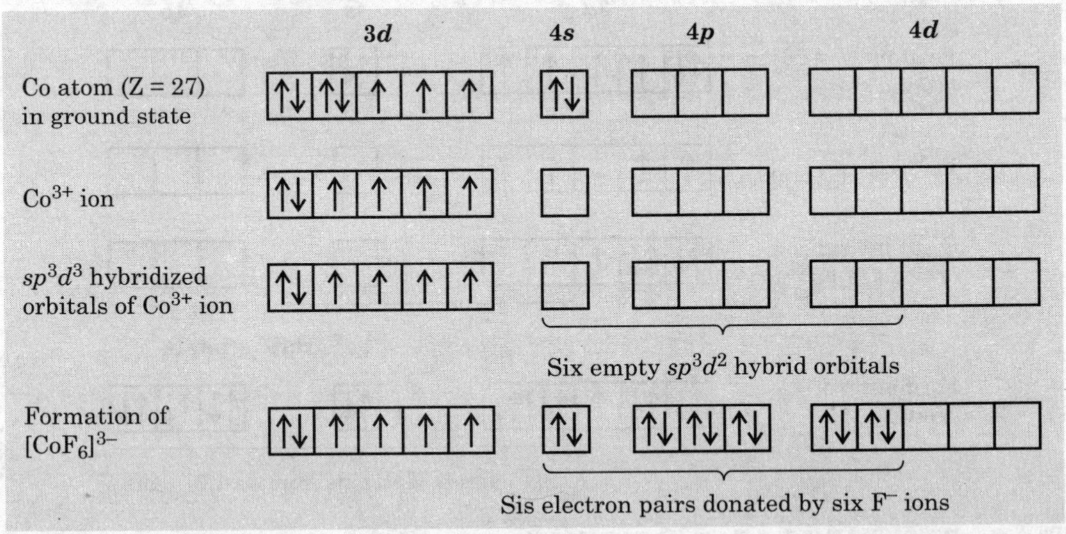

It may be pointed out here that F^- ion provides a weak ligand field and is unable to pair up the unpaired electrons of the 3d orbitals. Hence, six equivalent hybrid orbitals are obtained by mixing up one 4*s*, three 4*p* and two 4*d* orbitals. The high paramagnetic nature of the complex further confirms the presence of four unpaired electrons

As *d*-orbitals of the outermost shell take part in hybridization, the complexes thus formed are called **outer orbital complexes.**

Inner and Outer orbital complexes. We have seen that in octahedral structure, the central metal atom either uses inner $(n - 1)$ *d*-orbitals or outer *nd*- orbitals for hybridization. Accordingly, the complexes are classified as follows:

(i) **Inner orbital complexes.** When the complex formed involves the inner $(n-1)$ *d*-orbitals for hybridization (d^2sp^3), the complex is called **inner orbital complex.** In this case, the electron of the metal are made to pair up, so the complex will be either diamagnetic or will have lesser number of unpaired electrons. This type of complex is also known as **low spin complex.** This first three examples discussed above are inner orbital complexes. Some other examples of inner orbital complexes are:

$$[Co(NH_3)_6]^{3+}, [Mn(CN)_6]^{3-}, [Co(CN)_6]^{4-}, [Co(CN)_6]^{3-}, [Co(C_2O_4)_3]^{3-}, [Cr(H_2O)_6]^{3+}$$

(ii) **Outer orbital complex.** When the complex formed involves the use of outer *nd*-orbitals for hybridization, (sp^3d^2), the complex is called **outer orbital complex.** The complex will have large number of unpaired electrons as the configuration of the metal remains unchanged. This type of complex is also called **high spin complex.** One example i.e., of $[CoF_6]^{3-}$ has been discussed above. Some other examples of outer orbital complexes are:

$$[MnCl_6]^{3-}, [FeF_6]^{3-}, [CoF_6]^{3-}, [Ni(H_2O)_6]^{2+}, [Ni(NH_3)_6]^{2+}, [Fe(H_2O)_6]^{2+}, [Fe(H_2O)_6]^{3+}$$

Predicting the type of hybridization from magnetic behaviour. The magnetic moments of the coordination compounds can be determined by measuring their magnetic susceptibilities experimentally. This helps us to know the number of unpaired electrons present $\left(\text{as } \mu = \sqrt{n(n+2)}\text{ BM}\right)$.

For metal ions having upto 3 unpaired electron in the d-orbitals, like $Ti^{3+}(d^1)$, $V^{3+}(d^2)$ and $Cr^{3+}(d^3)$, two vacant *d*-orbitals are available for octahedral hybridization with 4*s* and 4*p* orbitals. Their magnetic moments are found to be corresponding to 1, 2 and 3 unpaired electron respectively.

For metal ions containing more than 3 unpaired electron in the *d*-orbitals, two empty *d*-orbitals can be made available only if 3*d* electrons pair up, e.g., d^4 (Cr^{2+}, Mn^{3+}), d^5(Mn^{2+}, Fe^{3+}) and d^6(Fe^{2+}, Co^{3+}). The magnetic moment data shows that maximum spin pairing occurs in many cases, especially in compounds with d^6 ions, but there are complications with species having d^4 and d^5 ions. A few example are given below:

(i) $[Mn(CN_6]^{3-}$ has magnetic moment of two unpaired electron while $[MnCl_6]^{3-}$ has for four unpaired electrons.

(ii) $[Fe(CN)_6]^{3-}$ has magnetic moment of one unpaired electron while $[FeF_6]^{3-}$ has magnetic moment of five unpaired electrons.

(iii) $[CoF_6]^{3-}$ has magnetic moment of four unpaired electrons while $[Co(C_2O_4)_3]^{3-}$ is diamagnetic.

The above result help us to explain whether the hybridization involved will be d^2sp^3 or sp^3d^2. Thus, we have

(i) $[Mn(CN_6)]^{3-}$, $[Fe(CN)_6]^{3-}$ and $[Co(C_2O_4)_3]^{3-}$ are inner orbital complexes involving d^2sp^3 hybridization.

(ii) $[MnCl_6]^{3-}$, $[FeF_6]^{3-}$ and $[CoF_6]^{3-}$ are outer orbital complexes involving sp^3d^2 hybridization.

(B) Tetrahedral complexes (*Coordination Number* = 4)

(1) Formation of $[Ni(CO_4)_6]^0$. Oxidation state of nickel in this complex is '0'. Its electronic configuration is [Ar] $3d^84s^2$. Hence, we have

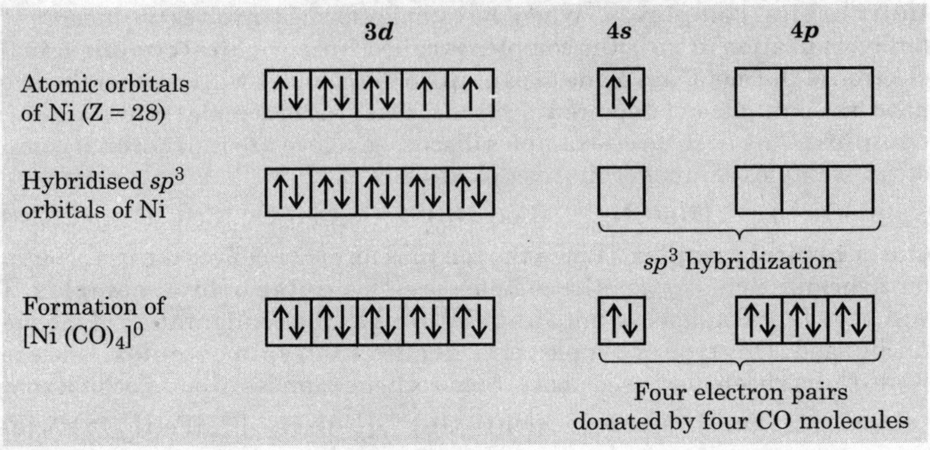

sp^3 hybrid orbital accommodate four pairs of electrons from four CO molecules and the resulting tetrahedral complex is diamagnetic due to absence of unpaired electrons.

(2) Formation of $[Zn(NH_3)_4]^{2+}$. Complexes of Zn^{2+} are invariably tetrahedral because they involve sp^3 hybridization as shown below:

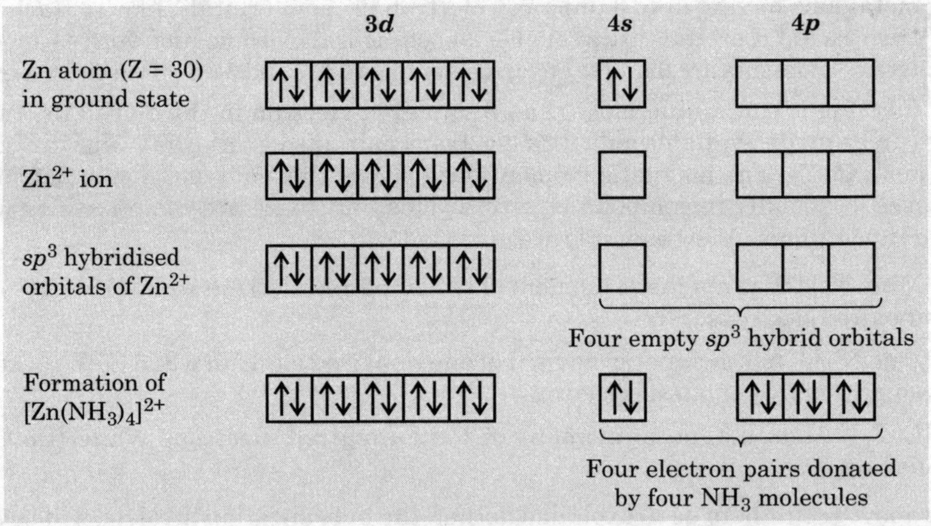

In the above case, 3*d*-orbitals are fully occupied, therefore, they cannot participate in hybridization. Four sp^3 orbitals can be formed by mixing of one 4*s* and there 4*p* orbitals. These four sp^3 hybrid orbitals accommodate four pairs of electron from four NH_3 molecules. The resulting $[Zn(NH_3)_4]^{2+}$ will be tetrahedral and will be diamagnetic as there are no unpaired electrons.

Some other examples of tetrahedral complexes are $[NiCl_4]^{2-}$, $[FeCl_4]^{2-}$.

(C) Square planer complexes. (*Coordination Number = 4*)

(1) Formation of $[Ni(CH)_4]^{2-}$. Nickel in $[Ni(CN)_4]^{2-}$ is in the +2 oxidation state, i.e., nickel is present as Ni^{2+} ion. The formation of $[Ni(CN)_4]^{2-}$ may be explained through hybridization as follows:

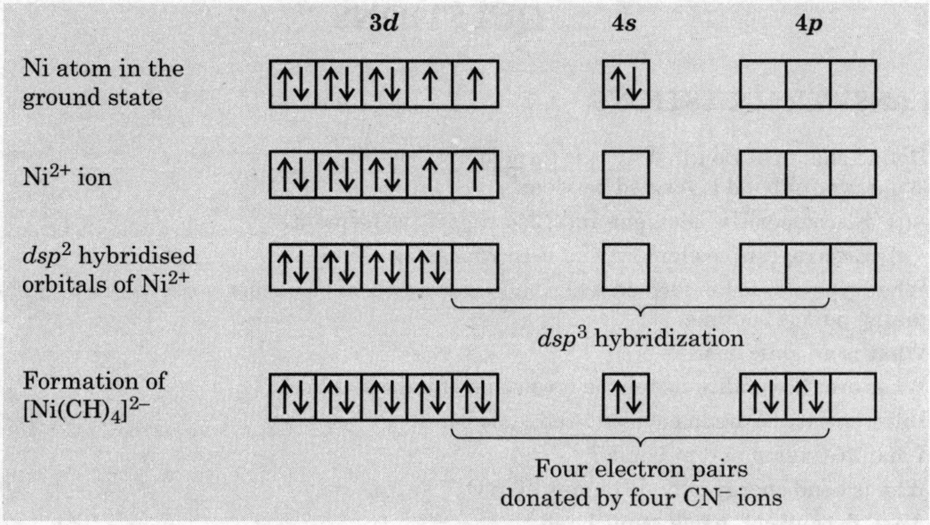

dsp^2-hybrid orbitals accommodate four pairs of electrons from four CN^- groups and the resulting square planar complex is diamagnetic since no unpaired electron is present.

It may be noted that from the configuration of Ni^{2+}, it appears at the first sight that the hybridization should be sp^3 but actually it is dsp^2, i.e., the two unpaired 3*d* electrons pair up against Hund's rule before hybridisation.

The is because CN^- is a strong ligand and when it approaches the metal ion, the electrons pair up resulting in a diamagnetic complex.

Note: In $[NiCl_4]^{2-}$, Cl^- provides a weal ligand field. It is, therefore, unable to pair up the unpaired electrons of the 3*d* orbital. Hence, the hybridization involved is sp^3 (involving one 4*s* and three 4*p* orbitals) and shape is tetrahedral. Moreover, it will be paramagnetic.

(2) Formation of $[Cu(NH_3)_4]^{2+}$. Oxidation state of Cu in $[Cu(NH_3)_4]^{2+} = +2$. Hence, we have

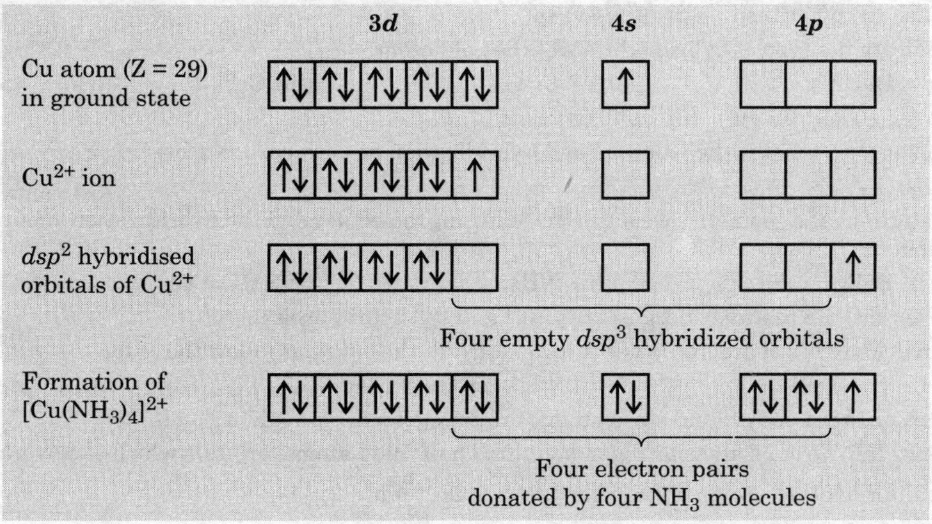

As it contains one unpaired electron, hence, it is paramagnetic.

QUESTIONS

SHORT ANSWERS QUESTIONS

1. Define chemical Bond? Why is it formed?
2. What type of bond is formed between
   (i) Electropositive element and Electropositive element.
   (ii) Electropositive element and Electronegative element.
3. What type of bond is formed when the shared pair of electrons is contributed by one of the atom taking part is bonding.
4. What is an ionic bond ?
5. What are the conditions for the formation of ionic bond ?
6. Differentiate between covalent and ionic bond?
7. What are sigma and pi bonds?
8. Why is bond energy of N_2 is higher than O_2?
   Ans: Because N_2 has a triple bond.
9. Arranges the covalent bonds C-F, C-Cl and C-I in increasing order of strength, giving reasons.
   Ans: The higher the electronegativity difference, the stronger is the bond. The electronegativity values increases in the order: $I < Br < CI < F$.
10. How many sigma and pi bonds are there between C-C atoms of $CaCO_2$?
    Ans: One sigma and two π bonds, because the triple $C \equiv C$ bond is a combination of these.
11. How many sigma and pi bonds are there between C-C atoms of CaC_2
12. CO_2, N_3^-, NO_2^+, are linear.
    Ans: All these have two bond pairs and hence are linear.
13. What do you mean by the term hybridization?
14. What is the type of hybridization of each carbon atom in the following molecule
    $\overset{4}{C}H_3 - \overset{3}{C}H = \overset{2}{C} = \overset{1}{C}H_2$
    Ans: (i) sp^2 (ii) sp (iii) sp^2 (iv) sp^3
15. Indicate the type of hybrid orbitals of the bold atom
    (i) **$BeCl_2$** (ii) **CCl_4** (iii) **C_2F_4** (iv) **SF6**
    Ans: (i) sp (ii) sp^3 (iii) sp^2 (iv) sp^3d^2
16. In benzene, what is the valence bond hybridization on each carbon atom?
    Ans: sp^2
17. Determine the geometry of each of the following molecules and the hybridization about the central atom.
    (i) BeF_2 (ii) NH_3 (iii) $HC \equiv CH$
18. What are the relative energies of sp, sp^2 and sp^3 hybrid orbitals.
    Ans: More is s character lower is the energy. so the energies follow the order
    $$sp^3 > sp^2 > sp$$
19. Can nitrogen use ground state atomic orbitals to form equivalent bonds.
    Ans: Yes. Ground state nitrogen has three half filled atomic orbitals which can overlap with the orbitals of other atoms to form bonds. ($2s^2$, $2p_x^1$, $2p_y^1$, $2p_3^1$)
20. Why do then nitrogen use hybrid orbitals for bonding.
    Ans: Hybrid orbitals give stronger bonds than atomic orbitals.
21. Draw outline sketches to show the formation of sp., sp^2 and sp^3 hybrid orbitals.

22. What type of hybridization is associated with the central atom when the atoms attached to it form
(a) an equilateral triangle
(b) a regular tetrahedron ?
23. Discuss the hybridization in PCl_5 and SF_6
24. What is the state of hybridization in CO^{-3} Ion?
25. Indicate the type of hybridization in the following molecules
(i) $CH - CH_3$ (ii) $H_3C - C \equiv C - CH_3$
(iii) $CH_2 = C = CH_2$ (iv) $CH_3 - C \equiv N$
26. What is the state of hybridization in SO_3 molecule.
Ans: sp^2
27. Differentiate between structural isomers and resonance structures.
Ans: Structural isomers are real molecules whose atoms are linked together in different ways. Whereas resonance structures are not real and are written when a single structure cannot explain all the properties of the molecule
28. What is the difference between a double salt and a coordination compound?
29. Explain the terms (i) Central metal atom (ii) Ligand (iii) Coordination sphere.
30. Discuss the outer and inner octahedral complexes.
31. Discuss the hybridization of $[Fe(CN)_6]^{3-}$, $[Ni(CO)_4]^0$, $[Ni(CN)_4]^{2-}$
32. What do you mean by coordination compounds?

LONG ANSWER QUESTIONS

1. What do you mean by a chemical bond? How do atoms combine? How many types of bonds are there?
2. How is covalent bond formed? Explain the formation of covalent bond between two hydrogen atoms. Why is such a bond not formed between two He atoms?
3. What do you understand by overlap of orbitals? How does it account for the stability of the bond between the two atoms? Give an example for each of the following types of orbital overlap. Also sketch the corresponding diagram leading to the bond formation.
(a) s–s overlap,
(b) s–p overlap
(c) Sideway p–p overlap and
(d) end to end p–p overlap.
4. What are sigma and pi bonds? Explain the different ways of their formation diagrammatically. Which one of them is stronger and why?
5. Explain the terms bond length, bond energy and bond angle.
6. Briefly explain orbital theory of covalent bond formation.
7. (a) What is the difference between coordinate bond and covalent bond?
(b) Under what circumstances can a fully filled orbital participate in bond formation? Discuss with an example.
8. (a) Flourine has seven valence electrons ($2s^2 2p^5$), so seven covalent bonds in principle could form around the atom. Such a compound might be FH_7 or FCl_7. These compounds have never been prepared. Why?
(b) Which of the following hydrogen halides has the most polar molecules and why?

HI, HBr, HCl, HF

(c) What type of chemical bonds are involved in the formation of the following compounds? Explain the bond formation on the basis of electronic structures of atoms involved in each case :

(i) NaCl
(ii) HCl
(iii) NH_4Cl

9. What is VSEPR theory? How does it explain the bond angles observed in CH_4, NH_3 and H_2O molecules?
10. Explain molecular orbital theory and the following observations.
    (i) CCl_4 and $SiCl_4$ are tetrahedral
    (ii) CO_2 molecule is linear and SO_2 is angular.
    (iii) BF_3 is planar but NH_3 is not
    (iv) H–S–H angle in H_2O is closer to 90° then H–O–H angle in H_2O.
11. What is hybridization of orbitals? Draw outline sketches to show the formation of sp, sp^2 and sp^3 hybrid orbitals.
12. (a) Explain why, carbon has valency of four and not two and why are four C–H bonds in methane are identical.
    (b) What type of hybridization is associated with the central atom when the atoms attached to it form
    (i) an equilateral triangle
    (ii) a regular tetrahedron.

❑❑❑

2

CHEMICAL BONDING-II

"The most incomprehensible thing about the world is that, it is comprehensible"

—Albert Einstein

CHAPTER OUTLINE

2.1 MOLECULAR ORBITAL THEORY

According to Valence Bond Theory discussed earlier, a covalent bond is formed by the overlapping of atomic orbitals and the bonded atoms in the molecule are held together by one or more pair of electrons. According to this theory, *the atoms retain their identity even after bond formation.*

On the other hand in the molecular orbital method, the valency electrons are considered to be associated with all the nuclei in the molecule i.e. electrons move in the field of more than one nucleus.

The molecular orbital theory given by R.S. Mulliken and F. Hund incorporates the following points.

1. When nuclei of two atoms come close to each other, their atomic orbitals interact, to form **molecular orbitals** in which the identity of both the atomic orbitals is lost.
2. The number of molecular orbitals is equal to the number of atomic orbitals involved in their formation.
3. Just as atomic orbitals in an atom are quantized, similarly molecular orbitals are also believed to be quantized.
4. The electrons in the atomic orbital are influenced by one nucleus, the electrons in the molecular orbitals are influenced by all the nuclei.
5. Electrons are filled in the molecular orbitals in the same way as they are filled in atomic orbitals, following Aufbau principle, Hund's rule of maximum multiplicity.

6. The molecular orbitals are represented by $\sigma\ \sigma^*\ \pi\pi^*$ etc.

A molecular orbital is constructed by combination of two atomic orbitals and this method is known as Linear Combination of Atomic orbitals (LCAO).

Table 2.1 Difference between Valence Bond Method and Molecular Orbital Method

Valence Bond Method	*Molecular Orbital Method*
1. In valence bond method, a molecule is composed of atoms which retain their individual identity even after bond formation.	In molecular orbital method, the combining atoms lose their individual identity.
2. An electron moves in the field of one nucleus only, i.e. atomic orbitals are monocentric.	An electron moves in the field of more than one nucleus i.e. molecular orbitals are polycentric.
3. Valence bond method does not explain the paramagnetism of O_2 successfully.	Molecular orbital theory explains the paramagnetism of oxygen successfully.
4. It does not involve the formation of bonding and antibonding orbitals.	It involves the formation of bonding and antibonding orbitals.
5. Atomic orbitals in valence bond method are represented by *s, p, d* and *f*.	Molecular orbitals in molecular orbital method are represented by $\sigma\ \sigma^*\ \pi\ \pi^*$ etc.

2.2 LINEAR COMBINATION OF ATOMIC ORBITALS

Electrons may be considered either as particle or waves. An electron in an atom may therefore be described by a wave function ψ, which is a solution to Schrodinger wave equation. According to Molecular Orbital Theory the electrons in a molecule are said to occupy molecular orbitals. The wave function describing a molecular orbital may be obtained by the method of **linear combination of atomic orbitals.**

Consider two atoms A and B which have atomic orbitals described by the wave functions Ψ_A and Ψ_B. The electron clouds of these two atoms overlap when the atoms approach each other. The wave function for the molecular orbital Ψ_{MO} can be obtained by a linear combination of the atomic orbitals Ψ_A and Ψ_B.

$$\Psi_{MO} = C_1\Psi_A + C_2\Psi_B$$

Where C_1 and C_2 are the coefficients chosen to give the lowest-energy for the molecular orbital. For homoatomic systems atoms A and B are same, and hence C_1 and C_2 are equal.

Now the probability of finding an electron in a volume of space dv is $\Psi^2 dv$, so the probability density for the combination of two atoms is given by:

$$\psi^2_{MO} = (C_1\Psi_A + C_2\Psi_B)^2$$

$$\psi^2_{MO} = (C_1^2\Psi_A^2 + 2C_1C_2\Psi_A\Psi_B + C_2^2\Psi_B^2)$$

The term $C_1^2\Psi_A^2$ gives the probability of finding an electron on atom A, if A is an isolated atom. The term $C_1^2\Psi_B^2$ gives the probability of finding an electron on atom B, if B is an isolated atom. The middle term $2C_1C_2\Psi_A\Psi_B$ is called the *overlap integral* as this term represents the main difference between the electron clouds in individual atoms and in the molecule. The larger this term the stronger the bond.

Overlap has a positive sign when the superimposed regions of the two orbitals have the same sign i.e. both (+) or both (-) (same phase). Overlap has a negative sign when the superimposed regions of the two orbitals have opposite signs (different phase). Zero overlap results when there are equal regions of overlap with opposite signs. Let us discuss this in more detail by considering the overlap of different orbitals.

Combination of s-s orbitals

Consider a simple case of combination of 1*s* atomic orbital of one hydrogen atom (A) with 1*s* atomic orbital of another hydrogen atom (B). Let wave function Ψ_A and Ψ_B represent the 1*s* atomic orbitals of the atoms A and B respectively. There are two possible ways of linearly combining the atomic orbitals.

(i) Additive combination-where the signs of the two wave functions are the same.*

(ii) Subtractive combination-where the signs of the two wave functions are different.* (Fig. 2.1)

Wave functions which have the same sign may be regarded as waves that are in phase, which when combined add up to give a larger resultant wave. Similarly wave functions of different signs correspond to waves that are completely out of phase and which cancel each other by destructive interfer-ence. Thus two combinations are —

$$\Psi_{(g)} = \Psi_A + \Psi_B$$

$$\Psi_{(u)} = \Psi_A - \Psi_B$$

The molecular orbital wave functions are designated by $\Psi_{(g)}$ and $\Psi_{(u)}$: *g* stands for *gerade* (even) and *u* for *ungerade* (odd). Actually the terms g and u refer to the symmetry of the orbital about its centre. If the sign of the lobe remains unchanged when the orbital is rotated about the line joining the two nuclei and then about a line perpendicular to this, then, the orbital is termed as gerade and if the sign changes, the orbital is termed as ungerade.

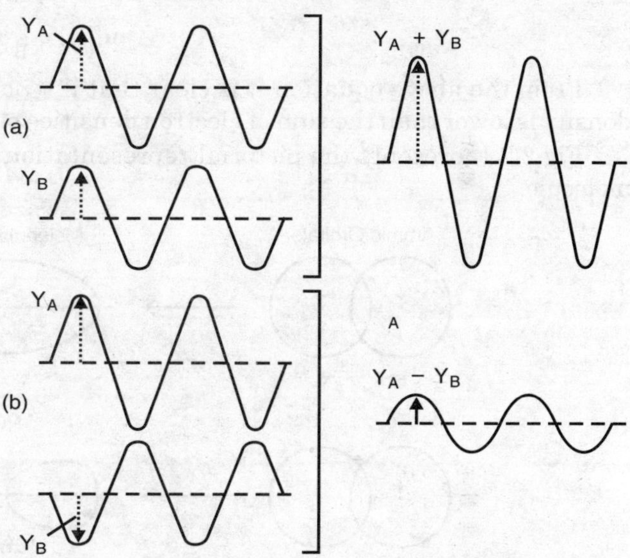

Fig. 2.1 (a) Additive effect of the electron waves in phase.
(b) Subtractive effect of the electron waves out of phase.

Bonding Molecular Orbital

The molecular wave function $\Psi_{(g)}$ leads to increased electron density in between the nuclei and is therefore a *Bonding Molecular Orbital*. The probability density of electron cloud can be obtained by squaring the wave function $\Psi_{(g)}$.

$$\Psi_g^2 = (\Psi_A + \Psi_B)^2$$

$$\Psi_g^2 = \Psi_A^2 + \Psi_B^2 + 2\Psi_A\Psi_B$$

---

* *One of the wave function Ψ_A is arbitrarily assigned a positive sign, the other may be either +ive or –ive. The signs + and – refer to signs of the wave functions, which determine their symmetry and have nothing to do with electrical charges.*

From the above equation, overlap criteria becomes clear. In a region where two orbitals Ψ_A and Ψ_B have positive overlap, the electron density is higher than the mere sum of the electron densities of the two separate orbitals i.e. $(\Psi_A + \Psi_B)^2$ is greater than $\psi^2{}_A + \psi^2{}_B$ by $2\,\psi_A\,\psi_B$. The attraction of both nuclei for these electrons is greater than the mutual repulsion of the nuclei. A net attractive force or bonding interaction results in the formation of stable chemical bond.

Antibonding Molecular Orbial

Conversely $\psi_{(u)}$ results in two lobes of opposite signs giving zero electron density in between the nuclei and hence is called the *antibonding molecular orbital*. The probability density of electron cloud in this case also is obtained by squaring the wave function ψ_u

$$\Psi_u^2 = (\Psi_A - \Psi_B)^2$$

$$= \Psi_A^2 + \Psi_B^2 - 2\Psi_A\Psi_B$$

From the above equation it is clear that if ψ_A and ψ_B have negative overlap, the electron density is lower than the sum of electron densities of the two separate orbitals.

Fig 2.2 represents the pictorial representation of bonding and antibonding orbitals of H_2 molecule.

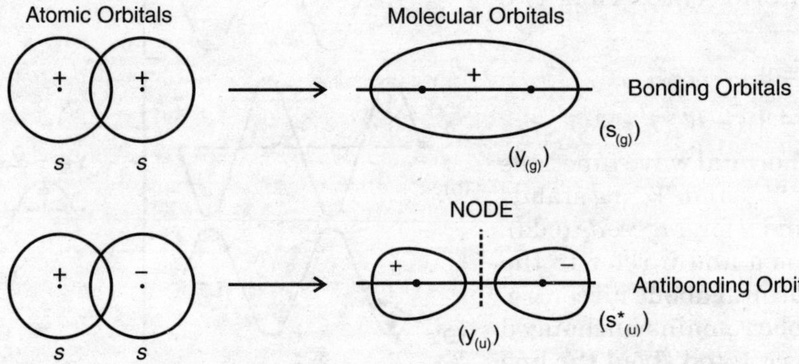

Fig 2.2 *s-s* combination of atomic orbitals

The energy of the bonding molecular orbital is lower than that of the atomic orbital by an amount Δ. This is known as **stabilization energy**. Similarly the energy of the antibonding molecular orbital is increased by Δ. The relative energies of the two molecular orbitals and the atomic orbitals are shown in Fig. 2.3.

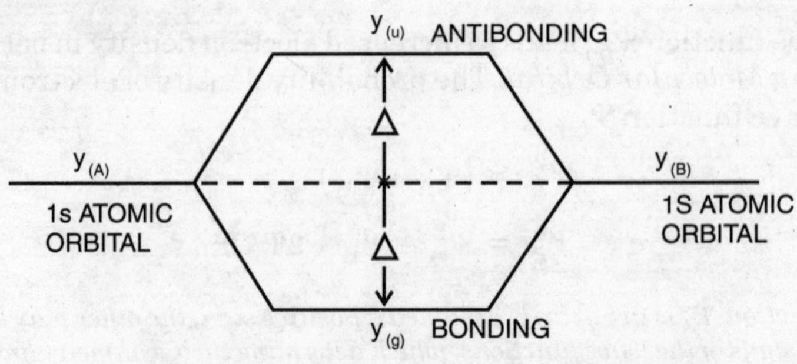

Fig. 2.3 Molecular Orbitals (Bonding and Antibonding)

Further, since the overlap of s-s orbital is along the axis joining the nuclei, the resulting molecular orbitals are designated as σ orbital. As a convention the antibonding molecular orbital is designated with an asterisk σ^*. Thus the two orbitals formed in the case of two combining hydrogen atoms, are designated as σ_{1s} and σ^*_{1s}.

Table 2.2 Difference between Bonding and Antibonding Molecular Orbitals

Bonding Molecular Orbital	*Antibonding Molecular Orbital*
1. They are formed by the additive effect of the atomic orbitals $\Psi_{MO} = \Psi_1 + \Psi_2$	They are formed by the subtractive effect of the atomic orbitals $\Psi_{MO} = \Psi_1 - \Psi_2$
2. It increases the electron density between the nuclei	It decreases the electron density between the nuclei
3. These orbitals have lower energy than the atomic orbitals from which they are formed.	These orbitals have higher energy than the atomic orbitals from which they are formed.
4. Bonding molecular orbitals stabilize the molecule.	Antibonding molecular orbitals destabilize the molecule.
5. These are represented by σ, π etc.	These are represented by σ^*, π^* etc.

Combination of s-p orbitals

An *s* orbital may combine with a *p* orbital provided the lobes of *p* orbital are directed along the internuclear axis. When the lobes, which are overlapping have the same sign, bonding molecular orbitals are formed with an increased electron density between the nuclei. The energy of this molecular orbital is lower than the energy of the atomic orbitals which have gone into its formation. The corresponding antibonding orbital is formed when the lobes of opposite signs overlap, with a reduced electron density in between the nuclei Fig. 2.4.

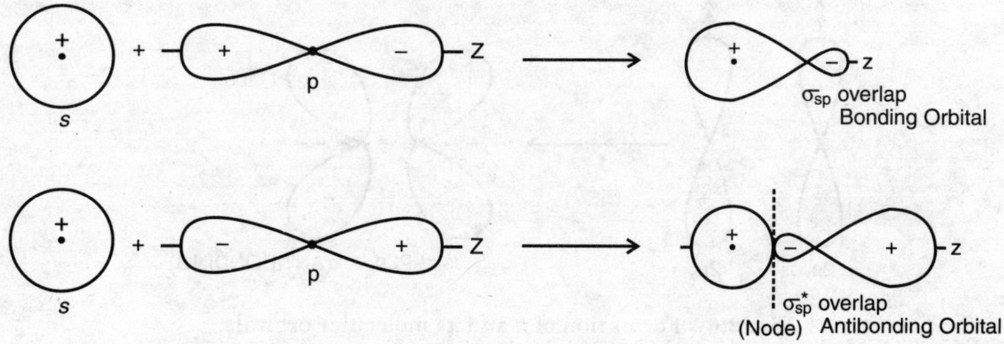

Fig. 2.4 Combination of *s-p* orbitals

Combination of p-p orbitals

The overlap of the two *p* orbitals, which have lobes pointing along the axis joining the nuclei results in σ_{2p} bonding σ^*_{2p} antibonding orbitals.

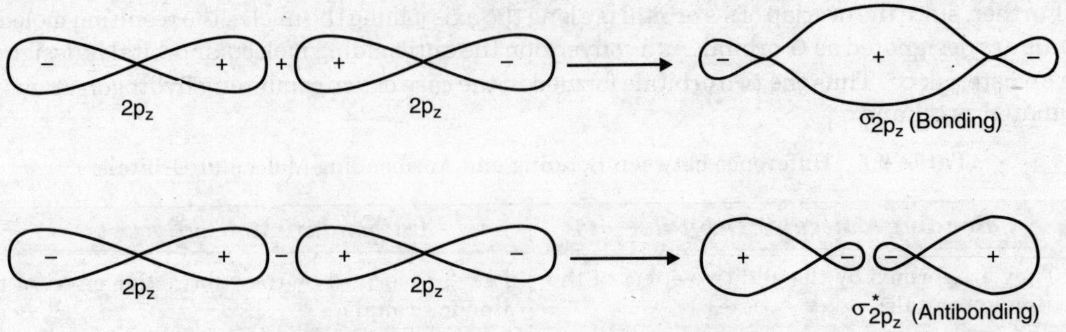

Fig. 2.5 Combination of *p-p* orbitals

Electron charge density between the two nuclei in σ^*_{2pz} is enhanced whereas the electron density in σ^*_{2pz} gets reduced.

The overlap between the two *p* orbitals, which both have lobes perpendicular to the axis joining the nuclei, is lateral. This results in the formation of **π bonding and π* antibonding molecular orbitals**.

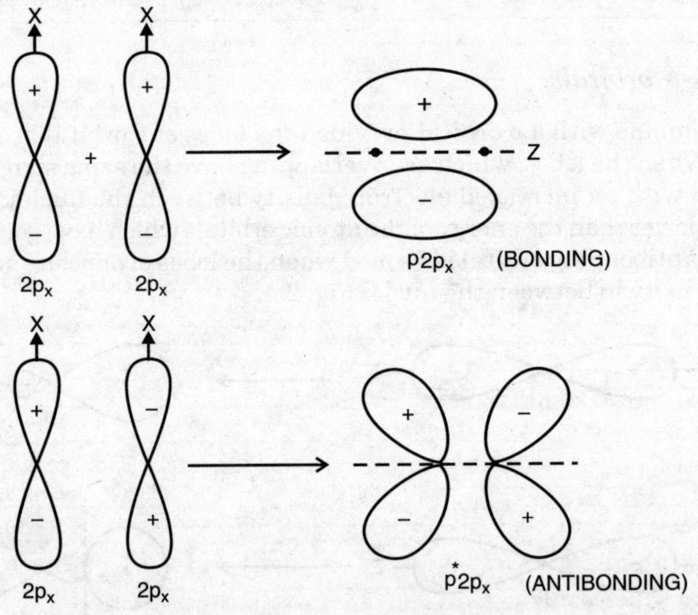

Fig. 2.6 Formation of π and π* molecular orbitals.

It is to be noted that for π molecular orbitals, ψ is zero along the internuclear line and consequently the electron density ψ^2 is also zero. This is in contrast to σ orbitals.

Combination of p-d Orbitals

A *p* orbital on one atom may overlap with a *d* orbital on another atom forming bonding and antibonding orbitals. Since the overlap is lateral and hence molecular orbitals are designated as π orbitals.

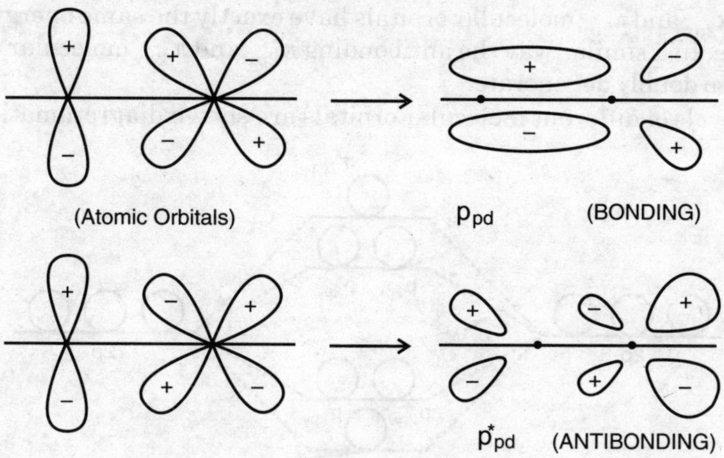

Fig. 2.7 *p-d* combination of atomic orbitals

Combination of d-d Orbitals

Two *dxy* or two dx^2-y^2 orbitals may overlap to form molecular orbitals.The molecular orbitals formed by the combination of *d-d* orbitals are designated as δ and δ* for bonding and antibonding molecular orbitals respectively.

Criteria for the Combination of Atomic Orbitals

Any two atomic orbitals will not essentially combine to form molecular orbitals. There are certain rules which are followed for their combination. The rules are:

1. The atomic orbitals which are involved in the formation of molecular orbitals should be approximately of the same energy.
2. The atomic orbitals must overlap to a considerable extent to form stable bonds. The overlap increases the electron charge density in between the two nuclei.
3. The atomic orbitals must have the same symmetry with respect to the molecular axis. Out of the three p orbitals which are designated as *px*, *py* and *pz* orbitals, *pz* orbitals are considered as pointing along the internuclear axis by convention. Thus *px* or *py* orbital of one atom does not combine with *s* orbital of the other atom. In fact ++ overlap is cancelled by +– overlap.

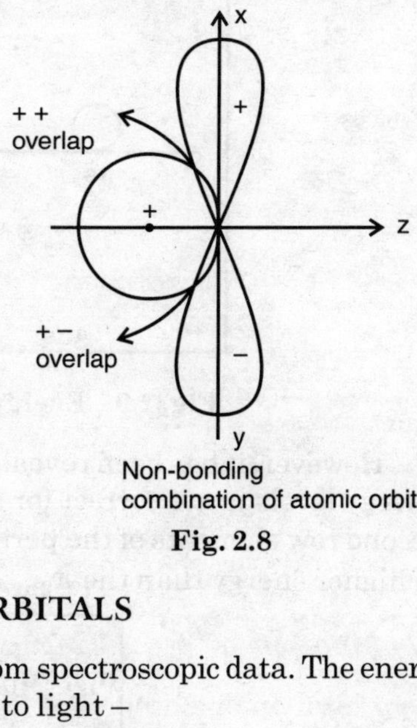

Non-bonding combination of atomic orbitals.

Fig. 2.8

2.3 ENERGY LEVEL DIAGRAM FOR MOLECULAR ORBITALS

The order of energy of molecular orbitals has been determined from spectroscopic data. The energy generally increases in the following order as we move from left to light –

$$\sigma_{1s},\ \overset{*}{\sigma}_{1s},\ \sigma_{2s},\ \overset{*}{\sigma}_{2s},\ \sigma_{2pz},\ \begin{Bmatrix}\pi_{2px}\\ \pi_{2py}\end{Bmatrix}\ \begin{Bmatrix}\overset{*}{\pi}_{2px}\\ \overset{*}{\pi}_{2py}\end{Bmatrix}\ \overset{*}{\sigma}_{2pz}$$

$\xrightarrow{\hspace{2em}}$ *increasing energy*

The bonding π_{2px} and π_{2py} molecular orbitals have exactly the same energy and are said to be double degenerate. In a similar way the antibonding π^*_{2px} and π^*_{2py} molecular orbitals have same energy and are also doubly degenerate.

The energy levels of different molecular orbitals are shown diagrammatically in Fig. 2.9.

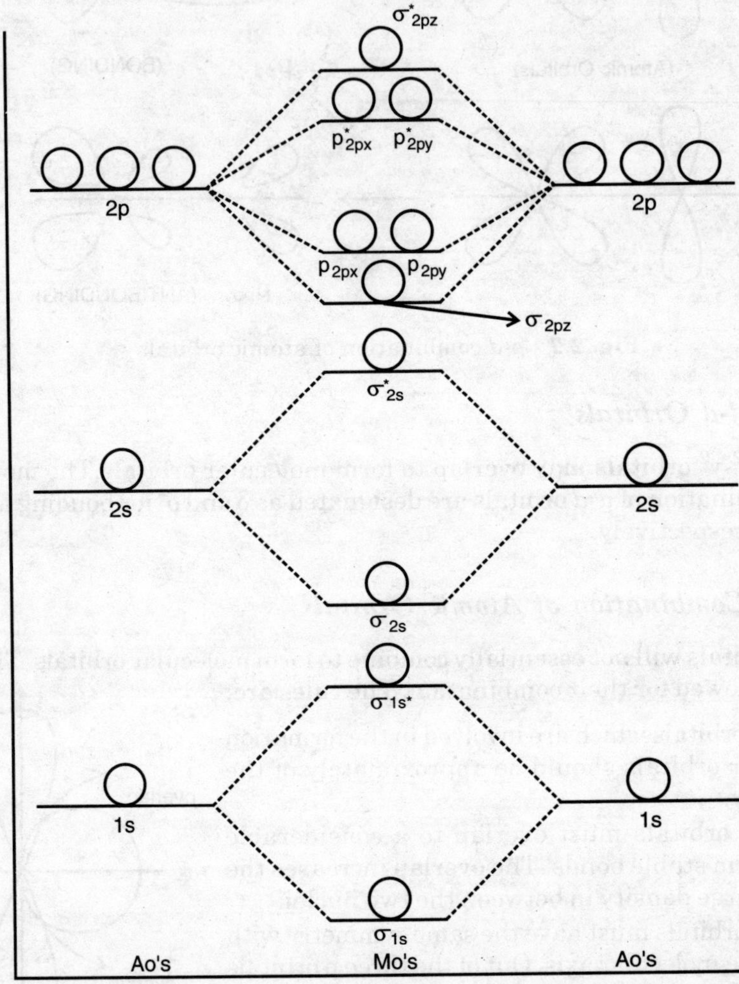

Fig. 2.9 Energy levels of different molecular orbitals of O_2, F_2, Ne_2.

However, it has been revealed by spectroscopic studies that the above mentioned sequence of energy levels is not correct for all molecules. In the case of homonuclear diatomic molecules of second row elements of the periodic table upto N_2 (viz. Li_2, Be_2, B_2, C_2, N_2), the σ_{2pz} orbitals are of higher energy than the π_{2px} and π_{2py} molecular orbitals. For these atoms the order is:

$$\sigma_{1s},\ \overset{*}{\sigma}_{1s},\ \sigma_{2s},\ \overset{*}{\sigma}_{2s},\ \left\{\begin{matrix}\pi_{2px}\\ \pi_{2py}\end{matrix}\right.\ \sigma_{2pz},\ \left\{\begin{matrix}\pi^*_{2px}\\ \pi^*_{2py}\end{matrix}\right.\ \sigma^*_{2pz}$$

$\longrightarrow$ *increasing energy*

The energy level diagram for these molecules is given in Fig. 2.10.

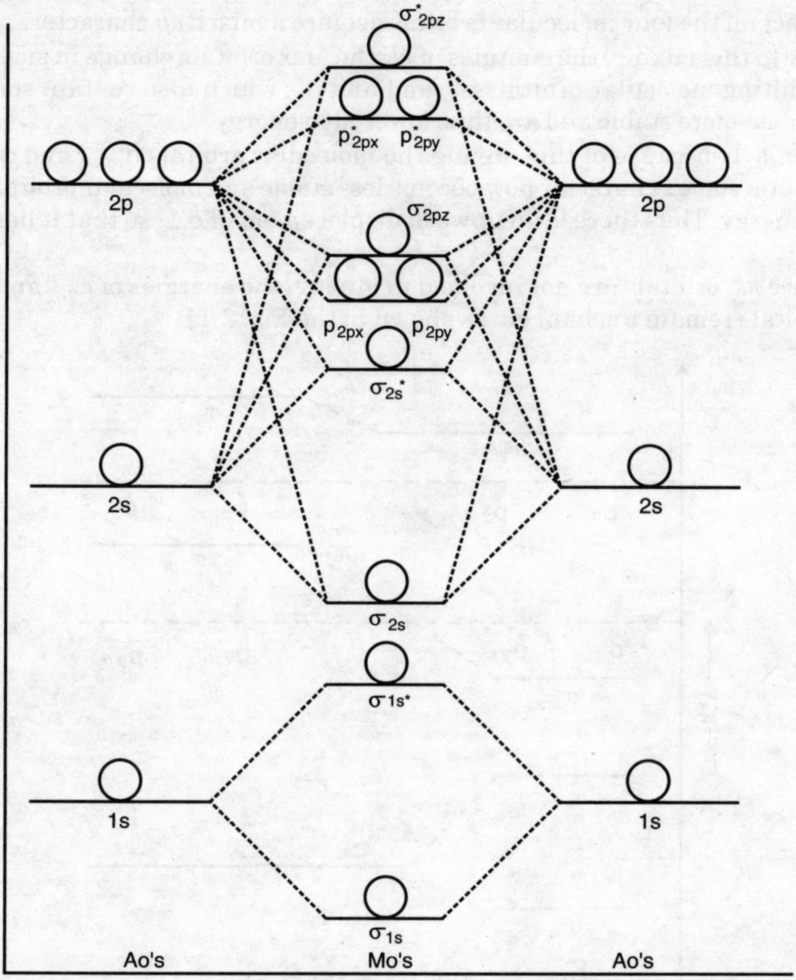

Fig. 2.10 Energy levels of different molecular orbitals in the case of lighter elements.

Now the question arises that why the order of energy of molecular orbitals is different for molecules lighter than F_2, viz., N_2, C_2, B_2 and Li_2 This can be explained as follows:

- In case of homonuclear diatomic molecules of second row elements, if the energy difference between 2s and 2p atomic orbital is **large**, then the 2s orbital of one atom interacts slightly with the 2pz orbital of the other atom. Though the symmetry is proper for overlap, but the energy difference between the two orbitals is so great that the overlap is not effective. Thus σ_s molecular orbitals are composed of the two 2s atomic orbitals and the order of energy levels is as shown in Fig. 2.11 This is true for diatomic molecules of O_2 and F_2.
- In case of elements upto nitrogen, in the second row, the energy difference between 2s and 2p atomic orbitals is **small** i.e. 2s orbital of one atom is close to 2pz orbital of second atom. As the two orbitals are of same symmetry their overlap cannot be ignored. As a result of interaction between *s* orbital of one atom with the *p* orbital of the other and vice versa, σ_{2s} and σ^*_{2s} molecular orbital **do not retain their pure *s* character** and σ_{2pz} and σ^*_{2pz} molecular orbitals **do not retain their pure *p* character**.

— Infact all the four molecular orbitals acquire a mixed *sp* character.
— Due to this mixing, the energies of all the four orbitals change in such a way that the resulting molecular orbitals σ_{2s} and and σ^*_{2s} which also contain some *p* character, become more stable and are thus lowered in energy.
— Similarly because of this mixing the molecular orbitals $\sigma_{(2pz)}$ and $\sigma^*_{(2px)}$ which also contain some *s* character now become less stable and molecular orbital are thus raised in energy. Thus there is an upward displacement of σ_{2pz} so that it lies above π_{2px} and π_{2py}.
— Since π_{2p} orbitals are not involved in mixing, the energies of π_{2px} and σ_{2py} molecular orbitals remain unchanged, as shown in the Fig. 2.11.

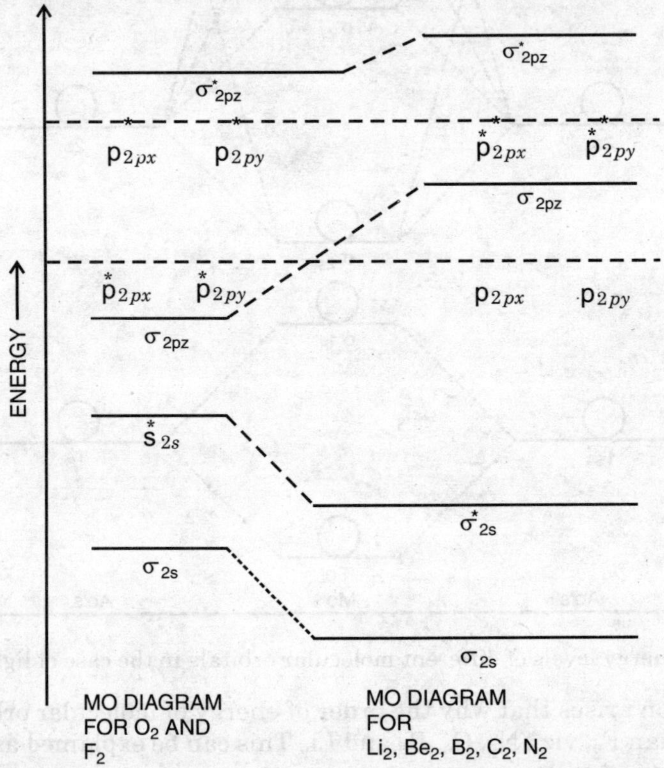

Fig. 2.11 The change in energies of MOs due to mixing of orbitals of the same symmetry.

- Considering the above points it can be concluded that the difference in 2*s* and 2*p* atomic orbitals of atoms up to nitrogen are small and hence these would overlap effectively resulting in the orbitals, the energies of which will be different than pure σ_{2s} and σ_{2pz} molecular orbitals. Same explanation is applicable for antibonding σ_{2s} and σ^*_{2px} orbitals. On the contrary, in the case of O_2 and F_2 the energy difference between 2*s* and 2*p* is quite large and hence no overlapping between these atomic orbitals is possible. Therefore energies of σ_{2s}, σ^*_{2s}, σ_{2pz}, σ^*_{2pz} remain unaltered and hence σ_{2pz} molecular orbital has lower energy than the energies of π_{2px} and π_{2py} molecular orbitals.

2.4 FILLING OF ELECTRONS IN MOLECULAR ORBITALS

For simple homoatomic molecules, the filling of electrons are carried out in a similar fashion as in Atomic Orbitals. Following rules are followed:

1. ***Aufbau Principle:*** The orbital with lowest energy is to be filled first of all i.e. electrons enter various molecular orbitals in the order of their increasing energies.
2. ***Pauli's Exclusion Principle:*** The maxmimum number of electrons in a molecular orbital cannot exceed two and these should have opposite spins.
3. ***Hund's Rule:*** Pairing of electrons takes place only when each molecular orbital of the same energy has at least one electron each.
4. A bond between two atoms is said to be formed when the total number of electrons in bonding orbitals is greater than the numbers in antibonding orbitals.
5. ***Bond order:*** Bond order is defined as half of the difference between the number of electrons present in the bonding and the antibonding orbitals i.e.

$$\text{Bond order} = \frac{1}{2}(\text{No. of bonding orbitals–No. of antibonding orbitals}).$$

A negative or zero bond order corresponds to an unstable molecule whereas positive bond order indicates a stable molecule.

2.5 HOMONUCLEAR DIATOMIC MOLECULES

1. ***Hydrogen molecule-ion H_2^+:*** This is the simplest molecular specie containing one electron only. Its existence has been detected spectroscopically when electric discharge is passed through hydrogen gas under reduced pressure. The only electron present in H_2^+ ion enters the σ_{1s} molecular orbital. Thus, electronic configuration of H_2^+ is written as $\sigma_{(1s)}{}^1$ and bond order = $\frac{1}{2}(1-0) = \frac{1}{2}$
2. ***Hydrogen molecule H_2:*** Each hydrogen atom contains 1 electron and hence there are two electrons in the molecule, occupying the lowest energy level. Thus, the electronic configuration of H_2 will be σ_{1s}^2 and bond order = $\frac{1}{2}(2-0) = 1$.
3. ***Hydrogen molecular negative ion H_2^-:*** This is formed by the combination of one hydrogen atom having one electron in 1s orbital with a hydrogen ion having 2 electrons. Thus H_2^- ion has 3 electrons. The electronic configuration will thus be $\sigma_{1s}^2\ \overset{*1}{\sigma_{1s}}$ and bond order = $\frac{1}{2}(2-1) = \frac{1}{2}$.

Table 2.3 Characteristics of H_2^+, H_2, H_2^-

Specie	*Electronic config.*	*Bond Order*	*Magnetic character*
H_2^+	σ_{1s}^1	$\frac{1}{2}$	Paramagnetic
H_2	σ_{1s}^2	1	Diamagnetic
H_2^-	$\sigma_{1s}^2\ \overset{*1}{\sigma_{1s}}$	$\frac{1}{2}$	Paramagnetic

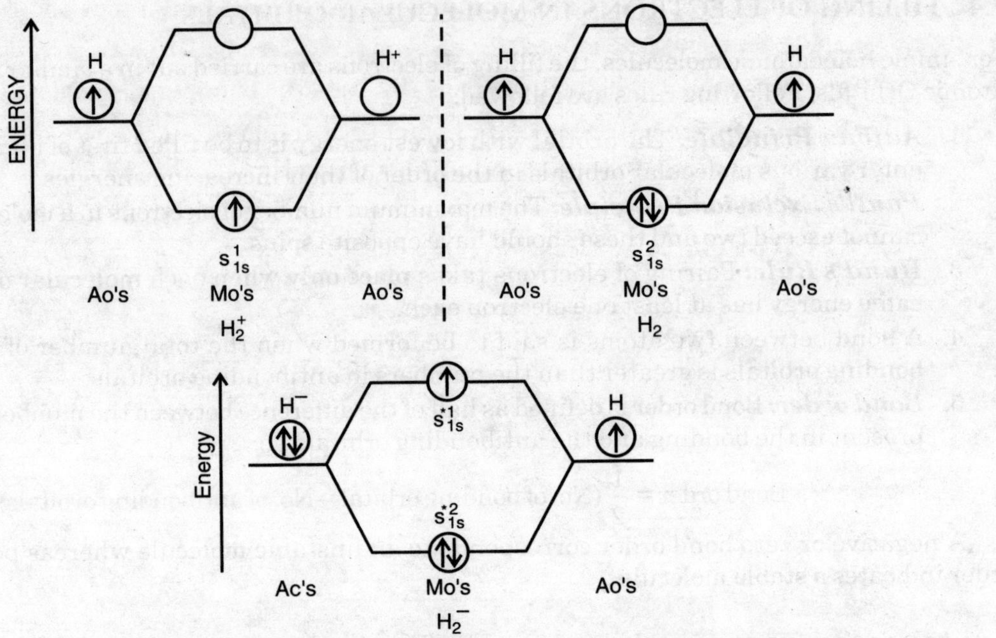

Fig. 2.12 Energy Level Diagrams for H_2^+ H_2 and H_2^-

4. ***Helium Molecule ion He_2^+*:** This may be considered to be formed from He atom and a He^+ ion. In the molecular ion there are three electrons which are arranged in the molecular orbitals as σ_{1s}^2, $\overset{*}{\sigma}{}_{1s}^{1}$. Thus helium molecular ion can exist, but it is not very stable. Bond order = $\frac{1}{2}(2-1) = \frac{1}{2}$.

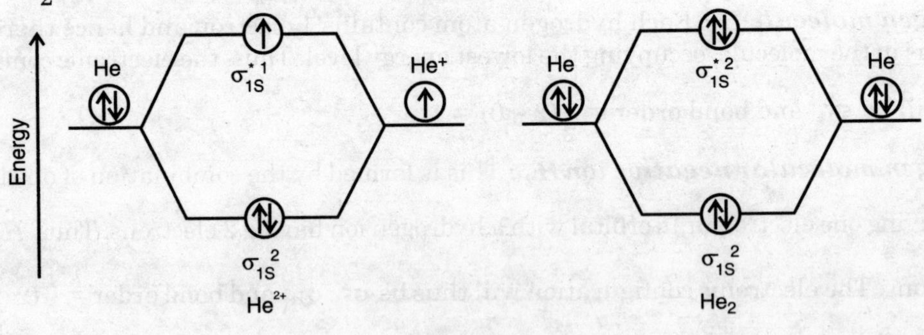

Fig. 2.13

5. ***Helium Molecule, He_2*:** The electronic configuration of the atom is $1s^2$. *He* molecule can be considered to form from two atomic orbitals. In the molecular He_2, there are four electrons which are arranged as σ_{1s}^2, $\overset{*}{\sigma}{}_{1s}^{2}$. Thus, the stabilizing effect of bonding orbitals is cancelled out by the destabilizing effect of antibonding orbitals and the molecule does not exist. Bond order = $\frac{1}{2}(2-2) = 0$.

6. ***Lithium Molecule Li_2:*** Each Li atom has two electrons in its inner shell and one in the valence shell. Thus in Li_2, a total of six electrons are present. These are arranged as σ_{1s}^2, $\overset{*}{\sigma}_{1s}^2$, σ_{2s}^2. The inner complete shells do not contribute to the bonding and molecular electron i.e. configuration can be written as $KK\sigma_{2s}^2$ where KK implies that K shell of the two atoms is complete and do not contribute to bonding much. Bond order of lithium molecule comes out to be $\frac{1}{2}(4-2)=1$

Lithium molecules do exist in the vapour state but solid state of lithium is energetically more favourable.

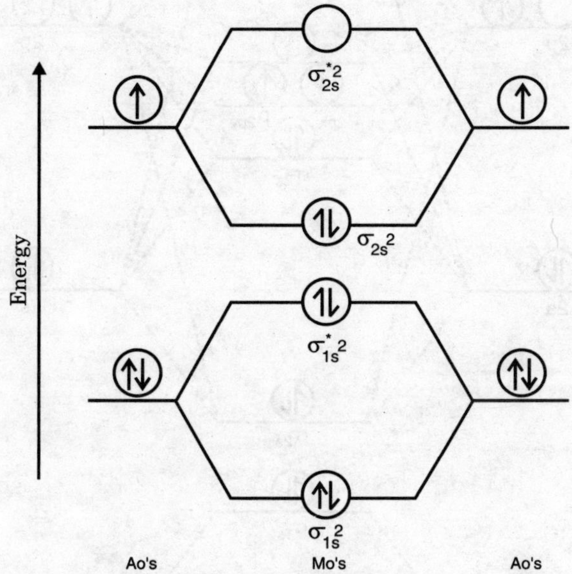

Fig. 2.14 Energy level diagram for Li_2.

7. ***Beryllium Molecule Be_2:*** Beryllium atom has 2 electrons in the valence shell. Hence in the formation of the diatomic molecule, electrons from the valence shell of each atom i.e. a total of four electrons are involved. These four electrons can be arranged as $KK\sigma_{2s}^2\ \overset{*}{\sigma}_{2s}^2$. Thus, it is apparent that the effects of the bonding σ_{2s} and antibonding $\overset{*}{\sigma}_{2s}$ levels cancel each other and hence the molecule does not exist. Bond order = $\frac{1}{2}(4-4)=0$. This zero bond order further suggests that existance of Be_2 molecule is not possible.

8. ***Boron Molecule B_2:*** The valence shell of each atom of boron contains 3 electrons (Z = 5). When two atoms of boron combine to form B_2 molecule the molecule contains a total of 10 electrons. These are arranged in the molecular orbitals as- $KK\,\sigma_{2s}^2\ \overset{*}{\sigma}_{2s}^2 \begin{cases}\pi 2p1_y \\ \pi 2p1_x\end{cases}$. It may again be pointed out here that boron is a light atom and order of energies of molecular orbits is different from the usual arrangement i.e π_{2px} and $2\pi_{2py}$ orbitals are at lower energy level than the σ_{2pz} orbital for the reason explained in section 2.3. Since π_{2px} and π_{2py} are degenerate,

the electrons are filled according to Hund's rule i.e each of these orbitals is singly occupied. The order of the energies of molecular orbitals as given above is supported by the experimental facts also. Boron molecule is found to be paramagnetic in nature which can be very well explained on the basis of this order, as there are two unpaired electrons one each in π_{2px} and π_{2py} molecular orbitals.

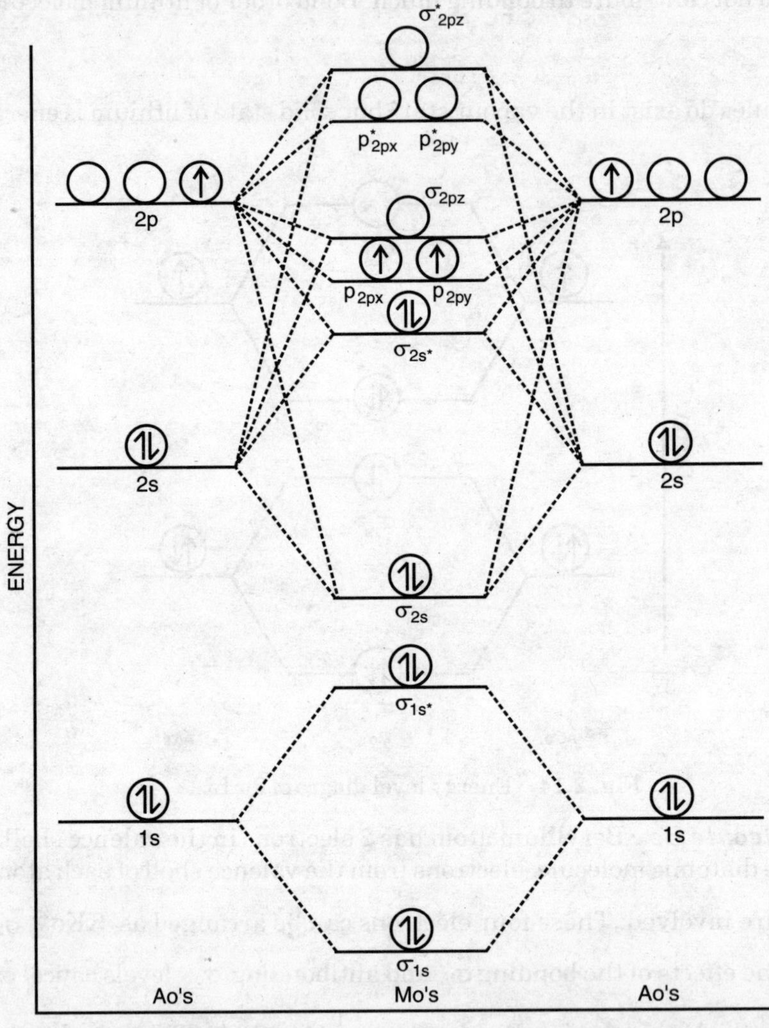

Fig. 2.15 Energy level diagram for Boron Molecule.

9. ***Carbon Molecule C_2:*** Carbon atom has 6 electrons. In the formation of C_2 molecule, a total of 12 electrons have to be accommodated in the molecular orbitals. These are arranged as:

KK $\sigma_{2s}^2\ \overset{*2}{\sigma}_{2s}\begin{cases}\pi_{2py}^2\\ \pi_{2px}^2\end{cases}$. Bond order $= \frac{1}{2}(8 - 4) = 2$. Thus the molecule should be stable and diamagnetic because all the electrons are paired. Similar to the case of B_2 molecule, in C_2 also reversal of energies of σ_{2pz} and π_{2px}, π_{2py} orbitals take place, due to the mixing of the 2s and

2p orbitals. Though it appears that C_2 molecule is stable, but carbon exists as macro molecule in graphite and diamond. These are even more stable than C_2, hence in preference to C_2, diamond and graphite are formed.

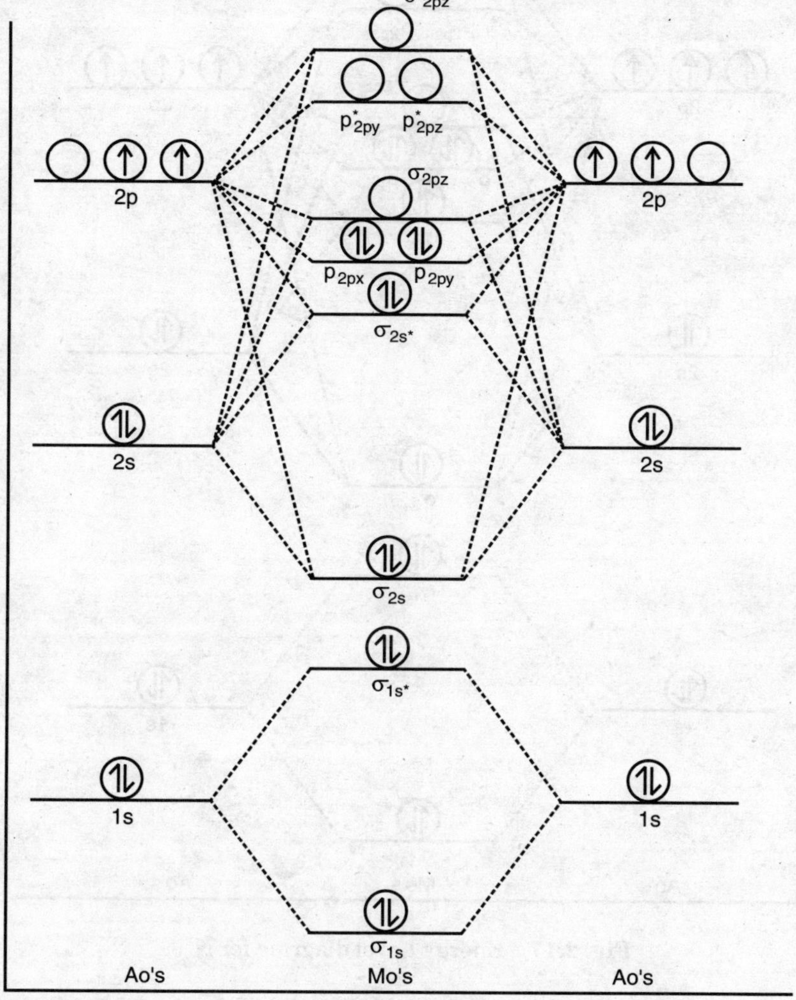

Fig. 2.16 Energy level diagram for C_2.

10. *Nitrogen Molecule N_2:* Nitrogen atom has seven electrons. In the formation of N_2 molecule, a total of 14 electrons are arranged in molecular orbitals as:

$$\text{KK}\ \sigma_{2s}^2\ \overset{*2}{\sigma}_{2s} \begin{cases} \pi_{2py}^2 \\ \pi_{2px}^2 \end{cases},\ \sigma_{2pz}^2\ .\ \text{Bond order} = \frac{1}{2}(10-4) = 3.$$

Thus nitrogen molecule contains a triple bond and is highly stable therefore nitrogen exist as N_2 molecule. Moreover all the electrons are paired therefore it is diamagnetic.

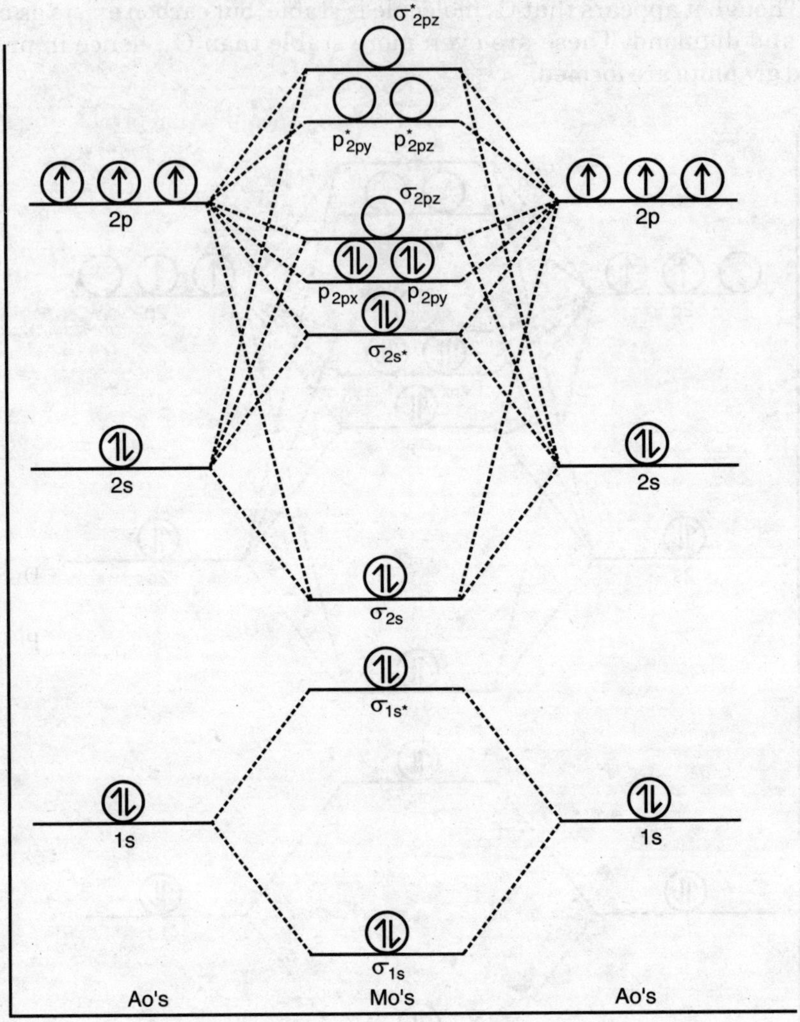

Fig. 2.17 Energy Level diagram for N_2.

N_2^+ ion: N_2^+ ion has one electron less than N_2 molecule. This electron will be lost from σ_{2pz} orbital. Hence the molecular electronic configuration will be –

$$N_2^+ = \text{KK}\ \sigma_{2s}^2\ \sigma_{2s}^2\ \pi_{2px}^2 = \pi_{2py}^2\ \sigma_{2pz}^1$$

$$\text{Bond order} = \frac{1}{2}(7-2) = 2\frac{1}{2}$$

The ion is paramagnetic due to the presence of unpaired electron.

N_2^- ion: N_2^+ ion has or one more electron than the nitrogen molecule. This extra electron is added in $\overset{*}{\pi}_{2px}\ \overset{*}{\pi}_{2py}$ orbital, thus molecular ion electronic configuration will be-

$$N_2^+ = \text{KK}\ \sigma_{2s}^2\ \overset{*2}{\sigma}_{2s}\ \pi_{2px}^2 = \pi_{2py}^2\ \sigma_{2pz}^2\ \overset{*1}{\pi}_{2px}$$

$$\text{Bond order} = \frac{1}{2}(8-3) = 2\frac{1}{2}$$

Moreover, as it contains one unpaired electron the ion is paramagnetic.

N_2^{2-} ion: N_2^{2-} ion has two more electrons than nitrogen molecule. These two extra electrons are added one each in $\overset{*}{\pi}_{2px}$ and $\overset{*}{\pi}_{2py}$ orbitals. Electronic configuration of N_2^{2-} ion can be written as:

$$N_2^{2-} = KK\ \sigma_{2s}^2\ \overset{*}{\sigma}{}_{2s}^{2} \begin{cases} \pi_{2px}^2 \\ \pi_{2py}^2 \end{cases} \sigma_{2pz}^2 \begin{cases} \overset{*}{\pi}{}_{2px}^{1} \\ \overset{*}{\pi}{}_{2py}^{1} \end{cases}$$

$$\text{Bond order} = \frac{1}{2}(8-4) = 2$$

Further it contains two unpaired electrons, it is paramagnetic.

Table 2.4 Comparision of N_2, N_2^+, N_2^- and N_2^{2-}

Molecular Specie	*Electronic Configuration*	*Bond Order*	*Magnetic Character*
N_2	KK $\sigma_{2s}^2\ \overset{*}{\sigma}{}_{2s}^{2}\ \pi_{2px}^2\ \pi_{2py}^2\ \sigma_{2pz}^2$	3	Diamagnetic
N_2^+	KK $\sigma_{2s}^2\ \overset{*}{\sigma}{}_{2s}^{2}\ \pi_{2px}^2\ \pi_{2py}^2\ \sigma_{2pz}^1$	$2\frac{1}{2}$	Paramagnetic
N_2^-	KK $\sigma_{2s}^2\ \overset{*}{\sigma}{}_{2s}^{2}\ \pi_{2px}^2\ \pi_{2py}^2\ \sigma_{2pz}^2\ \overset{*}{\pi}{}_{2px}^{1}$	$2\frac{1}{2}$	Paramagnetic
N_2^{2-}	*KK* $\sigma_{2s}^2\ \overset{*}{\sigma}{}_{2s}^{2}\ \pi_{2px}^2\ \pi_{2py}^2\ \sigma_{2pz}^2\ \overset{*}{\pi}{}_{2px}^{1}\ \overset{*}{\pi}{}_{2py}^{1}$	2	Paramagnetic

Stability of these species is in the order:

$$N_2 > N_2^+ = N_2^- N_2^{2-}$$

Oxygen Molecule O_2: Each oxygen atom has 6 valence electrons. Assuming that inner orbitals do not participate in bonding, altogether 12 electrons have to be accomodated in molecular orbitals. The molecular electronic configuration will be:

$$O_2 = KK\ \sigma_{2s}^2\ \overset{*}{\sigma}{}_{2s}^{2}\ \sigma_{2pz}^2 \begin{cases} \pi_{2py}^2 \\ \pi_{2px}^2 \end{cases} \begin{cases} \overset{*}{\pi}{}_{2py}^{1} \\ \overset{*}{\pi}{}_{2px}^{1} \end{cases}$$

In the case of oxygen, the difference between the energies of 2s and $2p_z$ atomic orbitals is quite large (1430 kJ/mol) and hence interaction between these orbitals and subsequent *changes in order of energies of molecular orbitals do not take place.* Thus, the usual order of energies as given in section 2.3 is followed. As is evident from the energy level diagram that two unpaired electrons are present, thus the molecule is paramagnetic. This is one of the great successes of molecular orbital theory, as it could predict the paramagnetism of oxygen molecule which could not be explained on the basis of valence bond theory.

Bond order = (10-6)/2 = 2. Thus a double bond is present in the molecule.

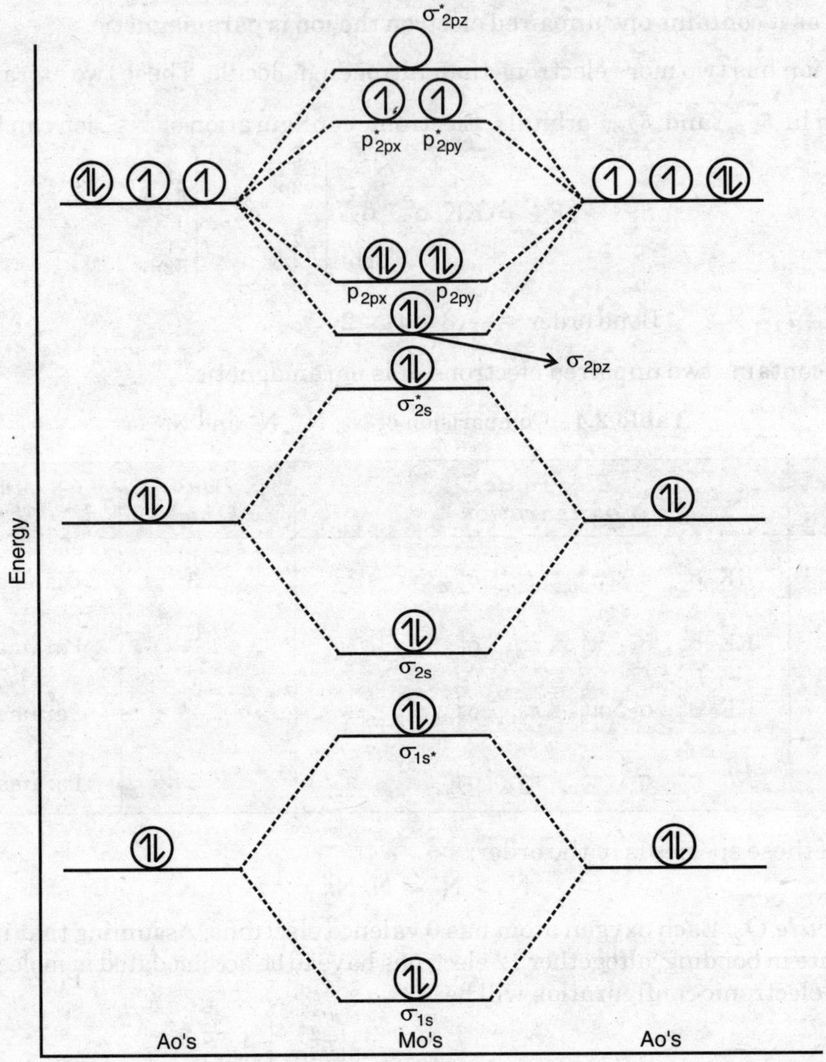

Fig. 2.18 Energy level diagram for O_2 molecule

O_2^- ion: These superoxide ions O_2^- has 17 electrons, i.e., one electron more than O_2 and hence this extra electron enters the π^*_{2px} or π^*_{2py} orbital. The molecular orbital electron configuration becomes:

$$= \text{KK}\ \sigma_{2s}^2\ \sigma_{2s}^{*2}\ \sigma_{2pz}^2 \begin{cases} \pi_{2y}^2 \\ \pi_{2px}^2 \end{cases} \begin{cases} \overset{*1}{\pi}_{2py} \\ \overset{*2}{\pi}_{2px} \end{cases}$$

$$\text{Bond order} = (10 - 7)/2 = 1\frac{1}{2}$$

The superoxide ion is paramagnetic.

O_2^{2-} ion: The peroxide ion O_2^{2-} has 18 electrons and has molecular orbital configuration as:

$$= KK\ \sigma_{2s}^2\ \overset{*}{\sigma}{}_{2s}^{2}\ \sigma_{2pz}^2 \begin{cases} \pi_{2py}^2 \\ \pi_{2px}^2 \end{cases} \begin{cases} \overset{*}{\pi}{}_{2py}^{2} \\ \overset{*}{\pi}{}_{2px}^{2} \end{cases}$$

Bond order = 10 – 8/2 = 1.

O_2^+ ion: This ion has 15 electrons. The molecular orbital configuration can be written as:

$$= KK\ \sigma_{2s}^2\ \overset{*}{\sigma}{}_{2s}^{2}\ \sigma_{2pz}^2 \begin{cases} \pi_{2py}^2 \\ \pi_{2px}^2 \end{cases} \begin{cases} \overset{*}{\pi}{}_{2py}^{0} \\ \overset{*}{\pi}{}_{2px}^{1} \end{cases}$$

$$\text{Bond Order} = \frac{1}{2}(8-3) = 2\frac{1}{2}.$$

Higher bond orders are associated with shorter bond lengths and higher bond strengths. Threfore we can conclude that bond strength of the above specie in decreasing order can be arranged as:

$$O_2^{2+} > O_2 > O_2^- > O_2^{2-}$$

Table 2.5 Comparison of O_2^+, O_2, O_2^-, O_2^{2-}

Molecular Species	*Electronic Configuration*	*Bond Order*	*Magnetic Character*
O_2^+	$KK\ \sigma_{2s}^2\ \overset{*}{\sigma}{}_{2s}^{2}\ \sigma_{2pz}^2\ \pi_{2px}^2\ \pi_{2py}^2\ \overset{*}{\pi}{}_{2px}^{1}$	$2\frac{1}{2}$	Paramagnatic.
O_2	$KK\ \sigma_{2s}^2\ \overset{*}{\sigma}{}_{2s}^{2}\ \sigma_{2pz}^2\ \pi_{2px}^2\ \pi_{2py}^2\ \overset{*}{\pi}{}_{2px}^{1}\ \overset{*}{\pi}{}_{2py}^{1}$	2	Paramagnatic.
O_2^-	$KK\ \sigma_{2s}^2\ \overset{*}{\sigma}{}_{2s}^{2}\ \sigma_{2pz}^2\ \pi_{2px}^2\ \pi_{2py}^2\ \overset{*}{\pi}{}_{2px}^{2}\ \overset{*}{\pi}{}_{2py}^{1}$	$1\frac{1}{2}$	Paramagnatic.
O_2^{2-}	$KK\ \sigma_{2s}^2\ \overset{*}{\sigma}{}_{2s}^{2}\ \overset{*}{\sigma}{}_{2pz}^{2}\ \pi_{2px}^2\ \pi_{2py}^2\ \overset{*}{\pi}{}_{2px}^{2}\ \overset{*}{\pi}{}_{2py}^{2}$	1	Diamagnetic

Flourine Molecule F_2: Flourine atom has 7 electrons in the outermost shell and hence 14 electrons are arranged in the molecular orbital as:

$$= KK\ \sigma_{2s}^2\ \overset{*}{\sigma}{}_{2s}^{2}\ \sigma_{2pz}^2 \begin{cases} \pi_{2py}^2 \\ \pi_{2px}^2 \end{cases} \begin{cases} \overset{*}{\pi}{}_{2py}^{2} \\ \overset{*}{\pi}{}_{2px}^{2} \end{cases}$$

$$\text{Bond Order} = \frac{1}{2}(8-6) = 1$$

The F-F bond is rather weak. Since there are no unpaired electrons, the molecule is diamagnatic.

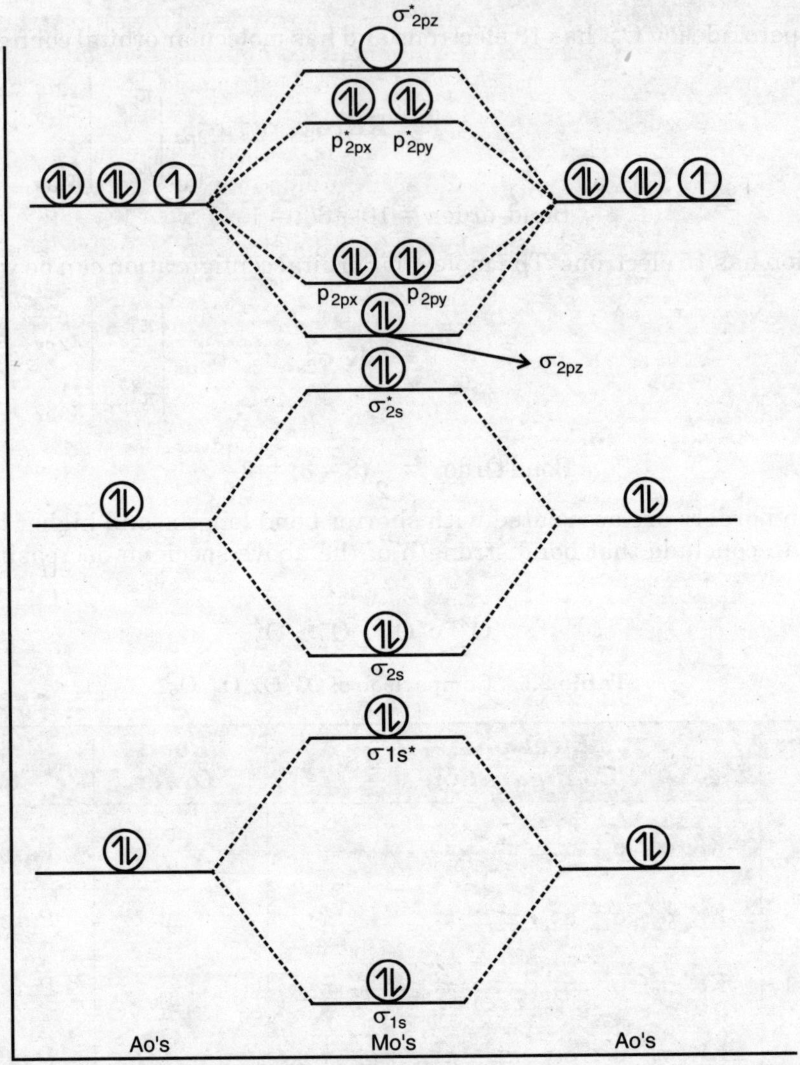

Fig. 2.19 Energy level diagram of flourine molecule.

Neon Molecule, Ne_2: Neon atom has 8 electrons in its valence shell. In the formation of Neon molecule, if possible, 16 electrons have to be accommodated in molecular orbitals. The molecular electronic configuration of Neon molecule would be

$$Ne_2 : KK\ \sigma_{2s}^2\ \sigma_{2s}^{*2}\ \sigma_{2pz}^2 \begin{cases} \pi_{2py}^2 \\ \pi_{2px}^2 \end{cases} \begin{cases} \pi 2py^{*2} \\ \pi 2px^{*2} \end{cases} \sigma_{2pz}^{*2}$$

It is apparant that the effects of the bonding orbitals and autibonding orbitals cancel, so the molecule does not exist.

2.6 HETERONUCLEAR DIATOMIC MOLECULES

The molecular orbitals of heteronuclear diatomic molecules differ from those of homonuclear diatomic molecules. Since the two atoms are different, the energies of their atomic orbitals are slightly different. Due to the difference in the electronegativities of the atoms, the electrons in a bonding molecular orbital will lie closer to the atom having high electronegativity. The electrons in an antibonding molecular orbitals are closer to the less electronegative atom.

Moreover, there is the energy difference between the two sets of atomic orbitals on the two different atoms. It means that two levels will interact less strongly and the energy lowering as a result of the overlap of atomic orbitals on different atoms in a heteronuclear molecule is less pronounced than in homonuclear molecule, when the orbitals have the same energies. Let us discuss the electronic structures and molecular orbitals diagrams for some heteronuclear diatomic molecules.

1. Hydrogen Fluoride molecule, HF: The electronic configuration of hydrogen and fluorine atoms are

$$H = 1s$$
$$F = 1s^2\, 2s^2\, 2p^5$$

The formation of HF molecule takes place by linear combination of the H (1s) atomic orbital with one of the F atomic orbitals. Spectroscopic evidence shows that energies of $1s^2$ and $2s^2$ electrons of fluorine are very low and these do not take part in the bonding. Out of the three 2p orbitals,only $2p_z$ orbital is able to combine with an s-orbital because of symmetry considerations.

Therefore, in the formation of HF molecule, the bonding results from the H (1s) electrons and $2p_z$ electrons of F and the other electrons of fluorine remains in atomicorbitals.

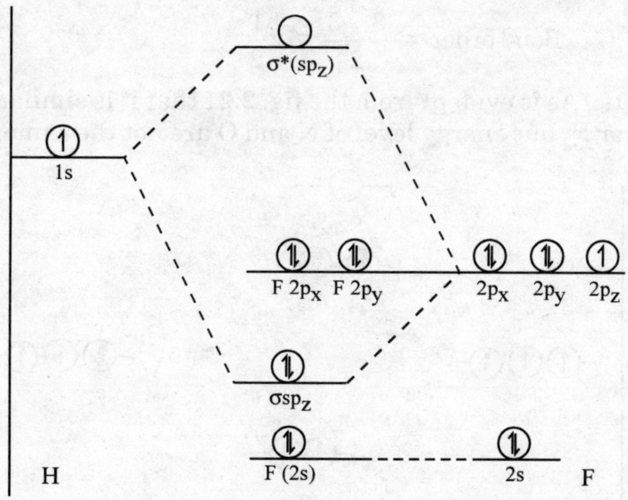

Fig. 2.20

The structure of HF molecule, therefore, may be written as

$$H\ (1s) + F\ (1s\ 2s^2\ 2p^5) \longrightarrow HF\ [K(2s)^2\ (\sigma_{spz})^2\ (2p_x)^2\ (2p_y)^2]$$

The bonding σ_{2pz} is the only molecular orbital formed in HF molecule.

The rest of the electrons remain in their atomic orbitals. The configuration of HF molecule, closely resembles that of the neon atom ($1s^2 2s^2\ 2p^6$). However, in neon, all the orbitals are atomic.

It may be noted that $\sigma_{(sp)}$ is the only MO formed in HF molecule and bonding MO contains two electrons.

In case of heteronuclear HF molecule, the electrons are no longer shared equally between the two nuclei. The $F(2p_z)$ atomic orbital is contributing more than the H (1s) orbital to the molecular orbitals. As seen in Fig. 2.20, the molecular orbital has an energy which lies closer to that of $2p_z$ orbital of fluorine than to that of 1s-orbital of hydrogen.

Similarly, the other HX molecules have single bond with an unequal sharing of the electrons between the nuclei. The chlorine, bromine and iodine atoms contribute respectively $3p_z$, $4p_z$ and $5p_z$ atomic orbitals in formation of HX molecules.

The molecular orbital diagram for HCl will be similar to HF.

2. Nitric Oxide Molecule (NO): The electronicconfigurations of N and O atoms are:

$$N = 1s^2\ 2s^2\ 2p_x^{\ 1}\ 2p_y^{\ 1}\ 2p_z^{\ 1}$$

$$O = 1s^2\ 2s^2\ 2p_x^{\ 2}\ 2p_y^{\ 1}\ 2p_z^{\ 1}$$

There are five electrons in the outer most shell of N atom and six electrons in outer shell of O atom.Thus, a total of 11 electrons are to be accommodated in the molecular orbitals of NO molecule.The MOdiagram for NO molecule is shown in Fig. 2.21. It is clear from the diagram that the bonding MO's are closer to oxygen and antibonding MO's are closer to N atom because oxygen is more electronegative than nitrogen.

The molecular orbital electronic configuration of NO molecule is:

$$NO = \left[\sigma_{2s}^2 \quad \sigma_{2s}^{*2} \quad \sigma_{2p_z}^2 \quad \begin{matrix} \pi_{2p_x}^2 & \pi_{2p_x}^{*1} \\ \pi_{2p_y}^2 & \pi_{2p_y}^{*o} \end{matrix} \right]$$

$$\text{Bond order} = \frac{8-3}{2} = 2\frac{1}{2}$$

NO is paramagnetic. As is evident from the fig. 2.21 that it is similar to as that of diatomic molecules such as O_2 or F_2 but energy level of N and O are not the same.

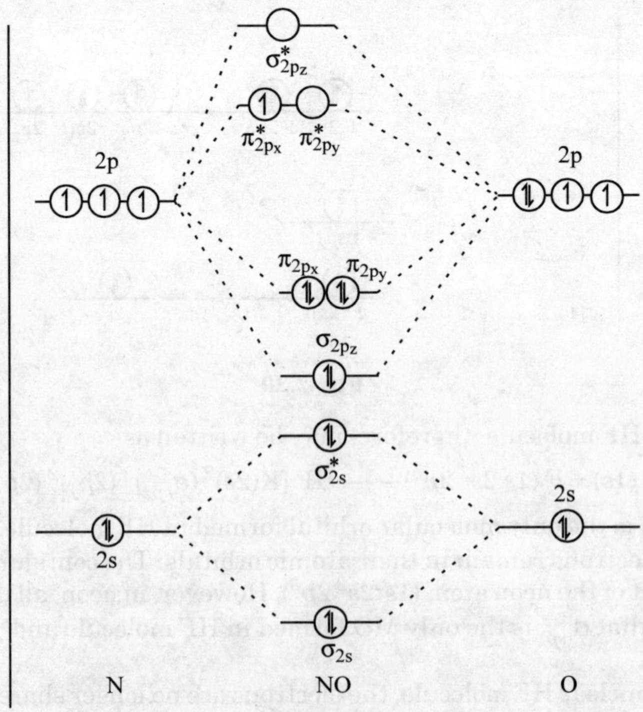

Fig. 2.21 Molecular orbital diagram for NO molecule.

QUESTIONS

SHORT ANSWER QUESTIONS

1. What do you understand by molecular orbital?
2. What is the difference between VBT and MOT?
3. What are the main point of MOT?
4. What do you understand by LCAO method
5. Sketch the shapes of molecular orbitals formed by the overlap of
   (i) Two s-orbitals.
   (ii) End on overlap of two p-orbitals.
   (iii) Side on overlap of two p-orbitals. Name the orbitals formed in each case.
6. What is the maximum number of electrons that can occupy a molecular orbital? How many MO.'s of H_2 originate from the hydrogen 1s atomic orbitals?
7. What is meant by bonding and antibonding molecular orbitals?
8. What are the conditions of combination of atomic orbitals?
9. Give molecular orbital description of hydrogen molecule and deduce the bond order.
10. Discuss the energy level diagram of F_2, O_2 and Ne_2.
11. What is the order of the energy of the molecular orbitals in case of diatomic molecular of second row elements of the periobic table up to N_2.
12. Explain why Ne_2 does not exist
13. Which species out of H_2, H_2^{2+} and H_2^{2-} are paramagnetic and why?
14. Use the energy level diagram to show that N_2 would be expected to have a triple bond.
15. What is meant by bond order? Calculate the bond order of H_2^+, O_2 and N_2 molecules.
16. Calculate the bond orders for He_2 and O_2^+.
17. Calculate the bond orders for N_2^+ and O_2. Calculate the bond order for H_2^+ ion. Is this ion expected to be paramagnetic or diamagnetic and why?
18. Draw M.O. diagram of oxygen molecule.
19. Arrange the following molecular species in increasing order of stability (giving bond orders):—

$$O_2,\ O_2^+,\ O_2^-,\ O_2^{2-}$$

20. Explain why the bond in H_2^+ is longer than that in H_2. Give the molecular orbital energy diagram for oxygen molecule and account for its paramagnetic property.
21. Explain why O_2 is paramagnetic while N_2 is diamagnetic.

LONG ANSWER QUESTIONS

1. (a) What is Molecular Orbital theory?
   (b) Give the electron charge density diagrams for bonding and antibonding molecular orbitals.
2. (a) What do you understand by the terms bonding and antibonding molecular orbitals? Why are they so called? Illustrate your answer by taking into consideration the structure of hydrogen ion H_2^+.
   (b) Is B_2 molecule paramagnetic or diamagnetic? Discuss.
3. Explain how the atomic orbitals combine to form bonding and antibonding molecular orbitals. What are the limitations to such combinations? Explain why in a homonuclear molecule, an s orbital of one atom cannot combine with a p_x or a p_y orbital of another atom of the same element.

4. With the help of molecular orbital diagrams. explain why
   (a) the bond order in N_2^+ ion is less than that in N_2 molecule whereas the bond order in O_2^+ is greater than the O_2 molecule.
   (b) Oxygen molecule is paramagnetic while nitrogen molecule is diamagentic.
   (c) Hydrogen forms diatomic molecule while helium remains monoatomic.
   (d) The bond order in O_2^- is less than that in O_2 which, in turn, is less than that in O_2^+
5. What is meant by bond order? Calculate the bond order of He_2^+, O_2, O_2^{2-} molecules.
6. Give the essential features of valence bond and molecular orbital theories.
7. Explain the following on the basis of MO theory
   (a) Ne_2 does not exist.
   (b) Why N_2 has greater bond dissociation energy than N_2^{2+} whereas O_2^{2+} has lesser bond dissociation energy than O_2
   (c) Why the bond in H_2 is shorter than that in H_2^-
8. What is meant by bond order? Calculate the bond order of He_2^+, O_2 and N_2 molecules.
9. Which species out of H_2, H_2^{2+} and H_2^{2-} are paramagnetic and why?
10. Explain LCAO. What are the conditions of combination of atomic orbitals.
11. Arrange the following molecular species in increasing order of their stability: O_2, O_2^+, O_2^-, O_2^{2-}
12. Discuss the shapes of molecular orbitals formed by combination of the following atomic orbitals:
   (a) $2p_z + 2p_z$ (along internuclear axis)
   (b) $2p_y + 2p_y$ (Lateral overlap)
   (c) $2s + 2pz$
13. Draw MO energy level diagram for the F_2 molecule. What is the order of stability among F_2^{2+}, F_2 and F_2^-.
14. State main points of difference between bonding and anti-bonding molecular orbitals.
15. (a) How are the bond order and bond length related? Explain with example.
   (b) Arrange the molecules H_2, O_2, F_2 and N_2 in the order of increasing bond length.
16. Differentiate between:
   (a) Valence bond theory and molecular orbital theory
   (b) Atomic orbital and molecular orbital.
17. Discuss briefly the molecular orbital theory. Discuss the formation of H_2 and O_2 molecule on the basis of this theory.
18. What do you understand by the term bonding and antibonding orbitals? Give the conditions for the combination of atomic orbitals to form molecular orbitals. How does molecular orbital theory help to explain the paramagnetic character of oxygen and diamagnetic character of nitrogen (O = 8, N = 7)
19. (a) Find out hybridizations and structures of any four of the following
   (i) BrF_3 (ii) SF_6 (iii) XeF_4
   (iv) IF_7 (v) SiF_4

❑❑❑

3 STRUCTURE AND STEREO STRUCTURE OF MOLECULES

"Imagination is more important than knowledge. Knowledge is limited, imagination encircles the world"

—Albert Einstein

CHAPTER OUTLINE

3.1 INTRODUCTION

One of the most distinctive feature of organic compounds is their ability to show isomerism. The term isomer (Greek 'iso-equal', 'mers'-parts) was coined by Berzelius to different compounds having the same molecular formula. The isomers are defined as—*two or more compounds having the same molecular formula but different chemical and physical properties and the phenomena is termed as isomerism.*

Isomerism is of two types:

1. Structural Isomerism
2. Stereoisomerism

Structural Isomerism: Isomers having the same molecular formula but different structures, i.e. different arrangement of atoms within the molecule are called structural isomers and the phenomena is called structural isomerism. Structural isomers can involve the carbon skeleton or nature and position of functional groups.

A simple example of structural isomers are the isomers ethanol and dimethyl ethers, as shown below:

$$CH_3 - CH_2 - OH \qquad\qquad CH_3 - O - CH_3$$
$$(C_2H_6O) \qquad\qquad (C_2H_6O)$$

Stereoisomerism: Stereoisomerism is also called space isomerism. It is due to the difference in spatial positions of atoms in a molecule, i.e. *the isomers are different from each other only in the way the atoms are oriented in space.* A small difference in their spatial arrangement may lead to a profound effect on their physical and chemical properties. The space relationship of atoms within organic molecules have assumed such an importance that a new branch of chemistry called "Stereochemistry" has been established. Stereochemistry is the branch of chemistry which deals with the structures of molecules in three dimensions and stereoisomerism is a part of this branch of chemistry.

Stereoisomers have 2 major classes:

1. Configurational isomers.
2. Conformational isomers.

Configurational isomers inturn are of two types:

(*i*) Enantiomers. (*ii*) Diastereomers

All these types of stereoisomers are discussed in the following sections.

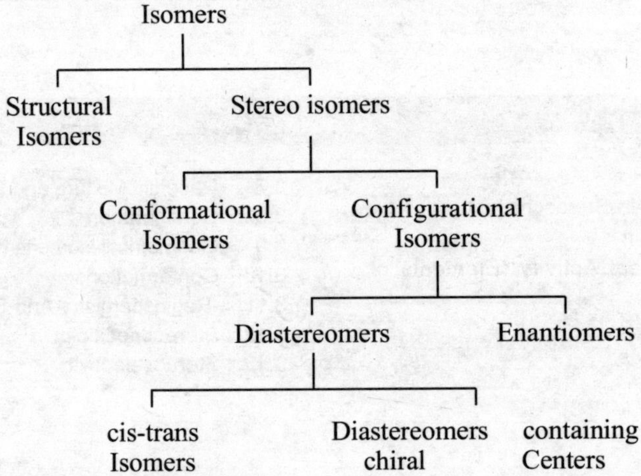

3.2 BASIC CONCEPTS IN STEREOCHEMISTRY

Before we discuss the details of the stereoisomers, it is worthwhile to be familiar with certain basic concepts involved.

1. Chirality (Stereogenicity)

Chirality (*cheir*, Greek for "hand") refers to objects which are not superimposable on (cannot be made to coincide with) their mirror image. ***A chiral molecule is one that is not superimposable on its mirror image.*** Achiral compounds (objects) can be superimposed on their mirror image. Our right and left hands are very similar, yet they are not identical. They are related to each other as mirror images, and as such they cannot be superimposed on top of each other. Hence, there are is said to be *chiral*. A more contemporary term for "chiral centre" (or chiral carbon) is stereogenic centre (or stereocenter).

> **Chiral**
>
> A molecule is *chiral* if its two mirror image forms are *not* superimposable upon one another.
>
> A molecule is *achiral* if it its two mirror image forms are superimposable

sp^3–hybridised carbon atoms possessing four different substituents display this property due to their tetrahedral geometry. A chiral carbon is one that has four different "group" attached to it.

The groups can be anything from a single H to functional groups to one more other carbons. A chiral carbon is often denote\ **by C*.**

An sp^3 hybridised with 1, 2 or 3 different atoms or groups attached can be superimposed on its mirror image and is, therefore *achiral.*

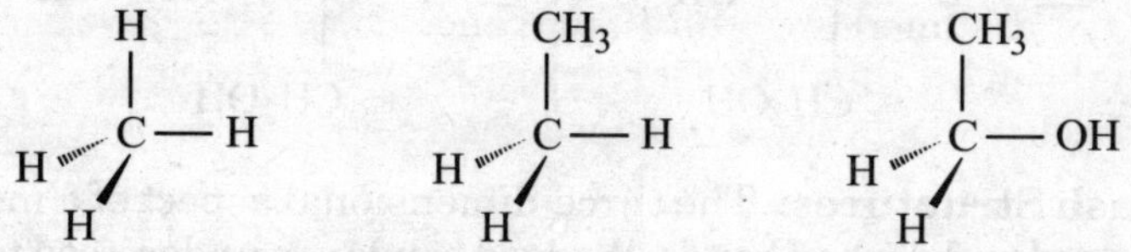

Achiral Carbons

However, if there are 4 different subsituents on carbon, then it is no longer superimposable on its mirror image and we say that carbon is chiral.

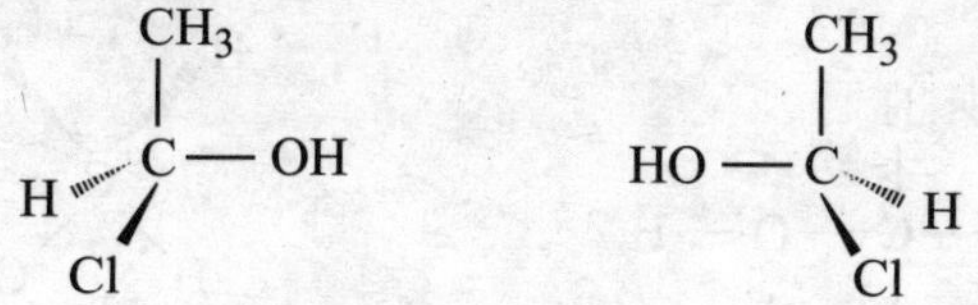

Chiral Carbons

For the carbon to be chiral, it follows that:

- the carbon atom is sp^3-hybridised
- there are four different groups attached to the carbon atom.

Consider 2-bromobutane

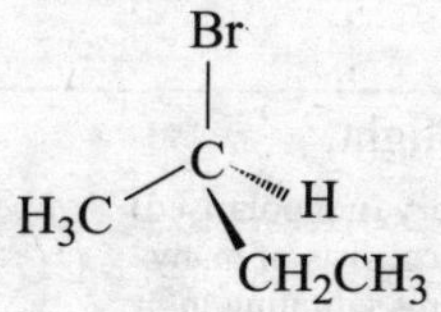

The carbon which is bound to the bromine is a stereogenic centre, it is sp^3 hybridised and it is bonded to four different groups:

1. a hydrogen
2. a bromine
3. a methyl group, and
4. an ethyl group.

2. Representing three dimensional structure:

(a) Fischer Projections Emil Fischer (late 1800's) introduced formulas depicting the spatial arrangement of groups around chiral carbon atoms. This convention represents molecular structure in a two-dimensional framework of vertical and horizontal bonds.

In Fischer projections, the vertical carbons are always viewed as being positioned behind the plane of the paper (away from the viewer) while the horizontal carbons are imagined as coming out of the plane (toward the viewer).

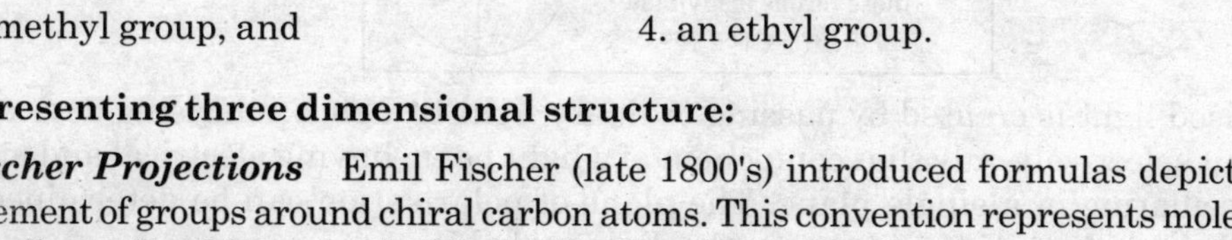

Fischer Projection of (S)-Lactic Acid

Similarly, we represent the isomers of glyceraldehyde as:

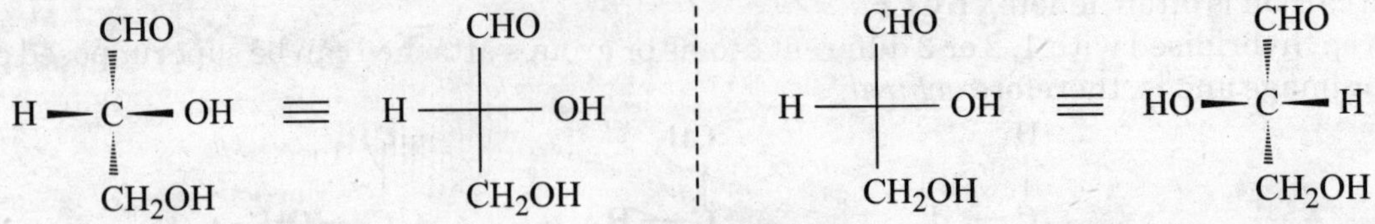

(b) Wedged and Dash Structures: The three-dimensional aspect of a molecule can be shown through the use of wedged and dashed bonds. Wedged bonds are understood to be coming out of the plane of the paper toward the reader, whereas dashed bonds are understood to be going away from the reader. Bonds that are drawn as plain lines in one of these structures are in the plane of the paper. This is shown for ethanol below.

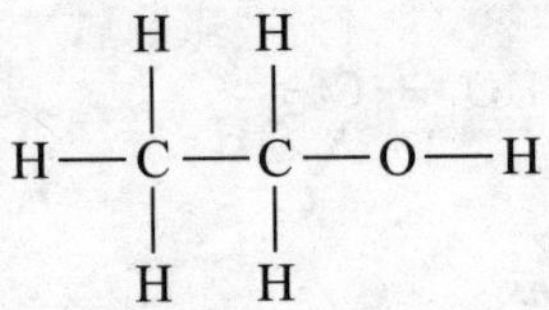

Kekule structure

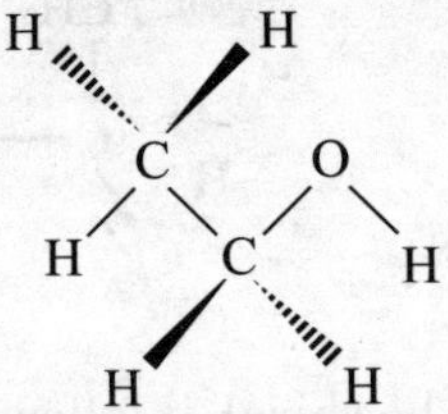

3-D structure

3. Plane Polarised Light

Ordinary light consists of waves vibrating in all planes perpendicular to the direction of propagation. Plane-polarised light is light in which all wave vibrations have been filtered out except for those in one plane.

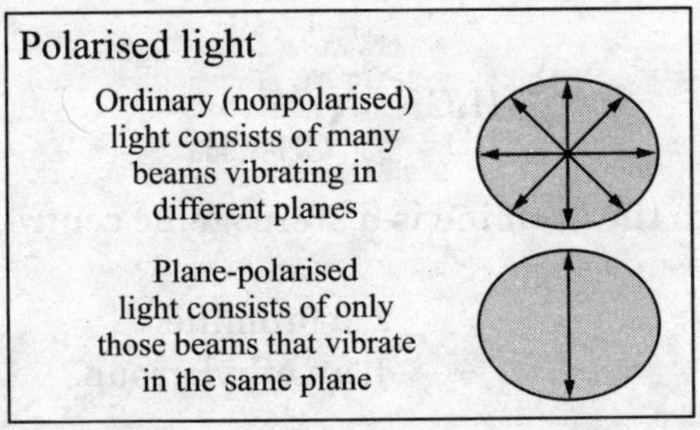

Plane-polarised light is created by passing ordinary light trough a polarising devices. Such devices transmit selectively only that component of a light beam having electrical and magnetic field vectors oscillating in a single plane. The plane of polarisation can be determined by an instrument called a **polarimeter,** shown in the diagram below.

Monochromatic (single wavelength) light, is polarised by a fixed polariser next to the light source. A sample cell holder is located in line with the light beam, followed by a movable polariser (the analyzer) and an eyepiece through which the light intensity can be observed. In modern instruments an electronic light detector takes the place of the human eye. In the absence of a sample, the light intensity at the detector is at a maximum when the second (movable) polariser is set parallel to the first polariser ($\alpha = 0°$). If the analyzer is turned 90° to the plane of initial polarisation, all the light will be blocked from reaching the detector.

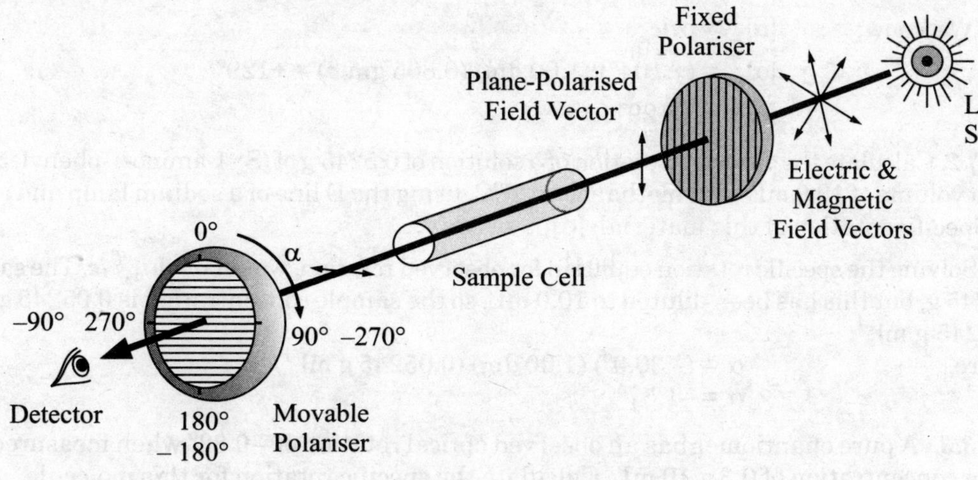

Fig. 3.1 Polarimeter

4. Optical Activity

Organic compounds are said to **optically active** because they rotate plane polarised light. Optically active compounds are chiral compounds which possess the ability to rotate plane-polarised light.

- **Polarimeter** is used to the measure the optical activity of compounds.
- If a solution containing an optically active substance is placed in the sample tube, it rotates the plane of the polarised light, and the degree of rotation is measured.
- If it rotates the light to the right (clockwise) it is called *dextrorotary* and is designated with a + sign. (+) Dextrorotatory (Latin *dexter* is "right").
- If it rotates light to the left (counterclockwise) it is called leavorotatory and it is designated – sign (–) Leavorotatory (Latin *laevus* is "left").

5. Specific Rotation

The magnitude of the rotation is an intrinsic property of the molecule. The observed rotation depends:

- on the concentration of the sample compound,
- the length of the sample tube,
- the temperature

Specific Rotation is the degree of rotation by 1.00 gram of compound in 1.00 mL of solution measured in a tube with a l decimeter path length. The amount of rotation is also dependent on the wavelength (typically the D line of sodium, 589.3 nm) and solvent.

$[\alpha]_D^{20} = \alpha/lc$ where α = observed rotation at 20°C

l = cell path length in decimeters

c = concentration in g/mL

D = wavelength (typically the D line of sodium, 589.3 nm)

or it can be expressed as : $[\alpha]_\lambda^T = \alpha/lc$

Example 3.1. A sample of pure (S)-2-butanol was placed in a 10.0 cm polarimeter tube. Using the *D* line of a sodium lamp, the observed rotation at 20°C was + 104°. The density of this compound is 0.805 gml^{-1}. What is the specific rotation of (S)-2. butanol?

Solution: We know;

$$[\alpha]_\lambda^T = \alpha/lc$$

$$[\alpha]_\lambda^T = (+104°)/(1.00 \text{ dm})(0.805 \text{ gm}^{-1}) = +129°$$

$$[\alpha]_\lambda^T = +129°$$

Example 3.2. Calculate the observed rotation of a solution of 0.5245 g of (S)-1-amino-1-phenylethane diluted to a volume of 10.0 mL with methanol at 20°C, using the D line of a sodium lamp and a 1.00 dm tube. Specific rotation of this material: $[\alpha]_\lambda^T = -30.0°$.

Solution: Solving the specific rotation equation for observed rotation, we get $\alpha = [\alpha]_\lambda^T lc$. The sample size is 0.5245 g, but this has been diluted to 10.0 mL, so the sample concentration is 0.05245 g/10.0 mL = 0.05245 g ml^{-1}.

Therefore,

$$\alpha = (-30.0°)(1.00 \text{ dm})(0.05245 \text{ g ml}^{-1}).$$

$$\alpha = -1.57°.$$

Example 3.3. A pure enantiomer has an observed optical rotation of – 0.82° when measured in a one dm at a concentration of 0.3 g/10 mL. Calculate the specific rotation for this molecule.

Solution:

The concentration of 0.3 g/10 mL is equivalent to 0.03 g/mL; c = 0.03 g/mL

The length of the analysis tube is one dm; 1 = 1.0 dm

The specific rotation is therefore: $[\alpha] = (-0.82°)/(0.03 \text{ g/mL} \times 1.0 \text{ dm})$

$$[\alpha] = -27.3° \text{ g}^{-1} \text{ mL}^{-1} \text{ dm}^{-1}$$

If the specific rotation of a pure enantiomer is known, the observed rotation can also be used to calculate **optical purity,** or the level of contamination of one compound with its enantiomer, using the simple convention:

Optical purity = (% of one enantiomer) – (% of the other enantiomer)

3.3 OPTICAL ISOMERISM

Optical isomers are compounds with the same molecular formula but differ in the way they rotate plane polarised light. These isomers are called "enantiomers" and are said to be optical active. *Enantiomers are stereoisomers which are related to each other as mirror images.* Such isomers will have one carbon that has four non-equivalent groups attached to the "chiral" carbon. The molecule will also be chiral. For example, CHClBrF would be a molecule in which its carbon would be chiral. Its mirror image would be non-superimposible, and the two molecules (the molecule and its mirror image) would be enantiomers.

Mirror-images–Non-superimposable
(enantiomers)

Whereas the molecule CH_2ClBr would not have a chiral carbon therefore the molecule is not chiral or "achiral", and the mirror image would be superimposible and therefore not enantiomers. There would be no optical activity in an achiral molecule.

Mirror-images–Non superimposable
Not Enantiomers

In enantiomers, one molecule will rotate plane polarised light to the left. The other enantiomer will rotate plane polarised light to the right by the same number of degrees. Both enantiomers are said to be optically active because of their ability to rotate plane polarised light. Hence, in a pair of optically active enantiomers, each enantiomer will rotate the plane of polarised light in equal and opposite directions; the enantiomers can therefore be referred to as (+) and (–) enantiomers.

Now, let us consider specific examples of the molecules which show enantiomerism.

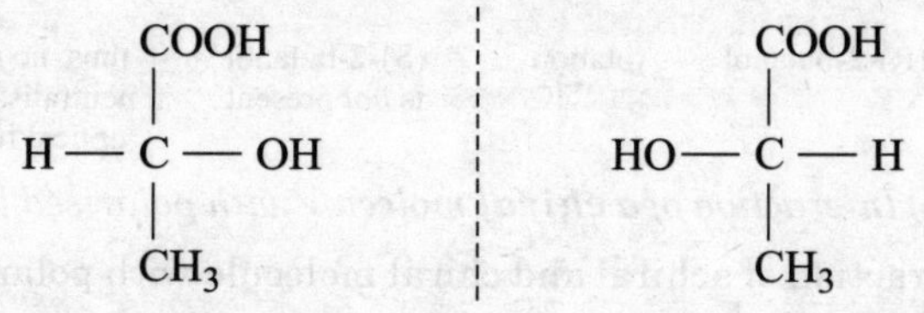

Enantiomers of Lactic acid

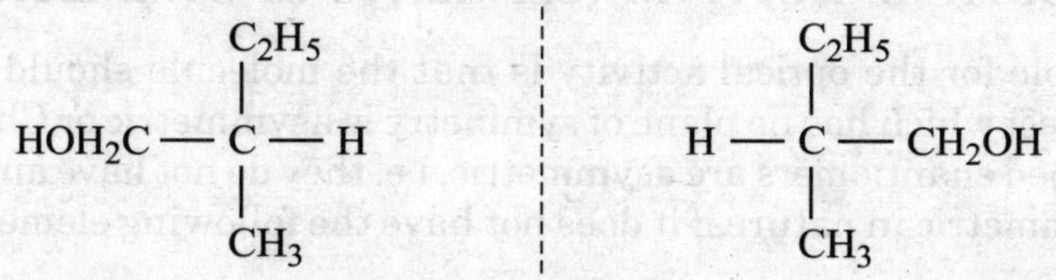

Enantiomers of 2 methyl-1-butanol

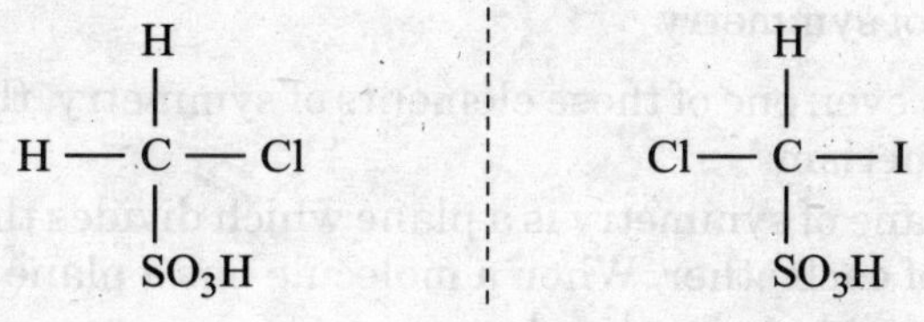

Enantiomers of chloroiodomethane sulf onic acid

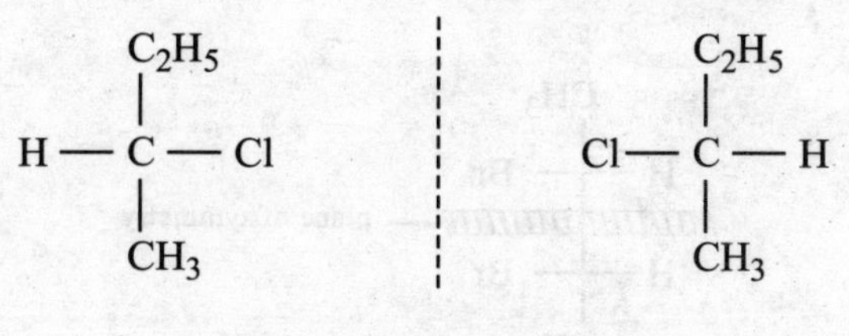

Enantiomers of sec-Butyl chloride

Most of the physical properties of a pair of enantiomers have identical values. For instance, enantiomers have the same boiling point, melting point, solubility, refraction index and many other properties. As mentined above enantiomers differ in only one physical properties, the interaction with plane-polarised light., as shown in the Figure below:

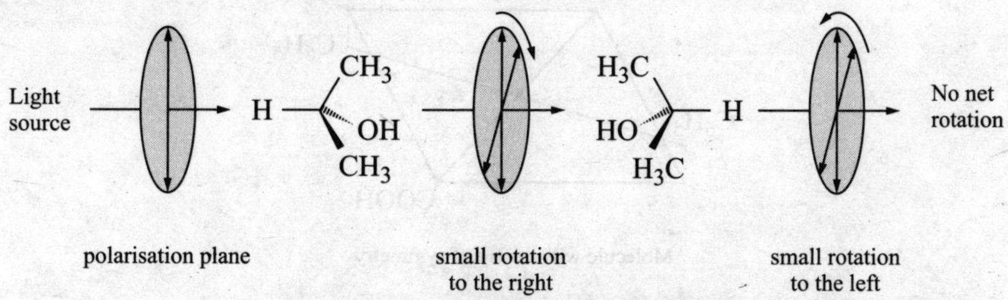

*(a) Interaction of an **achiral** molecule with polarised light.*

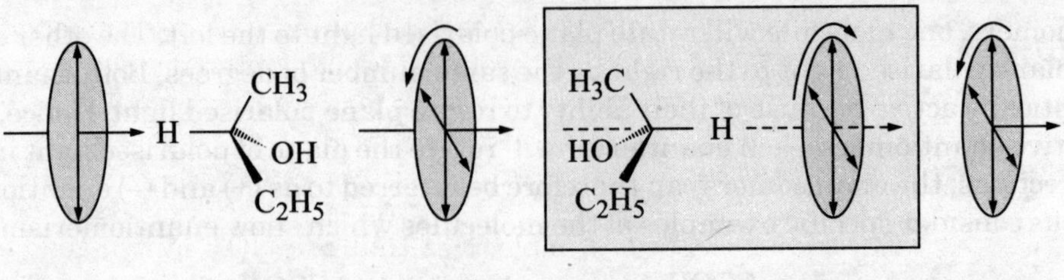

*(b) Interaction of a **chiral** molecule with polarised light.*

Fig. 3.2 Interaction of achiral and chiral molecule with polarised molecule.

3.4 CAUSE OF OPTICAL ACTIVITY (ELEMENTS OF SYMMETRY)

The factor responsible for the optical activity is that the molecule should have asymmetry or disymmetry. "An object which has no plane of symmetry is asymmetric or Chiral". Chiral carbons in the above-mentioned enantiomers are asymmetric, i.e. they do not have any plane of symmetry. A compound is disymmetric in nature if it does not have the following elements of symmetry.

(*i*) a plane of symmetry
(*ii*) a centre of symmetry
(*iii*) an alternating axis of symmetry

If the molecule contains even one of these elements of symmetry, the molecule is symmetrical and will not show enantiomerism.

(i) Plane of symmetry: A plane of symmetry is a plane which divides the molecule into two halves which are mirror images of each other. When a molecule has a plane of symmetry, it will not be optically active even though it has chiral carbon.

Example:

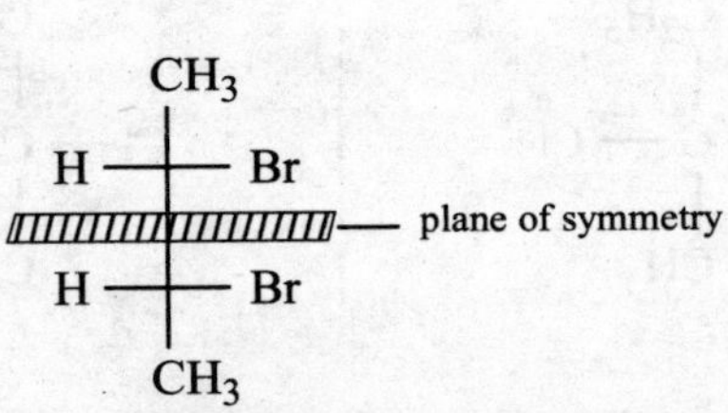

(ii) Centre of symmetry: Centre of symmetry is an imaginary point in a molecule from which if a line is drawn from any atom group to this point and extended beyond this point, another identical atom/group will be found just at the end of the line.

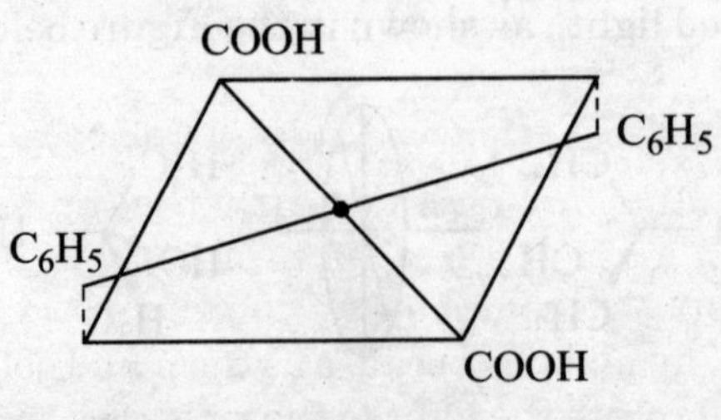

Molecule with centre of symmetry

(iii) Alternating axis of symmetry: The molecule has alternating axis of symmetry, when it is rotated through θ degrees about an axis passing through it and the rotated molecule is reflected in a mirror that is perpendicular to the axis of rotation.

3.5 DIASTEREOMERS

The other type of optical isomer are pair of molecules that are NOT mirror images but each molecule rotates plane polarised light. There are optically active. Such molecules are called "diastereoisomers".

When there are 2 or more chiral carbons in a molecule, it can have *more* optical isomers. For example, if there are 2 chiral carbons:

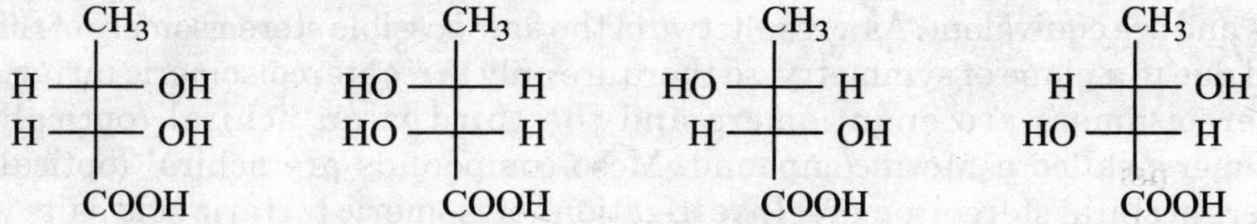

(S, S) (I)   (R, R) (II)   (R, S) (III)   (S, R) (IV)

This compound has 4 optical isomers and 2 pair of enantiomers. The pair of enantiomers are stereoisomers of one another. Stereoisomers which *are not mirror-image isomers* are called **diastereomers**. The structures I and III are not mirror images but are stereoisomers because they rotate the plane polarised light. Hence, I and III are a pair of diastereoisomers. Similarly, II and IV are diastereoisomers.

A compound with *n* chiral carbon atoms *can have maximum of* 2^n stereoisomers. The acid shown in the previous example has 2 chiral carbons and $2^2 = (= 4)$ stereoisomers.

3.6 MESO COMPOUNDS

Meso compounds are compounds that contain chiral carbons but *can be superimposed* on their mirror-image. Meso compounds are compounds that have chiral carbons (2) within the molecule, but the compound is *optically inactive.* Actually meso compounds have two chiral carbons and there is a plane of symmetry that will have half the molecule on one side of the plane of symmetry while the other half which is the mirror image of the first half on the other side of the plane.

Some compounds with 2 chiral carbons have *less than* 4 stereoisomers. For example,.

Achiral Diastereomers (meso-Compounds)

This molecule has an *internal plane of symmetry;* the top half is the mirror-image of the bottom half of the molecule. Although there are 2 chiral carbons in this molecule, this particular stereoisomers is **achiral**, or optically inactive.

The meso compound is optically inactive because as the plane polarised light passes through the molecule, half the molecule will rotate the light to the right and the other half of the molecule will rotate the light back in the other direction by the same number of degrees. This means that the light exits the sample with no apparent rotation.

The compound above is *meso*-tartaric acid. Since it has 2 chiral carbons, there is another pair of enantiomers possible,

(–) tartaric acid (+) tartaric acid

The chiral centers in the preceding examples are different, from one another. In this case of 2, 3-dihydroxybutanedioic acid, known as tartaric acid, the two chiral centers have the same four substituents and are equivalent. As a result, two of the four possible stereoisomers of this compound are identical due to a plane of symmetry, so there are only *three* stereoisomeric tartaric acids. Two of these stereoisomers are enantiomers and the third is an achiral (optically inactive) diastereoisomers, called a Meso compound. Meso compounds are achiral (optically inactive) diastereomers of chiral stereoisomers. Investigations of isomeric tartaric acid salts were carried out by *Louis Pasteur* in the mid 19th century.

3.7 RACEMIC MIXTURE AND ENANTIOMERIC EXCESS

A mixture consisting of equal parts of any pair of enantiomers is called a **racemic mixture** (or *racemic modification*) and is designated by (+/–). A racemic mixture does not rotate plane-polarised light because (+)-rotation caused by one enantiomer is cancelled by rotation in the opposite direction by the (–) enantiomer. A solution of a racemic mixture of enantiomers is optically inactive.

Clearly, unequal mixtures of two enantiomers will have a lower optical rotation than a pure enantiomer and the strength of this rotation will depend upon the ***enantiomeric excess (e.e.)*** of the mixture. Mixtures of unequal amounts of enantiomers are referred to as ***scalemic.***

Enantiomeric Excess = (% Enantiomer A – % Enantiomer B)%

Enantiomeric excess is a measure of the extent to which a particular enantiomer dominates the mixture .

Determining Optical Purity/Enantiomeric Excess:

The optical purity of a mixture of enantiomers is given by:

% Optical purity of sample = 100 * (specific rotation of sample)/(specific rotation of a pure enantiomer)

$$= 100[\alpha]_{mixture}/[\alpha]_{pure\ sample}$$

Example: (S)-bromobutane has a specific rotation of +23.1°, therefore, (R)-bromobutane has a specific rotation of –23.1°

Optical purity of a racemic mixture = 100 * (0°)/(+23.1°) = 0%

i.e. there is no one enantiomer present in excess.

A 50 : 50 mixture of the two enantiomers will not rotate plane polarised light because the effects of the two enantiomers cancel each other out, molecule for molecule. The specific rotation of a racemic mixture is zero.

Another way to express optical purity is as the "enantiomeric excess" or ee:

$$ee\% = 100\,([R] - [S])/([R] + [S])$$

where [R] = concentration of the R-isomer

[S] = concentration of the S-isomer

In a racemic mixture [R] = [S], therefore,

$$ee\% = 100 * 0/([R] + [S]) = 0\%$$

or, if we use the % as concentrations, then [R] = [S] = 50 % we get,

$$ee\% = 100 * (50 - 50)/(50 + 50) = 0\%$$

Example: What would be the ee of a mixture of the 2-bromobutanes of measured specific rotation = –9.2°?

The fact that the sign is negative tells us that in this case the R enantiomer is the dominant one.

Optical purity % = 100 * (–9.2°/–23.1°) = 40 % *ie there is a 40% excess of R over S.*

i.e R is in 40% excess. Out of remaining 60%, 30% is R and 30% is S.

Therefore, total R is 40 + 30 = 70% and S is 30%.

This corresponds to a mixture of 70% R and 30% S.

3.8 CONFIGURATION

The arrangement of atoms that characterizes a particular stereoisomer is called its configuration. Although enantiomers may be identified by their characteristic specific rotation, the assignment of a unique configuration to each has not yet been discussed. We have referred to the mirror-image configurations of enantiomers as "right-handed" and "left-handed". There are two conventions used to assign the configuration of enantiomers.

1. Relative Configuration: (D and L Configuration)

In 1960, a New York University chemist, M. A. Rosanoff, proposed the D.L. system. The glyceraldehyde was used as a standard and many thousands of compounds were related to it by chemical reactions that did not change any bond at a stereogenic unit.

In otherwords, this convention assigned a **D** prefix to enantiomers chemically related to a right-handed reference compound (glyceraldehyde) and a **L** prefix to a left-handed reference compound. Glyceraldehyde is chiral itself, and its two isomers are labeled *D* and *L.* In this system, compounds are named by analogy to glyceraldehyde.

The D/L labelling is unrelated to (+)/(–); it does not indicate which enantiomer is dextrorotatory and which is levorotatory. Rather, it says that the compound's stereochemistry is related to that of the dextrorotatory or levorotatory enantiomer of glyceraldehyde—the dextrorotatory isomer of glyceraldehyde is, in fact, the *D* isomer. A rule of thumb for determining the *D/L* isomeric form of an amino acid is the "CORN" rule. The groups:

COOH, R, NH_2 and H (where R is a variant carbon chain)

are arranged around the *chiral centre* carbon atom. Sighting with the hydrogen atom away from the viewer, if these groups are arranged clockwise around the carbon atom, then it is *D*-form. If counter-clockwise, it is *L*-form.

Although this notation is still applied to carbohydrates and amino acids, it required chemical transformations to establish group relationships, and proved to be doubtful in its general application.

2. R and S (Absolute) Configuration

In the early sixties, R. S. Cahn, C.K. Ingold, and V. Prelog introduced a new system, the *R, S* system, to name the absolute configuration because the *D, L* system is not unequivocal in all cases and cannot easily be applied to all families of compounds.

The resulting nomenclature system is sometimes called the **CIP** system or the **R-S** system. For chemists, the *R*/*S* system is the most important nomenclature system for denoting enantiomers, which does not involve a reference molecule such as glyceraldehyde.

In the *RS* system of nomenclature, each chiral centre in molecule is assigned a prefix (*R* or *S*), according to whether its configuration is right-or left-handed. No chemical reactions or interrelationship are required for this assignment. The symbol **R** comes from the Latin rectus for right, and L from the Latin *sinister* for left. The assignment of these prefixes depends on the application of following rules:

1. Rank the 4 atoms (groups) attached to the carbon according to the sequence rules or Cahn-Ingold-Prelog rules. (Given below)
2. Rotate the molecule until the lowest priority (number 4) substituent is in the back
3. Draw a curve from number 1 to number 2 to number 3 substituent.
4. If the curve is clockwise, the stereocenter is of *R* configuration. If the curve is counterclockwise, the stereocenter is of *S* configuration

The sequence Rules for R and S (Absolute) configuration:

(*i*) *Assigned sequence priorities to the four substituents by looking at the atoms attached directly to the chiral centre.* The higher the atomic number of the immediate substituent atom, the higher the priority.
For example, $H- < C- < O- < Cl-$ (Different isotopes of the same element are assigned a priority according to their atomic mass). In example below, iodine would be 1, bromine 2, chlorine 3, and fluorine 4, because iodine has the highest molecular weight (and is therefore highest priority) and fluorine has the lowest molecular weight (and is therefore the lowest priority).

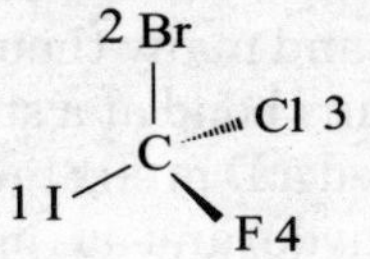

(*ii*) *If the first atom of two substituent happen to be identical in molecular weight, go to the next atom and make the molecular weight comparison* (e.g. and ethyl group would have higher priority over a methyl group). Evaluate atoms progressively further away from the chiral center until a difference is found.
For example, $CH_3- < C_2H_5- < ClCH_2- < BrCH_2- < CH_3O-$.

Thus, the four substituents on 2-bromobutane would be listed in the order. $Br > CH_2CH_3 > CH_3 > H$.

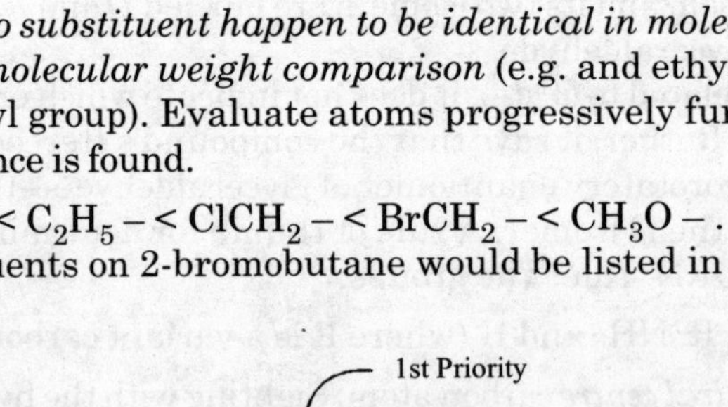

(*iii*) *If double or triple bonded groups are present as substituents, they are treated as an equivalent set of single-bonded atoms, i.e. a carbon with a double bond to another carbon is treated as a carbon singly bounded to two carbons, as shown below.* For example an ethylene substituent, $R–CH = CH_2$, will have a higher priority then an ethyl substituent ($R–CH_2CH_3$)

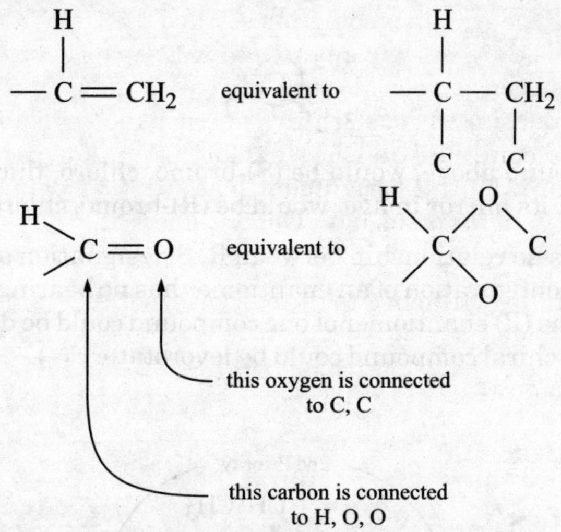

(*i*) Once the relative priorities of the four substituents have been determined, the chiral centre must be viewed from the side opposite the lowest priority group. If we number the substituent groups from 1 to 4, with 1 being the highest and 4 the lowest in priority sequence, the two enantiomeric configurations are shown in the following diagram along with a viewers eye on the side opposite substituent number 4.

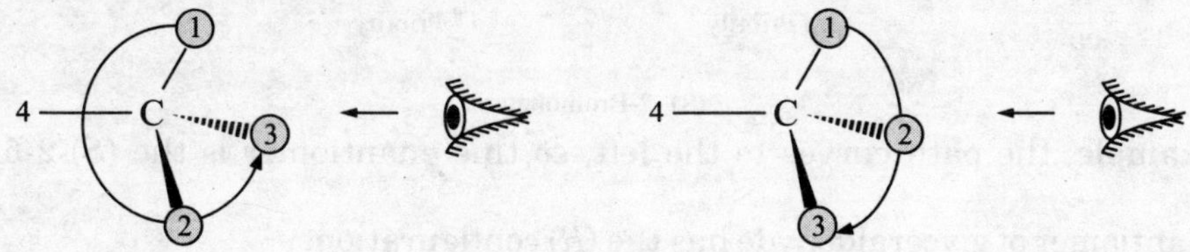

(S) (R)

Example: Let us assign the *R* or *S* configuration to the following molecule of bromochlorofluormethane based on above rules:

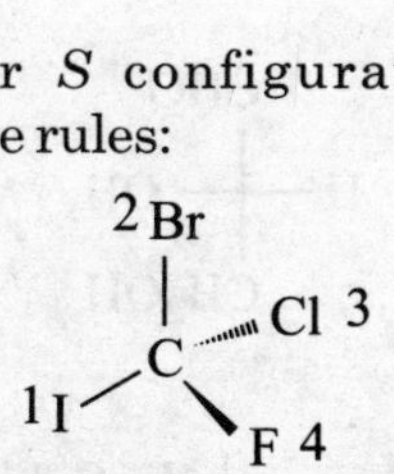

(*ii*) Rotate the molecule so the substituents can be arranged from 1 to 4. Rotate the molecule until the number 4 priority substituent is in the back, as shown below. (Remember that dashed lines mean a bond away from the viewer, and a solid wedged line indicates a bond towards the viewer).

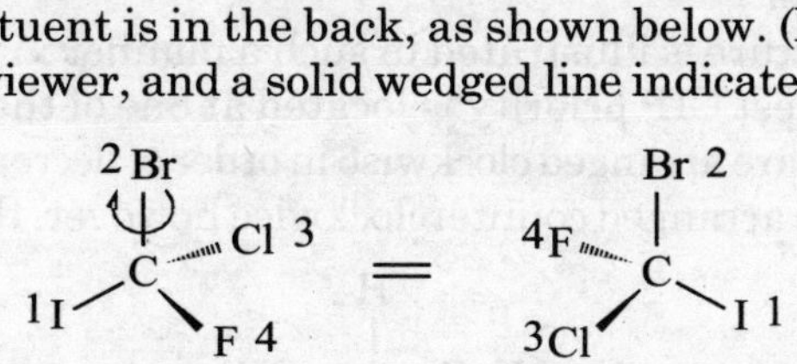

(*iii*) A curve is then drawn from the 1 to 2 to 3 priority substituents, ignoring the 4th priority substituent (as shown below). In the example below, the curve goes countercloskwise so the molecule is of *S* configuration.

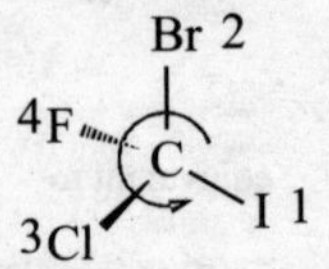

The name of the compound above, would be (S)-bromo, chloro, fluoro, iodomethane, and the name of its enantiomer, or its mirror image, would be (R)-bromo, chloro, fluro, iodomethane.

Note: Please note there is no relationship between R, S designation and the direction of optical rotation, i.e. the absolute configuration of an enantiomer has no bearing on its direction of rotation of plane-polarised light. The (R) enantiomer of one compound could be dextrorotatory (+) while the (R) enantiomer of another chiral compound could be levorotatory (–).

Some examples:

1.

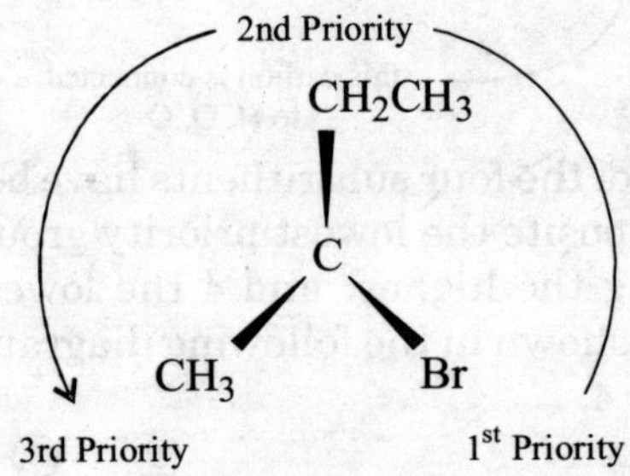

(S) -2-Bromobutane

In this example, the path curves to the left, so this enantiomer is the (*S*)-2-bromobutane stereoisomer

2. This enantiomer of glyceraldehyde has the (*R*) configuration.

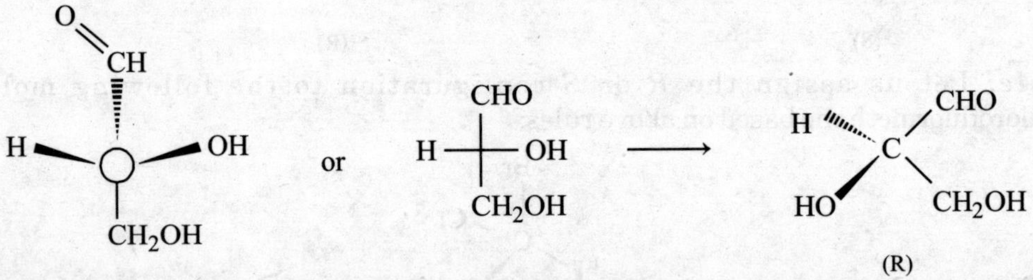

This naming of absolute configuration with *R*, *S* descriptors can be taken from the Fischer projection.

For this purpose, the structure is illustrated in such a manner in the Fischer projection, so that the substituent with the lowest CIP priority is located at one of the two vertical positions. If the remaining three substituents are arranged clockwise in order of decreasing CIP priority, the absolute configuration is *R*. If they are arranged counterclockwise however, the absolute configuration is *S*.

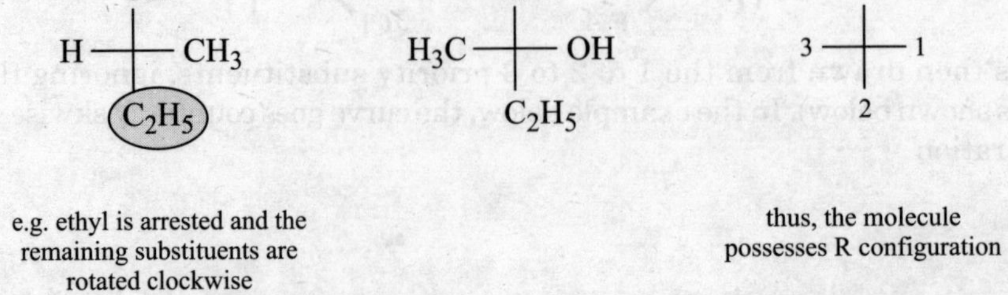

Isomers of Tartaric Acid

Fischer projection formulas provide a helpful view of the configurational relationships within the structures of these isomers. In the following illustration a mirror line is drawn between formulas that have a mirror-image relationship. In demonstrating the identity of the two meso-compound formulas, remember that a Fischer projection formula may be rotated 180° in the plane.

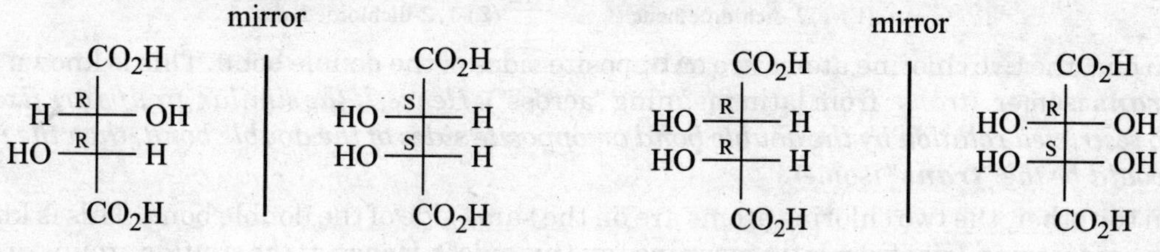

(+)-tartaric acid (–)-tartaric acid meso-tartaric acid

This enantiomeric pair do not possess an internal plane of symmetry and are optically active.

Tartaric acid, therefore, has only three stereoisomeric forms: one meso form and a pair of enantiomers. The meso form is a diastereomer of each of the enantiomers.

3.9 GEOMETRICAL ISOMERISM

Geometric isomerism (also known as cis-trans isomerism or E-Z isomerism) is a form of stereoisomerism. It was discussed in the previous section that *the stereoisomers which are not the mirror images of each other, are called Diastereomers.* Geometrical isomers are isomers that have the same molecular formula but the atoms are in different non-equivalent positions to one another. Geometrical isomers occur as a result of restricted rotation about a carbon-carbon bond.

How geometric isomers arise:

Single bonds between tow carbons in a non-cyclic structure may rotate around each other unhindered at room temperature because as the carbons rotate around each other the degree of overlap between the atomic orbitals is not changed and remain at maximum overlap as rotation occurs.

Consider two possible configurations of 1,2-dichloroethane. These two structures represent the same molecule

trans-1, 2-dichloroethene *cis*-1, 2-dichloroethene

However if there is a double bond between two carbons as in the case of alkenes, these sp^2 carbons have a sigma bond and a pi bond between two sp^2 hybrid orbitals. If the two sp^2 carbons attempt to rotate about each other this will reduce the p-p overlap of the Pi bond. This has a destabilising influence on the molecular system and so is resisted. Thus the Pi bond prevents the carbons from rotating abut each other (restricted rotation). As a result, the atoms attached to these sp^2 carbons will always remain fixed in three dimensional space. This gives rise to geometric isomers.

If the two sp^2 carbons have different atoms attached to themselves, then two configurations are possible: either two like groups (one on each sp^2 carbon) being on the SAME side of the double bond or two like groups being on the OPPOSITE sides of the double bond.

Example: as in 1,2-dichloroethene

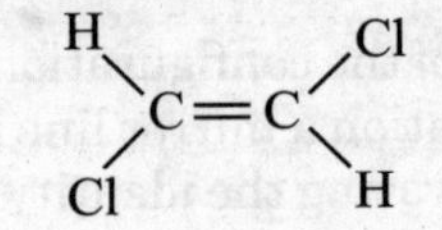

(E)-1, 2-dichloroethene

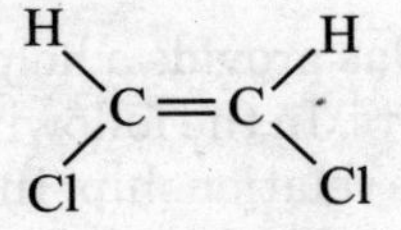

(Z)-1, 2-dichloroethene

- In one, the two chlorine atoms are on opposite sides of the double bond. This is known as the ***trans*** isomer. (*trans* : from latin meaning "across"). *Hence, if the similar groups are fixed due to restricted rotation by the double bond on opposite sides of the double bond, then the isomer would be the "**trans**" isomer.*
- In the other, the two chlorine atoms are on the same side of the double bond. This is know as the ***cis*** isomer. (*cis* : from latin meaning "on this side"). Hence, *if the similar groups are fixed on the same side as a result of this restricted rotation, then the isomer would be classified as "**cis**".*

Cis and trans isomers will differ in their physical and chemical properties. In order to identify them we prefix the cis or trans onto the front of the name of the isomer. For example, cis-1,2-dichloroethene would have two chlorines on the same side of the double bond. Trans-1,2-dichloroethene would have the two chlorines on opposite sides of the double bond.

Examples: The most likely example of geometric isomerism is but-2-ene. In one case, the CH_3 groups are on opposite sides of the double bond, and in the other case they are on the same side.

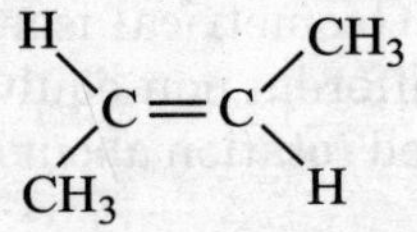

trans-but-2-ene

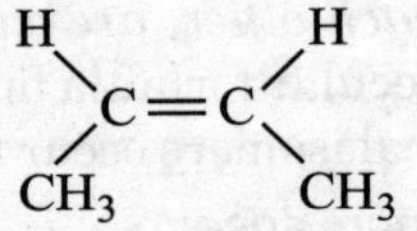

cis-but-2-ene

It must be emphasized that if the two atoms or groups of atoms attached to the same sp^2 carbon are the same, then **three cannot be geometrical isomerism** since the two like groups on the same sp^2 carbon could not differentiate between the two like groups.

On this basis we find that propylene, I-butene, and isobutylene should not show isomerism:

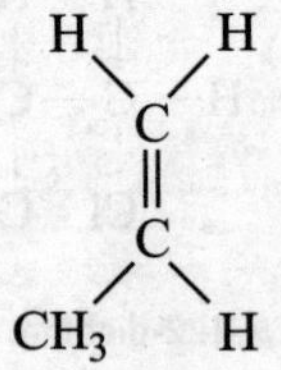

Propylene

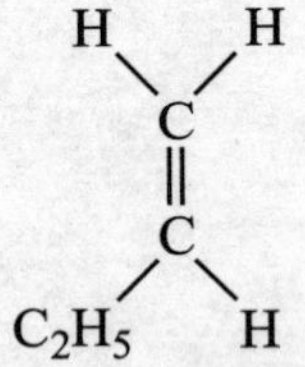

1-Butene

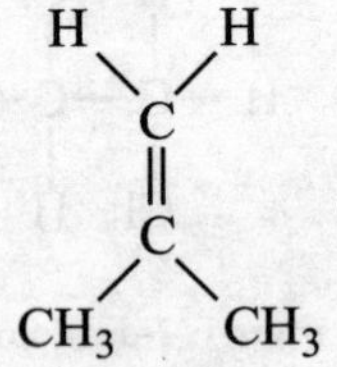

Isobutylene

Many higher alkenes may, of course, show geometric isomerism.

If we consider compounds other than hydrocarbons, we find that 1,1-dichloroethene and 1,1-dibromoethene should not show isomerism, whereas the 1,2-dichloroethenes and 1,2-dibromoethenes should. In every case these predictions have been found correct. Isomers of the following physical properties have been isolated.

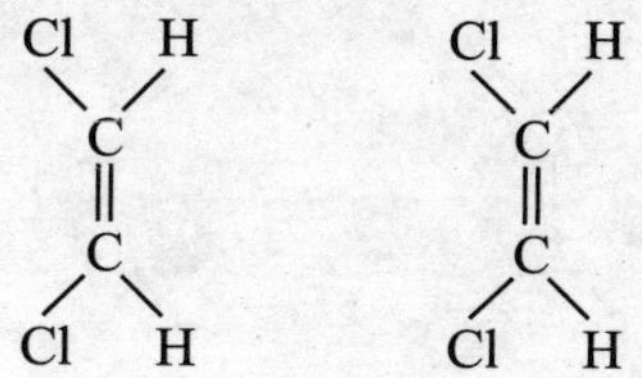

cis- *trans-*

1, 2-Dichloroethene

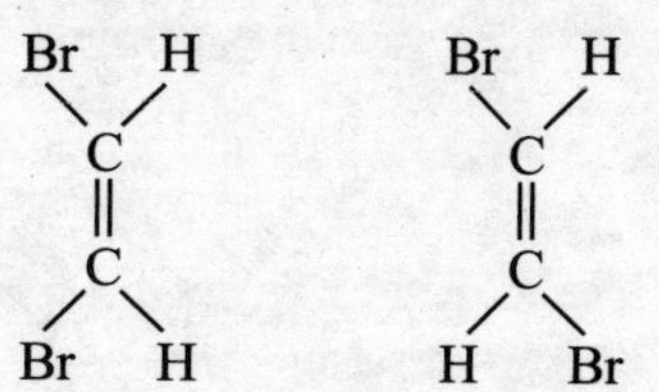

cis- *trans-*

1, 2-Dibromoethene

Geometric isomerism cannot exist if either carbon carries two identical groups. Some possible combinations are shown below.

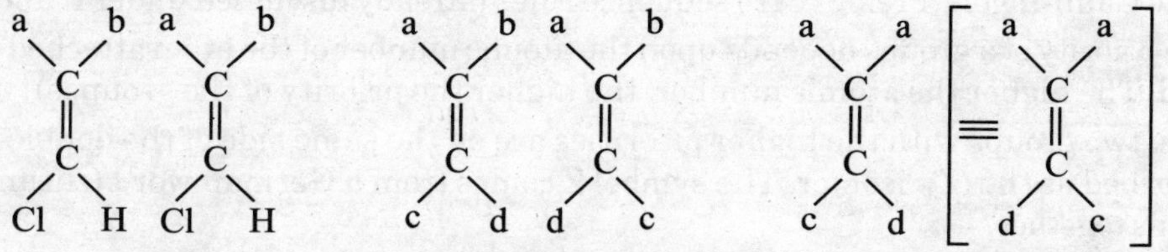

Isomerism *Isomerism* *No isomerism*

One might ask why there is no geometrical isomerism in Alkynes. That is because of the linear geometry around the sp hybrid carbons. The atoms attached to these carbons are in the same line as the carbons, so there is no difference in the geometrical orientation of these groups.

Geometrical Isomerism for Cyclic Hydrocarbon

The carbons found within the ring of a cyclic hydrocarbon also have restricted rotation. The carbons within the ring are sp^3 hybridised. If you attempt to rotate the carbons around each other again there will be a lessening of the overlap between two sp^3 orbitals. This, in turn, will cause instability if the overlap is decreased so there is a resistance to the rotation. Consequently, like the atoms attached to the sp^2 carbons in an Alkene, the atoms attached to the sp^3 carbons within the ring become fixed as long as the ring remains intact. If two similar atoms or groups of atoms find them selves on opposite sides of the ring, then the groups are trans and the isomer is the **trans** form. If the two similar groups are on the same side of the ring structure then we have the **cis** isomer. For example, Cis-1,2-dimethylcyclohexane would have the two methyl groups on the same side of the six-member ring system.

cis isomer **trans** isomer

diastereomers

Disubstituted cycloalkane can have two substituents on the same side of the ring (*cis* isomer, A) or on opposite sides of the ring (*trans* isomer, B). Note that these compounds are stereoisomers but not mirror images and hence are diastereomers.

E and Z nomenclature:

The expression cis and trans only applies to alkenes or cycloalkanes if one of the substituents on each of the carbons are the same. However, if there are three or four different substituents, it is not possible to

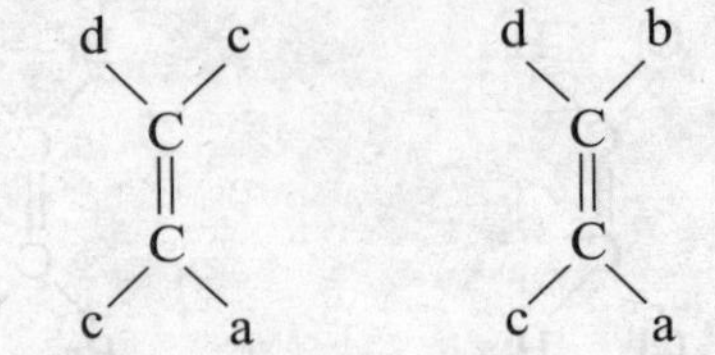

use the cis and trans system of naming the isomers. For such systems Cahn, Ingold and Prelog priority system or commonly known as E,Z nomenclature must be used.

Rules for E and Z nomenclature;

(1) Assign the priority to the groups attached to each carbon atom of the double bond according to the Cahn-Ingold-Prelog (CIP) sequence rules (already discussed under R and S system). (The priority of a groups depends upon the atomic number of the atom attached to the double bond. The higher the atomic number, the higher the priority of the group is)

(2) If the two groups with the higher priorities are on the same side of the double bond, that is described as the (Z)- isomer. The symbol Z comes from a German word (zusammen) which means together

(3) If the two groups with the higher priorities are on opposite sides of the double bond, then this is the (E)- isomer. E comes from the German entgegen which means opposite.

> (E)- Isomer: The higher priority groups are on opposite sides of the double bond.
> (Z)- Isomer: The higher priority groups are on the same side of the double bond

Let us explain the above rules with the help of following examples:

(1) 

- Compare the atomic numbers of the attached atoms to work out the various priorities.

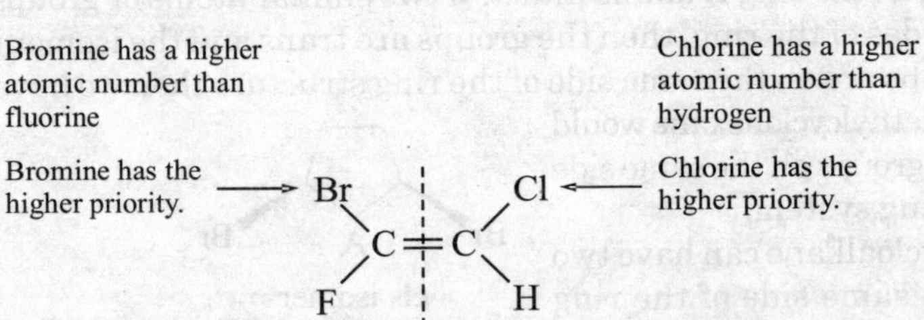

- Notice that the atoms with the higher priorities are both on the same side of the double bond. That counts as the (Z)- isomer.
- The second isomer obviously still has the same atoms at each end, but this time the higher priority atoms are on opposite sides of the double bond. That's the (E)- isomer.

(2) 1,2-dichloroethene.

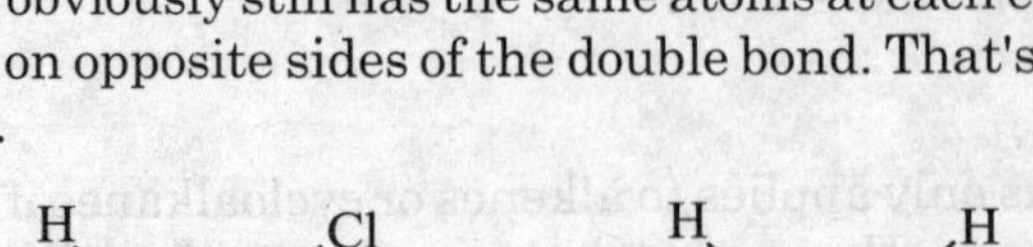

In the first isomer, the higher priority groups are on opposite sides of the bond. That must be the (E)- isomer. The other one, with the higher priority groups on the same side, is the (Z)- isomer.

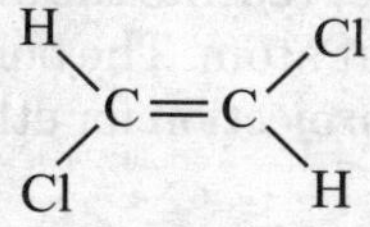

(E)-1, 2-dichloroethene

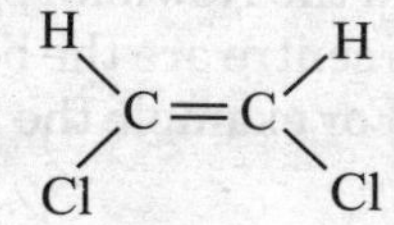

(Z)-1, 2-dichloroethene

(3) But-2-ene:

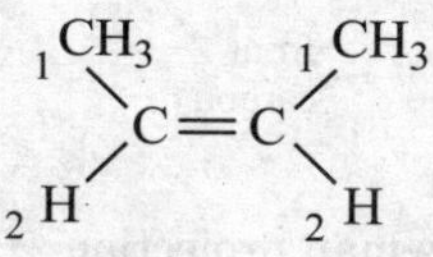

(Z) But-2-ene

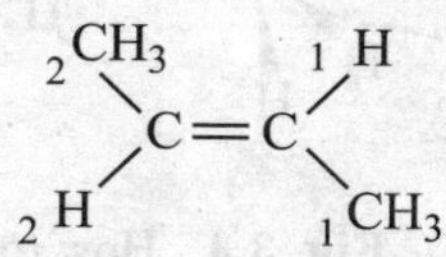

(E) But-2-ene

The CH_3 group has the higher priority because its carbon atom has an atomic number of 6 compared with an atomic number of 1 for the hydrogen also attached to the carbon-carbon double bond.

3.10 CONFORMATIONS

Conformations are different forms of a molecule, related by simple rotation about a single bond. For example alkanes of two or more carbons can be twisted into a number of different three-dimensional arrangements by rotation about a carbon-carbon single bond. The different spatial arrangements that a molecule can adopt due to rotation about σ bonds are called ***conformations*** and hence conformational isomers or ***conformers.***

Such interconversions are usually very rapid, so that a sample of a given molecule may exist in many different conformations.

Conformational isomerism is a form of stereoisomerism involving the phenomenon of molecules with the same structural formula existing as different conformational isomers or conformers due to atoms rotating about a bond. Different conformers can interconvert by rotation around single bonds, without breaking chemical bonds.

How Conformational Isomerism Arise:

The existence of more than one conformation is due to hindered rotation of sp^3 hybridised carbon bonds. Let us explain this by taking the simple example of ethane which has carbon- carbon sigma bond. Sigma bond is cylindrically symmetrical in nature and hence permits free rotation about its axis. In other words sigma bond permits *free rotation* about its axis. Hence, *the energetic barrier to rotation about sigma bonds is generally very low.*

The different spatial arrangements formed by rotations about a single bond are called conformations or conformers.

Difference between conformation and configuration : The term conformation should not be confused with the configuration which relates to those spatial arrangements of the atoms of a molecule that can be changed only by the breaking and making of bonds whereas the spatial arrangements in conformation are changed simply by rotation about a single bond.

Visualising Conformations:

(a) Newmann projection: A Newman projection can be used to specify the confirmation of a particular bond with clarity and detail. A Newman projection represents the head-on look down the bond of interest. The circle in the Newman projection represents the atom in front of the bond, and the lines radiating from the centre are the bonds of that atom. The bonds of the rear atom emerge from the sides of the circle. For example the Newman projection for ethane can be drawn like:

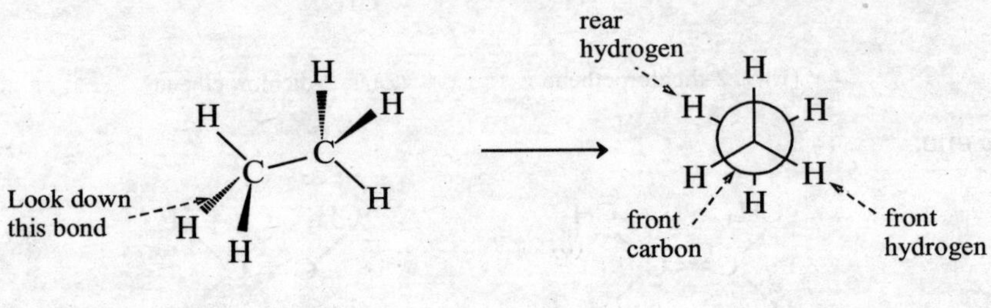

Fig. 3.3 How to draw a Newman projection

Newman projections can be characterised by the angles formed between bonds on the front atom and bonds on the rear atom. Such angles are called dihedral angles.

(b) Sawhorse Structures: Rotation around a bond in a linear structure can also be shown through the use of sawhorse structures. In this projection, the molecule is viewed along the axis of the model from an oblique angle. The central carbon-carbon bond (C—C) is draw as a straight line slightly tilted to right for the sake of clarity. The front carbon is shown as the lower left hand carbon and the rear carbon is shown as the upper right hand carbon. The three bonds around each carbon atom (C—H in ethane or C—C in higher alkanes) are shown by three lines

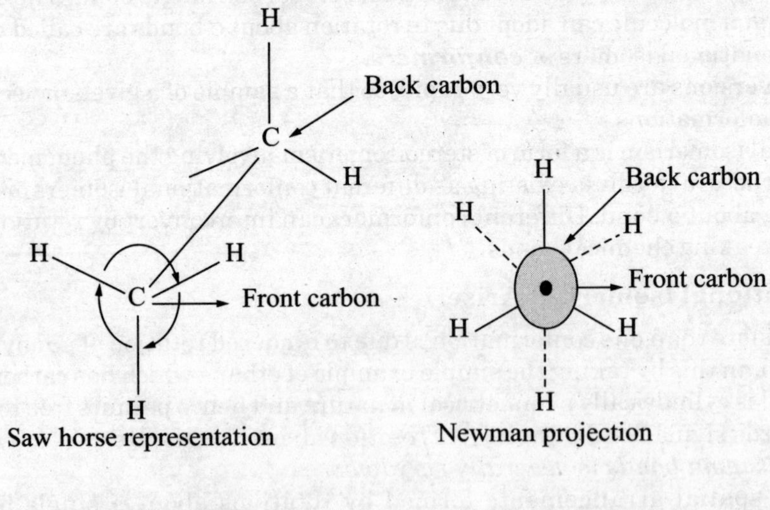

The Wedge and Dash, Sawhorse and Newmann Projection of ethane are shown in Figure Fig. 3.4:

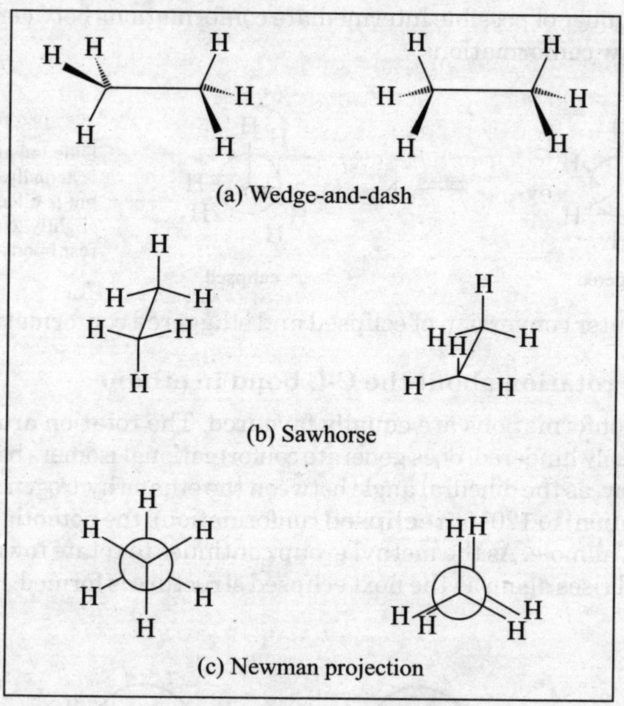

Fig. 3.4 *Wedge and Dash , Sawhorse and Newmann Projections of ethane*

Conformations of Ethane

While there are an infinite number of conformations about any sigma bond, in ethane two particular conformers represent the two extremes and have special names.

(i) Eclipsed: In the eclipsed conformation, the C-H bonds on the front and back carbons are aligned with each other with dihedral angles of 0 degrees.

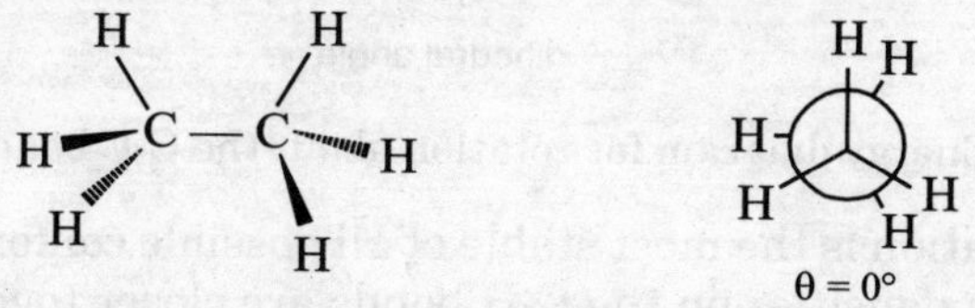

This conformation of ethane is the **eclipsed** form. The hydrogen atoms have a dihedral angle of 0°.

(ii) **Staggered:** In the staggered conformation, the C-H bonds on the rear carbon lie between those on the front carbon with dihedral angles of 60 degrees.

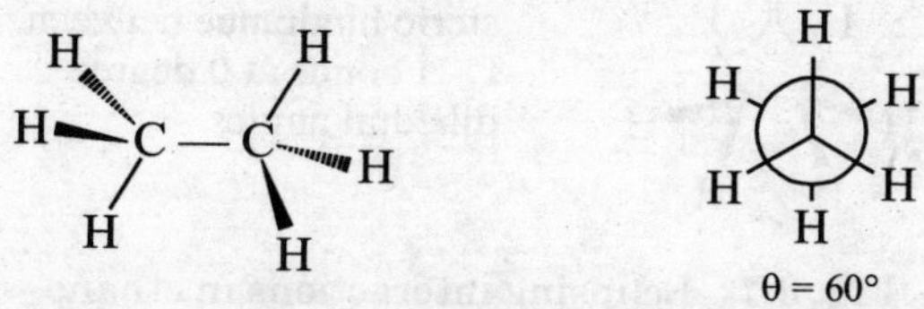

This conformation of ethane is the **staggered** from. The various hydrogen atoms on the two carbons form a dihedral angle of 60°.

(iii) The infinite number of possible intermediate conformations between these two extremes is referred to **skew** conformations

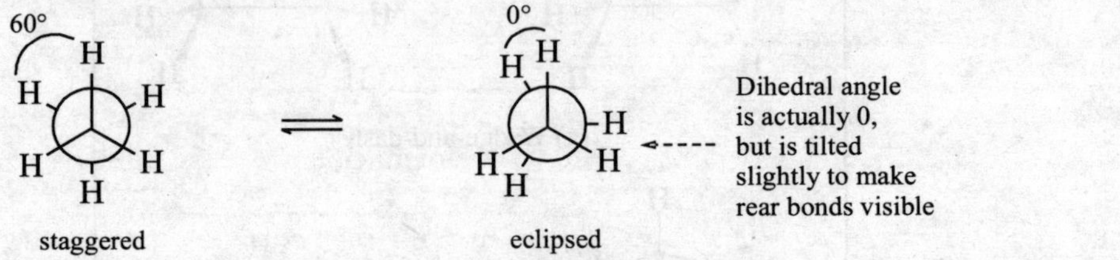

Fig. 3.5 The inter conversion of eclipsed and staggered conformations of ethane.

Energy diagram for rotation about the C-C bond in ethane:

Energetically, not all conformations are equally favoured. The rotation around the single bond in ethane, while not obviously hindered, does generate conformational isomers having different potential energies. As shown below, as the dihedral angle between the ethane hydrogen atoms changes from 60° (a **staggered** conformation) to 120° (an **eclipsed** conformation), the potential energy of the molecule increases by about 3 kCal/mole. As the methyl group continues to rotate towards 180°, the potential energy again drops and rises again as the next eclipsed structure is formed.

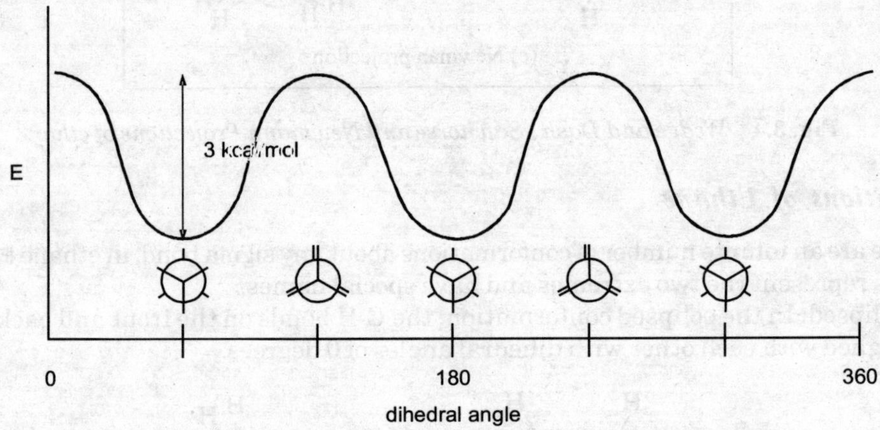

Fig. 3.6 Energy diagram for rotation about the C-C bond in ethane

The staggered conformation is the most stable of all possible conformations of ethane. In the eclipsed form, the electron densities on the C-H bonds are closer together than they are in the staggered form. When two C-H bonds are brought into a dihedral angle of zero degrees, their electron clouds experience repulsion, which raises the energy of the molecule. The eclipsed conformation of ethane has three such C-H eclipsing interactions.

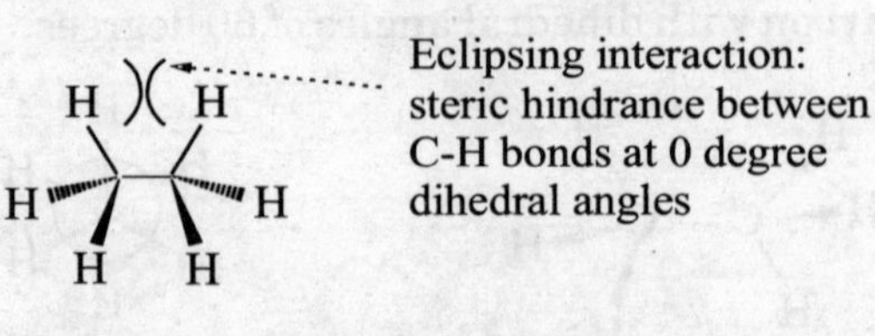

Fig. 3.7 Eclipsing interactions in ethane

Eclipsing interactions, which are due steric hindrance cause resistance to rotation, it is also called *torsional strain.* The energy (3 kCal/mol) needed to overcome this resistance is the torsional energy. At room temperature, ethane molecules have enough energy to be in a constant state of rotation. Although the term "conformational isomer" is sometimes used as a synonym for conformations, conformations of a molecule are not considered true isomers because of their rapid interconversion.

We can say that as one methyl group rotates 360° relative to the other, the compound passes through three stable staggered conformers via three unstable eclipsed forms. This barrier is small enough that at 25°C the compound changes conformation about 50 million times each second.

2. Conformations of Butane

In butane it is the rotation about the C2-C3 bond that is of most interest since the relative position of the two methyl groups is important. Butane has not only eclipsed and staggered conformations, but also forms that vary in the relative orientation of the methyl groups.

(i) **Fully Eclipsed (Syn):** The structure shown at 0° is fully **eclipsed**, that is, both methyl groups are aligned and are interacting.

(ii) **Gauche:** The two staggered forms with the methyl groups in closer proximity (60°) are the *gauche* conformations. As the front methyl group is rotated 60° in structure I, a **gauche** conformation is produced. The two staggered forms with the methyl groups in closer proximity (60°) (II and VI) are the *gauche* conformation

(iii) **Partially Eclipsed:** Another 60° rotation of conformation II produces an eclipsed version of the gauche conformation.

(iv) **Anti:** The staggered conformation where the two methyl groups are as for away from one another as possible (with a dihedral angle of 180°, seen in the Newman projection) is called the *anti* conformation, and is the lowest-energy arrangement. This is the most stable conformer in which no group is eclipsed.

Further rotations regenerate an equivalent eclipsed gauche conformer (at 240°), another gauche form (300°) and finally, the eclipsed form at 360°.

The illustration of butane below is represented by the Newman Projections below by designating the two middle carbons, one as the "front" and one as the "behind" carbon, and connecting the two end methyl groups accordingly.

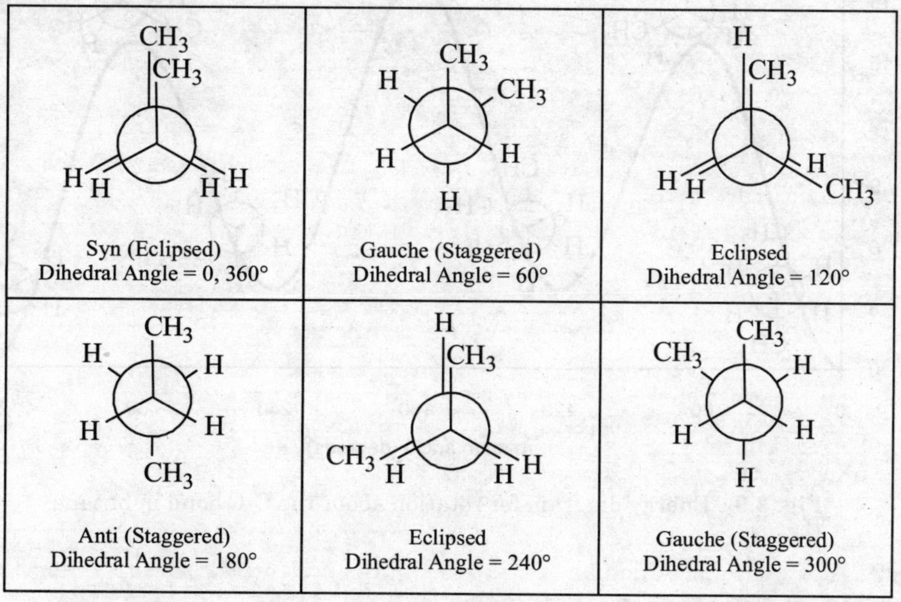

Energy diagram for rotation about the C-C bond in butane: (Conformational Analysis)

- As with ethane, the eclipsed conformations are higher energy than the staggered group are in close proximity with the molecule, therefore the potential energy is at its highest.

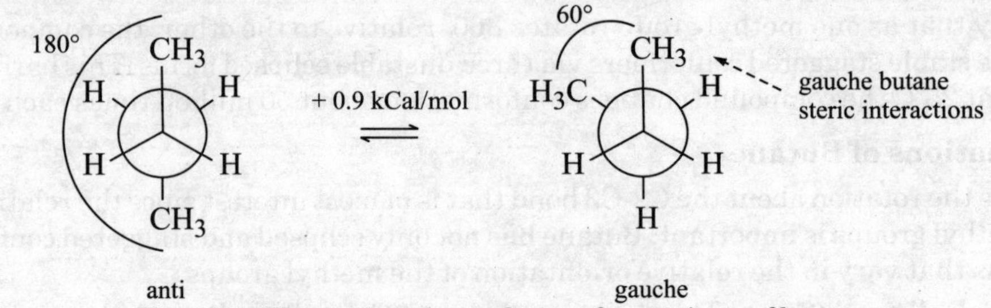

Fig. 3.8 The anti and gauche conformations of butane

- When the dihedral angle = 60° or 300°, (gauche conformation) in butane, the methyl groups are farther apart and therefore the potential energy drops by 4.1 kCal/mol. Gauche form is less stable than the anti form by 0.9 kCal/mol due to steric hindrance between the two methyl groups. Such an interaction is often referred to as a gauche-butane interaction because butane is the first alkane discovered to exhibit such an effect.
- When the dihedral angle = 120° or 240°, (eclipsed conformation) in butane, the methyl groups align themselves with hydrogens on the same plane, resulting in the butane having a higher potential energy than the gauche conformations though still having a lower potential energy than the syn conformation (dihedral angle - 0° or 360°) by 1.5 kCal/mol.
- The staggered conformation (when the dihedral angle = 180°) where the two methyl groups are as far away from one another as possible, i.e. the *anti* conformation, is the lowest-energy arrangement.

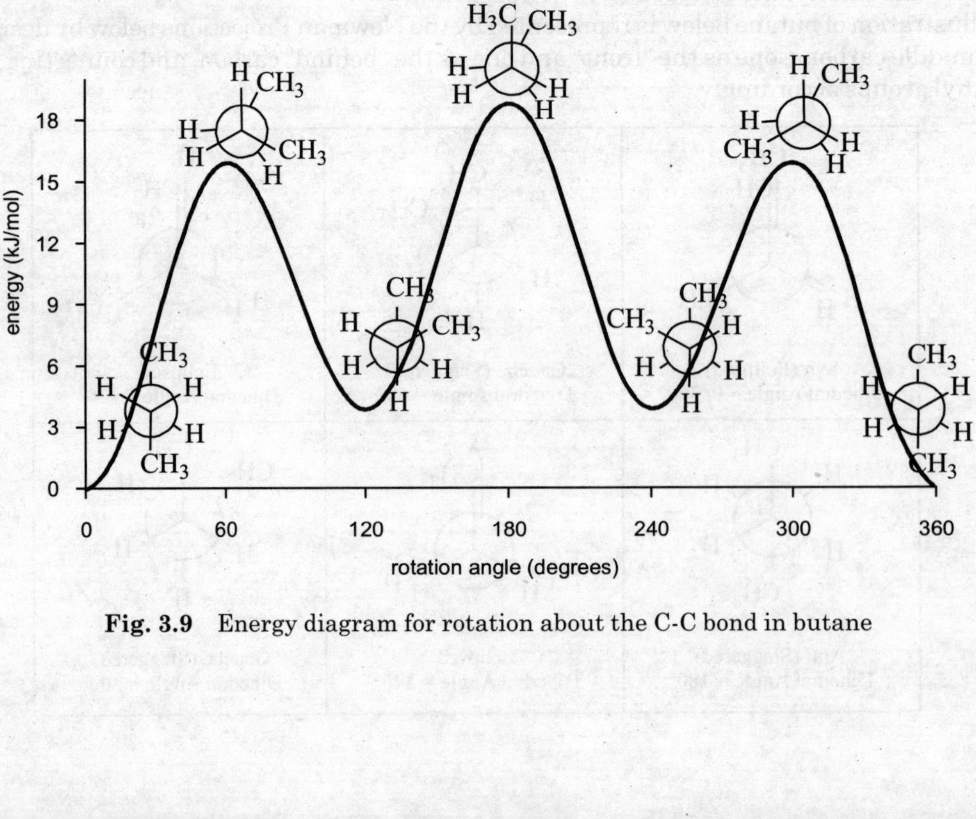

Fig. 3.9 Energy diagram for rotation about the C-C bond in butane

As with Newman Projections, **Sawhorse Projections** can be made for butane, such that its *eclipsed*, *gauche*, and *anti* conformations can be seen. Notice that it is much easier to determine the number of carbons in the longest chain using **Sawhorse Projections** than it is for Newman Projections because the carbon-carbon bond between the second and third carbons is drawn out.

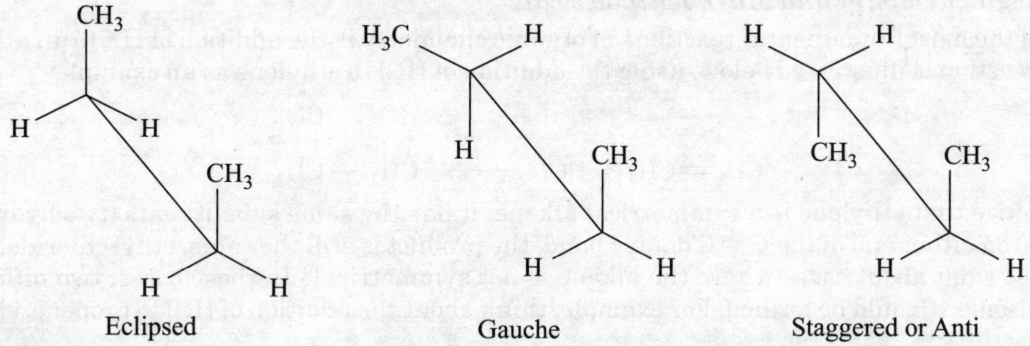

Fig. 3.10 *Saw horse Projections of Conformations of Butane*

KEY TERMS

Conformation - Different 3-dimensional shapes that can be taken by a molecule by rotating about single bonds.

Newman projection - A depiction of the conformation of a molecule where it is viewed down the atoms of a particular bond.

Dihedral angle - Angle formed between two bonds on neighbouring atoms in a Newman projection.

Staggered - A conformation where bonds have dihedral angles of sixty degrees. This typically minimizes the energy.

Eclipsed - A conformation where bonds have dihedral angles of zero degrees. This typically maximizes the energy.

Steric hindrance - Repulsion between the electron clouds on bulky groups of a molecule or between molecules.

Gauche - A staggered conformation where large groups at a dihedral angle of 60 degrees repel each other.

Anti - A staggered conformation where two large groups are placed in their most stable positions at a dihedral angle of 180 degrees. Also refers generally to any geometrically arrangement of groups in opposing positions

Syn - Steric hindrance between terminal methyl groups in a system of at least five carbon centers. Also a generic term describing groups oriented in the same direction; i.e. the opposite of anti.

Torsional strain - Resistance to rotation about a single bond caused by steric hindrance of eclipsing interactions.

Torsional energy - The energy it takes to overcome torsional strain, or the difference in energy between eclipsed and staggered conformations.

3.11 REGIOCHEMISTRY AND STEREOCHEMISTRY

Regiochemistry or regioselectivity refers to what region or area of two possible and similar regions will be involved in a reaction.

(a) Regiochemistry and Markovnikov's rule

One of the most fundamental reactions in organic chemistry is the addition of HX to an alkene. This reaction is illustrated below, using the addition of HCl to ethylene as an example:

$$CH_2{=}CH_2 + HCl \longrightarrow \underset{H}{CH_2} - \underset{Cl}{CH_2}$$

Notice that ethylene is a symmetrical alkene: it has the same substituents (two hydrogen atoms) on either end of the C=C double bond, the product is still the same, ethyl chloride:

But what about cases where the alkene is not symmetrical? In those cases, two different "regioisomers" could be formed. For example, think about the addition of HCl to propene, shown below:

$$H{-}Cl + H_2C{=}CHCH_3 \longrightarrow H_2\overset{H}{C}{-}\overset{Cl}{C}HCH_3 = H_3C{-}\overset{Cl}{CH}{-}CH_3$$

isoproply chloride

or

$$Cl{-}H + H_2C{=}CHCH_3 \longrightarrow H_2\overset{H\;\,Cl}{C}{-}CH{-}CH_3 = ClCH_2CH_2CH_3$$

isoproply chloride

In fact, the only product that is formed in the reaction of propene with HCl is isopropyl chloride. None of the regioisomer (n-propyl chloride) is formed. As such,we call this a "regiospecific" reaction, since only one of several possible regioisomers was formed.

(b) Regiochemistry in elimination reaction (Saytzef Product):

In an elimination reaction of 1-bromo-1- methylcyclohexane with sodium hydroxide, regioselectivity is the favoring of elimination between the carbons that would result in the more substituted alkene (the Saytzev product).

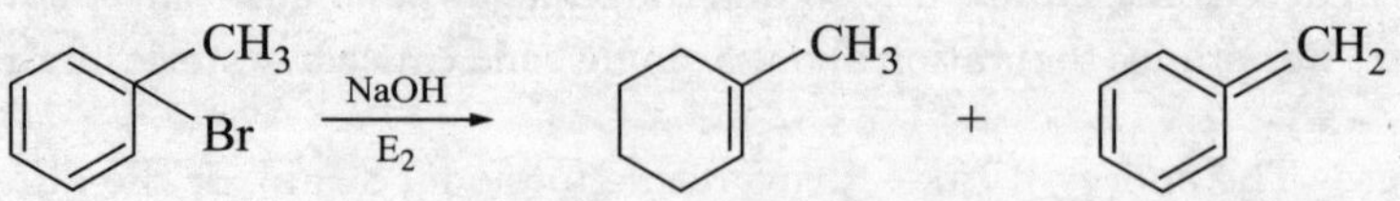

major product Saytzef monor product Hofmann

3.12 STEREOSPECIFICITY

A reaction that yields predominately one stereoisomer is called stereospecific reaction. Stereospecificity is where the mechanism of the reaction only allows formation of one stereoisomer.

Example:

2- butene has 2 geometrical isomers . On bromination it can give 3 sterioisomers as shown below:

$$CH_3CH{=}CHCH_3 + Br_2 \longrightarrow CH_3\underset{Br}{CH}\underset{Br}{CH}CH_3$$

2–butene → 2,3-dibromobutane

2 gemetric isomers *cis*- and *trans*- → 3 stereoisomers (*S, S*)-, (*R, R*)-, and (*R, S*)- *meso*-

Bromination of trans butene yield only meso -2,3–dibromobutane (stereospecific reaction)

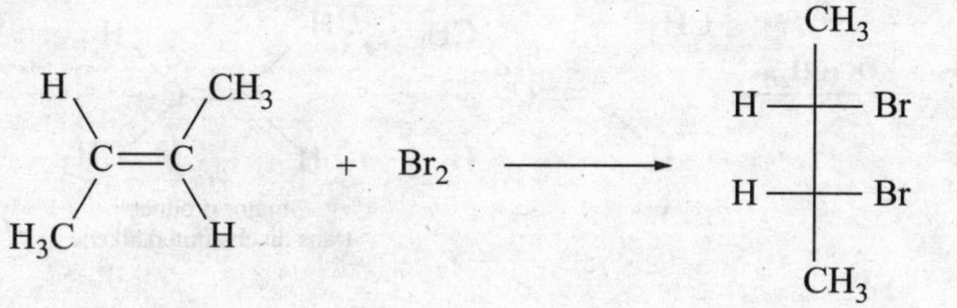

trans-2-butene → *meso*-2,3-dibromobutane *only product*

Bromination of cis butene yields racemic modification :

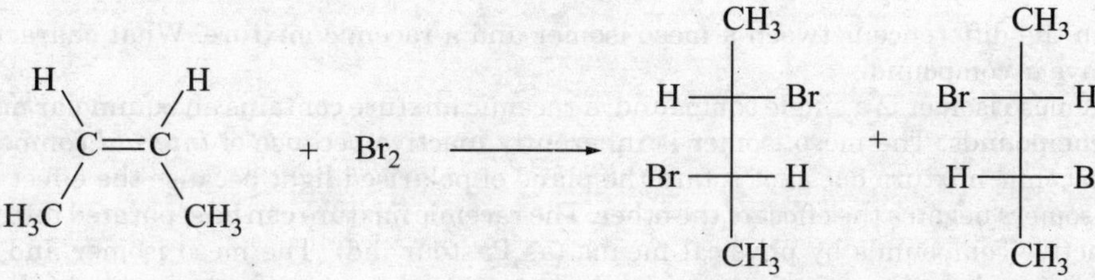

cis-2-butene → (*S, S*)- &(R, R)- 2, 3-dibromobutane *racemic modification only products*

A reaction in which stereochemically different molecules react differently is called a stereospecific reaction. In this case the *cis*- and *trans*- stereoisomers give different products.

3.13 STEREOSELECTIVITY

A stereoselective reaction is one where the mechanism allows formation of both products, but one product is preferred over the other.

According to this definition, a stereospecific process is necessarily stereoselective but not all stereoselective processes are stereospecific.

Recall that alkenes can be prepared from alkyl halides and alcohols via elimination reactions.

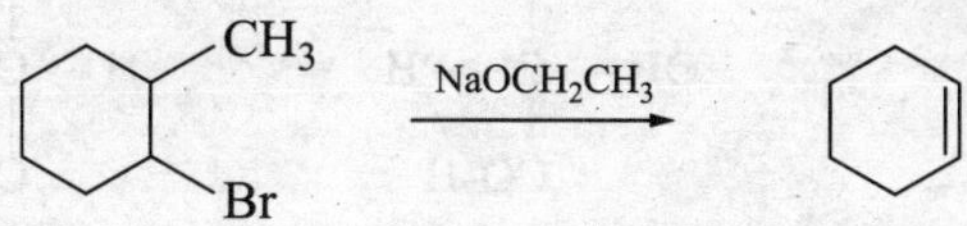

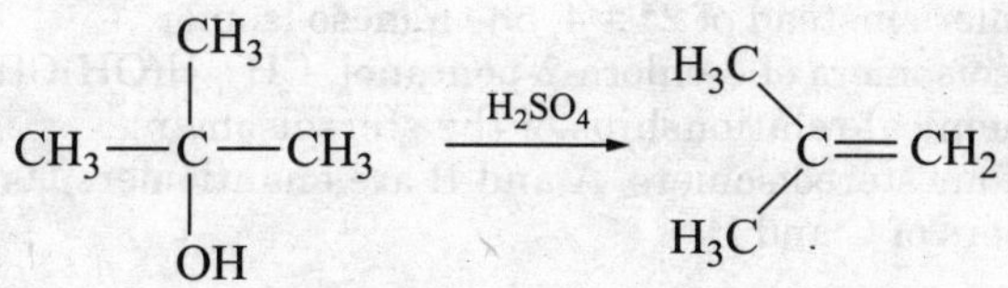

These elimination reactions are stereoselective and regioselective, the most stable alkene is usually formed as the major product.

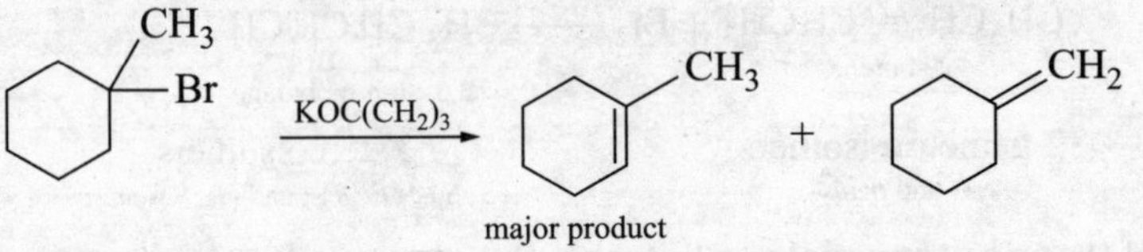

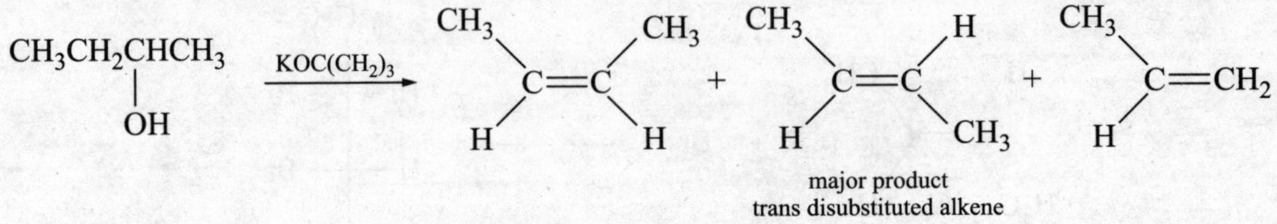

SOLVED EXERCISE

1. Explain the difference between a meso isomer and a racemic mixture. What characteristic(s) do they have in compound.

   Ans: A meso isomer is a single compound; a racemic mixture contains an equimolar mixture of two compounds. The meso isomer is inherently inactive because of *internal* compensation; the racemic mixture does not rotate the plane of polarised light because the effect of one of the isomers negates the effect of the other. The racemic mixture can be separated into two optically active compounds by physical means (as Pasteur did). The meso isomer and the racemic mixture have in common their inability to rotate the plane of polarised light. Both have two centres of (opposite) chirality.

2. How many optical isomers can exist for 2,3-butanediol? Would all of these be optically active?

   Ans: There are three optical isomers, one of which is inactive (the meso form) because of the two carbon atoms having the same groups arranged in opposite configurations.

3. Tartaric acid is a dihydroxydicarboxylic acid, $HOCOCHOHCHOHCO_2H$. Identify any chiral centers, draw all optical isomers, and explain the total number of such isomers of tartaric acid in terms of the 2^n rule.

   Ans: The two middle carbon atoms in each structure are chiral centers. They are starred, to indicate their chirality.

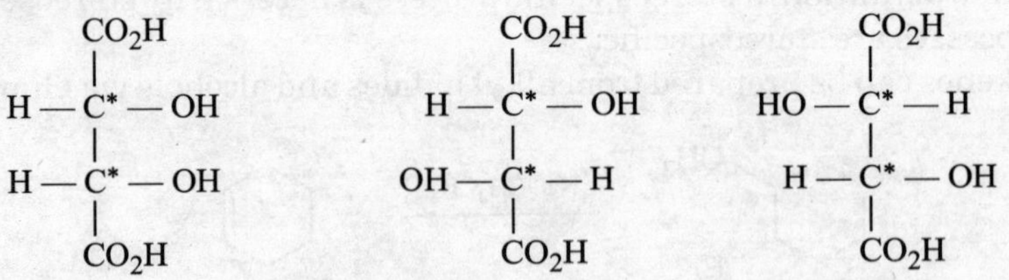

meso

   There are 3 isomers instead of $2^2 = 4$, one a meso isomer.

4. (a) Draw all the stereoisomers of 3-chloro-2-pentanol, $CH_{3C}H(OH)CHClCH_2CH_3$.

   (b) Give the stereochemical relationships of the stereoisomers.

   Ans: (a) There are four stereoisomers. A and B are enantiomers, as are C and D. (b) A and are diastereomers of C and D.

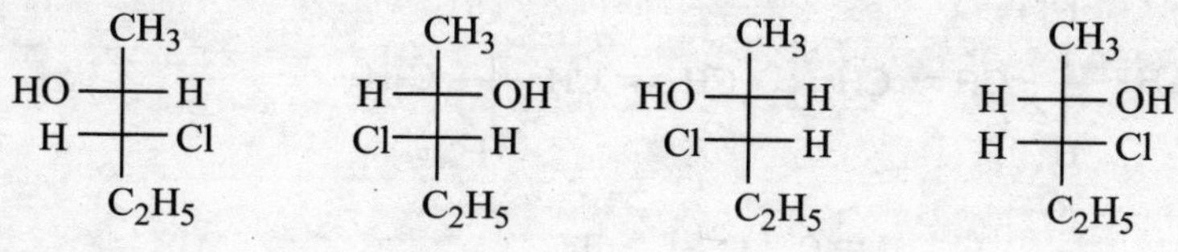

A B C D

5. Give the R/S designation for each of the stereoisomers of 1,3-dibromo-2-methylbutane.
Ans: There are two chiral C's (C^2 and C^3), and the four stereoisomers are shown below with their designation.

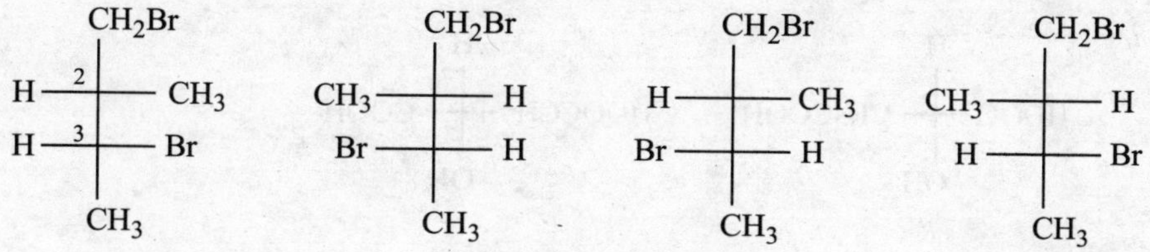

2S, 3R *2R, 3S* *2S, 3S* *2R, 3R*

6. (a) Theoretically, what is the maximum number of possible stereoisomers of a compound having *n* dissimilar chiral atoms?
(b) What is the maximum number of stereoisomers possible for a compound having (i) three (ii) four different chiral centers?
Ans: (a) A compound with *n* different chiral atoms can have 2^n stereoisomers.
(b) (i) $2^3 = 8$ (four racemic of enantiomers). (ii) $2^4 = 16$ (eight racemic pairs of enantiomers).
7. Draw the enantiomers of (a) chloroiodomethane sulphonic acid and label each as R or S.
Ans:

R *S*

I > Cl > S > H

8. What is the priority sequence of CH_3, 1°, 2° and 3° alkyls?
Ans: 3° > 2° > 1° > CH_3

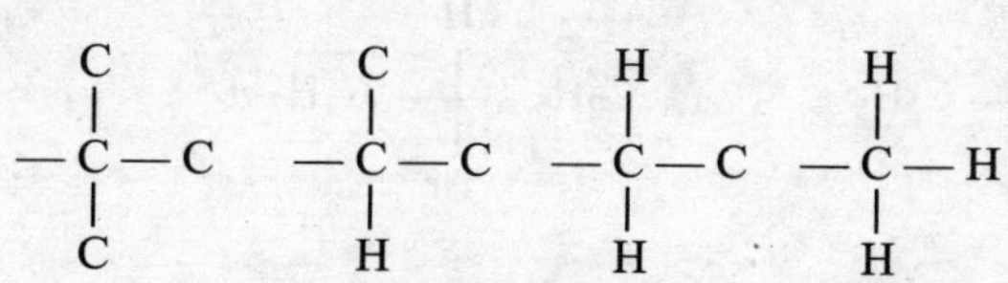

C, C, C	>	C, C, H	>	C, H, H	>	H, H, H
3°		2°		1°		CH_3

9. Draw and specify as R or S of the following:
(a) 3-chloro-l-pentene
(c) Malic acid
(c) $CH_3CH(NH_2)COOH$ (Alanine)
(d) $C_6H_5CH(CH_3)NH_2$

Ans: (a)

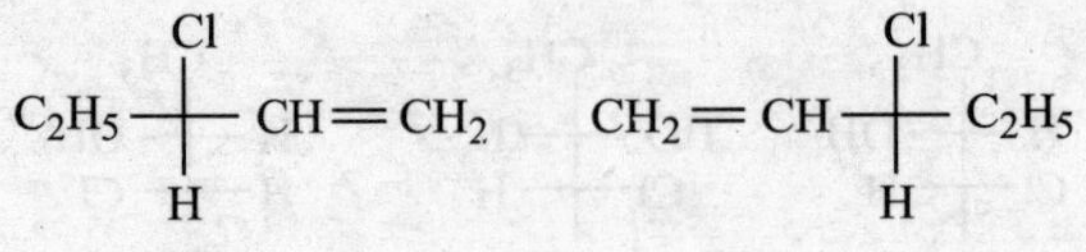

R *S*

Cl > C═C > C—C > H

(b)

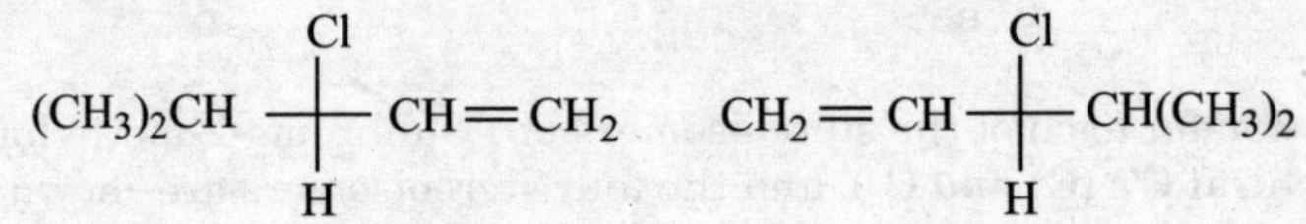

R *S*

Cl > C═C > C—C > H

(c)

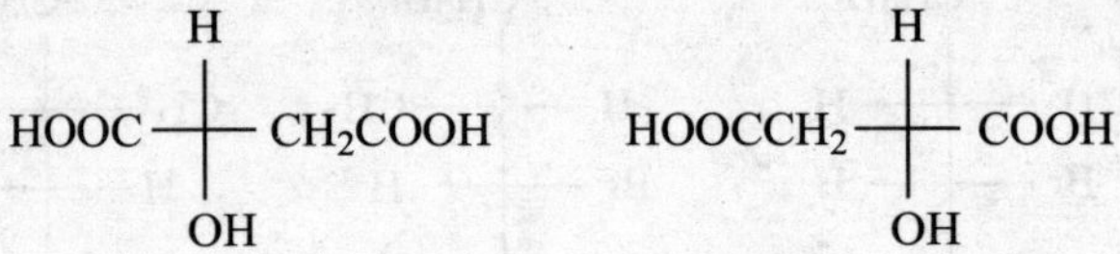

R *S*

O > C—O > C—C > H

(d)

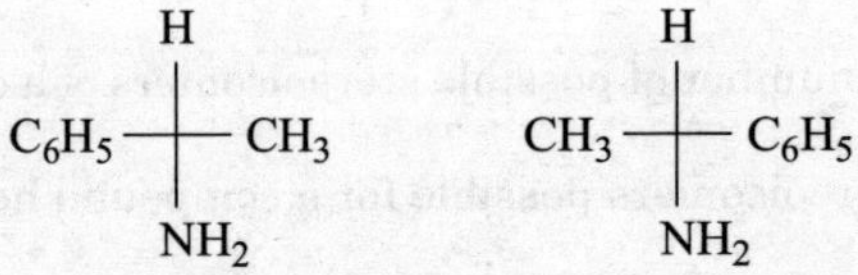

R *S*

N > C_{aryl} > C_{aliph} > H

(e)

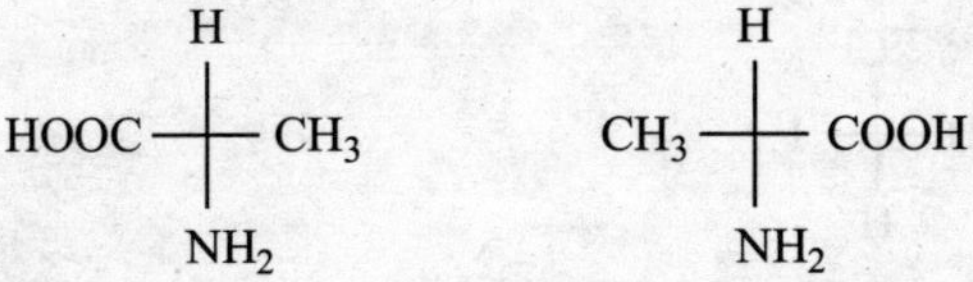

R *S*

N > C—O > C—H > H

10. Draw and specify as R or S enantiomer (if any) of:
(a) 3-bromohexane (b) 1-deuterio-1-chlorobutane

Ans: (a)

R *S*

Br > C,C,C > C,C,H > H

(b)

R *S*

Cl > C > D > H

11. Draw stereochemical formulas of the following compounds and give R/S designation (a) $CH_3CHBrCHOHCH_3$ (b) $CH_3CHBrCHBrCH_2Br$

Ans: (a)

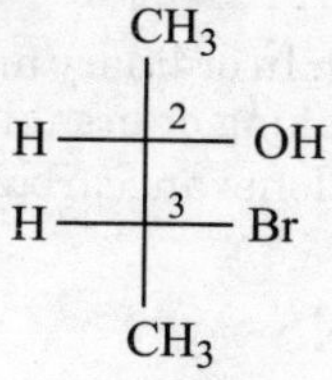

2*S*, 3*R*

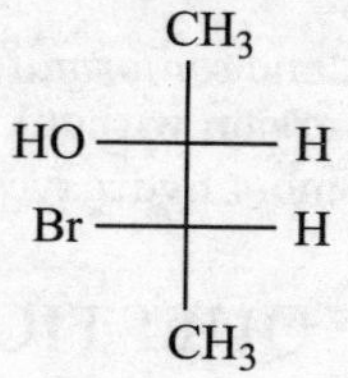

2*R*, 3*S*

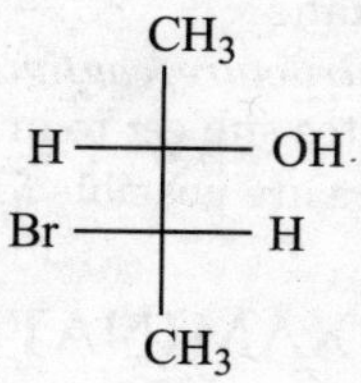

2*S*, 3*S*

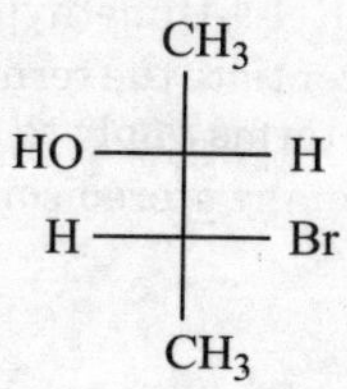

2*R*, 3*R*

Enantiomers *Enantiomers*

(b)

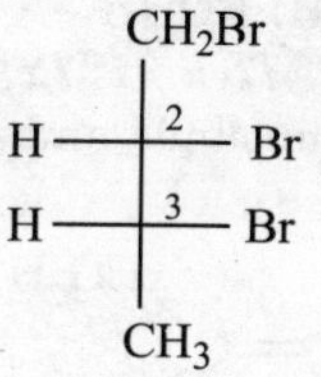

2*S*, 3*R*

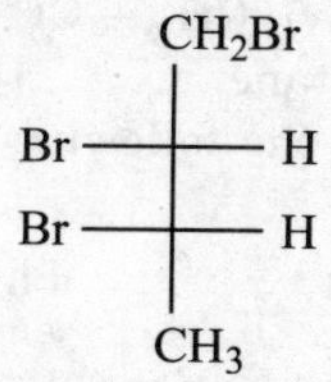

2*R*, 3*S*

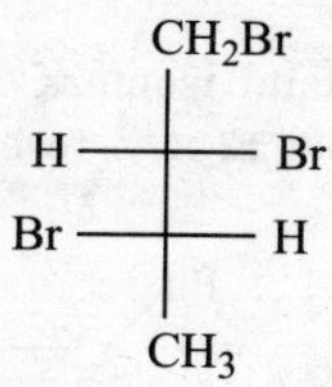

2*S*, 3*S*

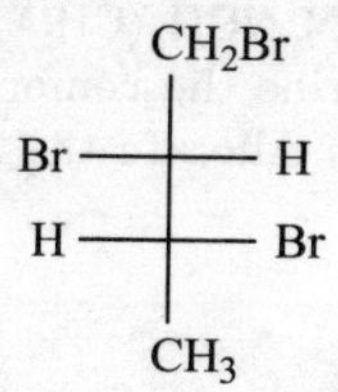

2*R*, 3*R*

Enantiomers *Enantiomers*

QUESTIONS

1. Explain optical isomerism of chiral organic compounds.
2. Write five examples of organic compounds showing optical isomerism without chirality.
3. Define the term; plane of symmetry, centre of symmetry and alternating axis of symmetry.
4. State differences between enantiomers and diastereoisomers with suitable examples.
5. An optical pure compound A, gave an $[\alpha]^{25} = +20°$. A mixture of A and its enantiomers B, gave $[\alpha]^{25} = +10°$. Calculate the ratio of A to B in the mixture.
6. An optically active compound A having molecular formula C_6H_{12} gives inactive compound B of molecular formula C_6H_{14} on hydrogenation. Write the structures of A and B.
7. What are the essential conditions for a compound to show geometrical isomerism?
8. Write three examples of geometrical isomerism.
9. Write out structures for the following substances.
   (a) The three conformations of 2-methylbutane with end-on formulas. Which formulas are superimposable?
   (b) Three conformations of 2-butanone with three-dimensional formulas (the groups attached to any unsaturated carbon atom all lie in one plane).
   (c) The two conformation of *cis*-4-isopropyl-1-methylcyclohexane.
   (d) The two conformations of *trans*-4-isopropyl-1-methylcyclohexane.
   (c) The two conformations of *cis*- and the two conformations of *trans*-1,3-dimethylcyclohexane.
10. Write out structural formulas for the following compounds.
    (a) *cis*-2,3-Dichloro-2-pentene
    (b) *trans*-4-Methyl-3-octene
    (c) *trans*-2-Phenyl-2-butane
    (d) *anti*-oxime of a-naphthaldehyde
11. Define the following terms:
    (a) Racemization
    (b) Mirror plane
    (c) Epimerization
    (d) Center of symmetry
    (e) Resolution
    (f) Diastereoisomerism
    (g) Equatorial bond

12. Write three-dimensional formulas for the following compounds:
    (a) CHClBrF (R configuration) (b) * (s configuration)
    (c) 3-R-4-s-Dimethylhexane
13. Differentiate the terms *structure, configuration,* and *conformation.* In ordinary molecules, which of these terms apply to states subject to interconversion without making or breaking bonds?
14. How many stereoisomers are possible for 2-bromo-4 hydroxycy-clohexanecarboxylic acid?

EXAMINATION QUESTIONS

1. Which of the following compounds would show optical isomerism :
   (i) H_2NCH_2 COOH (ii) $(CH_3)_2$ CHCHO
   (iii) N_2NCH $(CH_3)_2$ (iv) CH_3 – CH (OH) COOH
2. Describe the conformational isomers of *n*-butane. **[*U.P. Tech., 1*[st]*/2*[nd] *Sem., 2000-02*]**
3. (a) (i) Allocate the symbol E or Z to each of the following compounds and explain :

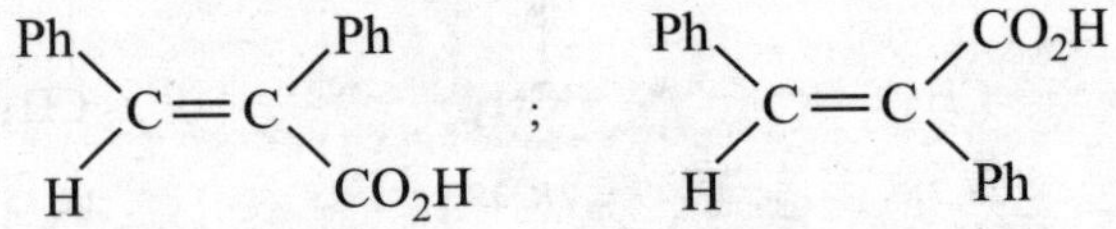

(Ph = Phenyl = C_6H_5)

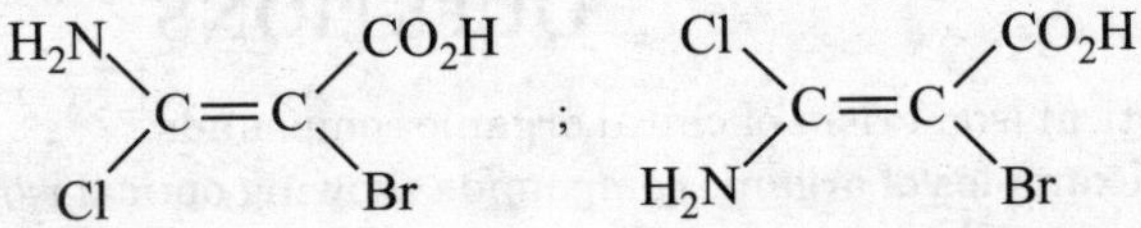

   (ii) Under what conditions can the terms cis and trans not be used ?
   (b) Discuss, in detail, the case of an organic compound with two Chiral centres.
   (c) Draw the potential energy diagram for the various conformations of *n*-butane.
   [U.P. Tech., B. Tech : 1st Sem., 2002-03]
4. Draw structures for the equilibrating chair conformes of trans- and cis-1, 2-dimethylcyclohexanes. Show all Me/H1, 3-diaxial interactions. Select the more stable conformation for each diastereoisomer. **[U.P. Tech., 2002-03]**
5. (a) (i) Indicate, with a suitable diagram, the potential energy changes during rotation about $C_{(2)} - C_{(3)}$ bond of *n*-butane. (ii) Comment on the relative stabilities of the conformers.
   (b) Name the elements of symmetry. Discuss, with the help of an example, an optically active compound without chirality. **[U.P. Tech., 2003-04]**
6. Attempt the following :
   (a) Write the two examples of optically active compounds without chirality.
   (b) Write short notes on: (i) Meso compound (ii) Racemisation
7. Write short notes on E, Z Nomenclature, Free Radicals, Electrophilic reagents and Nucleophilic reagents. **[U.P. Tech., 1st Sem., 2004-05]**
8. Define: (i) Enantiomers and Disastereoisoners.
   (ii) Which of the following compounds are optically active, and why, *n*-propanol, allenes, n-butanol and 2-chlorobutane.
   (iii) What is E-Z system of nomenclature ? In what way it is better than cis, trans-nomenclature?
   [U.P. Tech., 1st Sem., 2006-07]
9. What is optical activity ? Give the stereoisomers of Tartaric acid. How do you account for lack of optical activity in meso-forms and racemic mixtures. **[U.P. Tech., 2nd Sem., 2006-07]**

10. (a) Explain the mechanism of SN^1 and SN^2 organic reactions.

(b) (i) Differentiate between

– Enantiomers and diastereomers

– Racemic mixtures and meso compounds.

(ii) Draw the potential energy diagram for the various conformations of *n*-butane.

[*U.P. Tech., 2nd Sem., 2007-08*]

11. Which of the following compounds shows optical isomerism ?

(i) $CH_3 - CH(OH) - COOH$

(ii) $CH_3 - CHBr - CH_3$

(iii) $CH_3 - CH(CH_3) - CH_2CH_3$

(iv) $CH_3 - CH(OH) - CH_3$

[*U.P. Tech., 2nd Sem., 2007-08*]

12. (a) What do you understand by E-Z notations ? Assign E and Z configuration to the following compounds:

(i) 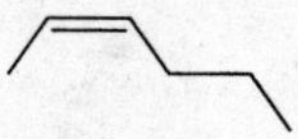

(ii) 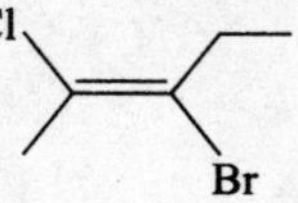

(iii) 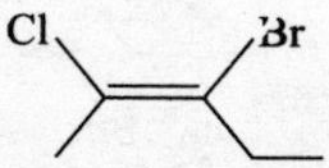

(iv) 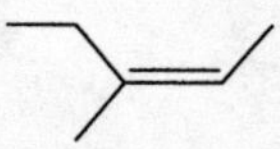

[*U.P. Tech., 2nd Sem., 2008-09*]

(b) Discuss the conformations of n-butane with the help of its potential energy diagram.

13. Explain with example

(i) Meso compounds

(ii) Enantiomers

(iii) Chiral compounds

(iv) E. Z nomenclature

[*U.P. Tech., 2nd Sem., 2008-09*]

❑❑❑

UNIT 2

Characterization and Analysis of Materials

4 Micro and Macro Properties of Molecules

Knowledge is a treasure, but practice is the key to it.

— ***Thomas Fullen***

CHAPTER OUTLINE

Materials constitute foundation of technology. Materials are so important in the development of human civilization that the historians have identified early periods of civilization by the name of most significantly used material, e.g.: Stone Age, Bronze Age.

Materials science is an applied science concerned with the relationship between the structure and properties of materials.

Chemists who work in the field study how different combinations of molecules and materials result in different properties. They use this knowledge to synthesize new materials with special properties. The central concept of materials science is relating the microstructure of a material to the properties you want it to have. By working with the microstructure, you can tailor the central properties of that material.

4.1 INTERMOLECULAR FORCES

The forces holding molecules together are generally called **intermolecular forces**. The term 'intermolecular forces' is used to describe all the forces, both attractive and repulsive, between groups of neutral molecules or atoms. (Forces between atoms within a molecule ('intramolecular forces') are usually called chemical bonds). The energy required to break molecules apart is much smaller than a typical bond-energy, but intermolecular forces play important roles in determining the properties of a substances. Intermolecular forces are particularly important in terms how molecules interact and form biological organisms or even life.

The various types of intermolecular forces acting between the molecules are:

4.1.1 Hydrogen bond

When a hydrogen atom is covalently bonded to a highly electronegative element such as flourine, oxygen and nitrogen, the resulting bond is highly polar. This is because highly electronegative element attracts the shared pair of electrons so strongly that it acquires partial negative charge and the hydrogen atom acquires a partial positive charge. The negative end of **one molecule** attracts the positive end of the **other molecule** forming a weak bond known as **hydrogen bond**. Thus hydrogen atom is able to act as a link between two electronegative atoms. This phenomenon was first considered by T.S. Moore T.F. Winwill and P.Pfeiffer. Its applicability was realized by W.M. Latimer and W.H. Rodebush.

"*The attractive force which binds hydrogen atom of one molecule with electronegative atom of another molecule is known as hydrogen bond.*" Hydrogen bond is generally represented by dotted line. This is illustrated in case of hydrogen flouride.

Thus hydrogen acts as a bridge between two strongly electronegative atoms holding one flourine atom by the covalent bond and other by electrostatic hydrogen bond.

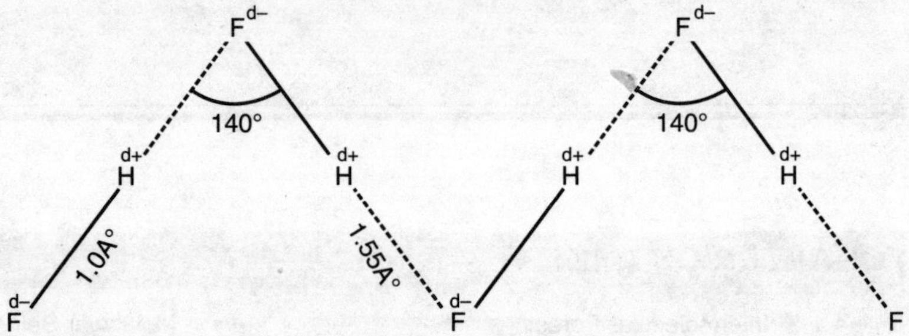

Fig. 4.1

Examples of hydrogen bonded compounds:

Ammonia

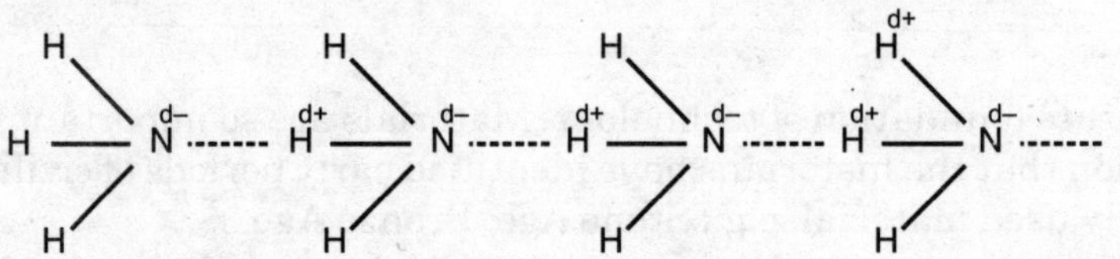

Water

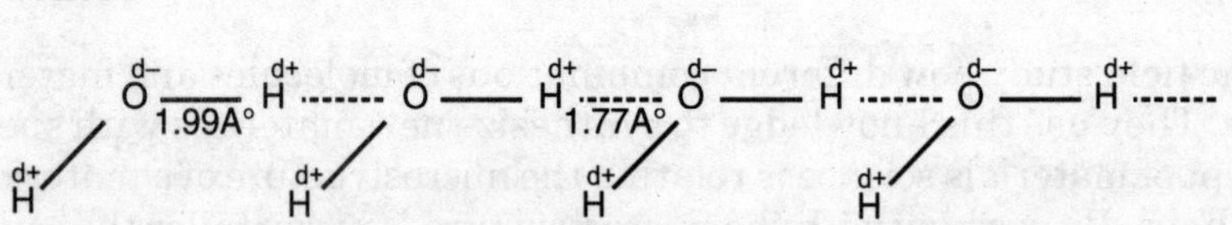

Alcohol

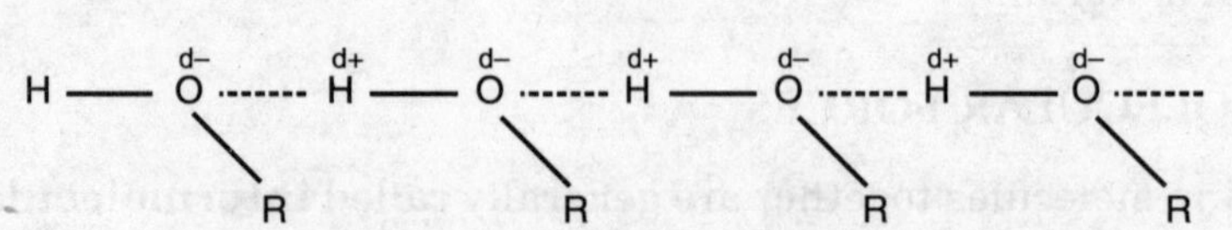

Carboxylic Acids

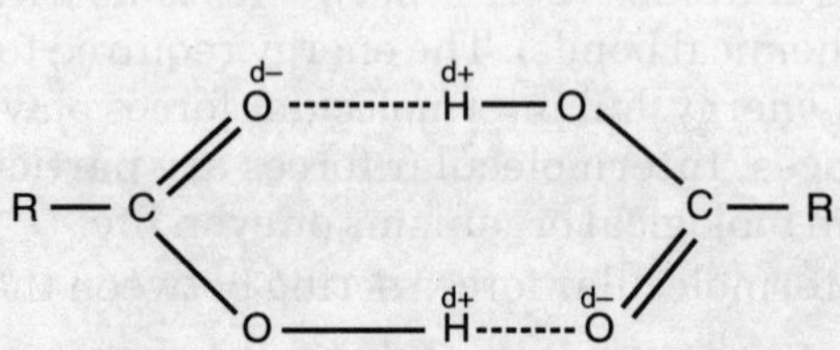

Conditions Necessary for the Formation of Hydrogen Bond: The formation of hydrogen bond is based on the following factors.

1. **High electronegativity:** The electronegativity of the atoms to which hydrogen atom is linked should be high. More is the electronegativity of the element to which hydrogen is linked, greater is the polar character and hence stronger is the hydrogen bond. The electronegativity of flourine is highest among flourine, oxygen and nitrogen. Hence strength of hydrogen bond is highest in HF (41.8 kJ mol^{-1}).
2. **Small size of the atom attached to hydrogen:** The smaller the size, greater will be the polarity of the covalent bond and hence greater will be the electrostatic attraction between the two molecules. When the size of the atom is large, its electron cloud is more diffused which results in less charge separation and hence hydrogen bond is not formed. Flourine, oxygen and nitrogen form strong hydrogen bonds but chlorine due to its large size, does not form hydrogen bond of appreciable strength.

Types of Hydrogen Bonds: There are two types of hydrogen bonds-
1. Intermolecular hydrogen bond.
2. Intramolecular hydrogen bond.

Intermolecular Hydrogen Bonding: When hydrogen bonding takes place between different molecules of the same or different compounds, it is called intermolecular hydrogen bonding. Examples include HF, H_2O and ROH where the hydrogen bonding is in between the molecules of the same compound. Hydrogen bonding in between molecules of different compounds is exhibited by alcohol and water or ammonia and water;

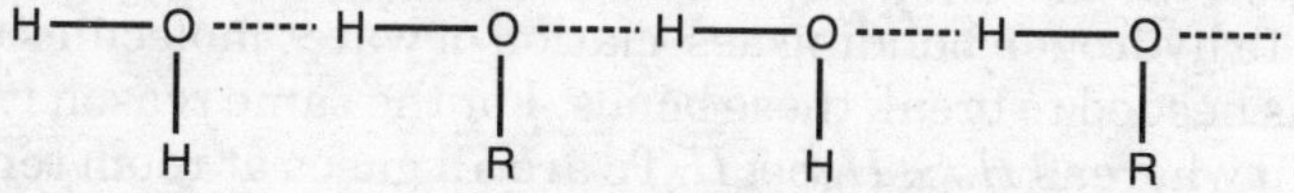

Intermolecular Hydrogen Bonding in water and alcohol molecules.

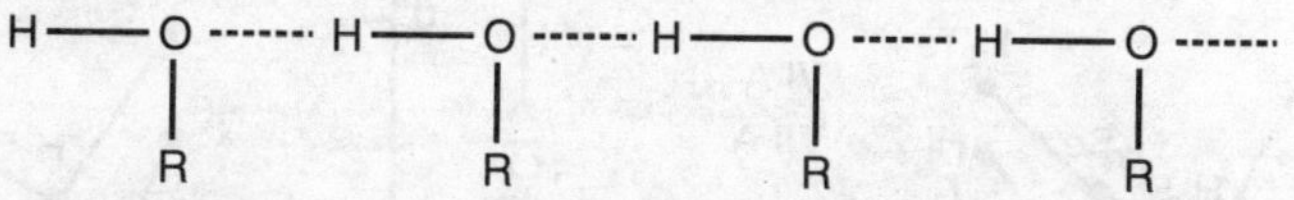

Intermolecular Hydrogen Bonding in similar molecules of alcohol.

Intramolecular Hydrogen Bonding: The hydrogen bonding which takes place within a molecule itself is called intramolecular hydrogen bonding. A few examples involving intramolecular hydrogen bonding include–

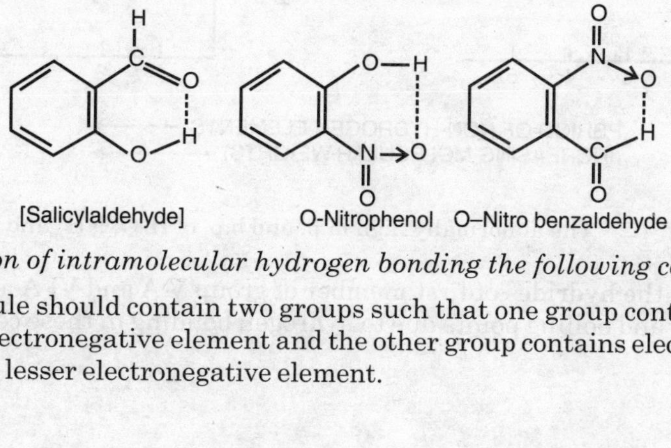

[Salicylaldehyde] O-Nitrophenol O–Nitro benzaldehyde

For the formation of intramolecular hydrogen bonding the following conditions apply–

(i) The molecule should contain two groups such that one group contains hydrogen atom linked to electronegative element and the other group contains electronegative element attached to lesser electronegative element.

(ii) The molecule should be planar.

(iii) The hydrogen bonding should lead to the formation of a six membered ring including hydrogen atom.

Strength of Hydrogen Bond: Hydrogen bond is a weak bond with the strength intermediate between the weak Vander Waals forces and strong covalent bonds. The bond dissociation energy of a covalent bond is 209-418 $kJ\ mol^{-1}$ and that for hydrogen bond it is only 12.6-41.8 $kJ\ mol^{-1}$. The hydrogen bond energies in case of three elements, F, O and N are-

Hydrogen Bond	*Bond Dissociation Energy* $kJ\ mol^{-1}$
H.........F	41.8
H.........O	29.3
H.........N	12.6

Consequences of Hydrogen Bonding

Though hydrogen bond is a weak bond, but still it affects a large number of physical properties of compounds-

1. ***High Melting and Boiling points :*** Substances in which molecules have stronger inter-molecular hydrogen bonding, show abnormally high melting and boiling points. To illustrate the point let us take the example of hydrides of group VI A. As we move down the group (Table 4.7), the boiling point increases from H_2S to H_2Te. But hydride of oxygen i.e H_2O has abnormally high boiling point. This can only be explained on the basis of hydrogen bonding. Due to hydrogen bonding association of water molecules takes place and hence extra energy is needed to break these bonds. For the same reason it can be explained that H_2O is a liquid whereas H_2S, H_2Se, H_2Te are all gases at room temperature.

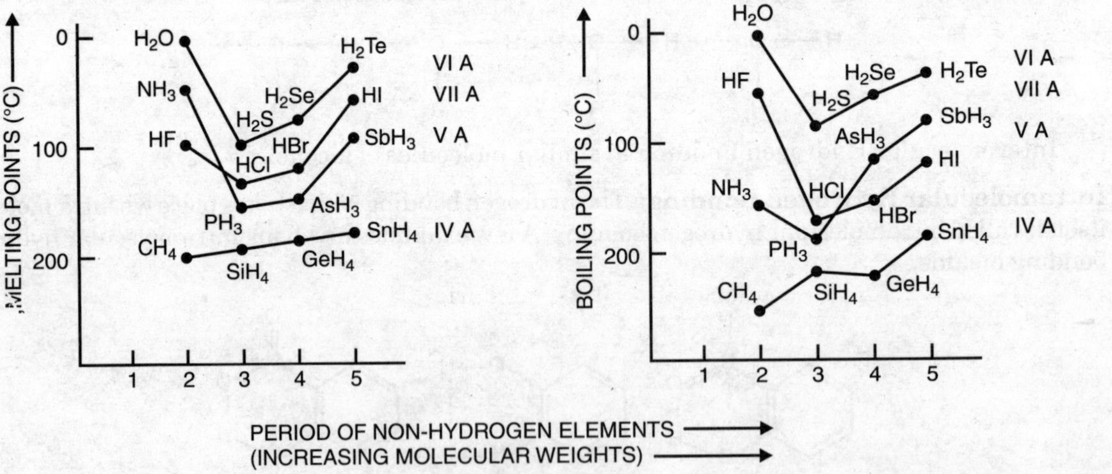

Fig. 4.2 The abnormally high m.p. and b.p. of H_2O, NH_3 and HF.

Similarly the hydrides of first member of group V A and VI A also have abnormally high melting and boiling points due to hydrogen bonding in these compounds. Fig. 4.2.

Table 4.4 Boiling points of hydrides of *V* group

Formula *(°C)*	*B. Pt*
H_2O	+100
H_2S	–59.6
H_2Se	–42.0
H_2Te	–1.8

2. **Solubility:** The compounds which can form hydrogen bonds with the solvent molecules are soluble in such solvents. The organic compounds generally are insoluble in polar solvent, water. But those organic compounds, which can form hydrogen bond with water molecules are soluble. Example, alcohols can form hydrogen bonding with water and hence are soluble.

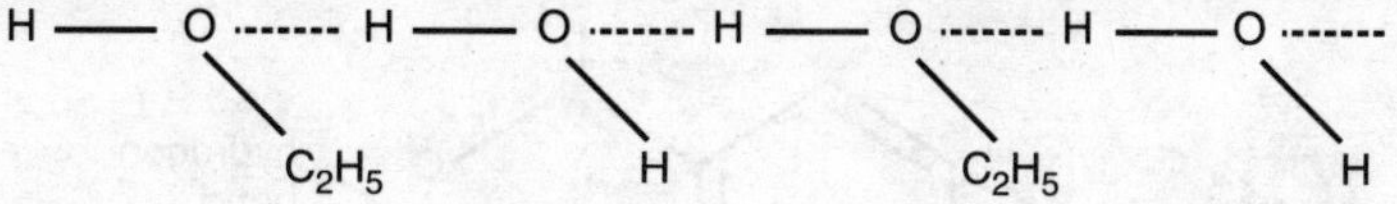

Similarly dimethyl ether is soluble in water because it is capable of forming hydrogen bonds with water.

3. **Viscosity:** Hydrogen bonded molecules have higher viscosity. Example glycerol has three -OH groups per molecule. There is extensive hydrogen bonding in glycerol and hence it is highly viscous.
4. **Association:** Hydrogen bond links up two or more molecules of a compound to form bigger units as in the case of carboxylic acids. Carboxylic acids exist as dimers even in the vapour state due to hydrogen bonding
 However in aqueous solutions, the molecules of these acids associate with water molecules rather than existing as dimers.

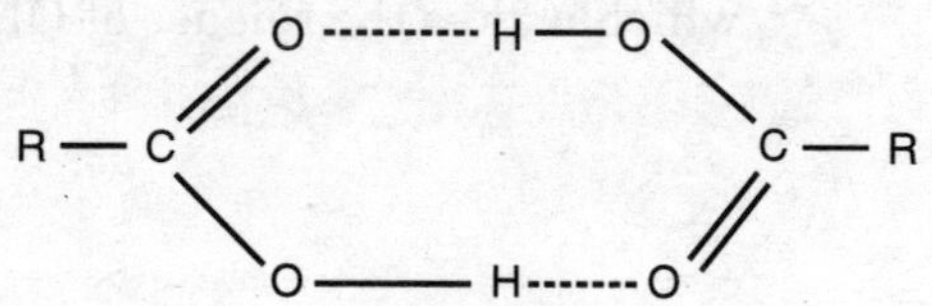

Fig. 4.3

5. **Some unique properties of water and Ice:** Due to hydrogen bonding water has some unique properties.
 (a) *Density of ice is less than that of water* : Gen-erally density of a substance in solid state is more as compared to liquid state. Actually, in ice the hydrogen bonding between water molecule is more extensive than in liquid water. The molecules are tetrahedrally oriented with respect to one another. Oxygen atom is surrounded by four hydrogen atoms, two of these are bonded covalently and the other two by hydrogen bonds. The hydrogen bonds are weaker and hence longer than the covalent bonds and thus an open cage like structure is formed (Fig.4.3). As ice melts, a large number of hydrogen bonds are broken and the molecules move into these open spaces and come closer thereby increasing the density of water. The density of liquid water at 4°C is hence greater than that of ice.
 (b) *Water contracts when heated between 0°C and 4°C:* Most of the substances expand when heated but as

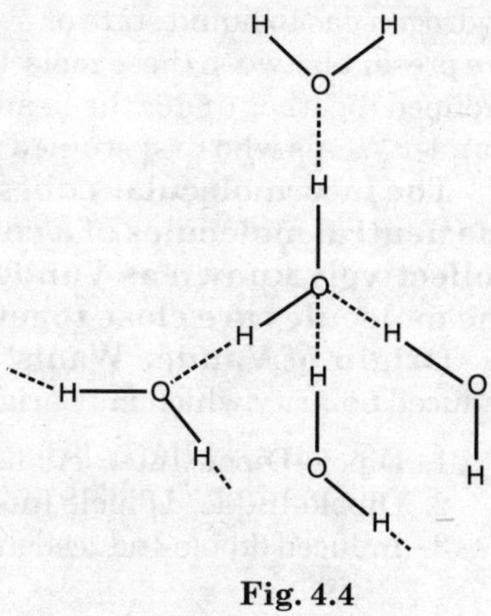

Fig. 4.4

the temperature of water (ice) is increased from 0°C to 4°C, it contracts in volume. This can also be explained on the basis of hydrogen bonding. As the temperature is raised hydrogen bond in ice starts breaking, thereby bringing the water molecules closer and closer together. This leads to contraction. Above 4°C the normal expansion as in the case of other substances takes place.

6. **Boiling point and intramolecular hydrogen bonding:** Intramolecular hydrogen bonding leads to lower boiling points than expected. Example, in O-nitrophenol, hydrogen bonding prevents the association of molecules and therefore the compound has low boiling point.
7. **Acidity and Basicity:** The acidity and basicity of some compounds exhibiting hydrogen bonding can be explained. For example, para-nitrophenol has higher acidic character as compared to ortho-nitrophenol. The presence of intramolecular hydrogen bonding in ortho nitrophenol makes the release of H^+ difficult decreasing its acidic strength.

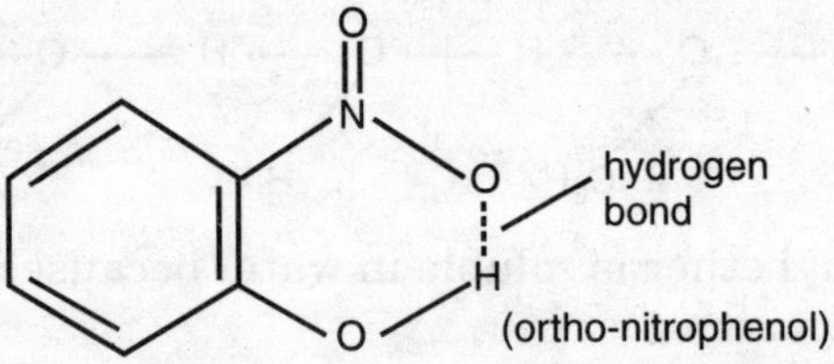

Similarly the basic strength of $(CH_3)_4N^+OH^-$ is higher as compared to $(CH_3)_3NH^+\overline{O}H$. This is again due to the presence of intramolecular hydrogen bonding in $(CH_3)_3NH^+\overline{O}H$, which makes the release of $\overline{O}H$ difficult.

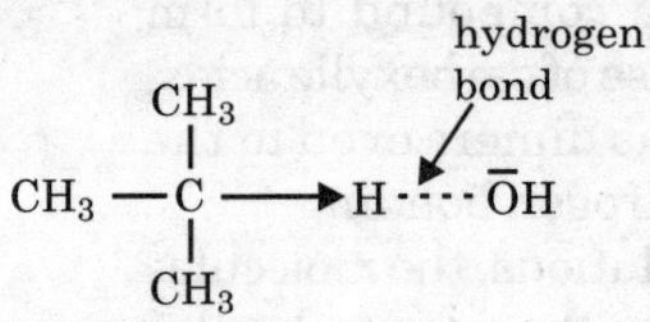

4.1.2 Vander Waals forces

Colvalent molecules like H_2, Cl_2, CH_4 etc cannot form additional bonds because the bonding capacity of the atoms involved in these molecules is saturated. But then the question is, what type of force holds these covalent molecules together? The condensation of covalent substances like hydrogen gas to liquid state or even solid state at low temperatures, implies that attractive forces are present between these molecules. These intermolecular attractions between the molecules are grouped together under the heading of Vander Waals' forces in honour of the Dutch chemist J.H. Vander Waals who first studied their nature.

The intermolecular cohesive forces operative between the molecules, which cause the neutral molecules of a covalent compound to condense into the liquid or solid are collectively known as Vander Waals' forces. These forces are significant only when the molecules are close together.

Origin of Vander Waals' Forces: These forces are electrical in nature and are due to induced polarity which may originate from three types of interactions :

1. Dipole-Dipole Interaction.
2. Dipole-Induced dipole Interaction.
3. Induced dipole-Induced dipole Interaction.

1. **Dipole-Dipole Interaction:** (Keesom forces). This effect was studied by Keesom in 1912 and is also referred to as the orientation effect. Gases such as HF, HCl, NH_3 etc have permanent dipoles. These dipoles orient themselves in such a way that the positive end of the molecule will be attracted to the negative end of the other molecule, when these are in close proximity. As a result, there is appreciable dipole-dipole interactions between the molecules. Because of interactive interactions these gases can easily be liquefied.

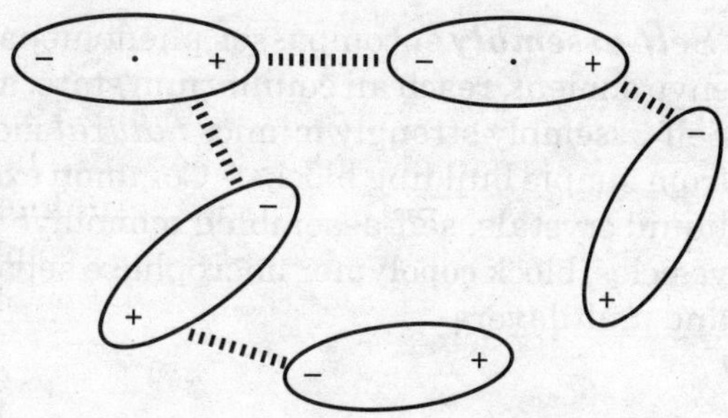

Fig. 4.5 Dipole Dipole Interaction

2. **Dipole-Induced dipole interaction:** (Debye forces) This effect was studied by Debye (1920). A polar molecule or a permanent dipole induces or polarizes a nonpolar molecule, which lies appreciably close to it. The induced dipole interacts with the permanent dipole. The attraction between the two will largely depend upon the polarizing power of the permanent dipole and polarizability of the non-polar molecule.

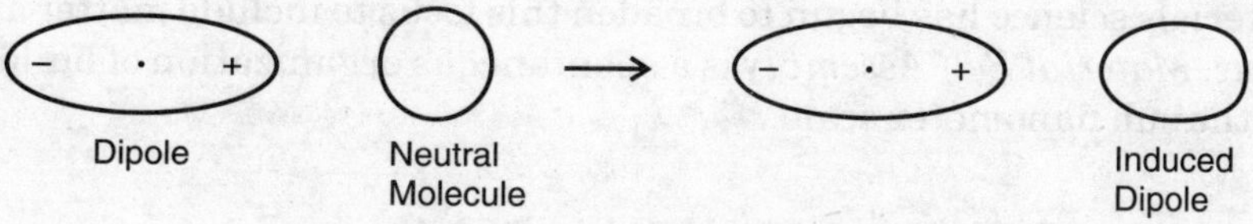

Fig. 4.6 Dipole-Induced dipole interaction

3. ***Induced dipole-Induced dipole Interaction:*** (London forces or Dispersive forces)-Vander Waals' forces exist in non-polar molecules also. F. London (1930) provided a satisfactory explaination for such type of forces. In non-polar substances, such as H_2, O_2, N_2 and also inert gases, there are no initial dipoles. The molecules are electrically neutral. A non-polar molecule may be considered as a positive centre surrounded by symmetrical electron cloud. But electron clouds keep on oscillating with respect to nuclei of the atoms. Therefore at a given instant the electron cloud may be concentrated on one side and a temporary dipole exists in a molecule i.e a non-polar molecule may become temporarily polarized. Now this self-polarized molecule may induce a dipole moment in a neighbouring molecule and in this way a large number of non-polar molecules become temporarily polar which are mutually attracted by weak forces. These forces are also known as dispersive forces.

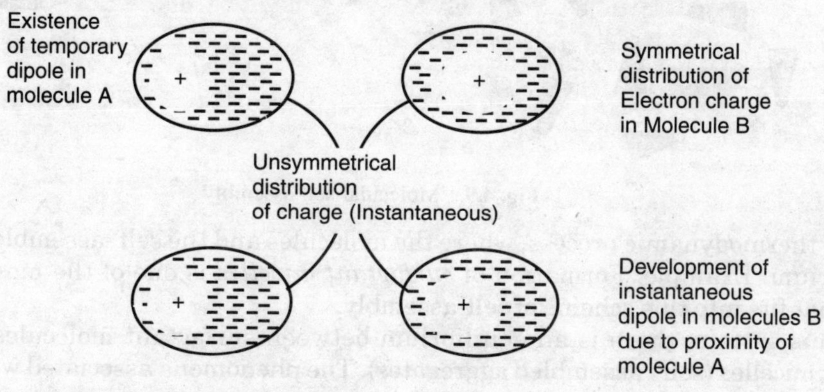

Fig. 4.7 Induced dipole-Induced dipole interaction

4.2 MATERIAL SELF ASSEMBLY (MOLECULAR AGGREGATION)

"***Self-assembly*** encompasses phenomena whereby small chemical units are introduced into an environment, reach an equilibrium state, and produce much larger-scale structures and patterns. Self-assembly strongly mimics *natural* mechanisms that form stable and complex architectures from simple building blocks." Common examples of self-assembled chemical structures include liquid crystals, self-assembled monolayers, organization of lipids into micelles and membrane vesicles, block copolymer microphase separation, and polymers adsorbed onto surfaces in layers and multilayers.

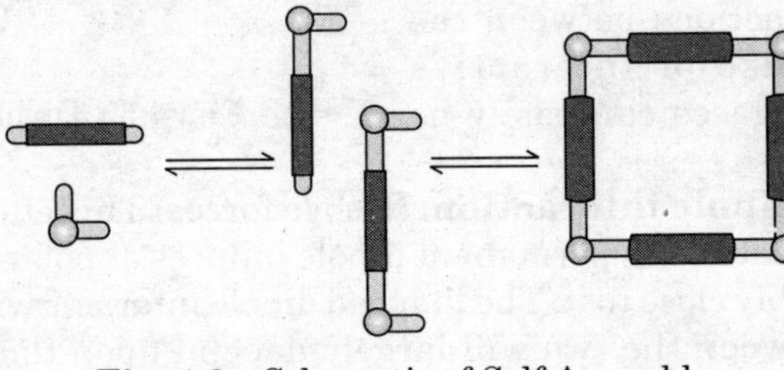

Fig. 4.8 Schematic of Self Assembly

The materials science has begun to broaden this focus to include matter at scales larger than the molecular. *Material Self Assembly* is a spontaneous organization of building blocks that are larger than the sub nanometer scale.

4.2.1 Molecular Self Assembly: Molecular Aggregation:

Molecular self-assembly is the assembly of molecules without guidance or management from an outside source. Traditionally, *self - assembly* has been defined as *spontaneous* association of molecules into defined three - dimensional geometry under a defined condition.

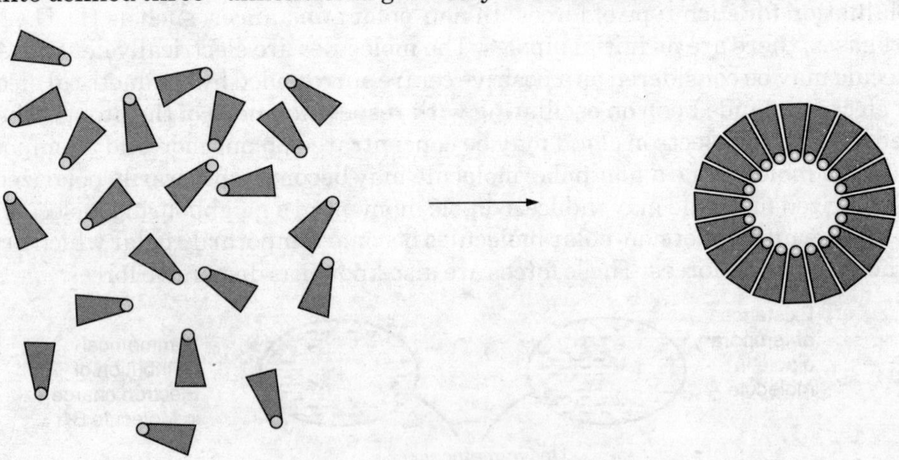

Fig. 4.9 Molecular self assembly

It is a thermodynamic process, where the molecules and the self -assembled aggregates are in equilibrium. Example: Formation of *surfactant micelles* is one of the most widely studied systems that fits into this scheme of self-assembly.

For this system, there is an equilibrium between surfactant molecules (monomer) and surfactant micelles (self - assembled aggregates). The phenomena associated with this scheme of *spontaneous* association are abundant in nature.

Polymeric self-assembly can be classified as molecular self -assembly as the sense of the building unit is polymer molecules. The biological system has long been known as a treasure house of self- assembly processes. Many biological systems use self-assembly to assemble various molecules and structures. Self-assembly forms the basis for many natural processes including protein folding, DNA transcribing and hybridization, and the formation of cell membranes.

Self-assembly is emerging as an elegant, "bottom-up" approach to fabricating nanostructured materials

4.2.2 General Scheme for the Formation of Self - Assembled Aggregates

Self-assembly is the force balance process between three classes of forces:

- attractive
- repulsive
- directional force.

Aggregation occurs when there is a net attraction and an equilibrium separation between the components. The equilibrium separation normally represents a balance between attraction and repulsion. These two interactions are fixed in molecular self-assembly but can be engineered independently in macroscopic self-assembly.

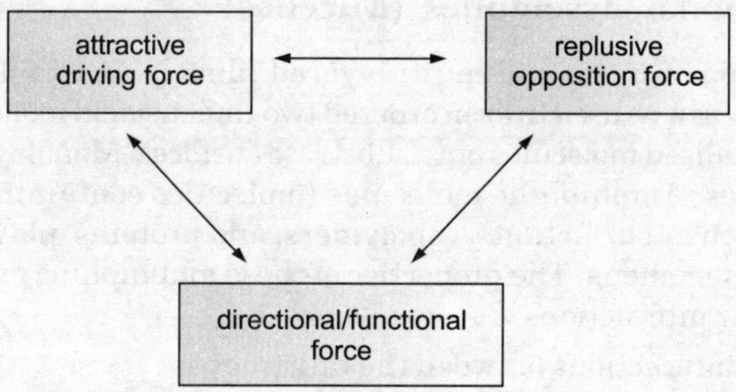

Fig. 4.10

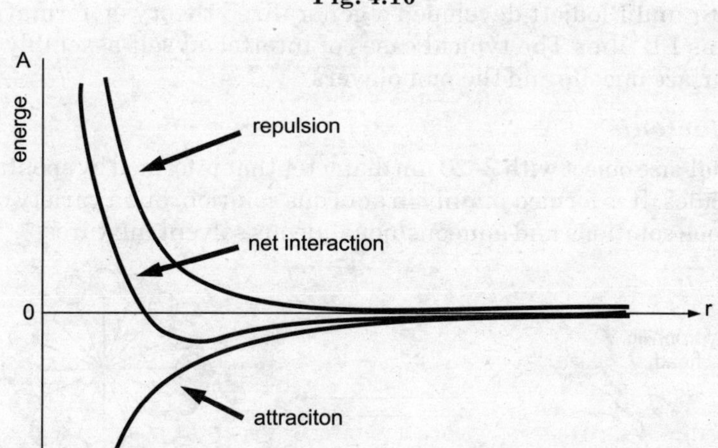

Fig. 4.11

The forces involved in self-assembly are generally weaker than the bonding forces that hold molecules together. They correspond to weaker aspects of Coulombic interactions and are found in many places throughout nature. For example, weak interactions called hydrogen bonds hold the hydrogen atom in one molecule of liquid water together with the oxygen atom of the next and prevent the molecules from becoming water vapor at room temperature. Hydrogen bonding also helps to hold proteins into particular three-dimensional structures that are necessary for their biological function.

4.2.3 Self-Assembly Materials

Self-assembly has become a very effective and promising approach to synthesize a wide range of novel nanoscale materials. In the self-assembly processes, atoms, molecules, particles, and other building blocks organize themselves into functional structures as driven by the energetics of the system. An additional aspect of nanotechnology is that often the formation of nanostructured systems involves self-assembly with multiple types of building units.

The most important driving force for self-assembly is the interaction energies between the subunits, whether they are atoms, molecules, or particles. Some of the significant Self Assembled Materials are discussed in the following sections.

4.2.4 Two Dimensional Assemblies (Micelles)

Layer by layer deposition of mono and multi layered films is an example of two dimensional assemblies. Self-assembly can be used to form ordered two-dimensional monolayers by chemisorption of appropriately functionalised molecules onto substrate surfaces. Monolayers are typically formed by amphiphilic molecules. Amphiphilic molecules (molecules containing both hydrophilic and hydrophobic groups), such as surfactants, copolymers, and proteins, play a critical role in a wide range of self-assembly phenomena. The properties of these amphiphilic molecules are determined by unique intermolecular interactions:

- the hydrophobic interactions between the tail groups,
- and the hydrophilic and/or the electrostatic interactions between the head groups.

Irving Langmuir and Blodjett developed a generalized theory of formation of monolayers sometimes known as LB films.The typical cases of interfacial self-assembly at the gas–liquid interface are the surface micelle and the monolayers

Micelles and Surfactants

A micelle is a colloidal-size object with 2–20 nm diameter that is formed by spontaneous association of surfactant molecules. It is formed mainly in aqueous solution, but a variety of micelles also are formed in nonaqueous solutions and aqueous/nonaqueous solvent mixtures

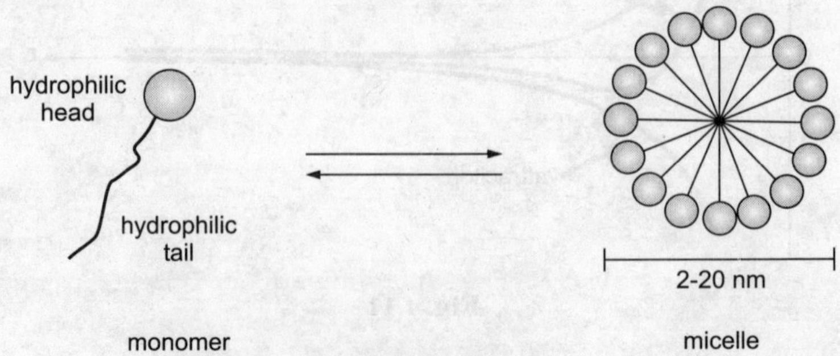

Fig. 4.12 Micelle formation: micellization.

Figure 4.13 shows the schematic illustration of the typical micellization process of surfactant molecules in aqueous solution. The surfactant molecules are soluble in aqueous solution. Part of the initially added molecules is adsorbed onto the air – liquid interface and forms an adsorbed monolayer. Since the space for the monolayer formation is limited at the surface, the rest of the molecules remain in the solution as a free form of molecules (monomers). For most of the cases, the main attractive driving force for the micellization is hydrophobic interaction.

Micelles can be inverted by placing amphiphiles into organic solvents. The hydrophilic head groups coagulate within the centre of the micelle, while the lipophilic portion of the amphiphile points outward in the organic solvent. In addition to micelles, *bilayers* and *vesicles* can also form. Bilayers are commonly found in biological cells as phospholipid bilayers .A vesicle is rather like a simple biological cell membrane in that it comprises a bilayer arrangement of amphiphiles encasing an aqueous inner region.

SAMs have a wide range of applications including surface wetting, non-fouling property, electrochemistry, surface passivation, protein binding, DNA assembly, corrosion resistance, biological arrays and cell interactions. SAMs also are contained, in large numbers, in materials for the semiconductor electronics industry such as nanowires, nanotransistors and nanosensors

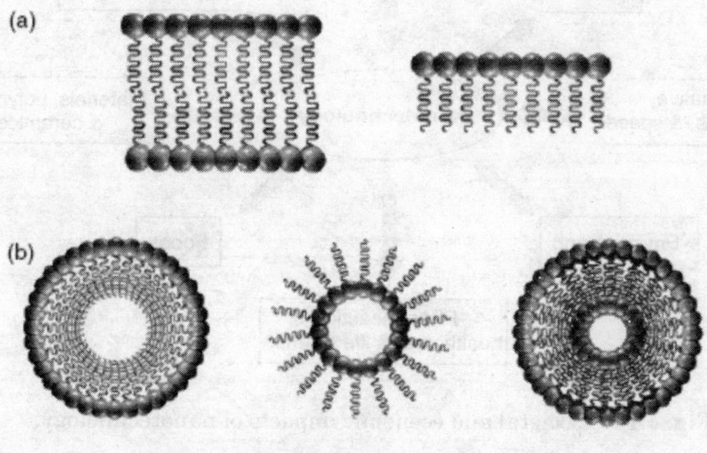

Fig. 4.13 Schematics of various amphiphile aggregates: (a) bilayer and monolayer; (b) micelle, reverse micelle and vesicle.

4.3 NANOTECHNOLOGY

Nano, derived from the Greek word *nanos*, designates a billionth fraction of a unit, e.g., of a meter. *Nanoscience* is the study of the fundamental principles of molecules and structures with at least one dimension roughly between 1 and 100 nanometers. These structures are known as *nanostructures*.

Nanostructures may be compared to a human hair which is ~50,000 nm thick whereas the diameters of nanostructures are ~0.3 nm for a water molecule, 1.2 nm for a single-wall carbon nanotube, and 20 nm for a small transistor. DNA molecules are 2.5 nm wide and proteins about 10 nm

Materials with relevant physical dimensions on the nanometer scales possess unique chemical and physical properties compared to their corresponding bulk or isolated atoms and molecules.

Their thermodynamic, dynamic, mechanical, optical, electronic, magnetic, and chemical properties can be significantly altered relative to their bulk counterparts

Nanotechnology is the application of these nanostructures into useful *nanoscale* devices. The widely accepted definition of nanotechnology is "the manipulation of atoms or molecules on a nanometer scale of ~ 1 – ~100 nm". A more detailed description could be "the ability to manipulate/ assemble nanometer - scale building units into integrated systems and the subsequent finding of novel properties/phenomena on a nanometer scale."

Nanotechnology literally means any technology done on a nanoscale that has applications in the real world. Nanotechnology is likely to have a profound impact on our economy and society perhaps comparable to that of information technology or cellular and molecular biology. Science and engineering research in nanotechnology promises breakthroughs in areas such as materials and manufacturing, electronics, medicine and healthcare, energy and the environment, biotechnology, information technology and national security. It is widely felt that nanotechnology will be the next industrial revolution.

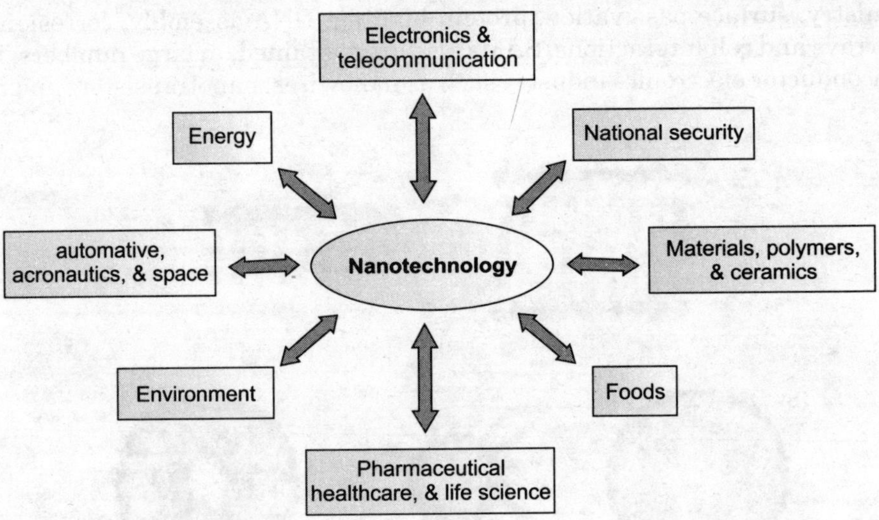

Fig. 4.14 Societal and economic impacts of nanotechnology.

Nanotechnology is the ability to manipulate individual atoms and molecules to produce nanostructured materials and submicron objects that have applications in the real world.

On December 29, 1959 at the California Institute of Technology, Nobel Laureate Richard P. Feynman gave a talk at the Annual meeting of the American Physical Society that has become one classic science lecture of the 20th century, titled "There's Plenty of Room at the Bottom." He presented a technological vision of extreme miniaturization in 1959, several years before the word "chip" became part of the lexicon. He talked about the problem of manipulating and controlling things on a small scale. Extrapolating from known physical laws, Feynman envisioned a technology using the ultimate toolbox of nature, building nanoobjects atom by atom or molecule by molecule. Since the 1980s,many inventions and discoveries in fabrication of nanoobjects have been a testament to his vision.

The two mainstream approaches to the issues of nanotechnology are *top -down* and *bottom-up*.

Top Down Approach: Traditionally, nanotechnological devices have been prepared by the break-down of materials using techniques developed by solid state physicists. For example, a large block of silicon wafer can be reduced to smaller components by cutting, etching and slicing down to a desired size or shape. This is known as the ***top-down approach***. By removing the excess portion of bulk materials by physical, chemical, and mechanical means, the desired nanostructured systems can be constructed. Hard lithography using optical laser-light, electron-beam, or *x*-ray are typical examples.

Bottom up Approach: The bottom-up approach, on the other hand, starts with the individual building units that are the objects of the nanometer scale, such as atoms, molecules, polymers, and colloids. By assembling these building units with the needed controllability, the desired nanostructured systems can be obtained. Nanomaterials synthesised from the bottom-up method have novel physico-chemical proprieties that differ from the bulk material.

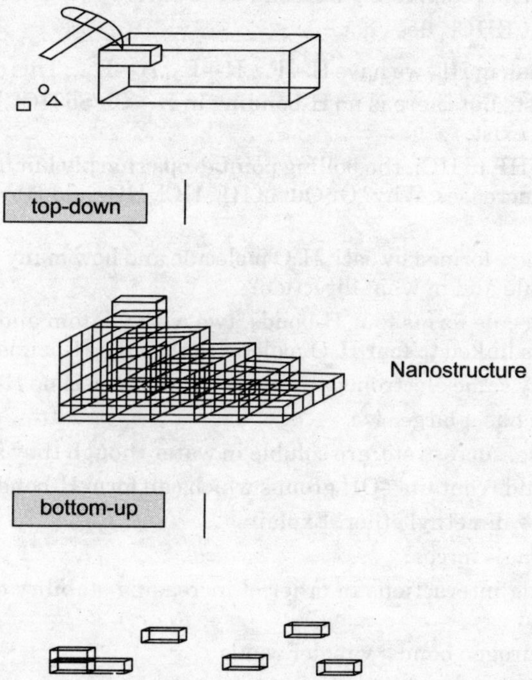

Fig. 4.15 Bottom-up versus top-down approach for the preparation of nanostructures

Molecular self assembly is widely used as bottom-up approach to synthesize nanomaterials and is discussed in detail in the following section.

QUESTIONS

1. Which one of the following is expected to show hydrogen bonding?
   (i) CH_3CH_2OH and CH_3OCH_3
   (ii) CH_3NH_2 and CH_3SH
   (iii) CH_3OH and $(CH_3)_3N$
   Ans: (i) CH_3CH_2OH (ii) CH_3NH_2 (iii) CH_3OH
2. Why the boiling point of $H_2O > CH_3OH$
   Ans: Because extent of hydrogen bonding is more in H_2O
3. Why CH_3OCH_3 is soluble in water and CH_3OCH_3 is not?

4. Why water contracts when heated between 0°C and 4°C ?
5. Why melting or boiling point of para isomer is more than ortho isomer?
6. Why o-nitrophenol can be steam distilled and not p-nitrophenol?

Ans: In p-nitrophenol intermolecular hydrogen bonding is present decreasing its volatility whereas in o-nitrophenol interamolecular hydrogen bonding is present hence the single molecules have high volatility and can be steam distilled.

7. Explain why the acidic strength of 0-nitrophenol < p-nitrophenol
8. Account for the following
   (i) Why H_2O is a liquid while H_2S is a gas?
   (ii) Why HF has higher boiling point than HCl?
   (iii) Why ethyl alcohol is completely miscible with water?
9. Why KHF_2 exists but $KHCl_2$ does not ?

Ans: Due to H-bonding in HF, we have H—F....H—F....H—F..... This can dissociate to give HF_2^- ion and hence KHF_2 exists but there is no H-bonding in H—Cl. So HCl_2^- ion does not exist and hence $KHCl_2$ also does not exist.

10. When we move from HF to HCl, the boiling point drops sharply but on moving further to HBr and HI the boiling point increases. Why? Or Out of HF, HCl, HBr and HI which has lowest boiling point and why?
11. How many H- bonds are formed by each H_2O molecule and how many water molecules are attached to each water molecule and in what direction?

Ans: Each H_2O molecule forms four H-bonds, two with O-atom and two with H-atoms. Further, each H_2O molecule is linked to four H_2O molecules through H-bonds tetrahedrally.

12. Though Cl has nearly same electronegativity as N, yet there is no H-bonding in HCl. Why?

Ans: Chlorine atom has a large size.

13. Why glucose, fructose, sucrose etc. are soluble in water though they are covalent compounds?

Ans: These compounds contain—OH groups which can form H-bonds with water.

14. Viscosity of glycerol > dimethyl ether. Explain.
15. What are van der Waals forces?
16. Arrange the following interactions in order of increasing stability covalent bond, van der waals forces, hydrogen bond

Ans: Covalent > hydrogen bond > van der waals

17. Why boiling point of halogens follow the order:

$$I_2 > Br_2 > Cl_2 > F_2$$

Ans: More is the molecular mass, more will be the van der waals forces between the molecules.

18. Why the boiling point of following hydrides follow the order:

$$GeH_4 > SiH_4 > CH_4$$

Ans: Strength of the van der waals forces depend upon size. The size decreases from Ge to C and hence the boiling point also decreases in the same order.

19. In which molecule the van der waals forces are likely to be the most important in determining the melting and boiling point : ICl, HCl, Br_2, H_2S, CO

Ans: Br_2, because in Br_2 only van der waals forces are present.

20. Define nanochemistry.
21. Discuss two approaches to prepare nanomaterials.
22. Molecular self-assembly is an important Technique to build nanomaterials. Comment.
23. Write short notes on (i) Molecular self assemblies (ii) Micelles.
24. Crystal growth is another sort of self-assembly. Explain.

❑❑❑

5 PHYSICAL METHODS OF PURIFICATION

Knowledge is a treasure, but practice is the key to it.

— Thomas Fullen

CHAPTER OUTLINE

5.1 INTRODUCTION

Once the compound is extracted from a natural source or synthesised in the laboratory, it is essential to purify it. Various methods used for the purification of compounds are based on the nature of the compound and the impurity present in it.

The common techniques used for purification are as follows:

(i) Sublimation
(ii) Crystallisation
(iii) Distillation
(iv) Differential extraction and
(v) Chromatography

Finally, the purity of a compound is ascertained by determining its melting or boiling point. Most of the pure compounds have sharp melting points and boiling points.

Various purification techniques are discussed in brief in the following sections;

5.2 RECRYSTALLISATION

Recrystallization is a separation and purification technique employed to produce a wide variety of materials. Crystallization may be defined as a phase change in which a crystalline product is obtained from a solution. It is based on the difference in the solubilities of the compound and the impurities in a suitable solvent. A suitable solvent is one which readily dissolves the solid (solute) when the solvent is hot but not when it is cold. The best solvents exhibit a large difference in solubility over a reasonable range of temperatures.

The purification of solids by crystallisation is based upon differences in their solubility in a given solvent or mixture of solvents.

In its simplest form, the crystallisation process consists of following steps:

(a) The impure material is dissolved in a suitable solvent, by shaking or vigorous stirring, at or near the boiling point, to form a near-saturated solution.
(b) The hot solution is filtered to remove any insoluble particles. To prevent crystallisation during this filtration, a heated (jacketed) filter funnel can be used or the solution can be somewhat diluted with more of the solvent.
(c) The solution is then allowed to cool so that the dissolved substance crystallises out.
(d) The crystals are separated from the mother liquor, either by centrifuging or by filtering, under suction, through a sintered glass, a Hirsch or a Buchner, funnel. Usually, centrifuging is much preferred because of the much greater ease and efficiency of separating crystals and mother liquor, and also because of the saving of time and effort, particularly when very small crystals are formed or when there is entrainment of solvent.
(e) The crystals are washed free from mother liquor with a little fresh cold solvent, then dried.

The resulting solid, after drying, is tested for purity (usually by a melting point determination. The process is repeated until the pure compound is obtained which is confirmed by the melting point.

If the solution contains extraneous coloured material likely to contaminate the crystals, this can often be removed by adding some activated charcoal (decolorising carbon) to the hot solution which is then shaken frequently for several minutes before being filtered. (The large active surface of the carbon makes it a good adsorbent for this purpose.)

5.2.1 Choice of Solvents

The best solvents for recrystallisation have the following properties:

(a) The material is much more soluble at higher temperatures than it is at room temperature or below.
(b) Well-formed (but not large) crystals are produced.
(c) Impurities are either very soluble or only sparingly soluble.
(d) The solvent must be readily removed from the purified material.
(e) There must be no reaction between the solvent and the substance being purified.
(d) The solvent must not be inconveniently volatile or too highly flammable.

The following generalisations provide a rough guide to the selection of a suitable solvent:

(a) Substances usually dissolve best in solvents to which they are most closely related in chemical and physical characteristics. Thus, hydroxylic compounds are likely to be most soluble in water, methanol, ethanol, acetic acid or acetone. Similarly, petroleum ether might be used with water insoluble substances. However, if the resemblance is too close, solubilities may become excessive.
(b) Higher members of homologous series approximate more and more closely to their parent hydrocarbon.
(c) Polar substances are more soluble in polar, than in non-polar, solvents.

5.2.2 Fractional Crystallisation

In fractional crystallization the compound is mixed with a solvent, heated, and then gradually cooled so that, as each of its constituent components crystallizes.It can be removed in its pure form from the solution. Fractional crystallization is a stagewise separation technique that relies upon liquid-solid phase transition and enables multicomponent mixtures to be split into narrow fractions, ultimately leading to top purities of selected components.

5.3 DISTILLATION

One of the most widely applicable and most commonly used methods of purification of liquids or low melting solids (especially of organic chemicals) is fractional distillation at atmospheric, or at lower pressure. Almost without exception, this method can be assumed to be suitable for all organic liquids and most of the low-melting organic solids.

Distillation is the process of heating a liquid until it boils, then condensing and collecting the resultant hot vapors. The boiling point of a liquid varies with the atmospheric pressure to which it is exposed. A liquid boils when its vapour pressure is the same as the external pressure on its surface.Its normal boiling point is the temperature at which its vapour pressure is equal to that of a standard atmosphere (760 mm Hg). Lowering the external pressure lowers the boiling point.

For most substances, boiling point and vapour pressure are related **by** the following equation,

$$\log \boldsymbol{p} = A + B/(t + 273),$$

where $\boldsymbol{p}$ is the pressure, t is in °C, and A and B are constants. Hence, if the boiling points at two different pressures are known the boiling point at another pressure can be calculated from a simple plot of log $\boldsymbol{p}$ ***versus*** $1/(t + 273)$.

When different compounds in a mixture have different boiling points, they separate into individual components when the mixture is carefully distilled.Distillation can be broght by the following methods:

5.3.1 Simple Distillation

Simple distillation is a procedure by which two liquids with different boiling points can be separated. Simple distillation can be used effectively to separate liquids that have at least fifty degrees difference in their boiling points. As the liquid being distilled is heated, the vapors that form will be richest in the component of the mixture that boils at the lowest temperature. Purified compounds will boil, and thus turn into vapors, over a relatively small temperature range (2 or 3°C); by carefully watching the temperature in the distillation flask, it is possible to affect a reasonably good separation. As distillation progresses, the concentration of the lowest boiling component will steadily decrease. The temperature will continue to increase

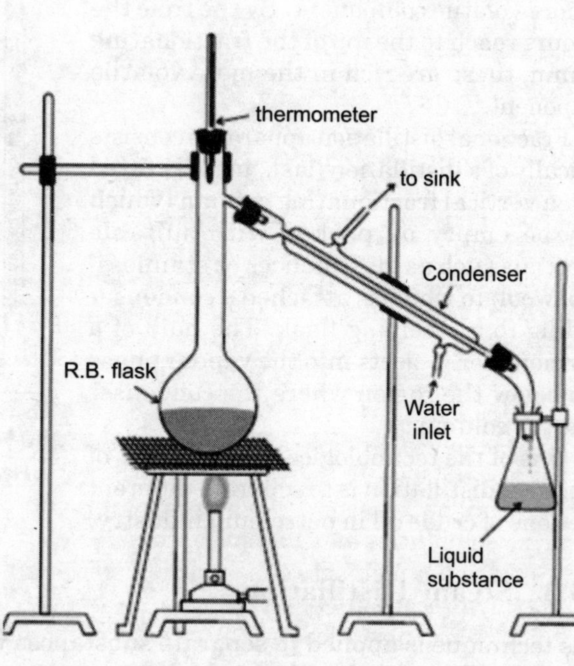

Figure 5.1 Simple Distillation Apparatus

until the boiling point of the next-lowest-boiling compound is approached. When the temperature again stabilizes, another pure fraction of the distillate can be collected. This fraction of distillate will be primarily the compound that boils at the second lowest temperature. This process can be repeated until all the fractions of the original mixture have been separated (Figure 5.1).

This important method is used to separate

(i) volatile liquids from nonvolatile impurities and
(ii) the liquids having sufficient difference in their boiling points. Liquids having different boiling points vaporise at different temperatures.

5.3.2 Fractional Distillation

Mixtures of liquids whose boiling points are similar (separated by less than 70°C) cannot be separated by a single simple distillation. In these situations, a fractional distillation is used.

Since the procedures of simple distillation are so similar to those involved in fractional distillation, the apparatus that are used in the procedures are also very similar. The only difference between the equipment used in fractional distillation and that used in simple distillation is that with fractional distillation, a packed fractionating column is attached to the top of the distillation flask and beneath the condenser. This provides the surface area on which rising vapors condense, and subsequently revaporize.The fractionating column is packed with glass beads (or something similar) to give the maximum possible surface area for vapour to condense on.

In this technique, vapours of a liquid mixture are passed through a fractionating column before condensation. The fractionating column is fitted over the mouth of the round bottom flask (Fig. 5.2).

Vapours of the liquid with higher boiling point condense before the vapours of the liquid with lower boiling point. The vapours rising up in the fractionating column become richer in more volatile component. By the time the vapours reach to the top of the fractionating column, these are rich in the more volatile component.

Fractional distillation apparatus consists basically of a distillation flask, usually fitted with a vertical fractionating column (which may be empty **or** packed with suitable materials such as glass helices or stainless-steel wool) to which is attached a condenser leading to a receiving flask. The bulb of a thermometer projects into the vapour phase just below the region where the condenser joins the column.

One of the technological applications of fractional distillation is to separate different fractions of crude oil in petroleum industry.

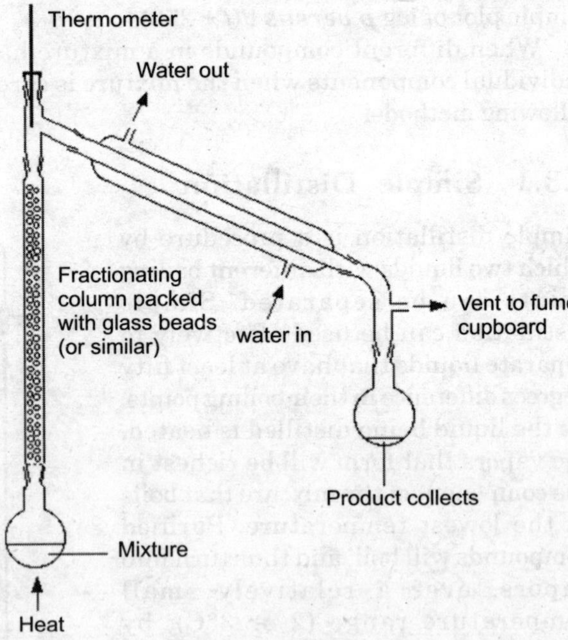

Fig. 5.2 Apparatus for Fractional Distillation.

5.3.3 Steam Distillation

This technique is applied to separate substances which are steam volatile and are immiscible with water. In steam distillation, steam from a steam generator is passed through a heated flask containing the liquid to be distilled. The mixture of steam and the volatile organic compound is condensed and collected. The compound is later separated from water using a separating funnel.

In steam distillation, the liquid boils when the sum of vapour pressures due to the organic liquid (p_1) and that due to water (p_2) becomes equal to the atmospheric pressure (p), i.e. $p = p_1 + p_2$. Since p_1 is lower than p, the organic liquid vaporises at lower temperature than its boiling point.

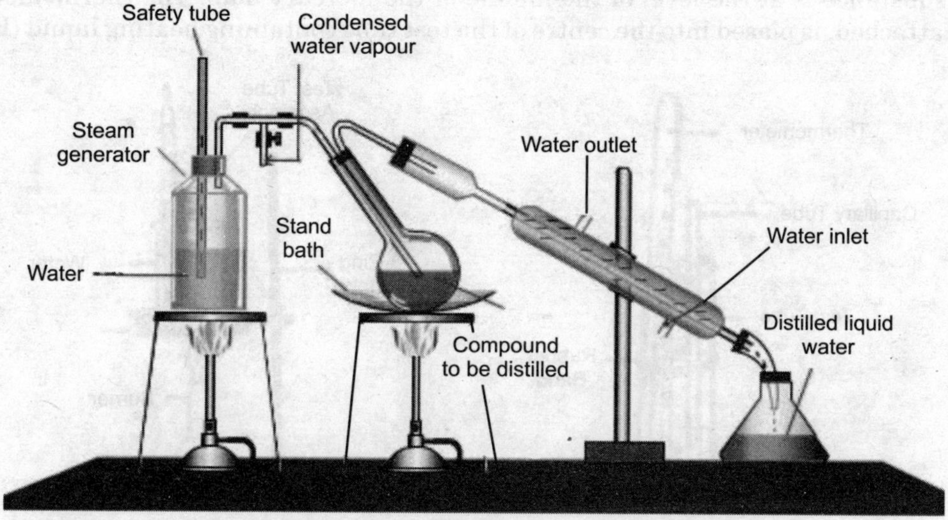

Fig. 5.3 Steam Distillation Apparatus

Thus, if one of the substances in the mixture is water and the other, a water insoluble substance, then the mixture will boil close to but below, 373K. A mixture of water and the substance is obtained which can be separated by using a separating funnel.

Its advantages are those of selectivity (because only some water-insoluble substances, such as naphthalene, nitrobenzene, phenol and aniline are volatile in steam) and of ability to distil certain high-boiling substances well below their boiling point. It also facilitates the recovery of a non-steam-volatile solid at a relatively low temperature from a high-boiling solvent such as nitrobenzene.

5.4 MELTING AND BOILING POINT: CRITERIA OF PURITY

Determining the melting point of a compound is one way to test if the substance is pure.A pure crystalline organic compound has, in general, a definite and sharp melting point; that is, the melting point range (the difference between the temperature at which the collapse of the crystals is first observed and the temperature at which the sample becomes completely liquid) does not exceed about 0.5 °C.

The presence of small quantities of miscible, or partially miscible, impurities will usually produce a marked increase in the melting point range and cause the commencement of melting to occur at a temperature lower than the melting point of the pure substance. The melting point is therefore a valuable criterion of purity for an organic compound.

The experimental method in most common use is to heat a small amount (about 1 mg) of the substance in a capillary tube inserted into a suitable melting point apparatus and to determine the temperature at which melting occurs.

About 25 mg of the dry substance is placed on a glass slide or upon a fragment of clean porous porcelain plate and finely powdered with a clean metal or glass spatula, and then formed into a small mound. The open end of the capillary tube is pushed into the powder, 'backing' the latter, if necessary, with a spatula. The solid is then shaken down the tube by tapping the closed end on the bench or by gently drawing the flat side of a triangular file (a pocket nail file is quite effective) along the upper end of the tube. The procedure is repeated until the length of lightly-packed material is 3-5 mm, and the outside of the tube is finally wiped clean.

The filled melting point tube is attached to the lower end of the thermometer in such a way that the substance is at the level of the middle of the mercury bulb. The thermometer, with capillary attached, is placed into the centre of the test tube containing heating liquid (Fig. 5.4).

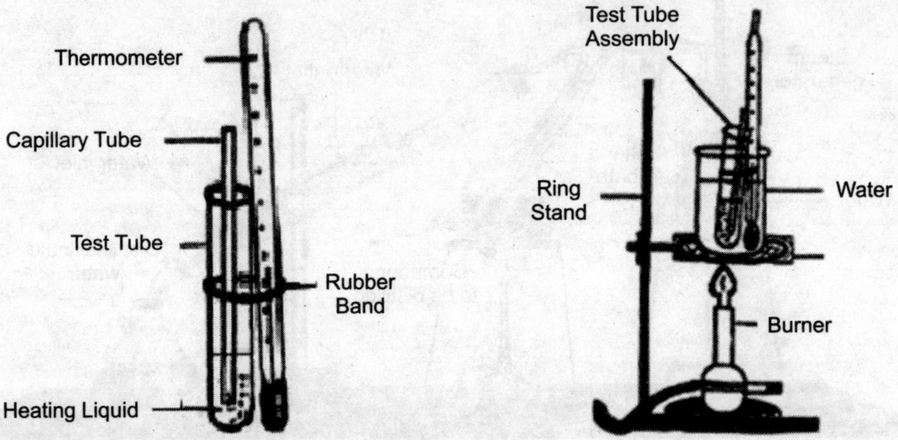

Fig. 5.4. (a) Heating Assembly (b) Determination of Melting Point

The two principal types of melting point apparatus in common use are those in which heating of the capillary tube is by means of a heated liquid bath (Fig 5.4(b)) .

A convenient form of bath is the Thiele apparatus (Fig. 5.5) in which the liquid (1) is contained in a tube with a closed bent side-arm.

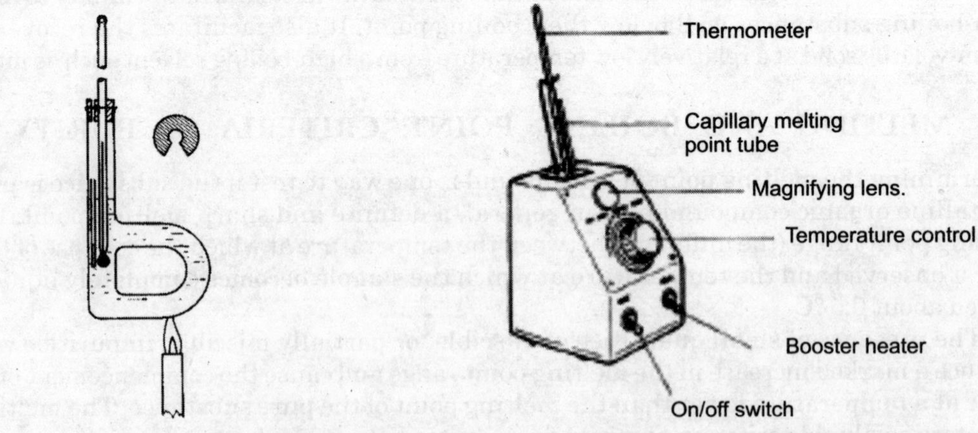

(a) Thiele Apparatus (b) Electric melting point apparatus

Fig. 5.5

The thermometer is located as shown, being inserted through a cork in such a way that the thermometer scale is visible, and also to allow for expansion of the air in the apparatus. The filled melting point tube is attached to the lower end of the thermometer in such a way that the substance is at the level of the middle of the mercury bulb. The thermometer, with capillary attached, is inserted into the centre of the main tube of the Thiele apparatus. On heating the bent side-arm, the heated liquid circulates and raises the temperature of the sample in such a way that no stirring of the bath liquid is required.

The melting point apparatus is heated comparatively rapidly with a small flame until the temperature of the bath is within 15 °C of the melting point of the substance, and then slowly and regularly at the rate of about 2 °C per minute until the compound melts completely. The temperature at whch the substance commences to liquefy and the temperature at which the solid has disappeared, i.e. the melting point range, is observed. For a pure compound, the melting point range should not exceed 0.5-1 °C; it is usually less.

In electrical apparatus (Fig 5.5(b)) the sample is loaded in a thin glass capillary tube and placed in the apparatus. The sample is closely observed through the magnifying lens or viewer while the temperature is slowly increased. The melting point range is between the temperature the first liquid appears and the last solid melts.

5.5 DETERMINATION OF BOILING POINT

When reasonable amounts of liquid components are available (> 5 ml) the boiling point is readily determined by slowly distilling the material from a pear- shaped flask in an apparatus assembly shown in Fig. 5.6, and recording the temperature at which the bulk of the compound distils.

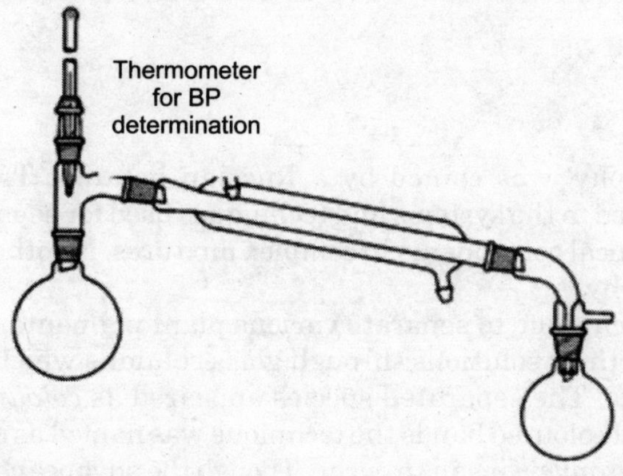

Fig. 5.6 Determination of Boiling Point during Distillation

QUESTIONS

1. Discuss the process of recrystallisation of solids.
2. What are the advantages of steam distillation over fractional distillation .
3. Write a short note on fractional distillation .
4. Why is it necessary to determine the melting point of a solid after crystallization

❑❑❑

6 CHROMATOGRAPHY

Even a mistake may turn out to be the one thing necessary to a worthwhile achievement.

— Henry Ford

CHAPTER OUTLINE

6.1 INTRODUCTION

The term 'Chromatography' was coined by a Russian botanist Tswett in 1906. The term 'Chromatography' is applied to the systems and techniques used for separation, identification, and determination of the chemical components in complex mixtures. No other separation method is as powerful as chromatography.

Twett employed the technique to separate various plant pigments, such as chlorophylls and xanthophylls, by passing their solutions through glass columns which were packed with finely divided calcium carbonate. The separated species appeared as *coloured* bands on the column. Because of the formation of coloured bands the technique was named as chromatography. In Greek *chroma* means colour and *graphein* means to write. Though the advance chromatographic techniques employed today for the separation of complex mixtures frequently involves colourless substances, but the original term *chromatography* is still retained.

All the chromatographic methods have one thing in common i.e. the use of two mutually immiscible phases, stationary phase and a mobile phase.

Stationary Phase: It is the phase that is fixed in a place, either packed in a column or on a fixed planar surface.

Mobile Phase: The mobile phase carriers the mixture of the compounds to be separated along with it and is moved through the stationary phase. The mobile phase can be a liquid or a gas. The components of the mixture are separated depending upon the rates at which they are carried along with the mobile phase through the stationary phase.

Chromatography involves a sample (or sample extract) being dissolved in a *mobile phase* (which may be a gas, a liquid or a supercritical fluid). The mobile phase is then passed through a immobile, immiscible *stationary phase*. The phases are so chosen such that components of the sample have differing solubilities in each phase. A component which is quite soluble in the stationary phase will take longer to travel through it, than a component which is soluble in the mobile phase. As a result of these differences in mobilities, sample components will become separated from each other as they travel through the stationary phase.

Some of the general terms which are used in chromatography are:

(*i*) *Elution:* It is a process in which solutes are carried through a stationary phase by the movement of mobile phase.

(*ii*) *Eluent:* It is a solvent used to carry the components of a mixture through a stationary phase.

6.2 CLASSIFICATION

Chromatographic techniques can be classified on the basis of nature of mobile phase or the stationary phase or the mechanism of separation.

(*i*) ***Based upon the nature of the mobile phase, chromatography is of two types:***

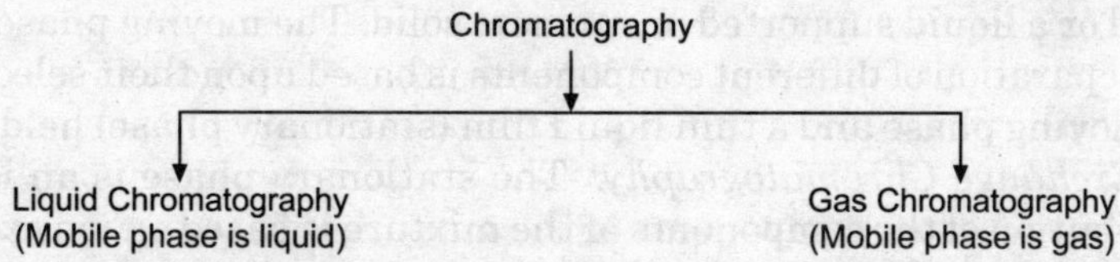

(*ii*) ***Based upon the nature of the stationary phase, these chromatography techniques are further divided into various types:***

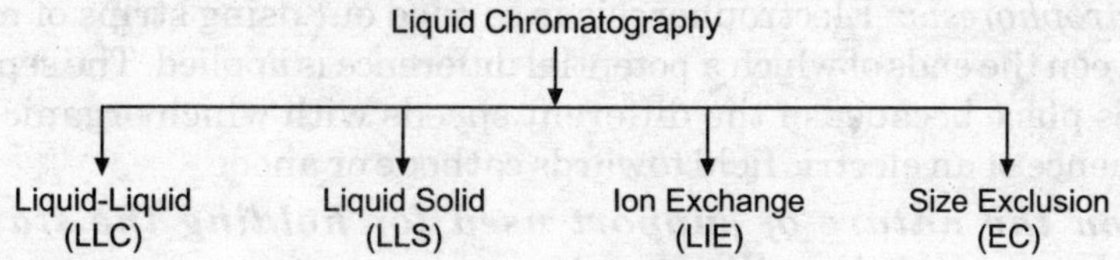

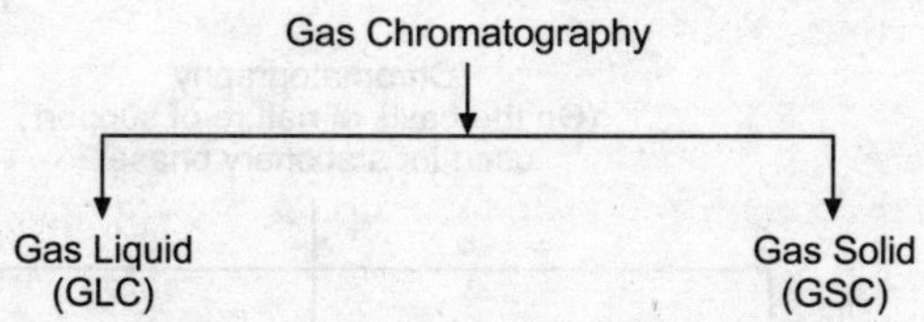

(a) *Liquid-Liquid Chromatography:* In this technique both the phases (stationary and mobile) are in the form of liquids. The liquid which forms the stationary phase is supported over a stationary support such as paper, plate etc.

(b) *Liquid-Solid Chromatography:* The stationary phase is solid (e.g. alumina, silica gel etc.). The solid phase can be placed in a column through which the mobile phase is passed or it can be thin layer supported on a glass plate.

(c) *Ion-Exchange Chromatography:* The ions present in the stationary phase are stoichiometrically exchanged for the ions in the mobile liquid phase.

(d) *Size-Exclusion Chromatography:* Separation is based mainly on exclusion effects, such as differences in molecular size and/or shape or in charge. The term *Size-Exclusion Chromatography* may also be used when separation is based on molecular size. The terms *Gel Filtration* and *Gel-Permeation Chromatography* (GPC) were used earlier to describe this process when the stationary phase is a swollen gel. The term *Ion-Exclusion Chromatography* is specifically used for the separation of ions in an aqueous phase.

(*iii*) ***Based on the mechanism of separation or the interacting phenomenon between the solute molecules and the stationary phase, the various chromatographic techniques have also been classified into the following types:***

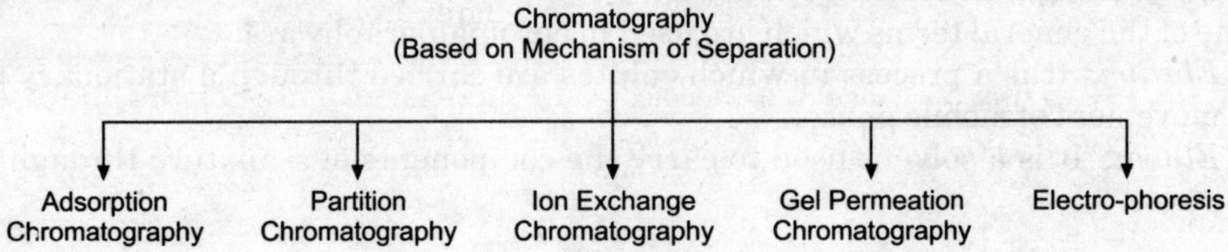

(*a*) *Adsorption Chromatography:* In this type of chromatography, the stationary phase consists of a solid adsorbent and the mobile phase is liquid or gas containing the mixture. The selective adsorption of the components of the mixture occurs on the surface of the solid.

(*b*) *Partition Chromatography:* In this type of chromatography, the stationary phase is a liquid or a liquid supported on an inert solid. The moving phase can be liquid or a gas. The separation of different components is based upon their selective partition between the moving phase and a thin liquid film (stationary phase) held on an inert solid.

(*c*) *Ion Exchange Chromatography:* The stationary phase is an ion exchanger and the distribution of the components of the mixture is based on ion exchange principle. This type of chromatography is applicable to ionic species.

(*d*) *Gel Permeation Chromatography:* In gel chromatography, the stationary phase is gel and the separation is based on its sieving action.

(*e*) *Electrophoresis:* Electrophoresis is carried out using strips of moistened filter paper between the ends of which a potential difference is applied. The separation of the mixture takes place because of the different speeds with which organic ions move under the influence of an electric field towards cathode or anode.

(*iv*) ***Based on the nature of support used for holding the stationary phase, the chromatography is broadly classified as:***

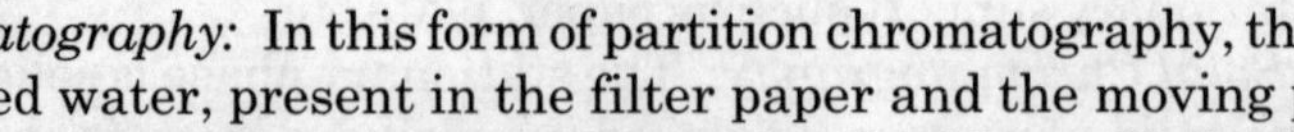

(*a*) *Paper Chromatography:* In this form of partition chromatography, the stationary phase is the absorbed water, present in the filter paper and the moving phase is a solvent saturated with water. Filter paper used is made of highly purified cellulose $(C_6H_{10}O_5)_n$.

(*b*) *Thin Layer Chromatography:* In this, the stationary phase is thin layers of certain fine solid particles, such as silica gel and alumina, supported on glass plates or plastic plates.

(*c*) *Column Chromatography:* In this type of chromatography, the stationary phase is formed into a column in a cylindrical tube.

6.3 LIQUID CHROMATOGRAPHY

As is discussed in section 6.2, that the term 'Liquid chromatography' has been used to cover a range of chromatographic techniques, all of which employ a *mobile liquid phase*.

Earlier liquid chromatography was performed in glass columns having diameters between 10 to 50 mm. These columns were packed with solid particles which constitute the stationary phase and through which the mobile phase percolates under gravity. The particle size in these earlier techniques were kept larger than 150 to 200 μm. The main problem in these techniques was slow speed which was increased by application of vacuum or pressure. But these were not effective because increase in flow rates were accompanied by increase in plate heights and also decreased the column efficiency.

In late 1960s, the technology for using packings with particle diameters as small as 5 to 10 μm was developed. New sophisticated instruments were developed which could withstand high pressures employed to increase the efficiency of the process.

Thus, the name High Performance Liquid Chromatography (HPLC) is often employed for these newer techniques to distinguish them from the older methods.

High Performance Liquid Chromatography (HPLC) is the chromatography is which liquid is the mobile phase and very finely divided particles comprise the stationary phase. To increase the efficiency or to obtain increased flow rates, the liquid is pressurised to several hundred pounds per square inch. HPLC have led to very successful separations of complex mixtures and is more versatile than gas chromatography.

6.4 INSTRUMENTATION FOR HPLC

The instrument used for chromatography is known as chromatograph. The basic HPLC equipment consists of the following components:

(*i*) Solvent reservoir
(*ii*) Pumping systems
(*iii*) Sample injection system
(*iv*) Column
(*v*) Detector
(*vi*) Data control and display

Fig. 6.1 represents the flow diagram of a typical liquid chromatograph.

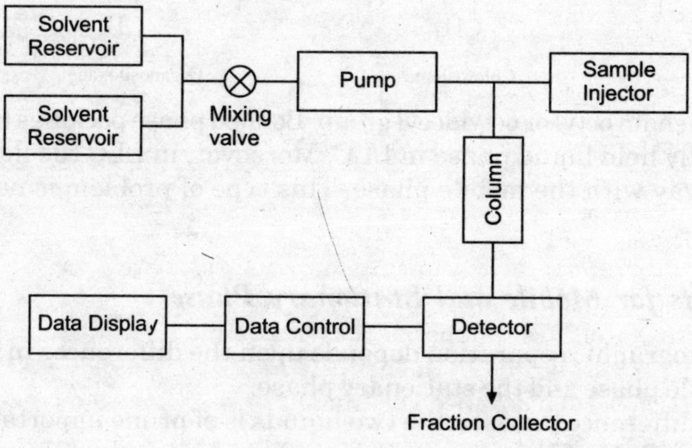

Fig. 6.1 Flow diagram of modern liquid chromatograph.

6.5 TYPES OF LIQUID—CHROMATOGRAPHY (BASED ON MECHANISM OF SEPARATION)

The most commonly and widely used types of HPLC techniques are:

(*i*) Partition or liquid-liquid chromatography
(*ii*) Absorption or liquid-solid chromatography
(*iii*) Ion exchange chromatography
(*iv*) Gel permeation chromatography
(*v*) Gel filteration chromatography

It should be emphasized here that in each chromatography technique, though one mechanism predominates but two or more mechanisms may be involved simultaneously.

6.5.1 Partition or Liquid-Liquid Chromatography

Partition chromatography is based upon the distribution of the solute particles between two immiscible liquid phases depending upon their relative solubilities.

One of the liquid is held on a suitable solid support and acts as a stationary 'phase'. The other liquid is used as the mobile phase. The stationary phase may be in the form of a packed column, a thin layer on glass plate or a strip of paper. The two liquids which are used must be immiscible and differ in the degree of polarity.

Stationary Phase Packings

The partition chromatography can be subdivided into liquid-liquid and liquid bonded phase chromatography. In liquid-liquid partition chromatography the stationary phase is the liquid held in place by *adsorption* on the surface of packing particles. In liquid-bonded partition chromatography, the stationary phase is an organic specie that is attached to the surface of support particles by chemical bonds.

The most common bonded hydrocarbon packings are prepared by reaction of an organochlorosilane with the —OH groups formed on the surface of silica particles by hydrolysis. An organosiloxane is formed as the product.

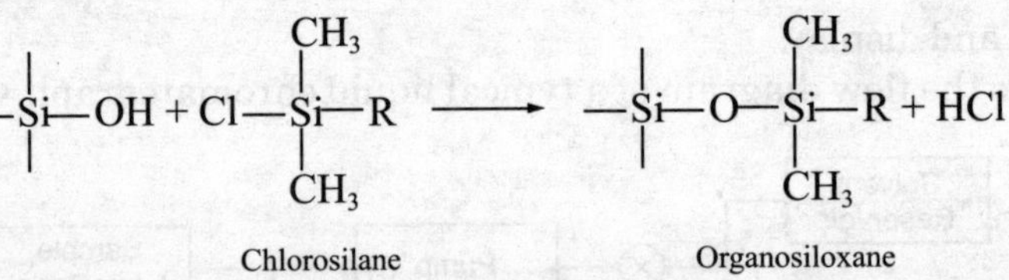

R is often a straight chain octyl or octyldecyl group. Bonded phase packings have greater stabilities compared to physically held liquid phase in LLC. Moreover, in LLC the liquid stationary phase may get dissolved away with the mobile phase. This type of problem is not met within bonded phase packing.

Selection of Solvents for Mobile and Stationary Phase

A successful chromatographic separation depends upon the differences in the interaction of the solutes with the mobile phase and the stationary phase.

Thus the polarity difference between the two liquids is of prime importance in order to obtain efficient separations, and should be properly balanced.

- If the polarity difference between the two phases (stationary and mobile) is too large then the solutes will either be retained too strongly or not be retained.

- If the polarity difference between the two phases is low then the solute will be soluble in both the solvents to some extent and hence separations would not be effective.

Thus the choice of a suitable mobile phase is vital in HPLC. The eluting power of mobile phase is determined by its overall polarity, the polarity of the stationary phase and the nature of the sample components. In general, the polarities of common organic functional groups in increasing order are:

Aliphatic hydrocarbons < olefins < aromatic hydrocarbons < halides < sulphides < ethers < nitro compounds < esters ≈ aldehydes ≈ ketones < alcohols ≈ amines < sulphones < amides < carboxylic acids < water.

Currently there is no one system which adequately ranks all of the common solvents in terms of polarity. However, the solvents are split into eight groups with similar characteristics.

6.5.2 Adsorption or Liquid Solid Chromatography

In adsorption or liquid solid chromatography the stationary phase is a finely divided solid adsorbent and the mobile phase is the organic liquid. The components of the sample are adsorbed onto the surface of the solid adsorbent. The mobile phase carries the sample through the stationary phase and the component which has greater affinity for stationary phase is retained by the stationary phase and the component with relatively higher affinity of the mobile phase travels at a much faster rate as it is carried along the mobile phase and hence separation results.

The solid adsorbent is packed in a column or spread on a plate. Finely divided silica and alumina are the stationary phases which are used commonly in adsorption chromatography. These solids have high surface areas. Silica is generally preferred for most of the applications because it is available in various useful forms.

It should be noted that highly active adsorbents usually give rise to irreversible solute adsorption. For example, silica gel which is slightly acidic may strongly retain basic components whereas alumina (non acid washed) is basic and, therefore, retains the acidic components strongly. Thus silica should not be used for acid sensitivie compounds and alumina should not be used for base sensitive compounds.

The role of mobile phase (solvent) in adsorption chromatography is very important because it is the only variable that affects the distribution of solute between two phases. This is in contrast to partition chromatography where the polarity of the stationary liquid phase can also be varied. Moreover, solvent molecules also compete with the solute molecules for polar adsorption sites on stationary phase.

Liquid solid HPLC is used for the compounds which are soluble in organic solvents and are non-ionic. The water soluble non-ionic compounds are separated using bonded phase chromatography.

6.5.3 Ion Exchange Chromatography

Ion exchange may be defined as the reversible interchange of like ions between a solid phase (ion-exchange material) and a liquid phase in which there is no permanent change in structure. Thus, Ion Exchange Chromatography may be regarded as a special example of liquid-solid chromatography wherein strong ionic attractions replace relatively weak polar adsorption forces. In this technique, substances to be separated are previously adsorbed on the ion exchangers and then released by eluting or washing the column with a solution containing a solute which is more tightly held by the adsorbent than the component to be separated.

Ion Exchange Resins: Ion exchange resins consist of cross linked, long chain organic polymers with a microporous structure. The functional groups attached to the polymeric chains are responsible for ion exchange.

Resins of suitable chemical compositions and physical properties may be synthesized for specific ion-exchange applications. The resins usually consist of styrene-divinylbenzene copolymer. Other polymers like methacrylic acid-divinylbenzene or phenol formaldehyde polymers are also used.

The functional groups attached to the resin may be acidic or basic and hence resins can be classified as:

(*a*) ***Cation exchange Resins:*** Resins containing acidic groups like –COOH or $–SO_3H$ group are termed as cation exchange resins. These are usually styrene divinyl benzene copolymers which on sulphonation or carboxylation become capable of exchanging their H^+ ions with the cations of the solution. A number of synthetic-resin cation exchangers which contain sulphonated phenolic or aromatic hydrocarbons have been prepared. These resins are called H-form cation exchangers. It is also possible to convert the H-form cation exchanger to Na-form exchanger by using NaCl. Schematically a cation exchanger can be represented as:

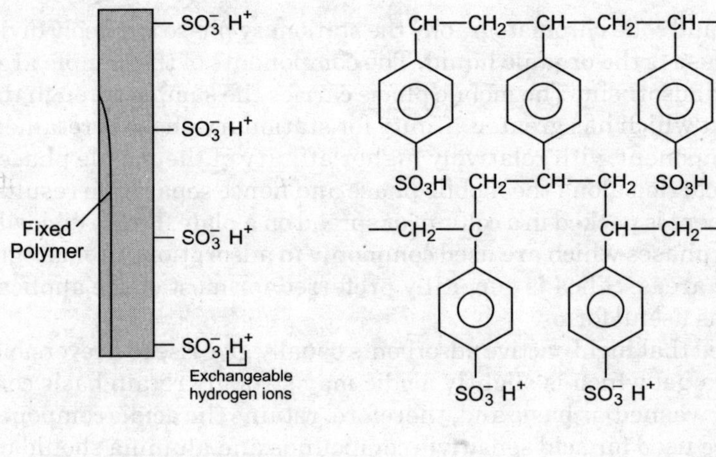

Fig. 6.2 Representation of Cation Exchanger

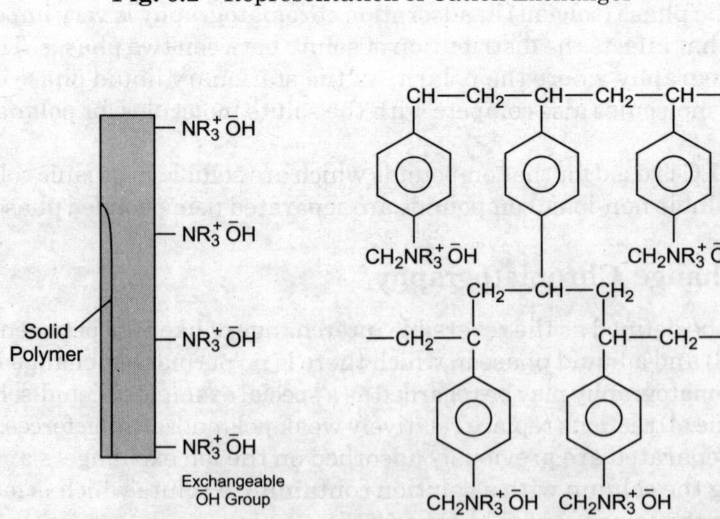

Fig. 6.3 Representation of Anion Exchanger

Resins having SO_3H group are termed as strongly acidic resins whereas having COOH group are termed as weakly acidic resins. The cation exchange resins can be represented as RH^+.

(*b*) ***Anion exchange Resins:*** Resins containing basic functional groups like amine, substituted amine or quaternary ammonium groups as their hydroxide salts are termed as anion exchange resins. These are styrene divinyl benzene complexes which, because of their basic functional groups become capable of exchanging their anions with those present in the solution.

Resins having $-NH_2$, =NH group are weakly basic whereas those having quaternary ammonium salts are strongly basic. Anion exchange resins are represented as $R\overline{O}H$.

Action of Ion Exchangers: The ion exchange resins are insoluble in aqueous media. When a cation exchanger is immersed in an aqueous solution of the sample containing M^{x+} ions, the following exchange equilibrium is quickly established.

$$\underset{\text{Resin}}{xRS\bar{O}_3H^+} + \underset{\text{Solution}}{M^{x+}} \rightleftharpoons \underset{\text{Resin}}{(RSO_3^-)_x M^{x+}} + \underset{\text{solution}}{xH^+}$$

Similarly the anion exchanger can exchange anions (A^{x-}) in a similar way:

$$\underset{\text{Resin}}{xRN(CH_3)_3^+ \overline{O}H} + \underset{\text{Solution}}{A^{x-}} \rightleftharpoons \underset{\text{Resin}}{[RN(CH_3)_3^+]_x Ax^-} + \underset{\text{Solution}}{x\overline{O}H}$$

The ion exchangers have been found to have greater affinities for polyvalent ions as compared to univalent ions i.e. they retain polyvalent ions more strongly than singly charged ions. For a sulphonated cation exchange resin, the affinity for univalent ions decrease in the order:

$$Ag^+ > Cs^+ > Rb^+ > K^+ > NH_4^+ > Na^+ > H^+ > Li^+$$

For divalent the order is

$Ba^{2+} > Pb^{2+} > Sr^{2+} > Ca^{2+} > Ni^{2+} > Cd^{2+} > Cu^{2+} > Co^{2+} > Zn^{2+} > Mg^{2+} > UO_2^{2+}$

6.6 TYPES OF CHROMATOGRAPHY BASED ON THE NATURE OF SUPPORT

6.6.1 Thin Layer Chromatography

Thin layer chromatography (TLC) is an efficient, fast and simple method of separating the mixtures. In this technique a thin layer of silica gel or alumina is spread on a glass or plastic plate and hence the name. The basic principle of separation is the *selective absorption* of the components of the mixture on the stationary phase, when they are carried through it by the flow of mobile phase. The steps involved are as follows:

(*i*) Preparation of thin layer plates.
(*ii*) Application of the material on the plate.
(*iii*) Development of the plate.
(*iv*) Visualization
(*v*) Calculation of R_f values.

(*i*) *Preparation of thin layer plates.*

The plates are usually prepared using glass plates or plastic plates of the size 20 × 20 cm or 5 × 20 cm. A thin layer of silica or alumina is spread on the glass or the plastic plate. The prepared thin layer on glass is often called a *chromaplate*. Chromaplates are prepared by applying a uniform layer of silica or alumina in the form of a thin aqueous paste. The solid adsorbent (silica or alumina) is usually mixed with a small amount of 'binder' such as calcium sulphate.

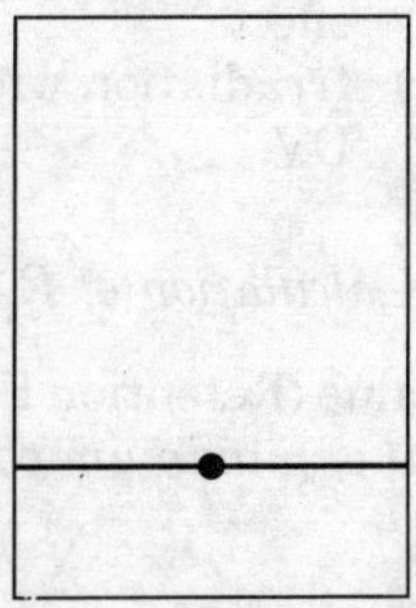

Fig. 6.4 Thin layer plate with the spot of sample

The binder helps in proper adherence of the adsorbent onto the plate. Special applicators are available commercially to spread the adsorbent. The thickness of the adsorbent layer is generally 0.25 mm. The paste or slurry of the adsorbent is prepared in water.

(ii) Application of the material on the plate.

The material is dissolved in an appropriate solvent and deposited on the thin layer with the help of a capillary. The capillary is first dipped in the sample solution, the solution goes up the capillary. The sample is applied on the plate by touching the capillary to the adsorbent surface at about 2 cm from the edge. Position of spot is marked with a pencil so that the Rf values may be calculated after separation.

(iii) Development of the plates

When the spot is dry, the chromaplate is placed in a jar (developing chamber) containing small amount of solvent. The jar must be saturated with solvent vapour before the plate is placed in it. This can be achieved by lining the jar with filter paper dipping into the solvent. Wide-mouthed, capped bottles are used as developing chambers in the laboratories. The plate is kept in such a way that lower end of the plate dips 1–2 cm below the surface of the solvent. The spot should be above this level. The top of the jar is firmly closed. The solvent rises through the adsorbent layer, *and the components of the mixture ascend at different rates depending on their affinities for the adsorbent.* When the solvent has nearly reached the top of the adsorbent layer (3/4 of the TLC plate), the components should be well separated. The relative distance travelled by the components can be increased by using a solvent of higher or lower polarity. Typical solvents in order of increasing elutive power:

Petroleum < benzene < chloroform < ether < ethanol

(iv) Visualization

The plate is removed from the tank, the position of the solvent front is marked with a pencil and the solvent is allowed to evaporate from the plate. If the components of the mixture are coloured, they can be detected by simple visual examination. However, if they are colourless, they are located by

(a) keeping the plate in a closed dry tank containing crystals of iodine. The spots show up as brown spots
(b) spraying the plate with concentrated H_2SO_4 and then heating the plate to about 200°C. The separated spots of organic compounds will render visible due to highly sensitive charring effect
(c) Irradiation with UV light to locate the substance which give flourescence on exposure to UV.

(v) Calculation of R_f Values

R_f value (Retention Factor) is defined as the *ratio of the distance travelled by the component at its point of maximum concentration to the distance travelled by the solvent.*

$$R_f \text{ Value} = \frac{\text{Distance moved by component}}{\text{Distance moved by the solvent front}}$$

The R_f value is characteristic of a substance for a given solvent.

6.6.2 Paper Chromatography

Paper chromatography (PC) is a simple technique to separate complex mixtures of metal ions, amino acids, sugar, dyes and drugs. It is a form of partition chromatography in which the stationary phase is the absorbed water, always present in filter paper. The support is the paper itself and the mobile phase is a solvent previously saturated with water.

Paper chromatography is a form of partition chromatography but adsorption phenomenon is also involved. Filter paper used is made of highly purified cellulose which, being a polyhydroxy compound of high molecular weight, absorbs and retains water molecules. The stationary phase is not the filter paper, but a very thin film of the solvent, usually water adhering to the surface of the filter paper. Thus the paper chromatography will not work in dry climates unless the paper is exposed to air of high humidity, before being placed in the developing solvent.

A drop of a solution of the mixture (or drops of individual components of the mixture) is applied to the paper by means of the capillary tube and the paper dried. The paper is then suspended in a suitable, closed container so that it can be irrigated with an organic liquid or a mixture of liquids The bottom of the paper should immerse in the solvent. The capillary action of the paper will cause the solvent to rise steadily up the paper. During this process, the solvent which now contains the mixture in solution, is continuously extracted by the retained water molecules in the paper.

The extent to which a particular solute has moved, relative to the extent to which the solvent has moved, known as R_f value, will be a function of the partition coefficient of the solute between the solvent and the film of retained water.

When the ascending solvent has travelled the required distance, the paper is removed, the position of solvent-front marked and the paper is dried. The position of various solutes, if they are coloured, appear as coloured spots. If the spots are not coloured, the location of the spots is determined by (i) spraying the paper with a chemical reagent which produces insoluble coloured derivatives with the solutes (e.g. ninhydrin for amino acids) (ii) or by UV light in which the solute exhibit flourescence. The R_f value is calculated.

$$R_f = \frac{\text{Distance moved by the substance}}{\text{Distance moved by the solvent}}$$

In paper chromatography, two types of techniques are used.

(*i*) *Ascending technique:* In this the solvent ascends through the filter paper as described above.

(*ii*) *Descending technique:* In this the solvent is contained in a small trough at the top of developing tank and travels downwards. The mixture is applied at a point external to the trough.

6.6.3 Column Chromatography

Column chromatography is a technique based on the use of a column of a solid absorbent (stationary phase), hence the name column chromatography. The separation of the components of the mixture may involve any physical principle like adsorption, partition or ion exchange.

The success of a separation by column chromatography depends on the choice of the stationary and mobile phases.

(i) **Selection of the stationary phase (Adsorbent)** There are number of adsorbents which can be used in column chromatography. The choice of the adsorbent is dependent upon the following factors:

- The adsorbent must not react with the substance under test.
- It should be insoluble in the solvent to be used.
- It should not catalyze the decomposition of the substances.
- The particle size of the adsorbent should be uniform.

(ii) **Selection of the mobile phase (Solvent)** The separation in a column involves adsorption, partition and ion exchange phenomenon, the choice of solvent depends upon these properties and other factors like polarity and solubility.

Different solvents are used for placing the solute in the column, developing the chromatogram and eluting the adsorbed material. For example, a relatively non polar solvent is used to place the mixture to be separated on the column, whereas polar solvents are used for eluting strongly adsorbed components and non polar solvents are used for weakly adsorbed components.

The solvents in increasing order of polarity are Petroleum ether < carbon tetrachloride < cyclo hexane < carbon disulphide < ether < acetone < benzene < esters < chloroform < ethanol < methanol < water < pyridine < organic acids etc.

The basic steps involved in the column chromatography are as follows:

1. Preparation of the column
2. Application of the sample
3. Development of the column.
4. Recovery of the pure constituents.

1. **Preparation of the column:** A glass column or a burette of 20 cm in length and 7-8 mm diameter is used as a column. The column is packed with the required adsorbent (alumina Al_2O_3; activated charcoal, calcium carbonate, starch etc.) as the stationary phase. For packing the adsorbent, the glass tube is first cleaned and dried. It is clamped to a stand. A small wad of cotton wool is used as the column support. Half of it is filled with the eluting solvent and a thin slurry of powdered packing material is poured into the tube. The slurry can be prepared by mixing the absorbent (silica, alumina etc.) with the requisite amount of the solvent. The stopper is opened and solvent is allowed to run slowly from the column so that the adsorbent particles settle. The column is tapped gently with a glass rod. The slurry is again added and the process is continued till all the adsorbent is added into the tube and the column of desired length is obtained. The column is covered with the layer of the solvent. The column can also be packed by dry powdered form of the adsorbent with continual tapping and then the solvent is added. But in this method air bubbles are liable to be trapped in the column.
2. **Application of the sample.** The mixture or the sample is dissolved in a minimum amount of the solvent (usually the same solvent used for preparing the column is taken). Liquid mixtures may be applied directly to the column. The solution of the sample is added in one lot at the top of the column.
3. **Development of the column.** As the solution of the sample penetrates into the column, these solutes are adsorbed. When the solution has just completely entered the column, fresh solvent is added at the top. The solvent flows down and redissolves the solutes and carries them to lower sections of the column where they are readsorbed in amounts according to their adsorption coefficients. The least tightly held material tends to move ahead more rapidly. The various components will pass through with different rates and separate out into coloured bands or zones. This is called the development of the column and the whole column with bands of separated mixture is known as chromatogram.
4. **Recovery of the pure constituents.** If the components are coloured their positions on the column is readily apparent as a series of coloured bands. Sometimes the positions of the

---

Note. It is customary to use a relatively non-polar solvent in the beginning. The eluting solvent must be significantly less polar than the components of the mixture. If it is more polar, then the components will remain in the mobile phase and little separation will result. Moreover, the mixture should be soluble in the solvent, otherwise it will remain permanently adsorbed on the adsorbent and no separation will take place.

bands can be detected by their fluorescence in ultraviolet light or by the use of an indicator or marking substance. In other cases, the addition of the solvent is continued. By successively changing the solvents in order of their increasing polarity, the components are allowed to pass through the column with the solvent and collected separately. This is called elution. The solvent used for washing a particular band is known as eluent.

6.7 GENERAL AND FUNDAMENTAL CONCEPTS OF CHROMATOGRAPHY

Chromatography involves a number of separation methods which have many similarities in experimental technique, basic principles and operating conditions. In all the chromatographic techniques various components in a mixture get separated during their passage through a porous media due to the differences in their migration rates. The separation is due to the difference in their distribution behaviour between stationary and mobile phase.

Fundamental Concepts

Some of the fundamental concepts used in all types of chromatographic techniques are:

(*i*) ***Distribution Ratio:*** This is one of the main fundamental concept in chromatographic technique. It is defined as the ratio of concentration of a solute in two phases (stationary and mobile) at equilibrium.

$$\text{Distribution ratio, } K_D = \frac{[\text{Solute}]_{\text{stationary phase}}}{[\text{Solute}]_{\text{mobile phase}}}$$

The value of K_D depends upon the temperature, pressure and is independent of concentration. If K_D is large the concentration of solute is more in stationary phase i.e. greater is the retention of the solute in stationary phase.

(*ii*) ***Retention Time:*** The retention time t_R is the time between injection of a sample and the appearance of a solute peak at the detector:

$$t_R = \frac{\text{Length of the column packing } (L)}{\text{Average linear rate of solute migration } (\bar{V})}$$

$$t_R = \frac{L}{\bar{V}}$$

$$\therefore \quad \bar{V} = \frac{L}{t_R}$$

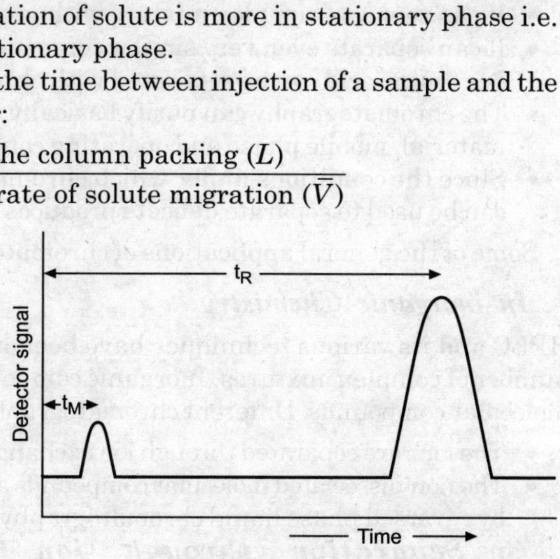

Fig. 6.5

The time taken for mobile phase to pass through the column is called t_m.

A term called the *retention factor*, k', is often used to describe the migration rate of an analyte on a column. It is also called the *capacity factor*. The retention factor for analyte A is defined as:

$$k'_A = \frac{t_R - t_M}{t_M}$$

tR and tM are easily obtained from a chromatogram. When an analytes retention factor is less than one, elution is so fast that accurate determination of the retention time is very difficult. High retention factors (greater than 20) mean that elution takes a very long time. Ideally, the retention factor for an analyte is between one and five.

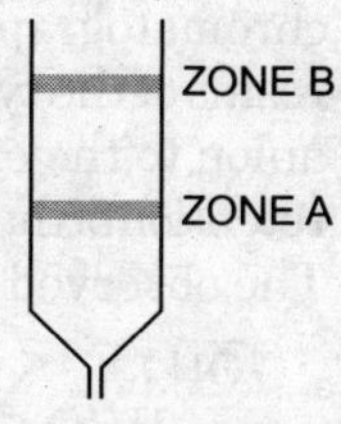

Fig. 6.6 Column showing different zones

(iii) ***Retardation Factor (R_f):*** The retention of the solute in partition or adsorption chromatography is described by its migration relative to the solvent of the mobile phase.

$$R_f = \frac{\text{Distance travelled by solute}}{\text{Distance travelled by mobile phase}}$$

The R_f values change with the type of migration, adsorbent and solvent.

(iv) ***Resolution:*** The effectiveness of the separation is determined by two general factors, the distance between the separated zones and the compactness of the zones.

As the distance between the zone centres increases, the degree of separation increases. The measure of this difference is called resolution. Sometimes as zones travel with the mobile phase, they tend to spread and broaden and overlapping of the bands take place.

(v) ***Selectivity: (Relative migration rates):*** The selectivity factor α for solutes *A* and *B* is defined as the ratio of partition coefficient of more strongly solute (say *B*) to the partition coefficient of the less strongly held solute (say *A*)

$$\alpha = \frac{K_A}{K_B}$$

Thus α is always greater than unity.

6.8 APPLICATIONS OF CHROMATOGRAPHY

In any chemical or bioprocessing industry we need to purify a product from a complex mixture, which is an important step. There are number of techniques available for separating the components of a mixture but chromatography offers a very special technique for number of reasons:

- It can separate complex mixture with great precision.
- It can separate even very similar components. For example, the proteins which differ only in one amino acid can be separated with chromatography.
- The chromatography can purify basically any soluble or volatile substance, if right adsorbent material, mobile phase and operating conditions are employed.
- Since the conditions under which chromatography is performed are not typically severe, it can be used to separate delicate produces.

Some of the general applications of chromatography are:

1. In inorganic Chemistry

HPLC and its various techniques have been applied in inorganic chemistry for separation of a number of complex mixtures. Inorganic compounds can be ionic compounds (composed of ions) or molecular compounds. Different chromatographic techniques have been used for these two classes.

- The ions are separated through ion exchange chromatography or ion exclusion chromatography.
- The non-dissociated molecular compounds are separated by reversed phase liquid chromatography.

(i) ***Separation through ion exchange chromatography:*** In ion exchange chromatography the anions are removed by using the resins of the type – NR^{3+}. Higher is the affinity of the anion to the exchange sites of the ion, higher is the retention of the respective ionic solutes.

The observed order of affinity for groups of ions is:

$$\overline{O}H, F^- < CH_3COO^- < HCOO^- < H_2PO_4^- < HCO_3^- < Cl^- < NO_2^- < HSO_3^- < Br^- < NO_3^- < HSO_4^- < I^-$$

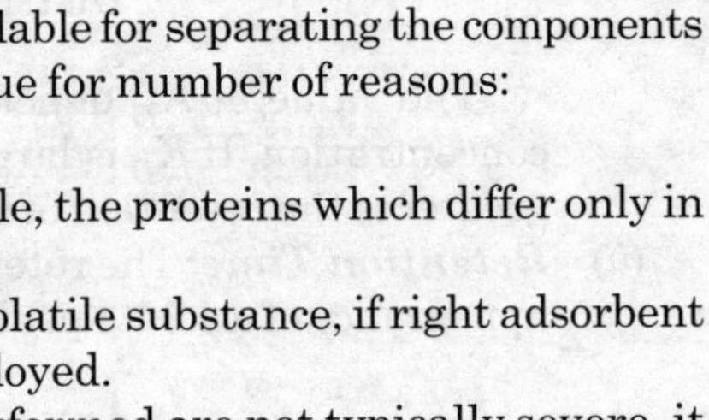

Fig. 6.7 Separation of F^-, Cl^-, PO_4^{3}, Br^- and SO_4^{2-} anions.

Thus, a mixture of ionic compounds containing anions such as F^-, Cl^-, PO_4^{3-}, Br^- and SO_4^{2-} can be separated easily. Fig. 6.7 shows the separation of these anions.

For the separation of anions such as I^-, IO_3^-, SCN^-, ClO_4, MoO_4^{2-}, CrO_4^{2-} and VO_4^{2-}, stronger eluents are generally employed.

(*ii*) ***Separation of strong acids from weak acids (organic acids):*** Using the stationary phase containing sulphonate type ion exchanger, strong acids are separated from weak acids. The strong acids are eluted with the mobile phase and weak acids are retained by the ion exchanger. The eluent is generally HCl.

(*iii*) ***Separation of the cations:*** For the separation of cations, sulphonated polymeric resins are used. These are used as thin coatings on the surface of the spherical glass beads (available under the name Zipax). Alkali metals like Li^{2+} to Cs^+ are eluted by using HCl. The exchange of cations depends upon the following factors:

- At low concentrations, the extent of exchange increases with increasing charge of the exchanging ion:

$$Na^+ < Ca^{2+} < Al^{3+} < Th^{4+}$$

- For singly charged ions, the extent of exchange increases with decrease in size of the hydrated cation:

$$Li^+ < Na^+ < NH_4^+ < K^+ < Rb^+ < Cs^+$$

- To exchange an ion of higher charge on the exchanger for one of lower charge in solution, exchange will be favoured by increasing the concentration of the solution. If the ion of lower charge is in exchanger and ion of higher charge is in solution, the exchange will be favoured by high dilutions. Fig. 6.8 shows applications of ion chromatography to separate a mixture of cations. Note that the amount of cations present in parts per million which would have been difficult to separate otherwise.

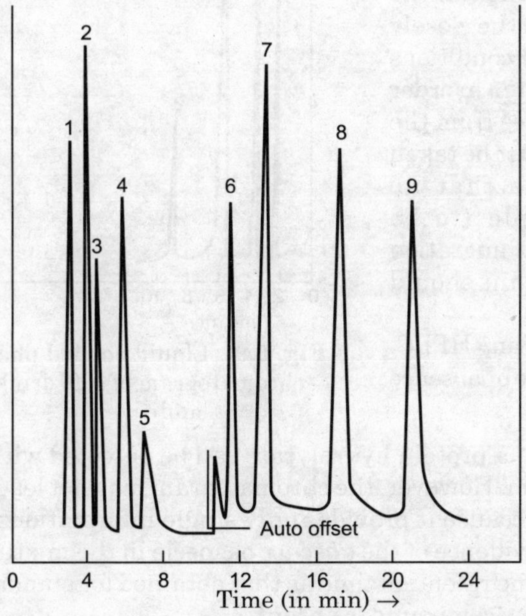

	ppm
1. Li^+	0.5
2. Na^+	2
3. NH_4^+	3
4. K^+	3
5. Morpholine	30
6. Cyclohexylamine	10
7. Mg^{2+}	1
8. Ca^{2+}	2
9. Sr^{2+}	10

Fig. 6.8 Ion chromatogram of a mixture of cations

2. *In Organic Chemistry*

Chromatographic techniques have wide range of applications in organic chemistry. The range of applications in organic chemistry include separation, preparation, purification and identification of the organic compounds.

(*i*) **Preparative HPLC:** It refers to the process of isolation and purification of the compounds. The compounds are isolated in high purity.

(*ii*) **Chemical Separations:** Almost all fields of organic chemistry utilize the chromatographic methods to separate the complex mixture of compounds due to its simplicity and relatively low cost.

Table 6.1 illustrates some of the fields and typical mixtures where this technique is applied extensively.

Table 6.1 Typical Applications of High-Performance Liquid Chromatography

Field	*Typical Mixtures*
Pharmaceuticals	Antibiotics, sedatives, steroids, analgesics
Biochemicals	Amino acids, proteins, carbohydrates, lipids
Food products	Artificial sweeteners, antioxidants, additives
Industrial chemicals	Condensed aromatics, surfactants, propellants, dyes
Pollutants	Pesticides, herbicides, phenols, PCBs
Forensic science	Drugs, poisons, blood alcohol, narcotics
Clinical Medicine	Bile acids, drug metabolites, urine extracts, estrogens
Petroleum and coal industries	Coal and oil products.

(*iii*) **Purification:** It refers to the process of separating or extracting the target compound from other structurally related compounds or contaminants. Under chromatographic conditions each compound produces a characteristic peak. Depending on the compound which has to be separated and the closely related samples, one can choose the proper conditions (such as mobile phase, stationary phase, etc.) in order to collect the desired compound as it elutes from the stationary phase. For purification, care must be taken to avoid overloading the column so that no contamination of the desired sample (to be collected) takes place. Moreover, the rate of migration of the desired sample and the contaminant should vary considerably.

(*iv*) **Identification:** This is the crucial part of any HPLC. It is widely used for recognizing the presence or absence of the components in mixtures.

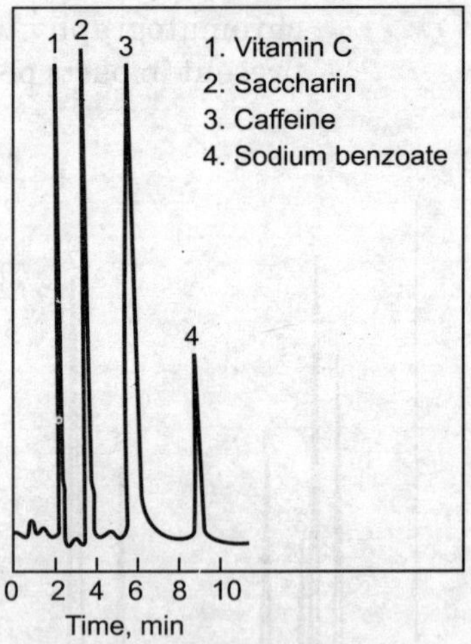

Fig. 6.9 Liquid bonded-phase chromatogram of soft-drink additives

For example, around 30 or more amino acids in a protein hydrolyrate can be detected with a reasonable degree of certainty with a chromatogram. However, the chromatogram may not lead to positive identification of the specie in a sample because it provides only single information i.e. retention time of the specie. But it provides sure evidence of the *absence* of specie in the mixture. The failure of a sample to produce a peak at the same retention time as that obtained for standard under identical condition is strong evidence that the compound is absent.

The technique has wide applicability in identifying the pollutants such as pesticides and carcinogens in air, water and food supplies. It has also been used in identifying the preservatives, antioxidants, additives in food products. Fig. 6.9 illustrates the identification of soft drink additives.

QUESTIONS

1. Define the term chromatography.
2. What are the different types of chromatographic techniques?
3. Define partition chromatography.
   [**Ans.** It is a technique in which mixtures of substances are separated by means of partition between a moving solvent and a stationary liquid which is held on a suitable support. The difference in partition coefficient of individual solutes in mobile and stationary phase is basis of separation.]

Based on thin layer chromatography

4. What is the principle of thin layer chromatography?
5. How the solid layers adhere to the glass plates?
   [**Ans.** With the help of binding agents like $CaSO_4$ etc.]
6. Define chromatogram, developer, adsorbent.
7. What is R_F value? How it is determined?
8. What are different materials used as adsorbent in thin layer chromatography?
9. What is the structure of ninhydrin.
10. What is silica Gel G?
11. In what ways TLC is superior to other chromatographic techniques?
    [**Ans.** It is simple, having short development time and is highly sensitive.]

Based on paper chromatography

12. What is the paper chromatography?
13. What is R_F value?
14. Differentiate between ascending and descending techniques?
15. Choice of solvent is an important criteria for separation in paper chromatography. Explain.
16. State the factors on which the rate of flow of substance depend?
    [**Ans.**The rate of flow of substance depend on the size, nature of the substance, solvent density and diffusion coefficient.]

Based on Column Chromatography

17. What is the principle of column chromatography?
18. How the column is prepared in the laboratory?
19. What do you mean by chromatogram?
20. How the pure constituents are recovered?
21. What is elution?
22. Why it is advisable to start elution with the non polar solvent in the beginning?

Based on ion exchange chromatography

23. Define ion exchange chromatography?
24. What are ion exchangers? How these are classified?
25. What is the principle of ion exchange chromatography?
26. What is ion exchange capacity?
27. Name some of the commercially available ion exchangers.
28. What is regeneration? What chemicals are used for regeneration for cationic and anionic exchangers respectively.

Mixed

29. Define the following:
    (a) Chromatography
    (b) Elution
    (c) Eluent

30. State the principle of column chromatography.
31. How are the different chromatographic techniques classified?
32. Give a brief outline of instrumentation for HPLC.
33. What are the various types of liquid-chromatography?
34. What are the main requirements of solid adsorbents used in column chromatography. Name some adsorbents commonly used.
35. Discuss ion exchange chromatography.
36. What are cation and anion exchange membranes.
37. Why do we change from less polar to more polar eluting solvents and not in the reverse order in column chromatography?
38. What is meant by R_f? What is its significance?
39. What are the main considerations in the selection of eluting solvent in column chromatography?
40. What is TLC? How will you test the purity of an organic compound using TLC?
41. What is paper chromatography? How is it better than column chromatography?
42. Compare HPLC with column chromatography.

❑❑❑

7 INSTRUMENTAL METHODS OF ANALYSIS

Knowledge is a treasure, but practice is the key to it.

— Thomas Fullen

CHAPTER OUTLINE

SPECTROSCOPY

Spectroscopy is the branch of science which deals with the determination of the structure of a compound through the interaction of electromagnetic radiation with the matter.

Though, a number of classical methods of structural analysis, such as examination of physical properties (melting point, boiling point, refractive index, optical rotation, solubility etc.) and chemical properties (functional group analysis etc.) are available, but spectroscopy has a number of advantages over these methods. *Some of these advantages are:*

- Spectroscopic methods require much less time as compared to classical methods.
- A very small amount of substance is required for spectroscopic analysis.
- Generally the substance remains unchanged during spectroscopic examination and can be used again for other tests.
- Spectroscopic methods are highly reliable in establishing the identity of two compounds.

Before going into the details of spectroscopy let us discuss the nature of electromagnetic radiations and how these interact with the matter.

7.1 ELECTROMAGNETIC RADIATIONS

An electromagnetic radiation is a form of energy that is transmitted through space at an enormous velocity. The common example of electromagnetic radiation is visible light. The quantum mechanics suggests that electromagnetic radiation has a dual character i.e. exhibiting wave as well as particle behaviour.

According to wave nature of radiations, the electromagnetic waves consist of oscillating electric and magnetic field and hence the name, electromagnetic radiations. The electric and magnetic components are mutually perpendicular to each other and are coplanar. (Fig. 7.1).

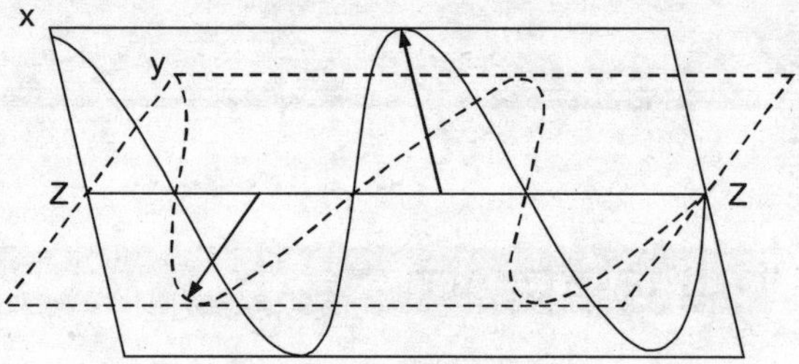

Fig. 7.1 Planes of electromagnetic waves

The electromagnetic radiations are characterized by different parameters, such as

(i) Wavelength: It is the distance between the two adjacent crests (points *C—C* in Fig. 7.2) or troughs (points *T—T* in Fig. 7.2) in particular wave. It is denoted by 'λ'. Mathematically,

$$\lambda = \frac{c}{v}$$

where *c* is the velocity of light and is equal to 3×10^{10} cm per sec and *v* is the frequency. Wavelengths are expressed in a number of units such as, cm, nm, Å etc.

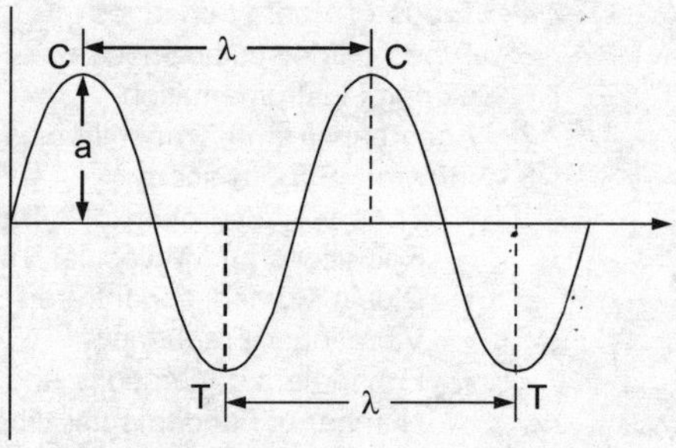

Fig. 7.2 Wave motion

(ii) Wave Number: Wave number is the reciprocal of the wavelength expressed in centimeters. It is the total number of waves which can pass through a space of one cm. It is denoted as $\bar{v}$ and is expressed in cm^{-1}.

$$\bar{v} = \frac{1}{\lambda \text{ in cm}}$$

(iii) Frequency: It is defined as the number of waves which can pass through a point in one second. It is denoted as '*v*' and is expressed in cycles per second or in Hertz where 1 Hz = 1 cycle sec^{-1}. Mathematically,

$$v = \frac{c}{\lambda}$$

where *c* is the velocity of light and λ is the wavelength. As λ and *v* are inversely related to each other, the shorter the wavelength, higher the frequency and vice versa.

(iv) Energy: According to quantum mechanics, an electromagnetic radiation consists of discrete packets of pure energy, called photons or quantas. Each photon is associated with a definite amount of energy. Energy of a wave of the particular radiation is given by Plank's relation,

$$E = hv$$

where h = Plank's constant and its value is 6.62×10^{-34} J-s and v is frequency of radiation in cycles sec^{-1}.

7.2 ELECTROMAGNETIC SPECTRUM

The radiation used in spectra is electromagnetic radiation. A wide range of electromagnetic radiation stretching from cosmic rays at one end to radiowaves at the other, constitutes the electromagnetic spectrum. The electromagnetic spectrum covers an enormous range of frequencies from radiowaves of lowest frequencies (or highest wavelengths) to cosmic rays of very high frequencies (or low wavelengths). In other words, electromagnetic spectrum is the arrangement of electromagnetic radiations in order of their increasing wavelengths or decreasing frequencies.

The visible light occupies a very small region in the electromagnetic spectrum extending from 4300 Å to 6900 Å. However, the complete spectrum may extend over very wide range from cosmic rays of wavelength .001 Å to radiowaves of wavelength 3×10^{13} Å. Table 7.1 gives an idea about the approximate wavelengths, frequencies and energies of different types of radiations.

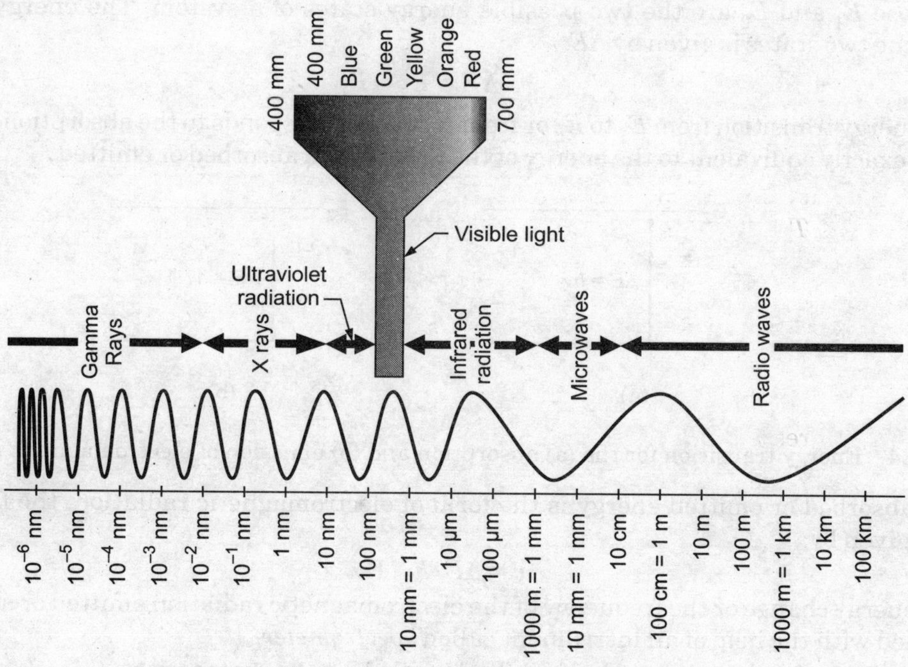

Fig. 7.3 Electromagnetic Spectrum

Table 7.1: The Electromagnetic Spectral Region.

Spectral Region	*Wavelength*	*Frequency, Hz*	*Energy J/Mole*	*Type of quantum change*
γ rays	100 to 1 pm	3×10^{18} to 3×10^{20}	10^9 to 10^{11}	Change of nuclear configuration
X rays	10 nm to 100 pm	3×10^{16} to 3×10^{18}	10^7 to 10^9	Change of electron distribution
U.V. and visible	1000 to 10 nm	3×10^{14} to 3×10^{16}	10^5 to 10^7	Change of electron distribution
I.R.	100 μm to 1000 pm	3×10^{12} to 3×10^{14}	10^3 to 10^5	Change of configuration
Microwave	1 cm to 100 μm	3×10^{10} to 3×10^{12}	10^1 to 10^3	Change of orientation
ESR	100 cm to 1 cm.	3×10^{8} to 3×10^{10}	10^{-1} to 10	Change of spin.
NMR	10 m to 100 cm	3×10^{6} to 3×10^{8}	10^{-3} to 10^{-1}	Change of spin

7.3 ORIGIN OF ELECTRONIC SPECTRA (INTRODUCTION OF RADIATION WITH MATTER)

If some light radiation is passed through a simple of an compound, then some of the wavelengths are absorbed while others remain unaffected. A molecule can only absorb radiation of certain frequency, if there exists within a molecule an energy difference of magnitude ΔE which is equal to hv. Suppose E_1 and E_2 are the two possible energy states of a system. The energy difference between the two states is given by ΔE.

$$\Delta E = E_2 - E_1$$

The energy transition from E_1 to E_2 or from E_2 to E_1 corresponds to the absorption or emission of energy exactly equivalent to the energy of the wavelength absorbed or emitted.

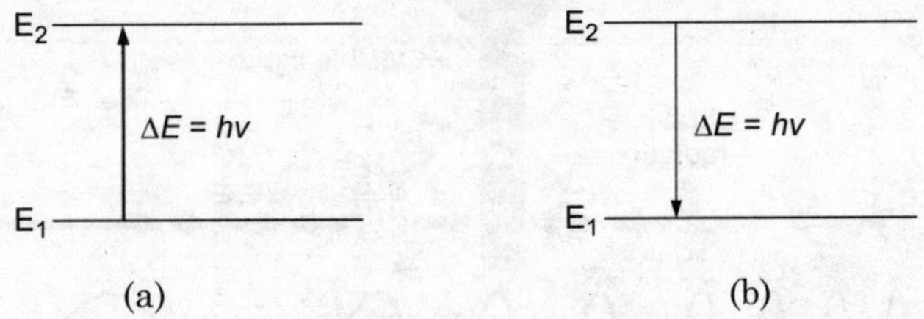

Fig. 7.4 Energy transition for the (a) absorption and (b) emission of electromagnetic radiation

The absorbed or emitted energy is the form of electromagnetic radiation, the frequency of which is given by:

$$v = \Delta E / h \quad \text{Hz}$$

The energy change or the frequency of the electromagnetic radiation emitted or absorbed can be recorded with the help of an instrument called *spectrometer*.

Thus, spectra so obtained can be broadly classified into two categories:

(i) **Emission Spectrum:** When there is a transition from a state of high energy to a stage of lower energy (Fig. 7.4b), the excess energy is emitted as a photon of energy hv provided, $\Delta E = hv$. The spectrum recorded will be *emission spectrum*.
Emission spectra are produced by heating a substance directly in a flame or heating it electrically and then passing the emitted radiation through a prism.

(ii) **Absorption Spectrum:** When there is a transition from a state of lower energy to a state of higher energy (Fig. 7.4a), the energy absorbed will be exactly equal to the energy difference ΔE. The spectrum recorded will be *absorption spectrum*.

Absorption spectrum is produced when the light from a source emitting a continuous spectrum is first passed through an absorbing substance, and then observed through a spectroscope. It will be found that certain wavelengths of light are absorbed by the substance and will leave dark lines or bands at their places giving rise to absorption spectrum.

Emission or absorption spectrum may be classified as Atomic spectra of Molecular spectra.

Atomic Spectra

Atomic spectra is concerned with the interaction of electromagnetic radiation with atoms which are commonly in their ground stage. In atomic spectra, the transition of electrons takes place from one *electronic energy* state to another i.e. between the atomic energy levels. The spectra consists of a series of sharp lines.

Molecular Spectra

Molecular spectroscopy deals with the interaction of electromagnetic radiation with molecules. The spectra by molecules are quite different from the atomic spectra because it involves the transition between *rotational* and *vibrational energy levels* in addition to *electronic transitions.*

The study of molecular spectra has given invaluable information regarding the structure of molecules, incorporating their size and shape, bond distance, bond strength and dissociation energy of the molecules.

In this chapter we will confine our discussion to *Molecular spectroscopy.*

7.4 TYPES OF ENERGY CHANGES

The total energy of the molecule is the sum of transational energy E_t, rotational energy E_r, vibrational energy E_v and electronic energy E_e.

$$E_{\text{total}} = E_t + E_r + E_v + E_r$$

The rotational and vibrational energies are also quantized as is the case with electronic energy. Also,

$$E_e >> E_v >> E_r > E_t$$

The transitional energy which is very small is dissipated as heat and is not involved in the molecular spectra.

The energy levels fall into widely spaced groups which are considered to correspond to different electronic states of the molecule. Associated with each electronic state are a series of vibrational energy levels designated by the vibrational quantum number v. Moreover, each vibrational energy level has a set of rotational sub-levels designated by the rotational quantum number J. The various levels can be represented by Fig. 7.5.

Electronic energy is associated with the transition of an electron from the ground state energy level to an excited state energy level of the molecule due to the absorption of a photon of suitable frequency.

Vibrational energy is due to the to-and-fro motion of the molecule such that centre of gravity does not change.

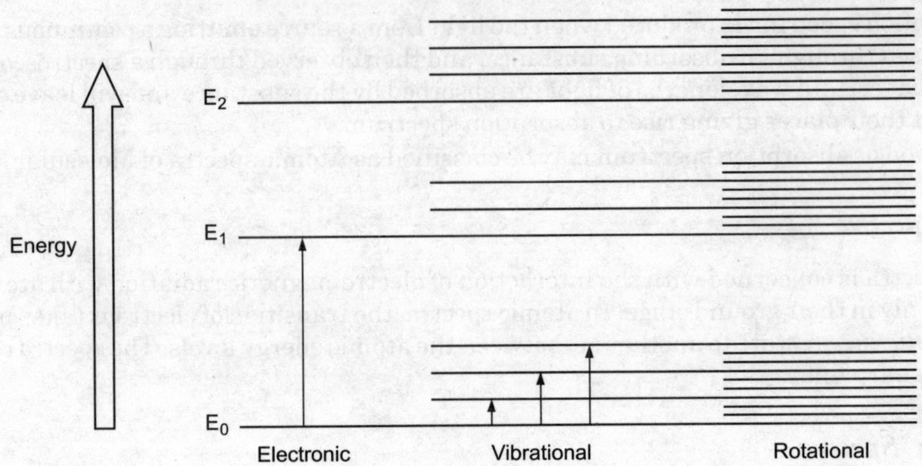

Fig. 7.5 Energy level diagram showing the various molecular energies

Rotational energy arises when the molecule rotates about an axis perpendicular to the internuclear axis and passing through the centre of gravity of the molecules.

Thus energy required for transition will be least for rotational and higher for vibrational and still higher for electronic transitions.

From the above discussion, it is clear that absorption of energy can cause three changes in a molecule.

(i) If the exciting energy is small so as to produce only *rotational transitions* within a vibrational energy level, pure rotation spectra will result. The rotational spectra of a molecule are observed in the *microwave region* or *far infrared region* (> 20 µm).

(ii) If the energy required for exitation are sufficiently large to cause *vibrational transitions* within the same electronic level, the energy changes observed correspond to the change in vibrational quantum numbers. But change in vibrational levels involves changes in rotational levels also, thus the resulting spectrum is called vibrational rotational spectrum. Such type of spectra are generally observed in the *near infrared region* (< 20 µm).

(iii) If the exciting energies are still higher, the *electronic transitions* take place. The electronic transitions are accompanied by a number of vibrational transitions and each vibrational transition in turn is associated with a number of rotational transition. Thus, the resulting spectrum will be complex, consisting of lines due to electronic, vibrational and rotational transitions. The set of rotational lines corresponding to a particular vibrational transition is called a band. Each vibrational transition gives rise to a series of such bands. The entire set of bands for a given electronic transition is known as *band group.* Such type of spectrum is called *electronic band spectra.* These type of transitions take place in *visible and ultraviolet spectral regions,* since considerably higher energies are involved.

7.5 TYPES OF MOLECULAR SPECTRA

Different parts of the electromagnetic spectrum have very different effects upon interaction with matter. Fig. 7.6 gives an idea of the transitions accompanying the absorption/emission of different parts of electromagnetic spectrum.

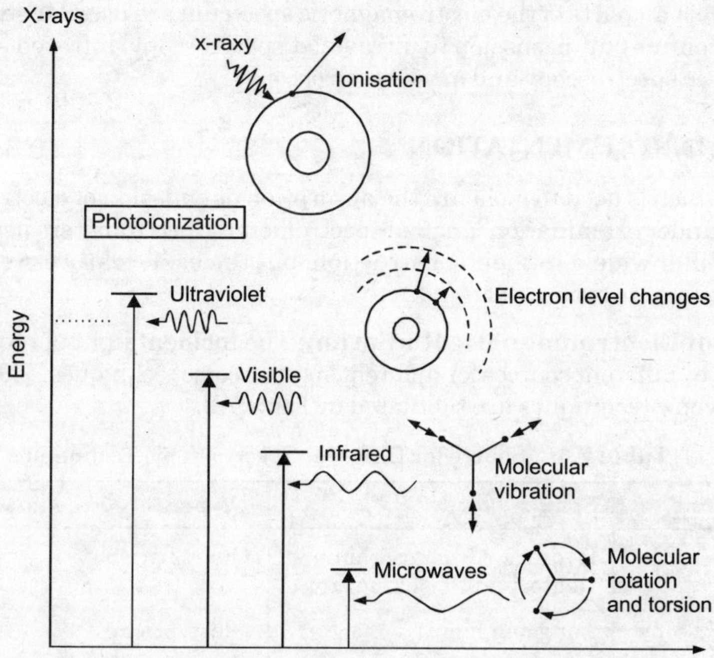

Fig. 7.6 Transitions accompanying the absorption of different parts of electromagnetic spectrum

Energy transitions within molecules are usually studied by the selective absorption of *radiations* passing through them. Molecules spectra are classified according to the region of electromagnetic spectrum in which they are observed or according to the transitions that take place.

Table 7.2 Various Types of Molecular Spectra and their Origin are Summarized

Various Types of Molecular Spectra			
Spectral region	***Wavelength involved***	***Transition***	***Spectroscopy***
1. Microwave	$1 - 100\ cm^{-1}$	Between rotational energy levels	Microwave spectroscopy (Rotational spectroscopy)
2. Infrared	$500 - 400\ cm^{-1}$	Between vibrational energy levels	Infrared spectroscopy (vibrational and vibrational–Rotational spectroscopy)
3. Visible and ultraviolet	$12500 - 25000\ cm^{-1}$ (visible) $25000 - 70{,}00\ cm^{-1}$ (uv)	Between energy levels	Ultraviolet-visible spectroscopy (Electronic spectroscopy)
4. Radio frequency $< 1\ cm^{-1}$ region		Between the nuclear spin energy of the molecules in applied magnetic field.	Nuclear Magnetic Resonance spectroscopy
5. Microwave region	$1 - 100\ cm^{-1}$ (2 – 9.6 GHz)	Between the electron spin energy of the molecules in applied magnetic field.	Electron Spin Resonance spectroscopy

Although almost all parts of the electromagnetic spectrum are used for studying matter but in this chapter will confine our discussion to ultraviolet spectroscopy, infrared spectroscopy nuclear magnetic resonance spectroscopy and mass spectroscopy.

7.6 GENERAL INSTRUMENTATION

The instrument which is used to measure the absorption or emission of electromagnetic radiation by the substance under examination is called spectrometer. Spectrometers used in various regions of the spectrum differ widely in their construction, but the basic features remain the same. The absorption spectrometers consists of :

(i) **Source of Electromagnetic Radiation:** The incident light of required wavelength is provided by different sources for different spectroscopic techniques. The sources for different spectroscopic techniques are tabulated in Table 7.3.

Table 7.3: Source for Different Spectroscopic Techniques

Spectral Region	*Source*	*Dispersing Elements*	*Detector*
1. Ultraviolet Spectroscopy	Hydrogen discharge lamp Deuterium Discharge lamp	Quartz prism	Photomultiplier tube
2. Visible Spectroscopy	Tungsten lamp	Glass prism	Photomultiplier
3. Infrared Spectroscopy	Electrically heated (~ 1500°C) rods of • Nernst glower (Sintered mixtures of oxides of Zr, Y Er etc.) • Globar (silicon carbide) • Ceramic element	Rock salt prism	Thermocouple
4. Microwave Spectroscopy	Klystron tube (A specially developed electron tube in which electrons generate electromagnetic radiations	—	Quartz crystal detector.

(ii) **Monochromator:** For conversion of polychromatic radiation from source into narrowbands or monochromatic radiations, the devices like filters or monochromators are used.

(iii) **Sample Holder:** The sample is placed in the sample container which are cuvettes or cells. The windows of the sample container must be chemically resistant to the sample. Sample holders for various techniques are given in Table 7.4.

Table 7.4

Technique	*Sample holder*
Visible-ultraviolet	Glass or quartz
Infrared	NaCl and KBr

(iv) **Detector Unit:** The main function of any detector is to absorb the energy of photons and convert it into measurable quantity such as electric current or thermal changes. Generally, the detectors generate an electrical signal which activates some type of meter or recorder. In UV-visible spectroscopy, photo cells or photomultipliers are used as detectors. In the infrared regions, thermocouples are used whereas in microwave region, the detector is a crystal rectifier.

(v) **Amplification and Recording Unit:** The main function of amplifier is to amplify a signal so that it can be measured. An amplifier takes an input signal from circuit and through electronic operations, produce on output signal which is several orders of magnitude than input. The signal is recorded on the graph or on the computer monitor. Most recording spectrophotometers record wavelength versus absorbance. The absorbance A or *optical density* is given by

$$A = \log \frac{I_0}{I}$$

where I_0 is the intensity of incident light and I is the intensity of transmitted light. Fig. 7.7 gives a schematic representation of a spectrometer in general.

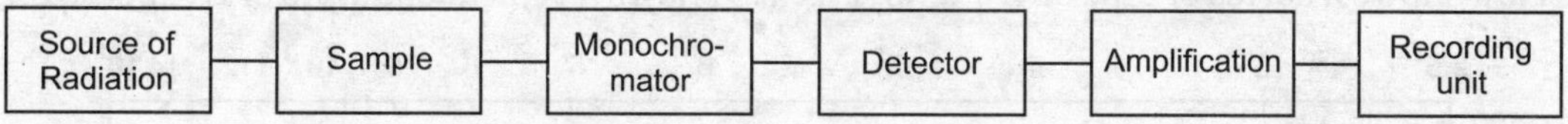

Fig. 7.7 Block Diagram of Spectrometer

Spectrometers for different spectroscopy units are discussed in their respective sections.

7.7 ABSORBANCE AND TRANSMITTANCE

Every molecular specie is capable of absorbing certain characteristic wavelength of electromagnetic radiation. When radiations are passed through a sample of an organic compound some of the frequencies are absorbed while other frequencies are transmitted through the sample, without being absorbed. If the percent Absorbance or percent Transmittance is plotted against frequency, the result is an *absorption spectra. Transmittance of a solution is defined as the ratio of intensity of radiation after passing through the solution to the intensity of intensity light i.e.,*

$$T = \frac{I}{I_0}$$

where I = intensity of light after passing through the solution
I_0 = intensity of the incident light

The transmittance can also be expressed in terms of absorbance of the solution i.e.,

$$A = \log \frac{I_0}{I}$$

where A is absorbance.

7.8 INFRARED SPECTROSCOPY

One of the most widely used tools for detection of functional groups in pure compounds and mixtures is Infrared Spectroscopy.

Infrared Spectroscopy, also known as vibrational spectroscopy, involves the transition between the vibrational energy levels or more precisely, the transition between vibrational and rotational energy levels of a molecule on absorption of radiation falling in the spectral range of 650 – 4000 cm^{-1} (infrared region). Infrared radiations do not have sufficient energy to cause the excitation of electrons. However, it causes atoms and groups of atoms of organic compounds to vibrate faster about the covalent bonds connecting them. As these vibrations are quantized and hence the compound absorbs the infrared energy in particular regions of the spectrum. *In other words, an infrared spectrum is obtained when the frequency of molecular vibration corresponds to the frequency of the infrared radiation absorbed.* In addition to the changes in vibrational energy, simultaneous changes in rotational energy can occur and hence some fine structures may be observed on the vibrational band.

Infrared Spectrum

The plot of percentage transmission of IR radiation versus wavelength is known as infrared spectrum. When a beam of IR radiation of varying frequency is passed through sample of a compound, then from time to time the energy of the beam is absorbed. This happens when the energy matches the difference between vibrational energy levels belonging to the bonds. Now the infrared spectrometer is so designed that the pen of the recorder moves downwards if radiation is absorbed, so vibrational spectrum shows a series of dips or troughs. Fig. 7.8 shows a typical example of IR spectrum.

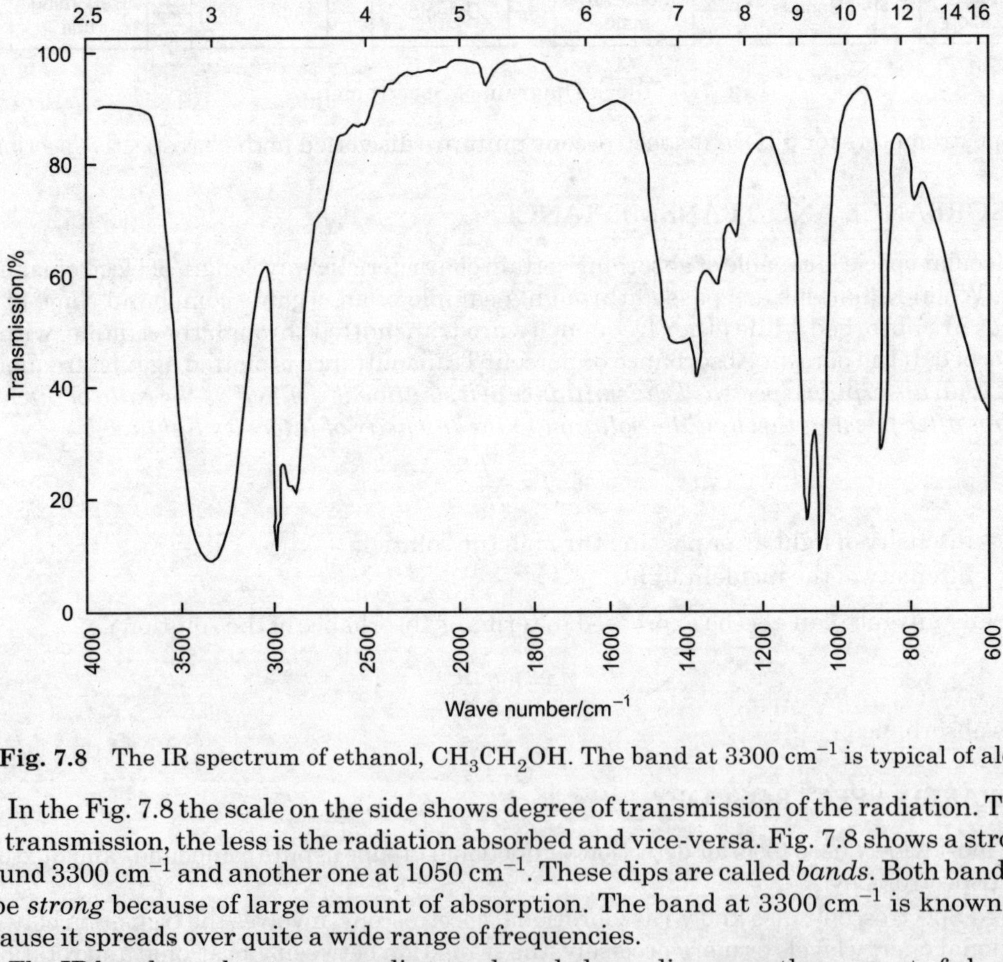

Fig. 7.8 The IR spectrum of ethanol, CH_3CH_2OH. The band at 3300 cm^{-1} is typical of alcohols.

In the Fig. 7.8 the scale on the side shows degree of transmission of the radiation. The higher the transmission, the less is the radiation absorbed and vice-versa. Fig. 7.8 shows a strong dip at around 3300 cm^{-1} and another one at 1050 cm^{-1}. These dips are called *bands*. Both bands are said to be *strong* because of large amount of absorption. The band at 3300 cm^{-1} is known as *broad* because it spreads over quite a wide range of frequencies.

The IR bands can be strong, medium and weak depending upon the amount of absorption. It must be noted again that infrared radiations can be expressed either in terms of wavelength (λ) or in wave number $(\bar{v})$. The relationship between the two is expressed as:

$$\text{Wave number } \bar{v} = \frac{1}{\text{wavelength } (\lambda) \text{ in cm}}$$

Hence the wave number has units of cm^{-1}. Further, IR spectroscopy is divided into three regions as shown below:

Near Infrared Region	Middle infrared (Fundamental vibrational Region)		Far Infrared
	Functional Group Region	Finger Print Region	Region (Pure Rotational Region)

12500 cm^{-1} 4000 cm^{-1} 1400 cm^{-1} 900 cm^{-1} 667 cm^{-1} 50 cm^{-1}

- The near infrared region corresponds to the wave number 12500 cm^{-1} to 4000 cm^{-1}. As there are few absorptions of organic molecules in this range, it is of little use to organic chemist.
- The middle or normal infrared region extends from 4000 cm^{-1} to 667 cm^{-1} and results in vibrational plus rotational transitions. This region is of paramount importance for an organic chemist as most absorptions occur in this region.
- The far infrared region deals with pure rotational motion of the molecules and only few absorptions occur here for organic molecule. It extends from 667 cm^{-1} to 50 cm^{-1}.

7.8.1 Theory—Absorption of Infrared Radiations and Molecular Vibrations

As is discussed earlier that due to relatively smaller amounts of energy associated with the infrared radiations, (2 – 10 k cal/mol) they cannot cause electronic excitations but can only induce transitions between vibrational and rotational energy levels of the molecule Fig. 7.9.

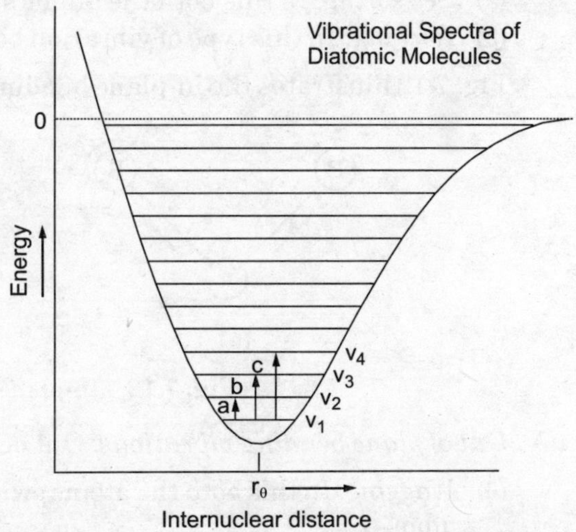

Fig. 7.9 Vibrational levels for vibrating bond.

Molecular Vibrations

Let us first understand what these molecular vibrations are. Actually, a molecule may be regarded as a system of balls (atoms) connected by tiny springs (bonds). The atoms in a molecule do not remain at fixed positions with respect to each other, but actually vibrate back and forth about an average value of interatomic distance (Fig. 7.10)

Now this vibrational motion is quantised as shown in Fig. 7.10. At room temperature most of the molecules in a given sample are in lowest vibrational state. However, when infrared light is passed through it, the molecule absorbs the radiation of appropriate energy and gets excited to higher vibrational level. This results in increasing the amplitude of the vibrational energy. *It may be mentioned here that IR light is absorbed only when oscillating dipole moment due to molecular vibrations interacts with the oscillating electric vector of an infrared beam.*

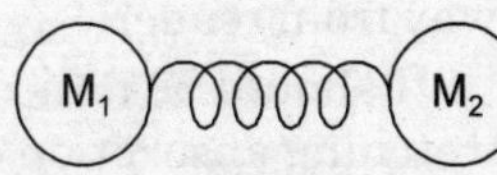

Fig. 7.10 M_1 and M_2 at equilibrium bond length

The simplest vibrational motions in molecules giving rise to absorption of infrared radiation are stretching and bending.

(1) Stretching Vibrations: In this type of vibrations, the distance between atoms increases or decreases but the atoms remain in the same bond axis. It is of two types: *symmetric* and *Antisymmetric*. For the molecule of the type AB_2 the vibrations can be illustrated as in Fig. 7.11.

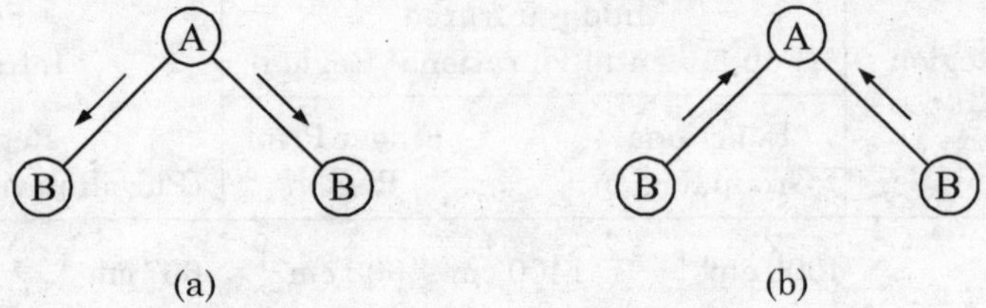

Fig. 7.11 Stretching Vibrations.

(i) *Symmetric:* In symmetric stretching the movement of the atoms with respect to the particular atom in a molecule is in the same direction. (Fig. 7.11a)

(ii) *Antisymmetric:* In this, one atom approaches the central atom while the other moves away. (Fig 7.11b)

(2) Bending Vibrations: In this type of vibrations, the positions of the atom change with respect to the original bond axis but the distance between the atoms remains constant.

Bending vibrations are of two types:

(i) *In plane bending vibrations:* In this type atoms remain in the same plane as the nodal plane of the system. It is further of two types:

(a) *Scissoring:* In this both the atoms swing in opposite direction.

(b) *Rocking:* In this type of vibration both the atoms move in the same direction.

Fig. 7.12 illustrates the in-plane bending vibrations.

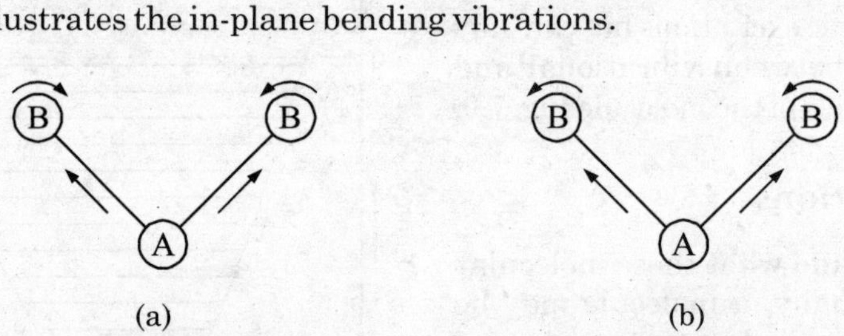

Fig. 7.12 In-plane bending vibrations

(ii) *Out of plane bending vibrations:* Out of plane vibration is of two types:

(a) *Wagging:* In this both the atoms swing up and below the plane with respect to central atom.

(b) *Twisting:* In this type, one atom moves up the plane and other moves down the plane with respect to the central atom.

Fig. 7.13 illustrates the out of plane bending. + and – sings indicate the motion towards and away from reader.

It should be noticed that as more energy is required to stretch a bond than to bend it hence stretching absorption of a band appears at higher frequency as compared to bending vibrations.

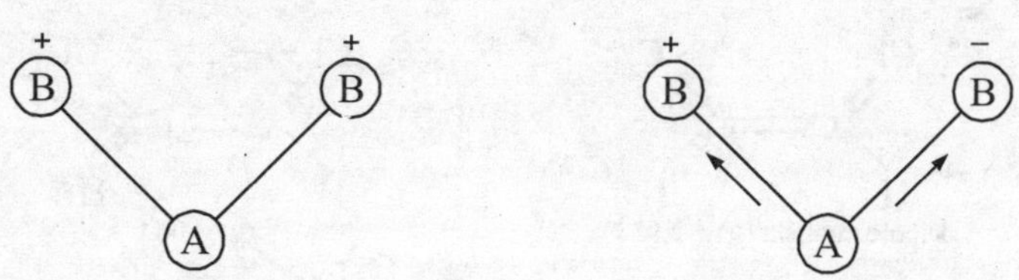

(a) Wagging (b) Twisting

Fig. 7.13 Out-of-plane bending motion.

7.8.2 Dipole Moment and Infrared Spectrum

Generally, the absorption of an infrared radiation of particular frequency by a molecule can occur only if the dipole moment of the molecule is different in the two vibrational levels. As stated earlier, the infrared light is absorbed when the oscillating dipole moment interacts with the oscillating electric vector of an infrared beam. Now, for this interaction to take place the dipole moment of the molecule in the ground state and the vibrational excited state should be different. The greater is the difference in these dipole moments, the more intense the absorption. If no change in dipole moment accompanies the vibrations, then that mode is infrared *inactive* or the molecule is said to be infrared inactive. For example, the symmetrical stretching vibration of — C = C — bond in ethylene can be represented as:

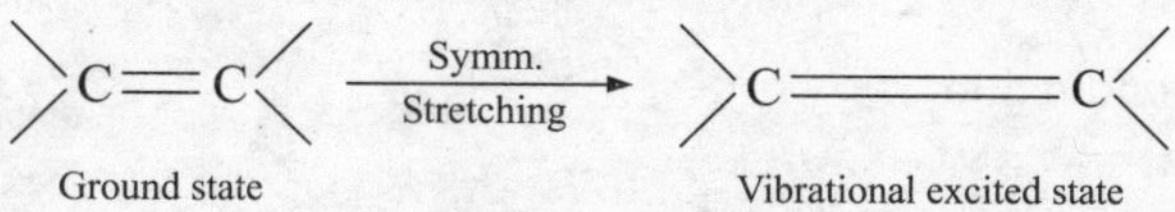

As $>C=C<$ is a nonpolar carbon-carbon double bond, on symmetrical stretching there is little or no change in dipole moment and hence infrared absorption is either absent or very weak. On the other hand the symmetrical stretching of polar carbonyl group can be represented as:

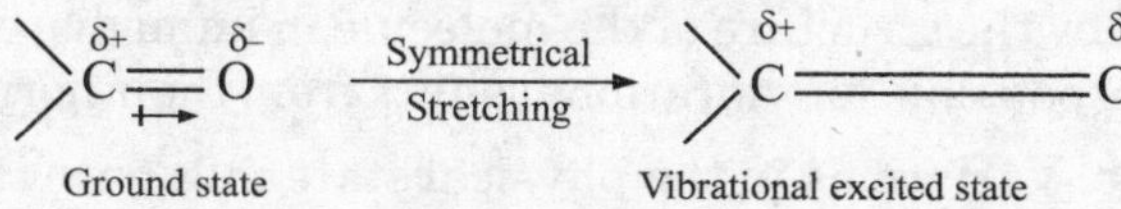

In carbonyl group ($>C=O$), the bond is connected by the atoms having different electronegativities and hence is polar. On stretching, the dipole moment increases considerably or there is a change in dipole moment on symmetrical stretching of the molecule. This change in dipole moment leads to the generation of an oscillating electrical field. If the frequency of this oscillating electrical field is equal to the frequency of fluctuating electric field of infrared radiation, the energy can be absorbed or emitted resulting in absorption infrared spectrum or emission infrared spectra respectively. This $>C=O$ is a infrared active molecule and will show an intense absorption.

Example 7.1: Why trans dichloroethylene shows no C═C stretching absorption while Cis isomer shows this bond.

Solution: The ($>C=C<$) bond in trans isomer is non-polar having zero dipole moment. On stretching, the dipole moment does not change and hence no absorption of infrared radiation takes place. Whereas Cis isomer has a dipole moment, the magnitude of which increases on stretching and hence C is isomer shows C═C stretching absorption band. However both the isomers show bands for C—H and C—Cl stretchings.

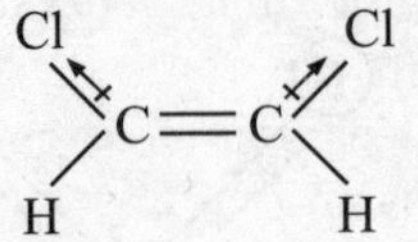

Dipole moment (μ) = 2.95 D

cis 1,2 dichloroethene

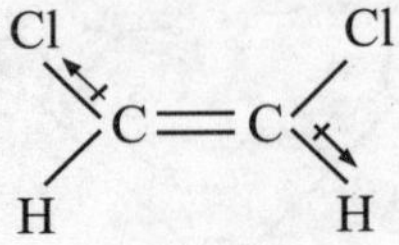

(μ) = 0 D

Trans 1,2 dichloroethene

For the same reason, it can be concluded that the molecules like H_2, N_2, Cl_2, O_2 are *infrared inactive molecules* as there occurs no change in dipole moment during vibrations. Whereas the molecules having permanent dipole moments and exhibiting a change in dipole moment during the vibration are *infrared active* molecules.

However, there are molecules which do not have permanent dipole moment but some of their vibrational modes are accompanied by change in dipole moment making these molecules infrared active. For example, the symmetrical stretching of C=O of CO_2 is infrared inactive while the asymmetrical stretching and bending vibrations are infrared active.

(a) Symmerical stretching (IR inactive) (b) Asymmerical stretching (IR active) (b) Bending (b) Bending (IR active)

Vibrations of Carbondioxide

7.8.3 Factors Influencing Vibrational Frequencies

As is discussed earlier that vibrational frequency can be calculated using Hooke's law. But the calculated value is never exactly equal to experimental value because vibrations of each group are significantly influenced by the structure of the molecule in immediate neighbourhood of bond. A number of factors are responsible for vibrational shifts from their normal values such as:

(i) **Physical state:** Variations in the physical state influence the infrared frequencies by means of association effects. In varpour state, a substance usually absorbs at higher frequency as compared to, when it is in liquid or solid state. For example, cyclopentanone shows the carbonyl absorption band in vapour phase at 1772 cm^{-1} and in liquid phase at 1746 cm^{-1}.

(ii) **Effect of solvent:** Change of solvent also shifts the IR band. For example, carbonyl stretching frequency of acetone occurs at 1742 cm^{-1} in the vapour phase whereas in the liquid phase dipole-dipole association of the type $(\text{---}\!>\!\overset{\delta^+}{C}=\overset{\delta^-}{0}\text{---}\!>\!\overset{\delta^+}{C}=\overset{\delta^-}{0}\text{---})$ lowers the frequency to 1718 cm^{-1}. In different solvents, the frequency of absorption is different viz.

Solvent	Hexane	Chloroform	Ethanol
Absorption band	1726 cm^{-1}	1713 cm^{-1}	1709 cm^{-1}

(*iii*) **Coupled Vibration and Fermi Resonance:** Vibrational coupling may take place between two bonds, through a common atom, vibrating with similar frequency. When the coupling is between two fundamental bands then it is called **coupled vibration** and when it is between fundamental vibration and overtone it is called **Fermi resonance**. In aldehydes, C—H stretching absorption appears at doublet at 2820 cm^{-1} and 2720 cm^{-1}

due to interaction between C—H stretching (fundamental band) and the overtone of C—H bending.

(*iv*) **Electronic Effects:** Under the influence of electronic effects (inductive effect, mesomeric effect) force constant or bond strength changes and the absorption shifts from the normal value. For example the absorption of carbonyl group in the following compounds occur at:

Compound	HCHO	CH_3CHO	CH_3COCH_3
Absorption (cm^{-1})	1750	1745	1715

The introduction of —CH_3 group cause +*I* effect in CH_3 CHO and CH_3 CO CH_3 which results in lengthening or weakening of the bond and hence force constant is lowered and wave number of absorption decreases.

(*v*) **Hydrogen Bonding:** Hydrogen bonding brings about downward frequency shifts. Stronger the hydrogen bonding, greater is the absorption shift towards lower wave number than normal value. Moreover inter and intramolecular hydrogen bonding can be easily distinguished by IR spectroscopy. Generally, intermolecular hydrogen bonds give rise to broad bands whereas intramolecular hydrogen bonds given rise to sharp and well defined bands. Moreover intermolecular hydrogen bonds are concentration dependent and intensities of such bonds decrease and finally disappear on dilution whereas intramolecular hydrogen bonds are independent of concentration.

7.8.4 Applications of IR Spectroscopy

Infrared spectrum of a compound gives the maximum information about its structure. It is highly characteristic property of an organic compound. The applications of IR spectrum are discussed below:

(1) Comparison of Two Compounds—Fingerprint Region

The most powerful function of infrared spectroscopy is establishing conclusively the identity of two samples. The possibility of two different compounds having the same IR spectra is negligible. That is why IR spectra is also called the "fingerprint" of a molecule.

- Thus, if two pure samples have *different* IR spectra, the compounds must be *different*.
- If they have exactly *same* spectra, when run in the same medium under similar conditions, the two samples are of the *same* compound..

Actually, the region from 4000–1500 cm^{-1} in an infrared spectrum, is useful for the identification of functional groups.

This region shows the absorption due to stretching mode, whereas, the region below 1500 cm (1500 – 650 cm^{-1}) contains number of vibrations caused by bending and stretching vibrations. As there are many more bending vibrations in a molecule than stretching vibrations, this region shows a number of vibrations which makes it complex. However, each organic compound has its own unique absorption pattern in this region. This part of spectrum is therefore known as **"Finger Print Region".**

The molecules containing the same functional group, show similar absorptions above 1500 cm^{-1}, but if the molecules are diferent than the spectra differ in finger print region, on the other hand if along with high frequency region (above 1500 cm^{-1}), the finger print region also matches, the two molecules are of the same compound.

This spectral comparison thus provides confirmation of their identity.

(2) Detection of Impurities in a Compound

The infrared spectrum cannot commonly distinguish a pure sample from an impure sample. However, spectrum of pure sample will have fairly sharp and well-resolved absorption bands. The spectrum of a impure sample that contains many different kinds of molecules will display broad and poorly resolved absorption bands.

(3) Detection of a Functional Group in a Compound

Infrared spectrum helps to reveal the functional group present in an unknown sample. In this technique almost all functional groups have their characteristic absorption bonds. Fig. 7.14 shows some of the most common groups with their absorption wavelength.

For example, if the spectrum contains a strong bond at 1718 cm^{-1}, the compound almost certainly contains a carbonyl group (refer Fig. 7.14) However, the spectrum by itself does not always provide further information as to the nature of the group i.e. whether the compound having the carbonyl group is aldehyde, a ketone, an acid or an amide, all of which have carbonyl group. For this, interpretation of infrared spectra for other diagnostic absorption bands has to be carried out. On the other hand, if the spectrum does not contain an absorption band characteristic of a particular functional group, this suggests the molecule does not contain that functional group. This negative evidence is sometimes more useful in confirming the presence or absence of functional group.

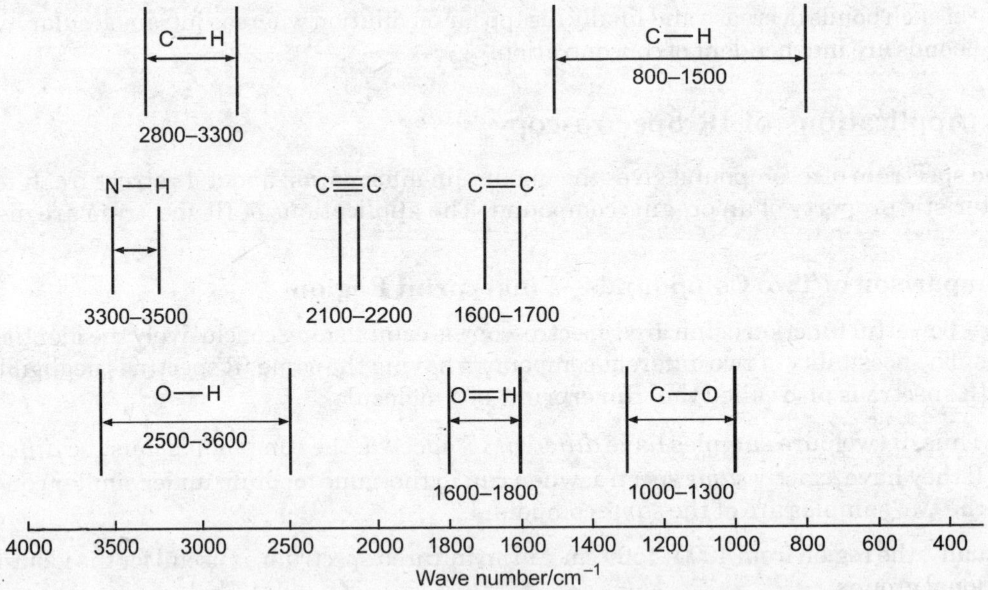

Fig. 7.14 The diagram shows where some of the most common groups appear on an infrared spectrum

(4) Elucidation of Structure

Structural elucidation of an unknown sample by infrared spectroscopy is one of the most significant application of infrared spectroscopy. When the spectrum of unknown material is obtained, a number of questions arise in the mind of a chemist, such as

- What is the type of carbon skeleton present in the compound?
- Which is the functional group present?
- What other environments are influencing it?

These questions can be answered fairly satisfactorily by systematic interpretation of the infrared spectrum. Next section discusses some of the important families giving characteristic absorptions and the environmental effects on them. *However, it is indeed not necessary to memorize all the absorption ranges but only a few values for general position can be remembered to interpret the data.*

7.8.5 Interpretation of Infrared Spectra of Some Representive Compounds

(1) IR Spectra of Hydrocarbons

As hydrocarbons are made up of carbon and hydrogen only, thus in infrared spectrum of these compounds, we shall see absorption bands due to vibrations of *carbon-hydrogen* and *carbon-carbon* bonds. Hydrocarbons may be saturated (alkanes), unsaturated (alkenes) cyclics or aromatics. The characteristic bonds due to stretching and bending vibrations in hydrocarbons are as follows:

(*a*) **C—H stretching Vibrations:**

Groups	***Infrared Absorption Wavelength (cm^{-1})***
Alkanes	2800 – 3000 cm^{-1}
Alkenes and Aromatics	3000 – 3100 cm^{-1}
Alkanes	3300 cm^{-1}

(*b*) **C—H bending Vibration:** These vibrations occur at lower frequency than stretching vibration.

Groups	***Infrared Absorption Wavelength (cm^{-1})***
Alkanes, C—H group	1340 cm^{-1}
Alkene, —CH_2— (Methylene group)	1485 – 1445 cm^{-1}
Alkane, —CH_3	1380 – 1370 cm^{-1}

Alkenes and Aromatic rings: In these compounds both in-plane and out-of-plane bending is present. However out of plane bending is more useful, which gives strong bonds in 800 –1000 cm^{-1}.

Groups	***IR Absorption***
Alkane, monosubstituted RCH = CH_2	910 – 920 cm^{-1} and 990 – 1000 cm^{-1}
Alkene, disubstituted, Cis RCH = CHR	675 – 730 cm^{-1}
Alkene, disubstituted, trans RCH = RHR	965 – 975 cm^{-1}
Alkyne ≡ C—H	630 cm^{-1}
Aromatic rings	730 – 770 cm^{-1}

(c) C—C stretching Vibration:

Groups	*IR absorption*
Aromatic	1500 – 1600 cm^{-1}
Alkenes, nonconjugated	1680 – 1620 cm^{-1}
Alkene, monosubstituted	1645 cm^{-1}
Alkene, disubstituted, cis	1658 cm^{-1}
alkene, disubstituted, trans	1675 cm^{-1}
Alkyne monosubstituted	2140 – 2100 cm^{-1}

Example 7.2: Explain the fact that ethylene and ethyne unlike propene and propyne, have no carbon to carbon multiple bond stretching bonds.

Solution: Ethyne and ethylene are ethylene and are *symmetrical* molecules and hence their dipole moment does not change when excited, and hence no absorption in IR region is observed. On the other hand, when propane and propyne are excited, a change in dipole moment occurs and hence a band is observed.

Example 7.3: The stretching frequencies for C—C, C = C and C ≡ C bonds increase in the order,

C—C	C=C	C≡C
1200 cm^{-1}	1680 – 1620 cm^{-1}	2140 – 2100 cm^{-1}

Explain.

Solution: The frequency of vibration is directly proportional to the force constant of the bond, which increase in the order,

$$C{-}C < C{=}C\ C < C \equiv C$$

and hence the stretching frequencies increase in the same order.

Example 7.4: How can you distinguish between an aliphatic and aromatic compound?

Solution: For aromatic compounds apart from other absorption peaks the most characteristic C=C stretching bands are at 1600 cm^{-1}, 1580 cm^{-1}, 1500 cm^{-1} and 1450 cm^{-1}. If there is no absorption in this region, it is a fair proof that the compound is not aromatic.

Example 7.5: How can you distinguish among hexane, 1-hexene and 1-hexyne using infrared spectroscopy.

$CH_3\ CH_2\ CH_2\ CH_2\ CH_2\ CH_3$
hexane

$CH_3\ CH_2\ CH_2\ CH_2\ CH = CH_2$
1-hexene

$CH_3\ CH_2\ CH_2\ CH_2\ C\ °\ CH$
1-hexyne

Solution:

(i) The IR spectrum of 1-hexyne will exhibit a strong≡C—H stretching absorption at 3300 cm^{-1} and a weaker C≡C absorption at 2120 cm^{-1}.

(ii) The absorption of 1-hexene show =C—H stretching at 3095 cm^{-1} and C=C stretching absorption at 1640 cm^{-1}. The additional bands due to terminal C—H out-of-plane bending will also appear at 990 and 910 cm^{-1}.

(iii) The C—H stretching absorptions in hexane will appear below 3000 cm^{-1}.

Example 7.6: How can you distinguish between the C—H stretching absorption of an alkane and alkene/aromatic compound

Solution: The distinction can usually be made by the fact that aromatic and alkene C—H stretching is just above 3000 cm^{-1} while alkane C—H stretching is just below 3000 cm^{-1}.

Example 7.7: How can you distinguish between an terminal alkyne and non-terminal alkyne.

Solution: Terminal alkynes are easily detected by the presence of strong C—H stretching band near 3300 cm^{-1} and a weaker C≡C stretching bond near 2200 cm^{-1}. However, nonterminal alkynes lack the valuable alkyne C—H stretching absorption and C≡C stretching is extremely weak..

Example 7.8: Mark the various bonds in the infrared spectrum which are most characteristic of benzene in the IR spectrum given below:

Infrared spectrum of benzene:

Solution: Position of characteristic absorptions,

(A) C—H Aromatic stretching vibration 3100 – 3000 cm^{-1}
(B) C═C stretching vibrations 1600, 1500 cm^{-1}
(C) C—H bending vibrations 900 – 700 cm^{-1}

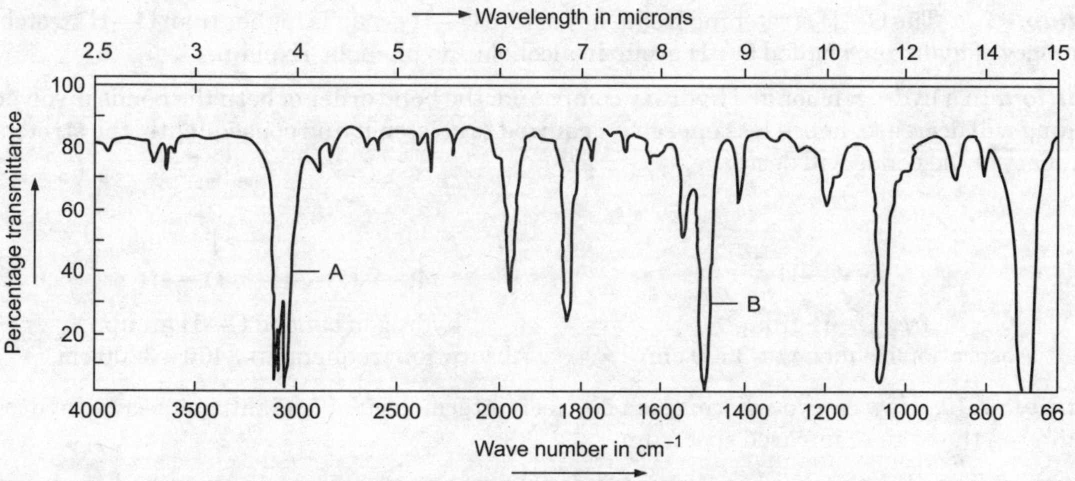

Fig. 7.15 Infra red spectrum of benzene

(2) Infrared spectra of O—H Compounds

Alcohols and phenols exhibit the property of hydrogen bonding and display the following characteristic absorptions due to vibrations and interactions of O—H and C—O bonds.

O—H Stretching vibrations: Alcohols and phenols, in varpour state or in dilute solution in non-polar solvents exhibit a weak O—H stretching absorption due to free OH group. When the spectrum is taken in concentrated solutions, the sharp band disappears and a broad band appears at lower frequencies

Group	*Type of Vibration*	*Region*	*Remarks*
Alcohols and Phenols			
Free OH group	O—H str	3700 – 3500 cm^{-1}	—
Intermolecular hydrogen bonded	O—H str	3400 – 3200 cm^{-1}	Dependent on concentration
Intramolecular hydrogen bonded	O—H str.	3570 – 3450 cm^{-1}	Independent of concentration

C—O stretching vibrations: The C—O stretching vibrations in alcohols and phenols produce a strong band near 1260 – 1000 cm^{-1}. The exact frequency depends on the nature of the alcohol.

Group	*Type of Vibration*	*Region*
Primary alcohols	C—O str. and O—H bending	1050 cm^{-1} and 1350 – 1260 cm^{-1}
Secondary alcohols	C —O str. and O—H bending	1100 cm^{-1} and 1350 – 1260 cm^{-1}
Tertiary alcohols	C—O str and O—H bending	1150 cm^{-1} and 1410 – 1310 cm^{-1}
Phenols	C—O str and O—H bending	1200 cm^{-1} and 1410 – 1310 cm^{-1}

Example 7.9: The O—H stretching frequency of free O—H group is higher than O—H stretching frequency of hydrogen bonded O—H group in alcohols and phenols. Explain.

Solution: In a hydrogen bonded hydroxy compound, the bond order of both the bonds involved in bonding will decrease, hence less energy is required to stretch it and consequently the stretching frequency of the bonds will decrease.

R—C—H

free O—H group
Absorption frequency = 3600 cm^{-1}

R—O—H·····O(R)—H

hydrogen bonded O—H group
Absorption frequency = 3400 – 3200 cm^{-1}

Example 7.10: How can you distinguish between intermolecular and intramolecular hydrogen bonding on the basis of infrared spectrum.

Solution: The IR spectra of the compounds in non polar solvents is examined in different concentrations. Intermolecular hydrogen bonded molecules will show absorption shifts on changing concentration whereas for intramolecular hydrogen bonded compounds there is no absorption shift on changing the concentration.

Example 7.11: The solutions of ethanol and 1, 2 – glycol in carbon tetrachloride show a broad O—H str. absorption at 3350 cm^{-1} in their IR spectra. On dilution with CCl_4, the spectrum of 1, 2 glycol does not change but the spectrum of ethanol show a sharp band at 3600 cm^{-1} in addition to the band at 3350 cm^{-1}. Explain.

Solution: No change in the spectra of 1, 2 glycol on dilution indicates the presence of intramolecular hydrogen bonding. Whereas intermolecular hydrogen bonding is present in ethanol and accounts for the appearance of higher frequency band on dilution.

(3) Infrared Spectra of Carbonyl Compounds (Aldehydes and Ketones)

The carbonyl group ($>C=O$) is the most important in infrared spectrum because it absorbs strongly in this region and exhibit high sensitivity towards minor changes in its environment.

The appearance of a strong band due to C=O stretching in the spectrum between **1650 – 1950 cm^{-1}** shows the presence of carbonyl group in the compound. The frequency of absorption due to the carbonyl group depends mainly on the force constant which intrun depends upon—

(*a*) Inductive effect
(*b*) Conjugative effect
(*c*) Ring Strain
(*d*) Field effects.

(*a*) Inductive Effect: Carbonyl group can be written in the form

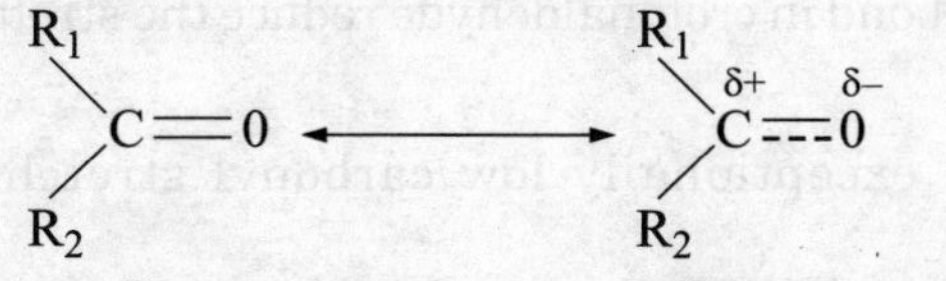

(I) (II)

- If R_1 and R_2 are electron repelling then structure II (polar) is predominant, the force constant for which is lowered as compared to structure I and hence absorption occurs at lower frequency.
- If R_1 and R_2 are electron withdrawing then structure I (double bond structure) is predominant, the force constant is high and hence absorption occurs at high frequency.

Due to large +*I* effect operating in ketones, the ketones absorb at lower frequency as compared to aldehydes

Aldehyde	HCHO	1750 cm^{-1}
	CH_3CHO	1745 cm^{-1}
Ketone	CH_3COCH_3	1718 cm^{-1}

(*b*) **Conjugative Effect:** Conjugation with an olefinic group or benzene ring results in delocalisation of the II electrons of both the unsaturated groups thus reducing the double bond character of both. Due to the introduction of single bond character, the carbonyl absorption frequency decreases from 1718 cm^{-1} to about 1690 cm^{-1}.

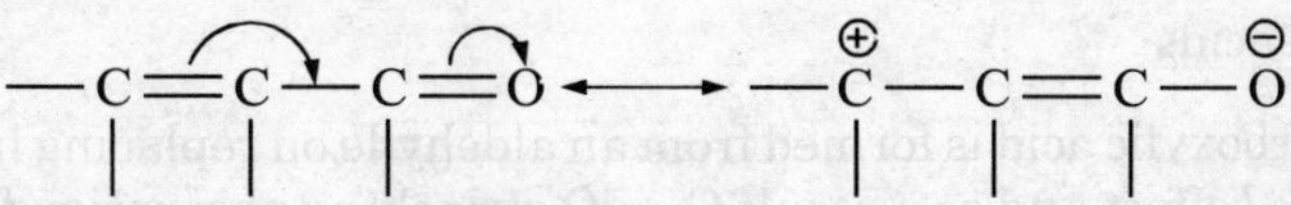

Example 7.12: Compare the C=O stretching absorption in the following compounds:

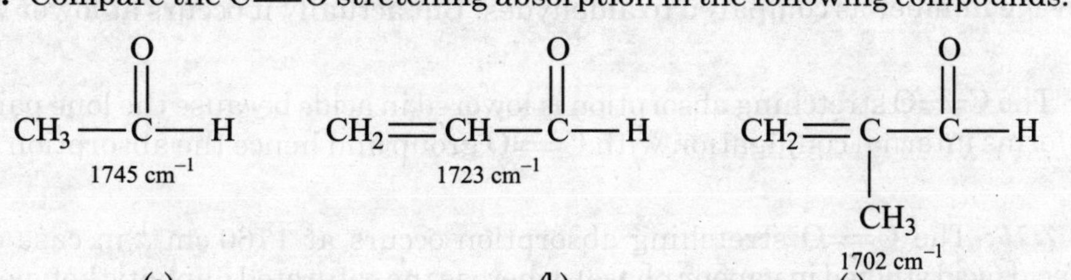

(a) (b) (c)

In compound (*b*) conjugative effect overpowers the –*I* effect and absorption occurs at lower frequency as compared to (*a*) In compound (*c*) both conjugative effect as well as +*I* effect due to CH_3 group lowers the absorption frequency as compared to the compound (*b*).

Example 7.13: Crotonaldehyde exhibit C=0 and C=C stretching at 1700 cm^{-1} and 1630 cm^{-1} whereas an aliphatic aldehyde containing unconjugated double bond exhibits C=O and C=C stretching at 1720 –1740 cm^{-1} and near 1650 cm^{-1} respectively. Explain.

Solution: Crotonaldeyhyde can be represented by the following two resonating forms:

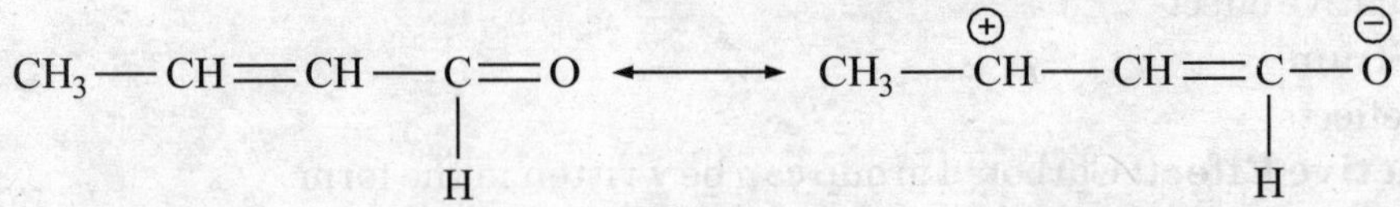

The presence of conjugated bond in crotonaldehyde reduce the stretching frequencies of C═O and C═C bond.

Example 7.14: Explain the exceptionally low carbonyl stretching frequency in 2, 4, 6 cycloheptatrienone.

Solution: Cycloheptatrienone can be represented as:

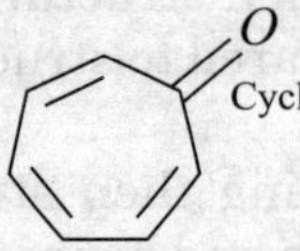

Cycloheptatrienone

The extremely low frequency of cyloheptatrienone is due to the presence of greater extent of conjugation because more is the conjugation more is the lowering of absorption frequency.

(4) Infrared Spectra of Carboxylic Acids

This is the easiest functional group to detect using infrared spectra as it can be considered as being formed from C═O and O—H units. Thus IR spectra of these compounds will show the absorption bands due to both the groups.

(i) Absorption due to *O—H stretching* appears as a broad bond near 3000 cm^{-1}, but extending from 3300 to 2500 cm^{-1}. (The carboxylic acids in liquid or solid state usually exist as dimers and hence the formation of bridges lowers the force constants of O—H and C═O bonds thus absorption is shifted to lower frequency)

(ii) Absorption due to C═O group occurs at the following positions for respective acids:

Aliphatic acid	1725 – 1700 cm^{-1}
Aromatic acid	1715 – 1680 cm^{-1}

or conjugated acids

Example 7.15: An carboxylic acid is formed from an aldehyde on replacing hydrogen atom by OH group. OH has strong *–I* effect and as a result C═O stretching absorption for acids should occur at higher wave number as compared to aldehydes. But actually it occurs at lower wave number. Explain.

Solution: The C═O stretching absorption is lowered in acids because the lone pair on oxygen of OH group forms internal conjugation with C═O group and hence the absorption wave number decreases.

Example 7.16: The C═O stretching absorption occurs at 1760 cm^{-1} in case of monomeric carboxylic acid (acid studied in vapour phase), whereas the saturated aliphatic ketones show C═O stretching absorption near 1720 cm^{-1}. Explain.

Solution: In carboxylic acid, two effects are predominant which work in opposite directions i.e. inductive effect and mesomeric effect, since in carboxylic acids the inductive effect is dominant over mesomeric effect and hence carbonyl stretching frequency in carboxylic acids is higher than ketones.

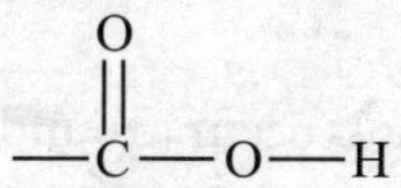

–*I* effect exerted by –OH group, increases absorption frequency

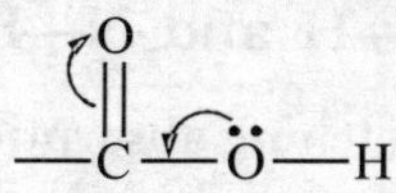

Mesomeric effect causes internal conjugation

Confirmation of the presence of carboxylic acid is done through conversion of acid to its salt. The spectrum of resulting carboxylate anion contains all the absorption pertaining to —COO^- group. Whereas absorptions due to –COOH group will be absent i.e. the absorption due to O—H stretching bond will disappear and C═O stretching frequency will be modified.

Example 7.17: Explain the difference between the carbonyl absorption of carboxylate anion and the carbonyl absorption in carboxylic acid.

Solution: When the acid is converted into its salt, the carboxylate anion formed is resonance hybridised having two CO bonds of equivalent strength, intermediate between C═O and C—O.

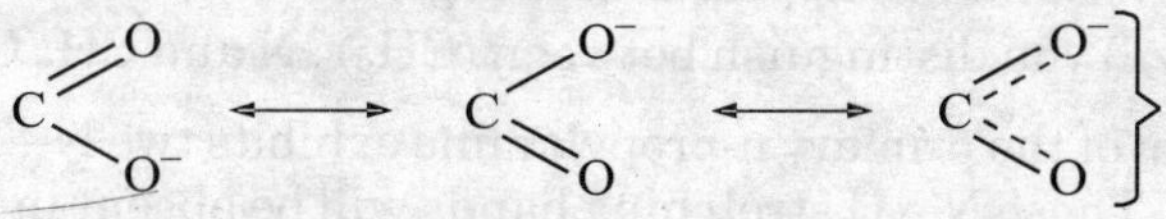

The coupling interaction between two equivalent C═O bonds of carboxylate anion will give rise to two bonds:

a strong asymmetic stretching bond at 1650 – 1550 cm^{-1}
weak symmetric stretching bond at 1400 cm^{-1}.

When the salt is converted to the free acid, the C═O stretching absorption appears at higher frequency than C═O stretching of carboxylate anion.

(6) Infrared Spectra of Amines

Amines are characterised by the appearance of the following absorptions in IR spectrum :

Group	*Type of Vibration*	*Region*	*Remarks*
(*a*) Primary amines	N—H str.	3500 – 3300 cm^{-1}	Two sharp bands
Secondary amines	N—H str	3500 – 3300 cm^{-1}	One band
Tertiary amines	—	—	Do not absorb in this region
(*b*) Primary amines	N—H in-plane bending	1650 – 1580 cm^{-1}	—
Secondary amines	N—H in-plane bending	1650 – 1550 cm^{-1}	—
(*c*) Aromatic primary amines	C—N vibration	1340 – 1250 cm^{-1}	
Aromatic secondary amine	C—N vibration	1350 – 1280 cm^{-1}	
Aromatic tertiary amine	C—N vibration	1360 – 1310 cm^{-1}	
Alphatic amines	C—N vibration	1220 – 1020 cm^{-1}	

Similarities in O—H and N—H Stretching Bands

The typical O—H stretching absorptions in alcohols and phenols occur around 3650 – 3200 cm^{-1} which is also the region in which N—H stretching vibrations of amines occur. Hence O—H str. and N—H str. regions can be sometimes confused with one another. However, since oxygen is more electronegative than nitrogen the O—H ... O hydrogen bonds are stronger than N—H ... N hydrogen bonds, hence frequency shifts due to hydrogen bonding in amines are smaller than alcohols and phenols. The N—H absorptions occur at lower absorption frequencies compared to O—H.

Example 7.18: How can you distinguish between N—H stretching absorption of a primary amine (I) and a secondary amine (II)

$$R\,NH_2, \quad R_2\,NH$$

Solution: These can be distinguished by N—H stretching absorptions. For primary amines N—H str. would appear as doublet (symmetric and antisymmetric stretching). For secondary amines generally a single band will appear in this region.

Example 7.19: How will you distinguish between $(CH_3)_3$ N and CH_3 CH_2 CH_2 NH_2.

Solution: The spectrum of the primary n-propylamine exhibits two N—H stretching bands in the region 3500-3300 cm^{-1}. These N—H stretching bands will be absent in tertiary butyl amine.

(7) Infrared Spectra of Nitro Compounds

In nitro compounds, the two oxygen atoms are equidistant from nitrogen atom and π electrons are distributed between the two N—O bonds:

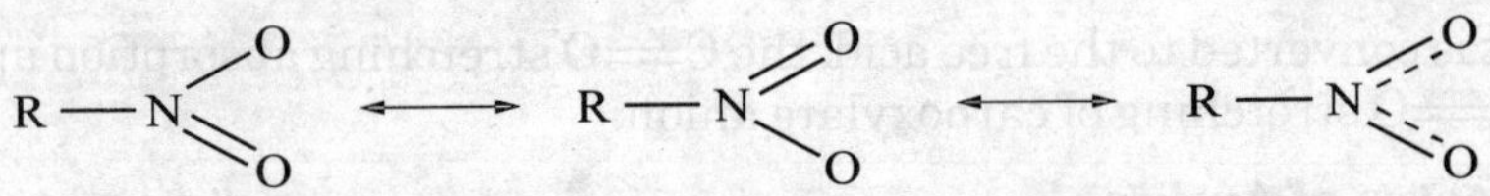

Thus in IR spectrum two strong bands arise from the symmetrical and asymmetrical stretching modes at 1620 – 1535 cm^{-1} and 1375 – 1275 cm^{-1} respectively.

Group	*Type of Vibration*	*Region*	*Remarks*
(i) Aliphatic nitro compounds	N=O str.	1620 – 1535 cm^{-1} 1375 – 1275 cm^{-1}	Symmetric str. Antisymmetric str.
(ii) Aromatic nitro compounds	N=O str	1570 – 1500 cm^{-1} 1370 – 1300 cm^{-1}	Electron Releasing groups at 'o' or *p* positions lower the N=O str. frequencies

(8) Infrared Spectrum of Ethers

Ethers, the derivatives of alcohols show characteristic C—O—C stretching absorptions in the range **1300 – 1000 cm^{-1}**. But a band in this region is also shown by alcohols, aldehydes, ketones and carboxylic acids. Hence the presence of an ether is confirmed only if the compounds shows no absorption band in the hydroxy (3650 – 3200 cm^{-1}) or carbonyl (1870 – 1650 cm^{-1}) region.

In case of dialkyl or diarylethers the symmetric C—O—C stretch is infrared inactive and only antisymmetric stretch is observed (one band). Whereas in case of highly unsymmetrical ethers i.e. alkyl aryl ethers, both the bands appear in infrared spectrum

Example 7.20: How can you distinguish between on ethyl alcohol and diethyl ether.

Solution: The absence of O—H str. band at 3200 – 3600 cm^{-1} in ether and presence of this band in alcohol distinguishes between the two compounds.

7.9 ULTRAVIOLET AND VISIBLE SPECTROSCOPY

Ultraviolet spectroscopy is used to measure the number of conjugated double bonds and also aromatic conjugation within the various molecules.

The electromagnetic radiations in UV-visible region extends from 200 – 750 nm (1 nm = 10^{-7} cm).

The light is of higher frequency and hence of greater energy than infrared light. When UV-visible light is absorbed by a molecule it produces the changes that require greater energy i.e. changes in electronic states. Hence UV-visible spectroscopy involves the transitions of electrons from the ground state to the higher energy states i.e. it leads to transitions among the electronic energy level of the molecule. It is also referred to as *electronic spectroscopy*.

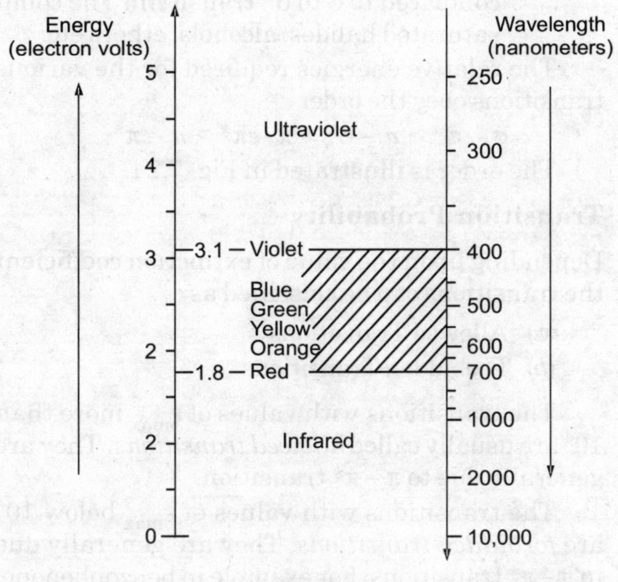

Fig. 7.16

7.9.1 Electronic Transitions: Types of Electronic Transitions

Absorption of ultraviolet radiation by an organic molecule leads to electronic transitions between various energy levels. These transitions consist of an excitation of an electron from an occupied molecular orbital to the next higher orbital.

Now, the electrons in organic molecules may be involved in bonding as strong σ bonds, weak π bonds or present as the non-bonding lone pair. Absorption of UV radiations can bring about the following transitions:

(*i*) **σ – σ* Transition:** The σ electrons are held firmly in the molecule, the transition from σ to σ* require large energies corresponding to absorptions in the far ultraviolet region in the range of 120 – 200 nm.

(*ii*) **π – π* Transitions:** This transition takes place in the compounds containing double or triple bonds. The excitation of π electron requires smaller energy and hence transition occurs at longer wavelengths.

In unconjugated alkenes, absorption bonds appear around 170 – 190 nm.

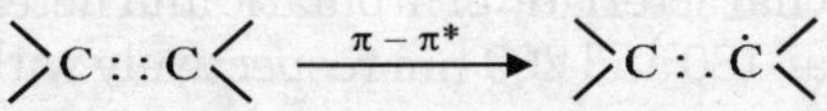

(*iii*) ***n* – π* Transitions:** In this transition, an electron of unshared electron pair on a hetero atom is excited to π* anti-bonding orbital. As these non-bonding electrons are held loosely, the transition requires longer wavelengths. These transitions occur with compound containing double bonds involving hetero atoms having unshared pair of electrons (e.g. *C* = *O*, *C* = *S*, and *N* = *O* etc.).

For example, aldehydes and ketones exhibit two bands at 180 – 200 nm ($\pi - \pi^*$ transition) and 280 nm ($n - \pi^*$). $n - \pi^*$ transition is always less intense because the electrons in the n-orbital are situated perpendicular to the plane of the π bond and hence the probability of the jump of an electron from n to π^* is low.

(*iv*) **$n - \sigma^*$ Transitions:** The excitation of an electron in an unshared pair on an atom to an antibonding σ orbital is called $n - \sigma^*$ transition. Such transitions require less energy compared to σ to σ^* transition. The compounds undergoing this type of transitions are saturated halides, alcohols, ethers etc.

The relative energies required for the various transitions obey the order

$$\sigma - \sigma^* > n - \sigma^* > \pi - \pi^* > n - \pi^*$$

The order is illustrated in Fig. 7.24.

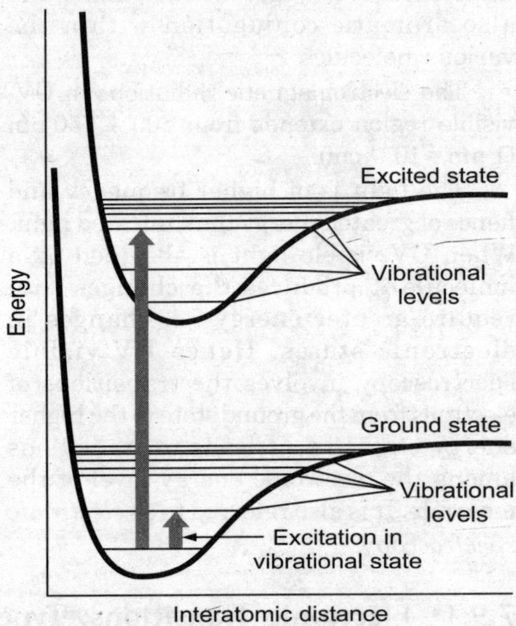

Fig. 7.17 Energy curves for a diatomic molecule

Transition Probability

Depending upon the value of extinction coefficient the transitions can be classified as :

(*a*) Allowed Transitions
(*b*) Forbidden Transitions

The transitions with values of ε_{max} more than 10^4 are usually called *allowed transitions.* They are generally due to $\pi - \pi^*$ transition.

The transitions with values of ε_{max} below 10^4 are *forbidden transitions.* They are generally due to $n - \pi^*$ transitions. For example in benzophenone, the two types of transitions involved are

(*i*) 252 nm ε_{max} 20,000 (Allowed)
(*ii*) 325 nm ε_{max} 180 (Forbidden)

Designation of UV Bands

The UV absorption bands are either designated by using electronic transitions such as ($\pi \rightarrow \pi^*$, $n \rightarrow \pi^*$ etc.) or by letter designation such as:

R Bands: The bands due to $n \rightarrow \pi^*$ transitions are called R-Bands (German-Radikalartig). They have low molar absorptivities ($\varepsilon_{max} < 100$) and are called forbidden bands.

K Bands: Such type of bands arise due to π–π^* transitions in conjugated π systems (German - Konjugierte). These type of bands are found in dienes, polyenes, enones etc. The ε_{max} is usually more than 10^4.

B Bands: B bonds are characteristic of aromatic and heteroaromatic compounds representing π to π^* transitions. The B-band occurs in the region from 250–255 nm as a broad band containing multiplet five structures.

E Bands: These are also characteristic of aromatic and heteroaromatic compounds. The E_1 and E_2 bands of benzene occur near 180 and 200 nm respectively with the ε value between 2000 – 14000.

7.9.2 Chromophores and Auxochromes

Chromophore: Originally, a chromophore was considered any system which is responsible for imparting colour to the compound. (Greek Chromophorus - colour carrier). For example, in nitro compounds the yellow colour is carried by — NO_2 group and hence NO_2 group is a chromophore.

The term has been extended these days and chromophore is defined as a covalently bonded unsaturated group that shows a characteristic absorption in UV or visible region. The absorption occurs irrespective of the fact whether the colour is produced or not. Typical examples are C = C, C ≡ C, C = N, C ≡ N, C = O etc.

Chromophores can further be classified into two types:

(*i*) Chromophores which contain π electrons and they undergo π – π* transitions. Example, C = C, C ≡ C etc.

(*ii*) Chromophores which contain both π electrons as well as non-bonding electrons and undergo π – π* and *n* – π* transitions. Examples are C = O, C ≡ N etc.

Auxochrome: Auxochrome can be defined as any group whose presence brings about a shift of the absorption band towards the longer wavelength. Actually auxochrome is the colour enhancing group. The effect of auxochrome is due to its ability to extend the conjugation of a chromophore by the sharing of non-bonding electrons. Thus, a new chromophore results which has different value of absorption maximum. The examples of auxochrome groups include —OR, —NH_2, —NR_2 etc.

Benzene shows an absorption maximum at 255 nm with ε_{max} 203 whereas aniline at 280 nm with ε_{max} 1430. Hence amino group (—NH_2) is an auxochrome.

Table 7.5 gives the approximate wavelength of absorption of organic chromophores.

Table 7.5: Simple Chromophores and the Approximate Wavelengths at which they Absorb.

Chromophore	*Transition*	*Absorption max. (mμ)*	ε_{max}	*Solvent*
C = C	π – π*	~ 175	~ 15000	Vapour
C ≡ C	π – π*	(*i*) ~ 175	~ 10000	Hexane
		(*ii*) 196	~ 2000	"
		(*iii*) 220	~ 150	"
C = O	*n* → σ*	160	18000	
	π – π*	180	10000	Hexane
	n → π*	285†	15	
R – NO_2	π – π*	~ 200	5000	Methanol
	n → π*	~ 274	15	
—C(OH) = O	*n* → π*	204	60	Methanol
—N = N—	*n* → π*	338	~ 5	Ethanol
— $CONH_2$	π → π*	178	9500	Hexane
	n → π*	220	63	Hexane

7.9.3 Absorption and Intensity Shifts

(*a*) **Bathochromic shift (Reds Shift):** Shift of an absorption maximum to longer wavelength is called Bathochromic shift or Red shift. It may be due to the presence of an auxochrome or change in the polarity of the solvent.

- The *n* – π* transition for carbonyl compounds experiences bathochromic shift when the polarity of the solvent is decreased.
- Conversion of phenol to phenolate anion causes bathochromic shift because the non-bonding electrons become available for interaction with the π-electron system of the ring.

(b) **Hypsochromic Shift (Blue Shift):** Shift of absorption maximum to shorter wavelength is called Hypsochromic shift or Blue shift. It may be due to increase in the polarity of the solvent or the removal of conjugation.

- Aniline absorbs at 280 nm because of the pair of electrons on nitrogen is in conjugation with the π bond system of benzene ring. However in its acidic solutions, a blue shift is caused and absorption occurs at ~ 203 nm because 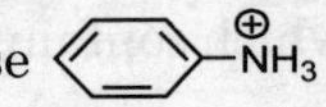 formed in acidic solution, the electron pair on nitrogen is not present and hence the conjugation is removed.

(c) **Hyperchromic Shift:** It is the effect leading to increase in intensity of the absorption maximum i.e. ε_{max} increases. The introduction of auxochrome usually increases the intensity of the absorption maximum.

- The B-bands for pyridine at 257 nm with ε_{max} 2750 shifts to 262 nm with ε_{max} 3560 for 2-methyl pyridine.
- Phenol absorbs at 210 nm with ε_{max} 6200 but the value of ε_{max} increases to 9400 in phenolate ion with λ_{max} 210 nm.

(d) **Hypochromic Shift:** It is the effect leading to decrease in intensity of the absorption maximum i.e. ε_{max} decreases. The groups which distort the geometry of the molecules cause hypochromic shift.

- Benzene absorbs at 230 nm with ε_{max} 11600 whereas benzoate absorbs at 224 nm with ε_{max} 8700. Hence the ε_{max} is decreased in benzoate ion.

Terminology for Absorption Shift

Nature of Shift	*Descriptive Term*
To Longer Wavelength	Bathochromic
To Shorter Wavelength	Hypsochromic
To Greater Absorbance	Hyperchromic
To Lower Absorbance	Hypochromic

Fig. 7.18(a)

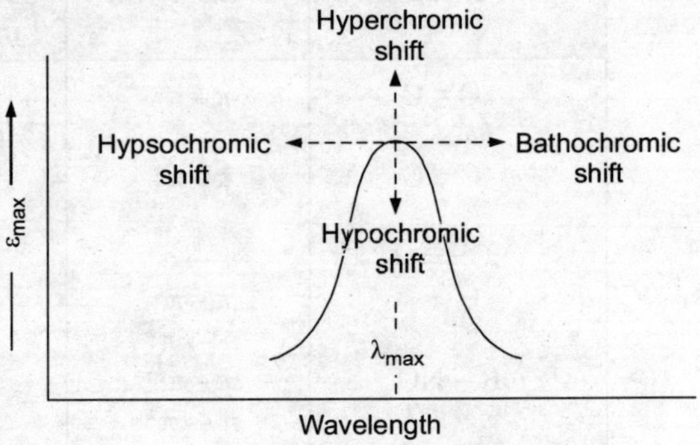

Fig. 7.18(b)

7.9.4 Solvent Effects

The most suitable solvent for UV spectroscopy is the one which does not absorb in the region under investigation. Most commonly useds solvents are ethanol, hexane, methanol, cyclohexane etc. They are transparent (do not absorb) upto above 210 nm.

The position and intensity of absorption bands is greatly affected by the polarity of the solvent used.

Following shifts are absorbed on changing the polarity of the solvents:

- Conjugated dienes and aromatic hydrocarbons experience very little shifts on changing the polarity of solvents.
- The polar compounds show a change in absorption maximum on changing the polarity of the solvent. α, β-unsaturated carbonyl compounds show two different shifts on increasing the polarity of the solvent:

(*a*) **The π-π* band** moves to longer wavelength. This is because the dipole-dipole interactions with the solvent molecules lower the energy of the excited state more than that of the ground state. In other words, π* orbitals are more stabilized by hydrogen bonding with polar solvents because of greater polarity of π* compared to π orbitals. Thus the energy required for π - π* transition is less and absorption moves to longer wavelengths.

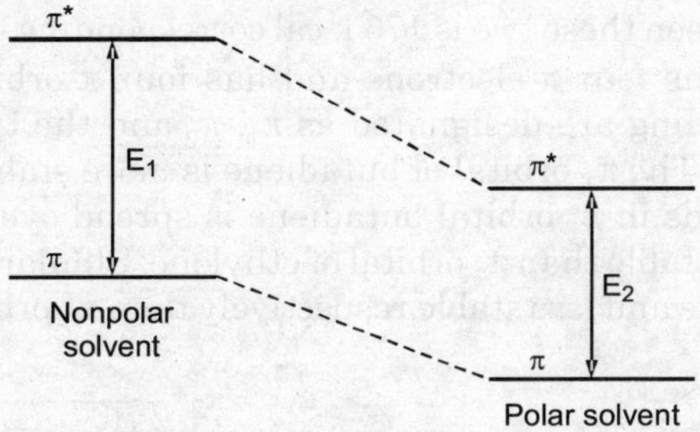

Fig. 7.19 Absorption shift with change in polarity in π-π* band.

(*b*) **The *n*-π* bond** moves to shorter wavelength by increasing the polarity of the solvent. This is because the polar solvent stabilizes the *n* orbitals more than the π* orbitals. This stabilization is more with hydrogen-bonding solvents (H_2O, ethanol etc.) Thus the net energy of *n* orbital is lowered, increasing the energy gap between ground and excited state. Hence absorption band moves to longer wavelengths.

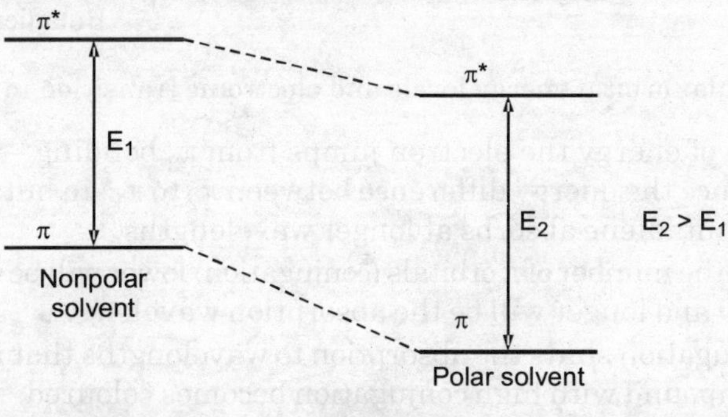

Fig. 7.20 Absorption shift with change in polarity in *n* – π* band

7.9.5 Ultra-violet Spectra of Organic Compounds

Organic applications of UV spectroscopy are mainly concerned with conjugated systems. Some of the most significant applications of UV spectroscopy in organic chemistry are discussed below:

(1) **Extent of Conjugation:** A conjugated system requires lower energy for the π-π* transition than an unconjugated system. Longer the conjugation, longer the absorption wavelength. For example, ethylene has one double band and absorbs at 170 nm due to π-π* transition whereas butadiene, having two double bonds in conjugation absorbs at 217 nm. Some more examples are

Compound	λ_{max}	ε_{max}
Ethylene	170	21,000
2,3 dimethyl butadiene	226	21,400
1,3,5 Hexatriene	254	21,400

How conjugation affects the λ_{max} value can be explained by considering the cases of ethylene and butadiene.

Ethylene has two π orbitals, one ground state π bonding orbital and one excited state π* antibonding orbital. When a molecule absorbs light at its longest-wavelength, an electron is excited from its highest occupied molecular orbital (HOMO) to the lowest unoccupied molecular orbital (LUMO). In ethylene or simple alkenes, the HOMO orbital is π orbital and LUMO is π* orbital. The energy difference between these two is 176 k cal corresponding to a wavelength of 170 nm.

Butadiene, however has four π electrons and has four π orbitals. Two bonding and two antibonding. The two bonding are designated as π_1, π_2 and the two antibonding orbitals are designated as π_1^* and π_2^*. The π_1 orbital of butadiene is more stable than π orbital of ethylene because the electrons clouds in π_1 orbital butadiene is spread over four carbon atoms. The π_2 orbital of butadiene is less stable than π_1 orbital of ethylene. Similarly the two π_1^* and π_2^* orbital of butadiene are more stable and less stable respectively than π* orbital of ethylene. (Fig. 7.21)

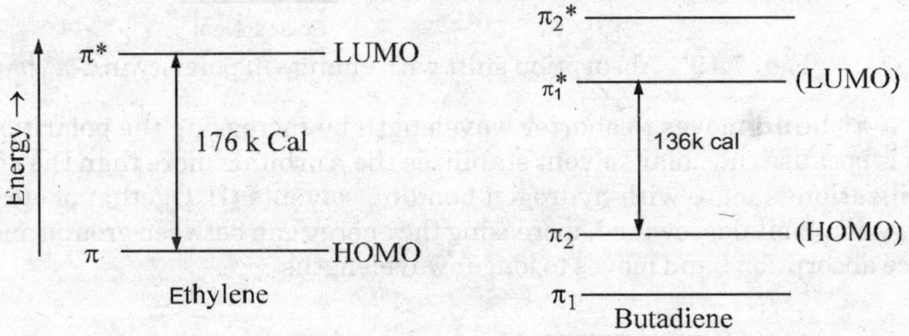

Fig. 7.21 Molecular orbital energy levels and electronic transition in ethylene and 1, 3 butadiene.

On absorption of energy the electron jumps from π_2 bonding → π_1^* antibonding (HOMO → LUMO) orbital. Since the energy difference between π_2 to π_1^* in butadiene is less as compared to π - π* in ethylene, butadiene absorbs at longer wavelengths.

Hence, greater the number of π orbitals (conjugation) lower will be the energy difference between HOMO and LUMO and longer will be the absorption wavelength.

Sufficient conjugation shifts the absorption to wavelengths that reach the visible region of the spectrum i.e. a compound with high conjugation becomes coloured.

For example, tomatoes are red in colour because of the presence of lycopene, the compound having eleven conjugated double bonds. Thus it absorbs at λ_{max} value of 454 nm which is the wavelength of blue colour - the complementary colour of red and hence the tomatoes appear red.

Similarly dyes owe their colours to conjugated systems.

Example 7.21: Which of the following will exhibit higher value of λ_{max} in the UV spectra ?

(*a*) CH_3COCH_3 and $CH_2 = CHCOCH_3$

(*b*) $CH_2 = CH - CH = CH - CH_2 - CH_3$ and $CH_2 = CH - CH_2 - CH = CH - CH_3$

Solution:(*a*) $CH_2 = CHCOCH_3$

(*b*) $CH_2 = CH - CH = CH - CH_2 - CH_3$

The UV absorption wavelength in conjugated dienes and trienes can be calculated using *Woodward- Feiser Rules* given in section 7.11:

(2) **Determination of Geometrical Isomers:** It has been found that cis-alkenes absorb at different wavelengths as compared to their corresponding trans isomers. Coplanarity is needed for

the most effective π orbital which eases the π–π* transition. The steric strain in cis-isomer prevents complete coplanarity. Thus in cis-isomer λ_{max} arises at shorter wavelength.

Example 7.22: Cis stilbene absorbs at 274 and trans stilbene absorbs at λ_{max} 294 nm. Explain.

Solution: The steric hindrance in cis-isomer prevent complete coplanarity required for full conjugation and hence it absorbs at shorter wavelength than trans isomer.

(3) **UV Spectra of Carbonyl Compounds:**

(*i*) **Saturated Carbonyl Compounds:** For a carbonyl compound, two types of transitions can take place $\pi \rightarrow \pi^*$ and $n \rightarrow \pi^*$.

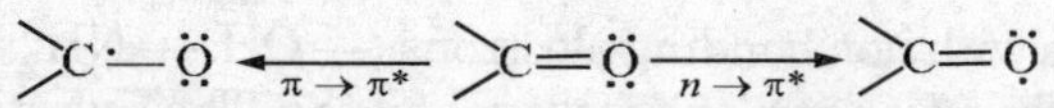

The $\pi \rightarrow \pi^*$ transition is very intense and corresponds to shorter wavelength. $n \rightarrow \pi^*$ transition is less intense and corresponds to longer wavelength. The λ_{max} values for isolated > C = 0 band are:

Transition	λ_{max}	ε_{max}
$\pi \rightarrow \pi^*$	170	10,000
$n \rightarrow \pi^*$	290	~ 20

Thus two bonds are observed for isolated > C = 0 group in UV spectra. When the carbonyl group is substituted by an auxochromic - as in ester, amide or acid, the π* orbital are raised and hence $n - \pi^*$ transitions occur at shorter wavelengths.

The presence of weak band in the 275 – 295 nm region is positive identification of a carbonyl group.

(*ii*) **α–β unsaturated Carbonyl Compounds:** When a carbonyl group is conjugated with an ethylenic linkage as in α – β unsaturated carbonyl compounds, then both the bands undergo bathochromic shift i.e. move towards longer wavelengths in the range of 300 – 350 nm.

The position and intensity is also imfluenced by solvent effects discussed in section 7.12.5. Increasing the polarity of the solvent induces red shift in $\pi - \pi^*$ transition and blue shift in $n - \pi^*$.

Moreover axially substituted isomers absorb at longer wavelengths than equitorially substituted isomers.

Example 7.23: Explain why a polar solvent usually shift the $\pi - \pi^*$ transition to longer wavelength and n - π* transition to shorter wavelength.

Solution: Refer to section 7.12.5

Example 7.24: Acetone absorbs at 279 nm in hexane whereas the values of λ_{max} in water is 264.5 nm. Explain.

Solution: The shift to shorter wavelength in water is due to the change in polarity of the solvent. Hydrogen bonding of acetone with water molecule lowers the energy of *n*-orbitals and hence *n*-π* transition requires more energy i.e. shorter wavelength.

(4) **UV Spectra of Benzene and its Derivatives:** Benzene shows three absorption bonds, all due to π–π* transitions. The three bonds occur at

λ_{max}	ε_{max}	***Remarks***
184 nm	60,000	Allowed transition
204 nm	7900	Forbidden transition
256 nm	200	Forbidden transition

The bond at 256 nm (B-bond) is easily identifiable and is called benzenoid band. It shows a great deal of fine structure in vapour state.

- When benzene is substituted by simple alkyl groups the absorptions are shifted to slightly longer wavelength. This bathochromic shift is due to hyperconjugation in which the σ-electrons of C—H of alkyl substituent takes part in resonance with the ring. For example, B band of benzene appears at 254 nm and is shifted to 261 nm in case of toluene.
- In case of disubstituted benzene the para isomer absorbs at the longest wavelength as compared to meta and ortho. Ortho absorbs at shortest wavelength. This is due to steric interaction between the ortho substituents, which reduce the effect of hyperconjugation.
- The substituents with non bonding electrons (—OH, —NH_2, —OR etc.) present in benzene ring, shift the absorptions to longer wavelengths. The fine structure of the B band is either diminished or eliminated, because of *n*-π conjugation.. Table 7.6 lists some of the absorption values of *K*-band of benzene and its derivations:

Table 7.6 Benzene and its Derivative

Compound	λ_{max} *in water*		*Intensity* ε_{max}	
	K Band	*B-Band*	*K Band*	*B-Band*
Benzene	204 nm	254 nm	7900	204
Chlorobenzene	290 nm	263.5 nm	7400	190
Aniline	230 nm	280 nm	8600	1430
Nitrobenzene	252 nm	280 nm	10,000	1000

- Absorption maximum for polynuclear aromatic hydrocarbons also move to longer wave-lengths.

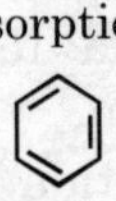

254 nm
Benzene

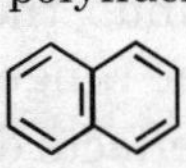

286 nm
Naphthalene

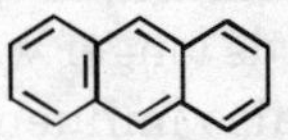

375 nm
Anthracene

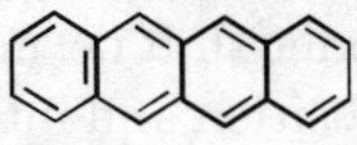

Tetracene

As might be expected from their absorption wavelengths that naphthalenes and anthracene are colourless but tetracene is coloured (orange).

APPLICATION OF ULTRA-VIOLET SPECTROSCOPY

(1) **Extent of Conjugation:** UV spectroscopy is widely used to determine the extent of conjugation in polymers. The value of λ_{max} increases with the extent of conjugation. (Refer to section - 7.13)

(2) **Distinction between Conjugated and Non-conjugated Compounds:** UV spectroscopy distinguishes between conjugated and non-conjugated compounds. For example in the following compounds,

$$(CH_3)_2C=CH-\underset{\underset{O}{\|}}{C}-CH_3 \qquad CH_2=\underset{\underset{CH_3}{|}}{C}-CH_2-\underset{\underset{O}{\|}}{C}-CH_3$$

(*a*) (*b*)

the compound (*a*) will have longer λ_{max} value because C = O group is in conjugation with double bond.

(3) **Characterization of Aromatic Compounds:** Anomatic compounds can be characterized by their characteristic absorption bands in UV spectroscopy. (Section 7.13).

(4) **Detection of Impurities:** Benzene is present as an impurity in ethanol. Because of toxic nature of benzene, it is important that ethanol for internal use should be freed from benzene. The presence of even minute amounts of benzene can be confirmed by UV spectrum of sample at 280 nm. At this wavelength ethanol is transparent but benzene exhibits an absorption band with $\varepsilon_{max} = 230$.

(5) **Quantitative Analysis:** This analysis is based on Beer-Lambert law.

$$A = \varepsilon cl$$

where A = absorbance at a particular wavelength

c = Concentration of the sample

ε = Constant known as molar absorptivity or extinction coefficient.

The absorbance of standard solution (A_1) and of that of the sample is measured (A_2) in the same sample cell (l = constant), concentration of standard solution, knowing the (c_1) the concentration of the sample (c_2) can be determined as

$$\frac{A_1}{A_2} = \frac{c_1}{c_2}$$

Example 7.25: For a solution of camphor in hexane in a 10 cm cell, the absorbance A was found to be 2.52 at 295 nm with $\varepsilon_{max} = 14$. What is the concentration of camphor.

Solution:

$$A = \varepsilon cl$$

$$A = 2.52, \quad \varepsilon = 14, \quad l = 10 \text{ cm}$$

$$\therefore \quad 2.52 = 14 \times c \times 10$$

$$c = \frac{2.52}{14 \times 10} = 1.8 \times 10^{-2} \text{ moles/litre}$$

(6) **Determination of strength of hydrogen bonding:** On the basis of shifts absorbed in polar solvents, it is possible to deduce the strength of hydrogen bonds. (Section 7.12.5)

(7) **Study of isomerism:** Refer to section 7.13 for geometric isomerism.

SOLVED PROBLEMS BASED ON UV SPECTROSCOPY

Example 7.26: Explain the variation in the value of λ_{max} for the following compounds;

CH_3Cl, λ_{max} = 173 nm

CH_3Br, λ_{max} = 204 nm

CH_3I, λ_{max} = 258 nm

Solution: These compounds undergo $n - \sigma^*$ transitions. As we move down the halogen group, the quantum numbers increase from Cl to I and hence the n electrons lie for away from the nucleus. Thus these are easily excited and λ_{max} occurs at higher values.

Example 7.27: Why λ_{max} for diene (I) is observed at lower nm than (II)

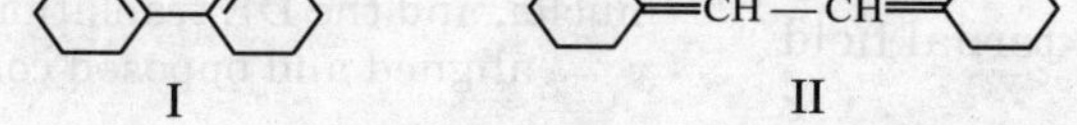

I II

Solution: Though there are four ring residues as substituents in I and II. But in diene (II), the two double bonds are exocyclic and hence according to Wood-Fieser rules, the λ_{max} will be higher by 2 × 10 nm.

Example 7.28: Why PABA (*p*-aminobenzoic acid) is used in sunscreen lotions.

Solution: PABA has λ_{max} at 289 nm with a high extinction coefficient of around 19,000. Due to this property it absorbs UV light and protect our skin from high energy UV radiation.

7.10 NMR SPECTROSCOPY

Nuclear magnetic resonance spectroscopy involves the transition of a nucleus from one spin state to another with the absorption of electromagnetic radiation by spin active nuclei, when they are placed in a magnetic field. Spin active nuclei are the nuclei having nuclear spin not equal to zero.

The electromagnetic radiations used in NMR spectroscopy have very long wavelengths ($\lambda = 10^7$ μ to 10^8 μ) and have extremely low energy. These radiations are *radio frequency waves* which are *not* energetic to effect the electronic, vibrational or rotational energy levels of the molecules. However they are able to interact with the nuclei of certain atoms which are exposed to a strong magnetic field, hence the name nuclear magnetic spectroscopy.

Before discussing the theory of NMR it is worth to review the magnetic properties of the nuclei of certain atoms.

7.10.1 Magnetic Properties of Nuclei

(*a*) **Nucleus—A Tiny Magnet:** The nucleus of certain isotopes (such as ^{1}H) behave as tiny spinning bar magnet because it possess both electric and magnetic spin. Hence like any other spinning charged body, the nucleus of hydrogen atom also generates a magnetic field. Thus the nucleus is equivalent to a minute magnet having magnetic moment μ.

(*b*) **Effect of External Magnetic Field:** Now if a magnetic nucleus such as ^{1}H is placed in a uniform external magnetic field, it is found that it can only adopt two orientations with respect to the direction of the applied external magnetic field —

- either aligned with the field (parallel)
- or opposed to the external field (antiparallel)

(*c*) **Precessional Motion :** The precessional motion of the nucleus can be understood by taking an example of a spinning top, the spinning axis of which moves slowly around the vertical axis of earth's gravitational field. This waltz like motion of the spinning top around the vertical axis ϕ is known as precessional motion.

Now as the proton is also a spinning magnet, it will also *precess* around the axis of the applied magnetic field. Either it can precess aligned with the field or opposed to the field. The aligned orientation is stable and is of low energy whereas the opposed orientation is unstable and is of higher energy.

The processional frequency υ is directly proportional to the strength of external field B_0 i.e.

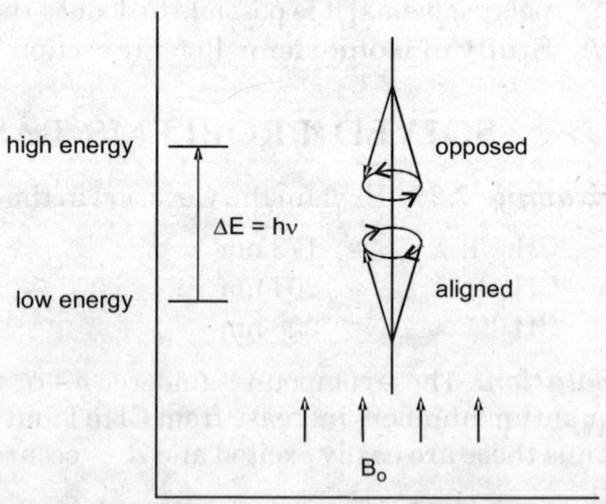

Fig. 7.22 Representation of precessing nuclei, and the DE transition between the aligned and opposed conditions.

$$\upsilon \propto B_0$$

This frequency of precession is extremely important in NMR spectroscopy.

(*d*) **Transitions between the Two States:** It is possible to induce transitions between these two transitions i.e. from lower energy state to higher energy state. The frequency of the radiation required to induce the transition is given by,

$$\upsilon = 2\,\mu B_0/h$$

where μ = magnetic moment
B_0 = strength of external magnetic field.
h = Planck's constant.

This frequency υ of electromagnetic radiation, required to induce transition from one nuclear spin state to another, is exactly equal to the processional frequency of the spinning nucleus. In other words, the processing proton will absorb energy from the radio frequency region only if the processing frequency is the same as the frequency of the radio frequency radiation. In such situation, the nucleus and the electromagnetic radiation are said to be in **resonance** and hence the name nuclear magnetic resonance.

(*e*) **Spin-active Nuclei:** Nuclei that exhibit the NMR phenomena are known as spin-active nuclei. As is evident from the above discussion that nuclear spin is the fundamental property of the nucleus to be involved in the NMR spectroscopy. Only those nuclei will exhibit the phenomenon of NMR for which the spin quantum number *I is greater than zero.* These nuclei are known as spin-active nuclei. For example, ^{1}H, ^{13}C and ^{19}F etc. ^{12}C, ^{16}O, ^{32}S etc. have $I = 0$ and are *not spin-active.* The spin active nuclei have permanent magnetic moments and quantized nuclear spin states.

The spin quantum number *I* is associated with mass number and atomic number of the nuclei

Mass Number	*Atomic Number*	*Spin Quantum Number I*
Odd	Odd or even	$\frac{1}{2}, \frac{3}{2}, \frac{5}{2}, \ldots$
Even	Even	0
Even	Odd	1, 2, 3

The number of possible orientations is given by $2I + 1$. In the nuclei with spin $\frac{1}{2}$ i.e. ^{1}H, ^{13}C etc. only two orientations are allowed.

7.10.2 Theory of Nuclear Magnetic Resonance

When a proton is placed in an external magnetic field, its magnetic moment, as discussed above, can be aligned in either of the two ways:

(*i*) Aligned with the field — parallel alignment.
(*ii*) Aligned against the field — antiparallel alignment.

It is possible to induce transitions between these two states i.e. a proton processing in the aligned orientation can pass into the opposed orientation by absorbing energy. Also, it can come back from the high energy, antiparallel alignment by losing energy. This transition from one energy state to the other is called *flipping* of the proton.

Now, how much energy is required to flip the proton from one state to another depends upon the strength of the external magnetic field applied. Stronger the field, greater will be the tendency of the nuclear magnet to remain lined up with it and higher will be the energy required to flip the proton to the higher energy state. If ΔE is the energy difference between the two states, it is related to the external field strength B_0 as,

$$\Delta E = 2\mu B_0 \qquad ...(ii)$$

or
$$\upsilon = \frac{2\mu B_0}{h} \qquad ...(iii)$$

where h is Planck's constant, μ is the magnetic moment.

For example, in the field of 14092 gauss, the energy required to cause flipping of the proton corresponds to the electromagnetic radiation of frequency 60 million cycles per second or 60 MHz. These radiations are in the radio frequency region. Hence, if we irradiate the processing nuclei with a beam of electromagnetic radiations of the required frequency, the flipping of proton from low energy state to higher energy state takes place. It should again be emphasized here that the processing proton will only absorb energy if the processing frequency of the spinning nucleus is same as the frequency of the radio frequency beam. When these occurs, the nucleus and the radio frequency beam are said to be in resonance.

Thus in NMR spectroscopy in principle, the substance can be placed in a magnetic field of constant strength and irradiated with electromagnetic radiations of varying frequencies and the frequency at which absorption occurs is observed. In practice however, it is generally more convenient to *keep the radio frequency constant and the strength of the magnetic field is constantly varied.* Now, at some value of the field strength, the energy required to flip the proton matches the energy of the radiation and absorption takes place. A signal is observed in the spectrophotometer. The spectrum so obtained is called nuclear magnetic resonance or NMR spectrum. When the nucleus involved is proton the spectrum is also called *PMR* or *proton magnetic resonance.*

7.10.3 The Chemical Shift

If the resonance frequencies for all the protons in a molecule were the same, then all the protons in the molecule would absorb at exactly the same field strength and the spectrum would consist of a single signal. Hence, NMR technique would be of little use to the organic chemist as it would tell us little about the structure of the molecule.

Fortunately, the precessional frequency of all the protons in the same external field is *not the same* and the precise value for any one proton depends upon a number of factors. The frequency at which a proton absorbs depends on the magnetic field which that proton *feels or experiences* and this effective field strength is different from the *applied field* strength. The magnetic field strength which the proton feels depends upon the environment of the proton, electron density at the proton, presence of other nearby protons and a number of other factors. For example, the surrounding electrons may shield the nucleus, so the effective magnetic field felt by the nucleus is not quite the same as the applied field.

Hence, each proton or more precisely, each set of equivalent protons will have a slightly different environment from every other set of protons. Thus each set will require a particular field strength at which absorption takes place. This particular field strength is measured and the result is that the spectrum show many absorption peaks.

Thus, the shift in the position of NMR spectrum as compared to a standard reference, due to chemical environment of the proton is called *Chemical Shift.*

Measurement of Chemical Shift: The measurement of precessional frequency of a group of nuclei in absolute frequency unit is extremely difficult and is not required also. However, relative absorption values can be obtained using a suitable standard. The chemical shift values of the protons in a particular compound are then determined with reference to this standard.

The universally accepted standard is NMR signal of tetramethyl silane (TMS)

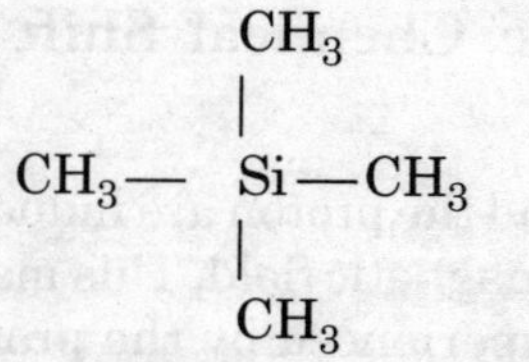

Tetramethyl silane (TMS)

TMS has 12 protons in magnetically equivalent positions and hence a sharp single signal is obtained. Moreover it is chemically inert and soluble in organic solvents used. It can be added to the sample solution as an internal standard.

As TMS is not soluble in water or D_2O, for solutions of the compounds in these solvents the sodium salt of 3-(trimethyl silyl)- propoanesulphonic acid is used.

The NMR signal value for a particular proton in a molecule will appear at different field strengths compared with a signal from TMS. This difference in the absorption position of the proton with respect to TMS signal is called *chemical-shift value.*

Units used in NMR: Chemical shift positions are frequently expressed in δ (delta) units which are defined as *the difference in parts per million from the TMS signal.* The δ unit is proportionality and is thus dimensionless.

The value of δ for a substance with respect to TMS can be obtained by measuring resonance frequency of the sample υ_s and resonance frequency of TMS

$$\delta = \frac{\upsilon_{\text{sample}} - \upsilon_{\text{TMS}}}{\text{operating frequency in magacycles}}$$

In majority of organic compounds, protons resonate at lower field than the protons of tetramethyl silane. Thus δ value of TMS is arbitrary fixed at zero and the protons absorbing at lower field than TMS are given positive value for δ. Most chemical shifts have δ values from 0 to 10.

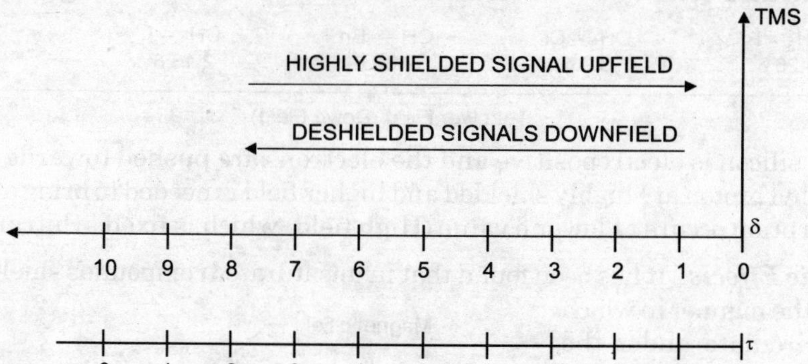

Fig. 7.23

NMR signal is usually plotted with magnetic field strength increasing to the right. Thus, the signal for TMS appears at extreme right of the spectrum with δ = 0 ppm. Greater the deshielding of the protons, larger will be the value of δ (Fig. 7.23). For most of the organic compoundes, signals appear downfield to the left of TMS signal.

Another scale used for the measurement of chemical shift is

τ scale: In τ (tau) scale, TMS is given an arbitrary value of 10 ppm. Thus

$$\tau = 10 - \delta$$

7.10.4 Factors Influencing the Chemical Shift Shielding and Deshielding Effects

In a magnetic field, the electrons around the proton are induced to circulate and thus they produce secondary magnetic field i.e. induced magnetic field. This magnetic field is opposed (diamagnetic) to the applied field. Thus the field experienced by the proton is diminished than that applied externally and the proton is said to be *shielded* from the influence of the applied field. Thus to bring a shielded proton to resonance, greater external field is required. Greater the electron density circulating around the proton, greater would be the shielding and greater the external field required to overcome the shielding effect.

On the other hand, rotation of electrons (π electrons) generates a field that can either oppose or reinforce the applied field at the proton. If the induced field reinforces the applied field the proton experiences a higher field strength as compared to the applied field and is said to be *deshielded.* Thus smaller external field is required to bring the deshielded proton to resonance.

Hence, shielding shifts the absorption peaks towards lower value of δ (higher values of external field B_0) and deshielding shifts the absorption peaks towards higher values of δ (lower value of external field, B_0)

Shielded proton → Higher Field (up field) → lower value of δ

Deshielded proton → lower field (down field) → Higher value of δ

Shielding and deshielding and hence chemical shift depends upon a number of factors discussed below:

(1) *Electronegativity:* There is a reasonable correlation between shielding and electronegativity of the substituent groups in simple saturated molecules. An electronegative group withdraws electron density from the neighbourhood of the proton, so that smaller field is needed to cause resonance. Hence absorption occurs downfield (higher δ values) farthest from TMS signal.

The chemical shift for the protons of halomethane is directly related to electronegativity of halo atom as is illustrated below:

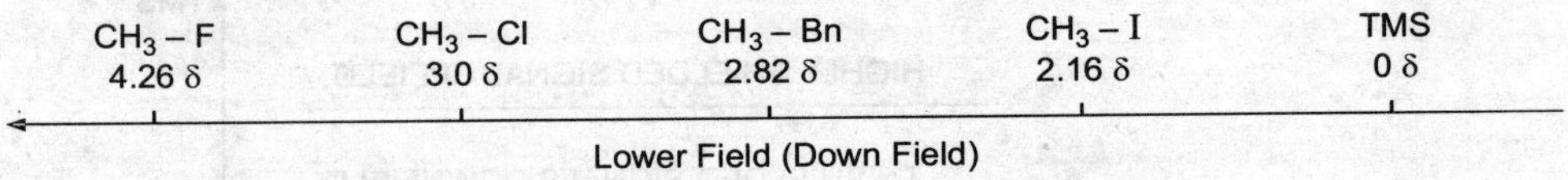

Similarly, silicon is electropositive and the electrons are pushed towards the methyl group. Hence the methyl proton are highly shielded and higher field is needed to bring about the resonance and thus absorption occurs at lower δ value (High field) which is fixed arbitrarily at 0.

(2) *Anisotropic Effects:* It has been found that in unsaturated compounds shielding or deshielding depends upon the manner in which π electrons circulate under the influence of applied field. The effects are diamagnetic (opposed to the field) in certain directions and paramagnetic (aligned with the field) in others, so these effects are called Anisotropic effects.

Alkene: Alkene is so oriented in space that the plane of the double bond is at right angles to the applied field. Circulation of π electrons generates a magnetic field

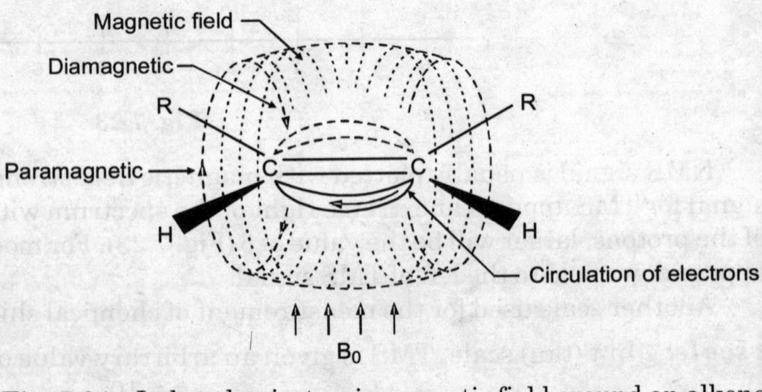

Fig. 7.24 Induced anisotropic magnetic field around an alkene group.

which is diamagnetic around the carbon atoms but paramagnetic in the region of alkene protons. Thus protons experience deshielding and require lower value of external magnetic field to exhibit resonance and hence absorption occurs at lower field (higher δ value) than expected.

Alkynes: In alkynes, electronic circulation around triple bond takes place in such a manner that the protons experience diamagnetic shielding effect. Thus protons feel smaller field strength and resonance require higher applied field (low δ values).

Aromatic Compounds: The movement of π electrons under the influence of applied magnetic field, around a benzene ring is called a ring current. Aromatic rings display strong magnetic anisotropy. Above and below the ring, there is strong shielding region (i.e. induced magnetic field is in opposition to the applied field) whereas in the plane of the ring there is strong deshielding region (induced magnetic field is in the same direction as the applied field). Consequently, the protons and groups attached to the ring are in the plane of the ring and are thus deshielded and give peaks at down field (higher δ values). The protons above and below the ring are shielded and resonate at low δ values.

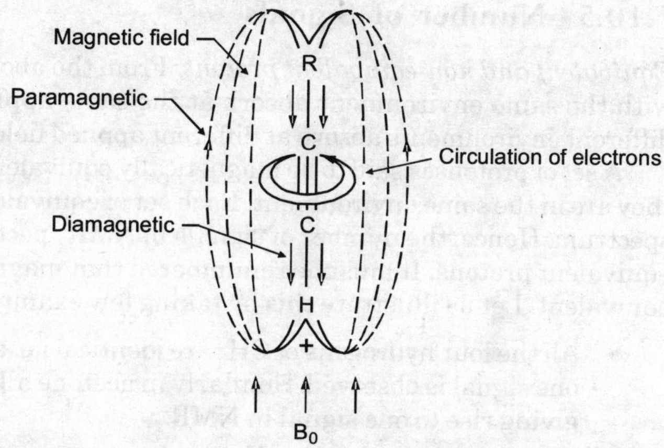

Fig. 7.25 Anisotropic shielding of a proton in an alkyne group.

Carbonyl Compounds: Similar situation arises for carbonyl compounds. The induced magnetic field by the π electrons in the region where protons are located, is oriented in the same direction as the applied field. Hence smaller field (higher δ value) is required to bring absorption due to deshielding.

Table 7.7 gives the approximate chemical shift ranges for protons attached to anisotropic groups.

Table 7.7: Chemical shifts due to anisotropic effects.

Compound	*Effect experienced by protons*	*Absorption*	*δ Value*
Alkene	Deshielding	Lower field	High (5 – 6)
Alkyne	Shielding	Higher field	Low (1.5 – 3.5)
Aromatic	Deshielding (for the protons attached to the ring)	Lower field	High (7 – 8)
Carbonyl	Deshielding	Lower field	High (9.5 – 10.0)

(3) *Hydrogen Bonding:* The hydrogen bonded proton being attached to a highly electronegative atom will have smaller electrons density around it and hence is deshielded. Thus absorption occurs at low field (higher δ). Also the downfield shift depends upon the strength of hydrogen bonding. The greater the degree of hydrogen bonding, greater would be the downfield shift.

Fig. 7.26 gives the approximate chemical shift positions for protons in organic molecules.

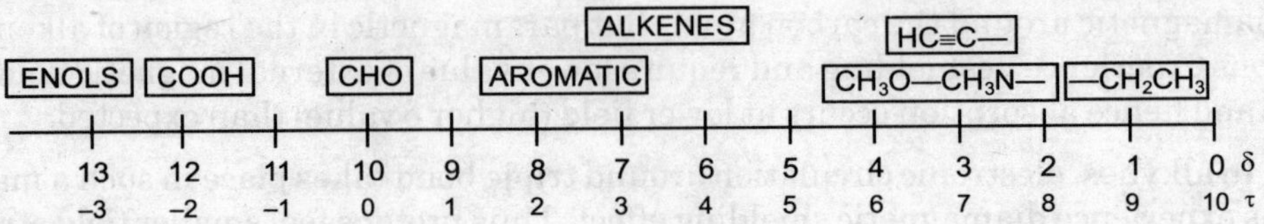

Fig. 7.26 Approximate chemical shift positions for protons in organic molecules.

7.10.5 Number of Signals

Equivalent and non-equivalent protons: From the above discussion it becomes clear that protons with the same environment absorb at the same applied field strength, while the protons with different environments absorb at different applied field strength.

A set of protons is said to be magnetically equivalent if they exhibit the same chemical shift i.e. they are in the same environment. Each set of equivalent protons will give rise to a signal in NMR spectrum. Hence, the number of signals in NMR spectrum tells us the number of different sets of equivalent protons. It must be renumbered that magnetically equivalent protons are chemically equivalent. Let us illustrate this by taking few examples.

- All the four hydrogens of CH_4 are identical i.e. are in the same environment and hence only one signal is observed. Similarly in acetone all the six protons are in similar environment giving rise to one signal in NMR.

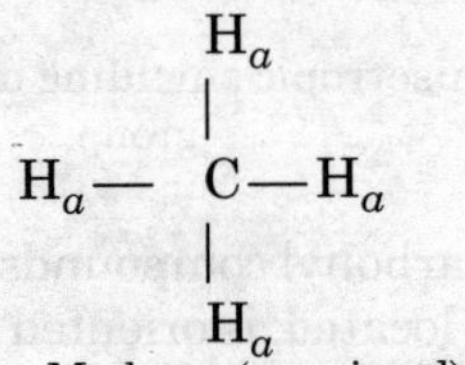

Methane (one signal)

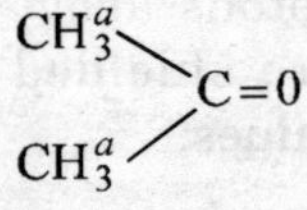

Acetone (One signal)

- In ethyl chloride the protons are in two different environments and hence two signals are observed. Same is true for other molecules such as methanol, propane, isobutylene etc.

$\underset{a}{CH_3}—\underset{b}{CH_2}—Cl$

Ethyl Chloride
(2 NMR signals)

$\underset{a}{CH_3}—\underset{b}{OH}$

Ethyl Chloride
(2 NMR signals)

$\overset{a}{CH_3}—\overset{b}{CH_2}—\overset{a}{CH_3}$

Propane
(2NMR signals)

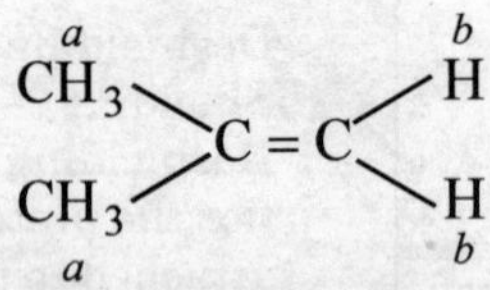

Isobutylene
(2 NMR signals)

- Some of the compounds having 3 NMR signals are as follows:

$\overset{c}{CH_3}—\overset{b}{CH_2}—\overset{a}{CH_2}—Cl$

3 NMR signals

$\overset{c}{CH_3}—\overset{b}{CH_2}—\overset{a}{NH_2}$

3 NMR signals

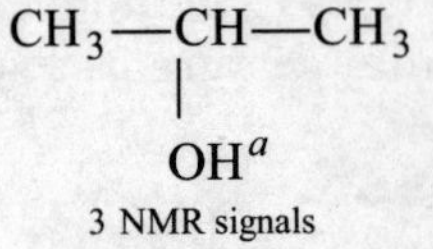

3 NMR signals

- It must be emphasized here that chemically equivalent protons must also be stereochemically equivalent also. Let us take the example of vinyl chloride.

$ClHC = CH_2$
Vinyl Chloride

or

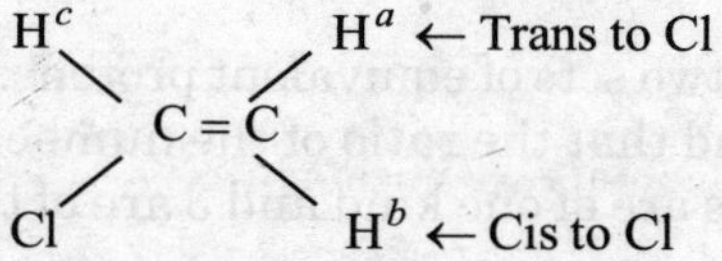

At first sight it appears that it has two sets of equivalent protons and would give rise to 2 NMR signals. But that is not true as one can see that out of two hydrogens attached to one of the carbon atom, one is cis to Cl atom while other is trans to Cl atom and hence are in different environment.

Moreover, these protons are in different environments than the protons on the C—Cl carbon. Thus all the three protons are in different environment, giving rise to 3 NMR signals. Based on similar arguments the number of signals in 2-bromopropene is 3

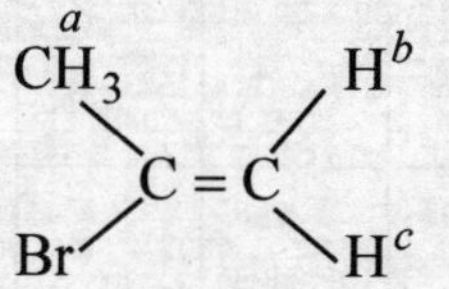

2 Bromopropene
3 NMR signals

Example 7.29: Predict the number of NMR signals in the following compounds: (*a*) Tert-butyl amine (*b*) Benzene (*c*) Propanol (*d*) benzene.

Solution: (*a*) $(CH_3^a)_3C—NH_2^b$ 2 NMR signals

Tert–butylamine

(*b*)  1 NMR signal

(*c*) $\overset{d}{CH_3}—\overset{c}{CH_2}—\overset{b}{CH_2}—\overset{a}{OH}$ 4 NMR signals

(*d*) 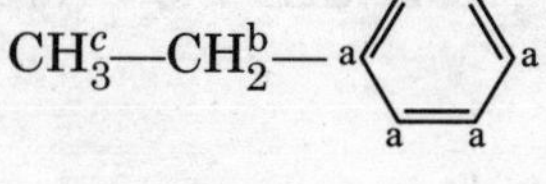

3 NMR signals

7.10.6 Peak Area and Proton Counting

The number of signals in NMR tells us how many different 'kinds' of protons are there in the molecule. The peak area tells us about *the number of protons in each equivalent set* present in the compound. The area under an NMR signal is directly proportional to the number of protons giving rise to the signal. Greater is the number of protons in a particular set, greater will be the energy absorbed.

In NMR, squares under each peak are counted and from this, the ratio between various kinds of protons is found out and converted into whole numbers. These whole numbers or some multiple of them tell the number of protons represented by each signal. Let us consider the NMR spectrum of toluene.

Two signals indicate two sets of equivalent protons. If the number of squares under each peak is counted, it will be found that the ratio of the number of squares under two peaks will be 5 : 3. Thus in toluene 5 protons are of one kind and 3 are of the another kind.

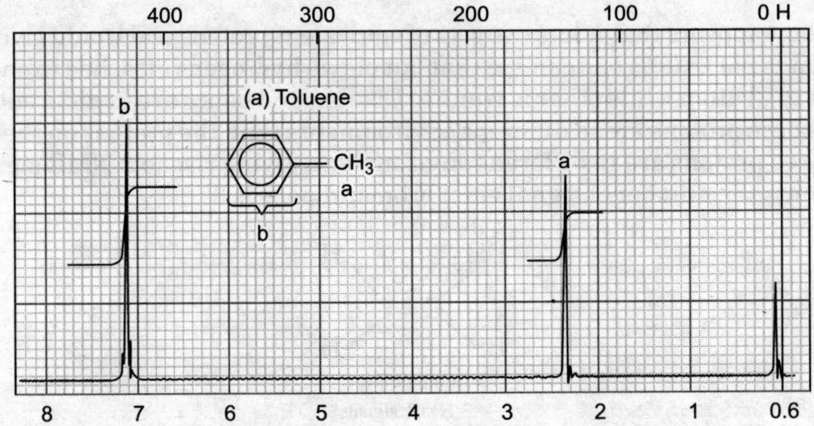

Fig. 7.27 NMR spectra of toluene.

SOLVED PROBLEMS BASED ON NMR SPECTROSCOPY

Example 7.30: Determine the equivalent and non-equivalent hydrogen in the following compounds and indicate the number of signals expected.

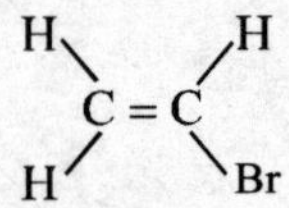

(*a*)

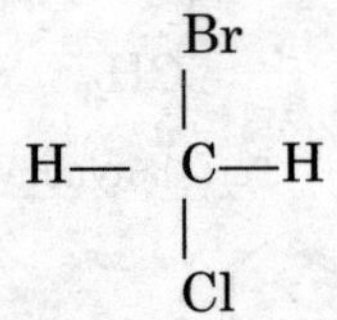

(*b*)

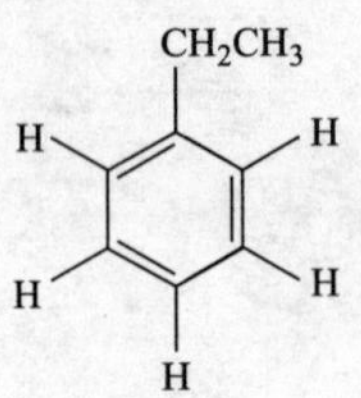

(*c*)

$CH_3CH_2—O\,CH_2CH_3$

(*d*)

Solution:

(*a*) 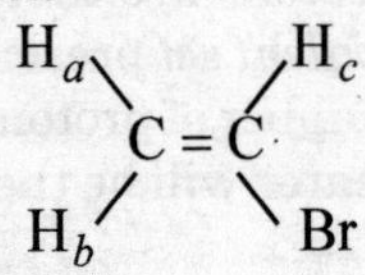

Three types of protons and hence number of signals = 3.

(b) 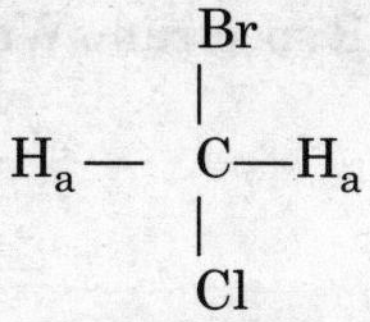

One set of equivalent protons hence number of signals = 1.

(c) 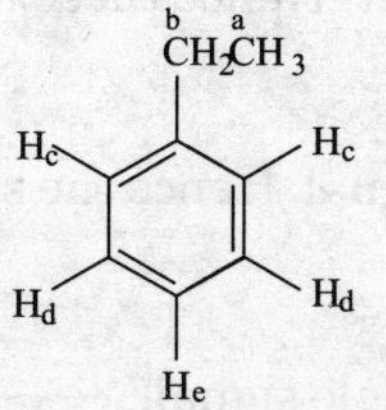

Five sets of equivalent protons and hence the number of signals = 5

(d) $\underset{b}{CH_3}\,\underset{a}{CH_2}—O—\underset{a}{CH_2}\,\underset{b}{CH_3}$

Two set of equivalent protons and hence number of signals = 2.

Example 7.31: **Which of the following compounds would be expected to show only a single peak in NMR or PMR spectrum**

(*i*) Acetone (*ii*) Methyl acetate
(*iii*) 1, 2-dibromo methane (*iv*) Methyl iodide
(*v*) 1-chloro-2-bromoethane

Solution:

(*i*) Acetone 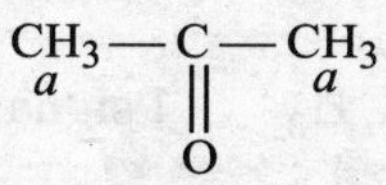

(*ii*) 1, 2-Dibromomethane $\overset{a}{CH_2}Br_2$

(*iii*) Methyl Iodide $\underset{a}{CH_3}I$

Example 7.32: **Methyl, methylene, methine and quaternary hydrogens resonate in successively lower fields in NMR spectroscopy, explain.**

Solution: The replacement of hydrogen by an alkyl group causes a downfield shift because hydrogen is more electropositive than carbon. Thus, methylene methine and quarternary hydrogens resonate at successively lower fields as shown below

	CH_4	CH_3CH_3	$(CH_3)_2\,CH_2$	$(CH_3)_3CH$
δ	0.23	0.86	1.33	1.68

Example 7.33: **A compound has molecular formula $C_{10}H_{14}$. It gives the following NMR data**

(*i*) 0.88 δ (9H, singlet),
(*ii*) 7.28 δ (5H, singlet, aromatic protons)

Assign the structural formula to the compound.

Solution:

(*i*) The nine proton singlet at 0.88 δ indicates the presence of tertiary butyl group.
(*ii*) Five proton singlet at 7.88 δ confirms the phenyl ring.

Hence the structural formula of the compound may be

Example 7.34: The following compounds show only one signal in the NMR spectrum. Write their structural formulae.

(*a*) C_5H_{12} (*b*) C_2H_6O

(*c*) $C_2H_4Br_2$ (*d*) C_4H_6

Solution:

(*a*) **C_5H_{12}:** Single signal means all the 12 hydrogens are equivalent. Hence these can be in four equivalent methyls. The structure is:

$$(CH_3)_4C$$

(*b*) **C_2H_6O:** Two equivalent methyl groups account for the single signal. Hence the structure is:

$$CH_3—O—CH_3$$

(*c*) **$C_2H_4Br_2$:** Two equivalent methylene groups account for the single signal.

$$Br—CH_2—CH_2—Br$$

(*d*) **C_4H_6:** Two equivalent methyl groups, with two carbons having no hydrogens account for the single signal

$$H_3C—C \equiv C—CH_3$$

Example 7.35: How many NMR signals are expected in each of the following compounds

(*a*) $CH_3OCH_2CH_2CH_3$ (*b*) CH_3OCH_3

(*c*) $CH_3CH_2CH_2CH_2OH$

(*d*) CH_3CHO (*e*) $CH_3—\underset{\underset{CH_3}{|}}{CH}—CH_2—Cl$

Solution:

(*a*) $\overset{a}{CH_3}\overset{b}{OCH_2}\overset{c}{CH_2}\overset{d}{CH_3}$ 4 signals (*b*) $\overset{a}{CH_3}O\overset{a}{CH_3}$ 1 signal

(*c*) $\overset{a}{CH_3}\overset{b}{CH_2}\overset{c}{CH_2}\overset{d}{CH_2}OH$ 5 signals (*d*) $\overset{a}{CH_3}\overset{b}{CHO}$ 2 signals

(*e*) $\overset{a}{CH_3}—\underset{\underset{\underset{a}{CH_3}}{|}}{\overset{b}{CH}}—\overset{c}{CH_2}—Cl$ 3 signals

Example 7.36: How will you distinguish between the following two compounds on the basis of NMR spectroscopy.

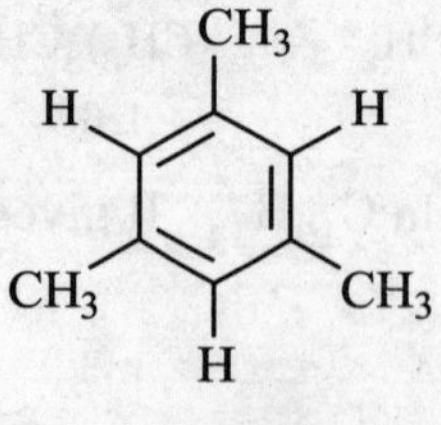

Mesitylene

2 sets of equiv. protons 2 signals

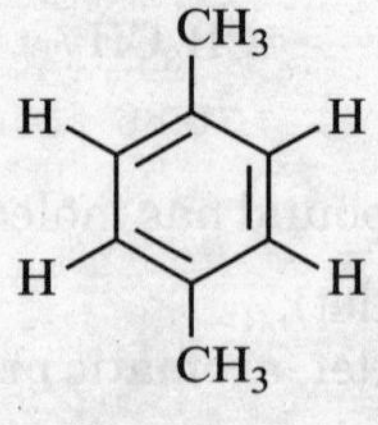

Xylene

2 sets of equiv protons 2 signals

Solution: Both the compounds would give two signals but the intensities of the peaks would be different in two compounds. In mesitylene the ratio of methyl protons to ring protons is 9 : 3 whereas in *p*-xylene it is 6 : 4.

QUESTIONS

Infra Red Spectroscopy

1. What is the difference between absorption and emission spectra.
2. Distinguish between atomic and molecular spectroscopy.
3. What do you understand by molecular spectroscopy. Name the different types of molecular spectra.
4. What are vibrational and rotational quantum numbers?
5. Explain Beer-Lambert's Law.
6. Why absorption spectroscopy is preferred? Give its experimental set-up.
7. (a) What are the factors that affect the width of spectral line?
   (b) What are the factors which affect the intensity of spectral line?
8. What are electromagnetic radiations? Give the regions of electromagnetic radiations in the increasing order of frequency.
9. Show vibrational, rotational and electronic energy levels on an energy level diagram. To which region of electromagnetic spectrum do rotational, vibrational and electronic spectra are concerned?
10. What type of molecules give pure rotational spectra?
11. Explain the principle and working of spectrophotometer, given its neat diagram.
12. Briefly explain, how a vibrational-rotational spectrum is obtained? What do you understand by fundamental bands and overtone bands?
13. What do you understand by force constant? How is the force constant of a molecule related to its vibrational frequency?
14. What do you understand by modes of vibration? Explain with suitable examples.
15. What type of molecules exhibit vibrational-rotational spectra and why?
16. Distinguish between the following, from their IR-spectra:
    (i) Ethanol and dimethylether (ii) Primary, secondary and tertiary amides.
17. What modes of vibrations are active in I.R. absorption spectra and why? Taking nitro group as an example show the various types of vibrations which will be observed in IR spectra.
18. Three test tubes contain three ketones one of them absorbs near 1785, one near 1750 and one near 1710 cm^{-1}. What types of ketones are they?
19. Explain, from the spectrum of acidified ethanol, it is seen that methyl absorption is split into a triplet (relative areas 1 : 2 : 1) and the methylene absorption is split into a quartet (relative areas 1 : 3 : 3 : 1).
20. (a)How many vibrational modes are there in a linear and a non linear molecules having n atoms? Explain with the help of degrees of freedom. Draw the diagram of vibrational modes for CO_2 molecules.
21. (a) What are the different kinds of electronic transitions.
    (b) Distinguish between the following pairs from their IR spectra.
    (i) 2-propanol and propanone (ii) Ethanol and dimethylether
    (iii) Ethylamine and acetamide

UV – Visible Spectroscopy

1. What is an absorption spectra? Write what do you know about $n \rightarrow \pi^*$ and $n \rightarrow \sigma^*$ transitions? Explain by giving examples.
2. Write short note on the electronic transitions caused by energy absorbed in the U.V. region.
3. Why molecules absorb in UV-visible region? What are the type of electronic transitions that can occur in a molecule? Discuss giving examples.
4. (a) Give one example each of $\sigma \rightarrow \sigma^*$, $n \rightarrow \sigma^*$, $n \rightarrow \sigma^*$, $n \rightarrow \pi$ and $\pi \rightarrow \pi^*$ transitions.
   (b) What is the effect of the presence of alpha-beta unsaturation on the carbonyl absorption in the U.V. spectrum.

5. List all the electronic transitions possible for (a) CH_4, (b) CH_3Cl, (c) $H_2C = 0$.

(**Ans.** (a) $\sigma \rightarrow \sigma^*$, (b) $\sigma \rightarrow \sigma^*$ and $n \rightarrow \sigma^*$ (there are no π or π^* MO's), (c) $\sigma \rightarrow \sigma^*$, $\sigma \rightarrow \pi^*$, $\pi \rightarrow \sigma^*$, $n \rightarrow \sigma^*$, $\pi \rightarrow \pi^*$ and $n \rightarrow \pi^*$)

6. The UV spectrum of acetone shows two peaks at λ_{max} = 280 nm. ε_{max} = 15, and λ_{max} = 190 nm, ε_{max} = 100. (a) Identify the electronic transition for each (b) Which is more intense ?

(**Ans.** (a) The longer wavelength (280 nm) is associated with smaller energy ($n \rightarrow \pi^*$) transition, $\pi \rightarrow \pi^*$ transition occurs at 190 nm. (b) $\pi \rightarrow \pi^*$ has the larger ε_{max} and is more intense.)

7. Identify the two geometric isomers of stilbene, C_6H_5 CH = CH C_6H_5, from their λ_{max} values, 294 nm and 278 nm.

(**Ans.** The higher energy *cis*-isomer has the shorter λ_{max}. Steric strain prevents full coplanarity of the *cis* phenyl groups and the conjugative effect is attenuated.)

8. Predict the kind of electronic transitions in (a) Cl_2 and (b) C = 0 group. Also give their intensity.

(**Ans.** (a) $\sigma \rightarrow \sigma^*$ (allowed transition, strong)
$n \rightarrow \sigma^*$ (forbidden transition, weak)
(b) $\pi \rightarrow \pi^*$ and $\sigma \rightarrow \sigma^*$ (both allowed, strong)
$n \rightarrow \pi^*$, and $n \rightarrow \sigma^*$ (both forbidden, weak)

9. Explain the effect of polar solvents on (1) $n \rightarrow \pi^*$ and $\pi \rightarrow \pi^*$ transitions.
10. (a) Define the following terms
    (i) Bathochromic shift (ii) hypochromic shift
    (iii) A chromophore (iv) Hyperchromic effect
11. Explain the various applications of ultra-violet spectroscopy.
12. "Increase in polarity of the solvent shifts $\pi \rightarrow \pi^*$ and $n \rightarrow \sigma^*$ band to shorter wavelength"
13. (a) Write short notes on K-band and R-band.
    (b) Which solvent in generally used in ultra-violet spectroscopy and why?
14. Draw a block diagram of a typical instrument for making UV/Visible absorption measurement. Briefly describe the various components of the instrument.
15. Explain the following in respect of UV-Visible spectroscopy:
    (b) Red shift or bathochromic effect. (b) Blue shift or hypsochromic effect.
    (c) Hyperchromic effect (d) Isobestic point

(NMR)

16. What do you mean by spin Active Nuclei?
17. Which of the following nuclei do not show nuclear magnetic resonance:

$$^{1}H,\ ^{2}H,\ ^{12}C,\ ^{13}C,\ ^{14}N,\ ^{15}N,\ ^{16}O,\ ^{19}F,\ ^{31}P$$

[**Ans.** Nuclei that contain odd number of electrons show NMR spectra. Thus, ^{13}C, ^{16}O, and ^{32}S having even number of electrons will not show NMR spectra.]

18. Describe briefly the theory of *nmr* spectrometry.
19. What do you understand by the positions of the signals in a *nmr* spectrum? How many signals are expected in each of the following compounds?
    (a) Propane (b) Isobutane (c) Ethanol (d) Cyclobutane (e) Ethylmethyl other
    (f) Ethyl acetate (g) Butanal (h) Glycol (i) α-alanine (f) Caprolactum.
20. (a) What is meant by the term chemical shift?
21. Write with examples, the shielding and the deshielding effects involved in *nmr* spectroscopy.
22. Describe with examples the various factors which affect the magnitude of the chemical shift.
23. Briefly describe how *nmr* spectrum of compound is canned?
24. Write brief notes on the following :
    (i) Chemical shift (ii) Spin-spin coupling
    (iii) Coupling constant.

25. How many kinds of H^1s are there in :

(i) $CH_3—CH_2—CH_3$? (ii) $CH_2=CH_2$?
(iii) $CH_3—CH=CH_2$? (iv) $C_6H_5—CH_3$?

26. Write the possible isomers corresponding to the following molecular formulae and draw their *nmr* structures.

(i) C_4H_{10} (ii) C_2H_6O
(iii) C_4H_8O (iv) $C_4H_6O_2$
(v) $C_4H_8O_2$.

27. (a) Write a short note on deshielding due to hydrogen bonding in NMR spectrum taking at least four examples.
(b) Name some important solvents, used in *nmr* spectroscopy. What are the important characteristics of the solvents used in this technique?

28. Why is TMS used as a standard reference ?

29. How many kinds of protons are there in

(i) CH_3CH_3 (ii) $CH_3CH_2CH_3$ (iii) $(CH_3)_2CHCH_2CH_3$
(iv) $C_6H_5CH_3$ (v) $C_6H_5CO_2$ (vi) $CH_2=CH_2$
(vii) $CH_3CH=CH_3$?

EXAMINATION QUESTIONS

1. What are the basic principles involved in the Nuclear-Magnetic Resonance (NMR).
[*U.P. Tech., 2nd Sem., 2003-04*]

2. (i) How many NMR signals do you expect from each of the following compounds. Indicate also the splitting pattern of the various signals:
(a) CH_3OCH_3 (b) $CH_3OCH_2CH_3$
(ii) Define Infra-red Spectroscopy. Describe the various molecular vibrations in the technique.
[*U.P. Tech., 2nd Sem., 2003-04*]

3. How will you distinguish vinyl acetate and methyl acrylate with the help of NMR spectroscopy?
[*U.P. Tech., 1st Sem. (C.O.), 2003*]

4. What are the basic principles involved in infrared and 1H-NMR spectroscopy?

5. A compound has molecular formula C_2H_3Cl. It can show geometral isomerism. The compound has two NMR-Signal. The splitting, under high resolution of NMR shows one doublet and one triplet. Identify the compound with the help of proper explanation. [*U.P. Tech., 1st Sem., 2006-07*]

6. Give the structure consistent with the following NMR data:
$C_3H_5Cl_3$ (a) Singlet, d 2.20, 3H
(b) Singlet, d 4.20, 2H [*U.P. Tech., 2nd Sem., 2006-07*]

7. (a) Write down the absorption positions of common functional groups found in organic compounds.
(b) What is 'Shielding' and 'Deshielding'? [*U.P. Tech., 1st Sem., 2007-08*]

8. Calculate the NMR signals and splitted signals in the following molecules :

(a) $CH_3 - O - CH_2 - CH_3$ (b) 

(c) 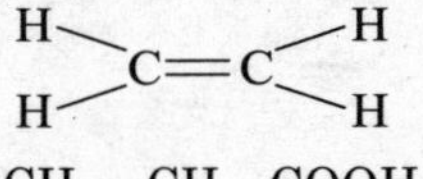 (d) $CH_3—CH_2—CH_2—Cl$

(e) $CH_3—CH—COOH$
|
Br

[*U.P. Tech., B.Tech., 1st Sem., 2007-08*]

9. (a) The absorbance is directly proportional to :
   (i) Wavelength (ii) Path length
   (iii) Concentration (iv) Concentration
   (b) IR active molecules are those which undergo a net change in ....
   (c) Number of NMR signals obtained in CH_3COCH_3 will be
   (i) 2 (ii) 6
   (iii) 1 (iv) 3 *[U.P. Tech, B. Tech., Ist Sen., 2008-09]*
10. (i) An organic compound having molecular formula C_7H_6O shows absorption peaks at 3010, 2700, 1600, 1580, 1520, 1480 and 1720 cm^{-1} in its IR spectrum Suggest its structure.
   (ii) Define chemical shift Show the expected NMR signals and their splitting in the following compounds
   $$CH_3\ CH_2\ CH_2\ OH \text{ and } C_6H_5\ CH_3.$$
   [U.P. Tech, B. Tech., Ist Sen., 2008-09]
11. Differential with one suitable example:
   (a) Shielding and deshielding of proton.
   (b) Chromophores and auxochromes
   (c) Bathochromic and hypsochromic shift. *[U.P. Tech, B. Tech., Ist Sen., 2008-09]*
12. "IR spectra is often characterized as molecular finger-prints". Comment on it.
   [U.P. Tech., B. Tech., 1st Sem., 2000-01]
13. Which of the following molecules will show infrared spectrum and why?
   H_2, HCl, CH_4, CO_2, H_2O *[U.P. Tech., B. Tech., 1st/2nd Sem., 2001-02]*
14. Point out the True and False statements out of the following:
   (i) Ultraviolet technique helps in predicting the number of conjugated double bonds in a molecule.
   (ii) The substance which absorbs in the visible region appears colored to the human eye.
   (iii) Spectroscopy involves the interaction between matter and electromagnetic radiation.
   (iv) In the nuclear magnetic resonance spectrum, radiation of radio frequency range is used.
   (v) In infra red, electron donating groups lower the wave number of absorption while the electron attraction groups raise the wave number. *[RGPV, B.E., Ist Sem., 2001]*
15. How can you distinguish $CH_3CH_2CH_2CHO$ from $(CH_3)_2CHCHO$ by proton NMR spectroscopy?
   [U.P. Tech., B. Tech., 2nd Sem., 2000-01]
16. Write a brief not on IR spectroscopy. *[U.P. Tech., 1st Sem. (C.O.), 2003]*
17. A gaseous hydrocarbon 'A' on passing through a quartz tube heated at 600°C, gave a liquid compound 'B' (Molecular weight : 78 amu). The latter compound was found to undergo electrophilic substitution reactions. It gave the following physical data on analysis. The IR spectrum showed a characteristic absorption band at 3040 cm^{-1} and a UV absorption, due to π–π^* transition, at 204 nm ($\log_e$, 3.84). The ^1H-NMR spectrum displayed a downfield, singlet (6H) at 7.3 τ. Identify the compound 'A' and 'B' and give your reasoning. *[U.P. Tech, 1st Sem., 2004-05]*
18. Define the terms chromophone and auxochrome, in UV spectroscopy. *[U.P. Tech., 2st Sem., 2006-07]*
19. What is Beer-Lambert law in UV-VIS absorption spectrospcopy ? A compound having concentration 10^{-3} g/l resulted absorbance value **0.20** at λ_{max} **510** nm using **1.0** cm cell. Calculate its absorptivity and molar absorptivity values. Molecular weight of compound is 400.
   [U.P. Tech, B.Tech. Ist Sem 2008-09]

❑❑❑

8 SURFACE TENSION, VISCOSITY AND CONDUCTIVITY

Knowledge is a treasure, but practice is the key to it.

— ***Thomas Fullen***

CHAPTER OUTLINE

8.1 SURFACE TENSION

The existence of strong intermolecular forces of attraction in liquids gives rise to a very important property known as surface tension. In the interior of a liquid a molecule is completely surrounded by the other molecules and hence experiences the force of attraction equally in all directions. (Fig. 8.1)

Fig. 8.1

On the other hand, a molecule at the surface is surrounded by liquid molecules on one side and that of vapour on the other. Thus the molecule at the surface experiences the attractive forces or inward pull perpendicular to the surface. (This is because of the difference of the number molecules per unit volume in the liquid and the vapour phase). Thus, under the influence of molecular forces of attraction from within, a liquid develops a tendency to contract its surface such that it acquires the smallest possible area for a given volume. This is due to the above reason that drops of a liquid acquire a spherical shape. As a result of the tendency to contract, a liquid surface behaves as if it were in a state of tension. This effect is known as *surface tension*

Definition. Surface tension is defined in SI units as the force in newtons (N) acting at right angle to any line of one meter length along the surface of the liquid. The surface tension is represented by a greek symbol g.

Units. In SI system Newton m^{-1}
In C.G.S system Dynes cm^{-1}

Surface Energy: The surface energy is a characteristic property of the surface and arises due to surface tension. The effect of surface tension is to reduce the area of the surface to minimum. It we wish to increase the area of the surface of a liquid, we have to work against the force of surface tension.

The work required to increase the surface area by one unit is called surface energy of the liquid

Units. Surface energy has units of work per unit area
In S.I System joule m^{-2}
C.G.S system erg cm^{-2}

It can easily be shown that,

$$\text{erg cm}^{-2} = \text{dyne} \times \text{cm cm}^{-2} = \text{dyne cm}^{-1}$$

In other words, the surface tension and surface energy are dimensionally equal.

8.1.1 Factors Affecting Surface Tension

1. **Effect of Temperature:** As temperature increases, the kinetic energy of the molecules increases. Thus, the intermolecular forces decrease and hence surface tension of the liquid decreases. The surface tension of a liquid is expected to disappear at its critical temperature but in most of cases it vanishes at $T_c - 6$.
2. **Effect of surfactants:** Surfactants are surface active materials and consist of molecules having polar and nonpolar parts. Presence of surfactants in aqueous solutions lowers the surface tension.

8.1.2 Some Effects of Surface Tension

1. Drops of liquid or bubbles of a gas are spherical in shape because of surface tension.
2. Water in the capillary tube is pulled upwards due to the force of surface tension and is pulled downwards due to force of gravity. Water continues to rise in the capillary till a balance between the two forces is reached.
3. When water rises in a glass, the shape of the meniscus is concave. This indicates that the adhesive forces between water and glass are stronger than the cohesive forces between water molecules. In some cases, e.g in case of mercury, the shape of the meniscus is convex . This is due to stronger cohesive forces as compared to adhesive forces.
4. In metallurgy, the froth floatation process for the concentration of ore is based on surface tension.

8.2 MEASUREMENT OF SURFACE TENSION

There are various methods for determining the surface tension of a liquid. The most common is stalagmometer method discussed in experimental section.

8.3 VISCOSITY

Different liquids flow with different speeds. For example, glycerine, castor oil, honey etc. flow slowly whereas water, ether alcohol etc. flow rapidly. There is some sort of internal friction operating in between the layers of the liquid which checks the flow of liquids and varies from one liquid to other. In case of liquids, the internal friction is primarily due to the forces of attraction between the molecules. This resistance to the flow of liquid is known as *viscosity*.

The phenomenon of viscosity is further elucidated by considering a laminar (stream lined) flow of the liquid through a narrow pipe. (The flow in which a liquid moves in layers is known as Laminar flow). In such a case, all parts of the liquid do not move with the same velocity. The layer immediately in contact with the surface of the inner wall of the pipe is stationary. The velocity of each successive layer increases as we move away from the surface towards the centre of the tube.

As a result, a velocity gradient is set up between different layers of the liquid. Moreover, the molecules moving in any one layer will tend to retard the movement of the molecules in the adjacent faster moving layer due to intermolecular attraction. Thus, there arises a frictional force or drag between the two layers which gives rise to viscosity which can be defined as *the resistance that one part of a liquid flowing with one velocity offers to another part of a liquid flowing with different velocity.*

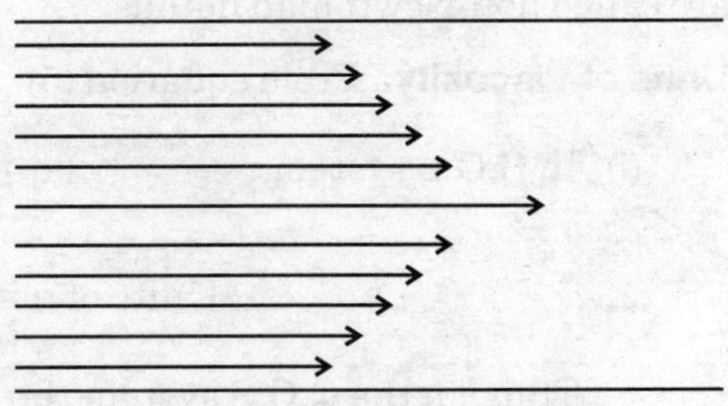

Fig. 8.2 Flow of liquid

The internal friction which resists the flow of a liquid can be measured in terms of the tangential force which is needed to keep the speeds of different layers constant. If A is the area of contact of two adjacent layers, *dx* the distance between two layers and *du* the velocity difference between two adjacent layers, then the tangential force required to maintain a constant velocity difference *du* will depend upon the following factors:

(i) It is directly proportional to A, the area of contact. Larger is A, greater will be the effect of intermolecular forces and hence larger would be the decrease in velocity. Therefore, larger force is required to maintain the velocity constant.

$$F \propto A \quad ...(1)$$

(ii) It is also directly proportional to *du*, the velocity difference between two adjacent Layers.

$$F \propto du \quad ...(2)$$

(iii) It is inversely proportional to *dx*. Larger is the distance, lesser would be intermolecular attractions and hence lesser force is required to maintain the velocity.

$$F \propto \frac{1}{dx} \quad ...(3)$$

Combining (1), (2) and (3)

$$F \propto A\frac{du}{dx}$$

or

$$F = \eta A\frac{du}{dx} \quad ...(4)$$

where η is known as *coefficient of viscosity* of a liquid or simply viscosity of the liquid. Rearranging (4)

Note. A streamline flow is that in which all the particles of the liquid move in an orderly manner along fixed paths. The velocity of particles of the liquid passing one after the other through a given point on a stream line remains unchanged in magnitude as well as in direction at that point.

The flow remains streamlined as long as the velocity is not large and remains below a certain velocity called critical velocity beyond which the flow becomes turbulent and highly irregular. In the turbulent flow the velocity of a water particle passing through a point is not the same as that of the particle which precede it.

$$\eta = \frac{F}{A\,(du/dx)} \qquad ...(5)$$

Thus η can be defined *as force per unit area required to maintain a velocity difference of unity between two parallel layers of liquid which are unit distance apart.*

Equation (4) is also known as *Newton's law of viscosity*. The fluid flow to which this equation is applicable is called laminar (streamlined) flow. The liquids which do not obey the above equation are called non-Newtonian liquids.

Units of Viscosity: From equation (5)

(i) In C.G.S system,

$$\eta = \frac{F}{A\,(du/dx)} = \frac{F}{A}\cdot\frac{dx}{du}$$

$$\text{Units of } \eta = \frac{\text{dyne cm}}{\text{cm}^2\ \text{cm s}^{-1}} = \text{dyne cm}^{-2}\ \text{s} = \text{poise}$$

Thus, in the C.G.S system the unit of viscosity is **dyne cm^{-2} s** and is known as poise.

$$\text{poise} = \text{dyne cm}^{-2}\ \text{s}$$
$$= \text{gcm s}^{-2}.\ \text{cm}^{-2}\ \text{s} = \text{g cm}^{-1}\text{s}^{-1}$$

In other words 1 poise = 1g cm^{-1} s^{-1}

(ii) In S.I system

$$\text{Units of } \eta = \frac{\text{Newton } m}{m^2\ ms^{-1}}$$
$$= \text{Nm}^{-2}\ \text{s}$$

Since Nm^{-2} = pascal (Pa) (unit of pressure)

∴ Units of η = Pas

Thus in S.I system the unit of viscosity is Pascal second (Pas).

Note.

$$1 \text{ poise} = 1\text{g cm}^{-1}\ \text{s}^{-1} = 1\ (10^{-3}\text{kg})\ (10^{-2}\ \text{m})^{-1}\ \text{s}^{-1}$$
$$= 10^{-1}\ \text{kg m}^{-1}\ \text{s}^{-1}$$
$$= 10^{-1}\ \text{Pas}$$

Hence 1 poise = 10^{-1} Pas.

Since the viscosities of most of the liquids are in small magnitude, therefore these are usually expressed in the units of centipoise and millipoise.

1 Centipoise = 10^{-2} poise

1 millipoise = 10^{-3} poise.

The reciprocal of viscosity is called fluidity and is represented by the symbol f so that,

$$\phi = \frac{1}{\eta}$$

Fluidity is the measurement of the ease with which the liquid flows. Its units are *Poise^{-1}*.

8.4 FACTOR ON WHICH THE VISOSITY OF THE LIQUID DEPENDS

1. Temperature. The viscosity of a liquid decreases with increase of temperature. In liquids, as the temperature rises, the kinetic energy of the molecules increases and the intermolecular forces of attraction become weak, resulting in the subsequent decrease in the viscosity. It has been found that the decrease is about 2% per degree rise of temperature.

The molecule taking part in liquid flow should have sufficient energy to overcome the forces of intermolecular attraction due to surrounding molecules. This energy is the activation energy for viscous flow and is designated as E_a. The variation of viscosity with temperature is best expressed by empirical relation given by Arrhenius

$$\eta \; \alpha \; e^{E_a/RT}$$

$$\eta = Ae^{E_a/RT} \quad ...(6)$$

where A is a constant of proportionality, T is the temperature and R is gas constant.

It is clear from equation (6) that as the temperature is increased the resistance to flow or the viscosity may be expected to decrease in a reciprocal manner.

Taking log on both sides of equations (6)

$$\text{In } \eta = \text{In } A + \frac{E_a}{RT}$$

$$\log \eta = \log A + \frac{E_a}{2.303\text{R}} \times \frac{1}{T} \quad ...(7)$$

The plot of log h against $\frac{1}{T}$, therefore, gives a straight line demonstrating the fact that the viscosity of a liquid rapidly decreases with rise in temperature.

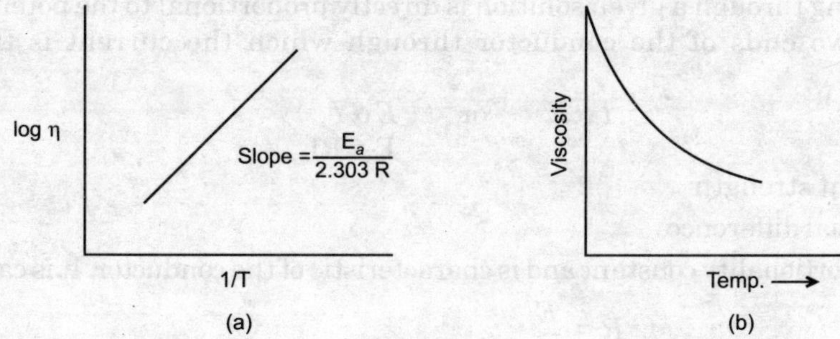

Fig. 8.3

In contrast to liquids, η increases with increasing temperature in the case of gases.

2. Effect of Pressure. The viscosity of the liquids increases with increase in pressure . As the pressure is increased the molecules of the liquid come closer and it becomes difficult for liquid molecules to move around and thus it becomes more difficult for them to flow.

3. Viscosity and Chemical Constitution. Though there is no significant relationship between viscosity and chemical constitution, however some general results have been observed.

(i) The viscosity increases as the size of the molecules increases. For example, in a homologous series, as we go up, the viscosity increases and increase per CH_2 group is almost constant.

(ii) In the compounds having chain isomerism, the n-isomers have generally greater viscosity than the branched chain. This is due to the fact that area of contact between the two layers in *n*-isomer is more and hence the molecules experience greater intermolecular attraction thereby increasing the viscosity.

However, molecules with flexible chains have very high viscosity due to entanglement of side chains.

(iii) Non-polar molecules have lower viscosities than polar molecules.

(iv) Presence of intermolecular hydrogen bonding increases the viscosity of the liquid. Glycerol and ethylene glycol have comparatively larger values of coefficients of viscosity since they have more than one hydroxyl group which results into hydrogen bonding.

(v) Presence of impurities increases the viscosity. Viscosity is greatly enhanced due to the presence of lyophillic colloids.

8.5 MEASUREMENT OF VISCOSITY

There are various experimental methods for the determination of viscosities of liquids and solutions.

(i) Ostwald's Viscometer method.

(ii) Fall of solid spheres through the liquids.

In laboratory the Ostwald's Viscometer method is commonly employed for the determination of viscosity, which is discussed in experimental section.

8.6 CONDUCTIVITY

Electrolytic solutions conduct electric current due to the movement of ions (cations and anions) towards oppositely charged electrodes.

Like metallic conductors, solutions of electrolytes also obey *Ohm's Law*. According to this law the current (I) passing through a given solution is directly proportional to the potential difference (E) between the two ends of the conductor through which the current is flowing. Thus, mathematically,

$$I \alpha E \quad \text{or} \quad E \alpha I$$

$$E = RI \qquad ...(1)$$

where I = Current strength

E = Potential difference

R = A proportionality constant and is characteristic of the conductor. It is called *resistance*.

$$R = \frac{E}{I} \qquad ...(2)$$

when E is measured in *volts*, current I is measured in *amperes*, the unit of resistance is known as *ohm* (Ω).

Thus the ratio E/I or R is the resistance of the conductor to the passage of electric current. The reciprocal of resistance R is called the *conductance* (*C*).

$$C = \frac{1}{R} \qquad ...(3)$$

The conductance is the ease of the flow of current through a solution of electrolyte. Since conductance is reciprocal of ohm its units are ohm^{-1} or *mho*. Now a days this unit of conductance is

$$IS = 1\ \Omega^{-1} \text{ known as } seimen\ (S).$$

8.7 SPECIFIC RESISTANCE AND SPECIFIC CONDUCTANCE

The resistance R of a conductor at a definite temperature, is directly proportional to its length (l cm) and inversely proportional to its area (a cm^2). Thus

$$R \alpha \frac{l}{a} \quad \text{or} \quad R = r \times \frac{l}{a} \qquad ...(4)$$

where ρ is a constant and is known as specific resistance or resistivity of the material. Its value is characteristic of the material at a specified temperature. If $l = 1$ cm and $a = 1\text{cm}^2$ then,

$$R = \rho \qquad ...(5)$$

Thus specific resistance of a substance is defined as the resistance offered by a cube of unit length of it to the passage of electricity through it. The units of ρ are *ohm cm* or *ohm m*.

The reciprocal of specific resistance or resistivity is called specific conductance or conductivity. It is denoted by the symbol K (kappa). Thus

$$K = \frac{1}{\rho} \qquad ...(6)$$

The specific conductance of a conductor is defined as the conductance of a cube of unit length.

The specific conductance of a solution can be defined as the conductance of a solution contained between two parallel electrodes which have cross sectional area 1 cm² and which are kept 1cm apart.

Taking the inverse of equation (4)

$$\frac{1}{R} = \frac{1}{\rho} \cdot \frac{1}{l/a} \qquad ...(7)$$

Substituting the value of $\frac{1}{\rho}$ from equation (6) in equation (7)

$$\frac{1}{R} = K \frac{1}{(1/a)} \qquad ...(8)$$

or $$C = \frac{K}{(l/a)}, \text{ where } C \text{ is the conductance} \qquad ...(9)$$

or $$K = C.\,(l/a) \qquad ...(10)$$

Units of K:

$K = C.\,l/a$, there fore its units are

$$K = mho \times \frac{\text{cm}}{\text{cm}^2} = \text{mho cm}^{-1} \text{ or siemens cm}^{-1}$$

Cell Constant. The ratio *(l/a)* in equation (10) is known as cell constant. It is constant for a given cell because it depends on the geometry of the cell i.e on the distance between the electrodes (*l*) and the cross sectional area of the electrodes (a). Thus.

$$\text{Cell Constant} = \frac{l}{a} \qquad ...(11)$$

Equation (10) can be written as,

Specific conductance = Measured conductance × cell constant or

$$\text{Cell constant} = \frac{\text{Specific conductance}}{\text{Measured conductance}}$$

The units of cell constant are cm⁻¹

Equivalent Conductance. It is defined as the conductance of a solution containing one gram equivalent of an electrolyte at any particular concentration, when placed between two large electrodes 1cm apart. Equivalent conductance of a solution is obtained by multiplying the specific conductance of the solution and the *volume in millilitres* which contains one gram equivalent of the solute. It is denoted by Λ_v, where v is the volume of the solution in millilitres

Equivalent Conductance, Λ_v = specific conductance (K) × Volume in $ml(v)$ per gram equivalent of the electrolyte ...(12)

The volume per gram equivalent of electrolyte is called the dilution factor of the solution. It the concentration of the solution is c gram equivalent per litre then the volume containing 1 equivalent of the electrolyte will be

$$v = \frac{1000}{c} \text{ ml per gram equiv.}$$

As $$\Lambda_v = K \times v$$

$\therefore$ $$\Lambda_v = K \times \frac{1000}{c} \quad ...(13)$$

Units of Λ_v = $(ohm^{-1}\ cm^{-1}) \times cm^3\ g\text{-}equiv^{-1} = \mathbf{ohm^{-1}cm^2g - equiv^{-1}}$

Molar Conductance. Molar conductance of an electrolyte at the particular dilution is the conductivity of the solution containing 1 mole of the electrolyte when placed between two large electrodes one cm apart . It is represented by m.

$$\mu = K \times V = K \times \frac{1000}{molarity} \quad ...(14)$$

where V is the volume of the solution in ml. containing 1 mole of the electrolyte

$$Units\ of\ m = ohm^{-1}\ cm^{-1} \times cm^3 mol^{-1} = ohm^{-1}\ cm^2 gmol^{-1}$$

Effect of dilution on conductivity: The specific as well as equivalent conductance values of solutions of electrolytes vary with dilution. The *specific conductance values decrease* on diluting the solutions. This is mainly due to decrease in the concentrations of ions per unit volume. On the other hand, the equivalent or *molar conductance increases with dilution* and attains a limiting value at infinite dilution. This increase in conductance in case of weak electrolyte is attributed to the increased dissociation of the electrolyte on dilution. Whereas in case of strong electrolyte, this increase in conductance is due to the increased mobility of the ions. (Fig. 8.4)

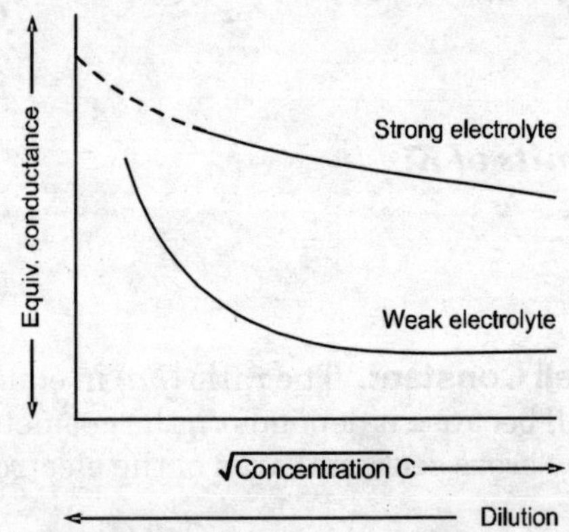

Fig. 8.4 Variation of equivalent conductance with dilution.

Effect of temperature on conductivity. The conductance of solution increases with increase in temperature. This is due to the.

- Increase in the velocity of ions,
- decrease in the viscosity of the medium
- decrease in the interaction between the ions.

In case of weak electrolytes, increase in the degree of ionization with increase in temperature also results in increased conductance. Generally, conductance of aqueous solutions increases by about 2% per degree rise in temperature.

Other factors influencing conductance

1. Number of free ions in the solution: More is the number of free ions, greater is the conductance.
2. Charge on the free ions: Greater the charge greater is the conductance.
3. Mobility of the ions: Greater the mobility greater is the conductance.

8.8 MEASUREMENT OF CONDUCTANCE

Conductance is the reciprocal of resistance which can be determined using the principle of wheat stone bridge circuit (Fig 8.5)

In the figure R_1 and R_2 represent fixed resistance, R_3 is variable resistance, and R_4 is unknown resistance. A small voltage is applied across the point AC and the resistance R_3 is varied till no current passes through the galvanometer connected across the points BD. Then according to wheat stone's principle,

$$\frac{R_1}{R_2} = \frac{R_4}{R_3}$$

Thus, the unknown resistance R_4 can be calculated , as all the other quantities are known. However, this method is faced with a number of difficulties when applied for determining the resistance of an electrolyte solution . The main difficulties are:

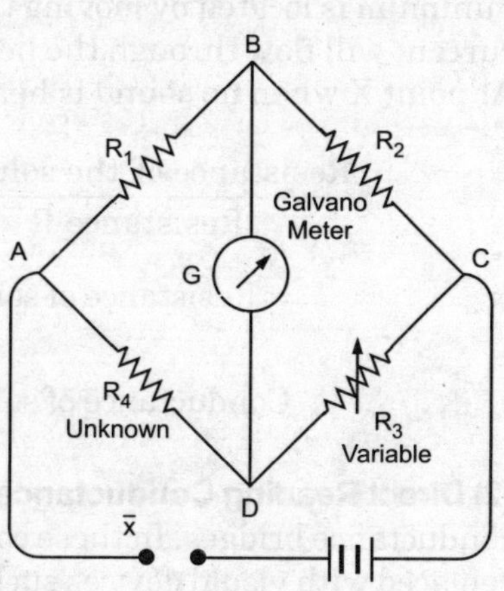

Fig. 8.5 A Simple Wheatstone bridge circuit.

(i) Direct current (DC) cause electrolysis of the solution and this changes the concentration of the solution.
(ii) Products of the electrolysis accumulate at the electrodes resulting is polarisation effects i.e a back e.m.f is set up which opposes the current being passed.

The above difficulties are removed by using alternating current of strength 500-2000 cycles per second. The galvanometer becomes useless while using alternating current and hence it is replaced by head phone.

A few conductance bridges based on the above principle, with modifications are:

(1) Kohlrauch's Conductance bridge. It is the widely used circuit in the measurement of conductance. The arrangement of the apparatus is shown in Fig. 8.6

The conductivity cell C is connected to the resistance box R on one side and to a long thin wire AB stretched along a scale on the other side.

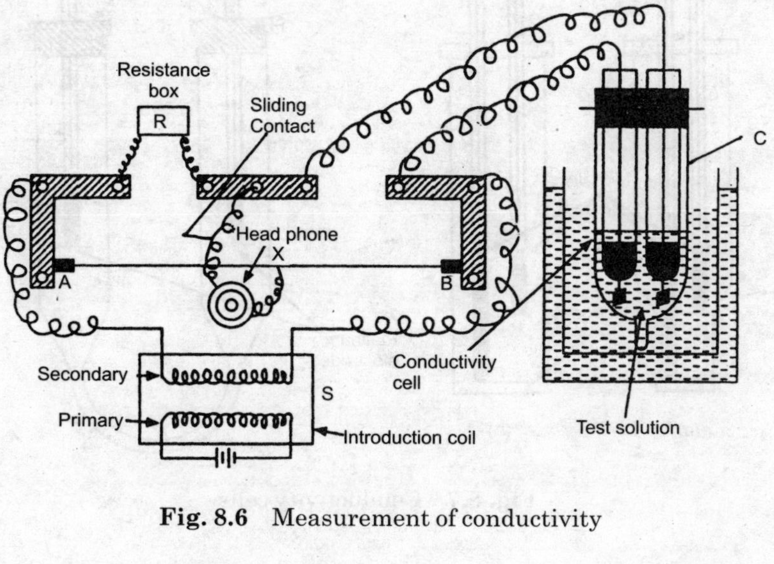

Fig. 8.6 Measurement of conductivity

An alternating current is passed and a point X at which the sound in the head phone P is minimum is located by moving the sliding contact along the wire AB. At every point other than X, current will flow through the head phone and a buzzing noise would be heard in the head phone. At point X when no sound is heard, no current flows through the head phone. At this point,

$$\frac{\text{Resistance of the solution}}{\text{Resistance R}} = \frac{\text{Resistance of wire AX}}{\text{Resistance of wire XB}} = \frac{\text{length of AX}}{\text{length of XB}}.$$

or
$$\text{Resistance of solution} = \frac{AX}{XB} \cdot \text{Resistance R}$$

$$\therefore \quad \text{Conductance of solution} = \frac{XB}{AX} \times \frac{1}{R}$$

(2) Direct Reading Conductance bridges. These days conductivity is determined by direct reading conductance bridges. In these conductivity bridges or conductometers, the head phone detector is replaced with visual devices such as electronic detector. The method of operation of an instrument depends on its make and the manual is supplied by the manufacturer. One such instrument is described under the experiment.

8.8.1 Conductivity Cells

The solution whose conductivity is to be determined is placed in a special type of cell known as conductivity cell. Conductivity cells are of various shapes and sizes. Two of these are shown is Fig. 8.7.

The conductivity cells are made up of pyrex glass (or some other resistant glass), which is fitted with a pair of platinum electrodes. These electrodes are platinum plates fused into the glass tubes. The tubes are firmly supported by an ebonite cover. The cover along with the platinum electrodes is fitted into the pyrex glass vessel so that the distance between the electrodes may not change.

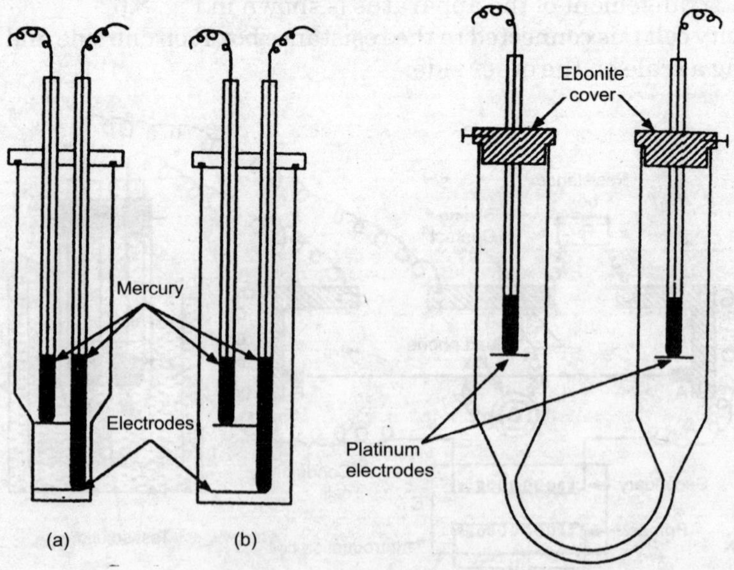

Fig. 8.7 Conductivity cells

Conductivity Water As ordinary distilled water contains impurities like CO_2, metal silicates etc., it cannot be used for preparing solutions for conductivity measurements. Hence, water should be specially purified before it can be used. Such water is known as *conductivity water*. Conductivity water is obtained by distilling good quality distilled water after adding potassium permanganate and KOH in a pyrex glass distillation unit. However, conductivity water is used when conductance values of high accuracy are required. For all other purposes simple distilled water can be used as long as the conductivity of water is not more than 10^{-5}S cm^{-1}

QUESTIONS

1. *Define surface tension.*

Ans. The surface tension γ is the magnitude F of the force exerted parallel to the surface of a liquid divided by the length L of the line over which the force acts:

$$\gamma = \frac{F}{L}$$

2. *What are the units of surface tension?*

Ans. **The common units for surface tension are dynes/cm or N/m. These units are equivalent.** Units of surface tension ($N \cdot m^{-1}$) are equivalent to joules per square metre ($J \cdot m^{-2}$). This means that surface tension can also be considered as **surface energy**. If a surface with surface tension γ is expanded by a unit area, then the increase in the surface's stored energy is also equal to γ.

3. *What are the forces responsible for surface tension?*

Ans. The **cohesive** forces between liquid molecules are responsible for the phenomenon known as surface tension.

4. *What is the surface tension of water?*

Ans. The **surface tension** of water is 72 dynes/cm at 25°C. It would take force of 72 dynes to break a surface film of water 1 cm long

5. *Why is Surface Tension important?*

Ans. If any formulation changes at the molecular level then the surface tension will change. If the formulation changes due to the addition of another chemical, the addition of a surfactant, or if anything contaminates the fluid in question, then the surface tension will change. Measuring surface tension is a direct indicator of the quality of any chemical and any formulation.

6. *Define Viscosity*

Ans. Viscosity is the resistance that one part of a liquid flowing with one velocity offers to another part of liquid flowing with different velocity.

7. *What are the units of viscosity?*

Ans. Units of Viscosity:

(i) In the C.G.S system the unit of viscosity is dyne cm^{-2} s and is known as poise.

$$\text{poise} = \text{dyne cm}^{-2}\,\text{s}$$
$$= \text{gcm s}^{-2}.\,\text{cm}^{-2}\,\text{s} = \text{g cm}^{-1}\,\text{s}^{-1}$$

In other word 1 poise = 1g cm^{-1} s^{-1}

(ii) In S.I system the unit of viscosity is Pascal second (Pas).

Note.
$$1\ \text{poise} = 1\text{g cm}^{-1}\,\text{s}^{-1} = 1\,(10^{-3}\,\text{kg})\,(10^{-2}\,\text{m})^{-1}\,\text{s}^{-1}$$
$$= 10^{-1}\,\text{kg m}^{-1}\,\text{s}^{-1}$$
$$= 10^{-1}\,\text{Pas}$$

Hence 1 poise = 10^{-1} Pas.

8. *What do you mean by fluidity?*

Ans. The reciprocal of viscosity is called fluidity and is represented by f so that,

$$\phi = \frac{1}{\eta}$$

Fluidity is the measurement of the ease with which the liquid flows. Its units are *Poise*$^{-1}$.

9. *How viscosity of a liquid varies with temperature?*

Ans. The viscosity of a liquid decreases with increase of temperature. As the temperature rises, the kinetic energy of the molecules increases and intermolecular forces of attraction become weak, resulting in subsequent decrease in the viscosity.

The variation of viscosity with temperature is best expressed by empirical relation given by Arrhenius

$$\eta \propto e^{Ea/RT}$$

$$\eta = Ae^{Ea/RT} \quad \text{....(6)}$$

where A is a constant of proportionality, T is the temperature and R is gas constant.

In contrast to liquids, h increases with increasing temperature in the case of gases.

10. *What is the effect of pressure on viscosity of a liquid?*

Ans. The viscosity of the liquids increases with increase in pressure. As the pressure is increased the molecules of the liquid come closer and it becomes difficult for liquid molecules to move around and thus it becomes more difficult for them to flow.

11. *How viscosity is related to chemical constitution?*

Ans. Though there is no significant relationship between viscosity and chemical constitution, however some general results have been observed.

(i) The viscosity increases as the size of the molecules increases.

(ii) In the compounds having chain isomerism, the n-isomers have generally greater viscosity than the branched chain.

However, molecules with flexible chains have very high viscosity due to entanglement of side chains.

(iii) Non-polar molecules have lower viscosities than polar molecules.

(iv) Presence of intermolecular hydrogen bonding increases the viscosity of the liquid.

(v) Presence of impurities increases the viscosity. Viscosity is greatly enhanced due to the presence of lyophilic colloids.

12. *What are the applications of viscosity measurements?*

Ans.
- Viscosity Measurements are useful in predicting the behaviour of lubricants and oils in hot and cold conditions.
- Viscosity is also used for determination of molar masses of polymers.
- For distinguishing between branched and chain isomers. Chain isomers generally have high viscosities.

13. Define conductance, specific conductivity and equivalent conductivity. What are their units?

14. How specific conductance and equivalent conductance vary with dilution?

15. What is the effect of temperature on conductance ?

16. Define cell constant. How is it determined ?

17. What is conductivity water ?

18. What are the advantages of conductometric titrations ?

19. Why KCl is used as a standard for verification of cell constant ?

20. What is the relation between specific conductivity, conductance and cell constant?

□□□

UNIT 3

Stability and Reactivity of Molecules: Unit Processes

9 Reactivity of Molecules: Fundamentals of Organic Mechanism

"The winds and waves are always on the side of the ablest navigators."

Edward Gibbon

CHAPTER OUTLINE

The term 'organic chemistry' was used to describe the compounds obtained from living organisms, while *inorganic chemistry* was used for the study of compounds from non living sources. Today, however we know that the common characteristic of organic compounds from living sources is that, all contain the element 'carbon'. Thus, organic chemistry is now defined as the chemistry of carbon compounds or more precisely hydrocarbons and their derivatives.

Most of the inorganic reactions are very fast and it is extremely difficult to know the detailed pathway of such reactions. However, the reactions of *organic compounds* are generally slow and proceed through well defined steps to form the products. In organic reactions, the reactant change into the *intermediate* substance and these intermediates then change into the final product.

The step-by-step sequence of elementary reactions by which the reactants change into the products is called reaction mechanism. In other words, the mechanism of an organic reaction is a detailed description of all the steps involved in that reaction. The study of mechanism of organic reactions is a complex area.

The general concepts and principles described in the following sections, help better understanding of the organic reaction mechanisms.

9.1 NATURE OF ORGANIC MOLECULES

Why do chemists treat organic chemistry as a separate branch of science? This is because of the ability of carbon atoms to bond together, forming chains or rings. Of all the elements, only carbon has this property and can form compounds from one carbon atom (such as methane) to tens of billions of carbon atoms (such as deoxyribonucleic acid, *DNA*).

Organic molecules have '*covalent bonds*' formed by sharing of electrons between the bonded atoms. Carbon can form multiple covalent bonds also. These covalent bonds are directional and hence the organic molecules have specific three dimensional shapes. Covalent bonding gives organic compounds, properties that are quite different form those of inorganic compounds. The nature of covalent bonds determine the path of the reaction in these compounds.

9.2 CLEAVAGE OF COVALENT BONDS

Organic reactions involve breaking and formation of covalent bonds. The covalent bonds can break by two ways. Different products will be formed depending upon the point of cleavage as discussed below:

(*i*) **Homolytic cleavage (Symmetrical cleavage):** We know that covalent bond is formed by the sharing of electrons between the two atoms. Let us represent the organic molecule as: "$A : B$". In homolytic cleavage the bond breaks in such a way that each atom departs with one electron of the shared pair.

$$\mathrm{A} \vdots \mathrm{B} \longrightarrow \underbrace{\mathrm{A}\cdot + \mathrm{B}\cdot}_{\text{Free radicals}}$$

Example:

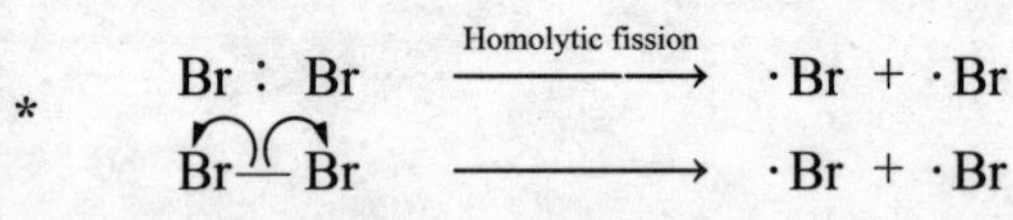

Homolytic cleavage or fission leads to the formation of highly reactive neutral species containing unpaired electrons which are known as *Free Radicals*.

Homolytic cleavage is possible only when the electronegativity ** of both the bonded atoms is same, *i.e.* both the atoms have similar attraction for the shared electron pair.

(*ii*) **Heterolytic cleavage (Unsymmetrical Cleavage):** In this case, the covalent bond is broken in such a way that one of the atoms carries away both the electrons of the shared pair. This result in the formation of a cation and an anion.

Note:* A partial head (fish hook) on the arrow indicates the shift of single electron. A full head arrow indicates the movement of electron pair.

** Electronegativity: It is the measure of the tendency of an atom in a molecule to attract the bonded pair of electrons.

$$A \vdots \cdot\cdot B \longrightarrow A^{\oplus} + B^{\ominus}$$ (When B is more electro negative than A)

or

$$A \cdot\cdot \vdots B \longrightarrow :A^{\ominus} + B^{\oplus}$$ (When A is more electro negative than B)

or

$$A - B \longrightarrow A^{+} + B^{-}$$

Heterolytic fission is possible when the electronegativities of the bonded atoms are different. The more electonegative atom retains both the electrons and acquires a negative charge, *i.e.* an anion is formed and the other atom which departs without the electrons acquires the positive charge, *i.e.* a cation is formed.

Example:

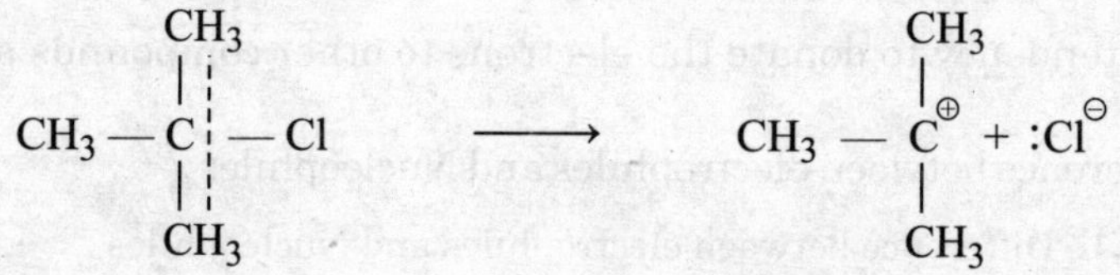

9.3 ATTACKING REAGENTS (ELECTROPHILES AND NUCLEOPHILES)

In a chemical reaction, the substrate (reactant) changes into the product under the influence of attacking reagent.

Substrate + Attacking Reagent → Product

The nature of attacking reagent determines the course of chemical reaction. The attacking reagent are of two types:

(*i*) Electrophiles or Electrophilic reagent
(*ii*) Nucleophiles or Nucleophile reagent.

(*i*) **Electrophiles:** (Electro = electron, philic = loving) Electrophiles are electron deficient species which can accept a pair of electrons. There are electron loving species and hence always attack at the centre of high electron density. Electrophiles may be either positively charged or electrically neutral chemical species.

(*a*) *Positively charged electrophiles:* These are electron deficient species and carry a positive charge. Examples,.

$$H^+, H_3O^+, Cl^+, Br^+, I^+, \overset{+}{N}O_2, \overset{+}{N}O, R^+ \text{ etc.}$$

(*b*) *Neutral Electrophiles:* These are electron deficient species but carry no charge. The central atom in neutral electrophiles has six electrons and can accept two electrons readily.

Examples: BF_3, $AlCl_3$, $FeCl_3$, $SnCl_4$, $\dot{R}$. (free radicals), : CR_2 (Carbenes), : NR (Nitrencs) etc.

Since, both the positively charged and neutral electrophiles are short from by a pair of electrons (except free radicals which are short by one electron), they have strong tendency to attract electron from other compounds and hence behave as *Lewis acids*.

(*ii*) **Nucleophiles:** (Nucleo = nucleus, philic = loving) Nucleophiles are electron rich species and have a tendency to donate a pair of electrons. Nucleophiles always attack at electron deficient (low electron density) centres. These may be either negatively charged or neutral chemical species.

(*a*) *Negatively charged nucleophiles:* These are electron rich species having negative charge. Examples: H^- (hydride ion), $\bar{Cl}$, $\bar{Br}$, $\bar{I}$, $\bar{R}$ (Carbanion), $\bar{O}H$, $\bar{O}R$, $S\bar{R}$, $\bar{NH_2}$, $\bar{CN}$, $RCO\bar{O}$ etc.

(*b*) *Neutral Nucleophile:* These are electron rich, having lone pair of electrons but carry no charge.

Examples: $H_2\ddot{\underset{..}{O}}$, $\ddot{N}H_3$, $R\ddot{N}H_2$, $R\ddot{\underset{..}{O}}H$, $R\ddot{\underset{..}{S}}H$, $R\ddot{\underset{..}{O}}R$ etc.

(c) *Ambident Nucleophile*: Sometimes more than one atom in a nucleophile is electron rich and hence it can attack the substrate through either of the atom. Such nucleophiles are called ambident nucleophile.

Examples: $\bar{C}N$, NO_2^-.

Nucleophiles have strong tendency to donate the electrons to other compounds and hence and behave as Lewis bases.

Table 9.1 gives the difference between electrophiles and Nucleophiles

Table 9.1: Difference between electrophiles and Nucleophiles.

Electrophiles	*Nucleophiles*
1. These are electron deficient species which can accept a pair of electrons.	These are electron rich species which can donate a pair of electrons.
2. They are positively charged or neutral species.	These are negatively charged or neutral species.
3. Attack on electron rich centre.	Attack on electron deficient centre.
4. These are Lewis acids as these can accept an electron pair.	There are Lewis bases as these can donate electron pair.
5. Example NO_2^+, Br^+, BF_3, $AlCl_3$, SO_3 etc.	Example : $\bar{O}H$, $\bar{O}R$, $\bar{C}N$, NH_3, H_2O etc.

9.4 ELECTRON DISPLACEMENTS IN ORGANIC MOLECULES

Covalent bonds in organic molecules undergo electronic displacements, sometimes on their own or under the influence of attacking reagent (nucleophile or electrophile). For a chemical reaction to take place in organic compounds, the substrate must develop some polarity so that charged centres are created. This is made possible by the displacement of electrons of the bonding electrons. The electron displacement influences the mechanism involving such bonds. Now, how do these electron displacements take place? These displacements are due to certain effects, some of which are permanent and others are temporary. Thus, *the permanent or temporary effects involving the displacement of electron in the substrate molecule are termed as electron displacement effects.* Some of the important displacement effects are:

(*i*) Inductive Effect
(*ii*) Mesomeric Effect
] Permanent Effects

(*i*) Electromeric Effect
(*ii*) Inductomeric Effect
] Temporary Effects

Some of these effects are discussed below.

9.4.1 Inductive Effect

A covalent bond is formed by the sharing of electron pair between two atoms. If the two atoms are dissimilar, the shared electrons will be attracted towards more electronegative atom. Thus an atom which is more electronegative acquires partial negative charge (δ^-) where as the other acquires partial positive charge (δ^+). In a carbon chlorine bond, for example, the electron pair between carbon and chlorine is displaced towards more electronegative chlorine atom as compared to carbon. Thus, chlorine atom acquires a small negative charge and the carbon atom a small positive charge.

$$-\overset{|}{\underset{|}{C}}{}^{\delta^+}\rightarrow Cl^{\delta^-}$$

This type of permanent displacement of electrons of a covalent bond towards the more electronegative element or group is called inductive effect.

If this carbon bond forms the terminal of carbon chain, the effect of this electron displacement is passed to other carbon atoms in the chain, *i.e.* the positive charge on C_1 attracts the electrons between C_1 and C_2 and hence C_2 will also acquire slight positive charge. But this positive charge will be less as compared to the positive charge on C_1. Similarly, effect can be passed further to C_3 but as the distance between carbon atom and chlorine increases, the effect decreases rapidly.

$$-\overset{|}{\underset{|}{C}}{}^{\delta^{+''}}-\overset{|}{\underset{|}{C}}{}^{\delta^{+'}}-\overset{|}{\underset{|}{C}}{}^{\delta^{+'}}-Cl^{\delta^-} \qquad \text{where } \delta'^{+} > \delta^{+''} > \delta^{+'''}$$

From the practical point of view, the effect is ignored after the second carbon atom. Hence, inductive effect may be defined as '*the process of electron displacement along a chain of atoms due to the presence of polar covalent bond*'. This is a permanent effect.

Inductive effect is represented by the symbol →, the arrow pointing towards the more electronegative element or group of atoms.

We can see that the inductive effect causes a certain degree of polarity in the bond which makes the bond much more liable to be attacked by other charged atoms or group.

In the foregoing example, the hetero atom, C_l is more electronegative than carbon. Inductive effect also takes place if the carbon is bonded to less electronegative atom. In that case, the electron displacement will take place in the opposite direction, *i.e.* the electrons are pushed away form the heteroatom.

(*i*) *– I Effect or electron withdrawing effect:* This effect is exerted by the groups attached to the carbon chain having high electronegativity than carbon. Some of the examples of groups having – *I* effect are:

$$NO_2 > F > Cl > Br > I > OCH_3 > OH$$

(*ii*) *+ I Effect or electron releasing inductive effect:* It is exerted by the atoms or groups attached to the carbon chain having lesser electronegativity than carbon.

Some of the examples of groups having + I effect are:

$$C_6H_5\bar{O} > CO\bar{O} > CR_3 > CHR_2 > CH_2R > CH_3 > H.$$

Characteristics of Inductive Effect

(*i*) Inductive effect does not involve an actual transfer of electrons from one atom to the other. It merely helps in displacing it permanently.
(*ii*) This effect is observed in saturated compounds involving the electrons of sigma bond.
(*iii*) Inductive effect is relayed along the carbon chain but decreases rapidly with increasing distance from the hetero atom.

Application of Inductive Effect

(*i*) **Dipole Moment:** Inductive effect leads to dipolar character in the molecule. More is the inductive effect, greater will be the dipole moment. Example,

$$CH_3,\ CH_3Br,\ CH_3Cl$$

$$\xrightarrow[\text{– I increases}]{}$$

$$\xrightarrow[\text{Dipole Moment increases}]{}$$

(*ii*) **Bond Length:** Usually the bond length of organic molecules decreases with increase in inductive effect. Example,

$$CH_3,\ CH_3Br,\ CH_3Cl$$

$$\xrightarrow[\text{– I increases}]{}$$

$$\xrightarrow[\text{Bond length decreases}]{}$$

(*iii*) **Reactivity of Alkyl halides:** Inductive effect has a great influence an reactivity of alkyl halides. Due to – I effect of halogen atom in the molecule of alkyl halides, the carbon attached to the halogen acquires a partial positive charge and hence is readily attacked by a nucleophile (negatively charged ion).

$$H_3C \rightarrow Cl \quad \text{or} \quad H_3C^{\delta+} — Cl^{\delta-}$$

↑ Attack by Nucleophile

Thus alkyl halides undergo '*nucleophilic substitution reactions*' The influence of inductive effect on alkyl halides is explained in detail under the section 'Nucleophilic Reactions'.

(*iv*) **Relative strengths of the Organic acids:** Carboxylic acids have the tendency to release a proton and change into carboxylate anion.

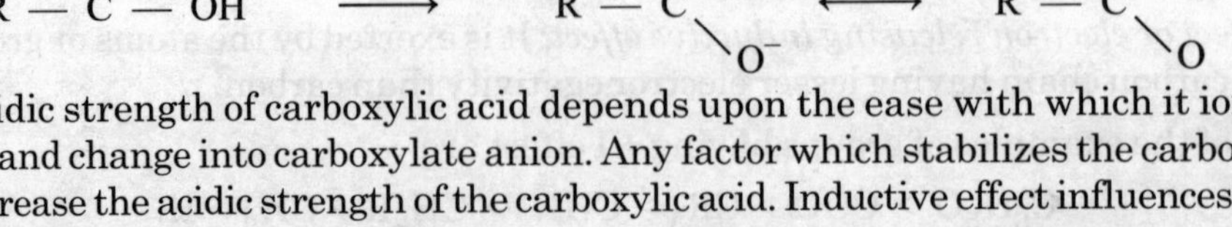

The acidic strength of carboxylic acid depends upon the ease with which it ionizes to give proton and change into carboxylate anion. Any factor which stabilizes the carboxylate anion will increase the acidic strength of the carboxylic acid. Inductive effect influences the strength of carboxylic acids as follows.

(*a*) Effect of +I groups: To explain the effect of +I groups, let us take the example of formic acid (HCOOH) and acetic acid (CH_3COOH)

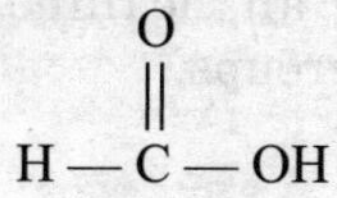

and

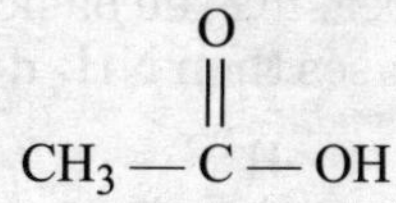

Formic acid (I) Acetic acid (II)

CH_3 group in (II) has +I effect (electron donating effect) and it will *destabilise* the acetate ion formed by release of proton because it enhances the negative charge on it. Whereas *no such destabilisation* is present in formate ion formed from (I). Hence formic acid is stronger acid than acetic acid.

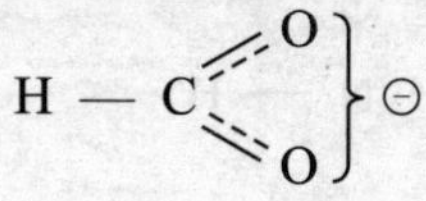

Formate ion

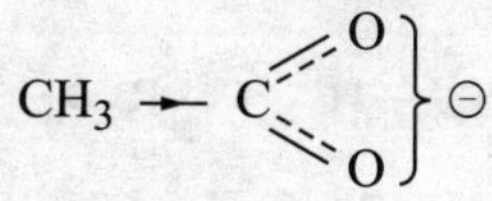

Acetate ion
(Destabilised by electron donating CH_3 group)

Further, the +I effect of alkyl groups increases with size,

$$CH_3- < CH_3\,CH_2- < (CH_3)_2\,CH- < (CH_3)_3\,C-$$

Therefore the acidic strength decreases in the same order. For example,

$$HCOOH > CH_3\,COOH > CH_3CH_2COOH > (CH_3)_2\,CHCOOH > (CH_3)_3\,CCOOH$$

(*b*) Effect of –I groups: The groups having –I effect will *stabilise* the carboxylate ion by dispersing the negative charge on it. Example,

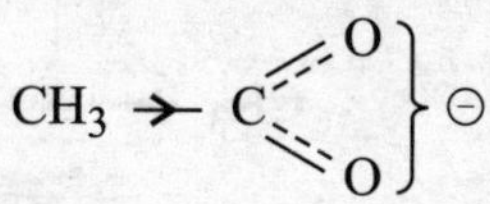

+ I effect of – CH_3 group intensifies the negative charge, destabilise the carboxylate ion

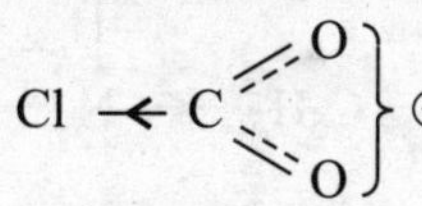

– I effect of halogen atom disperse the negative charge, stabilise the carboxylate ion

the acidic strength of chloroacetic acid is more than the acetic acid because the halogen attached to carbon atom stabilises the carboxylate ion formed after the release of proton as shown above.

Therefore, $Cl-CH_2\,COOH > CH_3\,COOH$ (Acidic strength)

More is the number of halogens on α – carbon, stronger is the acid. Example,

$$Cl_3C-COOH > Cl_2\,CH-COOH > ClCH_2-COOH > CH_3COOH.$$

Further, since – I effect of halogens decrease in the order, F > Cl > Br > I, the strength of α – halogen acids follow the same order.

$$FCH_2\,COOH > Cl\,CH_2\,COOH > Br\,CH_2\,COOH > I\,CH_2\,COOH$$

(*v*) **Relative basic strength of amines:** Amines contain a lone pair of electrons on the nitrogen atom and therefore have a strong tendency to donate the electron pair. Hence, amines behave as bases. The relative strength of aliphatic amines depends upon inductive effect, steric factors, etc. The inductive effect influences the strength as follows:

(*a*) +I Effect: The electron donating (+I) groups present on the nitrogen in amines increase the electron density on nitrogen. As a result nitrogen release the electron pair more easily, increasing the basic strength. For example, all the three classes of amines are stronger bases then NH_3 due to +I effect of alkyl groups.

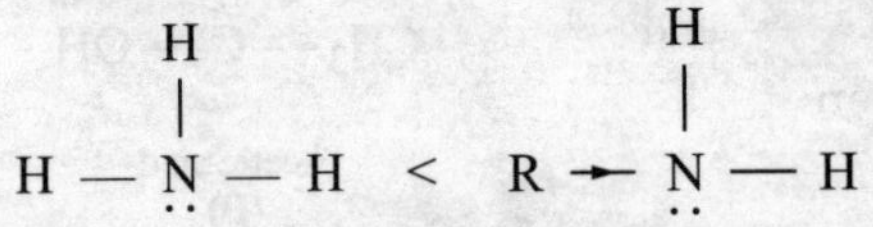

(Basic strength)

For the same reason, one should conclude that more the number of alkyl groups attached to nitrogen, more stronger the base is. On the basis of inductive effect alone, the strength of the amines should follow the order:

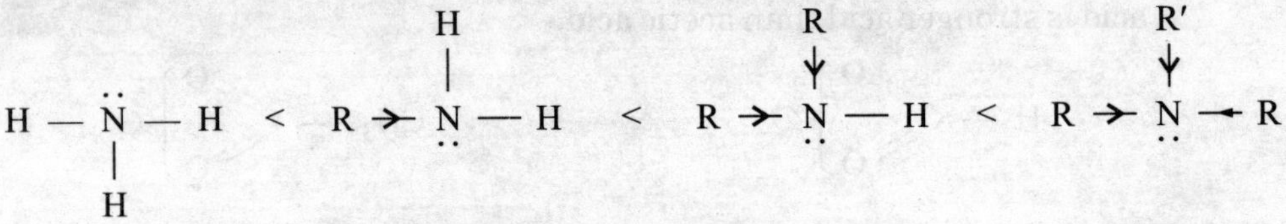

Primary amine Secondary amine Tertiary amine

However, the *actual order* of basicity in aqueous solution is:

2° Amine > 1° Amine > 3° Amine.

Thus, apart from inductive effect, other factors such as steric factors, stability of conjugate acid formed also play their role in determining the basicity.

(*b*) – I Effect: The electron withdrawing groups present on the nitrogen in amines decrease the electron density an nitrogen, making the electron release difficult. Hence – I group decrease the basic strength of amines Example:

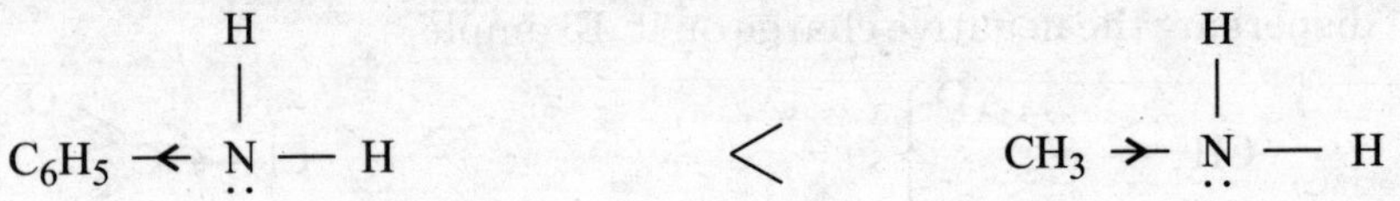

Electron withdrawing phenyl group decreases the electron density on Nitrogen. (Weaker Base)

Electron releasing alkyl group increases the electron density on Nitrogen (Stronger Base)

Similarly, the basic strength of aromatic amines follow the order:

$$C_6H_5\,NH_2 \;>\; (C_6H_5)_2\,NH \;>\; (C_6H_5)_3\,N$$

Aniline Diphenyl amine Triphenyl amine (Base strength)

9.4.2 Electromeric Effect

Electromeric effect involves the complete transfer of electrons of a multiple bond (double or triple bond) to one of the bonded atoms under the influence of an attacking agent.

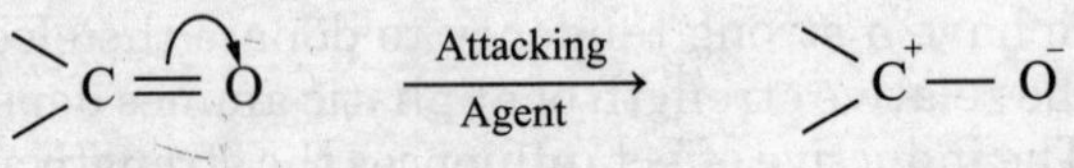

Characteristics:

(i) This is a *temporary effect* operating only under the influence of attacking reagent. As soon as the attacking agent is removed the electromeric effect disappears.

(ii) It involves complete transfer of π electrons of a multiple bond. For example, during the attack of a carbonyl group by a nucleophile reagent, the π electrons pair bonding the carbon and oxygen are transferred to oxygen.

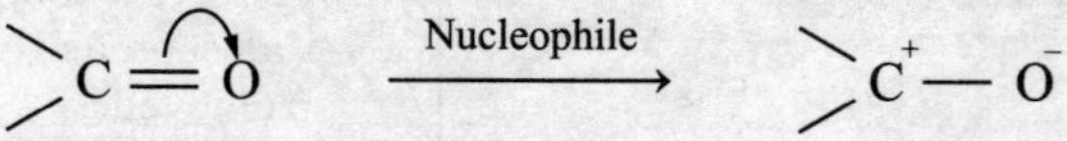

It causes the polarisation of the molecule leading to the formation positive and negative charges.

(iii) The displacement of electrons can take place in either direction in case of symmertical alkenes.

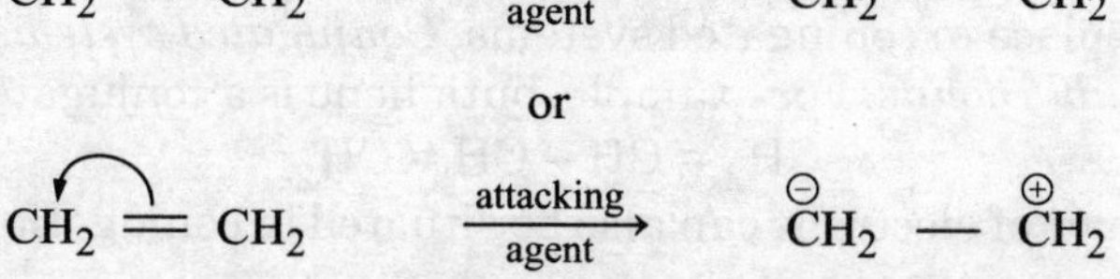

However in case of unsymmetrical alkenes the displacement of electrons is goverened by inductive effect.

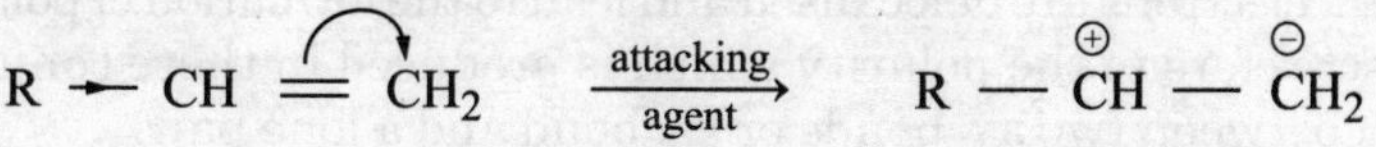

It the π bond is between the dissimilar atoms the shift occurs towards the more electronegative atom.

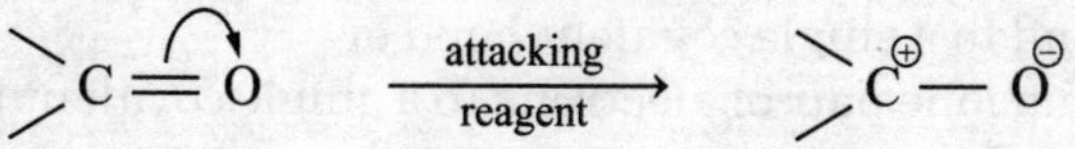

Types of Electromeric Effect

Electromeric effect is of 2 types:

(i) $+E$ Effect: The electrons of the π bond are transferred to that atom of the double bond to which the reagent gets finally attached, the effect is called $+E$ effect. For examples, during the addition of acids to alkenes,

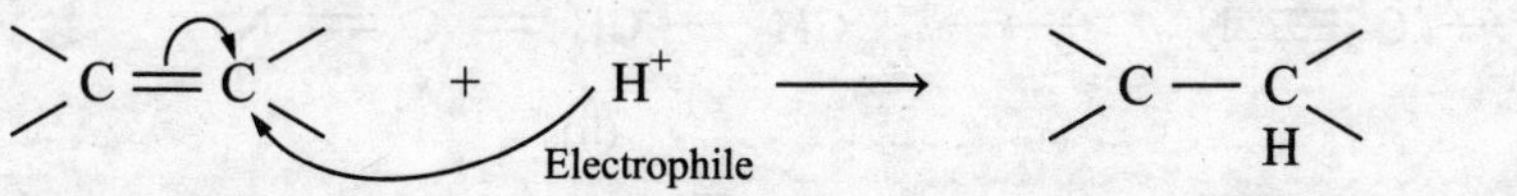

(+E Effect)

(ii) $-E$ Effect: If the electrons of this double bond are transferred to an atom of the double bond *other than* the one to which the attacking reagent get finally attached, then it is called – E effect. Example,

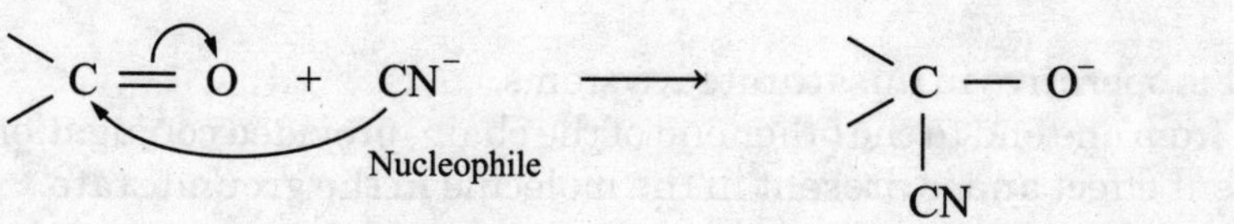

(–E Effect)

Difference between Electromeric and Inductive Effect.

Inductive effect	*Electromeric effect*
(i) Takes place due to the presence of electron donating or electron withdrawing group linked to the carbon chain.	(i) Takes place under the influence of an attacking reagent (electrophile or nucleophile).
(ii) This is a permanent effect	(ii) This is a temporary effect and disappears as soon as the attacking reagent is removed.
(iii) The electron pair is only displaced towards more electronegative atom	(iii) The electron pair is completely transferred to one of the multiple bonded atoms.
(iv) Shown by the substrate usually having single bond.	(iv) Shown by the substrate having multiple bonds.

9.4.3 Mesomeric Effect or Resonance Effect

Mesomeric effect takes place in conjugated systems. *Conjugated systems are the systems having alternate single and double bonds.* For example, butadiene is a conjugated system.

$$CH_2 = CH - CH = CH_2.$$

A group with lone pair of electrons can also be situated in conjugation with double bond. For example,

$$CH_2 = CH - \ddot{C}l:$$

In such systems, π electrons are delocalised and lead to the formation of polarity in the molecule. The mesomeric effect refers to the polarity which is produced in there conjugated systems as a result of interaction between two π – bonds or a π bond and a lone pair.

Mesomeric effect can be defined as " *the transference of electrons in conjugated systems,*

- from a multiple bond to an atom or
- from a multiple bond to a single covalent bond or
- from an atom having lone pair of electrons to a single covalent bond.

In other words, mesomeric effect is flow of electrons form one part of the molecule to the other creating centres of low and high electron density.

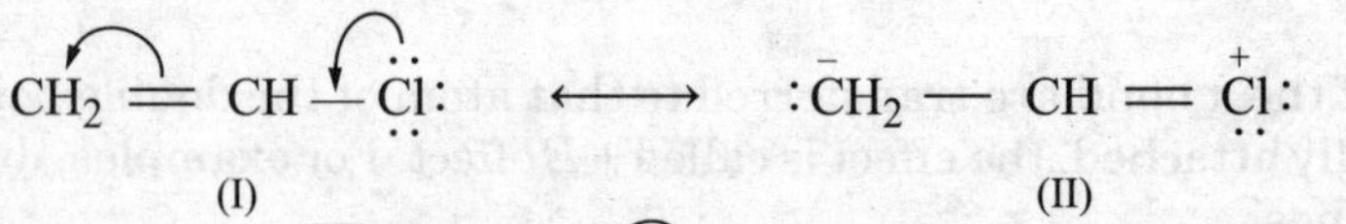

[The movement of the electrons from atom having lone pair to the single covalent bond]

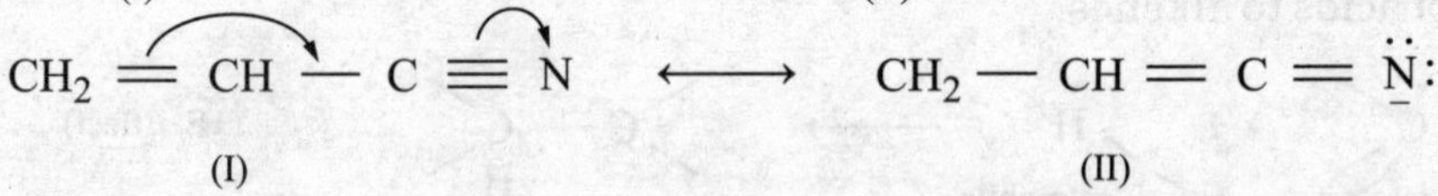

[The movement of the electrons from multiple bond to the single covalent bond and also from multiple bond to the atom]

Characteristics:

(i) Mesomeric effect is operative in unsaturated systems.
(ii) It is transmitted from one end to the other end of the chain, provided conjugation is present.
(iii) It is the permanent effect and is present in the molecule in the ground state.
(iv) The actual structure of the molecule lies in between the resonance (mesomeric) or contributing structures.

Types: Mesomeric effect is of two types:

(*i*) + M or + R Effect: A group is said to exhibit + M or + R effect, if it donates the electrons to a conjugated system. In such cases we can say that the transference of electrons is away form the atom or group in conjugated system. Some of the groups having + M effect are:

$$-\mathrm{Cl},\ -\mathrm{Br},\ -\mathrm{I},\ \mathrm{NH_2},\ -\mathrm{NR_2},\ -\mathrm{OH},\ -\mathrm{OCH_3}.$$

Example:

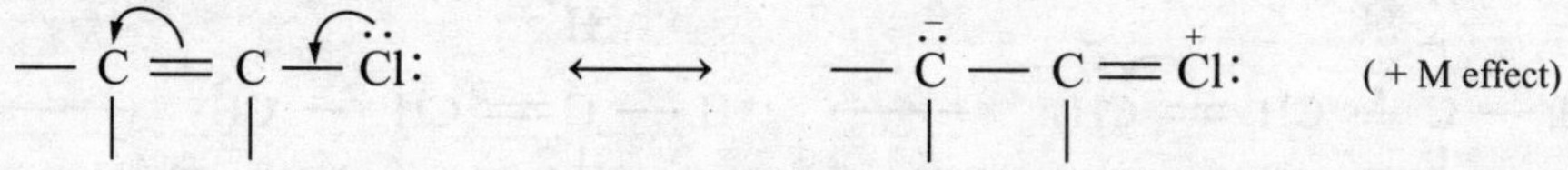

(I) (II)

(*ii*) – M or – R effect: A group or atom exhibit – M or – R effect if it withdraws electrons from the double bond or from a conjugated system. In such cases, electrons move towards the atom or group in conjugated system. Some of the groups having – M or – R effect are:

Example: $-\mathrm{NO_2}$, $-\mathrm{CN}$, $-\mathrm{C(=O)}-$, $-\mathrm{SO_3H}$

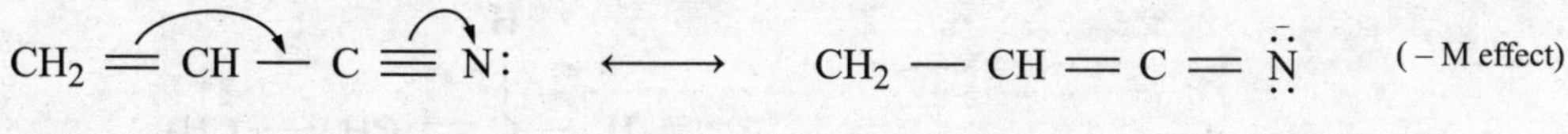

(I) (II)

9.4.4 Hyperconjugation

Hyperconjugation, also knows as Baker Nathan effect takes place in the organic molecules in which an alkyl group is attached to the unsaturated systems such as multiple bonds or benzene ring.

Baker and Nathan suggested that the alkyl groups with at least one hydrogen atom on the α carbon atom, attached to unsaturated carbon atom are able to transfer electrons by a mechanism similar to electromeric effect. Example,

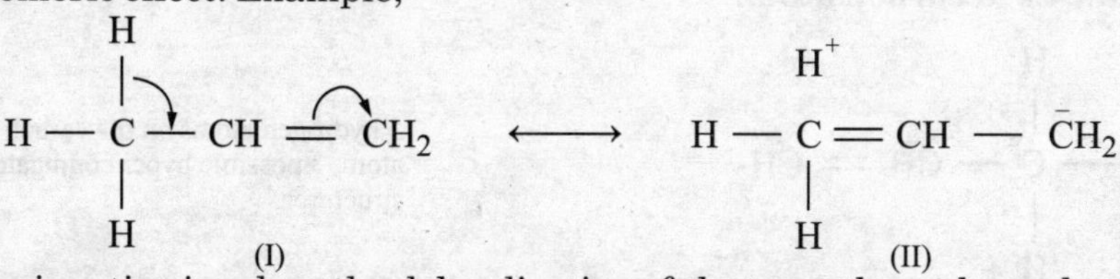

(I) (II)

Thus, hyperconjugation involves the delocalisation of electrons through overlapping of π orbitals of C = C double bond with σ orbital of the adjacent C – H bond in unsaturated system. This is also known as σ – π conjugation.

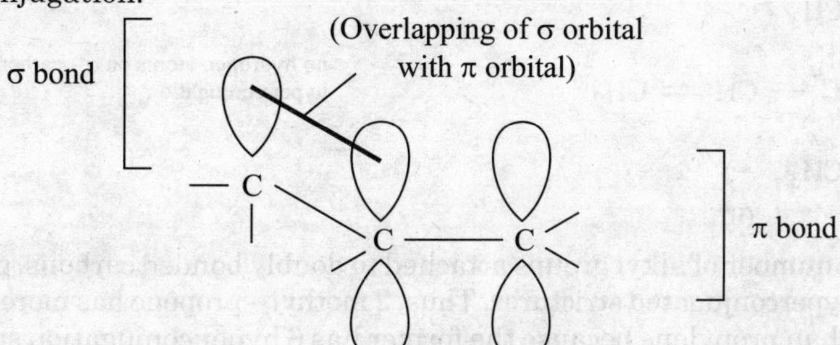

Orbital Representation of Hyperconjugation showing overlap of σ orbital with π orbital.

In fact, hyperconjugation effect is an extension of the mesomeric effect. The difference is that in mesomeric effect the delocalisation of π – electron of two adjacent double bonds or π electrons of a double bond and lone pair of heteroatom takes place, whereas in hyperconjugation the delocalisation of σ electrons takes place through overlapping of *p-orbitals of double bond* with σ – *orbital of the adjacent single bond.*

The polypropylene can have the following hyperconjugated structures in all:

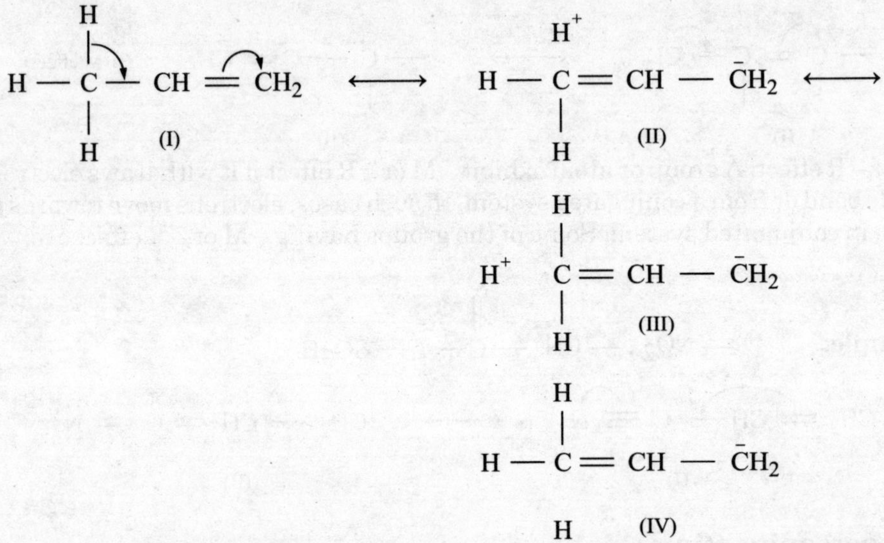

Structures II, III and IV are called contributing hyper conjugative structures. Since there is no real bond between C and one of hydrogens in these structures, hyperconjugation is also called *no bond resonance.* However, this proton is still firmly bond to the π-cloud and is not free to move.

Characteristics

(*i*) Hyperconjugation occurs through the H-atoms present on the carbon atom next to the double bond, *i.e.* α carbon atom.

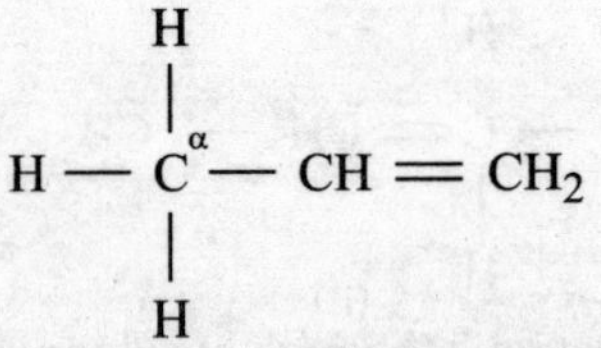

3 hydrogen atoms on α - carbon atom, 3 possible hyper conjugated structures.

(I)

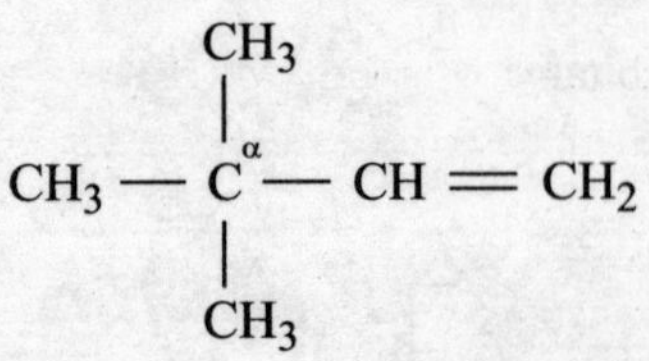

no hydrogen atoms on α - carbon, hence no hyper conjugation

(II)

(*ii*) Greater the number of alkyl groups attached to doubly bonded carbons, greater will be the number of hyperconjugated strictures. Thus, 2 methyl – propene has more hyper conjugated structures than propylene because the former has 6 hyper conjugation structures.

Compound	*No. of H on α carbon*	*Hyper conjugative structure*
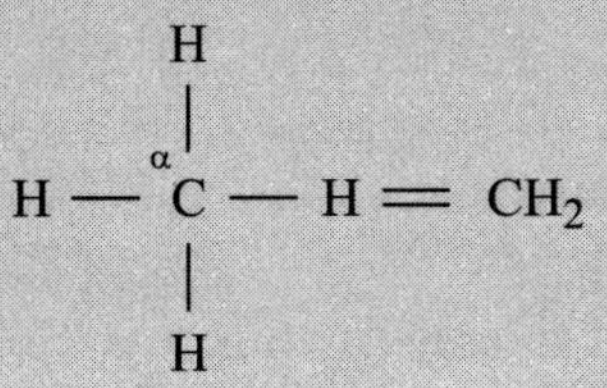 Propylene	3	3
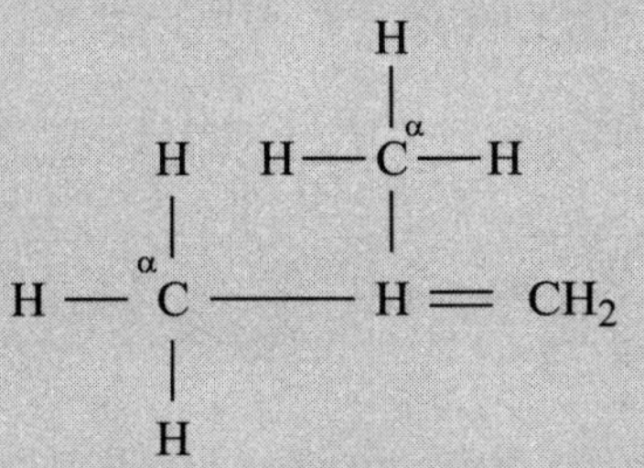 2 - methyl propylene	6	6

(*iii*) Larger is the number of hyper conjugative structures, greater will be stability of that substance. Thus, 2 methyl propene and butene – 2 ($CH_3CH = CH\ CH_3$) should be more stable than propene and this is actually so.

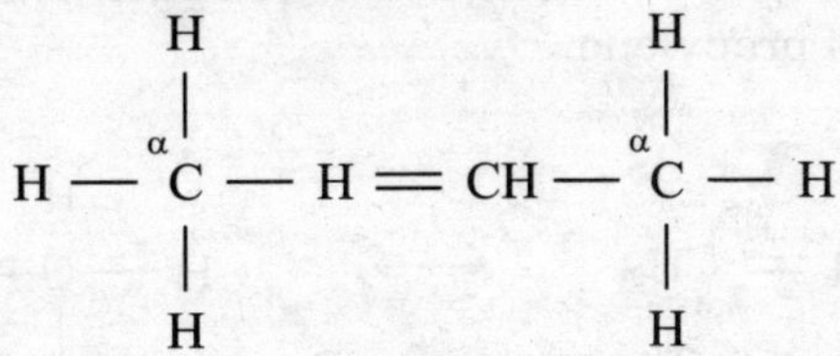

(6 hydrogen atoms on a carbon and hence 6 hyperconjugative structures)

Butene - 2

It may be mentioned here that the stabilising influence in case of hyper conjugative structures is not as large as in case of resonating structures.

Applications of Hyperconjugation

(*i*) *Stability of Alkenes:* Stability of certain alkenes over others can be explained by hyperconjugation. It has been found experimentally that greater is the number of alkyl groups attached to the α – carbon greater is the stability of alkene. It can be explained as follows: More is the number of hydrogen attached to α – carbon, more will be the hyperconjugative structures and more will be the stability. This is illustrated by taking following examples,

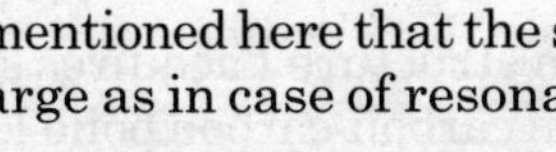

But - 2 - ene
6 α hydrogens
6 hyper conjugatine

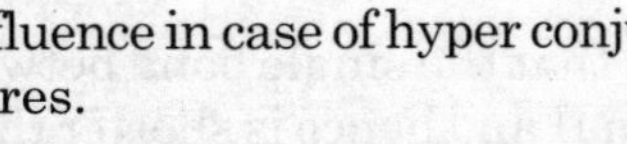

2, 3 Dimethyl but -2-ene
12 α-hydragens
12 hyper conjugatine structure

Thus 2, 3 Dimethyl but – 2 – ene is more stable than But – 2 – ene.
Similarly, propene is more stable than ethene.

$H_2C = CH_2$ $H_3C - CH = CH_2$ (Propene)
No α hydrogen 3 α hydrogen
No hyper conjugative structure 3 hyper conjugative structures
(more stable)

(ii) **Directive influence of Alkyl benzenes:** In electrophilic substitution reactions of alkylbenzene, the presence of alkyl group direct the incoming electrophile at ortho and para position. This can be explained by hyper conjugation.

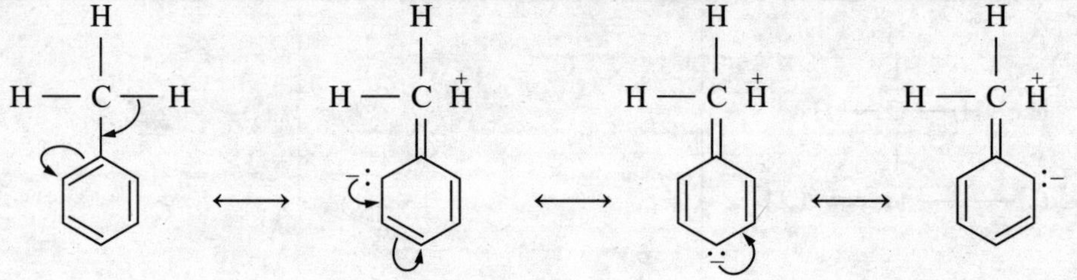

Similarly, six more structures can be drawn as two more α – *H* are present on alkyl group. From all there structures, it is evident that electron density is maximum either in ortho or para position.

Thus at a result of hyper conjugation, the electron density at o and p-position with respect or alkyl group increases and hence electrophile substitution reactions in alkyl substituted benzene will occur at o-and p-position. *Thus alkyl group are ortho and para directing.*

(iii) **Shortening of C – C single bond adjacent to multiple bonds:**- Let us illustrate this by taking the example of propylene.

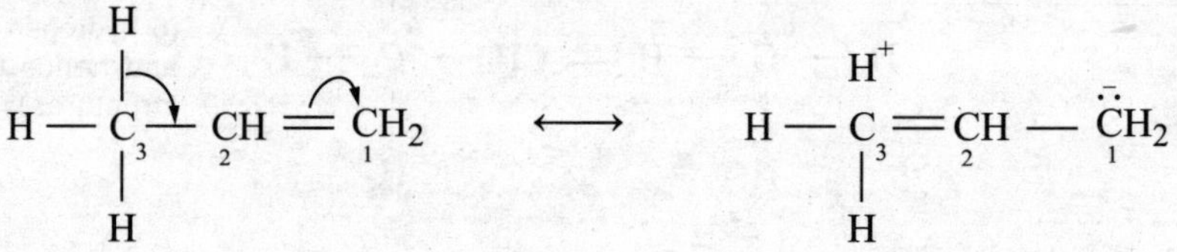

(I) (II)

We can see that the single bond between $C_2 - C_3$ in structure I acquires some double bond character in II and hence is shorter than the normal carbon-carbon bond length in propane.

(iv) **Stability of carbocations:** This is discussed in detail under the section reactive intermediates.

9.5 REACTIVE INTERMEDIATES

As discussed in previous section, homolytic or heterocyclic cleavage bonds leads to the formation of reactive organic species such as free radicals, carbocations, carbanions, etc. Most of the organic reactions occur through the involvement of these chemical species. These are generally short lived (10^{-6} seconds to a few seconds) and highly reactive and are called *reactive intermediate.*

Note: Reactive intermediates may not be confused with transition states. Reaction intermediates are stable enough to be isolated for a short period of time (few second) under appropriate condition whereas transition state refers to an imaginary state and cannot be isolated.

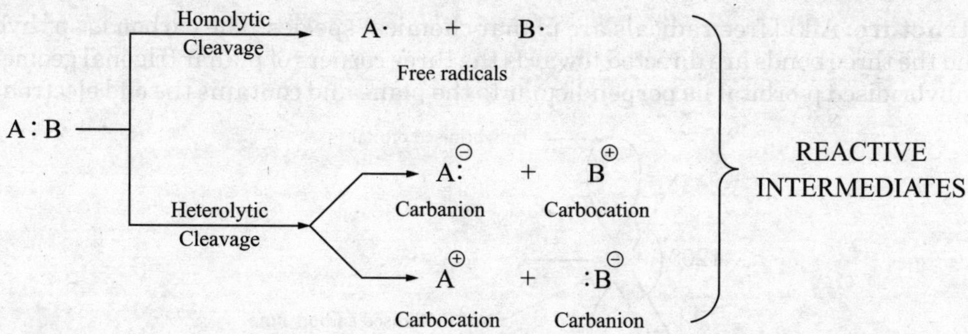

The three types of intermediates *viz*; free radicals, carbocations and carbanions are discussed below.

9.5.1 Free Radicals

A free radical may be defined as an atom or group of atoms having an odd or unpaired electron. Free radicals are formed due to homolytic cleavage of a covalent bond.

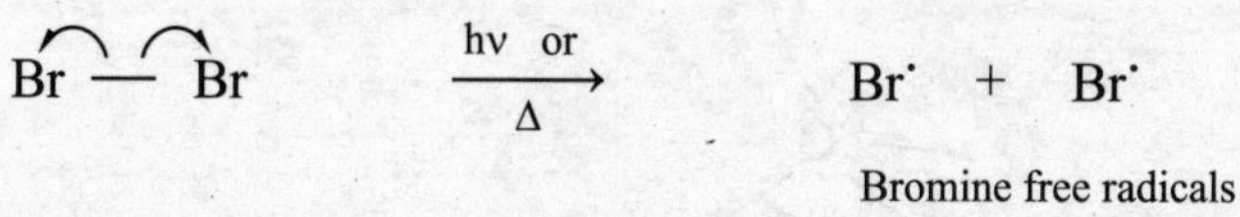

Free radicals are electrically neutral species. Due to the presence of odd electron these are *paramagnetic* in nature. Free radicals are highly reactive specie because of its tendency to gain an additional electron and complete its octet.

(*i*) **Generation:** Free radicals are generally formed in the reactions carried out at high temperature or under the influence of *high energy radiation* Moreover, the presence of weak bond also help in the formation of free radicals.

- The molecules containing bonds such as: O—O, —C—N, —C—S etc produce free radicals at high temperatures. Examples,

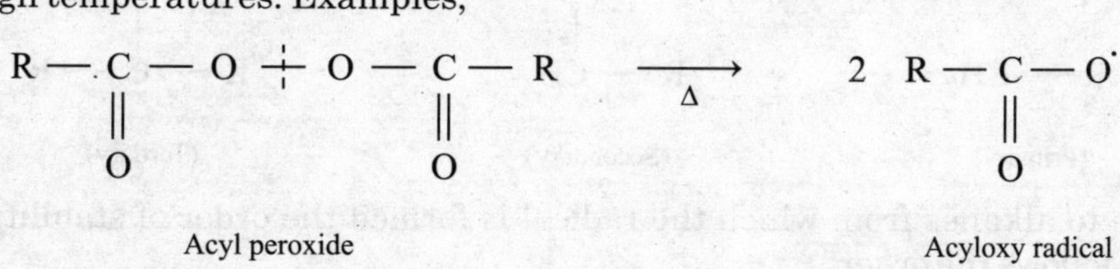

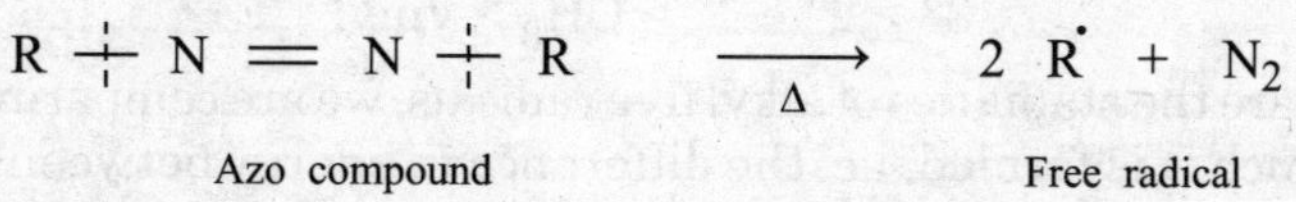

- Free radicals are produced by *high energy radiations* in photo chemical reactions. Example,

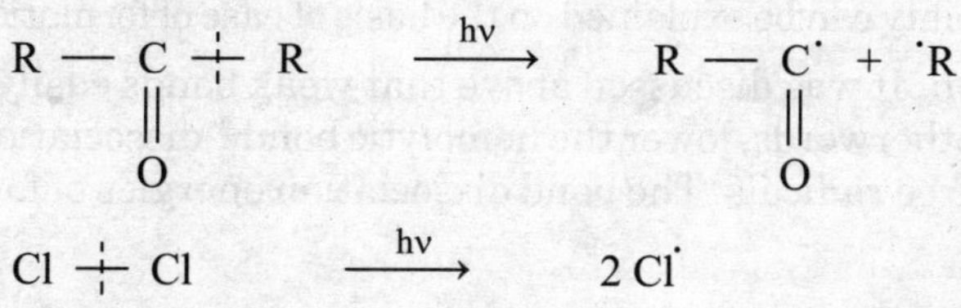

(*ii*) **Structure:** Alkyl free radicals are planar chemical species. The carbon is sp^2 hybridised and the three bonds are directed towards the three corners of planar trigonal geometry. The unhybridised p-orbital lie perpendicular to the plane and contains the odd electron.

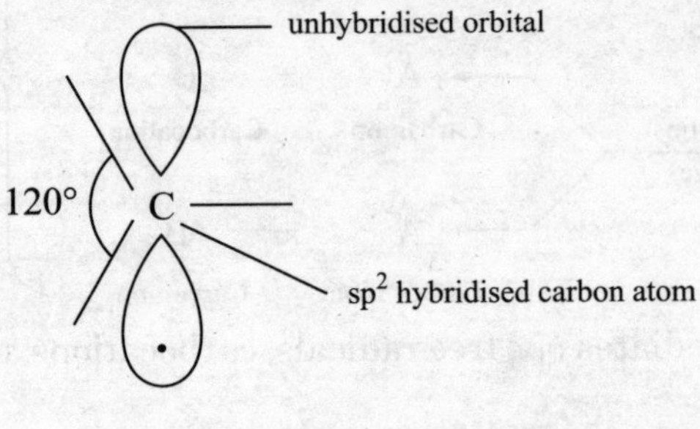

Fig. 9.1 Orbital structure of free radical planar triagonal

Sometimes, free radical may have pyramidal geometry also with sp^3 hybridised carbon.

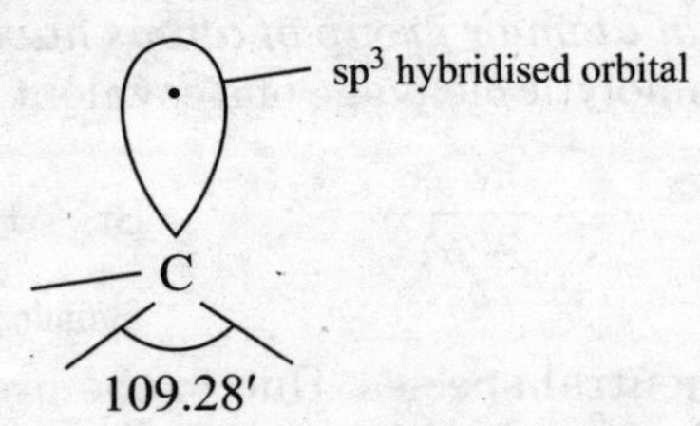

Fig. 9.2 Orbital structure of free radical – pyramidal

However, CH_3 and other simple alkyl radicals predeminantly have planar geometry which is supported by experimental proofs.

(*iii*) **Stability:** Free radicals are classified as primary, secondary or tertiary, according to the carbon atom carrying the free radical is primary, secondary and tertiary respectively.

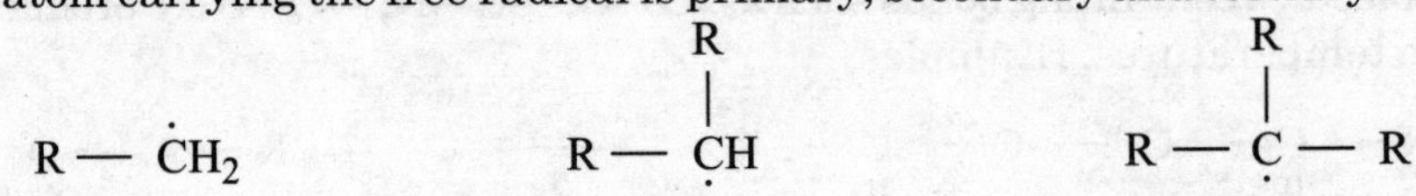

(Primary) 1°   (Secondary) 2°   (Tertiary) 3°

Relative to alkanes from which the radical is formed the order of stability of alkyl free radicals follows the order:

$$3° > 2° > 1° > CH_3 > \text{vinyl}$$

Note: When we compare the stabilities of alkyl free radicals, we are comparing it with the reference to the alkane form which it is formed, i.e. the difference in energy between methane and methyl radical is more than between ethane and ethyl radical (ref: Morrison and Boyd, Organic chemistry, Pearson).

The above order of stability can be explained on the basis of ease of formation and hyperconjugation.

(*a*) **Ease of formation:** It was discussed above that weak bonds easily cleave to homolytically to form free radicals. In otherwords, lower the homolytic bond * dissociation energy (ΔH) greater is the ease of formation of free radicals. The bond dissociation energies of formation for 1°, 2° and 3° alkyl radicals are:

***Note:** In this case, bond dissociation energy is the amount of energy that must be supplied to convert a mole of alkane into radicals and hydrogen atoms.

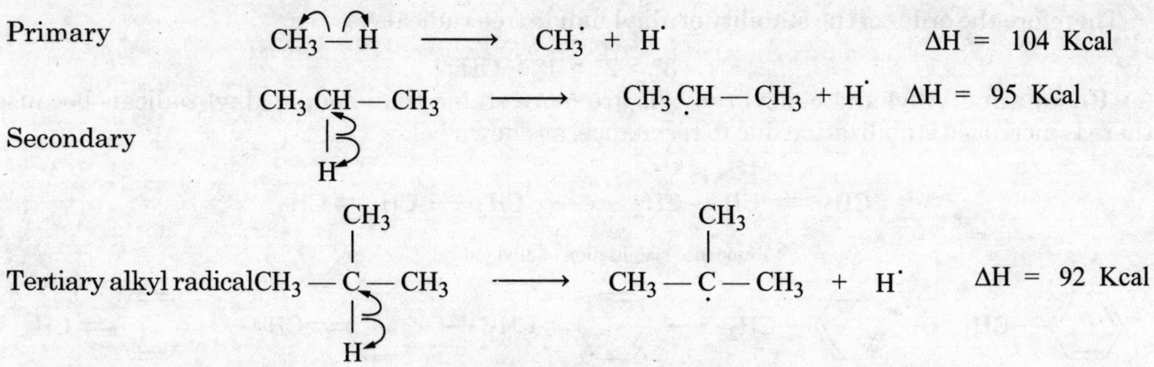

Therefore, the ease of formation of free radicals follows the order,

$$3° > 2° > 1° > CH_3°$$

More stable is the radical, the more rapidly it is formed. Hence the stability of alkyl radicals also follows the same order.

(*b*) **Hyperconjugation:-** The relative stabilities of 3°, 2° and 1° alkyl radicals can be accounted for on the basis of hyper conjugation. The was discussed in section 9.3 that more is the number of hyperconjugative (no-bond) structures, more will be the resonance energy and greater will be the stability of free radicals.

Ethyl radical has 3 hyperconjugative structures whereas isopropyl radical has 6 and t-butyl radical has 9 hyperconjugative structures as shown:

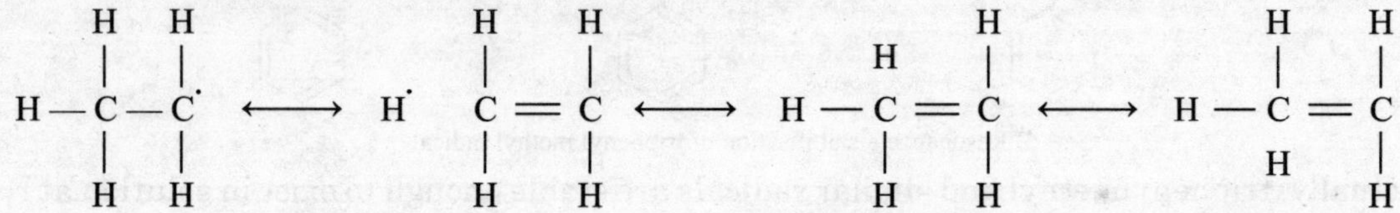

Ethyl radical — Hyperconjugative Structures (three)

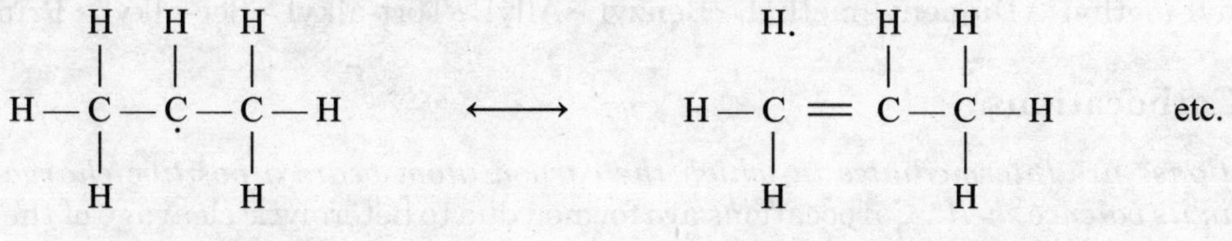

Isopropyl radical — 6 hyper conjugative structures are possible due to 6 – α hydrogen atoms.

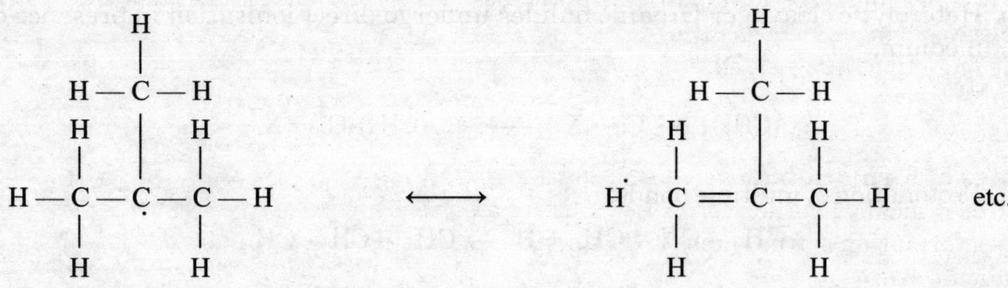

Ter - butyl radical — 9 - hyperconjugative structures possible due to 9 α - hydrogen atoms.

Therefore the order of the stability of alkyl halide free radicals is

$$3° > 2° > 1° > CH_3.$$

(*c*) **Resonance:** Allyl and benzyl radicals are more stable than simple alkyl radicals because there is increased stabilization due to resonance, as shown below:

Resonance-stabilisation of allyl radical

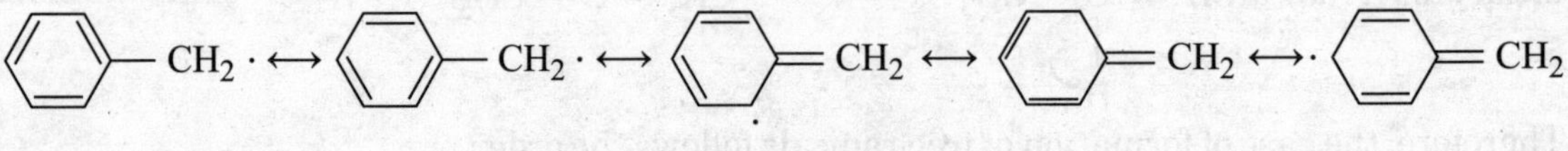

Resonance-stabilisation of benzyl radical

Diphenylmethyl and triphenylmethyl radicals, in which resonance stabilisation is much more marked should show very high stability.

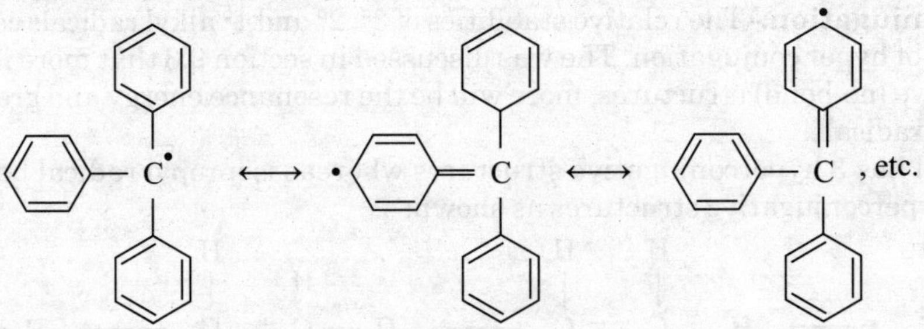

Resonance – stabilisation of triphenyl methyl radical

Actually, triphenylmethyl and similar radicals are stable enough to exist in solution at room temperature and hence can be isolated.

To sum up the decreasing order of stability of radicals mentioned above is:

Triphenyl-methyl > Diphenyl-methyl > Benzyl > Allyl > Tert-alkyl > Sec-alkyl > Prim-alkyl

9.5.2 Carbocations

"Carbocations are intermediates in which the carbon atom bears a positive charge with six electrons in its valence shell".* Carbocations are formed due to heterolytic cleavage of the covalent bond in which leaving group (more electronegative) takes away with it the shared pair of electrons.

(*i*) **Generation:** Carbocations are produced in one of the following ways:

(*a*) Heterolytic cleavage: Organic halides undergo direct ionisation in presence of polar medium.

$$(CH_3)_3 - C - X \longrightarrow (CH_3)_3C^+ + X^-$$

(*b*) Protonation of multiple bonds:

$$CH_3 - CH = CH_2 + H^+ \rightarrow CH_3 - \overset{+}{C}H - CH_3$$

---

*Note: Sometimes the alternative names such as carbonium ions or carbenium ion are used sepecially in older literature, for carbocations. But the term carbocation is most appropriate.

Carbocations are very reactive intermediates due to the electron deficient carbon atom which has only six electrons in its valence shell and thus have strong tendency to complete its octet.

(*ii*) **Structure:** Carbocations are planar chemical species. The carbon atom carrying positive charge is sp^2 hybridised and is bonded to three other atoms through its hybridised orbitals forming an angle of 120°. The unhybridised orbital, lying perpendicular to the plane of hybridised orbitals is empty.

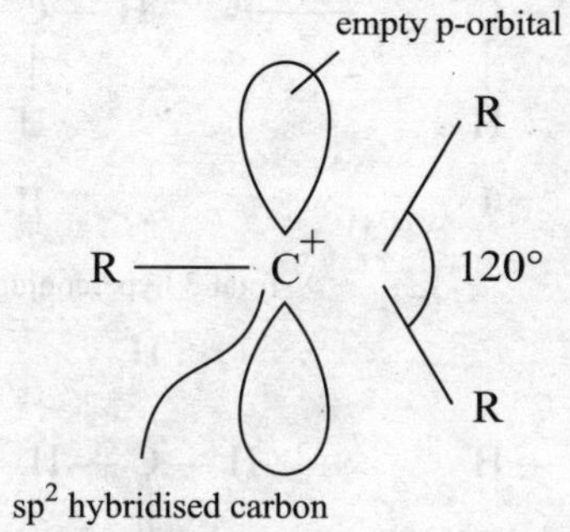

Fig. 9.3 Orbital Structure of Carbocation.

(*iii*) **Stability:** Depending upon the carbon which bears the positive charge, the carbocations can be classified as primary, secondary or tertiary carbocations.

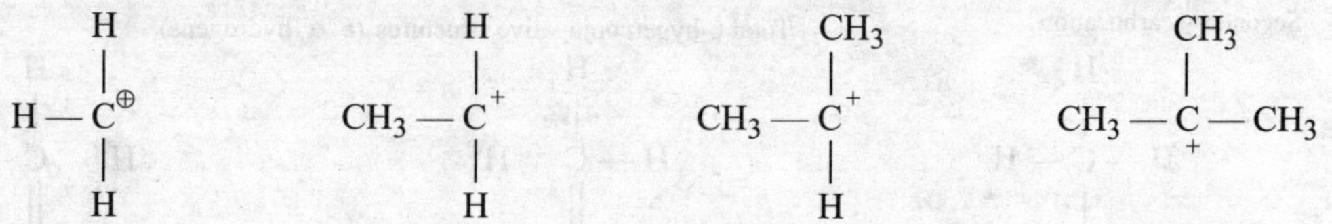

Methyl Carbocation | Primary carbocation | Secondary carbocation | Tertiary carbocation

The stability of carbocation generally follows the order:

$$3° > 2° > 1° > CH_3$$

This order of stability can easily be explained on the basis of inductive effect, hyerconjugation and resonance.

(*a*) *Inductive effect (+I):* Since carbocations are charged species, inductive effect has great influence on their stability The + I effect of alkyl groups decreases the net charge on the carbon and hence increases its stability. In otherwords, the positive charge on carbon is dispersed by alkyl groups. Greater is the number of alkyl groups on carbon atom bearing the positive charge, greater is the dispersal of charge and greater would be the stability.

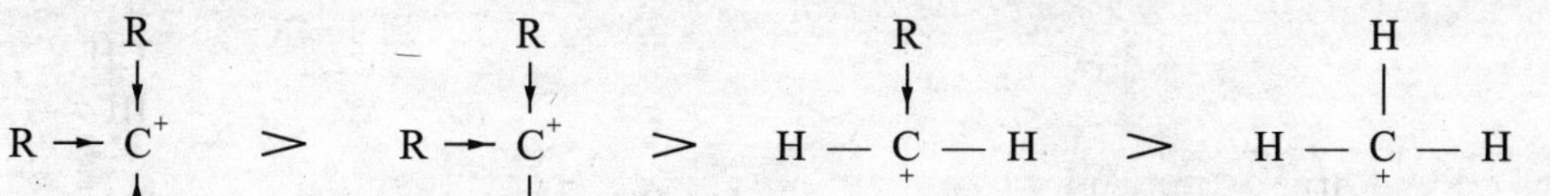

3° 2° 1° Methyl

(*b*) *Hyperconjugation:* Stability of alkyl carbocations can also be explained by hyperconjugation similar to the stability of alkyl free radicals. The + I effect and hyperconjugation in tertiary carbocations are responsible for its greater stability than secondary and primary.

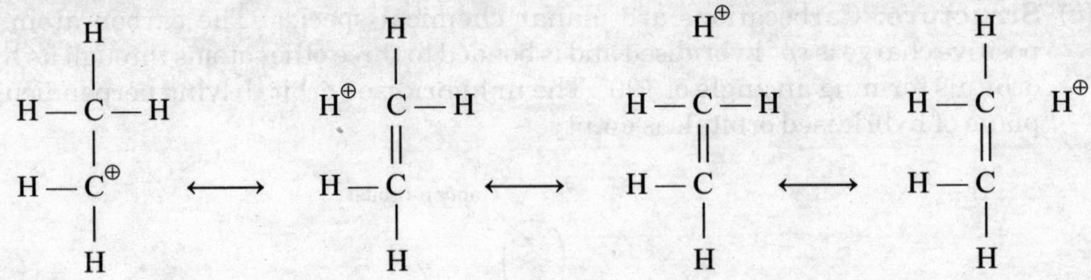

I II III

Primary carbocation

Total-3 hyperconjugative structures

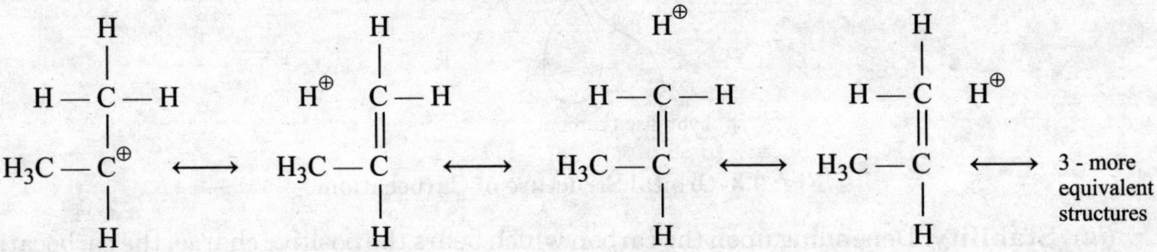

Isopropyl cation
Secondary carbocation

Total 6-hyperconjugative structures (6 α hydrogens)

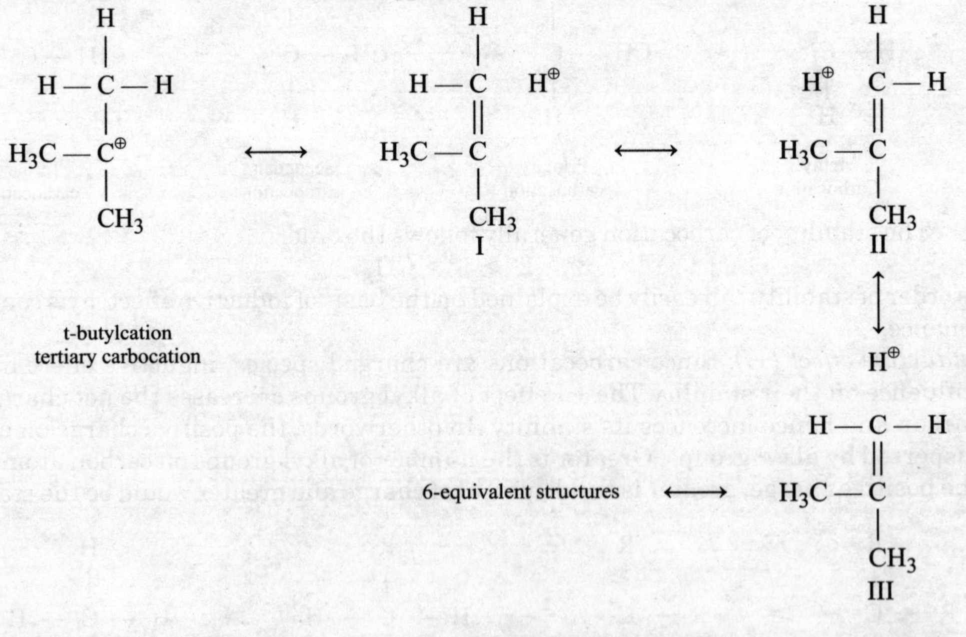

Total 9-hyperconjugative structures (9 α hydrogens)

Tertiary carbocation has 9, secondary has 6 and primary has 3 hyper conjugative structure. Hence, the order of stability is

$$3° > 2° > 1° > CH_3.$$

(*c*) *Resonance:* The stability of benzyl and allyl carbocation is explained on the basis of resonance as shown below:

Allyl cation: It is stabilised through resonance, as:

$$\overset{}{CH_2} = CH - \overset{\oplus}{CH_2} \longleftrightarrow \overset{\oplus}{CH_2} - CH = CH_2 \left(\equiv \overbrace{CH_2 \overset{---}{} CH \overset{---}{} CH_2}^{\oplus} \right)$$

Resonance - stabilisation of Allyl cation

Benzyl cation: This also shows increased stability due to resonance, as shown below:

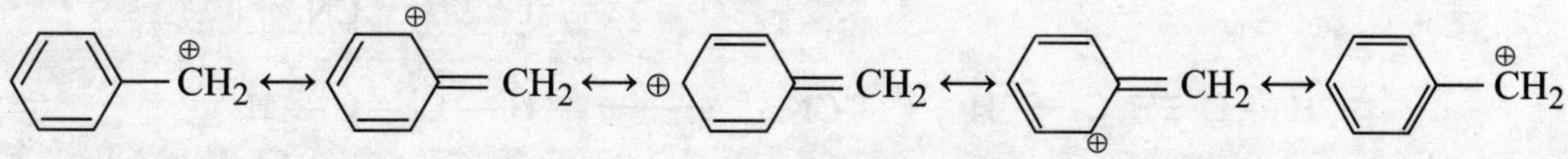

Resonance - stabilisation of Benzyl cation

Triphenyl methylcation: This is among the most stable carbonium ions known. Its remarkable stability is due to *extensive dispersal of the positive charge*, as depicted below:

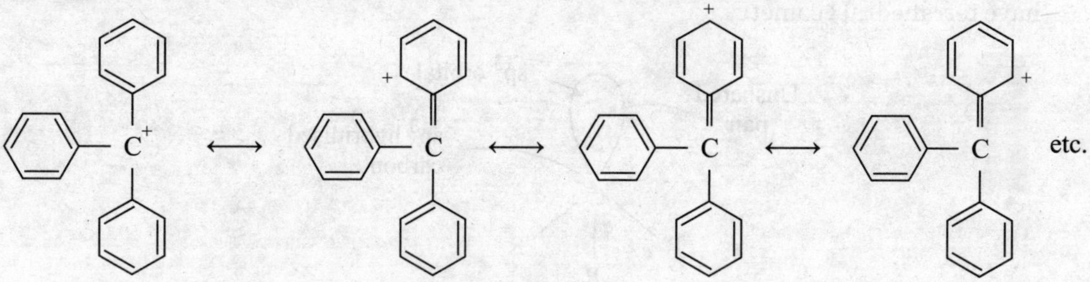

(Six more resonating structures are possible, which would disperse the positive change)

The order of stability of the above three ions is
Triphenyl methyl carbocation > Benzyl carbocation > allyl carbocation
These carbocations are more stable than alkyl carbocations.

9.5.3 Carbanions

"Carbanions are the reactive intermediates in which carbon bears a negative charge and possesses an unshared pair of electrons. It has eight electrons in its valence shell".

Carbanions are reactive intermediates and behave as Lewis bases. These are too unstable to be isolated.

(i) Generation:

- Heterolytic cleavage of covalent bond in which carbon is attached to less electronegative element.

$$CH_3 - \overset{O}{\overset{\|}{C}} - CH_2 - \vdots H \xrightarrow{NaO\,C_2H_5} CH_3 - \overset{O}{\overset{\|}{C}} - \overset{\bar{..}}{C}H_2 + H^+$$

$$H - \vdots CH_2 - CHO \xrightarrow{\bar{O}H} \overset{..}{C}H_2 - CHO + H_2O$$

Acetaldehyde

- Nucleophilic attack on a carbon multiple bond.

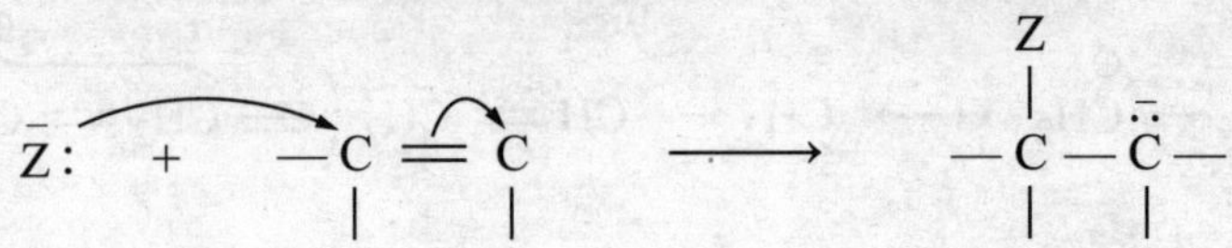

Nucleophile

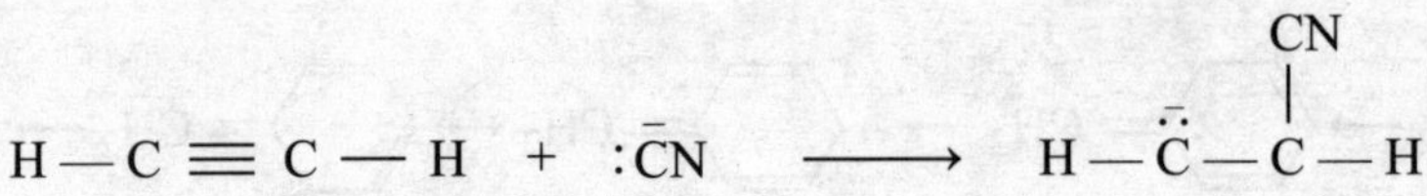

Eihyne

(*ii*) **Structure:** Carbocations have pyramidal structure. The carbon carrying the negative charge is sp^3 hybridised. The three of four sp^3 hybridised orbitals form three σ bonds with other groups whereas the fourth one occupies the lone pair of electrons. Thus carbanions have tetrahedral geometry.

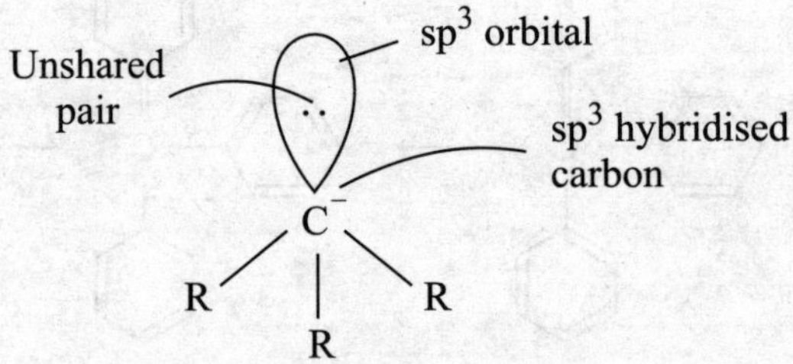

Fig. 9.4 *Orbital Structure of Carbanion*

(*iii*) **Stability:** The order of stability of carbanions is as follows:

$$\bar{C}H_3 > 1° > 2° > 3°.$$

The stability can be explained due to following effects:

(*a*) **Inductive effect:** The order is readily understood if we recall that alkyl groups are electron donating (+I effect) and hence they will intensify the negative charge on carbon, making the anion less stable. Greater the number of alkyl groups, greater would be the negative charge density at central carbon atom of carbanion and lesser would be the stability

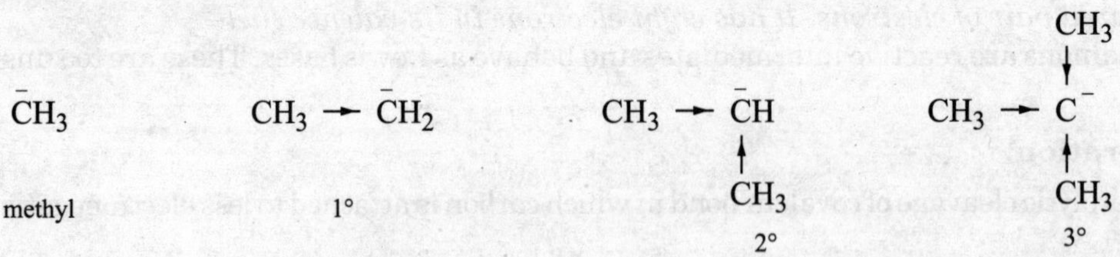

Hence $\bar{C}H_3$ is most stable followed by 1°, 2° and Tertiary.

(*b*) *s – Character:* The stability of carbanions derived from alkynes, alkenes and alkanes can be explained on the basis of their *s* character.

$HC \equiv \bar{C}$	$H_2C = \bar{C}H$	$H_3C - \bar{C}H_2$
(sp hybridised)	(sp^2 hybridised)	(sp^3 hybridised)
s character 50%	33%	25%

Higher the *s* character, closer the electrons to the nucleus and hence lower the energy, (more stability). Thus the order of relative stabilities for the above carbanions is:

$$HC \equiv \bar{C} > H_2C = \bar{C}H > H_3C - \bar{C}H_2$$

(*c*) *Resonance:* The stability of the carbanions in which the unshared pair is in conjugation with unsaturated bond is due to resonance. Example

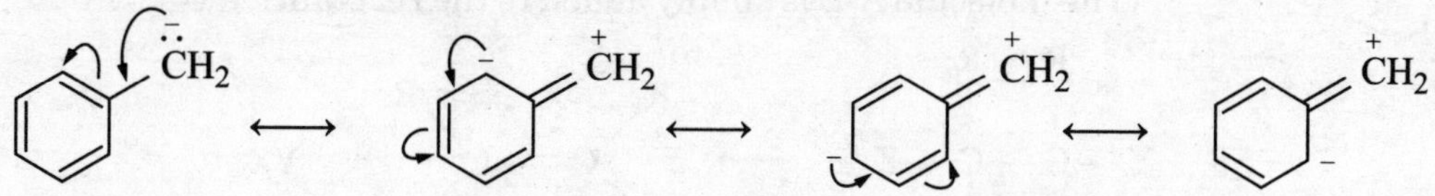

9.5.4 Carbenes

Carbenes are the reactive intermediates in which carbon has six electrons, two of which are free. These are neutral species. Electron deficient specie

The simplest example of carbene is methylene.

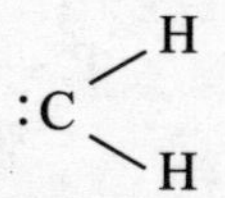

Methylene

(*i*) **Generation:-** Carbenes are produced by photolytic cleavage

$$CH_2{=}C{=}O \xrightarrow{h\nu} :CH_2 + CO$$

$$CH_2{=}\overset{+}{N}{=}\overset{-}{N} \xrightarrow{h\nu} :CH_2 + N_2$$

(*ii*) **Forms:** Carbene has two forms:

Singlet carbene: The unshared electron pair in carbene remains paired 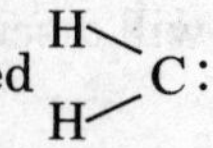

Triplet carbene: The unshared electron pair is unpaired, i.e. it is diradical

$$H - \dot{\underset{\cdot}{C}} - H\,.$$

9.6 TYPES OF ORGANIC REACTIONS

Organic reactions are broadly classified as:

(*i*) Substitution Reactions
(*ii*) Addition Reactions
(*iii*) Elimination Reactions
(*iv*) Rearrangement Reactions.

This classification is solely based on the structural changes occurring in the reactant molecule as shown below:

$$R-\underset{R}{\overset{R}{\underset{|}{\overset{|}{C}}}}-Y + Z \longrightarrow R-\underset{R}{\overset{R}{\underset{|}{\overset{|}{C}}}}-Z + Y$$

SUBSTITUTION

(The group/atom *Y* is substitued by *Z*)

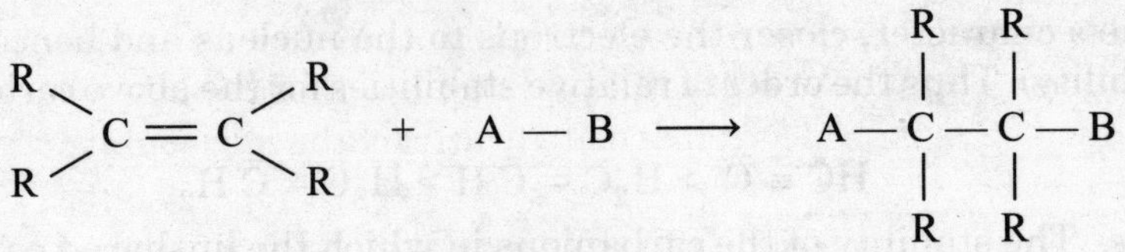

ADDITION

(The molecule A-B is simply added to the reactant)

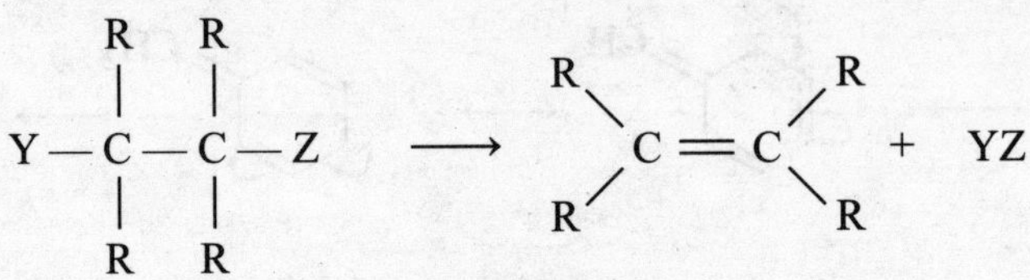

ELIMINATION

(The molecule/group Y-Z is eliminated from the reactant)

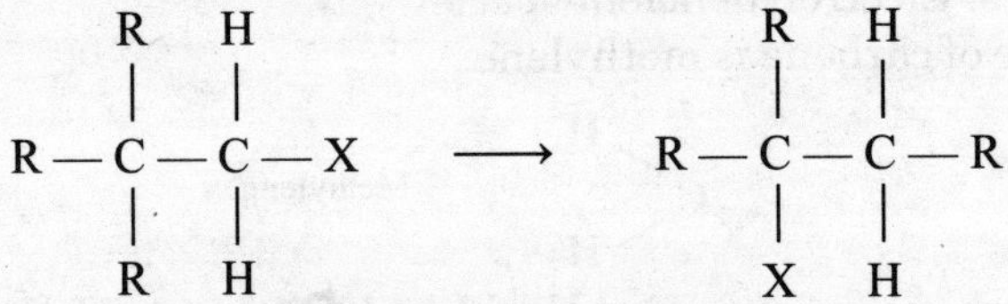

REARRANGEMENT

(The reactant molecule is simple rearranged)

Chemical reactions are commonly represented as equations such as:

$$\text{Reactant} + \text{Reagent} \xrightarrow[\text{Conditions}]{\text{Reaction}} \text{Products}$$

Now, we will discuss the above four classes briefly.

9.6.1 Substitution Reactions

The substitution reaction or displacement reactions are those reactions in which atoms or group of atoms attached to a carbon atom in a substrate molecule is replaced by another atoms or group of atoms.

The substitution reactions can take place through different attacking species (nucleophile, electrophile or a free radical). The substitution reactions are of the following types:

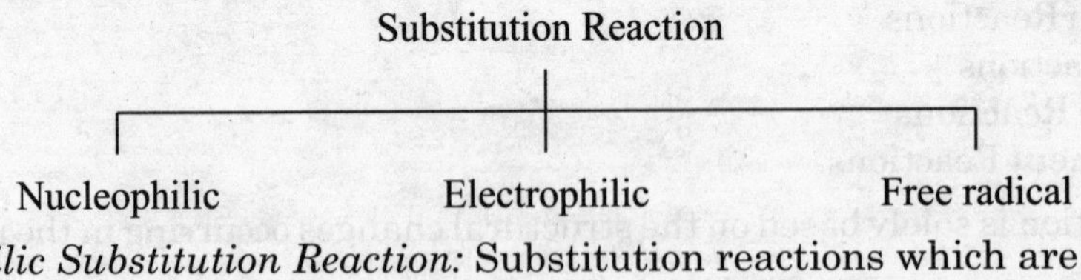

(i) Nucleophilic Substitution Reaction: Substitution reactions which are brought about by nucleophiles are called nucleophilic substitution reactions. There are denoted by SN reaction (S = substitution, N = Nucleophile). These reactions are typical of alkyl halides. Example

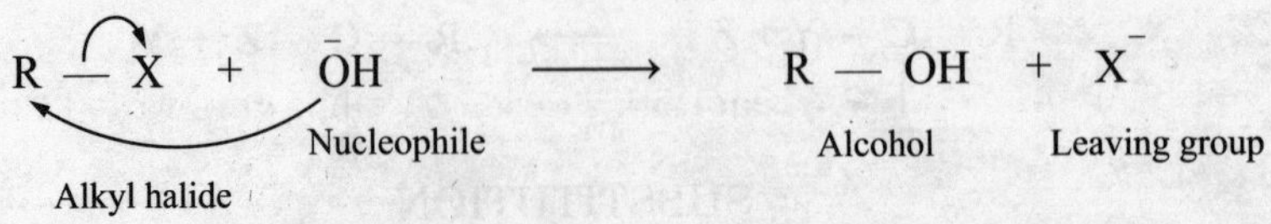

The mechanism of Nucleophilic substitution reactions is discussed in detail in the next section.

(ii) Electrophilic Substitution Reactions: Substitution reactions which are brought by electrophiles are called electrophilic substitution reaction. Electrophile systems are typical of aromatic systems which have high electron density to attract the nucleophile. Example, nitration of benzene ring

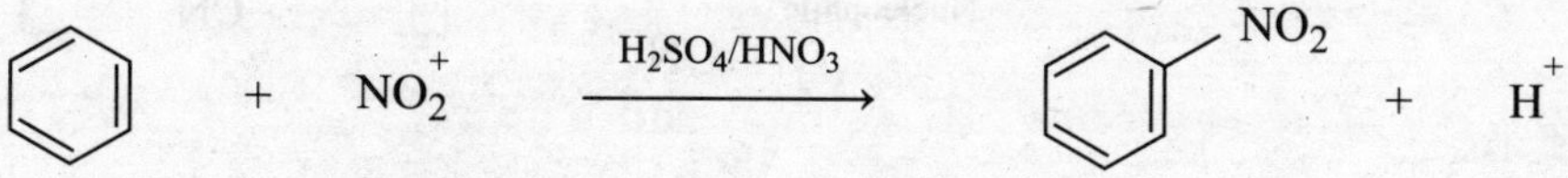

Electrophile

(iii) Free Radical Substitution Reaction: In this type of substitution reaction, the substitution is brought about by free radicals. For example, chlorination of alkanes in the presence of sunlight.

$$CH_4 + \dot{C}l \longrightarrow CH_3Cl + HCl$$

Free radical — Methyl chloride

The reaction occurs by free-radicals which are generated in presence of sunlight. The reaction involves three steps.

$$\text{Initiation}: Cl^\bullet + Cl^\bullet \rightarrow Cl_2$$
$$\text{Propagation}: CH_3^\bullet + Cl^\bullet \rightarrow CH_3Cl$$
$$CH_3^\bullet + Cl \rightarrow CH_3Cl$$

Propagation steps keep on producing reactive free radicals by chain propagation until the free radicals are terminated in termination step.

Termination. $Cl^\bullet + Cl^\bullet \rightarrow Cl_2$

$CH_3^\bullet + Cl^\bullet \rightarrow CH_3Cl$

$CH_3^\bullet + CH^\bullet \rightarrow CH_3 - CH_3$

9.6.2 Addition Reactions

Addition reactions are the reactions in which the attacking reagent adds up to the reactant (substrate) molecule to give the single molecule of the product. Addition reactions are typical of compounds containing double or triple bonds.

Like substitution reactions, addition reactions can also proceed through different attacking reagents and hence are of 3 types:

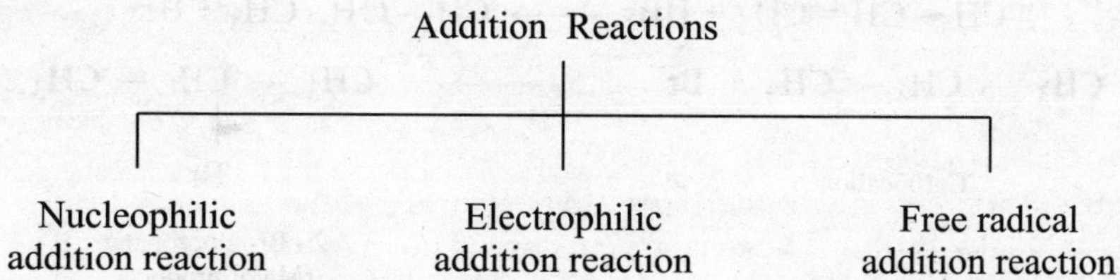

(i) Nucleophile Addition Reactions: When the addition reaction is brought about by nucleophiles, the addition is called nucleohilic addition reaction. These reactions generally take place in case of aldehydes and ketones. For example,

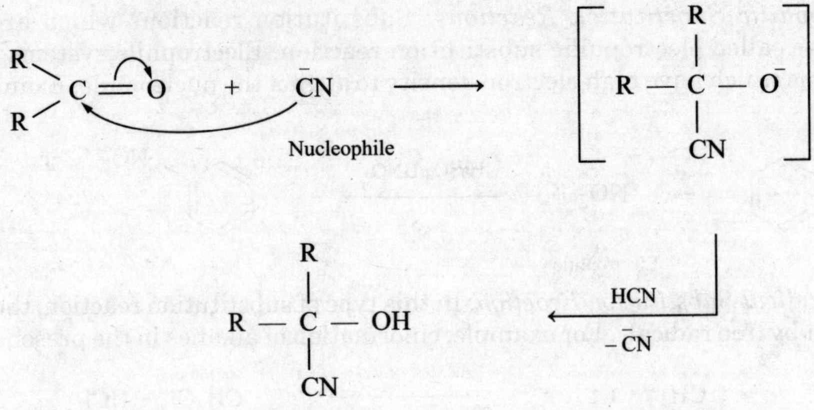

Addition product

(ii) Electrophilic Addition Reactions: Addition reaction brought about by electrophile is called electrophilic addition. Carbocations are formed as reaction intermediates. These reactions are typical of alkenes and alkynes.

$$CH_2 = CH_2 + HBr \longrightarrow CH_3\underset{}{\overset{Br}{\overset{|}{CH}}} - CH_3$$

The reaction takes place in two steps:

$$CH_2 = CH_2 + H^+ \xrightarrow{\text{slow}} \left[\overset{+}{CH} - CH_3\right]$$

Carbocation

$$\overset{+}{CH_2} - CH_3 + Br^- \xrightarrow{\text{fast}} \underset{\underset{Br}{|}}{CH} - CH_3$$

Bromoethane

Electrophilic addition to asymmetrical alkenes occurs according to the 'Markovnikav's Rule'.

Markovnikov's Rule:

If the two carbon atoms at the double bond are linked to a different number of hydrogen atoms, the halogen is found preferentially at the carbon with less hydrogen substituents; This observation is known as Markovnikov's rule.

$$CH_3 - CH = CH_2 + HBr \longrightarrow CH_3 - \overset{+}{C}H - CH_3 + Br^- \quad \text{...(i)}$$

$$\underset{\text{Carbocation}}{CH_3 - \overset{+}{C}H - CH_3} + Br^- \longrightarrow \underset{\text{2 - Bromopropane (Major product)}}{CH_3 - \underset{\underset{Br}{|}}{CH} - CH_3} \quad \text{...(ii)}$$

Markovnikov's rule or Markownikoff's rule is an observation formulated by the Russian chemist Markovnikov in 1870. In chemical reactions found particularly in organic chemisty involving the addition of H—X to the alkene, the rule states that,

"The addition of H-X to an unsymmertrical alkene, the acid hydrogen (H) becomes attached to the carbon with the greatest number of hydrogens, and the halide (X) group becomes attached to the carbon with the fewest hydrogens"

This is due to an intermediate which is produced by the abstraction of a hydrogen atom by the alkene from the acid (HBr) to form the most stable carbocation (i) (relative stability: 3° > 2° > 1° > methyl), as well as generating a Br anion.

(iii) Free Radical Addition Reaction: Addition reactions brought about by free radicals are called free radical addition reactions. Example, free radical addition of *Br_2* to alkene in presence of peroxides.

$$CH_3CH = CH_2 + HBr \xrightarrow{\text{Peroxide}} \underset{\text{n-propyl bromide}}{CH_3 - CH_2 - CH_2Br}$$

It is interesting to note that the free radical addition of Br_2 to propene takes place by *Anti-Markovnokov's rule.* In other words, the addition of hydrogen halide to the unsymmetrical alkene, in presence of peroxide, takes place in such a way that hydrogen gets attached to the carbon of the double bond having lesser number of hydrogen atom. This is also known as *peroxide effect.*

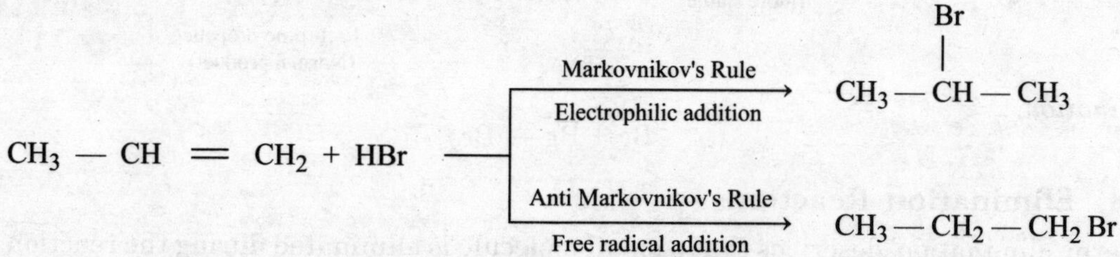

The above selectivity can be explained on the basis of stable intermediates formed. The free radical addition of HBr to unsymmetrical alkenes involves the following steps:

Initiation:

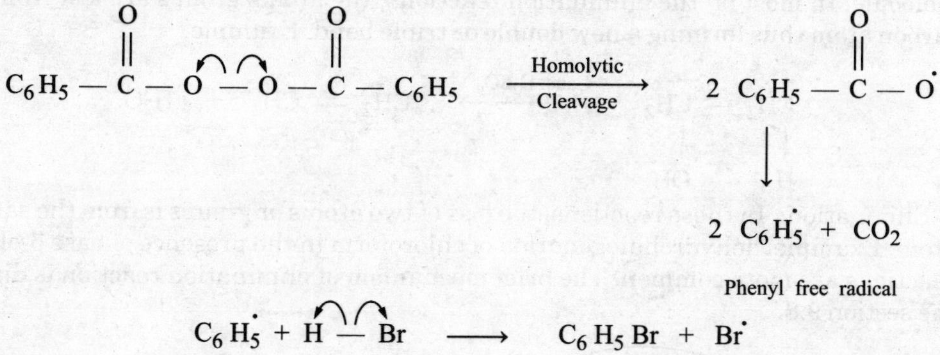

Propagation:

The $\dot{Br}$ free radical then attacks the propene to form primary and secondary free radicals which is chain propagating step.

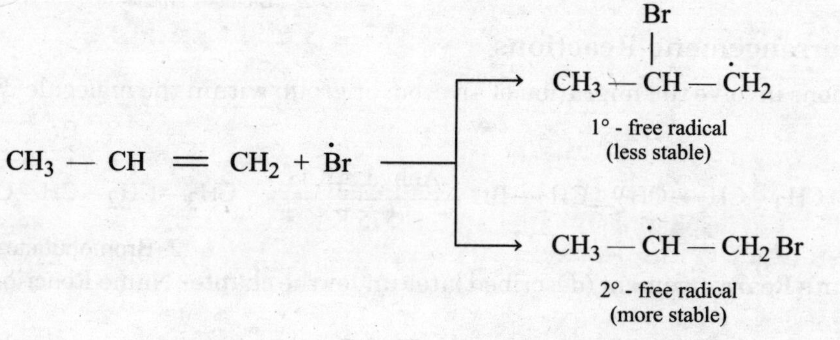

Free radical formed in step II is then attacked by HBr to form the product. Since secondary free radical is more stable than the primary free radical hence the final product is obtained form secondary free radical.

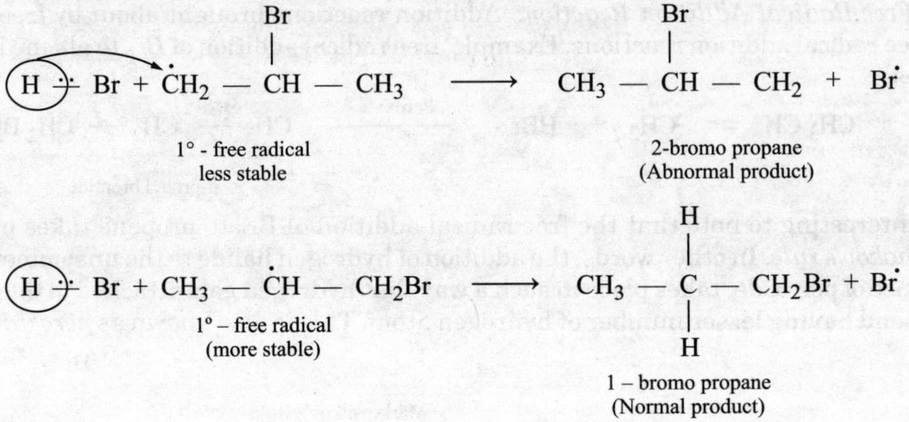

Termination.

$$\dot{Br} + \dot{Br} \rightarrow Br_2$$

9.6.3 Elimination Reactions

The term 'elimination' describes that a small molecule is eliminated during the reaction. These reactions are of two types:

(*i*) β-Elimination: In β elimination reactions, two atoms or groups are lost from the same molecule. In most of the elimination reactions, the atoms groups are lost from adjacent carbon atom thus forming a new double or triple bond. Example,

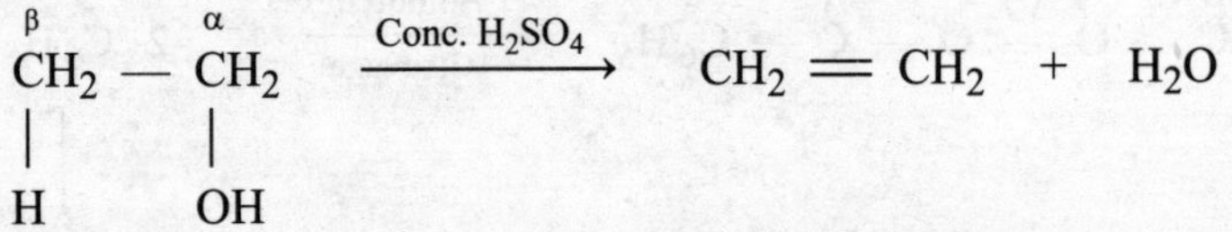

(*ii*) α-Elimination: In these reactions, the loss of two atoms or groups is from the same carbon atom. Example, dehydrohalogenation of chloroform in the presence of base β-elimination reactions are more common. The brief mechanism of elimination reaction is discussed in the section 9.8.

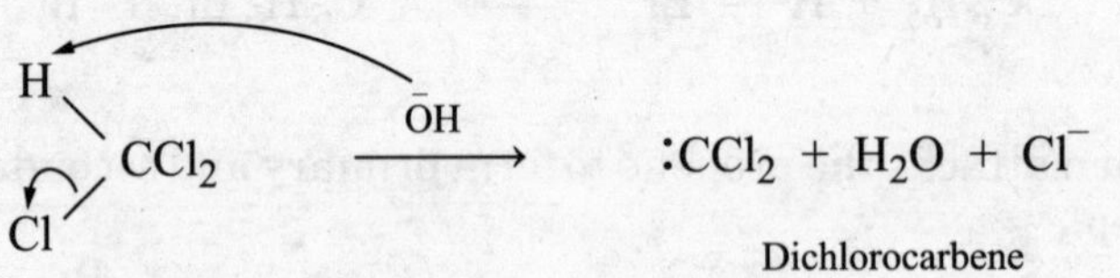

9.6.4 Rearrangement Reactions

These reactions involve the migration of an atom or group within the molecule. For example,

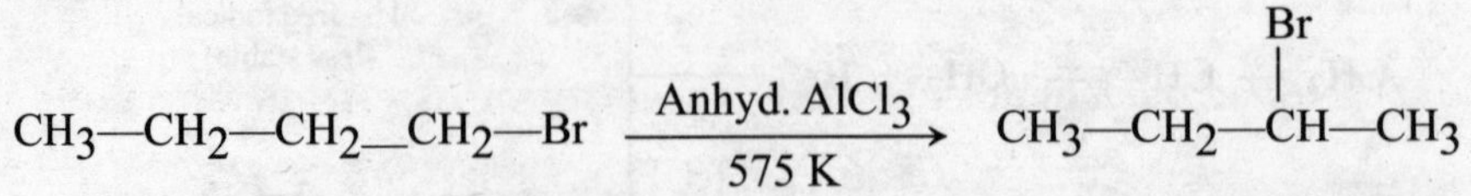

2–Bromobutane

Beckmann's Rearrangement (described later under the chapter Name Reaction)

9.7 NUCLEOPHILIC SUBSTITUTION REACTIONS

It was discussed in the previous section that the substitution reaction brought about by nucleophiles are known as nucleophilic substitution reactions and are denoted by S_N. Nucleophilic substitutions reactions are one of the most useful class of organic reactions.

A substitution implies that one group replace another,

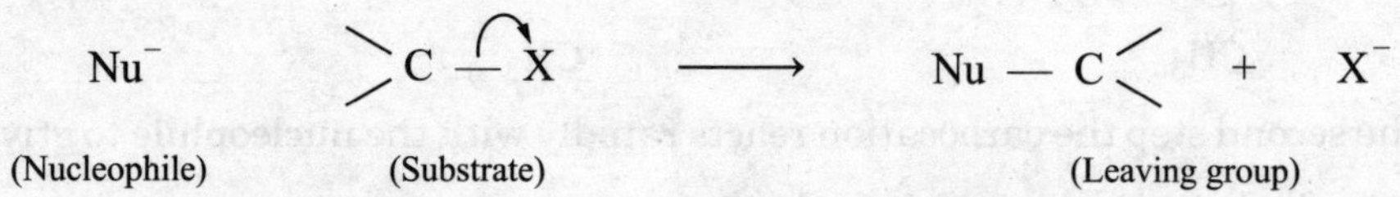

The carbon compound on which the substitution takes place is called the substrate. The substrate is characterised by the presence of leaving group, i.e. the group that leaves carbon taking electron pair with it. There are two fundamental steps taking place during the substitution reactions,

- Bond formation with nucleophile
- Bond breaking to release the leaving group.

Depending upon these two events, two types of mechanism are proposed for nucleophilic substitution reactions:

(*i*) SN_1 Mechanism
(*ii*) SN_2 Mechanism

Let us discues these one by one.

9.7.1 SN_1 Mechanism:

SN_1 indicates a *substitution, nucleophilic unimolecular reaction.*

The components required for nucleophilic substitution are: substrate, nucleophile and the solvent. In SN_1 type of reaction, the rate of the reaction depends upon the concentration of the substrate and is independent of the concentration of the nucleophile, i.e. it is unimolecular reaction. It shows first order kinetics and reaction rate is proportional to the concentration of the substrate

$$\text{Rate} \propto [\text{Substrate}].$$

Example: The reaction between tert-butyl bromide and hydroxide ion to yield tertiary butyl alcohol follows first order reaction.

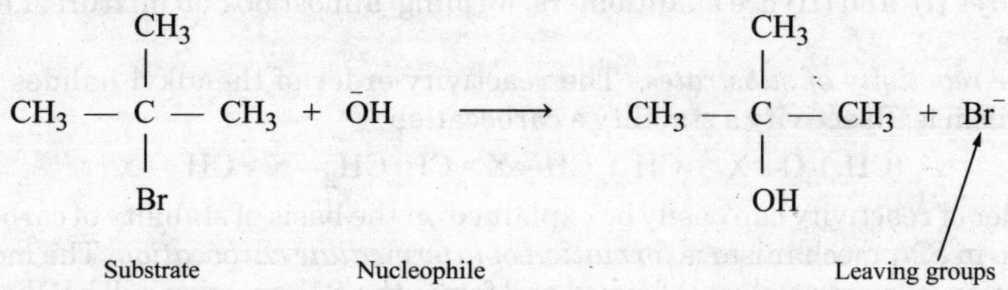

The rate of the reaction depends upon the concentration of tert-butyl bromide only.

$$\text{Rate} = k\,[(CH_3)_3\,C\text{-}Br]$$

Mechanism: The rate law suggest that reaction occurs in two steps:

Step I: In this step, the alkyl halide undergoes heterolytic cleavage to form a bromide ion and a carbocation.

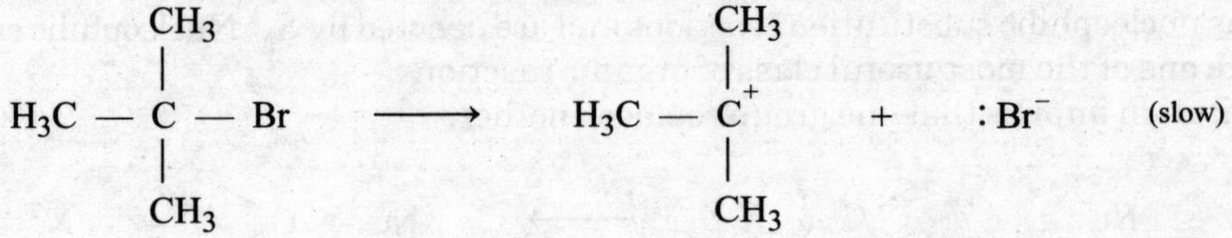

Step II: In the second step the carbocation reacts rapidly with the nucleophile to gtive the product

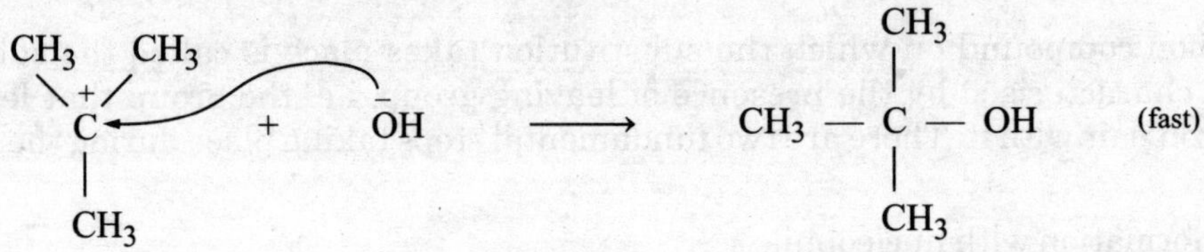

Subsitution product

So, the reaction rate is dependent upon the first step, which is the *slow* step. This mechanism is thus called SN_1 (Substitution, Nucleophilic Unimolecular). The term unimolecular is used here since rate determining step involvers only *one* molecule.

Characteristics of SN_1 Reaction:

(*i*) *Stereochemistry:* We can see that nucleophile can attach from either face of the carbocation, which is a planar molecule. Hence, if the alkyl halide is optically active, then the product will be *racemic mixture.*

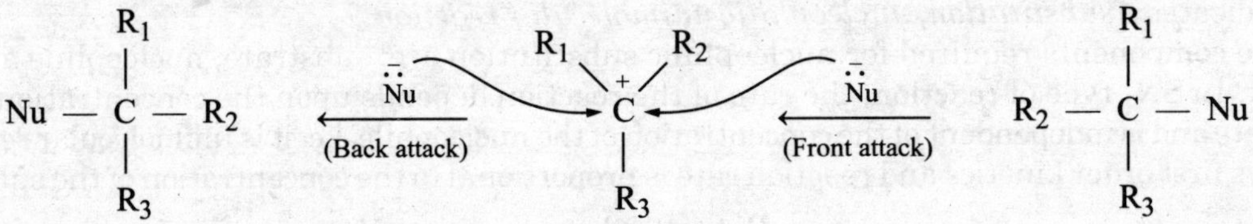

(I) 50% Carbocation (Planar) (II) 50%

Structures (I) and (II) are enantiomers, forming almost 50 : 50 mixture, i.e. the racemic mixture.

(*ii*) *Relative reactivity of substrates:* The reactivity order of the alkyl halides towards SN_1 mechanism is: Reactivity a stability a carbocation

$$(CH_3)_3C—X > (CH_3)_3CH—X > CH_3CH_2—X > CH_3—X$$

The order of reactivity can easily be explained on the basis of stability of carbocations. The key step in SN_1 mechanism is *formation of intermediate carbocation*. The more stable the carbocation is, more easily it is formed and faster the SN_1 reaction will be. We have studied in the previous section that the stability of carbocation follows the order:

$$3° > 2° > 1° > \overset{+}{C}H_3$$

due to inductive and hyperconjugation. Therefore tert-alkyl halides undergo SN_1 reaction mechanism predominantly.

(*iii*) *Energy diagram for SN_1 reaction:* Energy profile for SN_1 reactions:
Multi-step reactions have intermediates and a several transition states (*TS*).
In an SN_1 reaction there is loss of the leaving group which generates an intermediate carbocation which then undergoes a rapid reaction with the nucleophile
It can be seen that the intermediate carbocation has minimum energy w.r.t. transition states (1) and (2).

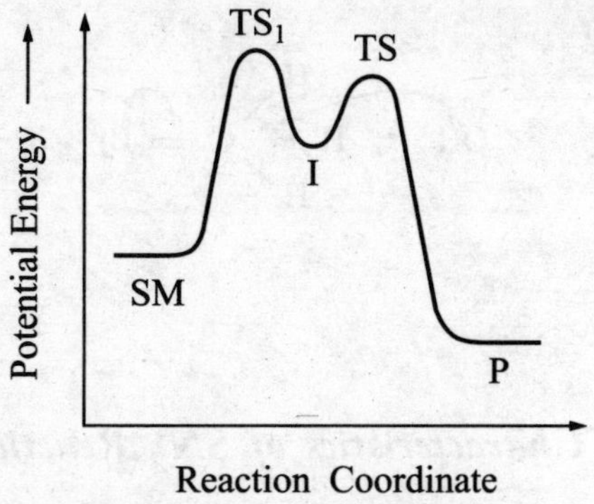

Fig. 9.5 Energy profile of the general reaction

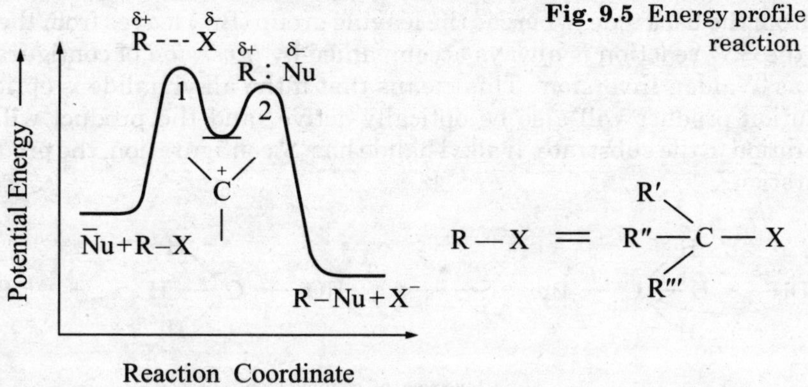

Fig. 9.6 Energy profile of the SN_1 reaction

9.7.2 SN_2 Mechanism

SN_2 indicates a substitution, nucleophile bimolecular reaction. In SN_2 type of bimolecular reactions, the rate of reaction *depends upon the concentration of both*, the substrate (alkyl halide) and the nucleophile, i.e. it is a bimolecular reaction. The alkyl halide and the nucleophile both are taking part in rate determining step of the reaction.

Thus SN_2 reactions are 'concerted reactions', i.e. they occur in one step. It shows second order i.e. kinetics the rate is proportional to substrate and the nucleophile. Hence,

$$\text{Rate} \ \alpha \ [\text{Substrate}] \ [\text{Nucleophile}]$$

Example: The reaction between CH_3Br and hydroxide ion to yield methanol,

$$CH_3Br + \bar{O}H \longrightarrow CH_3OH + Br^-$$

$$\text{Rate} = k[CH_3Br] \ [\bar{O}H]$$

Mechanism: The rate law suggests that the reaction requires a collision between alkyl halide and nucleophile. The breaking of bond and formation of bond takes place simultaneously.

SN_2 reactions proceed via a transition state in which carbon appear to be pentavalent. The transition state can be pictured as a structure in which carbon is partially bonded to both $\bar{O}H$ and Br^- *i.e.*, the C—OH is not completely formed and C—Br is not completely broken.

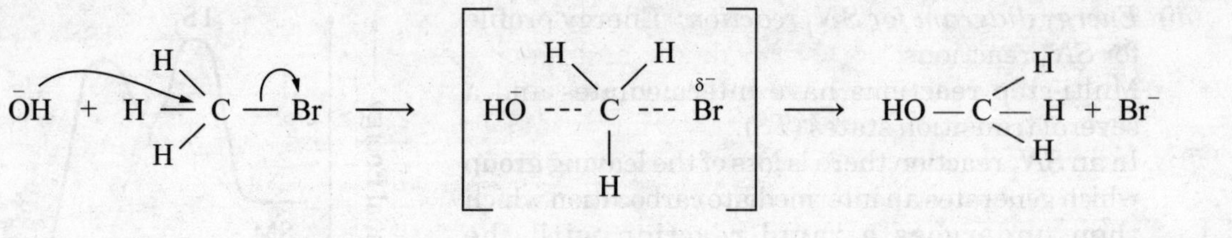

Transition state (I) Product

Characteristics of SN_2 Reactions

(i) *Stereochemistry:* We can see that in SN_2 reaction, the attack of the nucleophile ($\bar{O}H$) occurs from the back side, whereas the leaving group (Br^-) leaves from the front side. As a result, the SN_2 reaction is always accompanied by *inversion* of configuration commonly known as 'Walden Inversion'. This means that if the alkyl halide is optically active, the substitution product will also be optically active and the product will have opposite configuration to the substrate. If alkyl halide has '*S*' configuration, the product will have '*R*' configuration.

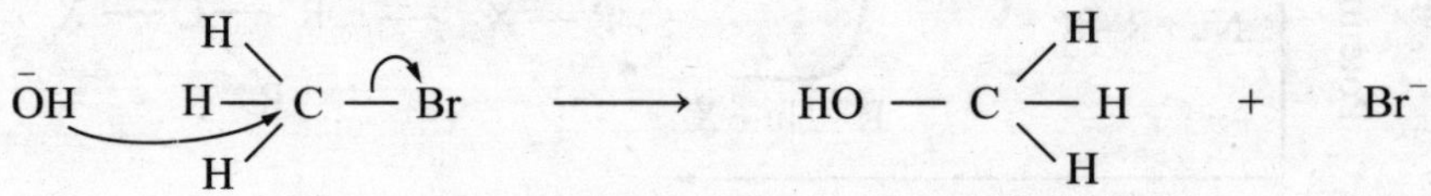

Inversion of configuration

(ii) *Relative reactivity of substrate:* The reactivity order of haloalkanes towards SN_2 mechanism is:

$$CH_3X > R–CH_2 – X > R_2\ CH – X > R_3CX$$

or simply, primary halides are more reactive towards SN_2 mechanism as compared to tertiary halides, which are least reactive.

This order of reactivity is explained on the basis of 'steric factors'. We can see from the mechanism that the nucleophile approaches from the opposite side of the leaving halide ion. If bulky groups are attached to the carbon, the approach of nucleophile would be quite difficult. We can say that nucleophile is 'sterically hindered' by the alkyl groups. Hence, as the number of substituents (alkyl groups) attached to the carbon bearing the halogen is *increased*, the reaction towards SN_2 *decreases*. Methyl halides or primary halides have much less sterical hinderance and hence undergo SN_2 type of reactions easily whereas, tertiary halides are sterically hindered and hence does not react through SN_2 mechanism.

(iii) *Effect of Nucleophile:* Since the nucleophile is involved in the rate determining step, the nature of the nucleophile is very important in an SN_2 reaction. SN_2 reactions are dependent upon the strength and concentration of the nucleophile. Strong nucleophiles will react more rapidly than nucleophiles that are described as poor or weak. Negative charged nucleophiles tend to be stronger than the neutral nucleophiles with the same central atom. For example, $\bar{O}H$ is stronger than H_2O and $R\bar{O}$ are better nucleophiles than R-O-H. Stronger nucleophiles will enhance the reaction rate of a reaction operating under SN_2 conditions.

Increasing the concentration of the nucleophile will enhance the rate of an SN_2 reaction since kinetically the reaction rate is also dependent upon the nucleophile concentration. This is in contrast to SN_1 where the concentration and nucleophilicity *have no effect* on the rate of an SN_1 reaction.

(iv) *Energy profile for SN_2 Reaction:* Single step reactions have no intermediates and have a single transition state (TS). In an SN_2 there is simultaneous formation of the carbon-nucleophile bond and breaking of the carbon-leaving group bond, hence the reaction proceeds via a TS in which the central C is partially bonded to five groups. No intermediate is present in the SN_2 reactions.

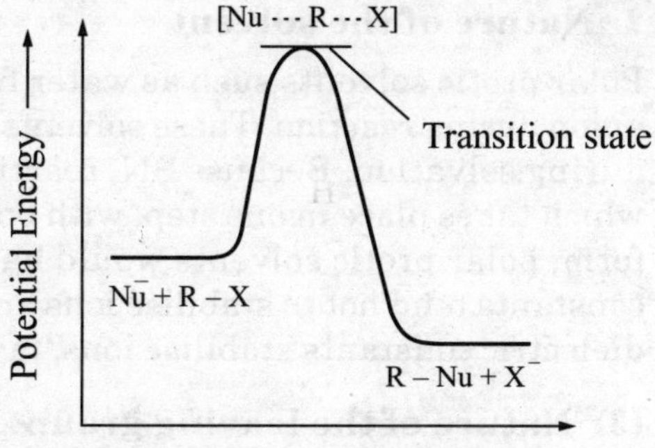

Fig. 9.7 Energy profile of SN_1 Reactions

(v) *Leaving Group Ability: Effect on SN_2 Reactions:* The leaving group ability (LGA) is defined as the ability of the leaving group to be displaced and remain stable. Leaving groups that are weak bases (coming form strong acids) are more stable and better leaving groups. After leaving the molecule these do not have any desire to return. However, the more basic the leaving group is (coming form weak acids), the poorer will be its ability to depart

Some examples of leaving groups that have excellent LGA are the halides except for fluoride ion. Chloride, Bromide, and Iodide all come form strong aides, HCl, HBr, and HI. Strong acids produce weak conjugate bases which make excellent leaving groups with high LGA. Flouride comes form HF which is a weak acid so its conjugate base, flouride ion would make a poor leaving group. Other groups with low LGA would include acetate ion, cyanide ion ($\bar{C}N$) and Nitrite ion, $N\bar{O}_2$.

The reaction rate of a reaction occurring via SN_2 path would increase with higher LGA for the leaving group. The same effect is observed for SN_1 reaction.

9.7.3 SN_1 Versus SN_2 Reactions

Whether an alkyl halide will undergo an SN_1 or an SN_2 reaction depends upon a number of factors. Some of the more common factors include the natures of the carbon skeleton, the solvent, the leaving group, and the nature of the nucleophile.

1. Nature of the carbon skeleton

Only those molecules that form extremely stable cations undergo SN_1 mechanisms. Normally, only compounds that yield 3° (tertiary) carbonications (or resonance-stabilised carbocations) undergo SN_1 mechanisms rather than SN_2 mechanisms. Carbocations of tertiary alkyl halides not only exhibit stability due to the inductive effect, but the original molecules exhibit steric hindrance of the rear lobe of the bonding orbital, which inhibits SN_2 mechanisms from occurring. Primary alkyl halides, which have little inductive stability of their cations and exhibit no steric hindrance of the rear lobe of the bonding orbital, generally undergo SN_2 mechanisms. Figure below illustrates the tendencies of alkyl halides toward the two types of substitution mechanisms.

Increasing SN_2 reactions ←

1° alkyl halide 2° alkyl halide 3° alkyl halide

→ Increasing SN_1 reactions

2. Nature of the solvent

Polar protic solvents such as water favor SN_1 reactions, which produce both a cation and an anion during reaction. These solvents are capable of stabilising the charges on the ions formed during solvation. Because SN_2 reactions occur via a **concerted mechanism** (a mechanism which takes place in one step, with bonds breaking and forming at the same time) and no ions form, polar protic solvents would have little effect upon them. Solvents with low dielectric constants tend not to stabilise ions and thus favor SN_2 reactions. Conversely, solvents of high dielectric constants stabilise ions, favoring SN_1 reactions.

(3) Nature of the leaving group

In general, good leaving groups are those capable of forming stable ions or molecules upon displacement from the original molecule. Conversely, poor leaving groups form ions of poor to moderate stability. Strong bases, such as OH^-, NH_2^-, and RO^-, make poor leaving groups. Water, which is less basic than a hydroxide ion, is a better leaving group. Poor bases usually make good leaving groups. A poor base is an ion or group in which the electrons are tightly bound to the molecule due to high electronegativity or resonance.

The following list ranks atoms and molecules in order of their stability as leaving groups, from most to least stable.

$$CH_3SO_3^-, \text{ tosyl}^- > I^- > Br^- > H_2O^+ > Cl^- > F^-$$

9.7.4 Nucleophilic Aromatic Substitution (SN_{Ar})

Even though the electophilic substitution is by far the most common mode of substitution in aromatic systems, the nucleophilic substitution is indeed possible and is a useful tool in certain cases. The early industrial syntheses of phenols and anilines in fact were based on the nucleophilic aromatic substitution reaction.

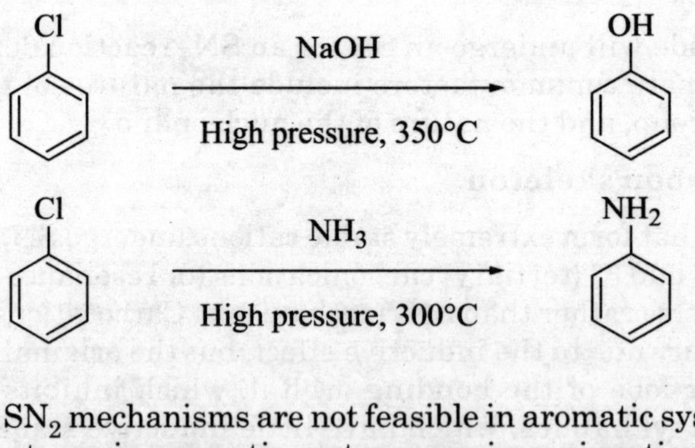

The typical SN_1 and SN_2 mechanisms are not feasible in aromatic systems. One of the major reasons is that the ∂ electrons in aromatic systems are in conjugation. More over, back side attack (as in SN_2) and inversion is precluded by the geometry of the ring. On the other hand, SN_1 type of substitution would require formation of the phenyl cation which is less stable than a primary carbocation. Obviously, the reaction has to proceed with a totally different type of mechanisms.

There are two mechanisms which are possible and they are:

(a) Addition-Elimination Mechanism
(b) Elimination-Addition Mechanism

Addition-Elimination Mechanism

Consider the reaction of an amine with chlorobenzene or nitro-substituted chlorobenzenes. Here, the reaction takes place with substituted chlorobenzene readily whereas under similar conditions, the chlorobenzene itself is unreactive. The nitro group clearly influences the rate of the reaction. In a similar manner, it was observed that the electron withdrawing groups such as –CN, $-CO^-$ etc. facilitated the substitution reaction.

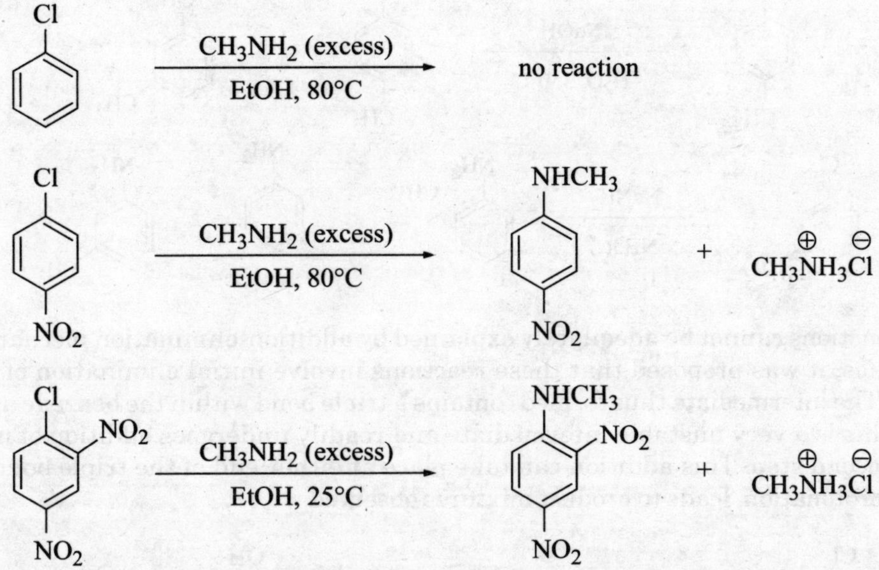

These observations suggest that electron withdrawing groups lower the energy of activation by stabilizing the negative charge developed as the nucleophile adds to aromatic ring. One can write a series of resonating structures for the cyclohexadienylanion generated after the addition of the nucleophile to the aromatic system. Finally, loss of chlorine completes the substitution process. This mechanism is parallel to the one which we studied for the electrophilic aromatic substitution reaction.

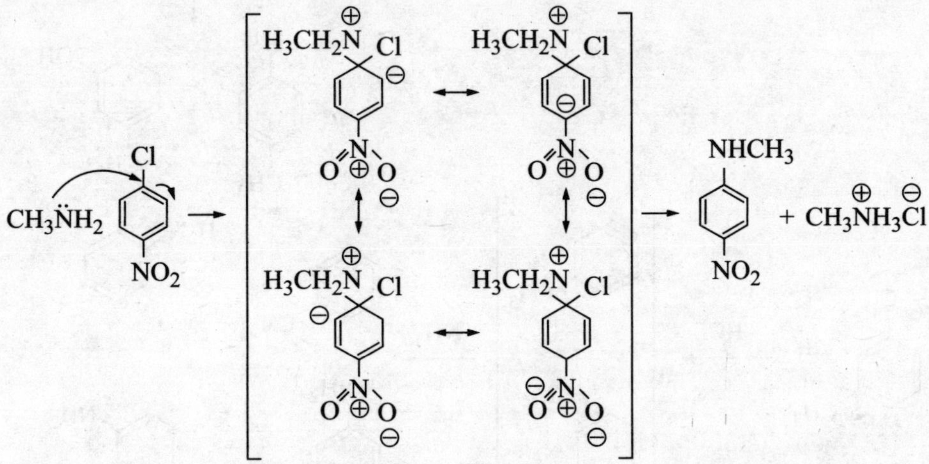

Elimination-Addition Mechanism – Benzyne Mechanism

As we have seen earlier, substituted halobenzenes on treatment with strong bases under harsh conditions lead to products formed by nuclephilic aromatic substitution reaction.

But position of substitution does not necessarily correspond to the carbon atom from which the leaving group departs. In other words, in cases, we see formation of regioisomeric products.

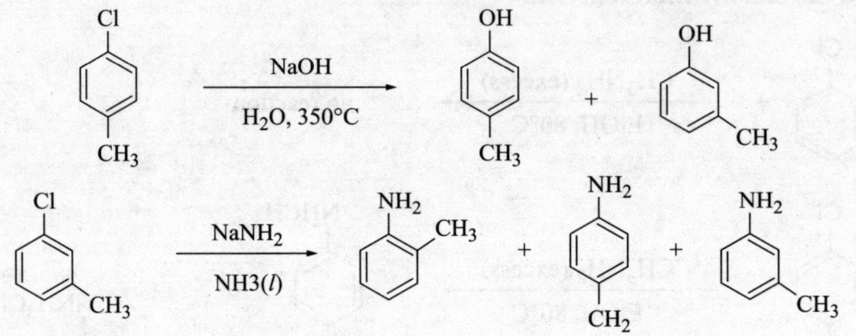

These reactions cannot be adequately explained by addition-elimination mechanism. After several studies, it was proposed that these reactions involve initial elimination of HX by the strong base. The intermediate thus formed contains a triple bond within the benzene and is called **benzyne**. This is a very unstable intermediate and readily undergoes addition of nucleophile (base) in a second step. This addition can take place at either end of the triple bond and after subsequent protonation, leads to product mixtures observed

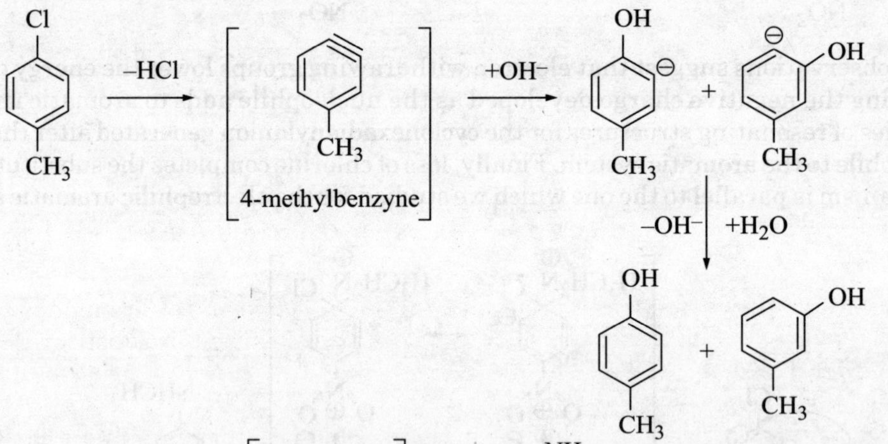

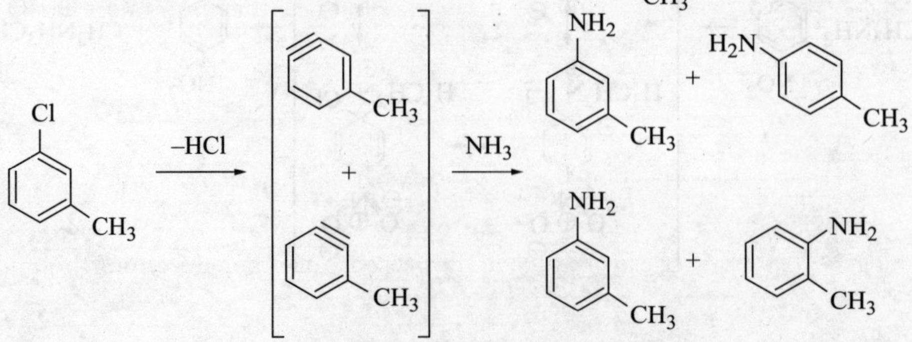

This is in line with what we would expect if elimination was to be the rate controlling step. Another evidence for this mechanism is reaction of 2,6-dimethylchlorobenzene with sodium amide in liquid ammonia. No substitution is observed in this reaction – obviously because there are no hydrogen atoms beta to the chlorine for the initial elimination step.

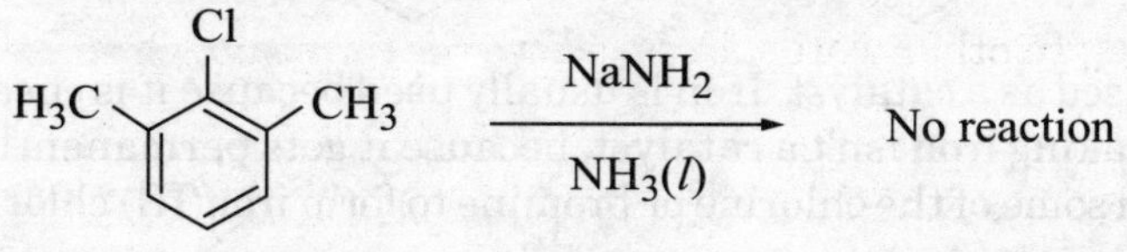

9.8 MECHANISM OF ELECTROPHILIC SUBSTITUTION REACTIONS (ELECTROPHILIC AROMATIC SUBSTITUTION REACTIONS)

It was discussed in the previous section that the substitution reactions brought about by the electrophilles are known as the Electrophilic substitution reactions. Electrophilic substitution is characteristic of aromatic compounds. Electrophilic aromatic substitution is an important way of introducing functional groups on benzene rings.

Although aromatic compounds have multiple double bonds, these compounds do not undergo addition reactions. Their lack of reactivity toward addition reactions is due to the great stability of the ring systems that result from complete π electron delocalisation (resonance). Aromatic compounds react by electrophilic aromatic substitution reactions, in which the aromaticity of the ring system is preserved.

Electrophilic Substitution Reactions of Benzene

A two-step mechanism has been proposed for these electrophilic substitution reactions. In the first, slow or rate-determining, step the electrophile forms a sigma-bond to the benzene ring, generating a positively charged **benzenonium intermediate**. In the second, fast step, a proton is removed from this intermediate, yielding a substituted benzene ring.

Every specific electrophilic aromatic substitution will contain these two steps plus whatever steps explain the initial generation of the electrophile required in step 1 above.

The mechanism is depicted in the following reactions , typical of benzene and its derivatives.

(1) Halogenation:

The electrophilic substitution reaction between benzene and chlorine or bromine is termed as halogenation. Benzene reacts with chlorine or bromine in an electrophilic substitution reaction, but only in the presence of a catalyst. The catalyst is either aluminium chloride (or aluminium bromide if benzene is reacted with bromine)

The reaction between benzene and chlorine in the presence of either aluminium chloride gives chlorobenzene.

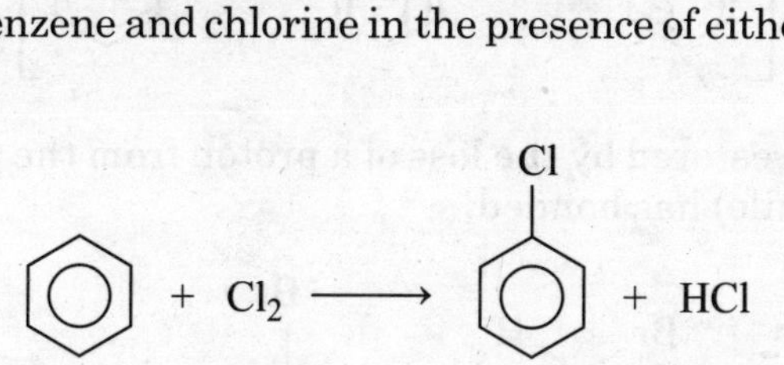

The reaction between benzene and bromine in the presence of either aluminium bromide gives bromobenzene.

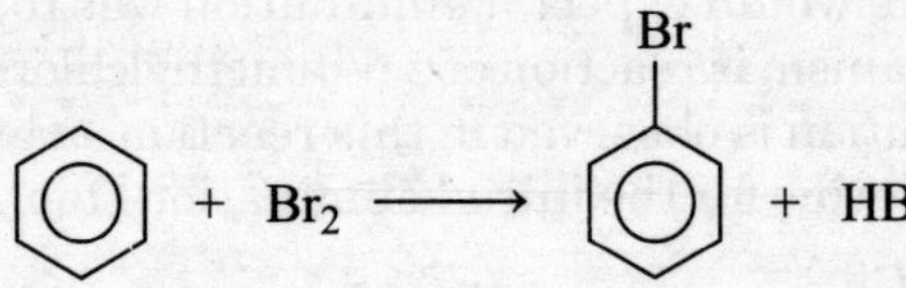

Iron may also be used as a catalyst. Iron is usually used because it is cheaper and more readily available. Strictly speaking iron isn't a catalyst, because it gets permanently changed during the reaction. It reacts with some of the chlorine or bromine to form iron(III) chloride, $FeCl_3$, or iron(III) bromide, $FeBr_3$.

$$2Fe + 3Cl_2 \longrightarrow 2FeCl_3$$
$$2Fe + 3Br_2 \longrightarrow 2FeBr_3$$

These compounds act as the catalyst and behave exactly like aluminium chloride in these reactions. All electrophilic aromatic substitution reactions share a common mechanism.

Mechanism: The bromination of benzene

This mechanism consists of a series of steps.

1. An **electrophile** is generated. For the bromination of benzene reaction, the electrophile is the Br^+ ion generated by the reaction of the bromine molecule with ferric bromide, a Lewis acid.

$$Br—Br + FeBr_3 \longrightarrow Br—Br^+—FeBr_3^- \longrightarrow \overset{+}{Br} + FeBr_4^-$$

2. The electrophile is attracted by the π electron system of the benzene ring to form a nonaromatic carbocation.

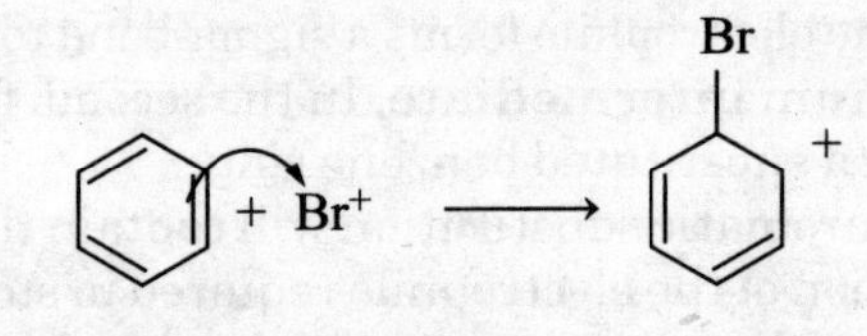

3. The positive charge on the carbocation that is formed is delocalised throughout the molecule, i.e., it is resonance hybridised.

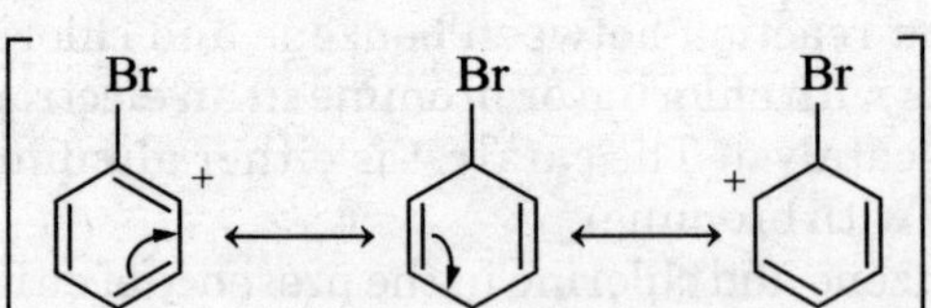

4. The aromaticity is restored by the loss of a proton from the atom to which the bromine atom (the electrophile) has bonded.

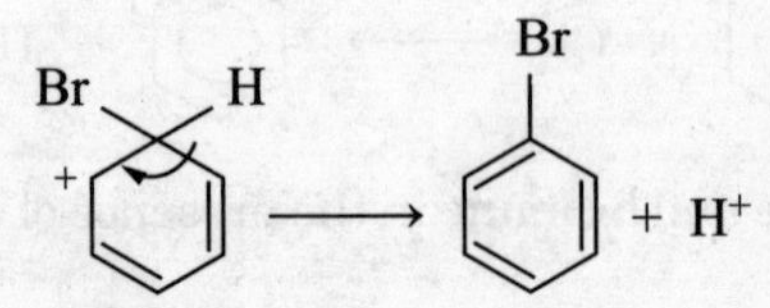

5. Finally, the proton reacts with the $FeBr_4^-$ to regenerate the $FeBr_3$ catalyst and form the product HBr.

$$H^+ + FeBr_4^- \longrightarrow FeBr_3 + HBr$$

2. The nitration of benzene

In another example of an electrophilic aromatic substitution reaction, benzene reacts with a mixture of concentrated nitric and sulfuric acids to create nitrobenzene.

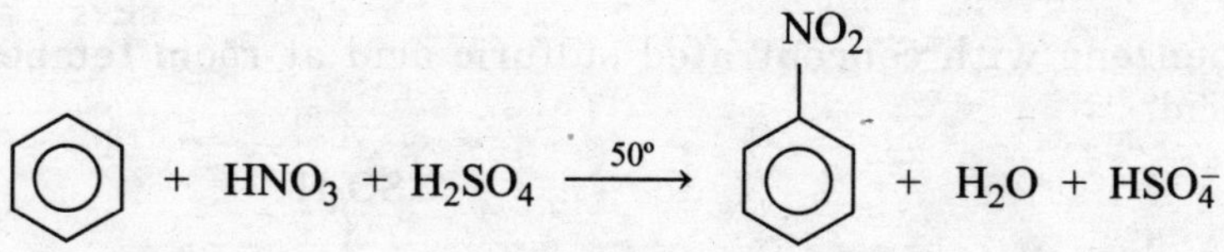

Mechanism: The mechanism for the nitrobenzene reaction is as follows:.

1. Sulfuric acid ionizes to produce a proton.

$$H_2SO_4 \rightleftharpoons H^+ + HSO_4^-$$

2. Nitric acid accepts the proton in an acid-base reaction.

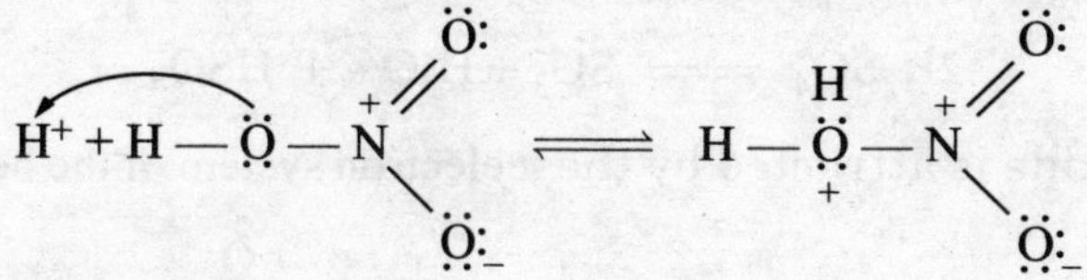

3. The protonated nitric acid dissociates to form a nitronium ion ($^+NO_2$).

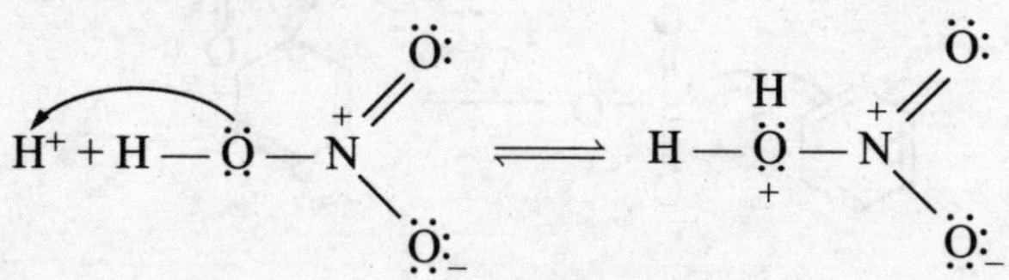

4. The nitronium ion acts as an electrophile and is attracted by the π electron system of the benzene ring.

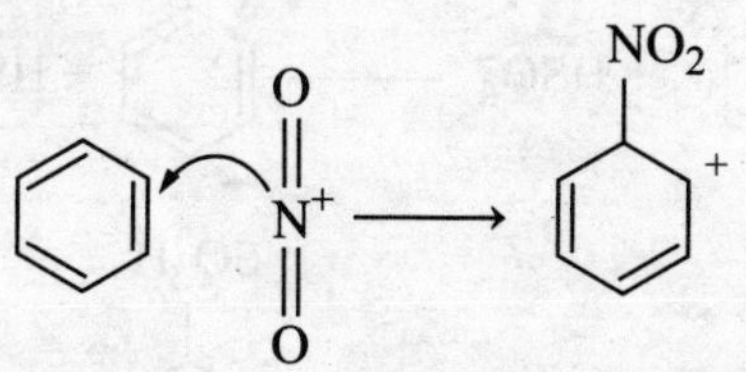

5. The nonaromatic carbocation that forms has resonating structures.

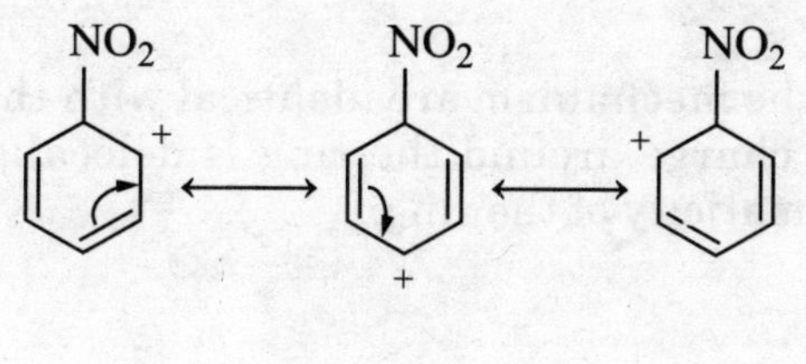

6. The aromaticity of the ring is reestablished by the loss of a proton from the carbon to which the nitro group is attached.

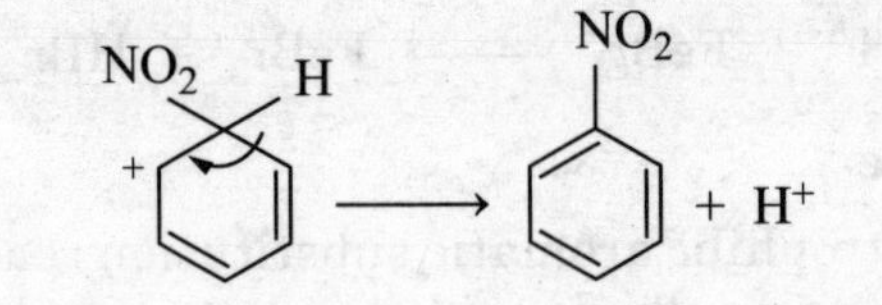

3. The sulfonation of benzene

The reaction of benzene with concentrated sulfuric acid at room temperature produces benzenesulfonic acid.

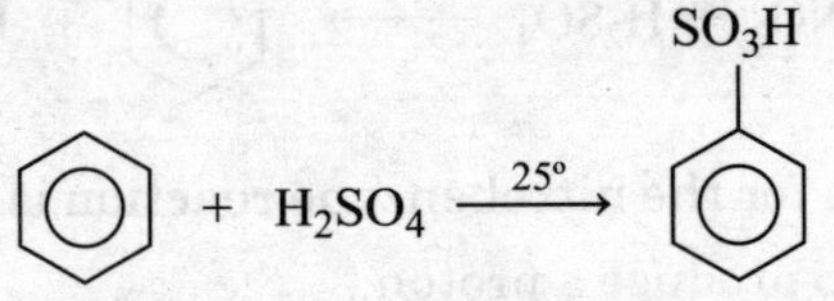

Mechanism: The mechanism for the reaction that produces benzenesulfonic acid occurs in the following steps:

1. The sulfuric acid reacts with itself to form sulfur trioxide, the electrophile.

$$2H_2SO_4 \rightleftharpoons SO_3 + H_3O^+ + HSO_4^-$$

2. The sulfur trioxide is attracted by the π electron system of the benzene molecule.

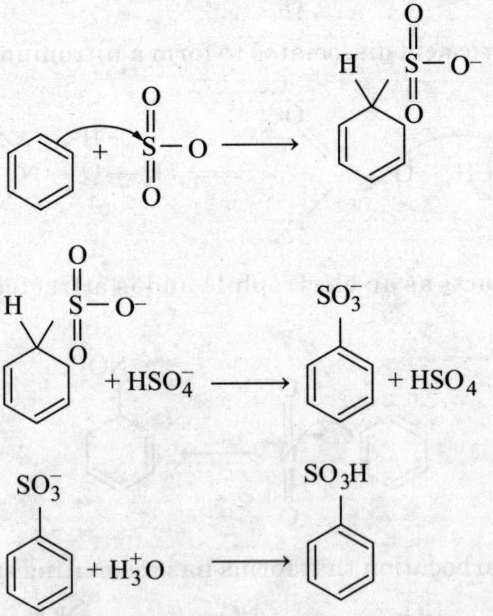

The remaining steps in the mechanism are identical with those in the bromination and nitration mechanisms: The charge around the ring is delocalised, and then the loss of a proton reestablishes the aromaticity of the ring.

3. Alkylation:

Alkylation of benzene means substituting an alkyl group into benzene ring. A hydrogen on the ring is replaced by a group like methyl or ethyl and so on. Benzene is treated with a chloroalkane (for example, chloromethane or chloroethane) in the presence of aluminium chloride as a catalyst

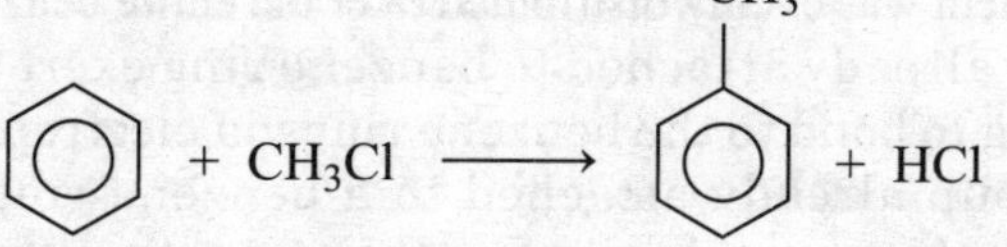

Mechanism:

1. The electrophile is CH_3^+. It is formed by reaction between the chloromethane and the aluminium chloride catalyst.

$$CH_3Cl + AlCl_3 \longrightarrow \overset{+}{C}H_3 + AlCl_4^-$$

2. The CH_3^+ is attracted by the π electron system of the benzene molecule. The nonaromatic carbocation that forms has resonating structures.

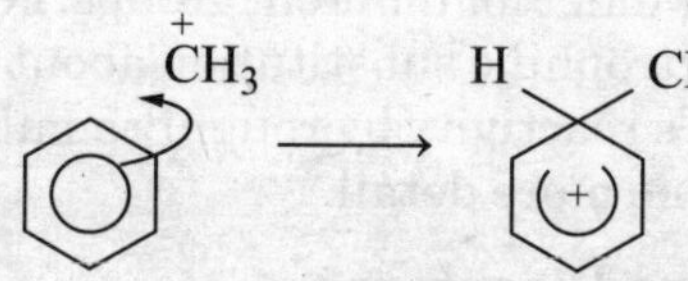

3. The aromaticity of the ring is reestablished by the loss of a proton from the carbon to which the methyl group is attached.

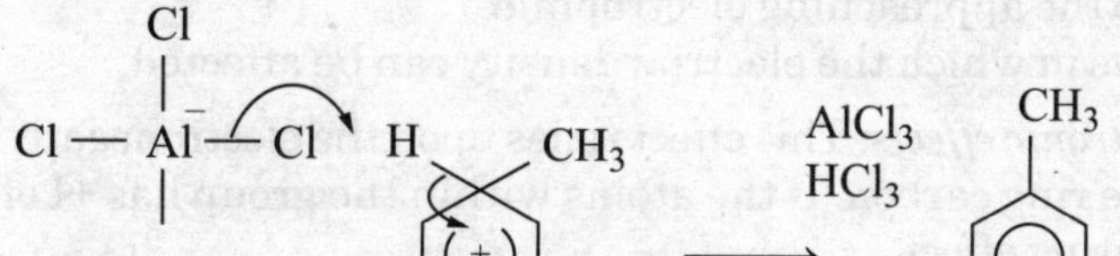

Summary of the aromatic electrophilic reactions is given in the table:

Reaction Type	*Typical Equation*	*Electrophile* $E^{(+)}$
Halogenation:	C_6H_6 + Cl_2 & $FeCl_3$ ⟶ $C_6H_5Cl + HCl$ heat catalyst Chlorobenzene	$Cl^{(+)}$ or $Br^{(+)}$
Nitration:	C_6H_6 + HNO_3 & H_2SO_4 ⟶ $C_6H_5NO_2 + H_2O$ heat catalyst Nitrobenzene	$NO_2^{(+)}$
Sulfonation:	C_6H_6 + $H_2SO_4 + SO_3$ ⟶ $C_6H_5SO_3H + H_2O$ heat Benzenesulfonic acid	$SO_3H^{(+)}$
Alkylation:	C_6H_6 + R-Cl & $AlCl_3$ ⟶ C_6H_5-R + HCl heat catalyst Benzenesulfonic acid	$R^{(+)}$
Friedel-Crafts		

9.8.1 Effects of Substituents (Directive influence)

We have only considered the substitution of unsubstituted benzene in the above reactions. In discussing these electrophilic aromatic substitutions on benzene, it was apparent that all six carbon atoms were equivalent in their ability to be the center of the substitution since the electron density exhibited by the π electron system was evenly distributed over the entire benzene ring.

However,substituents already attached to benzene ring exert an influence on additional atoms or groups attempting to bond to the benzene ring via electrophilic aromatic substitution reactions. An atom or group already attached to a benzene ring may direct an incoming electrophile to either the *ortho-para* positions or the *meta* position.

- Atoms or groups that make the benzene molecule more reactive by increasing the ring's electron density are called **activating groups.** Activating groups serve as ortho-para directors when they are attached to a benzene ring, meaning that they direct an incoming electrophile to the ortho or para positions.
- An atom or group that makes the benzene molecule less reactive by removing electron density from the ring acts as a **deactivating group.** Deactivating groups direct incoming electrophiles to the meta position.

These kinetic and directive effects differ for different groups. For example, a hydroxy or methoxy substituent increases the rate of electrophilic substitution about ten thousand fold. In contrast, a nitro substituent decreases the ring's reactivity by roughly a million.

Let us discuss this effects in some more detail.

1. Activating Groups: Ortho-Para Directors:

There are some groups which when attached to benzene ring will *activate* the ring toward further substitution. They do this by increasing the negative electron density in the ring thus making the ring more attractive to the approaching electrophile.

There are two ways in which the electron density can be affected:

(a) *Inductive Electronic effects*: This effect relies upon the electronegativity of the atom directly attached to the ring carbon. If the atoms within the group has +I effect than the ring will have an activating effect.

(b) *Resonance Electronic Effect*: This effect is the strongest of the two effects and accounts for some groups being powerful activators. Generally speaking any atom having at least one lone pair that is directly connected to a ring carbon will increase the electron density of the ring by lending that lone pair of electrons to the π electron system making the ring more attractive to the electrophile.

Most ring activators have atoms with unshared electron pairs directly attached to a carbon atom of the benzene ring. For example, the — OH group has two pairs of unshared electrons on the oxygen atom, which will form a bond to a carbon atom of the benzene ring. Thus, the — OH group will be an activating group. The following illustration shows why this group will act as an ortho-para director.

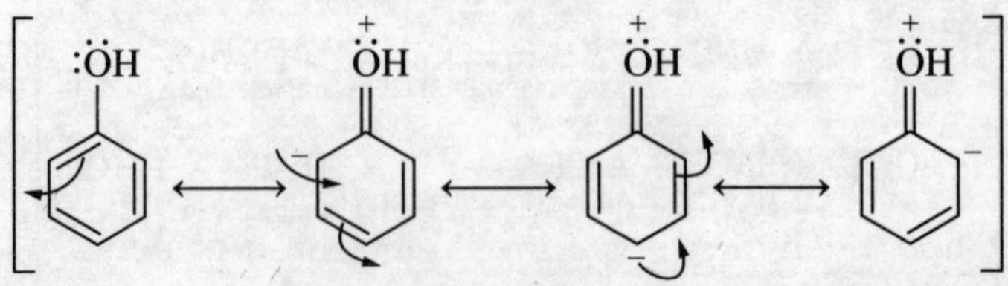

Notice that three of the four resonance structures show a negative charge residing on the positions ortho and para to the —OH group. These electron-rich positions should attract an electrophile more strongly than the less electron-rich meta positions do. Therefore, any group that possesses unshared electron pairs on the atom directly attached to a carbon atom of the benzene ring will be an ortho-para (activating) group. Groups that do not have unshared electron pairs on the atom directly attached to the benzene ring may also supply electrons to the benzene ring. This situation occurs if the atom in a group has weakly bonded π electrons attached to it or if the group has an inductive effect associated with it. The following diagram shows an example of the π electron movement giving ring activation

The most powerful and strongest activators are shown in Figure below.

Activating Substituents

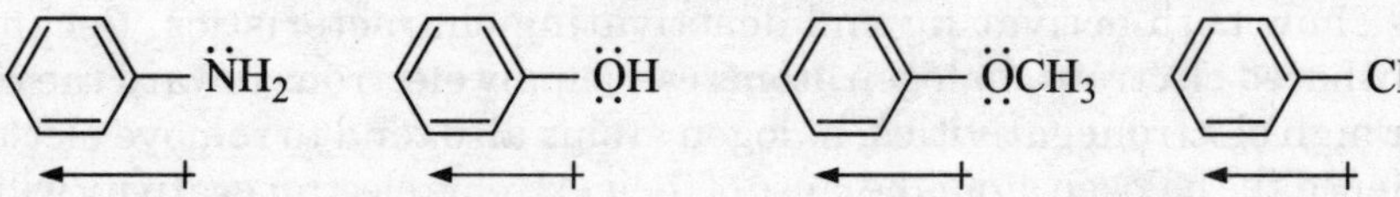

$$-NR_2 > -NHR > -NH_2 > -OH > OR > NH-\overset{O}{\overset{||}{C}}-CH_3 > CH_3 > Ten > C_6H_5$$

2. Deactivating Groups and meta direction

Most of these groups either have atoms attached to a ring carbon with a π bond or highly electronegative atoms attached as in the case of —CF_3 and —CCl_3 groups which are strong deactivators .These deactivators reduce the negative electron density of the ring by withdrawing it out of the ring either by resonance (Those atoms directly attached to the ring that have at least one π bond) or inductive effect as in the case of the —CCl_3, CF_3, and —NR_3^+ groups. Consequently, substituents in which nitrogen, oxygen and halogen atoms form sigma-bonds to the aromatic ring exert an inductive electron withdrawal, which deactivates the ring.

All deactivators with the exception of the weakly deactivating groups spoken of earlier will be meta directors. This is because when the electron density is dimenished in the ring , it occurs more at the ortho and para positions leaving the meta position relatively more negative.Therefore, if the electrophile does manage to be attracted to the deactivated ring, it will find the meta position that will be more negative and more attractive to the incoming electrophile.

Groups capable of doing this usually contain an atom that is directly attached to a carbon atom of the benzene ring and that bears a positive or partially positive charge. A typical example is the nitro group —NO_2. The structure of the nitro group is:

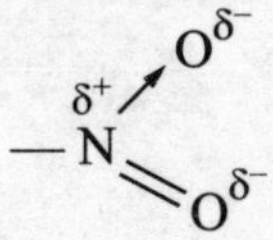

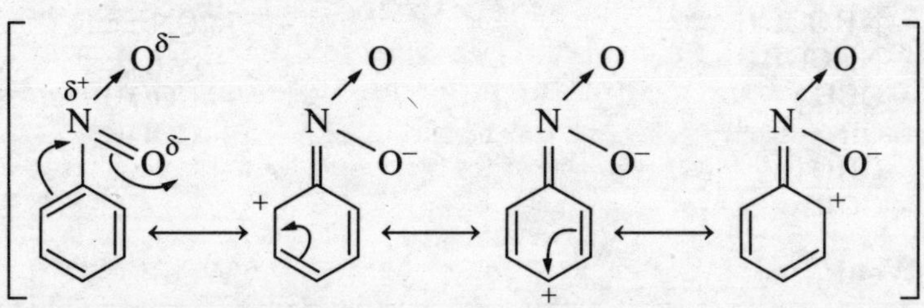

Notice that in three of the four resonance structures, a positive charge exists on the ortho and para positions. Thus, the hybrid structure is electron-poor in these areas, meaning that an electrophile generally attaches to the more electron-rich meta position. Some examples include:

Deactivating Substituents

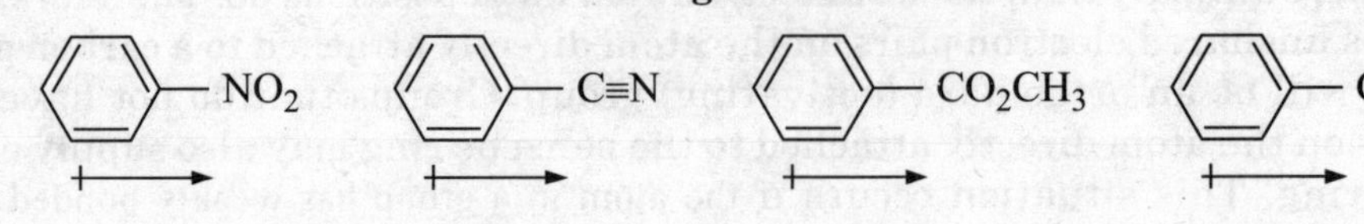

$$-NO_2 > -SO_3H > -C \equiv N - \overset{O}{\overset{||}{C}} - R > - \overset{O}{\overset{||}{C}} - OH$$

3. Halogens Deactivating but ortho and para directing

Halogen atoms show both activating and deactivating characteristics. Because they have three pairs of unshared electrons, halogen atoms can supply electrons toward the ring. However, because of their high electronegativities, halogen atoms also tend to remove electrons from the benzene ring. Hence,the halogen atoms because of their extreme electronegativity will pull electrons out of the ring leaving the ring more positive ,and therefore, less attractive to the incoming electrophile but the three lone pairs of non-bonding electrons found on the halogen atom will provide increased electron density at the ortho and para positions. Thus the benzene ring is deactivated ,but the relative electron density is greater at the ortho and para positions due to the resonance effect of the lone pairs. So, if the substitution occurs it will be slow but occur either at the ortho or para positions

There are three generalisations concerning this effect:

1. all meta-directing groups are deactivating
2. all ortho, para-directing groups are activating except the **halogens**
3. the halogens are deactivating

We can further classify activating and deactivating groups or atoms as strong, moderate, or weak in their directing influence. This table lists some typical activating and deactivating groups by the order of their strength.

The Influence of Various Directing Groups and Atoms

Ortho-Para Directions	Meta Directors
Strong	**Strong**
$—NH_2$	$—NO_2$
—NHR	$—NH_3^+$
$—NR_2$	$—NR_3^+$
—OH	$—CF_3$
$—O^-$	$—CCl_3$
Moderate	**Moderate**
$—NHCOCH_3$	$—CN\equiv N$
—NHCOR	$—SO_3H$
$—OCH_3$	—COOH
—OR	—COOR
—CHO	
—COR	
Weak	**Weak**
$—CH_3$	
$—CH_2CH_3$	
—R	
$—C_6H_5$	

SOLVED EXAMPLES

1. Which among acetic acid, ethanol, p-nitrophenol, p-methoxyhenol are *more acidic* than phenol ?

   Ans: *Ethanol and p-methoxyphenol.* This is due to the following *reasons* :
   Phenoxide ion ($C_6H_5O^-$) is *stabilised by resonance;* whereas ethoixde ion ($C_2H_5O^-$) is *not* resonance stabilised. Hence, pheonol is *more acidic* than ethanol. Moreover, resonance in p-methoxyphenoxide ion ($CH_3O.C_6H_4.O^-$) is *less* than that in phenol, because electron-release by O– group in phenoxide ion is opposed by electriondonating nature of $-OCH_3$ group. Hence, phenol is *more acidic* than p-methoxyphenol.

2. Why is phenol *less acidic* than carbonic acid ?

   Ans: Due to the presence of two O-atoms, the bicarbonate ion $HO-\overset{\overset{O}{\|}}{C}-O$ is *more resonance stabilised* than phenoxide ion ($C_6H_5O^-$) having having only one O-atom. Consequently, carbonic acid is a stronger acid than pheonol. Alternatively, phenol is a *weaker* acid than carbonic acid.

3. Arrange CH_3OH, CH_3CH_2OH, $(CH_3)_2CHOH$, $(CH_3)_3COH$ in decreasing order of their *acidity.*

   Ans: $CH_3OH > CH_3CH_2OH > (CH_3)_2CHOH > (CH_3)_3COH$, since the electronreleasing methyl groups *decrease* the acidic nature of alcohol.

4. Arrange $\overset{+}{C}H_3$, $R\overset{+}{C}H_2$, $R_2\overset{+}{C}H$, $R_3\overset{+}{C}$ carbonium ions in order of increasing *stability.*

   Ans: $\overset{+}{C}H_3 < R\overset{+}{C}H_2 < R_2\overset{+}{C}H < R_3\overset{+}{C}$

5. Which among $\dot{C}H_3$, $CH_3\dot{C}H_2$, $(CH_3)_2\dot{C}H$, $(CH_3)_3\dot{C}$ is *most stable* free radical?

   Ans: $(CH_3)_3\dot{C}$, since the stability of carbon free radical *increases* with the number of alkyl groups attached to the C-atom width odd electron.

6. Which among $C_6H_5CH_2Br$, $(CH_3)_2CHBr$, $(CH_3)_3CBr$, CH_3Br undergoes nucleophilic substitution reaction by SN_2 mechanism ?

   Ans: CH_3Br, since SN_2 reactivity of haloalkance in the order : $CH_3 > 1^o < 2^o > 3^o$. The rate determining factor is *steric hindrance.* There is minimum steric hindrance offered by H's of CH_3Br to the approach of nucleophile attacking at the C so C—Br bond.

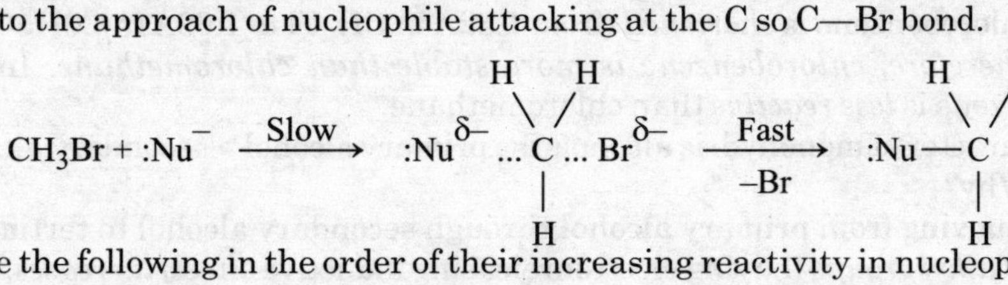

7. Arrange the following in the order of their increasing reactivity in nucleophili substitution reactions : CH_3F, CH_3I, CH_3Br, CH_3Cl.

   Ans: $CH_3I < CH_3Cl < CH_3Br < CH_3F$.

   Reason : The bond desiccation energices of : C—F > C—Cl > C—Br > C—I. Thus, C—I bond is not readily cleaveraged; while the C—F bond is cleaved with most difficulty. Now ***lesser*** *there bond dislocation energy,* ***more*** *is the reactivity of the haloalkance.* This explains the above order of reactivity.

8. Why is the halogen atom of an alkyl halide *readily substituted* by an other group?

   Ans: Alkyl halides (RX) are *partially polar* in nature with halogen atom carrying a fractional negative charge. Thus :

$$\overset{\delta+}{R—CH_2}—\overset{\delta-}{Cl}$$

Folar molecule

The electron-releasing inductive effect of an alkyl group facilitates the ionisation of halogen atom and its substitution by a nucleophile. Thus :

9. The treatment of alkyl chloride with aqueous KOH leads to the formation of alcohols; whereas in the presence of alc. KOH, alkences are formed as the major products. Explain.

Ans: *Aqueous KOH* contains *only* OH– *ions,* which act as nucleophile and these brings about *hydrolysis* of alkyl chloride to the corresponding alcohol.

$$RCl + \underset{\text{(Only nucleophile present)}}{OH^-(aq)} \xrightarrow{\text{Hydrolysis}} ROH + Cl^-(aq)$$

On the other hand, *alcoholic KOH* contains *ethoxide ions* (C2H5O–), which are *more basic than* OH– *ions.* Consequently, ethoxide ions preferentially bring about *dehydrohalgenatur* to form alkenes, e.g.

$$\underset{\boxed{H\quad Cl}}{H_2C—CH_2} + KOH(alc.) \longrightarrow H_2C{=}CH_2 + H_2O + KBr$$

10. Why are haloakanes more reactive than haloraenes?

Ans: Resonance structure of haloalkance (e.g., chlorobenzene) is :

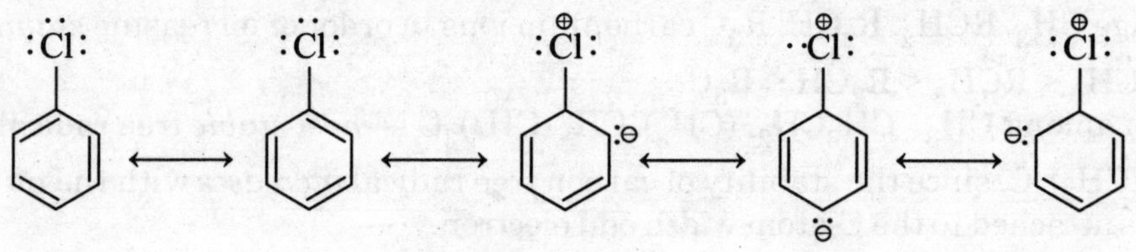

(I) (II) (III) (IV) (V)

Chlorobenzene is *resonance hybrid* of five structures; while haloalkance (e.g., chloromethane) is not resonance stabilised. As a result of contribution of structures III, IV and V, the C—Cl bond in chlorobenzene acquires *a partial double bond character*. Consequently, the C—Cl bond in chlorobenzene is *shortened* (from 1.82Å in CH_3Cl to 1.69Å in C6H5Cl) and is *more, stable. Therefore, chlorobenzene is more stable than chloromethane.* In other words, chlorobenzene is *less reactive* than chloromethane.

11. Acidic character of monohydric alcohols is : primary alcohol > secondary alcohol > tertiary alcohol. Why?

Ans: On moving from primary alcohol through secondary alcohol to tertiary alcohol, the *number of alkyl groups* (having electronreleasing inductive effect) *increases.* Consequently, the electronreleasing inductive effect of three alkyl groups in tertiary alcohols s maximum and it is minimum in primary alcohols. In other words, the primary alcohols are the *strongest* acids; while tertiary alcohols are *weakest* acids. Hence, the acidic character of alcohols is :

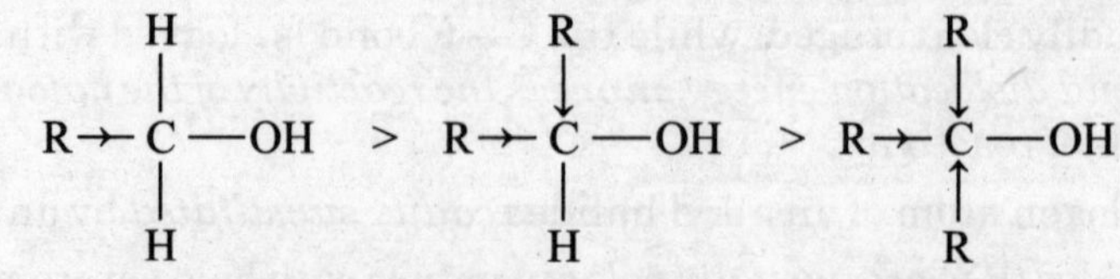

Primary alcohol Secondary alcohol Tertiary alcohol

12. Alcohols are *weaker acids* than water. Why?

Ans: The reaction : 2 ROH + 2Na ⟶ 2RONa + H_2 suggests that *alcohols are acidic in nature.* This is due to fact that the *highly electronegative* O-atom withdraws partially the electrons of the O—H bond towards itself, thereby the O—H bond becomes *weak* and splits off a proton (H+). However, alcohols are *weaker acids* than water, because *the electronreleasing effect of alkyl group (R–>–O—H) from the O—H bond becomes difficult.* Hence, alcoholds are *weaker acids* than water, due to *electronreleasing inductive effect of the alkyl group.*

13. Why phenol is *more acidic* than alcohols (or ethanol)?

Ans: Phenol is *acid and more acid* than alcohols (or ethanol) because of the following facts:

(i) The phenoxide anion ($C_6H_5O^-$) is *stabilised dby resonance;* whereas the alkoxide ion (or ethoxide ion) is *not* resonance stabilised.

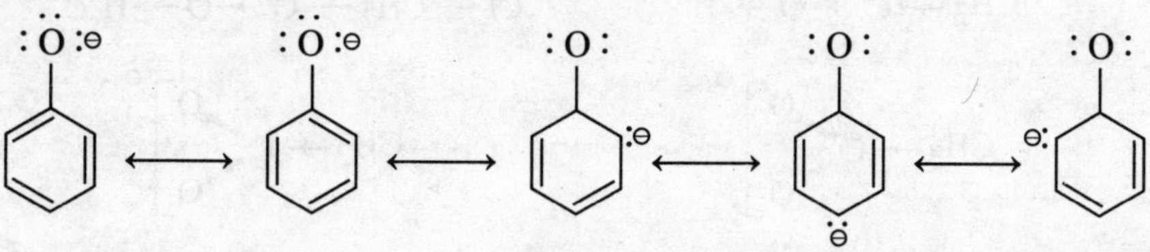

As a result of resonance, the oxygen atom of phenoxide ion acquires a *partial positive charge.* This is-turn *weakens* the O—H bond, thereby facilitating the release of a proton.

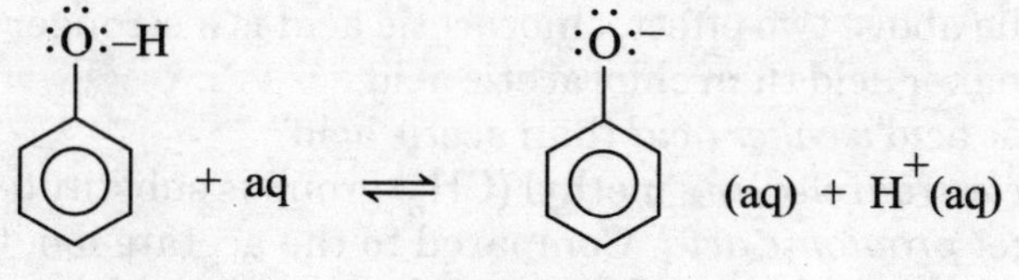

Phenol *(Less stable)* Phenoxide ion *(More stable)*

Thus, phenol has a strong tendency to form more stable phenoxide ion by releasing a proton. *Hence, phenol is acidic in nature.*

On the other hand, the alkoxide (or ethoxide) ion is *less stable* than alcohol (ethanol) and is *not* resonance stabilised. So alcohol (or ethanol) possesses *negligible tendency* to form *less stable* alkoxide (or ethoxide) ion by releasing a proton.

$$R—\ddot{\underset{..}{O}}—H \rightleftharpoons R—\ddot{\underset{..}{O}}:^{\ominus} + H^+$$

Alcohol *(More stable)* Alkoxide ion *(Less stable)*

14. Arrange the following compounds in the *increasing order* of the acid strength. Give reasons for you answers.

CH_3CH_2OH, CF_3CH_2OH, CCl_3CH_2OH

Ans: $CH_3CH_2OH < CCl_3CH_2OH < CF_3CH_2OH$.

Reason : (a) Acid strength of CF_3CH_2OH and CCl_3CH_2OH is *higher* than that of CH_3CH_2OH, due to the *inductive effect* (–I effect) of halogen atoms. This causes the electron-deficiency on the oxygen atom of '–OH' group, thereby *facilitating* the removal of proton (H+) from OH-bond.

(b) Electronegativeity of F of *higher* than Cl, so F exerts *higher* – I effect than Cl. Hence, CF_3CH_2OH is *stronger acid* than CCl_3CH_2OH.

15. Explain why chloroacetic acid is a *stronger acid* than acetic and.

Ans: (1) The chlorine atom in chloroacetic acid has *electron-withdrawing effect* (–I effect), so the chlorine atom withdraws itself the electrons from the O—H bond *decreases.* This in-turn *facilitates the release of H^+ ions.*

(2) Moreover, due to -I effect of chlorine atom, the chloroacetate ion (of chloroacetic acid) has *less electron-density* over the chlorocaboxylate anion than the carboxylate anion (of acetic acid). In other words, the chloroacetate anion is *more stabilised* relative to acetate anion. Consequently, the release of H+ ions from chloroacetic acid is *facilitated.*

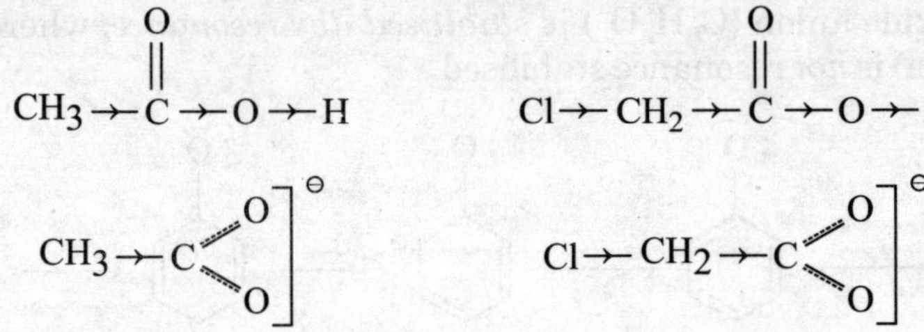

+I effect of the CH_3 group *intensifies* the negative charge on the carboxylate anion, thereby making it *less stable.*

–I effect of the Cl atom *disperses* the negative charge on the carboxylate anion, there making it *more stable.*

As a result of the above two effect, chloracetic acid is a stronger acid than acetic acid, i.e. acetic acid is weaker acid than chloracetic acid.

16. Why is propionic acid *weaker acid* than acetic acid?

Ans: When an *electronreleasing* methyl (CH_3) group is substituted for one H-atom of acetic acid, then we get *propionic acid.* Compared to the acetate ion, the *propionate ion is* ***less stable****. So propionic acid is* ***weaker acid*** *than acetic acid.*

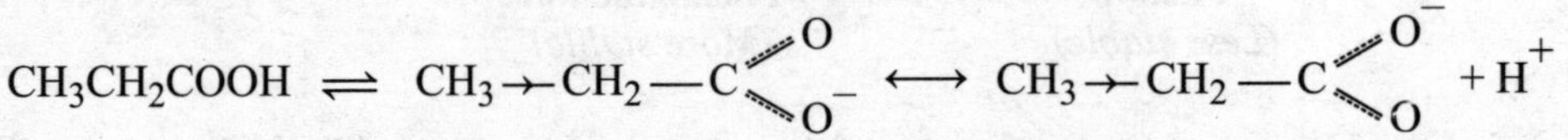

Electron-releasing CH_3 group *destabilises* the propionate anion relative to the acetate anion.

17. Explain the fact that aniline is *less basic* than ammonia.

Ans: Aniline is a *resonance hybrid* and as a result of this *the lone pair of electrons on the nitrogen atom of amino group in aniline get* ***delocalised*** *over the benzene ring,* thereby this long pair of electrons is ***less readily available*** for protonation. On the other hand, in case of ammonia, delocalisation of the lone pair of electrons on the nitrogen atom is *not possible* through resonance. Therefore, the long pair of electron on nitrogen atom of NH_3 is ***easily available*** for protonation. *Hence, aniline is less basic than ammonia.*

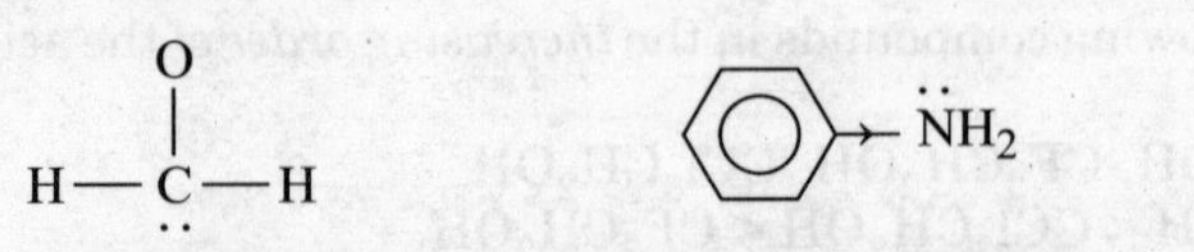

Electron pair on N atom is *easily* available for protonation

Due to resonance effect of phenyl radical, the electron pair on nitrogen atom is *less readily* available for protonation

QUESTIONS

1. What are the different types of organic reactions? Explain them with one example each.
2. What are: (a) Electrophilic reagents (b) Nucleophilic reagents? Give two examples of each.
3. Write notes on: (a) Electronegativity (b) Polar and non-polar covalent bonds (c) Inductive effect (d) Electromeric effect
4. Classify the following into electrophilic and nucleophilic reagents.
   (a) H^+ (b) Cl^- (c) NO_2^+ (d) OH^-
5. What is an addition reaction? Explain its reaction mechanism with an example.
6. What is an elimination reaction? Explain its reaction mechanism with an example.
7. Explain SN_1 reactions with examples.
8. Give a detailed account stereochemistry of SN_1 reaction.
9. How do bulky groups affect SN_1 reaction?
   (a) Comment on the stability of carbocations
   (b) Comment on elimination versus substitution.
10. What is the difference between a free radical and an ion?
11. What are different intermediate organic species? Discuss in detail their stabilities.
12. Distinguish between electromeric and inductive effects, giving some suitable examples.

❑❑❑

10 REACTION DYNAMICS CHEMICAL KINETICS

Don't go through life, grow through life

— ***Eric Butterworth***

CHAPTER OUTLINE

10.1 INTRODUCTION

Chemical kinetics is the branch of chemistry which deals with the study of the rate with which a chemical reaction occurs and the factors that affect this rate. This information is especially useful for determining *how* a reaction occurs.

What is meant by the *rate* of a reaction? The rate of a reaction is the speed at which the products are formed and the reactants are consumed.

The study of chemical kinetics has been highly useful in determining the factors that influence the rate, maximum yield and conversion in industrial processes. The mechanism or the sequence of steps by which the reaction occurs can be known. It is also useful in selecting the optimum conditions for maximum rate and yield of the chemical process.

Generally, chemical reactions can be put into three categories:

1. Instantaneous Reactions: These reactions take place instantaneously and the period of completion of such reactions is about 10^{-14} to 10^{-16} seconds. The ionic reactions are the examples of such reactions.

(*i*) $AgNO_3 + NaCl \longrightarrow AgCl + NaNO_3$

2. Very Slow Reactions: These reactions take a very long time for completion and it may appear that the reactions have not occurred at all. For example, the rusting of iron

$$Fe(s) + H_2O(g) + CO_2(g) + \frac{1}{2}O_2(g) \longrightarrow \underset{\text{Rust}}{Fe(OH)HCO_3}$$

The reactions proceed so slowly that is becomes difficult to determine the speed of the reactions.

3. Moderate Reactions: There are some reactions which occur in conveniently measurable time and their rates can be measured. These reactions are called moderate reactions. For example:

(*i*) Decomposition of hydrogen peroxide

$$2H_2O_2(l) \longrightarrow 2H_2O(l) + O_2(g)$$

(*ii*) Inversion of canesugar

$$C_{12}H_{22}O_{11}(s) + H_2O(l) \xrightarrow{H^+} \underset{\text{glucose}}{C_6H_{12}O_6(s)} + \underset{\text{fructose}}{C_6H_{12}O_6(s)}$$

(*iii*) Hydrolysis of ethyl acetate

$$CH_3COOC_2H_5(l) + H_2O(l) \xrightarrow{H^+} CH_3COOH(l) + C_2H_5OH(l)$$

These reactions are the main category of reactions for the kinetic study.

10.2 RATE OF REACTION

The quantity of a reactant species consumed or the quantity of a product species formed in a unit time in a chemical reaction is called the rate or velocity of the reaction. Precisely, the *rate* is defined as the change in concentration of any reactant or the product with time.

$$\frac{\text{change in concentration } \Delta c \text{ in time } \Delta t}{\Delta t}$$

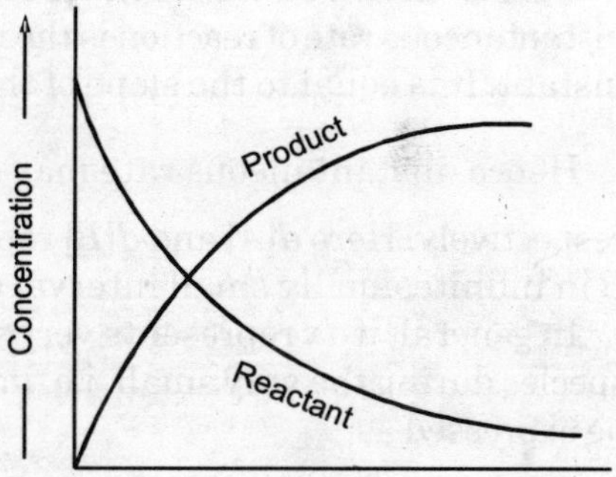

Fig. 10.1

Rate of the reaction can be measured by the

- the rate of formation of the product, or
- the rate of consumption of the reactant

Consider the reaction.

$$A \rightarrow B$$

where A is the reactant and B is the product

With the passage of time, the concentration of A goes on decreasing, whereas the concentration of B goes on increasing (Fig. 10.1).

The rate of reaction at any moment may be expressed as:

$$\text{Rate of reaction } (r) = \frac{\text{Amount of A consumed or decrease in concentration of A}}{\text{Time taken}}$$

$$\text{Rate of reaction } (r) = \frac{\text{Amount of B produced or increase in concentration of B}}{\text{Time taken}}$$

The decrease or increase in the concentration of reactants or products may also be expressed in terms of change in their concentration during the time interval, Δt, as

$$\text{Rate of reaction } (r) = \frac{-\Delta[A]}{\Delta t}$$

$$\text{Rate of reaction } (r) = \frac{+\Delta[B]}{\Delta t}$$

where $\Delta[A]$ gives the **decrease** in concentration of A and $\Delta[B]$ represents **increase** in concentration of B. The square brackets around the substances are used to express molar concentration (mole/litre). This rate is also known as **average rate** of reaction.

During a reaction, the concentration of reactants decreases with the passage of time and therefore, ΔA, (*i.e.*, final concentration – initial concentration) will be negative. The rate of reaction, then will be negative. But reaction rate is **always positive**, as also obtained from the rate of formation of the products. Therefore, the minus sign is put before $\frac{\Delta[A]}{\Delta t}$ so that the rate is positive.

As experiments show that for most reactions, the concentrations of all participants change most rapidly at the beginning of the reaction. That is, the [products] show the greatest rate of increase and the [reactants] show the greatest rate of decrease at the beginning. This means that the *rate of a reaction* **change with** *time,* so it is usual to define the rate over a very small time, Δt. Hence, we can express the rate as the *derivative* of concentration *with respect to time*

$$\text{rate} = \frac{d[\text{concentration}]}{dt}$$

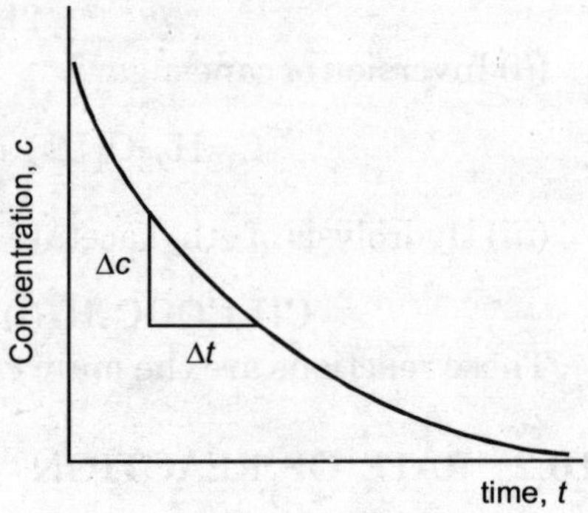

Fig. 10.2

This is known as **instantaneous rate** of reaction. Hence instantaneous rate of reaction is the reaction rate at a particular instant. It is equal to the slope of the line tangent to the concentration versus time graph.

Hence, instantaneous rate may be represented by $-\frac{d[A]}{dt}$ or $\frac{d[B]}{dt}$ instead of $\frac{\Delta[A]}{\Delta t}$ or $\frac{\Delta[B]}{\Delta t}$ respectively. Here $d[A]$ and $d[B]$ represent infinitesimally small changes in concentration of A and B in infinitesimally small interval of time, dt.

In general, if dx represents very small, i.e. infinitesimally small change in concentration of any species during the very small. i.e. infinitesimally small, change of time dt, the rate of reaction may be expressed as:

$$\text{Rate of reaction} = -\frac{dx}{dt}$$

(in terms of reactants; – sign indicates decrease in concentration)

or $$+\frac{dx}{dt}$$

(in terms of products; + sign indicates increase in concentration)

10.3 RELATION TO STOICHIOMETRIC EQUATIONS

Consider the reaction whose *stoichiometric equation* is

$$O_2 + 2H_2 \rightarrow 2H_2O$$

The stoichiometric equation shows how the number of moles of reactants and products are related. It must be balanced. This equation says that to form two moles of water, one mole of oxygen and two moles of hydrogen must react. It follows that the rate of consumption of hydrogen is twice that of oxygen.

$$\frac{d[H_2]}{dt} = 2\frac{d[O_2]}{dt}$$

Or the rate of formation of water is twice the rate of loss of oxygen, as two moles of water are formed from one mole of oxygen. As water is the product, the rate of change of its concentration is positive, i.e. the concentration is increasing with time; however, the rates of change of the concentrations of both hydrogen and oxygen are negative as these concentrations are decreasing with time. In derivative form:

$$\frac{d[H_2O]}{dt} = -2\frac{d[O_2]}{dt} \text{ or equivalently } -\frac{1}{2}\frac{d[H_2O]}{dt} = \frac{d[O_2]}{dt} \quad ...(1)$$

In order to get unique value of the reaction rate, it becomes necessary to divide each concentration change by its coefficient in the balanced equation of the reaction. These coefficients are the numbers in front of the species in the balanced chemical equation. So the stoichiometric coefficient is 1 for oxygen and 2 for both hydrogen and water. The rate of reaction, *r*, is defined for any species *A* as

$$r = \frac{1}{v_A}\frac{d[A]}{dt} \qquad ...(2)$$

where v_A is the stoichiometric coefficient of species *A* in the balanced chemical equation. In addition the stoichiometric coefficients are given negative signs for reactants and positive signs for products. This definition ensures that the rate is always *positive* and the *same* for a given reaction no matter which species is considered. For the above reaction rate is given by:

$$\text{Rate} = \frac{1}{2}\frac{d[H_2O]}{dt} = \frac{d[O_2]}{dt} \qquad ...(3)$$

For a general reaction,

$$aA + bB \rightarrow cC + dD$$

the reaction rate can be written in a number of different but equivalent ways,

$$\text{Rate} = -\frac{1}{a}\frac{d[A]}{dt} = \frac{1}{b}\frac{d[B]}{dt} = \frac{1}{c}\frac{d[C]}{dt} = \frac{1}{d}\frac{d[D]}{dt} \qquad ...(4)$$

As in our previous example, the negative signs account for material that is being consumed in the reaction and the positive signs account for material that is being formed in the reaction. The stoichiometry is preserved by dividing the rate of change of concentration of each substance by its stoichiometric coefficient.

Example. The rate of the reaction,

$2NO(g) + O_2(g) \rightarrow 2NO_2(g)$, is given by

$$\text{Rate} = \frac{1}{2}\frac{d[NO_2]}{dt} = -\frac{d[O_2]}{dt} = -\frac{1}{2}\frac{d[NO]}{dt}$$

10.4 RATE LAWS (RATE EQUATION)

The *rate law* is an expression indicating how the *rate* depends on the concentrations of the reactants and catalysts (if present).

A rate law of a reaction is a *mathematical expression* relating the rate of a reaction to the concentration of either reactants or products.

Often reactions are found to have rate laws of the form

$$r = k[A]^{\alpha}[B]^{\beta} ...$$

where *k* is the rate constant, characteristic of a particular reaction.

The rate law of a reaction must be proven *experimentally* by looking at either the appearance of products or disappearance of reactants.

Consider the following overall chemical reaction:

$$2NO(g) + O_2(g) \rightleftharpoons 2NO_2(g)$$

Here is an example of a rate law for the above reaction:

$$\text{Rate} = k[NO]^2[O_2]$$

In this rate law, the rate of disappearance of NO and O_2 is proportional to the concentrations of NO and O_2. Because this equation is an *experimental* law that describes the rate of the reaction, it is called the **rate law** for the reaction. The proportionality constant, *k,* is known as the **rate constant.**

10.5 ORDER AND MOLECULARITY

Law of Mass Action: According to the law of mass action, the rate of a chemical reaction is proportional to the product of the concentrations of the reactants.

For the reaction,

$$aA + bB \longrightarrow \text{Products}$$

$$\text{Rate} \propto [A]^a\,[B]^b$$

Low of mass Action is simply based upon the stoichiometry of the equation. On the other hand 'Rate Low' is based upon the experimental facts. Order of reaction is determined by Rate Law.

Order of **the reaction is defined as the sum of the powers of the concentration of the reactants in the rate law expression.** Or

Order of the reaction is the total number of reacting species whose concentration determine the rate of the reaction at a given temperature. The *power of the concentration* of a particular reactant in the rate law expression is called the order with respect to that reactant

For any general reaction

$$aA + bB \rightarrow cC + dD$$

If the rate law expression is;

$$\text{Rate} = r[A]^m\,[B]^n$$

where A and B represent the *molar concentrations* of A and B, m and n are the powers to which the concentrations must be raised, k is a constant of proportionality known as the rate constant. The values of 'm' and 'n' are *not* the stoichiometric numbers obtained from the balanced equation. These are obtained from the *experimental data.*

m is the *order* with respect to A,

n is the order with respect to B.

The overall order of the reaction will be given by *the sum of the powers of the concentration terms of the reactants in the rate law.*

The overall order of the reaction is $(m + n)$.

The rate law for a reaction **cannot** be predicted from the stoichiometry of the reaction; it must be determined **experimentally**. However, sometimes, the rate law is consistent with what we expect from the stoichiometry of the reaction.

$$2HI(g) \rightleftharpoons H_2(g) + I_2(g)$$

$$\text{Rate} = k[HI]^2$$

Often, however, it is not same as the stoichiometry of the reaction. Example:

$$2N_2O_5(g) \longrightarrow NO_2(g) + O_2(g)$$

$$\text{Rate} = k[N_2O_5]$$

Simple Rate Laws

A first order rate law is one in which the rate is proportional to the concentration raised to the power 1 (hence "first")

$$\text{rate} = k[A]^1$$

Here k is called a *first order rate constant.* The rate equation and the reaction it describes, is said to be *first order in A* or it is said that *the order with respect to A is one.*

Units of first order rate constant:

As the units of rate are concentration time^{-1} and the units of $[A]$ are concentration, the units of k are found as:

$$r = \frac{dx}{dt} = k\,[\text{conc.}]^n$$

$$k = \frac{dx}{dt} \times \frac{1}{[\text{conc.}]^n}$$

For first order,

$$k = \frac{r}{[\text{conc.}]^1} = \frac{dx/dt}{\text{conc.}} = \frac{\text{conc.}}{\text{Time}} \times \frac{1}{\text{conc}} = \text{Time}^{-1}$$

Hence first order rate constants have units of time^{-1}.

A second order rate law is the one in which the rate is proportional to the concentration raised to the power of 2.

$$r = k[A]^2$$

Here k is called a *second order rate constant.* The rate equation, and the reaction it describes, is said to be *second order in A,* or *the order with respect to A is two.*

Unit of second order rate constant:

The units of k are found to be conc^{-1} time^{-1}

$$k = \frac{dx}{dt} \times \frac{1}{[\text{conc}]^n}$$

For, second order,

$$k = \frac{dx}{dt} \times \frac{1}{[\text{conc}]^2} = \frac{\text{conc.}}{\text{Time}} \times \frac{1}{\text{conc.}^2} = \text{conc}^{-1}\ \text{time}^{-1}$$

Examples:

The decomposition of N_2O_5 is a **first-order reaction** because the rate of reaction depends on the concentration of N_2O_5 raised to the first power.

$$\text{Rate} = k[N_2O_5]$$

The decomposition of HI is a **second-order reaction** because the rate of reaction depends on the concentration of HI raised to the second power.

$$\text{Rate} = k[\text{HI}]^2$$

When the rate of a reaction depends on more that one reagent, we classify the reaction in terms of the order of each reagent.

For a zero-order reaction, the rate of reaction is a constant and does not depend on concentrations of the reactants. When the limiting reactant is completely consumed, the reaction abruptly stops.

$$\text{Rate} = k$$

Unit of zero order rate constant:

The rate constant, k, has units of conc. time^{-1}

$$k = \frac{dx}{dt} \times \frac{1}{[\text{conc.}]^0} = \frac{\text{conc}}{\text{Time}} = \text{conc. time}^{-1}$$

Example:

$$2\text{HI} \xrightarrow{\text{Au}} H_2 + I_2$$

$$\text{Rate} = k$$

$$H_{2(g)} + Cl_{2(g)} \xrightarrow{h\nu} 2HCl_{(g)}$$

$$\text{Rate} = k$$

Order	*Unit of k*
Zero	mol litre^{-1} time^{-1}
First	time^{-1}
Second	litre mol^{-1} time^{-1}

10.6 THE MOLECULARITY OF A REACTION

Molecularity is defined **as the number of atoms or molecules taking part in an elementary step leading to a chemical reaction.**

The overall chemical reaction may consist of many elementary steps. Each elementary reaction has its own molecularity which is equal to number of atoms or molecules. When a single molecule is consumed, the step is called **unimolecular**. When two molecules are consumed, it is **bimolecular**. When three molecules are colliding simultaneously then it is **ter molecular** reaction, although such reactions are very rare.

Elementary reactions are classified according to their molecularity

- In a **unimolecular Reaction**, a single molecule is involved in the reaction. Example, decomposition of NH_4NO_2 to N_2 and $2H_2O$

$$NH_4NO_2 \longrightarrow N_2 + 2H_2O$$

- **Bimolecular Reactions** involve the collision of pair of molecules. Bimolecular reactions are common. Example, oxidation of nitric oxide by ozone.

$$O_{3(g)} + NO_{(g)} \longrightarrow NO_{2(g)} + O_{2(g)}$$

In **Ter Molecular Reactions**, three molecules collide simultaneously. Molecularity rarely exceeds three. It is so because the occurrence of such reactions will involve the collision of four or more molecules simultaneously. However, chances of collisions of more than three molecules simultaneously, each having energy equal to or greater than the threshold energy, are rare.

$$2NO_{(g)} + O_{2(g)} \longrightarrow 2NO_{2(g)}$$

Thus, possible molecularities can be represented as:

Molecularity	**Elementary step**	**Rate law**
Unimolecular	$A \rightarrow$ product	$k[A]$
Bimolecular	$A + B \rightarrow$ product	$k[A]\,[B]$
	$A + A \rightarrow$ product	$k[A]^2$
Termolecular	$A + B + C \rightarrow$ product	$k[A]\,[B]\,[C]$
	$A + A + A \rightarrow$ product	$k[A]^3$
	$A + A + B \rightarrow$ product	$k[A]^2\,[B]$

Difference between Molecularity and Order

It is most important to distinguish **molecularity** from **reaction order:**

1. The **reaction order** (first order, etc.) is an empirical quantity, and obtained from the experimental rate law.
2. The **molecularity** (unimolecular/bimolecular) refers to an elementary reaction proposed as an individual step in a mechanism. In case of complex reactions, the overall molecularity has no significance because each step in the reaction has its own molecularity.

Molecularity	*Order*
It is the number or molecules of reactants taking part in elementary step of a reaction.	It is the sum of the powers of the concentration terms in the rate law expression.
It is theoretical concept	It is experimentally determined
It has integral values only	It can have fractional values also
It cannot be zero fractional or negative	It can be zero, fractional or negative
It provides no information on reaction mechanism.	The slowest step in the reaction can be judged by the order of the reaction and this gives further information about the mechanism
It is assigned for each elementary step of reaction.	It is assigned for overall reaction.

Let us illustrate the difference between the two with the following example:

Decomposition of nitrogen pentoxide can be represented as:

$$2N_2O_5 \longrightarrow 4NO_2 + O_2$$

From the stoichiometry it seems to be a **second order reaction**, but experimental verification reveals it to be **of first order**. There is however no correlation between the order and molecularity of a reaction, or between the stoichiometric representation and molecularity. The mechanism of the process consists of the following sequence of steps:

$N_2O_5 \rightarrow NO_2 \rightarrow NO_3$ (1) Unimolecular

$NO_2 + NO_3 \rightarrow NO_2 + O_2 + NO$ (2) Bimolecular

$NO + N_2O_5 \rightarrow 3NO_2$ (3) Bimolecular

$NO_2 + NO_3 \rightarrow N_2O_5$ (4) Bimolecular

Step (1) is unimolecular and the other three steps are bimolecular. The slowest step in this reaction is the decomposition of N_2O_5 to NO_2 and NO_3 and is the rate determining step and hence *first order* as only one molecule is involved.

Specific Rate Constant

In the rate equation k represents the rate constant. Rate constant or velocity constant (or) specific reaction rate is defined as the rate of the reaction when the concentration of each of the reactants is unity in the reaction.

When concentration of A and B is unity then, the rate constant is equal to the rate of the reaction.

The generalised rate equation for nth order reaction may be expressed as:

$$\text{Rate, } r = -\frac{dc}{dt} = k[c]^n \quad \text{...(1)}$$

When the concentration becomes unity, the specific reaction rate and the rate constant becomes equivalent.

$$r = k[1]^a = k = -\frac{dc}{dt}$$

Following are the important differences between rate and rate constant of a reaction.

Rate of Reaction	*Rate Constant of Reaction*
1. It represents the speed at which the reactants are converted into product at any instant.	1. It is the constant of proportionality in the rate law expression.
2. At any instant of time, the rate depends upon the concentration of reactants at that instant.	2. It refers to the rate of a reaction at the specific point when concentration of every reaction species is unity.
3. It decreases as the reaction proceeds.	3. It is constant and does not depend on the progress of the reaction.
4. Rate of rate determining step determines overall rate value.	4. It is an experimental value. It does not depend on the rate determining step.

10.7 PSEUDO-FIRST ORDER REACTION

The reactions which are not truly first order but which show first order kinetics under special conditions are called pseudo-first order reactions. In these reactions, one of the reactants is present in large excess. Since in elementary reactions, molecularity and order are identical, pseudo order reaction are also called pseudo-molecular reaction.

Let us consider a general reaction,

$$A + B \longrightarrow \text{Products}$$

Suppose the rate law for this reaction is,

$$\text{Rate} = k[A]\,[B] \qquad \text{...(1)}$$

Now, if $[B]$ is present in large excess, its concentration remains practically constant in the course of reaction and hence does not affect the rate. Therefore rate equation can be written in the form,

$$\text{Rate} = k'[A] \qquad \text{...(2)}$$

where $k' = k[B]$

Thus the actual order of the reaction is second order according to (1) but in practice it is first order (2). Thus the true law has been **forced into first-order form** by assuming the concentration of one the reactant B, constant, the effective rate law is called pseudo first order.

Many reactions in aqueous solutions that are reported as first or second order are actually pseudo first or second order. The water (solvent) participates in the reaction, but it is in such large excess that its concentration remains constant in the reaction.

Examples of Pseudo-Order Reactions

(1) Hydrolysis of an Ester. For example, ethyl acetate upon hydrolysis in aqueous solution using a mineral acid as catalyst forms acetic acid and ethyl alcohol.

$$\underset{\text{ethyl acetate}}{CH_3COOC_2H_5} + \underset{\text{(excess)}}{H_2O} \longrightarrow \underset{\text{acetic acid}}{CH_3COOH} + \underset{\text{ethyl alcohol}}{C_2H_5OH}$$

Here a large excess of water is used and any change in its concentration would be too small to be noticed, the rate law will be written as:

$$\text{Rate} = k[CH_3COOC_2H_5]\,[H_2O] = k'[CH_3COOC_2H_5]$$

The reaction is actually second-order but in practice it is found to be first order. Thus it is a pseudo-first-order reaction, the reaction appears to be dependent upon the concentration of ester only.

(2) Hydrolysis of Sucrose. Sucrose upon hydrolysis in the presence of a dilute mineral acid gives glucose and fructose.

$$\underset{\text{sucrose}}{C_{12}H_{22}O_{11}} + \underset{\text{(excess)}}{H_2O} \longrightarrow \underset{\text{glucose}}{C_6H_{12}O_6} + \underset{\text{fructose}}{C_6H_{12}O_6}$$

If a large excess of water is present, $[H_2O]$ is practically constant and the rate law may be written as:

$$\text{rate} = k[C_{12}H_{22}O_{11}]\,[H_2O] = k'[C_{12}H_{22}O_{11}]$$

The reaction though of second-order is experimentally found to be first-order. Thus it is pseudo-first-order reaction.

10.8 INTEGRATED RATE LAWS

The *differential rate law* describes how the rate of reaction varies with the concentrations of various species, usually reactants, in the system. The rate laws discussed above are differential rate laws and can be *integrated* in order to obtain an **integrated rate equation** that links concentrations of reactants or products with time.

- In other words differential rate law describes how the rate varies with change in concentration of reactants.
- Integrated rate law gives the variation of the concentration of a reactant with time as a chemical reaction proceeds.

Let us deduce the integrated rate laws for zero, first, second order reactions.

(1) Zero Orders Rate Law:

Consider the reaction,

$$A \longrightarrow P$$

For the zero order, the rate law takes the form

$$\text{Rate} = \frac{-d[A]}{dt} = k[A]^{\circ} = k \qquad ...(i)$$

The minus sign indicates the concentration of the reactant is decreasing with time.

Rearranging (*i*)

$$d[A] = -k\,dt \qquad ...(ii)$$

Integrating both sides, we get,

$$[A] = -kt + I \qquad ...(iii)$$

where I is the constant of integration.

The value of I, integration constant is determined from the initial conditions.

$$\text{At time } t = 0, [A] = [A]_0$$

where $[A]_0$ represents the initial concentration of A at time = 0. Therefore equation (*iii*) becomes,

$$[A]_0 = -k \times 0 + I$$

$$[A]_0 = I \qquad ...(iv)$$

Substituting, the value of I from (*iv*) in (*iii*).

We get,

$$[A] = -kt + [A]_0 \qquad ...(v)$$

$$kt = [A]_0 - [A]$$

or $$k = \frac{1}{t}\{[A]_0 - [A]\}$$ ...(*vi*)

This expression is the rate law expression for zero order reaction.

Half Life Period of Zero Order Reaction

Half life period of a reaction is defined as the time it takes for the concentration of the specified reagent to fall to half its initial value. This implies that, when

$$[A] = [A]_{1/2} \text{ at } t = t_{1/2}$$

Substituting, these values in equation (*vi*), we get,

$$t_{1/2} = \frac{1}{k}\{[A]_0 - [A]_0/2\}$$

$$= \frac{[A]_0}{2k}$$

$$\therefore t_{1/2} = \frac{[A]_0}{2k} \text{ for zero order reaction.}$$

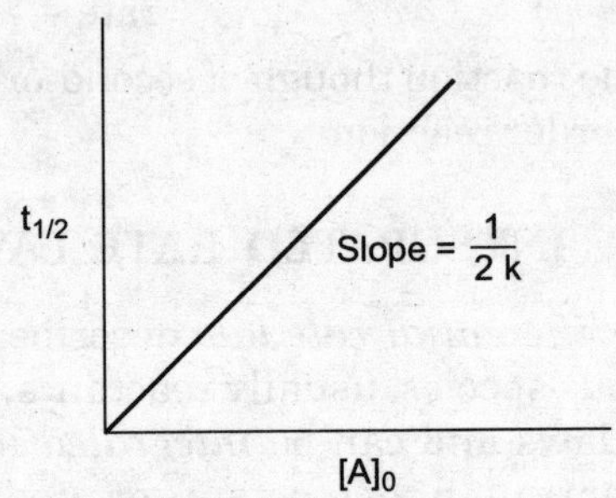

Fig. 10.3 Plot of half life ($t_{1/2}$) versus initial concentration for a zero order reaction

Thus half-life period of a zero order reaction is directly proportional to initial concentration. The plot of $t_{1/2}$ versus $[A]_0$ will be a straight line passing through the origin where slop = $\frac{1}{2k}$

(2) First Order Rate Law:

For a reaction,

$$A \longrightarrow P, \text{ a first order rate law takes the form}$$

$$\text{Rate} = -\frac{d[A]}{dt} = k[A]$$ ...(*i*)

where [*A*] represents the concentration of reactant at any time *t* and *k* is the rate constant.

Rearranging equation (*i*) we get,

$$\frac{d[A]}{[A]} = -k\,dt$$ ...(*ii*)

Integrating equation (*ii*) we get,

$$\int \frac{d[A]}{[A]} = -k \int dt$$

$$In[A] = -kt + I$$ ...(*iii*)

where *I* is the integration constant. The value of *I* can be determined as before from initial conditions.

At time $t = 0$, $[A] = [A]_0$, and equation (*iii*) becomes,

$$In[A]_0 = (-k \times 0) + I$$

$$In[A]_0 = I$$ ...(*iv*)

Substituting, the value of integration constant from (*iv*) into (*iii*) we get,

$$In[A] = -kt + In[A]_0$$ ...(*v*)

Rearranging (*v*),

$$In[A] - In[A]_0 = -kt$$

$$\frac{In[A]}{[A]_0} = -kt \qquad ...(vi)$$

or
$$In\frac{[A]_0}{[A]} = kt \qquad ...(vii)$$

Using logarithm to the base 10, we can write equation (*vii*) as

$$2.303 \log\frac{[A]_0}{[A]} = kt \qquad ...(viii)$$

or
$$\boxed{k = \frac{2.303}{t}\log\frac{[A]_0}{[A]}} \qquad ...(ix)$$

Equation (*ix*) is the first order rate equation. This equation is sometimes written in other form which is obtained as follows:

Suppose initial concentration of A is represented as 'a' moles per litre at time $t = 0$. After time t, suppose x moles per litre of it is decomposed. Therefore the concentration of A after time $t = (a - x)$ moles per litre.

So,
$$[A]_0 = a \text{ moles/L}$$
$$[A] = a - x \text{ moles/L}$$

Therefore, the equation (*ix*) becomes,

$$\boxed{k = \frac{2.303}{t}\log\frac{a}{a-x}} \qquad ...(x)$$

Equation (*vi*) can also be written as exponential form as:

$$In\frac{[A]}{[A]_0} = -kt \qquad ...(vi)$$

$$\boxed{[A] = [A]_0\, e^{-kt}} \qquad ...(xi)$$

Equation (*xi*) is the exponential form of the first order rate equation.

Characteristics of First Order Rate Equation

(*i*) Any reaction of first order must obey equations (*ix*), (*x*) and (*xi*).

(*ii*) Determination of rate constant of first order reaction:

From equation (*ix*),

$$\frac{k}{2.303}t = \log\frac{[A]_0}{[A]} = \log[A]_0 - \log[A]$$

or
$$\log[A] = \frac{-k}{2.303}t + \log[A]_0 \qquad ...(xii)$$

Equation (*xii*) is the equation of straight line ($y = mx + c$). Thus, if log $[A]$ is plotted against time t, the graph obtained should be a straight line. The rate constant k can be calculated from the slope of the graph.

(*iii*) Half life period of a first order reaction:

Equation (*x*) can be written as:

$$t = \frac{2.303}{k}\log\frac{a}{a-x} \qquad ...(xiii)$$

When half of the reaction is completed

$$x = \frac{a}{2} \text{ at } t_{1/2}.$$

Hence equation (*xiii*) becomes,

$$t_{1/2} = \frac{2.303}{k} \log \frac{a}{a - a/2}$$

$$= \frac{2.303}{k} \log 2 = \frac{0.693}{k}$$

$$\therefore t_{1/2} = \frac{0.693}{k} \qquad ...(xiv)$$

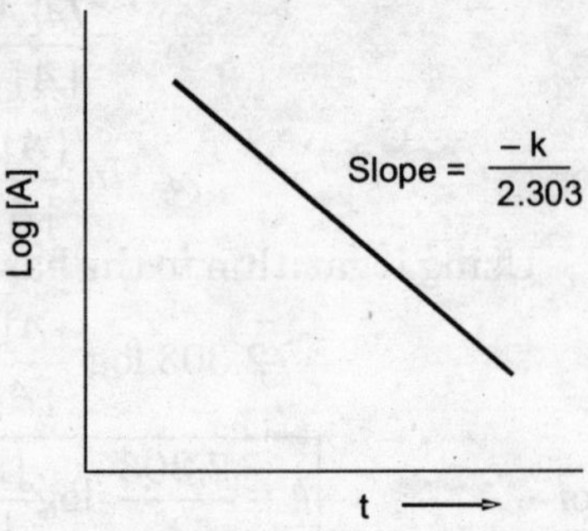

Fig. 10.4 Plot of log [A] vs t

Thus for *the first order reaction the half life period is independent of the initial concentration of the reactant.*

(3) Second Order Rate Law

Second order reactions may be represented as:

Type 1: When there is only one reactant involved

$$2A \longrightarrow \text{Products}$$

$$\frac{dx}{dt} = k[A]^2 \qquad \text{order} = 2$$

Type 2:

$$A + B \longrightarrow \text{Products}$$

$$\frac{dx}{dt} = k[A][B] \qquad \text{order} = 2$$

Now let us discuss the second order rate equation for these two types.

Type 1:

$$2A \longrightarrow \text{Products}$$

$$\frac{-d[A]}{dt} = k[A]^2 \qquad ...(i)$$

or

$$\frac{-d[A]}{[A]^2} = k\,dt \qquad ...(ii)$$

On integration,

$$\int \frac{-d[A]}{[A]^2} = \int k\,dt \quad \text{or} \quad \frac{1}{[A]} = kt + I \qquad ...(iii)$$

where I is the integration constant. I can be evaluated by putting $[A] = [A]_0$ at $t = 0$ in equation (*ii*)

$$\frac{1}{[A]_0} = k \times 0 + I \qquad ...(iv)$$

$$I = \frac{1}{[A]_0} \qquad ...(v)$$

substituting the value of I from equation (*v*) in equation (*iii*) we get,

$$\frac{1}{[A]} = kt + \frac{1}{[A]_0}$$

or

$$\boxed{\frac{1}{[A]} - \frac{1}{[A]_0} = kt} \qquad ...(vi)$$

Let if initial concentration of the reactant $[A]_0 = a$ at $t = 0$. After time t, let x moles of A have reacted. Then the concentration of A after time t, $[A] = (a - x)$. Putting these in equation (*vi*) we have,

$$\frac{1}{(a-x)} - \frac{1}{a} = kt \qquad ...(vii)$$

or
$$\frac{x}{a(a-x)} = kt$$

or
$$\boxed{k = \frac{1}{t}\cdot\frac{x}{a(a-x)}} \qquad ...(viii)$$

This is another form of integrated rate law for second order.

Characteristics of second order rate law when only one reactant is involved.

(i) Half Life Period

In equation (viii) if

$$x = \frac{1}{2} \text{ at } t = t_{1/2}\text{, then,}$$

$$k = \frac{1}{t_{1/2}}\cdot\frac{a/2}{a(a-a/2)} = \pm\frac{1}{t_{1/2}\cdot a}$$

or
$$\boxed{t_{1/2} = \frac{1}{k_a}} \qquad ...(ix)$$

From equation (*ix*) it is evident that half life period of second order reaction is inversely proportional to the concentration of the reactant.

(ii) Determination of rate constant of second order reaction.

From equation (*viii*), the rate constant of second order is given by,

$$k = \frac{x}{at(a-x)}$$

This equation can be written as,

$$\frac{1}{a-x} = kt + \frac{1}{a} \qquad ...(x)$$

$$y = mx + b$$

This is the equation of a straight line $y = mx + b$, where $y = \frac{1}{a-x}$, hence a plot of $\frac{1}{a-x}$ against time t will be a straight line (Fig. 10.5). The rate constant k = slope.

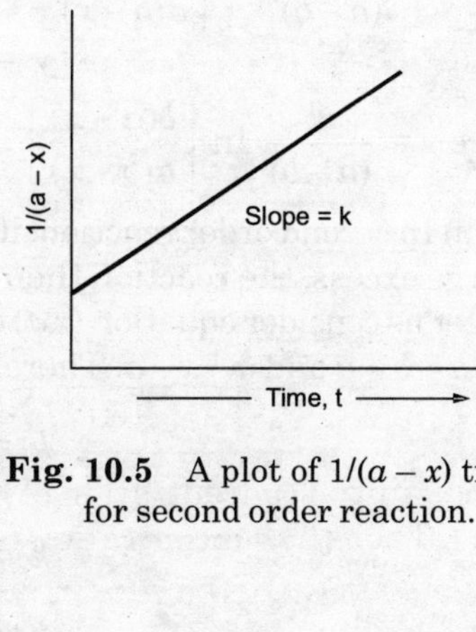

Fig. 10.5 A plot of $1/(a-x)$ time t for second order reaction.

Type 2: $A + B \longrightarrow$ Products

$$\frac{dx}{dt} = \frac{-d[A]}{dt} = \frac{-d[B]}{dt}$$

$$= k[A]\,[B] \qquad ...(xi)$$

Let the initial concentration of A and B be 'a' and 'b' moles respectively. After time 't' let 'x' moles of each been converted into products. The remaining concentration of A and B after time t will be $(a-x)$ and $(b-x)$ respectively

$$\therefore \qquad [A] = a - x \qquad \text{after time } t$$
$$[B] = b - x \qquad \text{after time } t$$

Putting these in (*xi*) we get,

$$\frac{dx}{dt} = k(a-x)(b-x) \qquad \text{...(xii)}$$

Separating the variables,

$$\frac{dx}{(a-x)(b-x)} = k\,dt \qquad \text{...(xiii)}$$

On integration,

$$\int_0^x \frac{dx}{(a-x)(b-x)} = \int_0^t k\,dt \qquad \text{...(xiv)}$$

$$= \frac{-1}{a-b}\int_0^x \left[\frac{1}{(a-x)} - \frac{1}{(b-x)}\right]dx = dt \qquad \text{...(xv)}$$

$$= \frac{1}{(a-b)}[\ln(a-x) - \ln(b-x)]_0^x = k \cdot t \qquad \text{...(xvi)}$$

$$= \frac{1}{(a-b)}\left[\ln\left[\frac{(a-x)}{a} - \ln\frac{(b-x)}{b}\right]\right] = k \cdot t \qquad \text{...(xvii)}$$

$$= \frac{1}{(a-b)} \ln \frac{b\,(a-x)}{a\,(b-x)} = k \cdot t \qquad \text{...(xviii)}$$

or

$$k = \frac{1}{t(a-b)} \ln \frac{b(a-x)}{b(b-x)} \qquad \text{...(xix)}$$

Changing the logarithm to the base 10, we get,

$$\boxed{k = \frac{2.303}{t(a-b)} \log \frac{b(a-x)}{a(b-x)}} \qquad \text{...(xx)}$$

Equation (*xvi*) is the rate order equation for the second order reaction, when there are two different reactants involved.

Characteristics:

(*i*) The above equation is the equation of a straight line

$$\frac{1}{(a-b)} \ln \left[\frac{b(a-x)}{a(b-x)}\right] = k \cdot t \qquad \text{...(xxi)}$$

$$y = mx$$

where $y = \dfrac{1}{(a-b)} \ln \left[\dfrac{b(a-x)}{a(b-x)}\right]$ and $x = t$, with slope = k)

(ii) In second order reaction, if one of the reactant is present in large excess, the reaction then becomes first order reaction.

Let us consider equation (*xxi*) once again and assume $a >> b$, then $a - b \approx a$ and $a - x \approx a$. Therefore equation (*xxi*) reduces to

$$k = \frac{2.303}{(a-b)t} \log \frac{b(a-x)}{a(b-x)}$$

$$k = \frac{2.303}{at} \log \frac{b}{(b-x)} \qquad \text{...(xxii)}$$

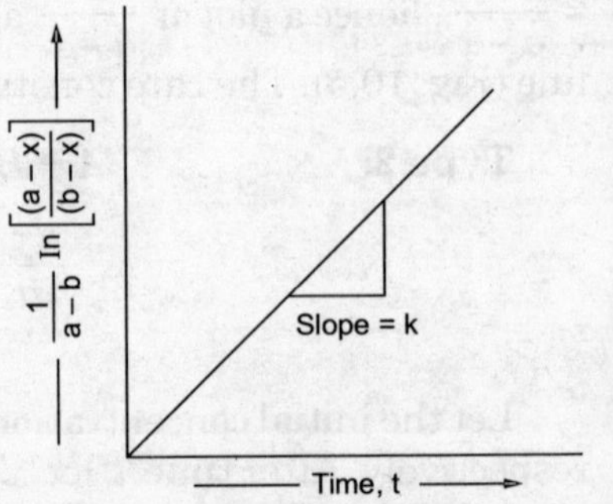

Fig. 10.6

where b = Initial concentration of reactant B

$b - x$ = Concentration of B after time t

a = The concentration of A, present in excess and hence constant.

$\therefore a \times k = k'$ (another constant)

Equation (xvii) becomes

$$k' = \frac{2.303}{t} \log \frac{b}{b-x} \quad \text{...(xxiii)}$$

The above equation is equivalent to that of first order rate equation.

A summary of the relationships for first and second order reactions is given below:

Order	*Differential rate law*	*Integrated rate law*	*Linear plot*	*Half life*	*Units*
Zero	$\frac{-d[A]}{dt} = k[A]^0$	$[A] = -kt + [A]_0$	$[A]$ vs t	$\frac{[A]_0}{2k}$	mol $L^{-1}s^{-1}$
First order	$\frac{-d[A]}{dt} = k[A]$	$[A] = [A]^0 e^{-kt}$	$\log [A]$ vs t slope = $\frac{-k}{2.303}$	$\frac{0.693}{k}$	s^{-1}
Second	$\frac{-d[A]}{dt} = k[A]^2$	$\frac{1}{[A]} - \frac{1}{[A]_0} = kt$	$\frac{1}{[A]}$ vs t slope = k	$\frac{1}{k[A]}$	L $mol^{-1}s^{-1}$

10.9 THEORIES OF REACTION RATES

(I) Collision Theory

According to this theory, for a reaction to occur, reactant molecules must collide with each other. But not sufficient in condition. The reaction can only takes place when these collisions are *"effective collisions"*. In order for a molecular collision to be effective it must meet two conditions:

- Energy barrier: The reactant molecules must collide with energy greater than some minimum value. This minimum value energy of collision required for two molecules to react is called threshold energy. Collisions with energy greater than this can result in the reaction. This implies that there is an energy barrier between reactants and products which must be crossed before products can be formed.
- Orientation barrier: The collision must occur with proper orientation

1. Energy barrier: Concept of Activation Energy:

The energy required for crossing the energy barrier comes from the kinetic energy of reacting molecules. *Hence, the minimum extra energy over and above the average potential energy of the reactants, which must be supplied so that the reactants are able to cross over the energy barrier is called the activation energy (E_a).*

Activation energy = (Threshold energy) – (Average energy of the reactants)

Each reaction has a characteristic value of energy of activation. At a given temperature, a reaction with low activation energy will proceed faster than a reaction with high energy of activation.

The effect of temperature on reaction rate can be explained from the energy distribution among molecules at different temperatures. Consider a reaction at two different temperatures T_1 and T_2, such that $T_2 > T_1$. This is represented in Fig. 10.8.

It is observed that the energy distribution curve is flatter at higher T and is shifted toward higher energy region. This is because, the kinetic energy of the molecules are higher at higher temperatures. Also, the fraction of molecules with energy treater than the threshold as indicated by the shaded portion is almost *double. It is very clear, that the rate of reaction increases with temperature, mainly due to increase in the number of collisions which occur with sufficient energy, i.e. with energy greater than the threshold.*

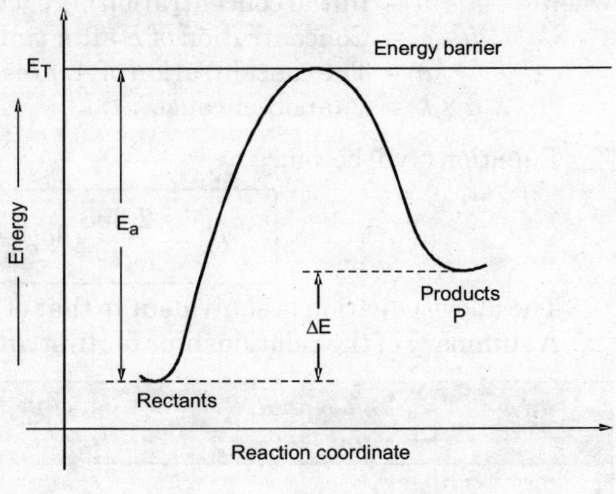

Fig 10.7 Energy profile of a chemical reaction.
E_a = Activation Energy
E_T = Threshold energy.

Quantitatively this relationship between the rate of a reaction proceeds and its temperature is determined by the *Arrhenius Equation.* At higher temperatures, the probability that two molecules will collide is higher. This higher collision rate results in a higher kinetic energy, which has an effect on the activation energy of the reaction.

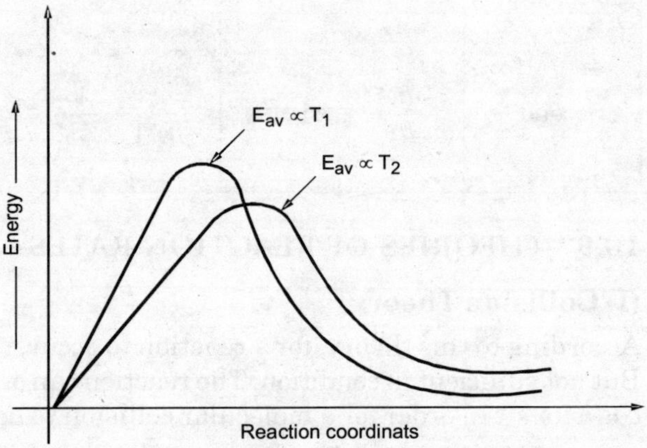

Fig. 10.8 Energy profile of a chemical reaction at two different temperatures

Arrhenius equation is given by:

$$k = Ae^{-E_a/RT}$$

where k is the rate coefficient, A is a constant (experimentally determined), E_a is the activation energy, R is the universal gas constant, and T is the temperature (in kelvin).

R has the value of 8.314×10^{-3} kJ mol^{-1} K^{-1}

In the logarithm form

$$\text{In } k = \text{In } A^{-E_a/RT}$$

If we plot In k against $\frac{1}{T}$, a straight line having a slope equal to $-E_a/R$ and an intercept In A can be obtained. From the slope and intercept, activation energy and frequency factor can be obtained.

E_a can be calculated by measuring the rate constant of a reaction at two different temperature. If k_1 and k_2 are the rate constants at temperature T_1 and T_2 respectively, then from Arrhenius equation, we have:

$$\text{In } k_1 = \text{In } A - \frac{E_a}{RT_1} \quad ...(i)$$

$$\text{In } k_2 = \text{In } A - \frac{E_a}{RT_2} \quad ....(ii)$$

Subtracting eq. (*i*) from eq. (*ii*), we have

$$\text{In } k_2 - \text{In } k_1 = \frac{E_a}{R}\left[\frac{1}{T_1} - \frac{1}{T_2}\right]$$

or,

$$\text{In } \frac{k_2}{k_1} = \frac{E_a}{R}\left[\frac{T_2 - T_1}{T_1 T_2}\right]$$

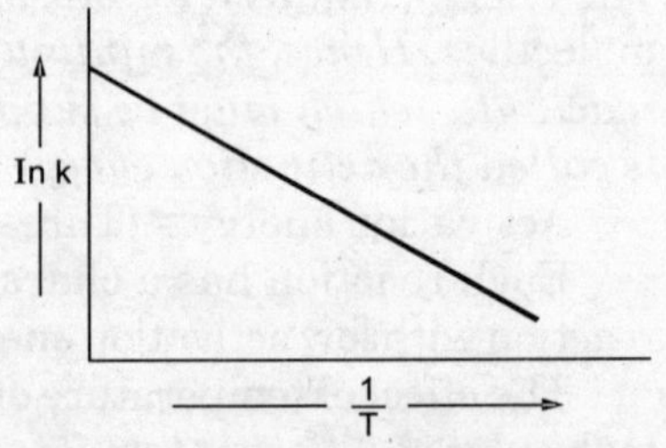

Fig. 10.9 Plot of In k against 1/T

or, $$E_a = \frac{RT_1T_2}{T_2 - T_1} \text{ In } \frac{k_2}{k_1}$$

Hence, by measuring rate constants at two different temperatures, E_a can easily be found out.

2. Orientation barrier

As discussed above, the rate of reaction also depends on the fraction of collisions that occur with proper orientation. This factor is *independent* of temperature. When two speeding objects collide, the extent of damage that occurs depends on the speeds of the project collide, the extent of damage that occurs depends on the speeds of the object and the position of impact.

Consider a simple reaction involving a collision between two molecules - ethene, $CH_2 = CH_2$, and hydrogen chloride, HCI, for example, These react to give chloroethane.

$CH_2 = CH_2 + HCI \longrightarrow CH_3\ CH_2CI$

As a result of the collision between the two molecules, the double between the two carbons is converted into a single bond. A hydrogen atom gets attached to one of the carbons and a chlorine atom to the other.

The reaction can only happen if the hydrogen end of the H–CI bond approaches the carbon-carbon double bond. Any other collision between the two molecules doesn't work. The two simply bounce off each other.

Collision theory explains the temperature dependence of rate of a reaction based specifically in the concept of effective collisions. Another theory that was proposed a few years later by Eyring, is the Transition State Theory.

(II) Transition State Theory

Transition state theory is also known as *activated-complex theory or theory of absolute reaction rates.* Transition state theory is a large improvement on simple collision theory,

According to this theory, the reactant molecules must come together to form an arrangement of atoms/molecules that facilitates the formation of products.

The theory was developed by Eyring and is also known as Absolute Reaction Rate Theory. The important points can be summarized as:

- Reactant molecules approach each other to form an activated complex by rearrangement of atoms and redistribution of energy.
- The activated complex is in equilibrium with reactants.
- The potential energy of the activated complex is very high and activated complex is very unstable.
- The activated complex although is unstable but it is considered to be a definite molecule with an independent identity.
- The activated complex subsequently breaks up to form products at a definite rate.

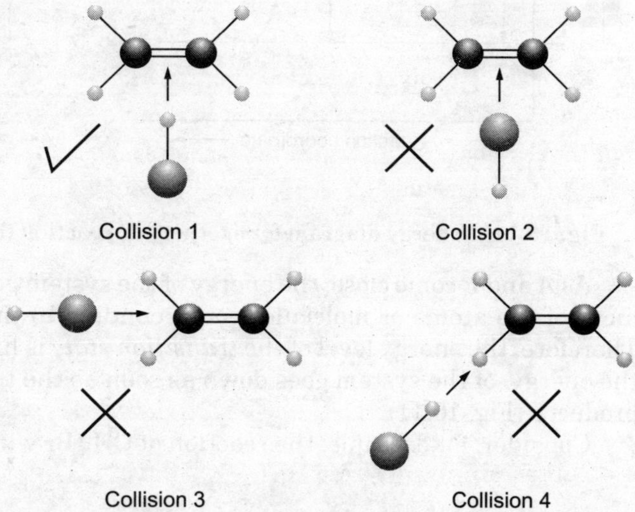

Fig. 10.10 Proper and improper orientations of colliding molecules

Schematically,

$$\text{Reactants} \rightleftharpoons \text{Activated Complex} \rightleftharpoons \text{Products}$$

A simple way of visualising the concept of the transition state is to think of a reaction proceeding along a reaction coordinate. For e.g., in a reaction where two reactants A and B react to form product

$$A + B \longrightarrow \text{Products},$$

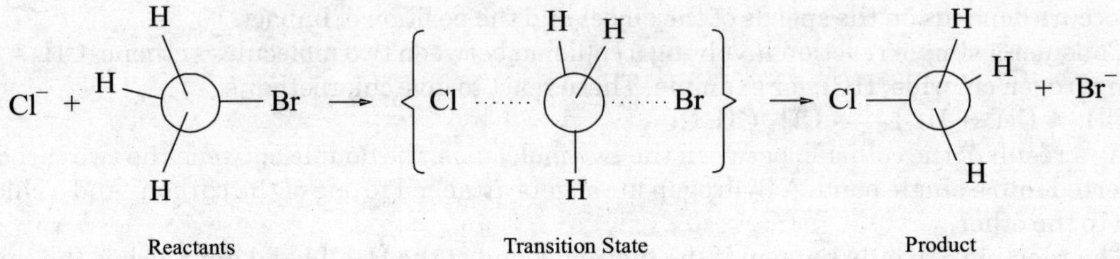

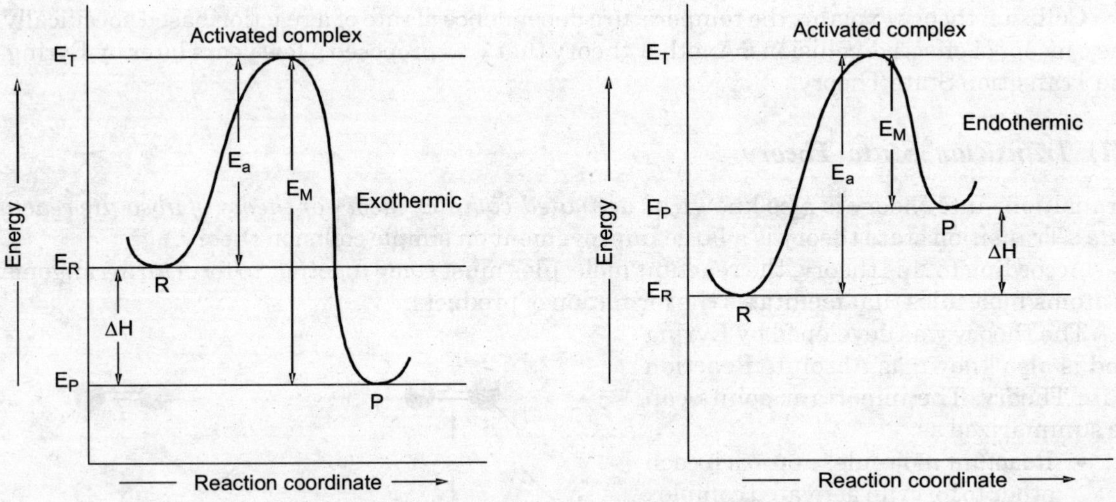

Fig. 10.11 Energy diagram for exothermic reaction (b) Energy diagram for endothermic reaction

As A and B come close, the energy of the system goes up and reaches a maxima. The arrangement of the atoms or molecules corresponding to this energy maxima is the transition state. Therefore, the energy level of the *transition state* is higher than that of the reactants. Therefore, the energy of the system goes down as soon as the transition state breaks and transforms into products. (Fig. 10.11)

Consider, for example, the reaction of CH_3 Br with CI in water at 298K.

SOLVED NUMERICALS

(I) Based on Rate of Reaction

Example 10.1. In the reaction $N_2(g) + 3H_2(g) \longrightarrow 2NH_3(g)$, the rate of formation of NH_3 is 5.0×10^{-7} mol^{-1} L^{-1} s^{-1}. Calculate

(*i*) the general rate of the reaction
(*ii*) the rate of consumption of N_2
(*iii*) the rate of consumption of H_2

Solution: In the reaction $N_2(g) + 3\,H_2(g) \longrightarrow 2\,NH_3(g)$

$$\text{Rate of formation of } NH_3 = \frac{d\,[NH_3]}{dt} = 5 \times 10^{-7} \text{ mol L}^{-1}\text{ s}^{-1}$$

$$(i) \textit{ General rate } = \frac{d[N_2]}{dt} = -\frac{1}{3}\frac{d[H_2]}{dt} = \frac{1}{2}\frac{d[NH_3]}{dt}$$

$$(ii) \text{ Rate of consumption of } N_2 = -\frac{d[N_2]}{dt} = \frac{1}{2}\frac{d[NH_3]}{dt}$$

$$= \frac{1}{2} \times 5 \times 10^{-7} \text{ mol L}^{-1}\text{s}^{-1} = 2.5 \times 10^{-7} \text{ mol L}^{-1}\text{ s}^{-1}$$

$$(iii) \text{ Rate of consumption of } H_2 = -\frac{d[H_2]}{dt} = \frac{3}{2}\frac{d[NH_3]}{dt} = 7.5 \times 10^{-7} \text{ mol L}^{-1}\text{ s}^{-1}$$

Example 10.2. The following reaction was studied in a closed vessel

$$2\,N_2O_5(g) \rightleftharpoons 4\,NO_2(g) + O_2(g)$$

It was found that the concentration of NO_2 Increases by 2.0×10^{-2} mol L^{-1} in five seconds. Calculate

(*i*) the rate of reaction
(*ii*) the rate of change of concentration of N_2O_5

Solution: (*i*) Rate of reaction $= \dfrac{1}{4}\dfrac{d[NO_2]}{dt}$

$$\text{But} \quad \frac{d[NO_2]}{dt} = \frac{2.0 \times 10^{-2} \text{ mol L}^{-1}}{5s} = 4 \times 10^{-3} \text{ mol L}^{-1}\text{ s}^{-1}$$

$$\Rightarrow \quad \text{Rate of reaction} = \frac{1}{4} \times 4 \times 10^{-3} = 10^{-3} \text{ mol L}^{-1}\text{s}^{-1}$$

(*ii*) Rate of change of concentration of N_2O_5

$$= -\frac{d[N_2O_5]}{dt} = -\frac{1}{2} \times \text{Rate of formation of } NO_2$$

$$= -\frac{1}{2}\frac{d[NO_2]}{dt} = -\frac{1}{2} \times (4 \times 10^{-3} \text{ mol L}^{-1}\text{ s}^{-1})$$

$$= 2 \times 10^{-3} \text{ mol L}^{-1}\text{s}^{-1}$$

Example 10.3. For the decomposition of dinitrogen pentoxide at 200 C,

$$N_2O_3\,(g) \longrightarrow N_2O_4\,(g) + \frac{1}{2}O_2\,(g),$$

if the initial pressure is 114 mm and after 25 minutes of the reaction, total pressure of the reaction mixture is 133 mm, calculate the average rate of reaction in (*a*) atm min^{-1} (*b*) mol L^{-1} s^{-1}.

Solution: (*a*) *If P* is the decrease in pressure of N_2O_3 in 25 minutes, then

$$N_2O_5\,(g) \longrightarrow N_2O_4\,(g) + \frac{1}{2}O_2\,(g)$$

Initial pressure	114 mm	0	0
After 25 min	$114 - p$	p	$\frac{1}{2}p$

$$\text{Total pressure} = 114 - p + p + \frac{1}{2}p = 114 + \frac{1}{2}p = 133 \text{ mm (Given)}$$

$$\therefore \qquad p = 2 \times (133 - 114) = 38 \text{ mm}$$

i.e., Decrease in pressure in 25 min = 38 mm = $\frac{38}{760}$ atm

$$\therefore \text{ Average rate of reaction} = \frac{38/760\text{atm}}{25 \text{ min}} = 0.002 \text{ atm min}^{-1}$$

(*b*) Applying $pV = nRT$, i.e. $\frac{n}{V} = \frac{P}{RT}$

$$\text{Decrease in molar concentration} = \frac{38/760 \text{ atm}}{0.0821 \text{ L atm K}^{-1} \times 473 \text{ K}} = 0.0012875 \text{ mol L}^{-1}$$

$$\therefore \qquad \text{Average rate of reaction} = \frac{0.0012875 \text{ mol L}^{-1}}{25 \times 60\, s} = 8.58 \times 10^{-7} \text{ mol L}^{-1} \text{ s}^{-1}$$

(II) Based on First Order Reactions

Example 10.4. What is the rate of a first order reaction with rate constant 3.0×10^{-10} s^{-1} when concentration of the reactant is 1.5 M.

Solution: Rate $= -\frac{d[A]}{dt} = k\,[A] = 3.0 \times 10^{-10}\text{ s}^{-1} \times 1.5 \text{ M} = 4.5 \times 10^{-10} \text{ Ms}^{-1}$

Example 10.5. The reaction $N_2O_3 \longrightarrow 2NO_2 + \frac{1}{2}O_2$ is of first order in N_2O_5. Its rate constant is 6.2×10^{-6} s^{-1}. If in the beginning $[N_2O_5]$ is 15 mol L^{-1}, calculate the rate of reaction in the beginning.

Solution: Rate of reaction $= k[A] = 6.2 \times 10^{-6} \text{ s}^{-1} \times 15 \text{ mol L}^{-1} = 9.3 \times 10^{-5} \text{ mol L}^{-1} \text{ s}^{-1}$

Example 10.6. The initial concentration of N_2O_5 in the following first order reaction

$$N_2O_5(g) \longrightarrow 2NO_2(g) + \frac{1}{2}O_2(g)$$

was 1.24×10^{-2} mol L^{-1} at 45°C. The concentration of N_2O_5 after 60 minutes was 0.20×10^{-2} mol L^{-1}. Calculate the rate constant of the reaction at 45° C.

Solution:

$$\log \frac{[A]_1}{[A]_2} = \frac{k(t_2 - t_1)}{2.303}$$

or

$$k = \frac{2.303}{t_2 - t_1} \log \frac{[A]_1}{[A]_2} = \frac{2.303}{60 \text{ min} - 0 \text{ min}} \log \frac{1.24 \times 10^{-2} \text{ mol L}^{-1}}{0.20 \times 10^{-2} \text{ mol L}^{-1}}$$

$$= \frac{2.303}{60} \log 6.2 \text{ min}^{-1} = \frac{2.303}{60} \times 0.7924 \text{ min}^{-1} = \mathbf{0.0304 \text{ min}^{-1}}$$

Example 10.7. The rate law for the decomposition of hydrogen peroxide at 70°C

$$2H_2O_2(aq) \longrightarrow 2H_2O(l) + O_2(g)$$

is first order in H_2O_2. The half-life of the reaction at 70 °C is 20 min. What is the concentration of hydrogen peroxide 60 minutes after the beginning of the reaction in which concentration of hydrogen peroxide is 0.30 mol L^{-1}?

Solution: Rate constant, $k = \dfrac{9.693}{t_{1/2}} = \dfrac{0.693}{20 \text{ min}} = 0.03465 \text{ min}^{-1}$

$$\log \frac{[A]_1}{[A]_2} = \frac{k(t_2 - t_1)}{2.303} = \frac{0.03465 \text{ min}^{-1} \times (60 \text{ min} - 0 \text{ min})}{2.303}$$

$$= \frac{0.03465 \times 60}{2.303} = 0.9027$$

$$\frac{[A]_1}{[A]_2} = \text{Antilog } 0.9027 = 7.993$$

$$[A]_2 = \frac{[A]_1}{7.993} = \frac{0.30 \text{ mol L}^{-1}}{7.993} = 0.0375 \text{ mol L}^{-1}$$

Example 10.8. A solution of H_2O_2 when titrated against $KMnO_4$ solution at different time intervals gave the following results:

t (minutes)	0	10	20
Vol $KMnO_4$ used for 10 ml H_2SO_2	23.8 ml	14.7 ml	9.1 ml

Show that the decomposition of H_2O_2 is a first order reaction.

Solution: The integrated rate equation for first order reaction is:

$$k = \frac{2.303}{t} \log \frac{a}{a - x}$$

Since, volume of $KMnO_4$ used in the titration is a measure of concentration of H_2O_2 in solution,

$a = 23.8$ ml
$(a - x) = 14.7$ when $t = 10$ mts
$(a - x) = 9.1$ when $t = 20$ mts

Substituting, these values in the rate equation above, we have

$$k = \frac{2.303}{10} \log \frac{23.8}{14.7} = 0.2303 (\log 23.8 - \log 14.7)$$

$$= 0.2303(1.3766 - 1.1673) = 0.04794$$

and

$$k = \frac{2.303}{20} \log \frac{23.8}{9.1} = 0.10165 (\log 23.8 - \log 9.1)$$

$$= 0.10165(1.366 - 0.9595) = 0.04806$$

Since, the value of *k* is constant, it is first order reaction.

Example 10.9. The rate constant of reaction with respect to reactant *A* is 6 min^{-1}. If we with [*A*] = 0.8 mol L^{-1}, when would [*A*] will reach this concentration of 0.08 mol L^{-1}.

Solution:

$$t = \frac{2.303}{k}\log\frac{[A]_o}{[A]_t} = \frac{2.303}{6\ \text{min}^{-1}}\log\frac{0.8\ \text{mol L}^{-1}}{0.08\ \text{mol L}^{-1}}$$

$$= \frac{2.303}{6\ \text{min}^{-1}}\log 10 = \frac{2.303}{6}\times 1\ \text{min} = 0.38\ \text{min}$$

Example 10.10. The following date was obtained on hydrolysis of methyl acetate at 25 °C in 0.35 N hydrochloric acid. Establish that it is a first order reaction.

t (secs) ml alkali	0	4500	7140	∞
used	24.36	29.32	31.72	47.15

Solution:

For a first order reaction, $k = \frac{2.303}{t}\log\frac{a}{a-x}$

At any time, the volume of alkali used is needed for the acid present as catalyst and the acid produced by hydrolysis.

The volume of alkali used for total change from t_0 to t_∞ gives the initial concentration of ester. Thus.

$$a = 47.15 - 24.36 = 22.79\ \text{ml}$$

$$(a-x)\ \text{after 4500 sec} = 47.15 - 29.32 = 17.83\ \text{ml}$$

$$(a-x)\ \text{after 7140 sec} = 47.15 - 31.72 = 15.43\ \text{ml}$$

Substituting, these values in the rate equation, we have,

$$k = \frac{2.303}{4500}\log\frac{22.79}{17.83} = 0.000\ 654$$

$$k = \frac{7.303}{7140}\log\frac{22.79}{15.43} = 0.000546$$

The value of '*k*' are fairly constant and hence it is first order reaction

Example 10.11. The decomposition of hydrogen Peroxide into water and oxygen was studied. The concentration of H_2O_2 was determined by titrating against standard $KMnO_4$ solution and the following data were obtained:

Time (min)	0	10	20	30	40	50
Conc. of H_2O_2	0.1297	0.1162	0.1045	0.0940	0.0845	0.0760

(*i*) Show that reaction is of first order.

(*ii*) Calculate the rate constant.

Solution: If the reaction is of first order, it must obey the equation.

$$k = \frac{2.303}{t}\log\frac{[A]_0}{[A]}$$

Here $[A]_0 = 0.1297$ and the values of *k* can be calculated by substituting the concentration of *A* at different times.

$$t = 10, [A] = 0.1162, k = \frac{2.303}{10} \log \frac{0.1297}{0.1162} = 0.01099$$

$$t = 20 \text{ min}, [A] = 0.1045, k = \frac{2.303}{20} \log \frac{0.1297}{0.1045} = 0.01080$$

$$t = 30 \text{ min}, [A] = 0.0940, k = \frac{2.303}{30} \log \frac{0.1297}{0.0940} = 0.01073$$

$$t = 40 \text{ min } [A] = 0.08445, k = \frac{2.303}{30} \log \frac{0.1297}{0.0940} = 0.01071$$

$$t = 50 \text{ min}, [A] = 0.0760, k = \frac{2.303}{50} \log \frac{0.1297}{0.0760} = 0.01069$$

Since, the value of k comes out to be almost constant, the reaction, therefore, is of first order. The mean value of $k = 0.01078 \text{ min}^{-1}$.

Example 10.12. A reaction that is of first order with respect to reactant A has a rate constant 6 min^{-1}, If we start with [A] = 5.0 mol L^{-1}, when would [A] reach the value of 0.05 mol L^{-1}?

Solution: For the first order reaction,

$$t = \frac{2.303}{k} \log \frac{[A]_0}{[A]}$$

$$[A]_0 = 5.0 \text{ mol L}^{-1}, k = 6 \text{ min}^{-1}.$$

$$t = \frac{2.303}{6} \log \frac{5.0}{0.05} = \frac{2.303}{6} \log 100 = 0.768 \text{ min.}$$

Example 10.13. What will be the initial rate of a reaction if its rate constant is 10^{-3} min^{-1} and the concentration of the reactant is 0.2 mol dm^{-3}? How much of the reactant will be converted into the products in 200 minutes?

Solution: Rate constant, $k = 10^{-3} \text{ min}^{-1}$.

$$\text{Initial rate} = k[A]$$

$$= (10^{-3} \text{ min}^{-1}) \times (0.2 \text{ mol dm}^{-3}) = 2 \times 10^{-4} \text{ mol dm}^{-3} \text{ min}^{-1}$$

Now, the integrated rate equation for the first order reaction is:

$$kt = 2.303 \log \frac{a}{a-x}$$

or
$$\log \frac{a}{a-x} = \frac{kt}{2.303}$$

$$t = 200 \text{ min}$$

$$k = 10^{-3} \text{ min}^{-1}$$

so that
$$\log \frac{a}{a-x} = \frac{10^{-3} \times 200}{2.303} = 8.68 \times 10^{-2}$$

or
$$\frac{a}{a-x} = 1.22 \qquad (\text{antilog } 8.68 \times 10^{-2} = 1.22)$$

or
$$a = 1.22a - 1.22x$$

or
$$1.22x = 1.22a - a - a = 0.22a$$

or
$$x = \frac{0.22a}{1.22} = 0.18a$$

$$\% \text{ conversion} = \frac{0.18a \times 100}{a} = 18\%$$

Example 10.14. For a first order reaction prove that the time required for the completion of 99.9% of reaction is about ten times the time taken for the completion of 50% of this reaction.

(*i*) When extent of reaction is 99.9%, then the amount of reactant unconsumed is

$$[A] = [A]_0 - 99.9\% \text{ of } [A]_0 = [A]_0 - 99.9\,[A]_0/100 = 0.1\,[A]_0/100 = 10^{-3}$$

$$\therefore \quad t\,(99.9\%) = \frac{2.303}{k}\log\frac{[A]_0}{[A]} = \frac{2.303}{k}\log\frac{[A]_0}{0.1[A]/100} = \frac{2.303}{k}\log 1000$$

$$= \frac{2.303}{k}\times 3 \qquad \text{...(1)} \because \log 1000 = 3$$

(ii) When extent of reaction is 50% then the amount of reactant unconsumed is

$$[A] = [A]0 - 0.50\% \text{ of } [A]_0 = [A]_0 - 50[A]_0/100 = 50[A]_0/100 = [A]_0/2$$

$$\therefore \quad t\,(50\%) = \frac{2.303}{k}\log\frac{[A]_0}{[A]} = \frac{2.303}{k}\log\frac{[A]_0}{[A]_0/2} = \frac{2.303}{k}\log 2$$

$$= \frac{2.303}{k}\times 0.3010 \qquad \text{...(2)} \because \log 2 = 0.3010$$

$$\text{Now} \quad \frac{t(99.9\%)}{t(50\%)} = \frac{2.303\times 3/k}{2.303\times 0.301/k} = \frac{3}{0.301} = 9.9966 = 10$$

That is, the time required for the completion of 99.9% of a first order reaction is about ten times the time taken for the completion of 50% of the same reaction.

(III) Based on Half Life Period

Example 10.15. Calculate the half life period of a first order reaction where the specific rate constant is (a) 200 s^{-1} (b) 2min^{-1} (c) 5 $year^{-1}$.

Solution: For the first order reaction,

$$t_{1/2} = \frac{0.693}{k}$$

(a) $$k = 200\ s^{-1}$$

$$\therefore \quad t_{1/2} = \frac{0.693}{200} = \mathbf{3.465 \times 10^{-3}\ s}$$

(b) $$k = 2\ min^{-1}$$

$$\therefore \quad t_{1/2} = \frac{0.693}{2} = \mathbf{0.3465\ min}$$

(c) $$k = 5\ year^{-1}$$

$$\therefore \quad t_{1/2} = \frac{0.693}{5} = \mathbf{0.1386\ year}$$

Example 10.16. The thermal decomposition of a compound is of first order. If 50% of the compound is decomposed in 120 minutes, how long will it take for 90% of the compound to decompose?

Solution: Let us first calculate, *k* as

$$k = \frac{0.693}{t_{1/2}}$$

$$\text{Here} \quad t_{1/2} = 120\ min$$

$$\therefore \quad k = \frac{0.693}{120\ min} = 5.77\times 10^{-3}\ min^{-1}$$

Now, for the first order reaction,

$$t = \frac{2.303}{k}\log\frac{[A]_0}{[A]}$$

If the initial concentration of *A* i.e. $[A]_0 = a$, then $[A] = a - a \times 0.90 = 0.10a$ because 90% of compound decomposes and only 10% is left.

$$\therefore \quad t = \frac{2.303}{5.77\times10^{-3}}\log\frac{a}{0.1a} = \frac{2.303}{5.77\times10^{-3}}\log 10 = \mathbf{399\ min}$$

Example 10.17. For a first order reaction, calculate the ratio between the time taken to complete three-fourth of the reaction and the time taken to complete half of the reaction.

Solution: For first order reaction,

$$t = \frac{2.303}{k}\log\frac{[A]_0}{[A]}$$

For 3/4 of a reaction to take place, $t = t_{3/4}$, $[A] = [A]_0 - 3/4[A]_0 = 1/4[A]_0$

Thus, $$t_{3/4} = \frac{2.303}{k}\log\frac{[A]_0}{1/4[A]_0} = \frac{2.303}{k}\log 4 \quad ...(i)$$

Now, for half of a reaction to take place,

$$t = t_{1/2};\ [A] = [A]_0 - 1/2[A]_0 = 1/2[A]_0$$

Thus,

$$t_{1/2} = \frac{2.303}{k}\log\frac{[A]_0}{[A]_0/2} = \frac{2.303}{k}\log 2$$

Dividing (i) by (ii)

$$\frac{t_{3/4}}{t_{1/2}} = \frac{\log 4}{\log 2} = \frac{0.6020}{0.3010} = 2$$

Thus, the time required for 3/4th of the reaction to occur is two times that required for half of the reaction.

(IV) Based on Second Order Reactions

Example 10.18. Hydrolysis of ethyl acetate by NaOH was studied by titrating 25 mL of the reaction mixture at different time intervals against standard acid. From the data given below, establish that this is a second order reaction.

t (min)	0	5	15	25
mL of acid used	16.00	10.24	6.13	4.32

Solution: The second order integrated rate equation is

$$k = \frac{1}{at}\cdot\frac{x}{a(a-x)}$$

The volume of acid used at any time is a measure of concentration of the unreacted substance at that them Therefore,

a, initial concentration = 16,00

after 5 mts $(a - x) = 10.24$ and $x = 5.76$

after 15 mts $(a - x) = 6.13$ and $x = 9.85$

after 25 mts $(a - x) = 4.32$ and $x = 11.68$

Substituting values in the rate equation (1), we have

$$k = \frac{1}{16 \times 5} \cdot \frac{5.76}{10.24} = 0.0070$$

$$k = \frac{1}{16 \times 5} \cdot \frac{9.85}{6.13} = 0.0067$$

$$k = \frac{1}{16 \times 25} \times \frac{11.68}{6.13} = 0.00675$$

The value of k being fairly constant, this reaction is of the second order.

(V) *Based on Arrhenius Equation*

Example 10.19. The rate of a reaction triples when temperature changes from 20° to 50°C. Calculate the energy of activation for such a reaction. (R = 8.314 JK^{-1} mol^{-1})

Solution: We know that

$$\log \frac{k_2}{k_1} = \frac{E_a}{2.303\,R}\left[\frac{T_2 - T_1}{T_1 T_2}\right]$$

$$\frac{k_2}{k_1} = 3,\ T_1 = 273 + 20 = 293 \text{ K}$$

$$T_2 = 273 + 50 = 323 \text{ K},\ R = 8.314 \text{ JK}^{-1}\text{ mol}^{-1}$$

$$\log 3 = \frac{E_a}{2.303 \times 8.314}\left[\frac{323 - 293}{323 \times 293}\right]$$

$$0.477 = \frac{E_a}{2.303 \times 8.314} \times \frac{30}{323 \times 293}$$

or

$$E_a = \frac{0.477 \times 2.303 \times 8.314 \times 323 \times 293}{30}$$

$$= 28.8 \times 10^3 \text{ J mol}^{-1} = 28.8 \text{ kJ mol}^{-1}$$

Example 10.20. The reaction,

$$2NO_2 \longrightarrow 2NO + O_2$$

has an activation energy of 110 kJ mol^{-1}. At 400°C, the rate constant is 7.8 mol^{-1}. $L^{-1}s^{-1}$. What is the value of rate constant at 430°C?

Solution: We know that

$$\log \frac{k_2}{k_1} = \frac{E_a}{2.303\,R}\left[\frac{1}{T_1} - \frac{1}{T_2}\right]$$

$E_a = 110$ kJ mol^{-1} = 110 × 103 J mol^{-1} $R = 8.314$ J mol^{-1} K^{-1}

$k_1 = 7.8$ mol^{-1} L s^{-1} $T_1 = 400 + 273 = 673$ K

$k_2 = ?$ $T_2 = 430 + 273 = 703$ K

Substituting the values,

$$\therefore \quad \log\frac{k_2}{7.8} = \frac{110\times10^3}{2.303\times8.314}\left[\frac{1}{673}-\frac{1}{703}\right]$$

$$\log\frac{k_2}{7.8} = \frac{110\times10^3\times(703-673)}{2.303\times8.314\times673\times703}$$

$$\log\frac{k_2}{7.8} = 0.3643 \text{ or } \frac{k_2}{7.8} = 2.31$$

$$k_2 = 2.31\times7.8 = 18.0 \text{ mol}^{-1}\text{ L s}^{-1}.$$

Example 10.21. For a first order reaction the rate constants are $3.20\times10^{-4}\text{s}^{-1}$ at 600 K and $1.60\times10^{-2}\text{s}^{-1}$ 1200 K. Calculate the energy of activation of the reaction. Take R = 8.314 $\text{JK}^{-1}\text{ mol}^{-1}$.

Solution: The energy of activation (E_a) of a reaction is given by Arrhenius equation.

$$E_a = \frac{2.303RT_1T_2}{T_2-T_1}\log\frac{k_2}{k_1}$$

On substituting the given values in equation (i) we get,

$$E_a = \frac{2.303\times8.314\text{ JK}^{-1}\text{mol}^{-1}\times600\text{ K}\times1200\text{ K}}{1200\text{ K}-600\text{ K}}\log\frac{1.60\times10^{-2}}{3.20\times10^{-4}}$$

$$= 39036\text{ J mol}^{-1} = 39.0\text{ kJ mol}^{-1}$$

Example 10.22. The rate of a reaction is doubled when the temperature is changed from 289 K to 308 K. Calculate the energy of activation of the reaction. R = 8.314 JK^{-1}.

Solution: The energy of activation is given by Arrhenius equation:

$$E_a = \frac{2.303\,RT_1T_2}{T_2-T_1}\log\left(\frac{k_2}{k_1}\right)$$

It is given that $\dfrac{k\text{ at }308\text{ K}}{k\text{ at }298\text{ K}} = 2 = \dfrac{k_2}{k_1}$

$$\therefore \quad E_a = \frac{2.303\times8.314\text{ JK}^{-1}\text{mol}^{-1}\times298\text{ K}\times308\text{ K}\log2}{308\text{ K}-298\text{ K}}$$

$$= 52897\text{ J mol}^{-1} = 52.9\text{ kJ mol}^{-1}$$

Example 10.23: The rate constant of a reaction is $2\times10^{-2}\text{ s}^{-1}$ at 300 K and $8\times10^{-2}\text{ s}^{-1}$ at 340 K. Calculate the energy of activation of the reaction.

Solution: The energy of activation (E_a) is given by:

$$E_a = \frac{2.303\,RT_1T_2}{T_2-T_1}\log\left(\frac{k_2}{k_1}\right)$$

It is given that : when $T_1 = 300$ K, $k_1 = 2\times10^{-2}\text{ s}^{-1}$

and when $T_2 = 340$ K, $k_2 = 8\times10^{-2}\text{ s}^{-1}$

Now, $$\log\frac{k_2}{k_1} = \log\frac{8\times10^{-2}s^{-1}}{2\times10^{-2}s^{-1}} = \log4 = 2\log2 = 0.6020$$

$$T_2 - T_1 = 340\text{ K} - 300\text{ K} = 40\text{ K}$$

$$T_2 - T_1 = 340\text{ K}\times300\text{ K} = 102000\text{ K}^2$$

$$2.303\text{ R} = 2.303\times8.314\text{ JK}^{-1}\text{ mol}^{-1} = 19.147\text{ JK}^{-1}\text{ mol}^{-1}$$

On substituting these values into equation (i) we get

$$E_a = \frac{19.147\ JK^{-1}mol{-1} \times 102000\ K^{-2} \times 0.602}{40\ K}$$

$$= 29392.6\ J\ mol^{-1} = 29.39\ kJ\ mol^{-1}.$$

QUESTIONS

1. What do you mean by chemical kinetics?
2. What is meant by rate of a reaction? What is the difference between the rate of reaction and rate constant?
3. What is the order of a reaction?
4. Can order of a reaction be fractional? Is it possible for a reaction to have identical values for molecularity and order? Given two examples of first order reaction.
5. A reaction is found to be zero order reaction, will its molecularity be also zero?
6. How is the half-life period of a first order reaction related to its rate constant?
7. What is the difference between order and molecularity of a reaction?
8. Show that the half-life of a first order reaction is independent of the initial concentration of the reactant.
9. What are the units of first order and second order reactions?
10. Given two differences between rate of a reaction and reaction rate constant.
11. What do you determine the order of a reaction by initial rate method?
12. Derive an integrated rate equation for a simple first order reaction
13. What is rate determining step? Show that the slowest step in the mechanism of the reaction largely determines the rate of the reaction.
14. What is meant by order of a chemical reaction? The reaction:

    $$2NO_2 + F_2 \longrightarrow 2NO_2F$$

    has been found to be first order with respect to NO_2 and with respect to F_2.

    (i) Write the rate law equation for the reaction.(ii) What is the overall order of the reaction?

    (iii) Suggest a suitable mechanism for the reaction.
15. Derive the integrated rate equation for rate constant for a first order reaction. What would be the units of (first order) rate constant if the concentration is expressed in moles per litre and time in seconds?
16. Explain with examples the meaning of the terms average rate and instantaneous rate of the reaction.
17. What is the effect of temperature upon the rates of chemical reactions? Explain how does a small rise in temperature appreciably increase the speed of many reactions?
18. How do we know that all collisions between the reactant molecules do not lead to chemical change?
19. What is pseudo unimolecular reaction? Discuss with a simple example.
20. What are the main features of Transition State Theory?
21. Define the rate of a reaction. For the reaction $X \rightarrow Y$, the rate of reaction can be denoted by:

    $$-\frac{dX}{dt} \quad \text{or} \quad +\frac{dY}{dt}$$

    State the significance of the plus and minus signs in these expressions.
22. Explain:

    (i) Energy of activation for a reaction. (ii) Half life of areaction.
23. The reaction $2A + B \rightarrow C + D$ has the rate law equation as:

    $$\text{rate} = k[A]^x\ [B]^y$$

    What is the

    (i) order of reaction (ii) apparent molecularity of reaction and

    (iii) order of reaction when B is present in large excess.

24. Write Arrhenius equation showing the effect of temperature in the reaction rates. How does it help to calculate the activation energy of a reaction?
25. Give important differences between
    (i) order and molecularity of a reaction
    (ii) reaction rate and rate constant
    (iii) activation energy and internal energy of a reaction.
26. Define the terms
    (i) Activated complex. (ii) Orientation barrier (iii) Effective Collision
27. Discuss the main points of collision theory of reaction rates.

PROBLEMS

1. In a reaction 5g of ethyl acetate is hydrolysed per litre in presence of dil HCl in 300 min. If the reaction is of first order and initial concentration of ethyl acetate is 22 g/L, calculate the rate constant for this reaction. **(8.73×10^{-4} min^{-1})**
2. A first order reaction has specific reaction rate of 2×10^{-3} s^{-1}. Calculate the half-life of this reaction. **(346.5 s)**
3. Calculate the half-life of a first order reaction when the specific rate constant is
   (i) 100 s^{-1} (ii) 2 min^{-1} **(0.00693 s and 0.3465 min)**
4. Time required to decompose SO_2Cl_2 to half its amount is 60 minutes. If decomposition is a first order reaction, find the rate constant of this reaction. **(0.0115 min^{-1})**
5. The reaction $SO_2Cl_2 \rightarrow SO_2 + Cl_2$ is a first order reaction with $k = 2.2 \times 10^{-5}$ s^{-1} at 320°C. What is its half-life? What percentage of SO_2Cl_2 would be decomposed on heating at 320°C? **(3.15×10^4 s, 11.27%)**
6. For a certain first order reaction, it takes 5 minutes for the initial concentration of 0.6 mol L^{-1} to become 0.2 mol L^{-1} to become 0.2 mol L^{-1}. What is the rate constant for this reaction? **(0.2198 min^{-1})**
7. A first order reaction is 20% complete in 5 minutes. Calculate the time taken for tyhe rectaion to be 60% complete. **(20.53 min)**
8. A first order reaction is found to have a rate constant k = 7.39×10^{-5} sec^{-1}. Find half-life of this reaction. **(0.1875×10^{-5} sec)**
9. The catalytic decomposition of hydrogen peroxide was studied by titrating it at different intervals with $KMnO_4$. Calculate velocity constant from the following data, assuming it to be a first order reaction.

t, sec	0	600	1200
$KMnO_4$ (mls)	22.8	13.8	8.2

(0.0008367 sec^{-1})

10. A reaction that is of first order with respect to reactant A, has rate constant of 6 min^{-1}. If we start with concentration of A equal to 0.5 mol L^{-1}, when would [A] reach the value of 0.05 mol L^{-1}. **(0.38 min)**
11. Show by using rate law, how much the rate of the reaction

$$2NO(g) + O_2(g) \rightleftharpoons 2NO_2(g)$$

will change if volume of the reaction vessel is diminished to 1/3 of its initial volume. The reaction is of first order.

12. From the following data for the decomposition of ammonium nitrate in aqueous solution, show that the reaction is of first order.

Time (minutes) :	10	20	25	∞
Volume of N_2 (ml) :	6.25	11.40	13.65	35.05

(0.0197 min^{-1})

13. A first order reaction is 20% complete in 10 minutes. Calculate:
    (i) the specific rate constant of the reaction and
    (ii) the time taken for the reaction to go to 75% completion. **[(i) 0.02231 min^{-1} (ii) 62.15 min^{-1}]**
14. Show that in case of a first order reaction, the time required for 99.9% of the reaction to take place is ten times that required for half of the reaction.

EXAMINATION QUESTIONS

1. (a) What do you mean by Energy of Activation ? Explain with the help of energy reaction coordinate diagram. Also explain how the nature of activated complex changes in the presence of a catalyst.
   (b) Differentiate between order and molecularity of a reaction.
   [U.P. Tech., B.Tech., 1st/2nd Sem., 2000-02]
2. Discuss the differences between order of a reaction and molecularity of a reaction.
   [U.P. Tech., March 2002]
3. For a first order reaction, the rate constant is found to be 7.0×10^{-7} at 7°C and 9×10^{-4} at 57°C. Calculate the energy of activation of the reaction.
   Ans. 1.098×105 mol^{-1} ($\log_{10} 7 = 0.8451$, $\log_{10} 9 = 0.9542$) ***[U.P. Tech., 2003-04]***
4. (i) What is meant by the rate of reaction ? Explain the factors which affect the rate of a reaction.
   (ii) Distinguish between Order and Molecularity of a Reaction. ***[U.P. Tech., 2002-03]***
5. (i) What meant by order of reaction ?
   (ii) Show that in case of a first order reaction, the time required for 99.9% of the reaction to take place is about 10 times that required for half the reaction. ***[U.P. Tech., 2004-05]***
6. (i) Define order of a reaction. What is condition that a second order reaction behaves like a first order?
   (ii) With the help of data given show that decomposition of H_2O_2 in aqueous solution is of first order.

Time (min)	:	0	10	20	30
Volume of KMnO4 required to decompose H_2O_2 (ml).	:	25.0	20.0	15.7	12.5

   [U.P. Tech., 1st Sem., 2006-07]
7. (i) The rate constant of a reaction is 1.5×10^7 Sec^{-1} at 50°C and 4.5×10^2 sec^{-1} at 100°C. Evaluate the Arrhenius parameters A and Ea.
   (ii) Derive the equation for half life of first order reaction. ***[U.P. Tech., 2nd-Sem., 2006-07]***
8. (a) What do you understand by the rate equation and the order of a reaction ?
   (b) Derive the rate Expression for Second-order reaction, when the reactants are different.
   (c) Discuss the effect of temperature on reaction rates. ***[U.P. Tech., 2007-08]***
9. (i) What is meant by rate of reaction ? Explain the factors which affect the rate of a reaction.
   (ii) What is activation energy ? How is the rate constant of a reaction related to its activation energy? ***[U.P. Tech., 2nd Sem., 2003-04]***
10. For the first gaseous reaction

$$X_{(g)} \longrightarrow 2Y_{(g)} + Z_{(g)}$$

   The initial pressure is 90 mm Hg. The pressure after 10 minute is 180 mm Hg. Calculate the rate constant of the reaction. ***[U.P. Tech., 1st Sem., (C.O.) 2003]***
11. Prove that for a second order reaction, the half life period is inversely proportional to its initial concentration. ***[U.P. Tech., 1st Sem., 2007-08]***
12. (a) Rate of reaction is directly proportional to the ________.
    (b) The reaction : $N_2O_5\,(g) \rightarrow N_2O_4\,(g)\ \frac{1}{2}O_2\,(g)$ is :
    (*i*) Zero order (*ii*) First order
    (*iii*) Second order (*iv*) Fractional order.
    (c) Threshold energy = Activation energy + ________. ***[U.P. Tech., 1st Sem., 2008-09]***
13. (a) Distinguish between order and molecularity of a reaction, calculate order and molecularity of the following reactions :

$$N_2O_5\,(g) \longrightarrow N_2O_4\,(g) + \frac{1}{2}O_2$$

$$CH_3COOCH_2H_5 + H_2)\ (excess) \xrightarrow{H^+} CH_3COOH + C_2H_5OH$$

$$H_2(g) + I_2\,(g) \rightarrow 2Hi(g)\text{ , and } CH_3CHO \xrightarrow{\Delta} CH_4 + CO$$

(b) What is activation energy ? Discuss its relationship with rate constant of a reaction.

[*U.P. Tech., 1st Sem., 2008-09*]

14. Derive an equation for half life period of a first order reaction.

A compound decomposes according to the first order rate low with a half life period of 30 min. Calculate the fraction of remaining compound after 120 min. **[*U.P. Tech., 1st Sem., 2008-09*]**

15. (a) Derive the rate equation for a second order chemical reaction. The specific rate constant for the decomposition of formic acid is 5.5×10^{-4} sec^{-1} at 413°K. Calculate the specific rate constant at 458°K if the energy of activation is 2.37×10^4 cal mol^{-1}.

(b) What is half life of a chemical reaction ? Show that time required to complete 99.9% of a first order reaction is about 10 times of its half life. **[*U.P. Tech., B. Tech. 1st Sem., 2008-09*]**

❑❑❑

11

CATALYSIS

"God grant me the serenity to accept the things I can not change, the courage to change the things I can, and the wisdom to know the difference."

— Reinhold Niebuhr

CHAPTER OUTLINE

It is a well known fact that all the chemical reactions take place at a definite rate under given experimental conditions. Some reactions are so rapid that they appear to be instantaneous, e.g. neutralization of strong acids and bases. But some reactions are so slow at ordinary temperatures, that over the years no detectable products are formed, e.g. combination of hydrogen and oxygen. Between these two extremes, lie a number of reactions which proceed with measurable velocities at ordinary temperatures. **The study of rates of reactions proceeding under given conditions of temperature, pressure and concentration is known as chemical kinetics.**

It has been a continuous effort to increase the rate of reaction, by working out different conditions for getting maximum yields of several products. This is done by varying the conditions on which rate of reaction depends. The most important of these include varying the concentration of the reactants, changing the temperature, use of radiation and presence of a catalyst. **The presence of a catalyst is the most wonderful aspect of chemical kinetics.** It has been observed that certain reactions which do not proceed to completion even when carried out for indefinite periods of time, are speeded up and get completed very fast when a small quantity of a catalyst is added.

The term catalyst was coined by Berzelius in 1836. **Catalyst is defined as a substance, which alters the rate of the reaction without undergoing any change in mass and composition at the end of the reaction.** This phenomenon of altering the velocity of a chemical reaction by the presence of catalyst is called **catalysis.**

How do catalysts increase reaction rates? The general answer is that catalysts provide a new path for the reaction to proceed. To understand this more clearly we must review the concept of activation energy and what difference the presence of catalyst make.

11.1 CONCEPT OF ACTIVATION ENERGY AND CATALYSIS

According to the collision theory of reaction rates, for a chemical reaction to occur, collisions between reactant molecules must take place. But only those collisions result in chemical reaction, in which the colliding molecules are associated with a certain minimum amount of energy. The minimum amount of energy which must be associated with molecules, so that the collisions between them result in chemical reaction is called **threshold energy**. The collisions between the molecules which are associated with energy less than the threshold energy do not result into chemical reaction. Most of the molecules, however have much less kinetic energy than the threshold energy. **The excess energy that the reactants molecules must acquire to change into products is known as activation energy.** Thus

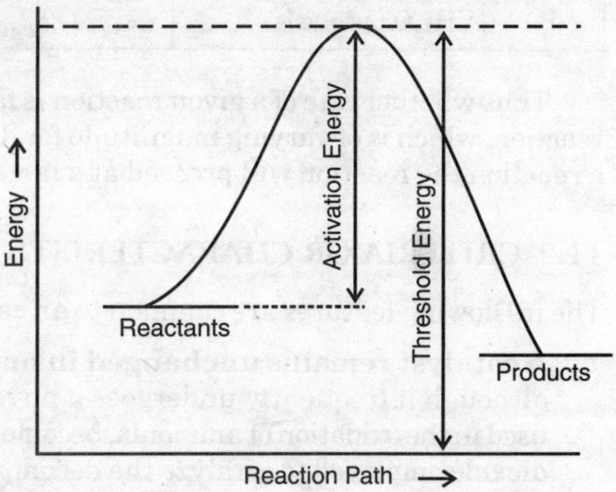

Fig. 11.1 The concept of activation energy

Activation Energy = Threshold Energy – Energy possessed by the molecules.

Activation energy is denoted by E_a.

Effect of Catalyst on activation energy

Catalyst provides an alternative route, with lower activation energy to the reactants to change into the products. In other words, a catalyst lowers the activation energy of the reaction by providing a new pathway. For instance, the activation energy for the decomposition of hydrogen peroxide in solution is 76 kJ mol^{-1} and the reaction is slow at room temperature. When a little iodine is added, the activation energy falls to 57 kJ mol^{-1} and the rate constant increases by a factor of 2000. Similarly enzymes, the biological active catalysts increase the reaction rate upto a factor of 10^{12} at body temperature. For example the enzyme saccharase, reduces the activation energy for acid hydrolysis of sucrose from 107 kJ mol^{-1} to 36 kJ mol^{-1} and hence reaction proceeds much faster at body temperature (310 K). Table (11.1) gives an idea about lowering of activation energy of catalyzed reaction.

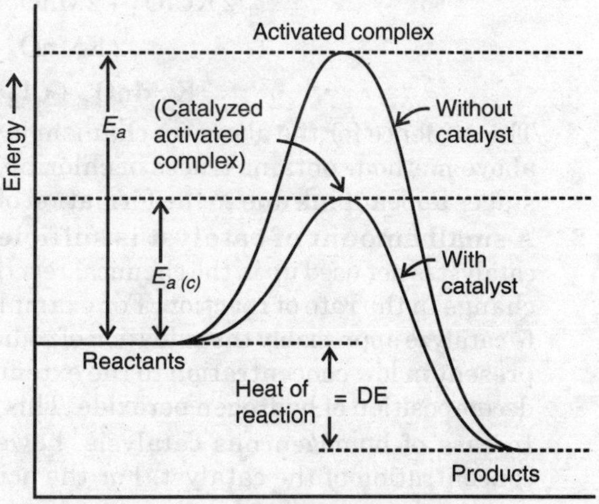

Fig. 11.2 Effect of catalyst in lowering the activation energy

Table 11.1

Reaction	*Catalyst*	*Activation Energy E_a (kJ mol^{-1})*	
		In Absence of Catalyst	*In Presence of Catalyst*
(i) $2HI \rightarrow H_2 + I_2$	$Au_{(s)}$	184	105
	$Pt_{(s)}$	184	59
(ii) $2HN_3 \rightarrow N_2 + 3H_2$	$W_{(s)}$	350	162
(iii) $2N_2O \rightarrow 2N_2 + O_2$	$Au_{(s)}$	245	121
	$Pt_{(s)}$	245	134
(iv) $(C_2H)_2O$ pyrolysis	$I_{2(g)}$	224	144

Thus whether rate of a given reaction is fast or slow depends upon the activation energy of the reaction, which is of varying magnitude for different reactions. If the activation energy is less for a reaction the reaction will proceed at a faster rate and vice versa.

11.2 CRITERIA OR CHARACTERISTICS OF CATALYSIS

The fo1llowing features are common to all catalytic reactions.

1. **A catalyst remains unchanged in mass and composition at the end of the reaction:** although it frequently undergoes a physical change. For example platinum gauze catalyst used in the oxidation of ammonia, becomes quite rough after some weeks. Similarly manganese dioxide employed to catalyze the decomposition of potassium chlorate, is in the form of large crystals in the beginning of the reaction but changes to a fine powder in the end.
   Since the catalyst undergoes a physical change, it implies that it is involved in the mechanism of the reaction but regenerated in the end. The participation of MnO_2 in the decomposition of $KClO_3$ is believed to be as explained below :
   $$2KClO_3 + 2MnO_2 \longrightarrow 2KMnO_4 + Cl_2 + O_2$$
   $$2KMnO_4 \longrightarrow K_2MnO_4 + MnO_2 + O_2$$
   $$K_2MnO_4 + Cl_2 \longrightarrow 2KCl + MnO_2 + O_2$$
   The evidence for the above mechanism is provided by the fact, that oxygen prepared by the above method contains traces of chlorine. Moreover the reaction mixture if cooled at early stages appear pink due to the formation of $KMnO_4$.
2. **A small amount of catalyst is sufficient to alter the rate of the reaction:** Since the catalyst is not used up in the chemical reaction, a small amount is sufficient to bring appreciable change in the rate of reaction. For example, 1 gm of copper (II) ions in 10^9 litres is sufficient to catalyze appreciably the oxidation of sodium sulphite by oxygen. Similarly colloidal platinum, present in low concentration to the extent of 1 mole in 10^8 litres, is enough to accelerate the decomposition of hydrogen peroxide. This is true in the case of heterogeneous catalysis.
   In case of homogeneous catalysis, however, the rate of reaction is proportional to the concentration of the catalyst. For the acid and alkaline hydrolysis of an ester, the rate of reaction is proportional to the concentration of the acid, which acts as an catalyst.
   $$RCOOR' + H_2O \xrightarrow[OH^-]{H^+ \text{ or}} RCOOH + R'OH$$

In Friedel Crafts reaction, anhydrous aluminium chloride function as a catalyst effectively, if present to the extent of 30% of the mass of benzene which is appreciably large amount.

$$C_6H_6 + C_2H_5\,Cl \xrightarrow{AlCl_3} C_6H_5C_2H_5 + HCl$$

In certain heterogeneous reactions, the rate of the reaction is proportional to the area of the catalytic surface. This explains why the finely divided state of a solid catalyst is much more effective in increasing the rate of reaction.

3. **A catalyst is specific in its action:** A particular substance can act as a catalyst only in particular reaction. For example MnO_2 may catalyze the decomposition of $KClO_3$, but not of $KClO_4$. Highly specific action of a catalyst can be compared to the specific use of a key which can open a particular lock. Enzymes are also highly specific in their action. But on the other hand, the transition metals like iron, cobalt, nickel, platinum and palladium can catalyze various reactions.
4. **Nature of the products is unaltered by the presence of the catalyst:** The combination of N_2 and H_2 under suitable conditions will give NH_3 in presence or absence of catalyst. Similarly SO_2 and O_2 will always combine to form SO_3 whether a catalyst is present or not. However there are some **exceptions** also. For example CO and H_2 combine to give three **different** products when different catalysts are used.

$$CO + 3H_2 \xrightarrow{Ni} CH_4 + H_2O$$
$$CO + H_2 \xrightarrow{Cu} HCHO$$
$$CO + 2H_2 \xrightarrow{ZnO + Cr_2O_3} CH_3OH$$

Similarly chlorination of toluene in the presence of a halogen carrier such as iron in the **absence of sunlight** takes place in the benzene ring.

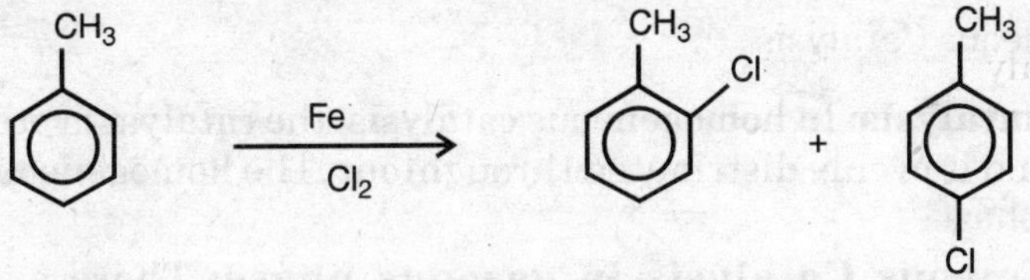

But chlorine is substituted in the side chain in the absence of a catalyst but in **presence of sunlight.**

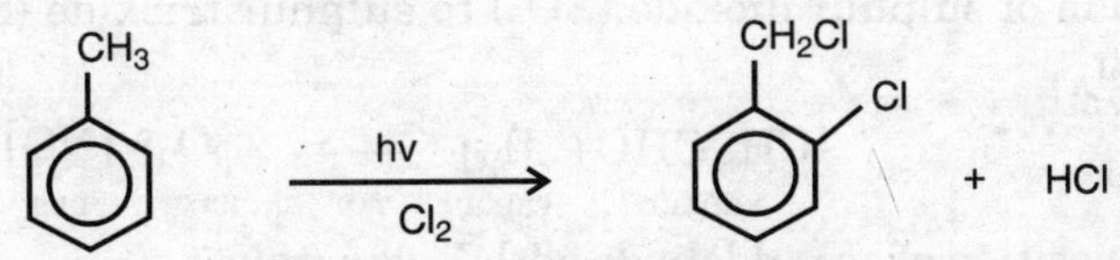

5. **The catalyst does not alter the position of equilibrium in a reversible reaction:** The catalyst helps in attaining the equilibrium more quickly by increasing the rates of both the forward and the reverse reactions to the same extent, but do not alter the concentration of the reactant and the products at equilibrium. Thus the equilibrium constant remains the same. It is also evident from the Fig. 11.2 that, the activation energy for the reverse reaction is also lowered by exactly the same amount as for the forward reaction, by the addition of the catalyst. Thus, if a catalyst doubles the rate of the forward reaction, it also doubles the rate of the backward reaction. In other words, the catalyst for the forward process in a reaction is also a catalyst for the reverse reaction.
6. **A catalyst does not intitiate the reaction:** It can only accelerate the rate of the reaction already taking place, even if very slowly. But this subject has been a matter of great controversy.

According to W. Ostwald (1895) catalyst cannot initiate the reaction but H.W. Armstrong (1885-1903) and T.M. Lowry (1925-1928) have suggested that there are certain reactions which can occur only if a catalyst is present and will not occur in its absence. For example, perfectly dry H_2 and O_2 do not combine to form water, even if they are left in contact for years, but in presence of a little water which act as a catalyst, the reaction proceeds quite rapidly. But still this does not prove that change did not occur at all, but that uncatalyzed combination may be so slow that millions of years would be required for a detectable amount of water to be formed.

7. **Optimum temperature:** There is a particular temperature at which the activity of the catalyst is maximum, this temperature is called optimum temperature.
8. **Catalytic promotors:** The addition of small amounts of foreign substances which themselves are not catalytically active, increases the activity of a catalyst. Such substances are called promoters. In the manufacture of NH_3 by Haber's process, finally divided iron acts as a catalyst, while molybdenum acts as a promotor.
9. **Catalytic poisons:** The activity of a catalyst is inhibited or completely destroyed by the presence of even minute traces of foriegn substances called catalytic poisons or anticatalysts. In the manufacture of H_2SO_4, a trace of As_2O_3 destroys the catalytic efficiency of spongy platinum.
10. **High surface area:** Catalysts have high surface areas. Even a small lump (1 cm^2) of a catalyst can have a surface area ranging to 100 m^2 because of the presence of pores.

11.3 TYPES OF CATALYSIS

Broadly two types of catalysis are known:

1. Homogeneous Catalysis
2. Heterogeneous Catalysis

Homogeneous Catalysis: In homogeneous catalysis, the catalyst is present in the same phase as the reactants and is evenly distributed throughtout. The homogeneous catalysis can occur in gaseous or liquid phase.

(i) Homogeneous Catalysis in gaseous phase: There are only a few cases of homogeneous catalysis in gaseous phase e.g.

(a) Oxidation of sulphur dioxide (SO_2) to sulphur trioxide (SO_3) with nitric oxide as catalyst.

$$\underset{\text{vapour}}{CH_3CHO} + \underset{\text{vapour}}{[I_2]} \longrightarrow \underset{\text{gas}}{2SO_3} + \underset{\text{gas}}{[NO]}$$

(b) Decomposition of acetaldehyde with I_2 as catalyst

$$\underset{\text{vapour}}{CH_3CHO} + \underset{\text{vapour}}{I_2} \longrightarrow \underset{\text{gas}}{CH_4} + \underset{\text{gas}}{CO} + I_2$$

(c) NO gas also acts as a catalyst in combination of CO with oxygen.

$$\underset{\text{gas}}{2CO} + \underset{\text{gas}}{O_2} \xrightarrow[\text{gas}]{NO} 2CO_2$$

(d) NO gas catalyses the combination of moist sulphur dioxide and oxygen in the lead chamber process for the manufacture of H_2SO_4.

$$\underset{\text{gas}}{2SO_2} + \underset{\text{vap}}{2H_2O} + \underset{\text{gas}}{O_2} \xrightarrow[\text{gas}]{[NO]} 2H_2SO_4$$

(ii) **Homogeneous catalysis in liquid phase:** Most of the reactions under this category involve the acid or base as the catalyst.

(a) Hydrolysis of an ester in the presence of acid or alkali.

$$CH_3COOC_2H_{5(l)} + H_2O_{(l)} \xrightarrow{H^+ \text{ or } \overline{OH}} CH_3COOH_{(l)} + C_2H_5OH_{(l)}$$

(b) Inversion of sucrose in presence of hydrogen ions in aqueous solution.

$$C_{12}H_{22}O_{11} + H_2O \xrightarrow{H^+} \underset{\text{Glucose}}{C_6H_{12}O_6} + \underset{\text{Fructose}}{C_6H_{12}O_6}$$

(c) The decomposition of hydrogen peroxide is catalyzed in the presence of chloride ions.

$$2H_2O_{2(l)} \xrightarrow{Cl_{(l)}} 2H_2O + O_2$$

(d) Conversion of acetone into diacetone alcohol is catalyzed by hydrogen ions

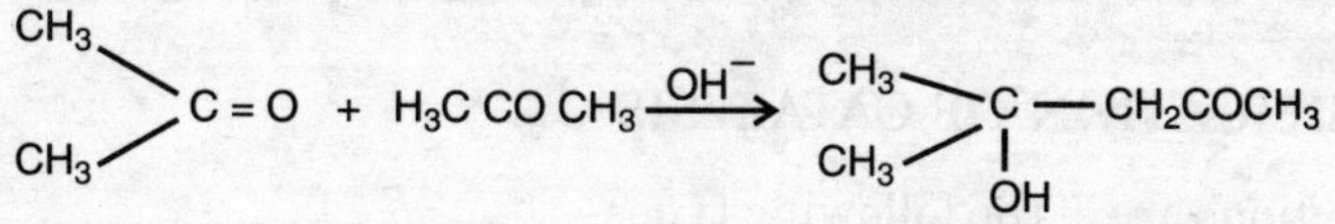

Diacetone alcohol

The mechanism of homogeneous catalysis can be explained on the basis of unstable intermediate compound formation theory. (discussed later).

Heterogeneous Catalysis: In heterogeneous catalysis the catalyst is in different phase from the reactants. In majority of the cases, the catalyst is solid while the reactants are either liquids or gases. Often the term "Contact Catalysis" is used for this type of catalysis as the reaction occurs by the contact of the reactants with the catalyst surface. This form of catalysis has a number of reactions of technical importance. The catalyst is usually a metal or metal oxide. Heterogeneous catalysis involve solid, liquid or gaseous reactants.

(i) **Heterogeneous Catalysis with gaseous reactants:** There are a large number of reactions which can be put under the heading of heterogeneous catalysis. Some of them are–

(a) Combination of SO_2 and oxygen in the presence of finely divided platinum or vanadium penta oxide V_2O_5 (Contact Process).

$$\underset{\text{gas}}{2SO_2} + \underset{\text{gas}}{O_2} \xrightarrow{Pt} \underset{\text{gas}}{2SO_3}$$

(b) Combination of hydrogen and oxygen in presence of finely divided platinum.

$$2H_2 + O_2 \xrightarrow{Pt} 2H_2O$$

(c) Hydrogenation of oils in presence of Nickel

$$\underset{\text{unsaturated oil}}{R-CH=CH-R'} + \underset{\text{gas}}{H_2} \xrightarrow{Ni} R-CH_2-CH_2-R'$$

(d) Synthesis of petroleum by the combination of CO and H_2 in the presence of iron, cobalt or nickel

$$xCO + yH_2 \xrightarrow[\text{or Ni}]{\text{Fe, Co}} \text{Petroleum} + H_2O.$$

(e) Manufacture of ammonia by Haber process

$$N_2 + 3H_2 \underset{Mo}{\overset{Fe}{\rightleftharpoons}} 2NH_3$$

(ii) **Heterogeneous catalysis with liquid reactants :** The common example under this category are–

(a) Benzene and ethanoyl chloride react in the presence of anhydrous aluminium chloride to form phenyl methyl ketone $(C_6H_5COCH_3)$

$$C_6H_{6(l)} + CH_3COCl_{(l)} \xrightarrow{AlCl_3} C_6H_5COCH_3 + HCl$$

(b) Decomposition of aqueous solution of hydrogen peroxide (H_2O_2) is catalysed by MnO_2 or platinum in colloidal form.

$$2H_2O_2 \xrightarrow{Pt} 2H_2O + O_2$$

(c) Decomposition of hypochlorites : The decomposition of hypochlorites in aqueous solutions is catalysed by oxides of cobalt and nickel.

$$Ca(ClO)_2 \xrightarrow{NiO} CaCl_2 + O_2$$

11.4 CLASSIFICATION OF CATALYSIS

Catalytic reactions are of the following types–

1. Positive Catalysis
2. Negative Catalysis
3. Auto Catalysis

1. **Positive Catalysis:** When the catalyst used accelerates the speed of a chemical reaction it is known as a positive catalyst. Initially the term 'catalyst' coined by Berzelius was used for the positive catalyst only, but later on this term was used for all the substances which **alter** the rate of reaction, and hence the substances which **increase** the rate of reaction are placed under positive catalysis. All the reactions discussed above are the example of positive catalysis.

2. **Negative Catalysis:** There are a few cases, where a catalyst instead of accelerating a reaction may actually **reduce** the rate of reaction. Such a catalyst used is termed as negative catalyst. The phenomenon is known as negative calatysis or inhibition. Negative Catalyst slows down or stops altogether an unwanted reaction. For example–

(i) $CHCl_3$ on oxidation by air forms carbonyl chloride $(COCl_2)$ which is a poisonous substance. The formation of $COCl_2$ is suppressed by the addition of ethanol which acts as a negative catalyst.

(ii) Small amount of glycerol or acetanilide retards the decomposition of hydrogen peroxide.

3. **Auto Catalysis:** There are certain reactions where one of the products of the reaction acts as the catalyst. Such reactions are known as Auto Catalyzed reactions. In auto catalysis the initial rate of the reaction rises as the catalytic product is formed thereby indicating that product accelerates the reaction rate. (Fig. 11.3).

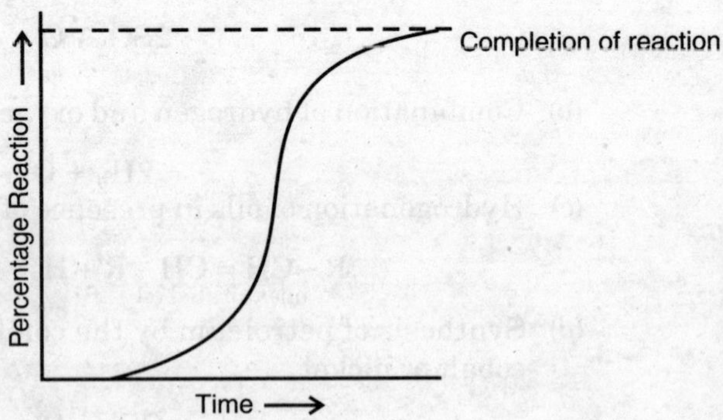

Fig. 11.3 Rise of Rate of reaction with the formation of product

Examples:

(i) During titration of warm solution of oxalic acid by acidified $KMnO_4$ solution, the first two drops take appreciable time to decolourize, since the reaction is initially very slow. But after some time, the decolourization takes place rapidly as Mn^{2+} ions formed during the reaction, catalyze the reaction.

$$2KMnO_4 + 5\begin{array}{c}COOH\\|\\COOH\end{array} + 3H_2SO_4 \xrightarrow[\text{Catalyst}]{2MnSO_4} K_2SO_4 + 8H_2O + 10CO_2$$

(ii) Hydrolysis of an ester is autocatalyzed by the acid which is formed during the reaction.

$$RCOOR' + H_2O \longrightarrow RCOOH + R'OH.$$

(iii) The free arsenic produced by the decomposition of arsine (AsH_3) auto catalyze the reaction.

$$2AsH_3 \longrightarrow \underset{\text{Catalyst}}{2As} + 3H_2$$

(iv) The action of HNO_3 on copper is very slow initially but NO_2^- ions produced during the reaction act as catalyst and the reaction proceeds rapidly.

$$3Cu + 8HNO_3 \longrightarrow 3Cu(NO_3)_2 + 4H_2O + 2NO$$

Nitric oxide formed above dissolves in H_2O to provide NO_2^- ions.

11.5 THEORIES OF CATALYSIS

Different theories have been put forward to explain the homogeneous and heterogeneous catalysis.

Theory of Homogeneous Catalysis (Intermediate Compound Formation Theory)

This theory was put forward by element and Desormes in 1806. This theory explains satisfactorily the mechanism of homogeneous catalysis. According to this theory, the catalyst combines with one of the reactants to form an intermediate compound which is unstable and hence react immediately with other reactants to form the products and regenerating the catalyst. Consider a reaction between the reactants X and Y to form the product XY in the presence of a catalyst C.

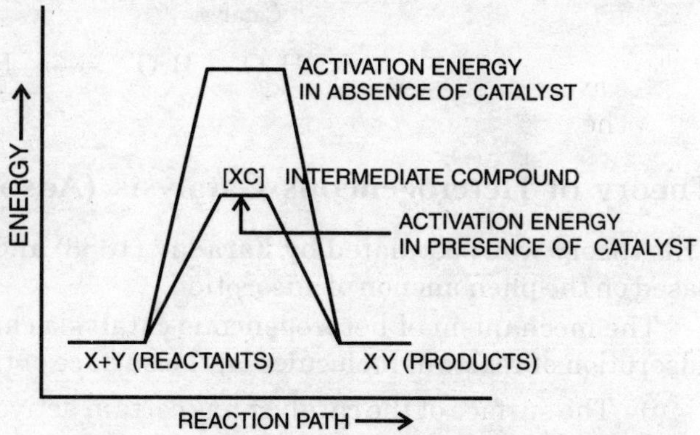

Fig. 11.4 Energy Profile for Homogeneous Catalysis

$$X + Y \xrightarrow{C} XY$$

According to the intermediate compound formation theory, the reaction may be split up into the following two steps:

(i) **First step** – Formation of intermediate

$$X + C \longrightarrow [XC] \text{ intermediate Compound}$$

(ii) **Second step** – Dissociation of intermediate.

$$[XC] + Y \longrightarrow \underset{\text{Product}}{XY} + \underset{\text{Catalyst}}{C}$$

Infact, the formation of intermediate compound provides an alternative reaction path having lower activation energy and hence reaction proceeds at an increased rate.

It is seen from the Fig. 11.4 that the total energies of the reactants *X* and *Y*, and those of the product *XY* are unaffected by the catalyst. The only difference between the two is lowering of the activation energy in the presence of a catalyst. A number of catalytic reactions can be explained on the basis of intermediate compound formation theory. A few of them are discussed below:

(i) Ordinarily, the gas phase reaction of sulphur dioxide with oxygen proceeds slowly:

$$2SO_2 + O_2 \longrightarrow 2SO_3$$

But if the same reaction is carried out in presence of NO as catalyst, the reactions proceeds relatively faster.

$$\underset{\text{Catalyst}}{NO} + \frac{1}{2}O_2 \longrightarrow \underset{\text{Intermediate}}{NO_2}$$

$$NO_2 + SO_2 \longrightarrow SO_3 + \underset{\text{Catalyst Regenerated}}{NO}$$

(ii) In Friedel Craft's reaction, anhydrous aluminium chloride is used as a catalyst for the preparation of benzene. The mechanism can be discussed on the basis of intermediate formation theory. For example, benzoylation of benzoyl chloride proceeds as follows–

$$\underset{\text{(Reactant)}}{C_6H_5COCl} + \underset{\text{(Catalyst)}}{[AlCl_3]} \longrightarrow \underset{\text{Intermediate}}{C_6H_5COCl.AlCl_3}$$

$$C_6H_5COCl.AlCl_3 + C_6H_6 \longrightarrow \underset{\text{Product}}{C_6H_5COC_6H_5} + HCl + \underset{\text{Catalyst}}{AlCl_3}$$

(iii) Decomposition of hydrogen peroxide into water and oxygen catalyzed by bromide ions is believed to take place as:

$$H_2O_{2(aq)} + \underset{\text{Catalyst}}{Br^-_{(aq)}} \longrightarrow H_2O_{(l)} + BrO^-_{(aq)}$$

$$H_2O_2 + BrO^- \longrightarrow H_2O_{(l)} + O_{2(g)} + \underset{\text{Catalyst}}{Br^-_{(aq)}}$$

Theory of Heterogeneous Catalysis (Adsorption Theory)

This theory was postulated by Faraday (1883) and later on modified by others. This theory is based on the phenomenon of adsorption.

The mechanism of heterogeneous catalysis can be explained satisfactorily on the basis of adsorption of reactant molecules on the surface of the catalysis. According to this theory,

(i) The surface of the catalyst has certain active centres due to the unsaturation of valencies at the surface.

(ii) The reactant molecules strike the surface and are chemisorbed on it.

(iii) The exothermic nature of the chemisorptive process is able to supply energy of the same order of magnitude as in chemical reactions and hence the dissociation of bonds of adsorbed molecules may take place

(iv) The bonds between adsorbate and adsorbent are usually covalent but sometimes may be ionic also, as in the case of metal oxygen bonds.

(v) Due to the close proximity between adsorbed molecules on the catalyst surface, they undergo interaction to form an activated complex.

(vi) The activated complex breaks to form the products and the free surface of the catalyst is regenerated.

(vii) The molecules of the products diffuse away from the surface of the catalyst.

Hydrogenation of ethene in presence of Nickel can be explained, to illustrate the above points.

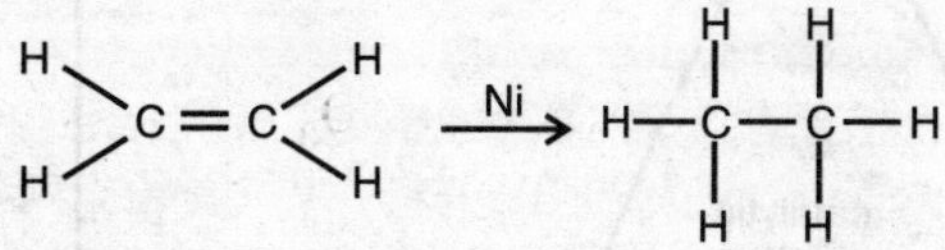

The catalyst operates by the following steps–

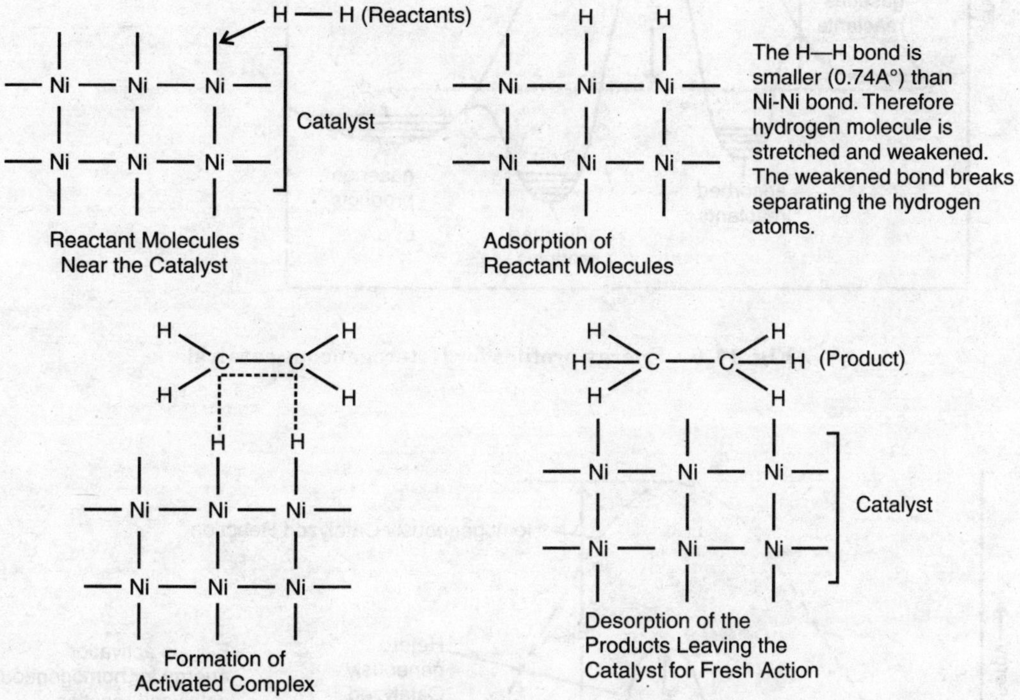

Fig. 11.5 Mechanism of Heterogeneous catalysis.

The process of heterogeneous catalysis also provides an alternative reaction path of much lower activation energy as compared to uncatalysed reaction. Fig. 11.5 give energy curves fon non-catalytic reaction, and heterogeneous catalysis. (Fig. 11.5)

A number of reactions have been studied both as homogeneous and heterogeneous processes. It has been found that lowering of activation energy is more in case of heterogeneous catalysis as compared to homogeneous catalysis (Fig. 11.6).

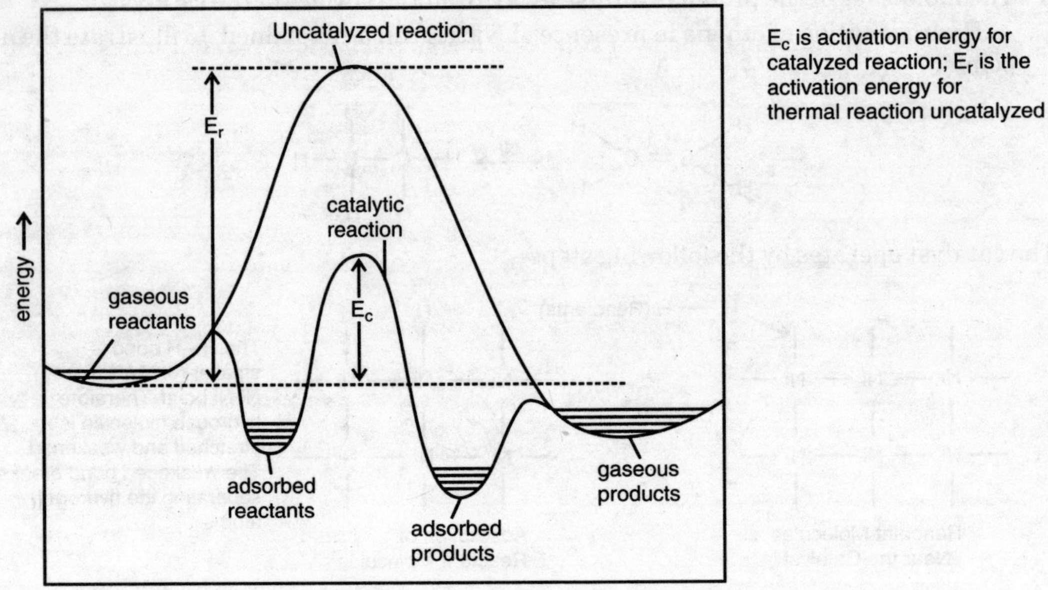

Fig. 11.6 Energy profiles for Heterogeneous catalysis.

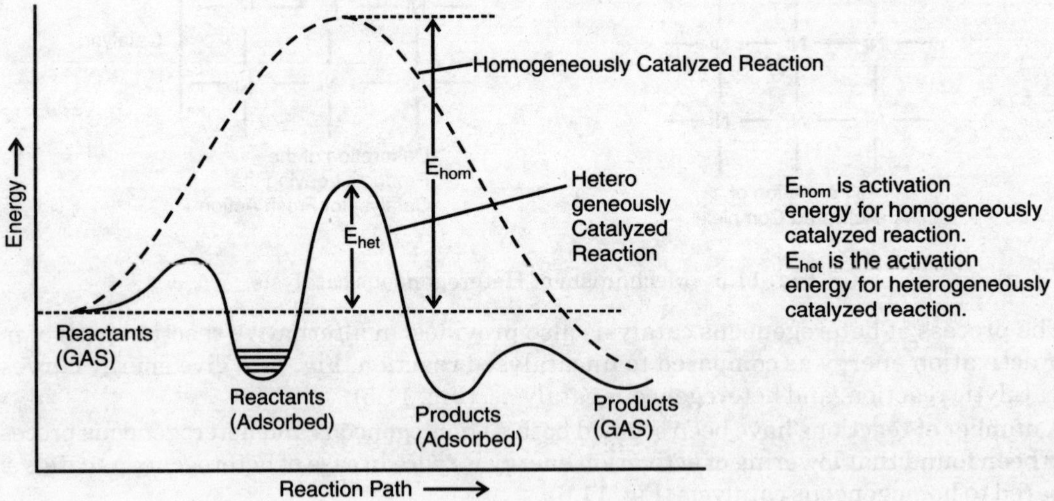

Fig. 11.7 Energy Profile for Homogeneously & Heterlogeneously Catalyzed Reaction

Adsorption theory of heterogeneous catalysis explains a number of the observed facts satisfactorily. Some of there are discussed below.

KINETIC OF SOME OF THE IMPORTANT CATALYTIC PROCESSES

Development, selection and application of catalysts forms the basis of a major part of chemical industries. To give a more clear idea of the concept, some of the catalysis processes are discussed in detail in the following sections.

11.6 ACID BASE CATALYSIS

The most important category of homogeneously catalyzed reactions in solution is acid-base catalysis.

Early investigations by Ostwald and Arrhenius on the study of acid base catalysis showed that the rate of the reaction was dependent on the concentration of H^+ ions in acid catalyzed reactions and on the concentration of $\bar{O}H$ ions in base catalyzed reactions. Thus H^+ or $\bar{O}H$ were thought to be effective catalytic agents. It was also observed the nature of anion or undissociated acid in acid catalyzed reaction and the nature of cation or undissociated base in base catalysed reaction does not influence the rate of reaction. But later on, the work of Taylor, Lowry and others showed the reactions catalysed by H^+ ions are influenced by the undissociated acids and the reactions catalyzed by $\bar{O}H$ ions are influenced by the undissociated bases. Many observations of similar nature were made by other workers including S. Arrhenius during the course of his work later. Acid-base catalysis can be divided into the following categories–

(a) Specific acid catalysis.
(b) Specific base catalysis.
(c) General acid catalysis.
(d) General base catalysis.

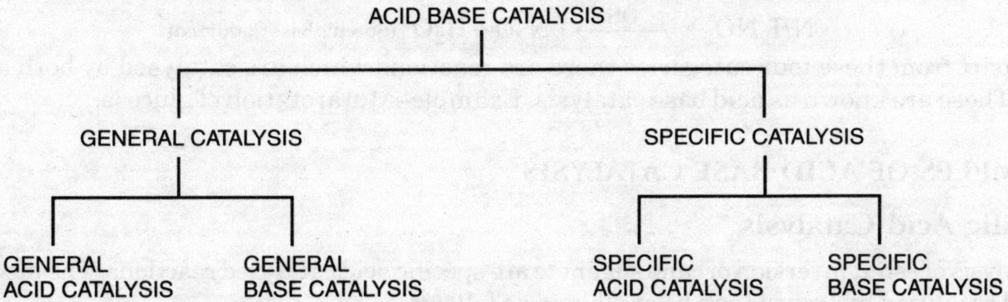

(a) **Specific Acid Catalysis:** These are the reactions which are catalysed by H^+ or H_3O^+ ions **only** or simply these are proton catalyzed reactions. Example – inversion of cane sugar in the presence of acid as a catalyst. The rate is influenced by the concentration of H^+ ions.

$$\underset{}{C_{12}H_{22}O_{11}} + H_2O \xrightarrow{H^+} \underset{\text{Glucose}}{C_6H_{12}O_6} + \underset{\text{Fructose}}{C_6H_{12}O_6}$$

Keto Enol transformation also involves specific acid catalysis.

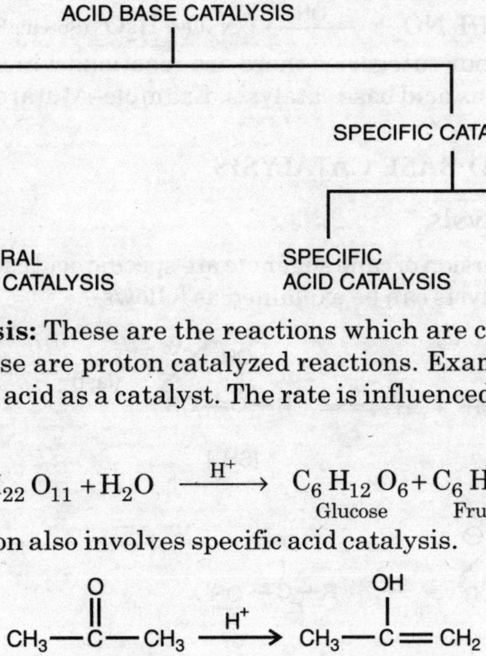

Keto form *Enol form*

The rate of reaction of these reactions is found to be proportional to the concentration of hydrogen ions present in the solution.

(b) **Specific base Catalysis:** These reactions are catalyzed by $\bar{O}H$ ions only and are not influenced by undissociated bases or proton acceptors. For example conversion of acetone into diacetone alcohol is catalyzed by $\bar{O}H$ ions only

$$CH_3-\overset{O}{\overset{\|}{C}}-CH_3 + CH_3-\overset{O}{\overset{\|}{C}}-CH_3 \xrightarrow{\bar{O}H} CH_3COCH_2C(CH_3)_2OH$$

The rate of reaction in this case is dependent upon the concentration of $\bar{O}H$ ions and is independent of any other basic specie present in the system.

(c) **General Acid Catalysis:** In this type of catalysis the rate is affected by not only H^+ ions present in the system, but also undissociated acid or cations of weak bases, influence the rate of reaction. In other words, all the substances which have a tendency to lose a proton (bronsted acid) can function as catalyst. Example–Iodination of acetone catalyzed by monochloroacetic acid and its salt.

$$CH_3COCH_3 + I_2 \xrightarrow[ClCH_2CO\bar{O}]{ClCH_2COOH} CH_3COCH_2I + HI$$

The rate of the reaction of these type of reactions depend upon the concentration of $\bar{O}H$ ions as well as all effective proton donors.

(d) **General Base Catalysis:** These reactions are catalyzed by hydroxyl ions, undissociated bases and even anions of weak acids. In other words, all the substances which have a tendency to gain a proton (bronsted bases) will act as catalyst. Example-decomposition of nitramide is catalyzed by hydroxyl ions, or by acetate ions.

$$NH_2NO_2 \xrightarrow{CH_3COO^-} N_2O + H_2O \text{ (General base catalyzed)}$$

$$NH_2NO_2 \xrightarrow{\bar{O}H} N_2O + H_2O \text{ (Specific base Catalyzed)}$$

Apart from these four categories there are reactions which are catalysed by both acid and base. These are known as acid base catalysis. Example–Mutarotation of glucose.

EXAMPLES OF ACID BASE CATALYSIS

Specific Acid Catalysis

Hydrolysis of ester, inversion of cane sugar etc are specific acid catalyzed reactions. The mechanism of acid catalyzed hydrolysis can be explained as follows–

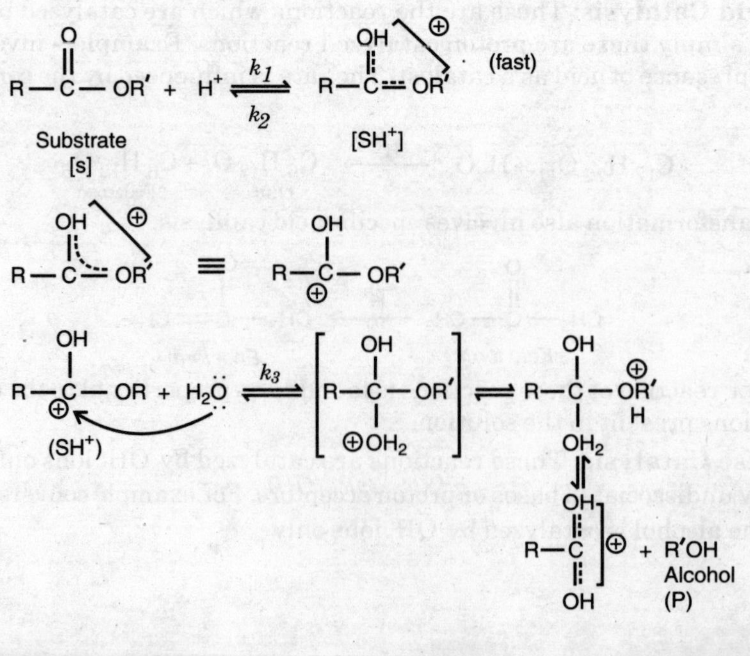

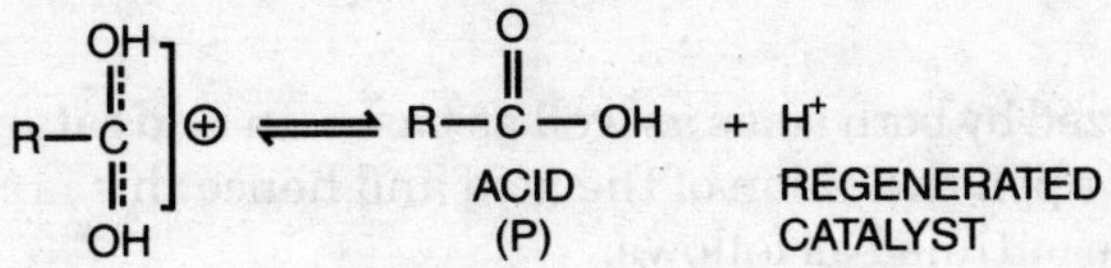

Acidic hydrolysis is a reversible reaction

$$RCOOR' + H_2O \underset{H^+}{\overset{H^+}{\rightleftharpoons}} RCOOH + R'OH$$

Thus in the reverse process i.e. esterification the mechanism is just the mechanism of hydrolysis taken in the opposite directions.

Rate of the reaction is given by–

$$r = k_3[SH^+]$$

$$\text{or } r = \frac{k_1 k_3 [RCOOR'][H^+]}{k_2 + k_3} \quad ...(20)$$

as is discussed in kinetics of the reaction. Thus rate is first order with respect to ester as well as H^+. At constant H^+ the reaction is first order with respect to ester, or can be termed as pseudo first order reaction.

Specific Base Catalysis

Hydrolysis of ester may also be accomplished by alkalies. But this is irreversible reaction, since the reverse reaction, esterification, cannot be brought about in the presence of alkalies. This is because carboxylic acid reacts with alkali to form a resonance stabilised carboxylate anion which has very little tendency to react with an alcohol.

Hydrolysis of ester depends upon the concentration of $\bar{O}H$ ion and hence is specific base catalysed reaction. The mechanism can be outlined as follows:

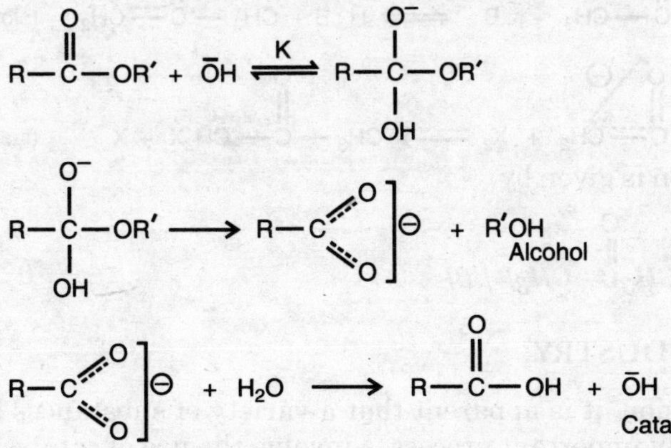

Thus the rate of the reaction depends upon the concentration of the ester as well as hydroxide ion.

$$r = k\,[R\overset{O}{\overset{\|}{C}} - OR]\,[\bar{O}H] \quad ...(21)$$

General Acid Catalysis

The halogenation of a ketone is catalyzed by both acids as well as bases. In acid catalyzed reaction, the rate of the reaction is dependent upon the anion of the acid and hence this is called general acid catalysis. The mechanism can be outlined as follows:

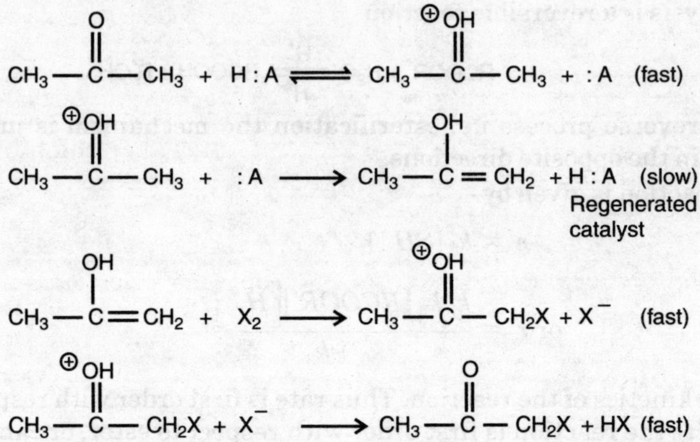

It is seen from the above mechanism that the rate determining step (slowest step) involves the anion :A, the anion of the acid.

General Base Catalysis

The halogenation of a ketone in the presence of a base. The rate determining step involves the bronsted base (the substance which can accept the proton) and not the hydroxide ion and hence this reaction is general base catalysed reaction.

The mechanism can be outlined as follows–

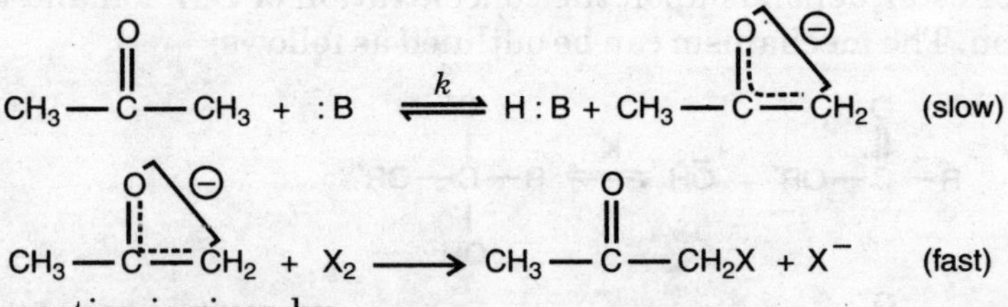

Thus rate of the reaction is given by

$$r = k\,[CH_3\overset{\overset{\displaystyle O}{\|}}{C} - CH_3]\ [:B] \qquad ...(22)$$

11.7 CATALYSTS IN INDUSTRY

From the foregoing discussion, it is apparent that a variety of substances have been used as catalysts. Many industrially important processes involve the use of catalysts. The criteria for choosing a catalyst for industrial applications are as follows:

(i) The catalyst must possess maximum possible activity so as to give maximum yield of the desired product in minimum time.

(ii) The cost of the catalyst should be low.

(iii) The catalyst must be easily available.

(iv) The catalyst should be such that it is regenerated easily without much difficulty.

(v) The catalyst should be highly selective so that the desired product should be formed in major amount.

Many of the primitive methods of preparing certain substances involved the unconscious use of catalysis. For example

- the fermentation of wine to acetic acid.
- the manufacture of soap from fats and alkalies.
- the formation of ether from alcohol and H_2SO_4.

All of these are catalytic reactions. But the application of catalyst to industrial processes was undertaken in 19th century. P. Phillips used platinum to oxidize sulphur dioxide to sulphur trioxide in presence of air. This process became a great success towards the close of 19th century. There after numerous catalysis were discovered and affected the growth and development of industrial processes. Catalyst manufacture is itself a rapidly growing industrial process.

Modern chemical industry depends on the development, selection and application of the catalysts. Some of the important industrial processes along with the corresponding catalyst are tabulated in Table 11.2.

Table 11.2

Process	***Catalyst***
1. Synthesis of NH_3 (Haber's Process)	*Fe*
$N_2 + 3H_2 \rightleftharpoons 2NH_3$	
2. Manufacture of sulpuric acid	
(a) Contact Process :	
$2SO_2 + O_2 \rightleftharpoons 2SO_3$	V_2O_5
(b) Chamber Process :	
$2SO_2 + O_2 \rightleftharpoons 2SO_3$	NO_2
3. Cracking of Petroleum	$Al_2(SiO_3)_3$ Zeolites
$C_{10}H_{22} \rightarrow C_5H_{12} + C_5H_{10}$	Silica-Alumina Catalyst.
4. Hydrogenation of Unsaturated Hydrocarbons.	Ni, Pt
$-CH = CH - + H_2 \rightarrow -CH_2 - CH_2-$	
5. Oxidation of Hydrocarbons in Automobile Exhausts	CuO, V_2O_5, Pt, Pd
6. Isomerization of n–butane to isobutane	$AlCl_3$, HCl
$n - CH_3 - CH_2 - CH_2 - CH_3 \xrightarrow{AlCl_3} CH_3 - CH - CH_3$	
7. Catalytic Reforming	
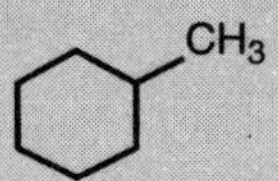 $\longrightarrow C_6H_5CH_3 + 3H_2$ Methylcyclohexane	$Mo_2O_3 - Al_2O_3$
8. Polymerization	Ziegler Natta Catalyst
$n\,CH_2 = CH_2 \rightarrow -(CH_2 - CH_2)_n$	$TiCl_3 + (C_2H_5)_3Al$
9. Manufacture of Chlorine (Deacon's process)	Cu_2Cl_2
$4HCl + O_2 \rightleftharpoons 2CH_3COOH$	
10. Oxidation of NH_3 to NO (Ostwald's Process)	Pt Gauze
$4NH_3 + 5O_2 \rightleftharpoons 4NO + 6H_2O$	

CATALYST AND AUTOMOBILE INDUSTRY

A recent area of research in the field of industrial catalysis is the development of catalysts which can convert the harmful auto exhaust gases into less harmful ones. For example-Catalytic converters in the cars. One such type of catalytic converter is discussed briefly.

Catalytic Converters

In running cars, the temperature inside the engine becomes high. At this high temperature nitrogen and oxygen gases react to form nitric oxide.

$$N_{2(g)} + O_{2(g)} \rightleftharpoons 2NO_{(g)}$$

$NO_{(g)}$ is released into the atmosphere where it combines with O_2 to form NO_2. Nitrogen dioxide and other gases which are also emitted by the automobile like CO and unburnt hydrocarbons are the major cause of air pollution.

To overcome this difficulty to some extent most of the cars are fitted with the catalytic converters. The catalyst employed are mostly Pt or Pd. Sometimes transition metal oxide such as CuO or Cr_2O_3 are also used.

An efficient catalytic converter serves the following purposes

(i) It oxidizes CO and unburnt hydrocarbons to CO_2 and H_2O.

(ii) It reduces NO and NO_2 to N_2 and O_2.

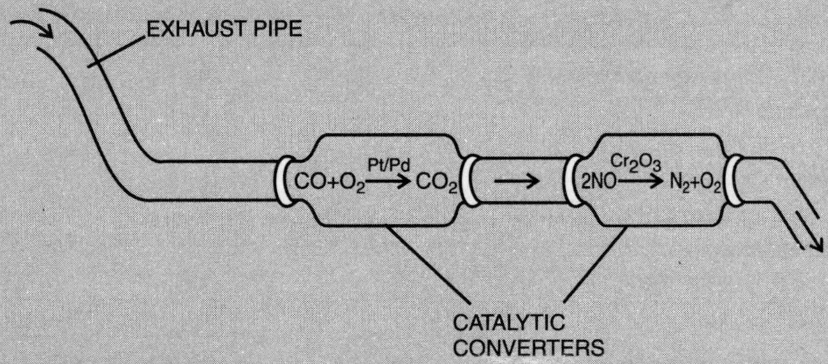

Fig. 11.8 Catalytic Converters for an Automobile.

Fig. 11.20 gives a simplified diagram of catalytic converter. Hot exhaust gases are first mixed with air and are passed through the first chamber where the complete burning of hydrocarbons take place. This chamber has Pt or Pd as catalyst. Then the exhaust gases are passed through the second chamber containing a different catalyst mostly a transition metal oxide (Cr_2O_3 or CuO), which operates at a low temperature. In this chamber NO is dissociated into N_2 and O_2 before the exhaust is allowed to escape through tail pipe.

QUESTIONS

Short Answer Questions

1. What do you mean by activation energy?
2. What is the role of the catalyst?
3. What are negative catalysts?
4. What is the advantage of catalyst over high temperature to promote chemical reactions.
5. What are catalytic promoters?
6. What are catalytic poisons?
7. Light speeds up many reactions and is sometimes spoken of as a catalyst. Criticize the statement.
8. What is the difference between intermediate and a catalyst?

9. What do you understand by the term homogeneous catalysis and heterogeneous catalysis.
10. Explain induced catalysis?

    Ans. When one reaction influences the speed of the other reaction, the phenomenon is termed as induced catalysis. For example, oxidation of sodium sulphite acts as a catalyst for oxidation of sodium arsenite solution.
11. Give three examples of homogeneous catalysis.
12. Give three examples of heterogeneous catalysis.
13. Define (*i*) Positive Catalyst (*ii*) Negative Catalyst (*iii*) Auto catalyst
14. What are the limitations of intermediate compound formation theory?
15. What is adsorption?
16. Differentiate between physical adsorption and chemisorption.
17. What are active centres.
18. What is catalytic poisoning?
19. Distinguish between temporary and permanent poisoning.
20. In catalytic cracking of hydrocarbons, the catalyst gets exhausted after some time. Explain?
21. What is general and specific acid catalysis?
22. Give some examples of specific acid catalysis.
23. Give some examples of specific base catalysis.

Long Answer Questions

1. What is meant by catalysis and catalyst? Give general characteristic of catalytic reactions.
2. Describe briefly homogeneous and heterogeneous catalysis giving one example of each. Explain Intermediate Compound Formation theory of catalysis.
3. How does a catalyst work?
4. What are promoters and inhibitors? Give suitable examples.
5. Explain the difference between homogeneous and heterogeneous catalysis. Discuss the theories of catalysis.
6. Give:
    (a) Mechanism of solid catalyst.
    (b) Action of Promoters and Poisons with examples.
    (c) Applications of catalysts.
7. Derive a general expression for the rate of reaction of an acid-catalysed reaction.
8. Explain with a suitable example that heterogeneous catalysts are highly specific in nature.
9. Write a short note on Activation Energy.

❑❑❑

UNIT 4

Water and Its Treatment

Chapter 12: Water and its Treatment

12 WATER AND ITS TREATMENT

"Science cannot solve the mystery of Nature. It is because in the analysis we ourselves are part of the mystery we are trying to solve"

— ***Max Plank***

CHAPTER OUTLINE

The planet earth is abundantly rich in water, which constitute 75% of its surface. Besides, being an essential ingredient of animal and plant life, water forms the basis of human life. It constitutes about 65% of human body. It is so common that its importance is overlooked many a times. Ben Franklin has rightly said, *"When the well is dry, we know the worth of water."* Water is essential not only for survival of human beings, animals, plants and all other living beings but is also utilized in a large amount in industries. Out of the total water consumed by human beings, more than 50% of it is consumed for industrial activity and only a small proportion is used for drinking purposes.

Most of the water which is available is located in oceans (97.2%). The high concentration of salts make it virtually unusable as a source of water for municipal, agricultural or most industrial needs. However, we do use the oceans for thermal cooling of power plants and as a sink for much of our pollution. Hence we are left with only 2.98% of water, out of which, considerable amount of water is present in the form of polar ice caps and may remain locked as such for thousands of years. A small portion of it is also present in rock crevices and minerals, lying 800 meters deep below the earth's crust, which is again very expensive to pump out. Thus, the conclusion is that, only 0.3 percent of world's water resources is available for possible human use.

Quantity of water is not the only concern. The second major problem is the *quality* of water. Overuse has resulted in the progressive deterioration of water quality. Our traditional confidence in the quality of our drinking water is lost. Even the water which is used for industries has to be treated prior to its use. The treatment to which it is subjected, depends upon the purpose, for which it would be used. Thus, both the problems, *water quantity* and *quality* have to be treated with equal importance.

The impurities present in water vary greatly from one place to another. Treatment of water obtained from *different sources* is therefore essential so that it may be used safely for industrial or municipal purposes.

Maintaining an adequate supply of water for municipal or industrial purposes requires the efforts of scientists and engineers. The following aspects are taken into consideration for treatment of water for various uses.

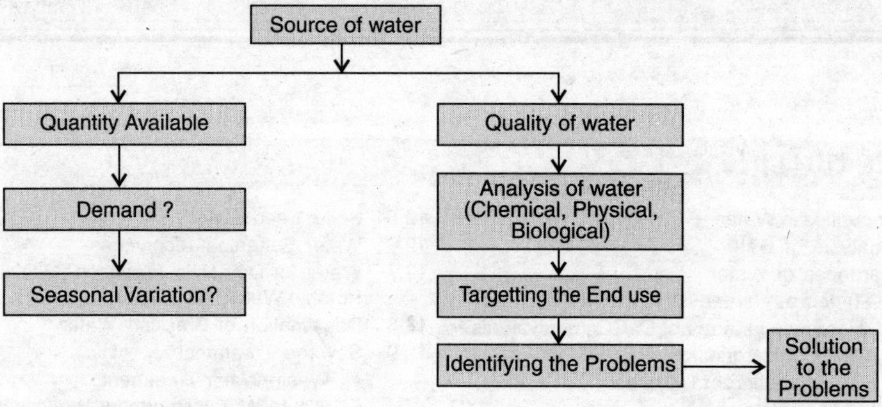

12.1 IMPURITIES IN WATER

Water from a water supply scheme may contain a number of impurities which it takes up from

- the ground or soil with which it comes in contact. For e.g. soil particles, minerals, garbage etc.
- the sewage and industrial waste. For eg. chemical wastes, acids, colours, heavy metals etc.
- the decomposition of plants and animals.
- the growth of bacteria, algae, viruses etc.
- the atmosphere in the form of dissolved gases.

Types of Impurities: The impurities present in water may be broadly classified as given below in Table 12.1.

Table 12.1

	Impurities
Dissolved Impurities	(a) Salts : Carbonates, bicarbonates, sulphates and chlorides of calcium, magnesium, iron, sodium and potassium.
	(b) Gases : CO_2, O_2, N_2, H_2S, NH_3 etc.
Suspended Impurities	(a) Inorganic – sand, clay
	(b) Organic – Vegetable and animal matter
Colloidal Impurities	Finely divided silica and clay, organic waste products, colouring matter etc.
Micro Organisms	Bacteria, Algae, Fungi, Virus etc.

Effect of Impurities: The various types of impurities present in water impart certain properties to water. Colour, decaying aquatic organisms and certain chemicals may impart odour, suspended matter may impart turbidity to the water, presence of dissolved salts make the water hard, whereas excess quantities of metals and dissolved gases makes it corrosive. Bacteriological impurities due to pathogenic bacteria make water unfit for drinking or domestic purposes.

Hence, the water from any source has to be treated before it is used. The treatment to which it is subjected depends upon its use. For e.g., if it has to be used for drinking purposes, the treatment would include removal of objectionable colour, taste and pathogenic microorganisms, where as the water for industrial use require removal of dissolved salts if it is used for steam generation, removal of Ca, Mg, Fe and Mn salts for its use in laundries. Similarly, the water used in alcoholic breweries requires the high purity and the presence of a few micro organisms.

From the foregoing discussion, it is clear that each industry has its special water treatment requirement. Moreover, the treatment for municipal water, used for drinking and domestic purposes is different from the treatment to which water for industrial use is subjected. Subsequent sections give an account of the specifications and treatment of water for its use in industries and for drinking purposes.

12.2 ANALYSIS OF WATER

Analysis of water includes the determination of different constituents in water to ascertain the suitability of water for a specific use, and the treatment to which it should be subjected prior to its use. The following characteristics of water are usually determined :

- Hardness
- Alkalinity
- Dissolved Oxygen (BOD, COD)
- Dissolved Carbondioxide
- Free Chlorine
- Chlorides
- Silica
- Total Solids.

Each of these is discussed in the foregoing sections.

12.3 HARDNESS OF WATER

Hardness is one of the most important characteristic of water. This term was applied to the waters which make cleansing action of the soap difficult. The water which **does not form lather with soap** and is **hard** to wash in, is termed as hard water. Hardness of a water sample is a measure of its capacity to precipitate soap or its soap consuming capacity.

On the other hand, the water which forms copius lather with soap is termed as soft water.

Cause of Hardness : Hardness of water is due to the presence of calcium and magnesium salts. Generally, soaps consist of sodium salts of higher fatty acids such as stearic, oleic, palmitic etc. which are highly soluble in water and thus exert their cleansing action. If calcium and magnesium salts are present in water, these react with soluble sodium soap to form insoluble salts of calcium and magnesium. For example,

$$\underset{\text{(Sodium Stearate)}}{2C_{17}H_{35}COONa} + CaSO_4 \longrightarrow \underset{\text{(Insoluble salt)}}{(C_{17}H_{35}COO)_2\,Ca} + Na_2SO_4$$

$$2C_{17}H_{35}COONa + MgCl_2 \longrightarrow (C_{17}H_{35}COO)_2\,Mg + 2NaCl$$

Thus, a white scum or precipitate of insoluble soap of calcium and magnesium is formed. Lather is not produced until the cations, Ca^{2+} and Mg^{2+} are completely precipitated out in the form of insoluble salts. As a result, a large quantity of soap is required to produce lather with hard water. Thus, hardness is rightly defined as *soap consuming capacity of water*.

12.3.1 Types of Hardness

Hardness of water is of two types

(a) Temporary Hardness
(b) Permanent Hardness

(a) **Temporary Hardness :** Hardness due to the presence of bicarbonate ions of calcium and magnesium is called temporary hardness or *carbonate hardness*. This is also known as alkaline hardness. This hardness can be removed by mere boiling of water, which converts the bicarbonates into insoluble carbonates or hydroxides. The insoluble carbonates or hydroxides can be removed by filteration.

$$Ca(HCO_3)_2 \xrightarrow{\text{boil}} CaCO_3\downarrow + H_2O + CO_2$$

$$Mg(HCO_3)_2 \xrightarrow{\text{boil}} MgCO_3 + H_2O + CO_2.$$

$$MgCO_3 + H_2O \longrightarrow Mg(OH)_2\downarrow + CO_2.$$

Temporary hardness can also be removed by addition of calculated amount of lime ($Ca(OH)_2$)

$$Ca(HCO_3)_2 + Ca(OH)_2 \longrightarrow 2CaCO_3\downarrow + 2H_2O$$

$$Mg(HCO_3)_2 + Ca(OH)_2 \longrightarrow MgCO_3\downarrow + CaCO_3\downarrow + 2H_2O$$

(b) **Permanent Hardness :** This type of hardness is also known as *non-carbonate* hardness. It is due to the presence of sulphates and chlorides of calcium and magnesium. Permanent hardness cannot be removed by boiling. Special methods (discussed later) are used for removal of permanent hardness.

12.3.2 Hardness in Terms of $CaCO_3$ Equivalents

From the foregoing discussion, it is clear that a number of salts can cause hardness such as bicarbonates, sulphates, chlorides of calcium and magnesium. To compare the hardness of different samples of water, it would be easier, if the hardness caused by different salts can be expressed in terms of single salt.

It is customary to express hardness in terms of equivalents of $CaCO_3$. The reason for choosing $CaCO_3$ as the standard for calculating hardness of water is due to–

- its molecular weight is exactly 100, which makes mathematical calculations easier.
- it is the most insoluble salt, thus can be easily precipitated in water treatment processes.

Now 1 mole of $CaCO_3$ (or 100 parts by weight) would react with the same amount of soap as 1 mole of $CaCl_2$ (111 parts by weight), or 1 mole of $MgCl_2$ (95 parts by weight) because 1 mole of all these salts will produce same number of calcium or magnesium ions (Avogadro's number).

Thus $CaCO_3$ equivalents for various salts are as follows:

100 g of $CaCO_3 \equiv$ 111 g of $CaCl_2 \equiv$ 136 g of $CaSO_4 \equiv$ 95 g of $MgCl_2 \equiv$ 120 g of $MgSO_4 \equiv$ 162 g of $Ca(HCO_3)_2 \equiv$ 146 g of $Mg(HCO_3)_2 \equiv$ 164 g of $Ca(NO_3)_2 \equiv$ 44 g of $CO_2 \equiv$ 148 g of $Mg(NO_3)_2$

If x g of $CaCl_2$ is present in a water sample, then the amount of $CaCl_2$ present in terms of its $CaCO_3$ equivalent will be –

1 g mole of $CaCl_2 \equiv$ 1 g mole of $CaCO_3$

111 g of $CaCl_2 \equiv$ 100 g of $CaCO_3$

55.5 g of $CaCl_2$ = 50 g of $CaCO_3$

($\because$ Eq. wt of $CaCO_3$ = 50 g Eq. wt of $CaCl_2$ = 55.5 g)

$$x \text{ g of } CaCl_2 = \frac{50}{55.5} \times x \text{g of } CaCO_3.$$

Thus to convert various hardness causing salts into $CaCO_3$ equivalents we may use–

$$\text{Equivalents of } CaCO_3 = \frac{\text{Mass of the substance}}{\text{Equivalent weight of the substance}} \times \text{Equivalent weight of } CaCO_3$$

As equivalent weight of most of the hardness causing substances and that of $CaCO_3$ = Molecular weight/2, hence to make mathematical calculations more easier, the above formula can be written in terms of molecular weights i.e.

$$\text{Equivalents of } CaCO_3 = \frac{\text{Mass of the substance}}{\text{Molecular mass of the substance}/2} \times 100/2$$

or $$\text{Equivalents of } CaCO_3 = \frac{\text{Mass of the substance}}{\text{Molecular mass}} \times 100$$

In otherwords, the weight of the substance has to be multiplied by a certain fraction i.e. 100 / *Molecular Mass*, to obtain equivalents of $CaCO_3$ in case of *divalent salts*. Similarly the multiplication factors for other hardness causing substances can be found out. (Table 12.2)

Table – 12.2

	Salt	***Molar Mass***	***Multiplication Factor***
Divalent Salts	$Ca(HCO_3)_2$	162	100/162
Divalent Salts	$Mg(HCO_3)_2$	146	100/146
Divalent Salts	$CaCO_3$	100	100/100
Divalent Salts	$CaSO_4$	136	100/136
Divalent Salts	$CaCl_2$	111	100/111
Divalent Salts	$MgCO_3$	84	100/84
Divalent Salts	$MgSO_4$	120	100/120
Divalent Salts	$MgCl_2$	95	100/95
Divalent Salts	$Mg(NO_3)_2$	148	100/148
*	$Al_2(SO_4)_3$	342	100/114
	$FeSO_4.7H_2O$	278	100/278
**	$NaAlO_2$	82	100/82×2
	HCO_3^-	61	100/61×2
	CO_3^{2-}	60	100/60
	OH^-	17	100/17×2
	H^+	1	100/1×2

Notes: *The multiplication factor for $Al_2(SO_4)_3$ is given by:

$$\frac{\text{Equivalent weight of } CaCO_3}{\text{Equivatent weight of } Al_2(SO_4)_3} = \frac{100/2}{342/6} = \frac{50}{57}$$

To maintain the uniformity of the figure 100 in the numerator, as in the case of other multiplication factors, we may take multiplication factor as,

$$\frac{2 \times \text{Equivalent weight of } CaCO_3}{2 \times \text{Equivalent weight of } Al_2(SO_4)_3} = \frac{100}{114}$$

** Similarly for monovalent salts also, we take multiplication factor as:

$$\frac{2 \times \text{Equivalent weight of } CaCO_3}{2 \times \text{Equivalent weight of monovalent substance}}$$

Now, Equivalent mass of monovalent substance = Molar mass.

$$\therefore \qquad \text{Multiplication factor} = \frac{100}{2 \times \text{Molar mass}}$$

12.3.3 Units of Hardness

The results of the analysis of hard water are expressed in one of the following units :

1. **Parts per million (ppm) :** ppm is the parts of calcium carbonate equivalent hardness per 10^6 parts of water.
2. **Milligrams per litre (mg/L) :** It is the number of milligrams of $CaCO_3$ equivalent hardness present per litre of water.

$$1 \text{ mg/L} = 1 \text{mg of } CaCO_3 \text{ equivalent hardness per L of water.}$$

But 1L of water weighs = 1000 gm.

= 1000 × 1000 mg

1mg/L = 1mg/10^6 mg = 1 ppm.

3. **Clarke's degree (°Cl) :** It is the number of grains (1 grain $= \frac{1}{7,000} lb$) of $CaCO_3$ equivalent hardness per gallon (10 lb) of water. Or simply, it is the parts of $CaCO_3$ equivalent hardness per 70,000 parts of water.

1^0 Clarke = 1 part of $CaCO_3$ equivalent hardness per 70,000 parts of water.

4. **Degree French (°Fr) :** It is the parts of $CaCO_3$ equivalent hardness per 10^5 parts of water.

Relationship between various units of hardness :

1 ppm = 1mg/L = 0.1°Fr = 0.07°Cl

1°Fr = 10 ppm = 10 mg/L = 0.7°Cl

1°Cl = 14.3 ppm = 14.3 mg/L = 1.43°Fr

NUMERICALS BASED ON INTERCONVERSION OF UNITS

Example 12.1: A water sample has the following analysis.

$Mg(HCO_3)_2$ = 83 mg/L; $Ca(HCO_3)_2$ = 134 mg/L; $CaSO_4$ = 124 mg/L; $MgCl_2$ = 84 mg/L; $CaCl_2$ = 94 mg/L; NaCl = 50 mg/L. Calculate the temporary and permanent hardness in degree Clark.

Solution: Calculation of $CaCO_3$ equivalents

Constituent	*Quantity mg/L*	*Multiplication Factor*	*$CaCO_3$ Equivalents (mg/L)*
$Mg(HCO_3)_2$	83	100/146	56.8
$Ca(HCO_3)_2$	134	100/162	82.7
$CaSO_4$	124	100/136	91.2
$MgCl_2$	84	100/95	88.4
$CaCl_2$	94	100/111	84.7
NaCl	50	Does not produce hardness	—

Temporary hardness = hardness due to ($Mg(HCO_3)_2 + Ca(HCO_3)_2$
= 139.5 mg/L = 139.5 ppm
= 139.5 × 0.07°Cl = 9.77°Cl

Permanent hardness = Hardness due to {$CaSO_4 + MgCl_2 + CaCl_2$}
= 264.3 mg/L = 264.3 ppm
= 264.3 × 0.07°Cl = 18.50°Cl

Example 12.2: Calculate the total hardness of a water sample having following composition in ppm.

$CaCO_3$ = 20°Cl; $MgSO_4$ = 4° Cl, $CaSO_4$ = 2° Cl; Fe_2O_3 = 0.04° Cl, $MgCl_2$ = 4° Cl ; SiO_2 = 0.09° Cl

Solution:

Constituent	*Amount (°Cl)*	*Multiplication Factor*	*$CaCO_3$ equivalents (°Cl)*
$CaCO_3$	20	100/100	20
$CaSO_4$	2	100/136	1.47
$MgCl_2$	4	100/95	4.21
$MgSO_4$	4	100/120	3.33
Fe_2O_3	0.04	—	—
SiO_2	0.09	—	—

Total hardness of water sample = 29.01°Cl
= 29.01 × 14.3 ppm = 414.84 ppm

Example 12.3: Calculate the total hardness of water in °Fr and °Cl from the following analysis of water sample.

$Mg(HCO_3)_2$ = 20.8 mg/L; $Mg(NO_3)_2$ = 28.8 mg/L; $MgCl_2$ = 21.0 mg/L; $MgSO_4$ = 20.0 mg/L; $CaCO_3$ = 21.2 mg/L; KCl= 73.2 mg/L

Solution: Calculation of $CaCO_3$ equivalents.

Salt	*Amount mg/L*	*Multiplication Factor*	*Equivalents of $CaCO_3$ mg/L*
$Mg(HCO_3)_2$	20.8	100/146	14.2
$Mg(NO_3)_2$	28.8	100/148	19.45
$MgCl_2$	21.0	100/95	22.10
$MgSO_4$	20.0	100/120	16.67
$CaCO_3$	21.2	100/100	21.2
KCl	73.2	—	—
			Total Hardness = 93.62 mg/L

$$\text{Total Hardness} = 93.62 \text{ mg/L} = 93.62 \times 0.1°\text{Fr} = 9.362°\text{Fr}$$

$$\text{Total hardness in °Cl} = 93.62 \times 0.07°\text{Cl} = 6.55°\text{Cl}$$

12.3.4 Determination of Hardness

Hardness of water can be determined by a number of methods :

(i) O.Hehner's Method
(ii) Soap Titration Method
(iii) EDTA Method

Out of the above three methods, the EDTA method is more accurate and is used widely for the estimation of hardness in water. This method is discussed in detail in the following section :

EDTA Method

This method makes use of complexometric titration of water sample with EDTA using Eriochrome Black *T* as the indicator.

EDTA, Ethylene diamine tetra acetic acid is a strong complexing agent having the structure,

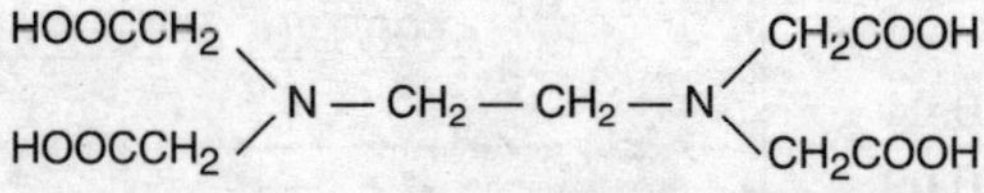

Ethylene diamine tetra acetic acid (EDTA)

When EDTA is dissolved in water, EDTA forms a zwitterion which has the structure,

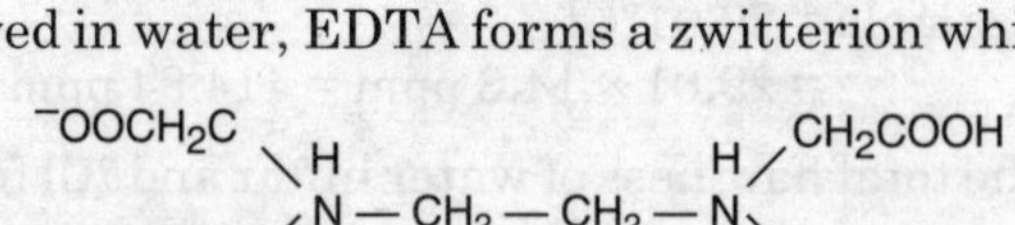

The various EDTA species are abbreviated as H_4Y, H_3Y^-, H_2Y^{2-}, HY^{3-} and Y^{4-} where H represent the four ionizable hydrogens and the rest of the remaining structure is represented by Y. The relative amounts of these species is a function of pH.

EDTA is an hexadentate ligand and can form 4 or 6 dative bonds with Ca^{2+} and Mg^{2+}. As such, it is not very soluble in water and it is used as its disodium salt which is quite soluble in water. It forms a very stable, colourless, 1:1 complex with calcium or magnesium ions. The complex formed is generally very soluble in water (Fig. 12.1).

Eriochrome Black T: Eriochrome black T is an azo dye which is capable of forming soluble wine-red coloured complexes with Ca^{2+} and Mg^{2+}.

The Eriochrome Black T is black in unionized form whereas its ionized form is red in colour. (Fig. 12.2)

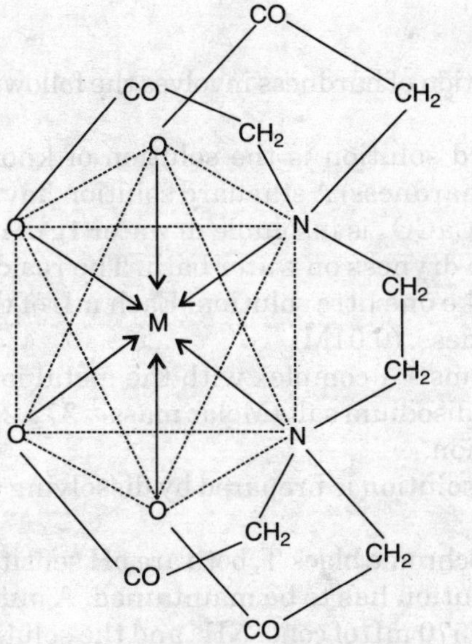

Fig. 12.1 EDTA Complex

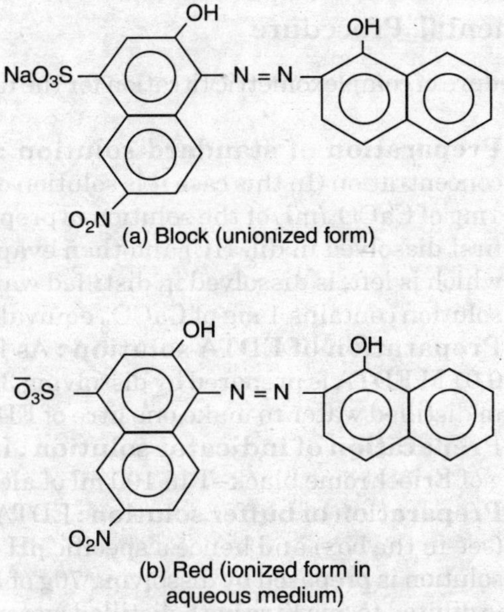

Fig. 12.2 Eriochrome Black T (sodium 1 – (1–hydroxy–2–naphthylazo) 6 – nitro –2 – naphthol – 4 – sulphonate

However, the phenolic groups on EBT may also get ionized at higher pH value (see box)

Principle of Complexometric Titration: Estimation of hardness by EDTA method is based on the following principle –

(i) First, the indicator Eriochrome Black-T, which is a blue coloured dye, forms an unstable complex with calcium or magnesium ions in water at a pH of 9 to 10. The complex is wine red in colour.

$$Mg^{2+}/Ca^{2+} + \text{Eriochrome Black T} \longrightarrow \underset{\text{Unstable Complex (Wine red)}}{[\text{Mg/Ca–Eriochrome Black T}]}$$

(ii) As this solution is titrated against EDTA, the free Ca^{2+} and Mg^{2+} ions in water form stable metal –ion EDTA complex

$$Mg^{2+}/Ca^{2+} + \text{EDTA} \longrightarrow \underset{\text{Stable Complex (Colourless)}}{[\text{Mg/Ca – EDTA}]}$$

Once the free metal ions are complexed, the EDTA replaces Ca and Mg ions from the unstable indicator complex also, to form a stable complex, with the result, the indicator is set free. Since the free indicator is blue in colour at the above mentioned pH, the end point is the appearance of blue colour.

$$\underset{\text{Wine red}}{[\text{Mg/Ca–Eriochrome Black T}]} + \text{EDTA} \longrightarrow [\text{Mg/Ca – EDTA}] + \underset{\text{(Blue)}}{\text{Eriochrome Black T}}$$

Thus the amount of EDTA used corresponds to the hardness of water.

The temporary hardness is removed by boiling and after the removal of precipitate by filteration, the permanent hardness in the filterate is determined by titration with EDTA as above.

Temporary hardness will be given by the difference of total hardness and permanent hardness.

Experimental Procedure

The procedure of complexometric titration for the determination of hardness involves the following steps:

(i) **Preparation of standard solution :** Standard solution is the solution of known concentration (In this case it is solution of known hardness) A standard solution having 1mg of $CaCO_3$/mL of the solution is prepared. As $CaCO_3$ is insoluble in water 1g of it is first dissolved in dil. HCl and then evaporated to dryness on water bath. The residue which is left, is dissolved in distilled water to make one litre solution. Each mL of this solution contains 1 mg of $CaCO_3$ equivalent hardness. (0.01M)

(ii) **Preparation of EDTA solution :** As EDTA forms 1:1 complex with the metal ions, 0.01 M EDTA is prepared by dissolving 3.7 g of its disodium salt (Molar mass = 372.25g) in distilled water to make one litre of EDTA solution.

(iii) **Preparation of indicator solution :** Indicator solution is prepared by dissolving 0.5 g of Eriochrome black–T in 100 ml of alcohol.

(iv) **Preparation of buffer solution :** EDTA and Eriochrome black T, both are pH sensitive (see in the box) and hence a specific pH of the solution has to be maintained. A buffer solution is prepared by dissolving 70g of NH_4Cl in 570 mL of conc. NH_3 and the solution is diluted to one litre with distilled water.

Ethylene diamine tetraacetic acid is highly pH sensitive ligand. Since it is tetraprotic acid, it can be written as H_4Y, where H represent the four ionizable hydrogens. Around

pH > 12 most of the EDTA exists as tetranion, Y^{4-}.

At pH ≥ 8, the EDTA exists mostly as HY^{3-}, whereas

At pH ≈ 5, the predominant specie is H_2Y^{2-}.

The reaction with Ca^{2+} or Mg^{2+} ions can be represented as

$$M^{2+} + H_2Y^{2-} \rightleftharpoons \underset{\text{Metal EDTA Complex}}{MY^{2-}} + 2H^+$$

$$M^{2+} + HY^{3-} \rightleftharpoons MY + HY^-$$

depending upon the pH of the solution. The protons liberated in these reactions would considerably alter the pH of the solution and may cause dissociation of metal–EDTA complex. However this is prevented by *adding an excess of buffer solution* to the titration system.

Erichrome Black T : The ionized form of Eriochrome black T is red in colour (Fig.12.2) The phenolic group on Eriochrome black T may also get ionized at higher pH values. If indicator is represented as H_3In, its dissociations will be represented as –

$$\underset{\text{Black}}{H_3In} + H_2O \longrightarrow H_3O^+ + \underset{\text{Red}}{H_2In^-}$$

$$\underset{\text{Red}}{H_3In^-} + H_2O \rightleftharpoons H_3O^+ + \underset{\text{Blue}}{H\,In^{2-}}$$

$$\underset{\text{Blue}}{H\,In^{2-}} + H_2O \longrightarrow H_3O^+ + \underset{\text{orange}}{In^{3-}}$$

Hence, it follows from the above discussion that the indicator would exhibit different colours, depending upon its dissociation, which in turn depends upon the pH of the solution i.e.

at $pH < 6$ the indicator exists as H_2In^-(red)
$pH > 7$ the indicator exists as HIn^{2-}(blue)

Thus, the pH of the titrating solution has to be maintained by adding a suitable buffer solution.

The EDTA titrations are performed in buffers of pH 8 – 10, the range in which the predominant form of the indictor is blue (HIn^{2-}). When metallic salts (Ca^{2+} or Mg^{2+}) are added, the colour of the indicator changes from blue to red as indicator complexes with metal ion. At the end point, the indicator is set free because of the formation of a stable metal EDTA complex and the colour of the solution changes to blue from wine red.

$$\underset{\text{(Indicator)}}{H\,In^{2-}} + M^{2+} \longrightarrow \underset{\substack{\text{Wine red}\\\text{comples}}}{M\,In^-} + H^+$$

$$M\,In^- + \underset{\text{(EDTA)}}{H_2Y^{2-}} \rightleftharpoons \underset{\text{(Colourless)}}{MY^{2-}} + \underset{\text{(Blue)}}{H\,In^{2-}} + H^+$$

(v) Standardization of EDTA solution : To find out the exact strength of EDTA solution, it is titrated against standard hard water prepared in step (i), using Eriochrome black T as indicator and in the presence of 10–12ml of buffer solution. The end point being wine red to blue. EDTA is taken in the burette and water is taken in the titration flask.

(vi) Titration of the sample water : In a similar way, the standard hard water is titrated against EDTA solution using Eriochrome black-T as indicator, in the presence of buffer solution.

(vii) Titration of water having permanent hardness : 250 ml of the water sample is taken in a 500 ml beaker and is boiled till the volume reduces to about 50 ml. Any temporary hardness present will thereby would be precipitated out as $CaCO_3$ or $MgCO_3$. The precipitates are filtered off and the volume of the filterate is made to 250 ml by adding distilled water. This is now titrated against EDTA as in step (v) and (vi).

Observations : Let volume of the water taken for each titration = 50mL

Volume of EDTA used when titrated against standard hard water = V_1mL

Volume of EDTA used when titrated against sample hard water = V_2mL

Volume of EDTA used when titrated against water having permanent hardness = V_3mL

Calculations :

(a) Strength of EDTA solution:

1 ml of standard Hard water (SHW) contains = 1 mg of $CaCO_3$

50 ml of SHW contains = 50 mg of $CaCO_3$

Volume of EDTA consumed for 50 ml of SHW = V_1 ml

∴ V_1 ml of EDTA is used for = 50 mg of $CaCO_3$

$$1 \text{ ml of EDTA is used for} = \frac{50}{V_1} \text{ mg of } CaCO_3$$

$$\text{or strength of EDTA solution} = \frac{50}{V_1} \text{ mg/ml of EDTA}$$

(b) Total Hardness :

Volume of EDTA used for 50 ml of sample hard water = V_2ml

$$\text{Since 1 ml of EDTA consumes} = \frac{50}{V_1} \text{ mg of } CaCO_3$$

$$V_2 \text{ ml of EDTA will consume} = \frac{50}{V_1} \times V_2 \text{ mg of } CaCO_3.$$

$$\text{Hence, 50 ml of sample water} = \frac{50}{V_1} \times V_2 \text{ mg of } CaCO_3.$$

$$\text{1000 ml of sample water} = \frac{50 \times V_2}{V_1} \times \frac{1000}{50} \text{ mg of } CaCO_3.$$

$$= \frac{V_2 \times 1000}{V_1} \text{ mg}$$

$$\therefore \quad \text{Total hardness} = \frac{V_2 \times 1000}{V_1} \text{ ppm.}$$

(c) Permanent Hardness:

Volume of EDTA used for 50ml of water, containing permanent hardness = V_3 mL

$$V_3 \text{ ml of EDTA will consume} = \frac{50}{V_1} \times V_3 \text{ mg of } CaCO_3.$$

$$\text{Hence, 50 ml of water contains pemanent hardness} = \frac{50}{V_1} \times V_3 \text{ mg of } CaCO_3.$$

$$\text{1000 ml of water contains permanent hardness} = \frac{50}{V_1} \times \frac{V_3}{50} \times 1000 \text{ mg of } CaCO_3.$$

$$= \frac{V_3 \times 1000}{V_1} \text{ mg}$$

$$\text{Permanent Hardness} = \frac{V_3}{V_1} \times 1000 \text{ ppm.}$$

(d) Temporary Hardness : Total hardness—Permanent hardness.

Table 12.3: *General Scale of Hardness*

Hardness of Water (mg/L)	*Description*
0–17	Soft
17–60	Slightly hard
60–120	Moderately hard
120–180	Hard
> 180	Very hard

Temporary hardness due to the presence of carbonates and bicarbonates is referred to as the carbonate hardness and the amount of hardness in excess carbonate (*i.e.*, permanent hardness) is referred to as the non-carbonate hardness.

∴ Total hardness = Carbonate + Non-carbonate hardness.

NUMERICALS BASED ON HARDNESS OF WATER

Example 12.4: One gram of $CaCO_3$ was dissolved in dil. HCl and the solution diluted to one litre. 100 mL of this solution required 90 ml of EDTA solution, while 100 mL of the sample water required 40 mL of EDTA. On the other hand, 100mL of the boiled sample water when titrated against EDTA consumed 20mL of solution. Calculate each type of hardness in ppm.

Solution: Strength of standard hard water = 1 mg/mL = 1g/L

Strength of EDTA

90 ml of EDTA is required for = 100ml of standard hard water

= 100 mg of $CaCO_3$

∴ 1 ml of EDTA is required for = $\frac{100}{90}$ mg of $CaCO_3$

Total hardness:

100 ml of sample required = 40 ml of EDTA

Since 1ml of EDTA is required for = 10/9mg of $CaCO_3$.

∴ 40ml of EDTA is required for = 10/9 × 40mg of $CaCO_3$.

Hence 100 ml of sample water contains 400/9 mg of $CaCO_3$.

1000ml of sample water contains = $\frac{400}{9} \times \frac{1000}{100}$ mg of $CaCO_3$ = 444.4mg/L

Total hardness = 444.4ppm.

Temporary hardness:

After boiling 100 ml of sample, water requires = 20 ml of EDTA

100 ml of sample water contains = $20 \times \frac{10}{9} = \frac{200}{9}$ mg of $CaCO_3$.

1000 ml of sample water contains = $\frac{200}{9} \times \frac{1000}{100}$ mg of $CaCO_3$.

Permanent hardness = 222.2mg/L

= 222.2ppm

Temporary hardness = 444.4–222.2ppm = 222.2ppm.

Example 12.5: 100ml of a sample of water required 20ml of 0.01M EDTA for titration using Eriochrome T as indicator. After boiling 100ml of the same sample, required 10ml of 0.01M EDTA. Calculate

(i) the total hardness
(ii) permanent hardness
(iii) and temporary hardness.

Solution: Since EDTA forms 1 : 1 complex with the metal ions hence,

1 molar solution of EDTA reacts with 1mole of $CaCO_3$.

or 1000ml of 1M EDTA reacts with = 100g of $CaCO_3$.

$$1 \text{ ml of 1M EDTA reacts with} = \frac{100}{1000} \times 1000 \text{ mg of } CaCO_3.$$

$$= 100\text{mg of } CaCO_3.$$

$$1\text{ml of .01M EDTA reacts with} = \frac{100}{100} = 1\text{mg of } CaCO_3.$$

Strength of the given EDTA solution = 1mg/mL.

(i) Total hardness:

Since 1ml of EDTA is required for = 1 mg of $CaCO_3$.

20ml of EDTA is required for = 20 mg of $CaCO_3$.

Hence 100ml of sample hard water contains = 20 mg of $CaCO_3$.

$$1000\text{ml of sample hard water contains} = \frac{20}{100} \times 1000 = 200\text{mg of } CaCO_3.$$

Total hardness = 200mg/L = 200ppm.

(ii) Permanent hardness:

10ml of EDTA is required for = 1 × 10mg of $CaCO_3$.

Hence 100ml of boiled sample hard water contains = 10mg of $CaCO_3$.

1000 ml of boiled sample hard water contains = 100mg of $CaCO_3$.

Permanent Hardness = 100mg/L = 100ppm.

(iii) Temporary hardness = Total hardness – permanent hardness
= 200 – 100 = 100ppm.

Example 12.6: 500 ml of a sample of water on EDTA titration with Eriochrome Black-T as indicator consumed 18ml of 0.045M EDTA till end point is reached. Calculate the hardness of water.

Solution: (1ml of 0.01M EDTA = 1mg of $CaCO_3$)

$$18\text{ml of 0.045M EDTA} = \frac{1 \times 18}{.01} \times 0.45 \text{ mg of } CaCO_3.$$

$$= 81\text{mg of } CaCO_3.$$

500 ml of water contains 81mg of $CaCO_3$, so hardness present in 1 litre

$$= \frac{81}{500} \times 1000 \text{ mg/L}$$

$$= 162 \text{ mg/L} = 162\text{ppm}.$$

Example 12.7: 0.50g of $CaCO_3$ was dissolved in dil. HCl and diluted to 500 ml. 100ml of this solution required 20ml of EDTA solution for titration. 100ml of a hard water sample required 30ml of the same EDTA solution for titration. On boiling, titration of 100 ml of the same water sample required 10ml of EDTA. Calculate total, permanent and temporary hardness. (1ml of 0.01M EDTA = 1mg of $CaCO_3$)

Solution: 500ml of $CaCl_2$ solution contains 0.50g of $CaCO_3$. eq. hardness
= 500mg of $CaCO_3$ eq. hardness
1ml of $CaCl_2$ solution = 1mg of $CaCO_3$ eq. hardness

Strength of EDTA:

100ml of $CaCl_2$ solution = 100mg of $CaCO_3$.

20ml of EDTA solution is required for 100ml of $CaCl_2$ solution.
= 100mg of $CaCO_3$

1ml of EDTA solution is required for 100/20 mg of $CaCO_3$
= 5mg of $CaCO_3$

Total hardness:

100ml of hard water = 30ml of EDTA solution
= 30 × 5mg of $CaCO_3$
= 150mg of $CaCO_3$.
1000ml of hard water = 1500mg of $CaCO_3$.
Total hardness = 1500 ppm

Permanent hardness:

100 ml of permanent hard water = 10 ml of EDTA solution
= 5 × 10mg of $CaCO_3$.
1000ml of permanent hard water = 500mg of $CaCO_3$
Temporary hardness = Total hardness – Permanent hardness
= 1500 – 500 = 1000mg of $CaCO_3$
Temporary hardness = 1000 ppm

Potential Health Effects of Hard Water

Hard water is not a health hazard. In fact, National Academy of Sciences states that hard drinking water generally contributes a small amount toward calcium and magnesium human dietary needs. They further state that in some instances, where dissolved calcium and magnesium are very high, water could be a major contributor of calcium and magnesium to the diet.

However, hard water has some advantages also.

Advantages of hard water:

1. It tastes better due to the presence of calcium and magnesium ions.
2. It is thought to reduce the number of heart illnesses.
3. It provides useful calcium ions for the healthy growth of bones and teeth.
4. The formation of lime scale in pipes due to hard water causes the inside of the pipe to be covered with insoluble carbonates. This layer of carbonate prevents the water in the pipe from coming into contact with the metal of the pipe and so prevents pipe corrosion and poisonous metal salts becoming dissolved in the water.

12.4 ALKALINITY OF WATER

Another important characteristic of water is its alkalinity, which is a measure of the ability of water to neutralize the acids. The alkalinity of water is normally due to the presence of bicarbonates, carbonates and hydroxides of sodium, potassium, calcium and magnesium. Some of the salts, not all, which cause alkalinity also cause hardness. However, presence of other bases such as borates, silicates, phosphates can also contribute to the total alkalinity, though to a small extent.

Depending upon the anion present, the alkalinity can be classified as:

(i) Caustic alkalinity (due to $\bar{O}H$ and CO_3^{2-} ions) ($\because$ CO_3^{2-} also produce $\bar{O}H$ on hydrolysis)

(ii) Bicarbonate alkalinity (due to HCO_3^- ions)

The alkalinity in water can be present due to the presence of the above mentioned ions present alone or in combination as

- hydroxide only ($\bar{O}H$)
- Carbonates only (CO_3^{2-})
- Bicarbonates only (HCO_3^-)
- Hydroxides and Carbonates ($\bar{O}H + CO_3^{2-}$)
- Carbonate and bicarbonates ($CO_3^{2-} + HCO_3^-$)

The presence of hydroxide and bicarbonates existing together is ruled out because they undergo the following reaction to form CO_3^{2-} ions.

$$\bar{O}H + HCO_3^- \longrightarrow CO_3^{2-} + H_2O$$

And hence the possibility of all the three ions existing together is also ruled out.

12.4.1 Determination of Alkalinity

Principle: Determination of alkalinity due to different ions is based on the titration of the water sample against a standard acid making *selective use of indicators*. The indicators used are - phenolphthalein and methyl orange. The reactions taking place may be represented by the following equations:

$$\left.\left.\begin{array}{ll}\text{(a)} & \bar{O}H + H^+ \longrightarrow H_2O \\ \text{(b)} & CO_3^{2-} + H^+ \longrightarrow HCO_3^- \end{array}\right] P \quad \begin{array}{ll} \text{(c)} & HCO_3^- + H^+ \longrightarrow H_2CO_3 \end{array}\right] M$$

- The volume of the acid used upto phenolphthalein end point corresponds to the reaction (a) and (b) i.e. complete neutralization of $\bar{O}H$ ions and neutralization of CO_3^{2-} ions upto HCO_3^- stage (half neutralization of CO_3^{2-} ions).
- The volume of the acid used upto methyl orange end point corresponds to the reaction (a), (b) and (c) i.e. complete neutralization of $\bar{O}H$, CO_3^{2-} and HCO_3^- ions.

Thus from the volume of the acid used, the respective strengths of various ion can be determined as discussed in the following section.

Procedure: A known volume of the sample water is titrated against a standard acid, using phenolphthalein as indicator. The end point is disappearance of pink colour. To the same solution is added the indicator, methyl orange.

This solution is then titrated with the standard acid solution till end point, from yellow to pink is reached.

Observations:

Let the volume of the water sample taken for each titration = 100ml

Volume of the acid used upto phenolphthalein end point = V_1ml

Extra volume of the acid used upto methyl orange end point = V_2ml

Total volume of the acid used upto methyl orange endpoint = $(V_1 + V_2)$ml

Normality of the acid used for titration = N/50

Calculations:

(i) Phenolphthalein alkalinity in terms of $CaCO_3$ equivalents

$$\underset{\text{(Acid)}}{N_1 V_1} = \underset{\text{(Water sample)}}{N_2 \times 100}$$

$$\frac{1}{50} \times V_1 = N_2 \times 100$$

$$N_2 = \frac{1}{50} \times \frac{V_1}{100} = \frac{V_1}{5000}$$

$$\text{Strength in terms of } CaCO_3 \text{ equivalents} = N_2 \times \text{Eq.wt.of } CaCO_3 = N_2 \times 50\text{g/L}$$

$$= \frac{V_1}{5000} \text{ g/L}$$

$$\text{Phenolphthalein alkalinity, P} = \frac{V_1}{100} \times 1000 \text{ mg/L} = 10V_1 \text{ ppm}$$

(ii) Methyl orange alkalinity in terms of $CaCO_3$ equivalents

$$\underset{\text{(Acid)}}{N_1(V_1 + V_2)} = \underset{\text{(Water sample)}}{N_2' \times 100}$$

$$\frac{1}{50} \times (V_1 + V_2) = N_2' \times 100$$

$$N_2' = \frac{1}{50} \times \frac{(V_1 + V_2)}{100}$$

$$\text{Strength in terms of } CaCO_3 \text{ equivalent} = N_2' \times \text{Eq.wt. of } CaCO_3$$

$$= \frac{V_1 + V_2}{5000} \times 50 \text{ g/L}$$

$$\text{Methyl orange alkalinity, M} = 10\,(V_1 + V_2) \text{ mg/L} = 10\,(V_1 + V_2) \text{ ppm}$$

On the basis of the analysis of water with respect to phenolphthalein alkalinity and methyl orange alkalinity, the alkalinity due to differents ions can be calculated. The results are summarized in the following table.

Table 12.4: *Computing the alkalinity due to different ions*

Alkalinity	$\bar{O}H$ *(ppm)*	CO_3^{2-} *(ppm)*	HCO_3^- *(ppm)*
P = O	NIL	NIL	M
$P = \frac{1}{2}M$	NIL	2P	NIL
$P < \frac{1}{2}M$	NIL	2P	(M-2P)
$P > \frac{1}{2}M$	(2P–M)	2 (M–P)	NIL
P = M	P = M	NIL	NIL

Discussion:

Case (i): When phenolphthalein alkalinity = 0, this implies that volume of the acid used till phenolphthalein end point is 0; This can only happen if CO_3^{2-} and $\bar{O}H$ both are absent. Whatever alkalinity is present is due to HCO_3^- ions and can be detected using methyl orange as indicator.

Case (ii): When $P = \frac{1}{2}M$, this implies that only CO_3^{2-} ions are present. Neutralization using phenolphthalein as indicator reaches upto HCO_3^- stage. Same amount of acid will be further used when methyl orange is used as indicator to neutralize HCO_3^- to H_2O and CO_2 as represented in Fig. 12.3 (a) The strength of CO_3^{2-} will be given by M or 2P

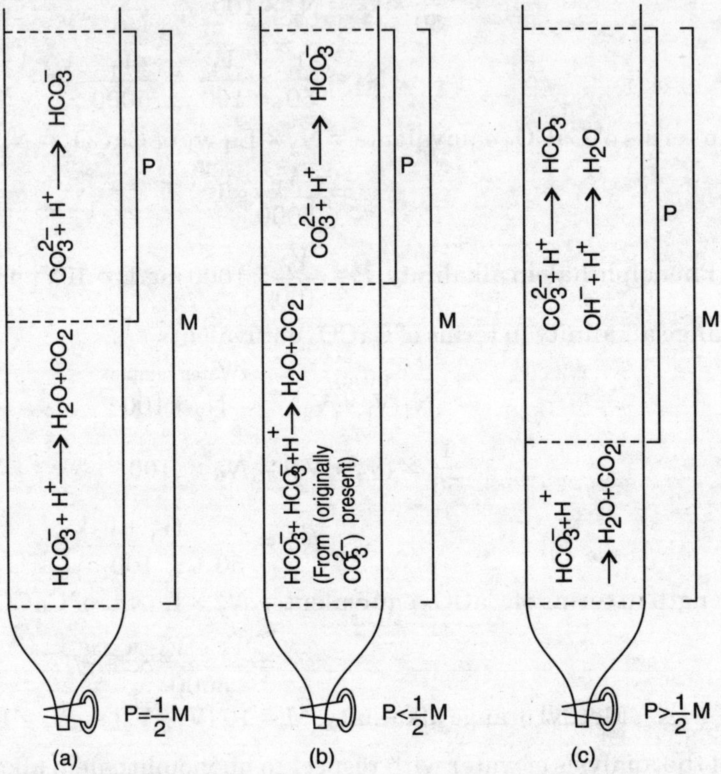

Fig. 12.3 *Relative amounts of Acid using phenolphthalein and methyl orange as indicators*

Case (iii): When $P < \frac{1}{2}M$, this implies that besides CO_3^{2-} ions, HCO_3^- ions are also present. The acid used uptill phenolphthalein end point corresponds to half neutralization of CO_3^{2-} ions. Further titration using methyl orange as indicator corresponds to neutralization of HCO_3^- obtained from CO_3^{2-} ions and HCO_3^- originally present in the reaction mixture as shown in Fig. 12.3(b)

Alkalinity due to $CO_3^{2-} = 2P$

Alkalinity due to $HCO_3^- = (M–2P)$

Case (iv) : When $P > \frac{1}{2}M$. This implies that besides CO_3^{2-} ions, $\overline{O}H$ ions are also present. $\overline{O}H$ will be completely neturalized by the acid whereas CO_3^{2-} will be neutralized up to the HCO_3^- stage, using phenolphthalein as indicator. On further titration with the acid using methyl orange as indicator, the neutralization of HCO_3^- takes place as shown in Fig. 12.3 (c).

Alkalinity due to $CO_3^{2-} = 2\,(M–P)$.

Alkalinity due to $\overline{O}H = M - 2\,(M–P) = (2P–M)$

Case (v) : When P = M implies that only $\overline{O}H$ ions are present

Alkalinity due to $\overline{O}H$ = P = M.

SOLVED NUMERICALS BASED ON ALKALINITY OF WATER

Example 12.8: 100ml of a water sample required 30ml of N/50 H_2SO_4 for neutralization to phenolphthalein end point. After this, methyl orange indicator was added to this and further acid required was again 30ml. Calculate the alkalinity of water as $CaCO_3$ in ppm.

Solution: (i) 100ml of water upto phenolphthalein end point

$$= 30\text{ml of N/50 } H_2SO_4.$$

$$100\text{ml} \times N_P = 30\text{mL} \times N/50$$

$$\text{Normality, } N_P = \frac{30}{100} \times \frac{1}{50} = \frac{3}{500} N$$

Strength of alkalinity upto phenolphthalein point (in terms of $CaCO_3$ equiv.)

$$(P) = \frac{3}{500} \times 50 \text{ g/L} = .3\text{g/L} = 300\text{mg/L}$$

(ii) 100ml of water upto methyl orange end point = 30mL + 30mL of N/50 H_2SO_4.

$$\text{Normality, } N_M = \frac{60}{100} \times \frac{1}{50} = \frac{6}{500} N$$

Strength in terms of $CaCO_3$ equivalents

$$(M) = \frac{6}{500} \times 50\text{g/L g/L} = .6\text{g/L}$$

$$= 600\text{mg/L}$$

From (i) and (ii) $P = \frac{1}{2}M$, so alkalinity is due to CO_3^{2-} ions only.

Hence total alkalinity = 2P = M = 600mg/L

= 600ppm.

Example 12.9: A water sample is alkaline to both phenolphthalein as well as methyl orange. 100ml of water sample on titration with N/50 HCl required 6.8 mL of the acid up to phenolphthalein end point. When a few drops of methyl orange are added to the same solution and the titration further continued, the yellow colour of the solution just turned red after the addition of another 12.6mL of the acid solution. Elucidate the type and extent of alkalinity present in water sample.

Solution: (i) 100ml of water upto phenolphthalein end point = 6.8 mL of N/50 HCl

$$\text{Normality, } N_P = \frac{6.8}{100} \times \frac{1}{50} = .00136\text{N}$$

Strength in terms of $CaCO_3$ eq.

$$(P) = 0.00136 \times 50\text{g/L} = .068\text{g/L} = 68\text{mg/L} = 68\text{ppm}$$

(ii) 100 ml of water upto methyl orange end point = 6.8 mL + 12.6 mL of N/50 HCl

$$\text{Normality, } N_M = \frac{19.4}{100} \times \frac{1}{50} = .0039 \text{ N}$$

Strength in terms of $CaCO_3$ eq.

$$(M) = 0.0039 \times 50\text{g/L} = .195\text{g/L} = 195\text{mg/L} = 195\text{ppm}$$

From (i) and (ii), $P < \frac{1}{2}M$, therefore the alkalinity is due to CO_3^{2-} and HCO_3^-

Alkalinity due to CO_3^{2-} = 2P = 2× 68ppm = 136ppm

Alkalinity due to HCO_3^- = (M–2P) = 195– 136ppm = 59ppm

Example 12.10: A sample of water was alkaline both to phenolphthalein and methyl orange. 100ml of this water sample required 20mL of N/50 H_2SO_4 for phenolphthalein end point and another 15mL for complete neutralization. Calculate the type of alkalinity in ppm.

Solution: (i) 100mL of water upto phenolphthalein end point = 20mL of N/50 H_2SO_4

$$\text{Normality } N_P = \frac{20}{100} \times \frac{1}{50} = .004\text{N}$$

Strength in terms of $CaCO_3$ eq.

$$(P) = 0.004 \times 50\text{g/L}$$

(ii) 100mL of water upto methyl orange end point

$$\text{Normality, } N_M \equiv 20\text{ml} + 15\text{mL of N/50 } H_2SO_4$$

$$\text{Normality, } N_M = \frac{35}{100} \times \frac{1}{50} = .007\text{N}$$

Strength in terms of $CaCO_3$ eq.

$$(M) = 0.007 \times 50 \text{ g/L} = 0.35\text{g/L} = 350\text{mg/L}$$

From (i) and (ii), $P > \frac{1}{2}M$, therefore the alkalinity is due to $\bar{O}H$ and CO_3^{2-} ions.

$$\text{Alkalinity due to } CO_3^{2-} \text{ ions} = 2\,(M - P) = 2\,(350\text{–}200)\text{mg/L}$$
$$= 300\text{mg/L} = 300\text{ppm}$$

$$\text{Alkalinity due to } \bar{O}H \text{ ions} = (2P - M) = [2\,(200)\text{–}350]\text{mg/L}$$
$$= 50\text{mg/L} = 50\text{ppm}$$

12.5 BOILER FEED WATER

In industries, water is mainly required for cooling and steam generation purposes. The water used for cooling purposes need not to be of high quality and hence requires very less treatment. Sometimes supply of water for cooling purposes are used without any treatment.

On the other hand water required for boiler feed purposes i.e., for steam generation should be of very high quality and thus requires a lot of treatment. However in treatment of boiler feed waters, total elimination of all the impurities is generally not required. Only those impurities which lead to certain problems are eliminated or kept within the prescribed range.

Untreated waters, containing impurities may lead to the following problems in boilers:

1. Scale and sludge formation.
2. Boiler Corrosion
3. Caustic Embrittlement
4. Priming and Foaming

All these problems are discussed below in detail-

1. Scale and Sludge Formation

In boilers water is converted into steam. When water containing impurities is used for generating steam, the concentration of the dissolved salts increases progressively because of evaporation of water. When the concentration reaches beyond the saturation point, the dissolved solids start precipitating out in order of their solubilities, the salt which is least soluble separating out first.

When precipitation of the dissolved salts takes place in the form of loose, non adherent, slimy precipitates, it is called **sludge**. On the otherhand if the precipitates form hard, adhering coating on the inner walls of the boilers, it is called **Scales**.

For the deposition of scale and sludge,

- the ionic product of the salt exceeds their solubility product and hence precipitation occurs.
- the increase in temperature leads to reactions that result in the formation of insoluble substances.

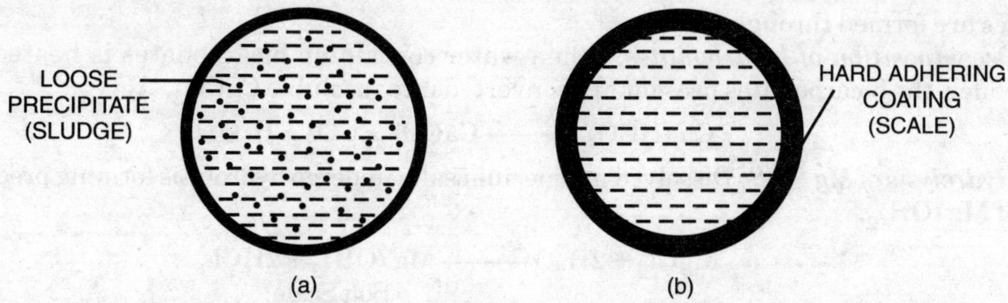

Fig. 12.4 *Sludge (a) and Scale (b) in boilers*

The salts responsible for the formation of scales and sludges are $CaCO_3$, $CaSO_4$, Mg $(OH)_2$ $MgCl_2$, and SiO_2 etc. The substances like $CaSO_4$, the solubility of which decrease with temperature forms scales on the hot surface and sludge in the cooler regions. Whereas if the solubility increases with increase in temperature, the material will form scale on the cooler surfaces and sludges in the hotter regions.

Though, formation of scales and sludges both is undersirable but the formation of scales is far more dangerous as compared to sludges as the latter can be removed easily.

Sludge Formation

Sludges are formed by the substances having greater solubilities in hot water than in the cold water e.g $MgCO_3$, $MgCl_2$, $CaCl_2$, $MgSO_4$ etc.

Disadvantages of Sludge Formation

- Wastage of Fuel: Sludges are poor conductor of heat so they tend to waste a lot of amount of heat. The boilers will have to be heated to higher temperatures in order to maintain the supply of heat.
- Decreased efficiency of the boilers: Sludge formation decrease the efficiency of the boilers.
- Choking of the pipes: The formation of sludges slows down the water circulation as it settles in the regions such as pipe connections, plug openings etc. Excessive formation may even cause choking of pipes.

Prevention of Sludge Formation

- Sludge formation can be prevented by using soft water for boiler feed purposes.
- By frequently carrying out the cleaning operations i.e. removal of a portion of concentrated water and replacing it by fresh water. This is called "*blow down operation*" and the soft water which is added to make up the volume is called "*make up water*".

Removal of Sludge

- Sludges can easily be removed by scrapping off with a brush.

Scale Formation

Scales stick very firmly to the inner walls of the boilers and hence are very difficult to remove. Some of them are so hard that they cannot be removed even with the help of hammer and chisel.

The reason for the deposition of some substances as scales and some substances as sludges have not been explained satisfactorily. According to Partridge, when a bubble of water is formed and strikes on a heated surface of the boiler, the salts contained in the solution get deposited on the surface and water leaves as steam. The bubbles are formed so rapidly that the solid deposited at the surface does not get any chance to get washed away and instead another bubble strikes it, depositing more solid over it and so on and subsquently a thick adhering coating is deposited at the inner walls.

Scales are formed through the

- *Decomposition of bicarbonates:* When water containing bicarbonates is heated in the boiler, the bicarbonates present are converted into insoluble $CaCO_3$.

$$Ca\,(HCO)_3 \longrightarrow CaCO_3 + CO_2 + H_2O.$$

- *Hydrolysis of Mg Salts:* Dissolved magnesium salts undergo hydrolysis forming precipitates of $Mg\,(OH)_2$.

$$MgCl_2 + 2H_2O \longrightarrow \underset{\text{Soft Scale}}{Mg\,(OH)_2} + 2HCl.$$

- *Calcium sulphate* is soluble in cold water but the solubility decreases with rise of temperature. Consequently, $CaSO_4$ gets precipitated as hard scale on the heated portions of the boiler. $CaSO_4$ is the main cause of scale formation in high-pressure boilers.
- *Presence of Silica:* Salts like calcium silicate, magnesium silicate are sparingly soluble in cold water, but are completely insoluble in hot water. These deposits stick very firmly on the inner side of the boiler surface and are very difficult to remove.

Disadvantages of Scale Formation

- *Wastage of Fuel :* Because of low thermal conductivity of the scales, the heat supplied to the boiler is not efficiently transferred to water. Hence more heat has to be supplied to maintain the supply of steam leading to wastage of fuels.
- *Bagging:* The distortion of boiler material is known as bagging. The superheating of the boiler leads to the distortion of the boiler material. Due to overheating, rapid reaction between water and iron occurs at high temperature, causing thinning of the boiler material.

$$3Fe + 4H_2O \rightarrow Fe_3O_4 + 4H_2$$

- *Danger of Explosion:* The uneven expansion of the scale may lead to the cracking of the scale. The super-heated steel plates come in contact with boiler water. This causes formation of a large amount of steam suddenly, creating high pressure inside the boiler. When the boiler material is not strong enough to withstand the preasure, it bursts.

Removal of Scales

- Soft scales are removed with the help of scrapper or wire brush.
- Brittle or hard scales can be removed by giving thermal shocks i.e. heating the boiler and suddenly cooling with cold water.
- Scales can also be dissolved in certain chemicals and hence can be removed along with water. For e.g $CaCO_3$ scales can be dissolved by using 5-10% HCl. Similarly EDTA can also be used to dissolve the salts of Ca, with which it forms soluble complexes.

Prevention of Scale Formation

The scale formation can be minimized by subjecting the boiler feed water to certain treatments which may be

(i) **External treatment**. which includes the water softening techniques. (discussed in the next section).

(ii) **Internal Treatment:** The internal treatment of boiler feed waters refers to the treatment of water in the boiler itself (discussed in the next section).

2. Boiler Corrosion

Boiler Corrosion is the decay of boiler material due to the attack of certain chemicals on its surface, resulting into the formation of compounds such as oxides, sulphides etc. The material of the boiler thus gets dissolved and rusted, thereby reducing the life of the boiler.

In general, corrosion occurs, if Fe of the boiler material enters water as Fe^{2+} ions and combines with the water molecules to form ferrous hydroxide, which reacts with oxygen to form Fe $(OH)_3$, or rust.

$$Fe^{2+} + 2H_2O \longrightarrow Fe\,(OH)_2 + 2H^+$$

$$4Fe\,(OH)_2 + O_2 \longrightarrow \underset{\text{Rust}}{2\,[Fe_2O_3.\,2H_2O]}$$

Main reasons for boiler corrosions are-

(a) Presence of dissolved oxygen.
(b) Presence of dissolved carbon dioxide.
(c) Presence of Mineral Acids.

(a) **Presence of dissolved oxygen:** Dissolved oxygen is the main cause of boiler corrosion. At room temperature about 8ml of dissolved oxygen is present in water. Oxygen enters the boiler through raw make up water. When water containing dissolved oxygen is heated in the boiler, the free gas is evolved under prevailing high temperature of boiler and attacks the boiler material.

$$2\,Fe + 2H_2O + O_2 \longrightarrow 2Fe\,(OH)_2$$

$$4Fe\,(OH)_2 + O_2 + 2H_2O \longrightarrow 2[Fe_2O_3.\,2H_2O]$$

Removal of dissolved Oxygen: Oxygen can be removed by chemical or mechanical means.

Chemical Means: Oxygen can be removed by adding calculated amounts of certain chemicals such as Na_2SO_3, Na_2S and N_2H_4.

$$2Na_2SO_3 + O_2 \longrightarrow 2Na_2SO_4$$

$$Na_2S + O_2 \longrightarrow Na_2SO_4.$$

However, the sodium sulphate so formed may decompose in high pressure to form sulphurous acid. Sulphurous acid is corrosive in nature. Therefore a very low concentration of 5 – 10 ppm of Na_2SO_3 is maintained in the boiler water.

Hydrazine is found to be an ideal compound for removing dissolved oxygen since the products are water and nitrogen gas which does not form harmful products.

$$\underset{\text{Hydrazine}}{N_2H_4} + O_2 \longrightarrow N_2 + 2H_2O$$

Thus hydrazine is now extensively used to remove dissolved oxygen in high pressure boilers. Though pure hydrazine is an explosive, inflammable liquid, but 40% aqueous solution is quite safe to handle and is used in water treatment. Calculated amount of hydrazine should be added. Any excess of it decompose to give NH_3, which may cause corrosion of some alloys like copper etc used in condensor tubes.

$$3N_2H_4 \longrightarrow 4NH_3 + N_2.$$

Mechanical Means: Mechanical means for the removal of oxygen make use of Henry's law and Dalton's Law i.e solubility of a gas is directly proportional to pressure and inversely proportional to temperature.

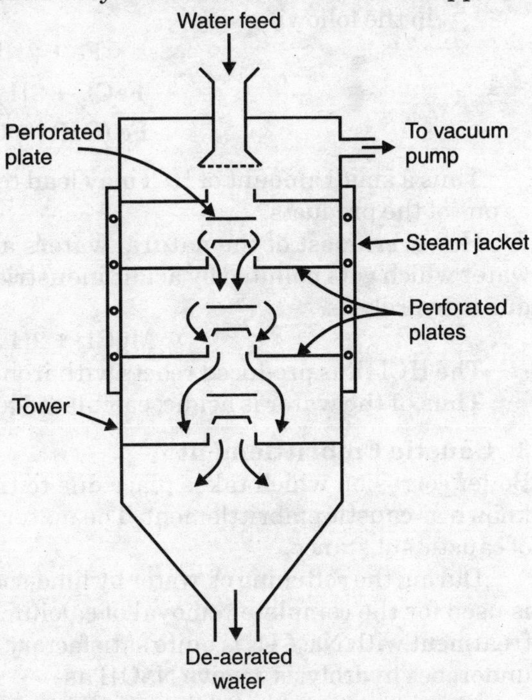

Fig. 12.5 *Mechanical deaeration of water*

Mechanical deaeration involves, injecting hot feed-water, as a fine spray, into a vaccum chamber heated externally by steam. The chamber is fitted with perforated plates to provide large surface area. Large exposed surface, high temperature, and low pressure reduces the solubility of the gas, the water free of dissolved oxygen is collected at the bottom. (Fig. 12.5)

(b) **Presence of dissolved CO_2:** Carbon dioxide is dissolved in raw water and also it is formed by the decomposition of dissolved bicarbonates under high temperature of boiler.

$$Mg(HCO_3)_2 \longrightarrow MgCO_3 + CO_2 + H_2O.$$

Carbondioxide forms carbonic acid in presence of water which has slow corrosive effect on boiler material.

$$CO_2 + H_2O \longrightarrow \underset{\text{Carbonic acid}}{H_2CO_3}.$$

Removal of CO_2 : Removal of CO_2 also involves chemical or mechanical means.

Chemical means: Chemical means for removal of CO_2 includes addition of calculated amount of NH_4OH.

$$2NH_4OH + CO_2 \longrightarrow (NH_4)_2CO_3 + H_2O.$$

However, if NH_4OH used is more than the calculated amount, it may attack the condeser tubes made of copper.

(ii) Filtering water through lime stone also removes CO_2

$$CaCO_3 + H_2O + CO_2 \longrightarrow Ca(HCO_3)_2$$

But this method leads ot increase in hardness content due to the formation of Ca $(HCO_3)_2$.

Mechanical means: Process is similar as discussed in removal of oxygen. CO_2 is also removed along with oxygen in mechanical deaeration.

(c) **Presence of Acids:** Free mineral acids, present in boiler water, reacts with iron material in the following way

$$Fe + 2HCl \longrightarrow FeCl_2 + H_2$$

$$FeCl_2 + 2H_2O \longrightarrow Fe(OH)_2 + 2HCl$$

$$Fe(OH)_2 + O_2 \longrightarrow 2[Fe_2O_3.2H_2O]$$

Thus a small amount of HCl may lead to corrosion to a great extent as HCl is further formed as one of the products.

However, most of the natural waters are alkaline excepting waters from mining areas or water which gets polluted by acidic industrial wastes. But some salts like $MgCl_2$ etc liberate HCl due to hydrolysis.

$$MgCl_2 + 2H_2O \longrightarrow Mg(OH)_2 + 2HCl.$$

The HCl thus produced reacts with iron and hence corrodes the boiler material.

Thus, if the water is acidic, calculated amount of alkali is added to neutralize it.

3. Caustic Embrittlement

Boiler corrosion which takes place due to the presence of highly alkaline water in the boiler is known as caustic embrittlement. The material of the boiler becomes brittle due to accumulation of caustic substance.

During the softening of water by lime-soda process (discussed later) a slight excess of Na_2CO_3 is used for the complete removal of calcium and magnesium salts. In low pressure boilers, the treatment with Na_2CO_3 is quite satisfactory but in high pressure boilers the free Na_2CO_3 present undergoes hydrolysis to give NaOH as-

$$Na_2CO_3 + H_2O \longrightarrow 2NaOH + CO_2.$$

NaOH so formed along with the natural alkalinity of the water make the boiler water caustic. The NaOH travels to the minute hair cracks, riveted seams, bends and joints. Water evaporates and the concentration of NaOH at these points keep on increasing with time due to poor circulation of water at these points. This caustic soda attacks the surrounding area e.g the intergranular cracks within the cracks, thereby corroding the boiler material or dissolving the iron of the boiler. This causes embrittlement of the boiler parts, particularly the stressed parts. The embrittlement appears to be favoured under the following conditions:

- Materials subjected to great stresses beyond their yield points.
- High conc. of NaOH.
- Presence of compounds such as sodium silicate etc.

The caustic embrittlement can be explained by the formation of the concentration cell as

Iron at Rivets	Conc. NaOH	Dil. NaOH	Iron at plane surfaces

The iron surrounded by concentrated NaOH becomes anodic side, which is corroded whereas the iron surrounded by dil. NaOH becomes cathodic side. (Fig. 12.6)

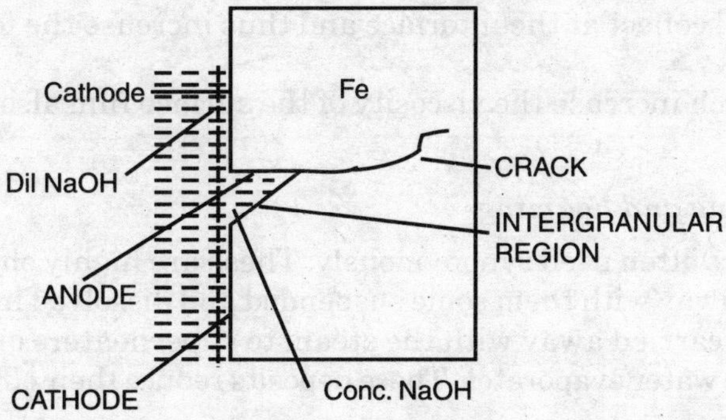

Fig 12.6 *Caustic Embrittlement at the Cracks*

The concentrated alkali dissolves the iron as sodium ferroate ($Na_2 FeO_2$). $Na_2 FeO_2$ decomposes a short distance away from its place of formation according to the reaction.

$$3Na_2FeO_2 + 4H_2O \longrightarrow 6NaOH + Fe_3O_4 + H_2.$$

Further dissolution of iron takes place because of the regeneration of NaOH and precipitation of Fe_3O_4 according to the above reaction.

Prevention of Caustic Embrittlement: The caustic embrittlement can be prevented by

- Using sodium phosphate as softening reagent instead of sodium carbonate.
- By adding liquids or tannin to boiler water. These block up the hair cracks thus prevent the infiltration of sodium hydroxide.
- It is observed that boiler waters containing sodium sulphate or sodium phosphate inhibit the caustic embrittlement by blocking the capillaries, thereby preventing the infiltration of caustic soda solution to these. The concentration of Na_2SO_4 : NaOH are maintained at 1 : 1, 2 : 1 or 3 : 1 for operating pressures of 10, 20 and > 20 atmospheres respectively to check the caustic embrittlement.

However, presence of Na_2SO_4 may lead to the formation of $CaSO_4$ scales. Thus it is preferable to use Na_3PO_4 instead of Na_2SO_4. It is over 300 times as effective as Na_2SO_4 in suppressing embrittlement.

4. Priming and Foaming

Due to the extremely rapid boiling of water in the boiler, the steam formed may be associated with small droplets of water. These droplets of water are carried away along with the steam. *The phenomenon of formation of wet steam i.e. steam associated with liquid water droplets is called priming.*

The liquid contamination in the steam is expressed in percentage by weight of steam. For e.g if steam contains 0.5% moisture, its steam quality will be reported as 99.5%.

Cause: Priming is caused by-

- The high steam velocities.
- Very high water level in the boiler.
- Sudden steam demands leading to sudden drop of pressure in the steam line.
- Faulty boiler design.
- Presence of suspended and dissolved impurities in water.

Foaming is the formation of small persistent bubbles at the surface of water in the boiler. Bubbles are formed when there is a difference in concentration of solute or suspended matter between the surface film and the bulk of the liquid. Any material which lowers the surface tension of the water will collect at the interface and thus increase the foaming tendency of the liquid e.g oil or grease.

The substances which increase the viscosity of the surface film also increase foam forming tendency.

Disadvantages of Priming and Foaming

Priming and foaming are often used synonymously. These are highly objectionable because the droplets of water carry away with them some suspended and dissolved impurities present in the boiler water. These are carried away with the steam to superheaters or turbine blades, where they get deposited as the water evaporates. These deposits reduce their efficiency by hindering the flow of steam.

- A part of the dried salts may be carried along with the steam farther and deposit on the engine valves thereby decreasing their life.
- Presence of water droplets in the steam may lead to corrosion in the steam-inlet ends of superheaters.
- Height of the water column cannot be judged properly due to foaming thereby making the maintenance of the boiler pressure difficult.

Prevention of Priming: Priming can be avoided by-

- Keeping the water level lower.
- Reducing steam velocities also decreases the moisture content of steam because of greater chances of condensing available for the drop.
- Efficient softening and filteration of boiler feed waters so as to minimize the dissolved and suspended impurities.
- Good boiler design fitted with mechanical steam purifiers.

Prevention of Foaming: Foaming can be avoided by-

- Adding antifoaming chemicals: Antifoaming chemicals act by counteracting the reduction in surface tension. For example polyamide antifoamers alters the surface tension and lead to the formation of only large unstable bubbles. Antifoaming agents may reduce foaming by simple mechanical action. For example, castor oil spreads on the surface of water and prevents foaming.

- Removing oil from boiler water also prevents foaming. Oil and grease can be removed by addition of coagulants such as sodium aluminate, ferrous sulphate etc. The coagulants like sodium aluminate etc form hydroxides [Al $(OH)_3$] which entraps oil drops.

5. Removal of Silica

Presence of silica in boiler feed water causes number of boiler problems such as-

- Silica forms thermally resistant scales with calcium and magnesium. With alumina it forms sodium alumino silicate or silica may form scale consisting of silica only.
- In high pressure boilers silica is carried over with steam and deposits on turbine blades.

Silica is removed from the boiler feed water by the following methods:

- By the addition of Magnesia (MgO) to the water, after the removal of temporary hardness. Magnesia acts as magnesium hydroxide. Silica is removed by magnesia by adsorption phenomenon or by the formation of magnesium silicate.
- By the use of ferrous sulphate or sodium aluminate as coagulants. They act as $Fe(OH)_2$ or $Mg(OH)_2$ and $Al(OH)_3$, which enmesh or trap finely suspended and colloidal impurities including oil and silica.
- By the use of highly basic anion exchanger. Water is first dimineralized by passing through anion and cation exchangers. After which it is again passed through a strongly basic special anion exchanger, which removes silica as H_2SiO_3. The silica content is usually reduced to a fraction of a ppm.
- Silicic acid (H_2SiO_3) is very feebly ionized and not removed by passing over weakly basic anion exchanger. Instead flourosilicic acid (H_2SiF_6) is strongly ionized and can be removed by passing over weakly basic anion exchanger. So silica is first converted into flourosilicic acid and then it is removed by passing it over weakly anion exchanger.

Silica in water can be converted into flourosilicic acid by adding sodium flouride and then passing through hydrogen-cation exchanger.

$$SiO_2 + 6NaF + 3H_2R \longrightarrow 3Na_2R + H_2SiF_6 + 2H_2O.$$

For each ppm of SiO_2, 4.2 ppm of NaF is used. Now finally when water is passed through weakly basic anion exchanger, silica is removed.

Silica can also be removed by distillation.

12.6 WATER SOFTENING TECHNIQUES

Internal Treatment of Boiler Feed Water

This involves adding of certain chemicals directly to the water to remove the scale forming salts, either by precipitation of these salts in some or the other form or by converting these salts into soluble complexes, which remain dissolved in water and thus do not form scales. The internal treatment is often referred to as "Sequestration". *Sequestration is a process in which an ion is prohibited to exhibit its usual properties, by either converting it to some other form or by forming a complex with some other compound.* The chemical which is added to bring about sequestration is called sequestering agent.

The methods used for internal treatment of water are-

(a) Colloidal Conditioning.
(b) Carbonate Conditioning.
(c) Phosphate Conditioning.
(d) Calgon Conditioning.

(a) *Colloidal Conditioning:* Scale formation can be reduced by introducing colloidal substances like kerosene, tannin, glue, agar etc. to the boiler water. These substances act as protective coatings. They surround the minute particles of scale forming salts and prevent their coalescence and coagulation. Hence, the salt remains as loose, non sticky deposits in the form of sludges which can be removed by simple blow down operation. This process is suitable for low pressure boilers.

(b) *Carbonate Conditioning:* For the precipitation of a salt, the ionic product of its ions should increase the solubility product. Thus, for the precipitation of $CaCO_3$, the product of its ions Ca^{2+} and CO_3^{2-}, must exceed the solubility product. Similarly, for the precipitation of $CaSO_4$, the ionic product of Ca^{2+} and SO_4^{2-} must exceed the solubility product of $CaSO_4$. The precipitation of $CaSO_4$ thus can be prevented by removing the Ca^{2+} from the solution so that ionic product does not exceed the solubility product. One way of doing it is, by carbonate conditioning, in which Na_2CO_3 is added to the boiler water. The ionization of $CaSO_4$ can be represented as:

$$CaSO_4 \rightleftharpoons Ca^{2+} + SO_4^{2-}$$

By the addition of Na_2CO_3, the concentration ratio of carbonate ions to sulphate ions is increased than the ratio of solubility product of $CaCO_3$ to that of solubility product of $CaSO_4$, i.e.

$$\frac{[CO_3^{\ 2-}]}{[SO_4^{\ 2-}]} > \frac{K_{CaCO_3}}{K_{CaSO_4}} = K'$$

where K_{CaCO_3} and K_{CaCO_4} represent the solubility product of $CaCO_3$ and $CaSO_4$ respectively. From the above discussion, it becomes clear that if $\left[CO_3^{2-}\right] > K'\left[SO^{2-}_4\right]$, then, only $CaCO_3$ will be precipitated in preference to $CaSO_4$, as solubility product of $CaSO_4$ will not be reached under these conditions. Thus the deposition of $CaSO_4$ scales is prevented. The sludge which is formed due to $CaCO_3$ can be removed by blow down operation.

$$Na_2CO_3 + CaSO_4 \rightleftharpoons Na_2SO_4 + CaCO_3$$

(c) *Phosphate Conditioning:* In low pressure boilers, the carbonate conditioning by adding Na_2CO_3 is applicable, while at high pressures, the CO_3^{2-} undergoes hydrolysis to form $\overline{O}H$

$$CO_3^{2-} + H_2O \rightleftharpoons HCO_3^- + \overline{O}H$$
$$HCO_3^- + H_2O \rightleftharpoons H_2O + CO_2 + \overline{O}H$$

Thus the concentration of CO_3^{2-} is decreased and $CaCO_3$ is not precipitated out. Moreover, this is likely to lead to caustic embrittlement. In such cases, it is preferable to precipitate Ca as *the most insoluble compounds of* Ca *i.e* $Ca_3(PO_4)_2$. This is done by the addition of Na_3PO_4

$$3CaSO_4 + 2Na_3PO_4 \rightleftharpoons Ca_3(PO_4)_2 + 3Na_2SO_4.$$

Here also, the excess of PO_4^{3-} is maintained, over that of SO_4^{2-}, to favour the forward reaction i.e the concentration ratio of phosphate ions to sulphate ions is increased than the solubility product of $Ca_3(PO_4)_2$ to $CaSO_4$ by the addition of sodium phosphate.

$$\frac{[PO_4{}^{3-}]^{2/3}}{[SO_4{}^{2-}]} > \frac{[K_{Ca_3(PO_4)_2}]^{2/3}}{K_{CaSO_4}} = K'[SO_4{}^{2-}]$$

or

$$[PO_4{}^{3-}]^{2/3} = K'[SO_4{}^{2-}]$$

Therefore, as long as $[PO_4{}^{3-}]^{2/3} > K'[SO_4{}^{2-}]$, only $Ca_3(PO_4)_2$ will be precipitated because the solubility of $CaSO_4$ cannot be exceeded by the ionic product under these conditions. The sludge, $Ca_3(PO_4)_2$ is removed by blow down operation.

Further, calcium can only be properly precipitated at pH of 9.5 or above. The main phosphates used for colloidal conditioning are-

(a) NaH_2PO_4 - Sodium dihydrogen phosphate (acidic)
(b) Na_2HPO_4 - Disodium hydrogen phosphate (weakly alkaline)
(c) Na_3PO_4 - Trisodium phosphate (alkaline)

The exact choice of phosphate depends upon the alkalinity of the boiler feed water. For instance, if the alkalinity of boiler water is low, trisodium phosphate is used as it is most alkaline. But if the alkalinity is too high then sodium dihydrogen phosphate is used.

(d) *Calgon Conditioning:* Calgon is the commercial name of sodium hexametaphosphate. It forms the complexes with the scale forming salt in which Ca^{2+} is trapped by the phosphate, even if present in very small amounts. The complex which is formed, is soluble in water and hence remains in the dissolved form without causing any harm to the boiler.

$$Na_2[Na_4(PO_3)_6] \longrightarrow 2Na^+ + [Na_4P_6O_{18}]^{2-}$$

$$2CaSO_4 + [Na_4P_6O_{18}]^{2-} \longrightarrow [Ca_2P_6O_{18}]^{2-} + 2Na_2SO_4$$

However, this treatment is not capable of preventing the formation of iron oxide and copper depositions, which can be prevented by adding EDTA or its sodium salt (trylon B) to the boiler water. EDTA binds the cations of scale forming salts into soluble complexes.

External Treatment of Boiler Feed Water

The process of removing the hardness causing salts from water is called softening of water. As is discussed earlier, water used for industrial purposes (boiler feed water, water for cooling purposes) must be sufficiently softened before put to use. The three most common methods adopted for softening of water are-

1. Lime Soda Process
2. Base Exchange or Zeolite Process
3. Demineralization or Ion Exchange Process.

1. Base Exchange or Zeolite Process

A reversible exchange of ions between a liquid phase and a solid phase is termed as ion exchange. A certain class of naturally occuring aluminosilicate minerals called zeolites act as solid phase, capable of exchanging its cations.

The name zeolite is derived from two greek words, Zein+lithos-which mean boiling stone. The term was first coined by Cronstedt, a Swedish geologist in 1756 for a group of minerals which released their water of hydration in the form of steam. The chemical structure of sodium zeolite may be represented as:

$$Na_2O.Al_2O_3.xSiO_2.yH_2O \text{ where } x = 2 - 10 \text{ and } y = 2 - 6,.$$

It may be regarded as hydrated sodium alumino-silicate, which is capable of exchanging its sodium ions for divalent ions of alkaline earth group and for the multivalent ions of some of the metals in water. Zeolites find application in softening of water for domestic and industrial purpose. Two types of zeolites in use are-

(i) **Natural Zeolites:** The naturally available zeolite is green in colour and is referred to as green sand. The common ones being Thomsonite (Na_2O, $Al_2O_3.3SiO_2.2H_2O$).

(ii) **Synthetic Zeolites:** Synthetic zeolites are porous and posses a gel structure. They can be prepared by heating together solutions of sodium silicate, aluminium sulphate and sodium aluminate. Synthetic Zeolites possess higher exchange capacity compared to that of natural zeolites. The most common artificial zeolite is a white coloured substance prepared from feldspar, kaolin, clay and soda and is called permutit. Hence Zeolite process is sometimes referred to as permutit process.

Structure: A zeolite crystal can be considered to be formed from the linking of several SiO_4 tetrahedra. In it, each oxygen of a tetrahedra is being shared with an adjacent one. The empirical formula is thus $(SiO_2)_n$. However some of the Si^{4+} ions are replaced by Al^{3+} ions. In order to balance the charges, extra positive ion Na^+ or K^+ must be incorporated for every Al^{3+} introduced. Thus the linking of tetrahedra results in an open structure with cavities in which Na^+ or K^+ ions are loosly held. (Fig. 12.7)

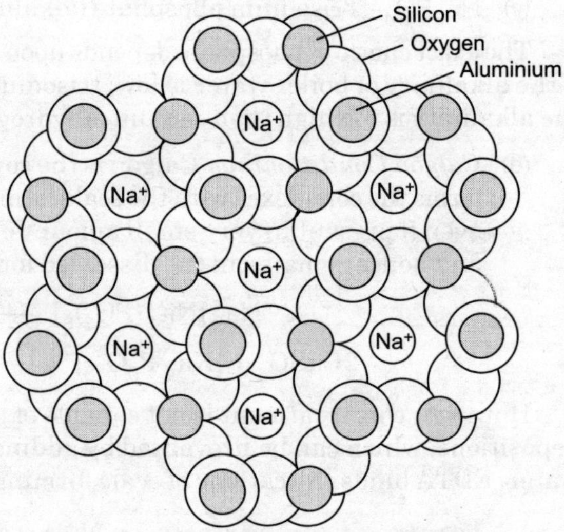

***Fig. 12.7** Structure of a naturally occurring zeolite; The sodium ions are loosely held in holes in the lattice.*

Principle of Water Softening

Zeolites can be represented as Na_2Z, where Z is the insoluble radical framework. When hard water is passed through a bed of active granular Na_2Z, the Ca^{2+} and Mg^{2+} ions of the hard water are exchanged for Na^+ ions of zeolite. The various reactions taking place may be indicated as follows:

$$Ca(HCO_3)_2 + Na_2Z \longrightarrow CaZ + 2NaHCO_3$$
$$Mg(HCO_3)_2 + Na_2Z \longrightarrow MgZ + 2NaHCO_3$$
$$CaSO_4 + Na_2Z \longrightarrow CaZ + Na_2SO_4$$
$$MgSO_4 + Na_2Z \longrightarrow MgZ + Na_2SO_4$$
$$CaCl_2 + Na_2Z \longrightarrow CaZ + 2NaCl$$
$$MgCl_2 + Na_2Z \longrightarrow MgZ + 2NaCl.$$

It is seen from the above reactions that sodium zeolite is converted to calcium and magnesium zeolites. This exchange takes place because the extent of exchange increases with increasing valency of the exchanging ions i.e monovalent ions are exchanged for divalent ions.

Regeneration: When the zeolite bed gets exhausted i.e nearly all its Na^+ are exchanged for Ca^{2+} and Mg^{2+} ions, it loses its sodium exchange capacity. The zeolite bed is regenerated by first backwashing it and then by passing a concentrated NaCl solution through it. Though other solutions like $NaNO_2$, Na_2SO_4, KCl etc can also be used, but brine solution (NaCl solution) is used because it is cheap and also because the products formed by regeneration reactions are $CaCl_2$ and $MgCl_2$ which are highly soluble and can be readily rinsed out from the zeolite bed.

The following reactions take place during regeneration process :

$$CaZ + 2NaCl \rightleftharpoons Na_2Z + CaCl_2$$
$$MgZ + 2NaCl \rightleftharpoons NaZ + MgCl$$
$$\underset{\text{zeolite}}{\text{Exhausted}} \qquad \underset{\text{zeolite}}{\text{Regenerated}}$$

Process: A zeolite softener consists of a steel tank packed with a thick layer of permutit or zeolite. The hard water is percolated at a specified rate through the bed of zeolite. Calcium and Magnesium ions present in it are retained by the zeolite in the form of CaZ and MgZ respectively. The outgoing water becomes rich in sodium salts. The softened water is collected from the bottom. (Fig. 12.8)

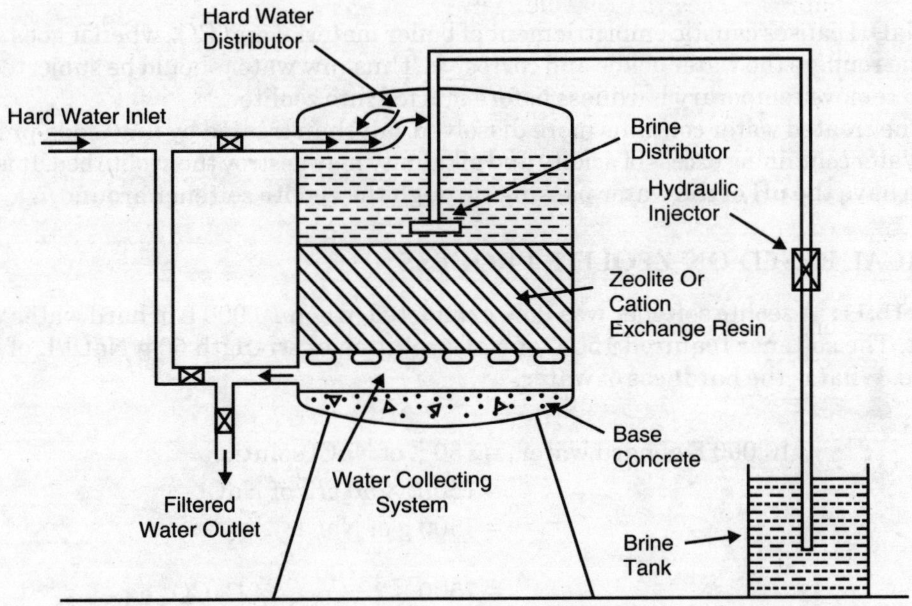

Fig. 12.8 *Sectional view of Zeolite process for Water Softening*

When a significant portion of the sodium in the zeolite has been replaced by calcium and magnesium, it is regenerated by first washing it with water by reversing the flow and then treating it with conc. NaCl solution. The soluble $CaCl_2$ and $MgCl_2$ so formed in regeneration process are washed with soft water and washings are led to sink. The zeolite bed is ready for use again for softening purpose. Theoretically each mg/L of $CaCO_3$ hardness requires 14.17 mg/L of NaCl solution. But thrice the amount of theoretically required amount of NaCl is used to make regeneration reaction more efficient.

Advantages of Zeolite Process

(i) Water having very low hardness of about 10ppm hardness is produced. Even water having zero hardness can be produced.

(ii) The plant is compact and occupies less space.
(iii) No sludge is formed and hence there is no problem of sludge disposal.
(iv) The running, maintenance and operation cost is quite less.
(v) The plant can be installed in the water supply line itself, avoiding double pumping.
(vi) It requires less skill for maintenance as well as operation.

Disadvantages

(i) This process is not suitable for treating highly turbid waters because the turbidity will clog the pores of zeolite bed and thus make zeolite less efficient. Therefore the turbidity must be removed before the water is fed into zeolite.
(ii) Water containing large amounts of Fe and Mn salts when passed through zeolite bed are converted to iron zeolite and manganese zeolite which cannot be regenerated easily. The zeolite is thus wasted.
(iii) The process leaves sodium bicarbonate in water, which causes priming and foaming in industrial or boiler feed waters. Moreover in boilers $NaHCO_3$ dissociates forming NaOH and CO_2.

$$NaHCO_3 \longrightarrow NaOH + CO_2.$$

NaOH causes caustic embrittlement of boiler material and CO_2 when it goes with steam the renders the water acidic and corrosive. Thus raw water should be subjected to boiling to remove temporary hardness before it is fed into zeolite.
(iv) The treated water contains more dissolved salt than treated by lime-soda process.
(v) Water containing excess of acidity and alkalinity may destroy the zeolite bed. It is preferable to have the pH of the water passing through the zeolite softener around 7.

NUMERICAL BASED ON ZEOLITE PROCESS

Example 12.11: A zeolite softener was 95% exhausted, when 10,000 L of hard water was passed through it. The softener required 150 L of NaCl solution of strength 50 g NaCl/L of solution to regenerate. What is the hardness of water.

Solution:

$$\begin{aligned} \text{10,000 L of hard water} &= \text{150 L of NaCl solution} \\ &= 150 \text{ L} \times 50 \text{ g/L of NaCl} \\ &= \text{7500 g of NaCl.} \\ &= 7500 \times \frac{100}{58.5 \times 2} \text{ of } CaCO_3 \text{ eq.} \\ &= 6410.26 \text{ g of } CaCO_3 \text{ eq.} \\ \text{1 L of hard water} &= 6410.26 \text{ g of } CaCO_3 \text{ eq/10,000.} \\ &= 0.6410 \text{ g of } CaCO_3 \text{ eq.} \\ &= 641 \text{ mg of } CaCO_3 \text{ eq.} \\ \text{Hence hardness of water} &= 641 \text{ mg/L.} \end{aligned}$$

Example 12.12: A Zeolite softener was completely exhausted and then regenerated by passing 200 litres of NaCl solution containing 100 g per litre of NaCl. How many litres of a sample of water of hardness 500 ppm can be softened by the softener.

Solution:

$$\text{200 litres of NaCl} = 100 \text{ L} \times 200 \text{ g/L} = 20000 \text{ g of NaCl}$$

$$= 20000 \times \frac{100}{58.5 \times 2} \text{ g of } CaCO_3 \text{ eq.}$$

$$= 17094 \text{ g of } CaCO_3 \text{ eq.}$$

Now 500 mg of hardness is present in = 1 litre of water

$$17094 \times 10^3 \text{ mg of } CaCO_3 \text{ eq. hardness is present in } = \frac{1}{500} \times 17094 \times 10^3 \text{ L}$$

$$= 34188 \text{ litres of hard water.}$$

Hence, the softener can soften 34188 litres of hard water.

2. Ion Exchange or Deionization Process (Demineralization of Water)

In Zeolite process for softening of water, the anions viz; SO_4^{2-} and HCO_3^- remain in the softened water combined with Na^+. The total content of dissolved solids in water remains the same. However with the development of organic synthetic cation exchangers the removal of even anions has been made possible.

The process of complete removal of all ions present in water is called dimineralization. Demineralization of water is achieved by using ion-exchange resins.

Ion Exchange Resins: Ion exchange resins consist of cross-linked, long chain organic polymers with a microporous structure. The functional groups attached to the polymeric chains are responsible for ion exchange.

Resins of suitable chemical compositions and physical properties may be synthesized for specific ion-exchange applications. The resins usually consist of styrene-divinylbenzene copolymer. Other polymers like methacrylic acid-divinylbenzene or phenol formaldehyde polymers are also used.

The functional groups attached to the resin may be acidic or basic and hence resins can be classified as:

(a) **Cation exchange Resins:** Resins containing acidic groups like –COOH or $-SO_3H$ group are termed as cation exchange resins. These are usually styrene divinyl benzene copolymers which on sulphonation or carboxylation become capable of exchanging their H^+ ions with the cations of the solution. A number of synthetic-resin cation exchangers which contain sulphonated phenolic or aromatic hydrocarbons have been prepared. These resins are called H-form cation exchangers. It is also possible to convert the H-form cation exchanger to Na-form exchanger by using NaCl. Schematically a cation exchanger can be represented as:

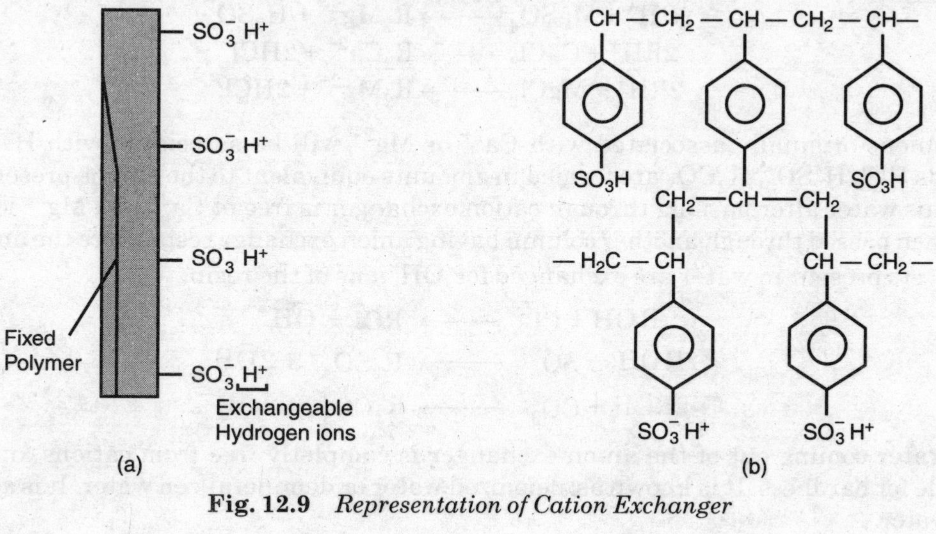

Fig. 12.9 *Representation of Cation Exchanger*

Resins having SO_3H group are termed as *strongly acidic resins* whereas having -COOH group are termed as *weakly acidic resins*. The cation exchange resins can be represented as RH^+.

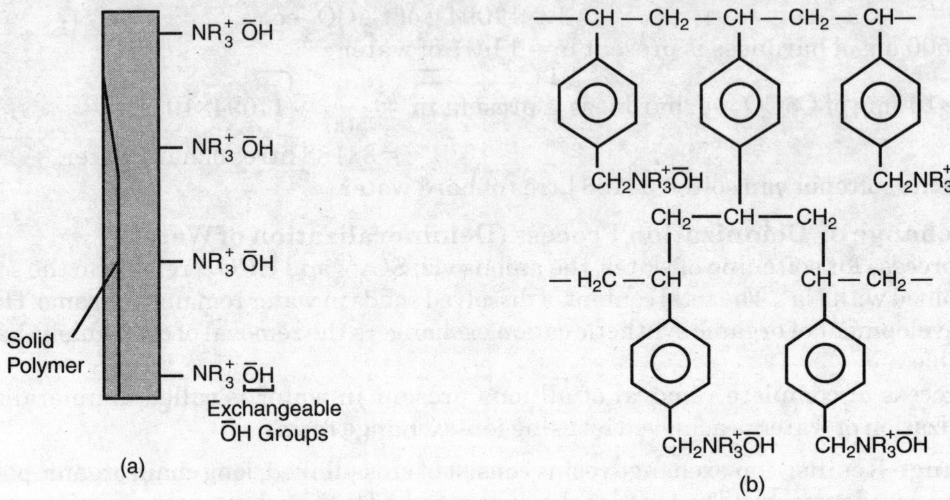

Fig. 12.10 *Representation of Anion Exchanger*

(b) **Anion exchange Resins:** Resin containing basic functional groups like amine, substituted amine or quaternary ammonium groups as their hydroxide salts are termed as anion exchange resins. There are styrene divinyl benzene complexes which, because of their basic functional groups become capable of exchanging their anions with those present in the solution.

Resins having $-NH_2$, = NH group are *weakly basic* whereas those having quaternary ammonium salts are *strongly basic*. Anion exchange resins are represented as $R\bar{O}H$.

Process: The ion exchange unit consists of two tanks. Cation exchange resins and anion exchange resins are placed in first and second tank respectively. First, hard water is made to pass through cation exchange resin which exchanges its H^+ ions for Ca^{2+} and Mg^{2+} ions in water.

$$2RH^+ + CaSO_4 \longrightarrow R_2Ca^{2+} + H_2SO_4$$
$$2RH^+ + MgSO_4 \longrightarrow R_2Mg^{2+} + H_2SO_4$$
$$2RH^+ + CaCl_2 \longrightarrow R_2Ca^{2+} + 2HCl$$
$$2RH^+ + MgCl_2 \longrightarrow R_2Mg^{2+} + 2HCl.$$

The anions originally associated with Ca^{2+} or Mg^{2+} will be associated with H^+ ions. In otherwords HCl, H_2SO_4, H_2CO_3 are formed in amounts equivalent to the anions present in raw water. Thus water after passing through cation exchanger is free of Ca^{2+} and Mg^{2+} ions. This water is then passed through another column having anion exchange resin. Here the anions like SO_4^{2-}, Cl^- etc present in water are exchanged for $\bar{O}H$ ions of the resin.

$$R\bar{O}H + Cl^- \longrightarrow R\bar{Cl} + \bar{O}H$$
$$2R\bar{O}H + SO_4^{2-} \longrightarrow R_2SO_4^{2-} + 2\bar{O}H$$
$$2ROH + CO_3^{2-} \longrightarrow R_2CO_3^{2-} + 2\bar{O}H$$

The water coming out of the anion exchanger is complelty free from cations and anions responsible for hardness. It is known as deionized water or demineralized water. It is as pure as distilled water.

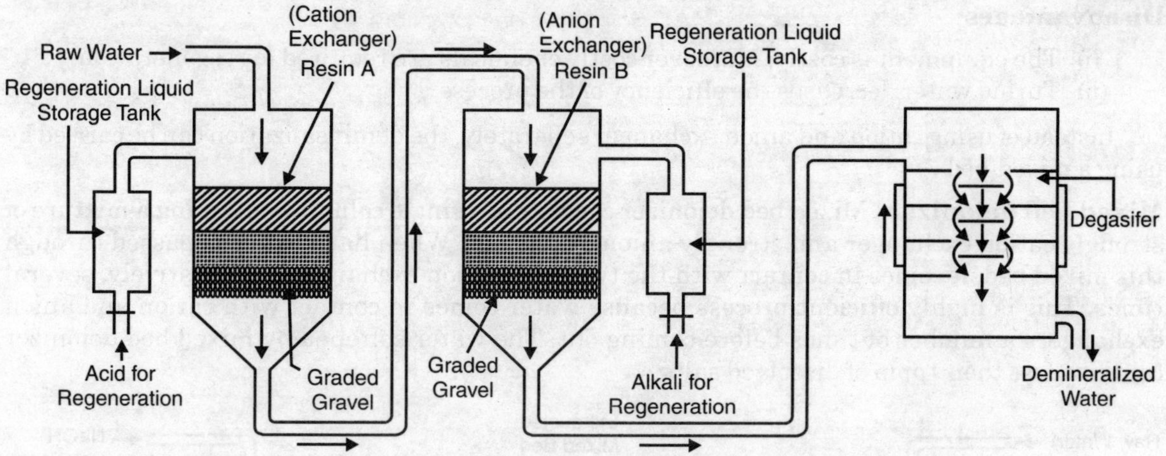

Fig.12.11 *Softener for Demineralization*

H^+ released from cation exchanger and $\bar{O}H$ released from anion exchanger combine to form unionized water.

$$H^+ + \bar{O}H \longrightarrow H_2O.$$

Regeneration: After some time the cation and the anion exchangers get exhausted and stop working.

The cation exchanger is regenerated by first back washing and then passing a 2% H_2SO_4 solution through it. The regeneration reactions can be represented as:

$$R_2Ca^{2+} + H_2SO_4 \rightarrow 2RH^+ + CaSO_4$$
$$R_2Mg^{2+} + H_2SO_4 \rightarrow 2RH^+ + MgSO_4.$$

The column is then rinsed with soft water to remove $CaSO_4$ and $MgSO_4$. However if HCl is used for regeneration, $MgCl_2$ and $CaCl_2$ are formed.

$$R_2Ca^{2+} + 2HCl \longrightarrow 2RH^+ + CaCl_2$$
$$R2Mg^{2+} + 2HCl \longrightarrow 2RH^+ + MgCl_2.$$

HCl is ideally suited for regeneration since $CaCl_2$ and $MgCl_2$ are more soluble.

The anion exchanger is regenerated by back washing and then passing a solution of NaOH. The Na_2SO_4 or NaCl so formed, are rinsed with soft water and washing are led to sink.

$$R_2SO_4{}^{2-} + 2NaOH \longrightarrow 2R\bar{O}H + Na_2SO_4$$
$$R_2Cl^- + NaOH \longrightarrow R\bar{O}H + NaCl.$$

The regenerated exchanger is ready for further use.

Ion exchanger do not remove CO_2 which is removed in degasifier in the deionization process.

Advantages

(i) Highly acidic or alkaline water can be softened.

(ii) Water of very low hardness is produced (about 2ppm)

(iii) Anions as well as cations are removed thereby problems like caustic embrittlement and corrosion are reduced when treated water is used for boiler feed purposes.

Disadvantages

(i) The equipment is costly. Moreover costly chemicals are required for regeneration.
(ii) Turbid water decreases the efficiency of the process.

Instead of using cation and anion exchanger separately, the deminsalization can be carried by using a mixed-bed.

Mixed bed deionizer: Mixed bed deionizer consists of a single column containing a mixture of strongly cation exchanger and strongly anion exchanger. When hard water is passed through this mixed bed, it comes in contact with the two types of ion exchangers alternatively, several times. This is highly efficient process because water comes in contact with cation and anion exchangers a number of times before coming out. The water softened by mixed bed deionizer contains less then 1ppm of dissolved salts.

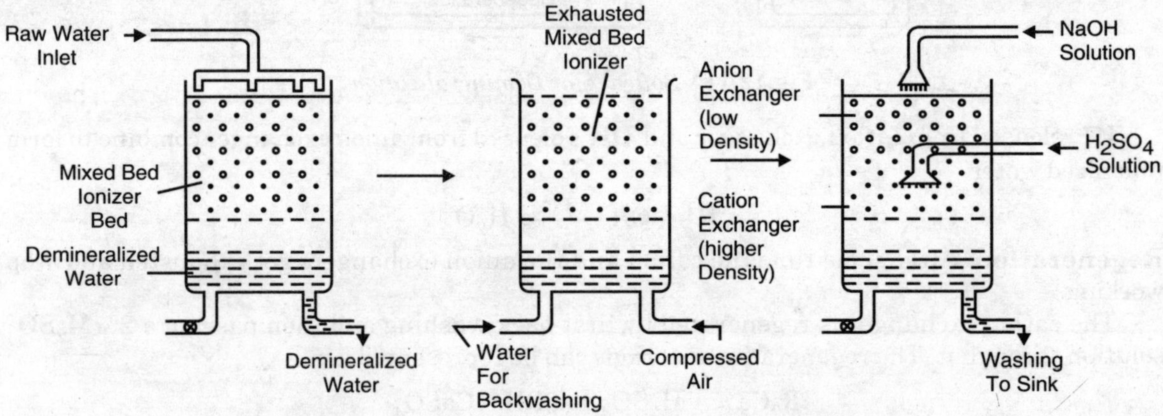

Fig. 12.12 *Regeneration of Mixed-Bed Ion Exchanger*

Regeneration: When the resins are exhausted, the bed is back washed. During washing the anion exchanger being lighter gets displaced above the cation exchanger. This hydraulic separation is highly useful as it makes the regeneration process much easier. The anion exchanger is regenerated by letting in caustic soda solution (NaOH) from the top. The cation exchange resin is regenerated by H_2SO_4 solution. The two beds are then rinsed with soft water. The beds are mixed again by sending a stream of compressed air from below and are used again.

12.7 WATER FOR DOMESTIC PURPOSES (Municipal Water/Potable Water)

One of the basic question a person asks on arriving at a place is, whether the water available, is safe to drink or not. About 80% of all sickness in the world is attributed to the lack of safe drinking water. It is unfortunate that a large portion of world's population lack a safe supply of water for drinking, washing and other domestic puposes.

As water flows over the surface of earth it picks up a number of impurities in the form of dissolved particles of soil, garbage, sewage, pesticides and other human, animal or chemical wastes. Some of these impurities may make water more useful and potable (water which is safe to drink) and some of these render it harmful and unfit. For example, certain minerals such as iron, calcium, magnesium, flourine etc may be useful in small quantities. But when these materials are present in large amounts or in combination with other impurities, the water may become unfit for human consumption. For example, the water may contain toxic or poisonous substances such as arsenic, barium, cadmium, lead etc which are very harmful even if present in low quantities.

It is clear from the above discussion that water has to be treated to remove various impurities which are objectionable, to make it safe for drinking purposes. Municipal water treatment aim at maintaining an adequate supply of domestic water, confirming to all the required specifications of drinking water.

Specification of Drinking Water

Drinking water standards fall into two categories:

(a) **Primary Standards:** Primary standards specify the maximum contaminant levels (MCL) of various dissolved minerals based on their effect on human health.

(b) **Secondary Standards:** Secondary standards are the standards, which may vary from place to place depending upon the taste, odour, colour and hardness of water etc and do not have any anticipated effect on the health.

The common standards for drinking water are

(i) It should be free from objectionable minerals which have deterimental effect on health. Table (12.5) gives the maximum permissible limit i.e. maximum contaminant levels for certain chemicals and their ill effect if these limits are exceeded.

Table 12.5: *Maximum contaminant levels for certain chemicals and their effect on human health*

Contaminant	*Permissible Limit (mg/L)*	*Principle health effect, if exceeded beyond permissible limit*
Arsenic	0.050	Dermal and nervous system toxicity effects.
Barium	1.000	Circulatory system effects.
Cadmium	0.010	Kidney is effected.
Chromium	0.050	Liver and kidney are effected
Flouride	1.000	Flourosis.
Lead	0.050	Central and peripheral nervous system damage, kidney effects, highly toxic to infants.
Mercury	0.001	Central nervous system disorders.
Nitrite and Nitrate	45.000	Methaemoglobinemia (blue baby syndrome)
Selenium	0.010	Gastrointestinal effects.
Silver	0.050	Skin discolouration.
Sulphate	200.000	Gastrointestinal irritation.

(ii) pH Value: It should have pH value ranging from 6.5 to 8.5. Beyond this range, the water will effect the mucous membrane.

(iii) Total Dissolved Solids: Total dissolved solids should be less than 500ppm.

(iv) Pathogenic bacteria: It should be free from disease causing bacteria. Coliform bacteria (Escherichia coli) are used as indicator organisms whose presence suggests that the water is contaminated. Water should contain one coliform per 100ml.

(v) It should be free of iron and manganese, the permissible limit being .3 mg/L and 0.05mg/L respectively. They change the taste of water and also are responsible for producing stains on clothes in laundry.

(vi) It should be non corrosive.

(vii) It should be free from objectionable dissolved gases like H_2S.

(viii) It should be sparkling clear, odourless and should have pleasant taste.

(ix) It should not be turbid. The permissible turbidity is less than 10 ppm.

WATER TREATMENT FOR DOMESTIC/MUNICIPAL PURPOSE

The treatment of water for domestic supply will depend to a large extent on the quantity of the source water. A typical treatment plant for surface water might include the following sequence of steps:

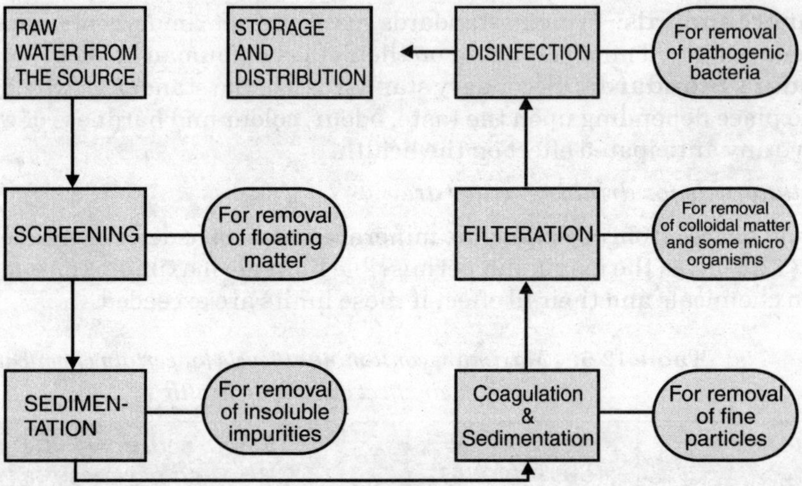

Fig. 12.13 *Schematic Representation of a Typical Water Treatment Plant for Surface Water*

1. **Screening:** Screening is the process by which the floating material like wood pieces, leaves etc are removed. Raw water is made to pass through large screens having number of holes which hold back the floating matter and allow the water to pass.

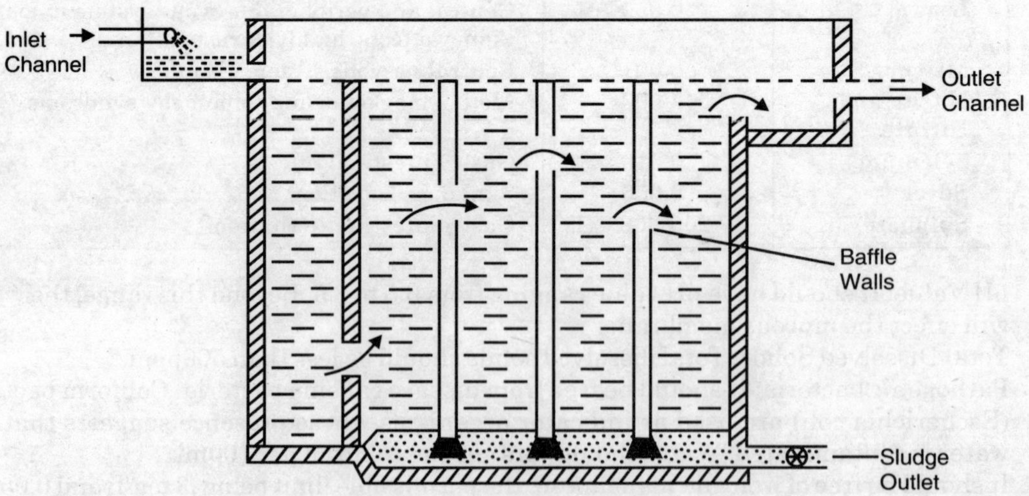

Fig. 12.14 *Horizontal Flow Sedimentation*

2. **Sedimentation:** Sedimentation is the process of allowing water to stand undisturbed in big tanks for sufficient period of time, so that the suspended particles settle down at the bottom. The detention period may vary from 2 to 8 hours. The process of sedimentation is

carried out in continous flow type tanks, in which the flow of the water is slowed down, so that substances held up by turbulence of fast moving water, settle at the bottom under the force of gravity. The basin in which the flow of water is retarded is called the settling tank or sedimentation tank. Thus in such tanks, the flow of water is reduced and water is not brought to complete rest. Fig (12.17) shows a horizontal flow sedimentation tank. It consists of a rectangular tank. It is provided with baffle walls to reduce the velocity of the incoming water. The settled sludge is taken out through the outlet at the bottom.

The working of continous flow type of tanks is simple. Water enters from one end and is collected from the other end. The velocity is sufficiently reduced by providing sufficient length of travel.

3. **Sedimentation Aided with Coagulation:** Finely divided silica, clay and organic matter does not settle down easily and hence cannot be removed by simple sedimentation. Most of these are in the form of colloids and are generally negatively charged. Because of their same charge, these particles do no coalesce together to form larger particles and hence do not settle down easily. They can however be removed easily by adding certain chemicals called **coagulants**. Coagulants when added to water, form insoluble gelatinous precipitates which entrap very fine suspended particles, forming bigger sized agglomerates which settle down easily. Cationic coagulants are generally added to bring about coagulation in water.

The cationic coagulants neutralize the negative charge on the colloids which there by come together and coalesce to form bigger particles and hence settle down.

The coagulants may also exhibit their action by adsorbing the fine colloidal particles over their surface and forming bigger flocs which settle down under the force of gravity. The process is called flocculation.

The most commonly used coagulants in water treatment plants are Alum, sodium aluminate and ferrous sulphate.

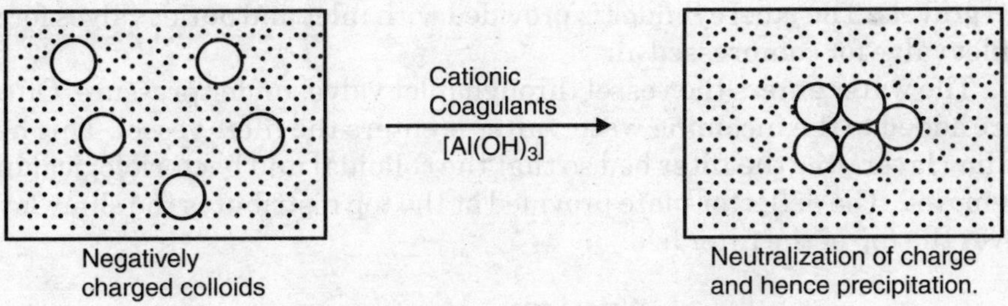

Fig. 12.15 *Action of a Cationic Coagulant*

Alum [$K_2(SO_4)_3$. $Al_2(SO_4)_3$. $24H_2O$]: Alum forms $Al(OH)_3$ on reacting with water in which alkalinity is also present. Some alkali is added, if the water is not sufficiently alkaline. It is mostly used in the pH range of 6.5 – 8.5

$$Al_2(SO_4)_3 + 3Ca(HCO_3)_2 \longrightarrow 2Al(OH)_3 + 3CaSO_4\downarrow + 6CO_2$$

The $Al(OH)_3$ so formed acts as a floc or coagulant, which removes the finely divided particles or colloidal impurities by neutralizing the charge on them as well as by adsorbing the colloidal particles forming heavier flocs which settle down.

Sodium Aluminate [$NaAlO_2$]: For treament of acidic waters sodium aluminate can be used. It works in the range of 5.5 – 8.0. It is costlier than alum. The reactions taking place are-

$$NaAlO_2 + 2H_2O \longrightarrow Al(OH)_3\downarrow + NaOH.$$

$Al(OH)_3$ acts in the same way as discussed above and NaOH react with $MgSO_4$ forming Mg $(OH)_2$.

$$2NaOH + MgSO_4 \longrightarrow Mg(OH)_2\downarrow + Na_2SO_4.$$

Ferrous Sulphate: [$FeSO_4$. $7H_2O$] This is used when water is slightly alkaline, in the pH range > 8.5. If water is not alkaline, the pH must be raised by adding [$Ca(OH)_2$].

$$FeSO_4 + Mg(HCO_3)_2 \longrightarrow Fe(OH)_2 + MgCO_3 + CO_2 + H_2O$$
$$4Fe(OH)_2 + O_2 + 2H_2O \longrightarrow 4Fe(OH)_3.$$

The raw water is thoroughly and rapidly mixed with the required chemicals. It is then led to the flocculation basin where it is gently agitated. During this, the precipitating Al $(OH)_3$ attracts colloidal particles, forming flocs.

After flocculation, the water flows through a sedimentation basin. Solids that collect on the bottom of the tank may be removed manually by periodically shutting down the tank and washing out the collected sludge or it may be mechanically cleaned using a bottom scraper.

4. **Filteration:** The effluent from the tank is then filtered. Filteration is the process of removing insoluble, colloidal and bacterial impurities from water by means of filters. One of the most widely used filteration units is the rapid-sand filter. Though slow-sand filters are also used but have become obsolete these days.

Rapid sand filters are of two types-

(a) Rapid gravity filters

(b) Rapid presssure filters.

(a) **Rapid Gravity Filters:** It consists of a large rectangular tank made up of concrete. The filtering medium is a layer of coarser sand which is supported by a bed of graded gravel. A layer of fine sand is carefully seived at the top of coarser sand. To collect the filtered water, steel cylinders are used as pressure vessel, containing filtering media consisting of the layers of fine sand followed by coarse sand, which inturn is supported on gravels. The steel cylinder is provided with inlet and outlet valves for water and an inlet valve for compressed air.

The water enters the vessel through inlet valve, under pressure. Often coagulants are added to the incoming water, after it enters the filter vessel. This helps to form a slimy layer over the filter bed so that the colloidal and bacteriological impurities are removed. The deflector plate provided at the top distributes the slimy layer uniformly over the top of the filter bed.

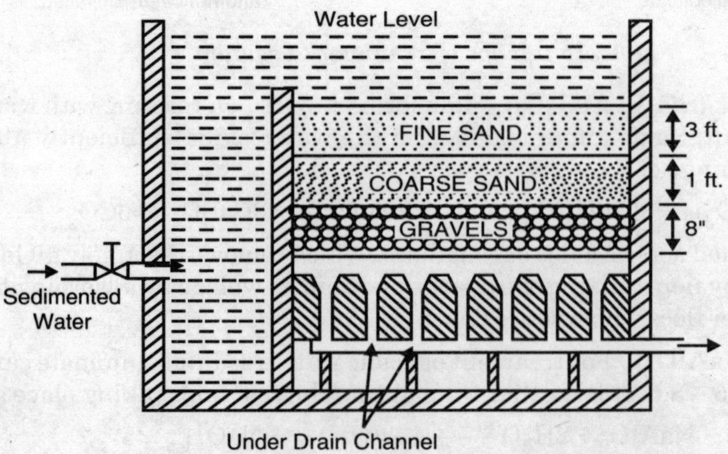

Fig. 12.16 *Gravity sand filter*

The cleaning of the filter is carried out by back washing as is done in gravity filter. Before backwashing, compressed air is passed to agitate the sand. Compact automatic pressure filters, in which cleaning is done at a specific interval of time are available now a days. Water, drains are provided at the bottom. (Fig. 12.19)

Sedimented water entering the sand filter is uniformly distributed over the entire fine sand bed. After passing through the filter beds, the filtered water is taken out from the drain system provided at the bottom. When the filter bed is clogged by the particles removed from the water, it has to be cleaned. When the rate of filteration becomes slow, the working of the filter is stopped. The filter bed is cleaned by backwashing i.e by reversing the flow of water through the filter. The top most layer of sand is scrapped off and replaced with clean sand. The rapid gravity filters get clogged very frequently and have to be washed every 24 to 48 hours.

(b) **Rapid Pressure Filters:** Pressure filters are similar to rapid gravity filters, placed in closed vessels. Water is passed through pressure filters at a pressure greater than the atmospheric pressure, it is necessary that these filters are located in air tight vessels. The raw water is pumped into the vessels by means of pumps.

5. **Disinfection:** During coagulation, settling and filteration, practically all of the suspended solids, most of the colour, and a few percent of the bacteria are removed. The final step is disinfection to kill the remaining pathogenic organisms, (disease causing organisms).

The process of destroying these pathogenic bacteria and micro organisms is known as sterilization or disinfection. The chemicals and the substances used for disinfecting are known as disinfectants.

The process of purification is the most important, because the presence of pathogenic bacteria may lead to spread of various diseases and their epidemics, thus causing disasters to public life.

The disinfection of water can be carried out by the following methods:

(i) **Chlorination:** Chlorination in its various forms is almost universally used for disinfecting domestic water supplies. This is because it is cheap, reliable and easy to handle. Moreover it is capable of providing residual disinfecting effects for long periods and thus prevents future recontamination of water.

Disinfecting Action of Chlorine

When chlorine is added to water, it forms hypochlorous acid or hypochlorite ions which have immediate disastrous effect on most forms of microscopic organisms.

$$Cl_2 + H_2O \xrightarrow{pH > 8} HOCl + HCl$$

The hypochlorous acid is unstable and may dissociate as

$$HOCl \underset{pH > 7}{\overset{pH > 8}{\rightleftharpoons}} \underset{\text{Hypochlorite ions}}{H^+ + \bar{O}Cl}$$

The above reaction is pH dependent. The dissociation of HOCl is most effective at higher pH values and vice versa. At pH < 7, HOCl exist without dissociating whereas at pH<5, Cl_2 does not react and remain as elemental chlorine. At pH > 10, $\bar{O}Cl$ ions are the predominant specie. (Table 12.6)

Table 12.6

pH	*Predominant specie*
< 5	Cl_2
> 5 < 7	HOCl
> 10	$\bar{O}Cl$

All the three forms viz. HOCl, $\bar{O}Cl$, Cl_2 existing in a sample water are termed as **free chlorine.** Out of these three forms the HOCl is found to be the most destructive according to Enzymatic hypothesis given by Gleen and stumpt.

For this reason, *the* pH *value of water during chlorination is generally maintained slightly less than* 7 *so as to prevent the dissociation of* HOCl.

According to the enzymatic hypothesis the HOCl reacts with the enzymes in the cells of the bacteria which are essential for their growth and other metabolic processes. Thus due to deactivation of enzymes, in the cells of organisms by HOCl, the bacteria get killed and no multiplication of micro organisms takes place. HOCl is about 80 times more destructive to bacteria than $\bar{O}Cl$ ions.

Moreover the HOCl will immediately react with ammonia likely to be present in water to form various chloramines as

$$NH_3 + HOCl \longrightarrow NH_2Cl + H_2O (pH > 7.5)$$
$$NH_2Cl + HOCl \longrightarrow NHCl_2 + H_2O (pH = 5 \text{ to } 6.5)$$
$$NHCl_2 + HOCl \longrightarrow NCl_3 + H_2O (pH < 4.5)$$

In the usual chlorination process, the pH is kept slightly less than 7 i.e. around 6.5 and hence dichloramine is the predominant specie.

These chloramines so formed are stable and are found to possess disinfecting properties. They act as chlorine reserves. The chlorine in this form is called combined chlorine. As compared to the free chlorine, combined chlorine is less effective.

Break Point Chlorination: The amount of chlorine required for disinfecting water depends upon the inorganic and organic impurities present in water. When chlorine is added to a sample of water and its amount is estimated after a few minutes, it is found that the available chlorine is not equal to the amount of chlorine added. A relationship between the amount of chlorine added to water and the free residual chlorine (determined experimentally) is shown in Fig.12.17. The curve shown in the fig can be divided into four stages.

I stage: It is seen from the figure that for lower doses of Cl_2 all the chlorine added is consumed and there is no residual chlorine. This is due to the fact that initially Cl_2 reacts with the inorganic impurities present in water forming chlorides. These chlorides do not have any residual oxidizing power.

II stage: As the amount of chlorine is increased, there is steady increase in the amount of residual chlorine. This stage

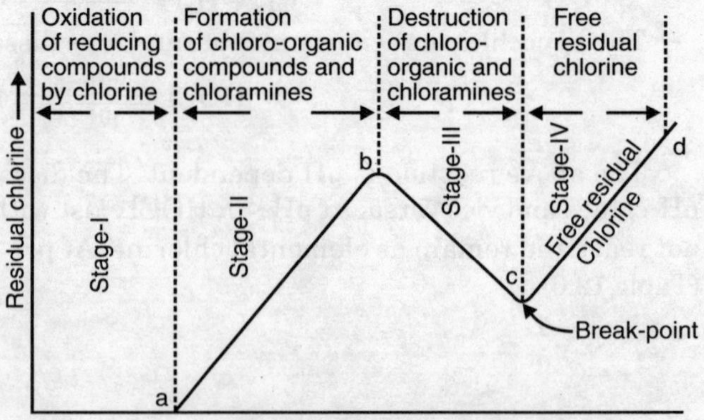

Fig. 12.17 Break-point chlorination curve

corresponds to the formation of chloramines i.e chlorine is consumed by NH_3. But chloramines respond to the test for estimation of chlorine in the same way as free chlorine and hence curve represents the chlorine available from combined residuals.

Chloramines remain in water body for longer periods which help to kill the micro organisms even in a distribution system. Simultaneously this stage also corresponds to the formation of chloro organic compounds without oxidizing them. These are responsible for bad odour and unpleasant taste in water.

III stage: When the dosage of chlorine is further increased, the complete oxidation of chloramines and chloro-organic compounds take place utilizing the chlorine and accordingly free residual chlorine also decreases and reaches a dip when the oxidative destruction is complete. This dip is known as break point. The chlorination upto breakpoint ensures complete destruction of organic compounds which give unpleasant taste and bad odour.

IV stage: After breakpoint, the amount of chlorine added is not used in any reaction and the residual chlorine agrees fairly well with the quantity of chlorine added. Thus free chlorine is available only if the demand for chlorine by other reacting substances present in water is met with.

Thus break point chlorination of water may be defined as the chlorination to such an extent that micro organisms as well as bad tastes and odours present in water are also destroyed.

Chlorination after break point increases the free residual chlorine (Cl_2, HOCl, $\overline{O}Cl$). The free chlorine as well as the combined chlorine (chloramines) will cause germicidal action on micro organisms. Free chlorine will instantaneously kill the pathogens, while the combined chlorine will provide long term germicidal effect. Hence to use chlorine as an effective disinfectant, the chlorine dosage has to be slightly more than the break point. In general, water is satisfactorily disinfected if the free chlorine residual is about 0.2ppm i.e. the chlorine added in the dosage upto break point + 0.2 ppm. Amount greater than 0.2ppm of free residual chlorine causes bad taste in water and is harmful as it irritates the membrane of intestine.

Various forms in which Chlorine can be Applied: Chlorine is generally added to water in the following forms:

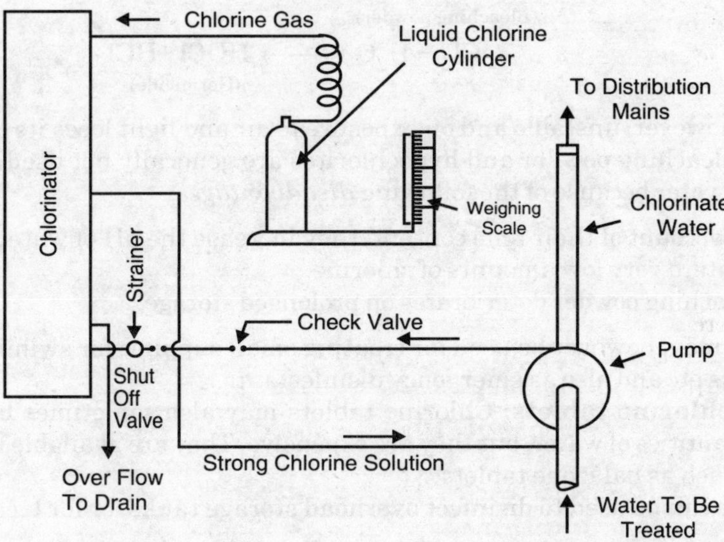

Fig. 12.18 *Line diagram of a typical solution feed chlorinator installation*

(a) Chlorine gas/liquid: The molecular chlorine can be applied to the water in liquid or gaseous form, though liquid form is generally preferred these days. The liquid chlorine is 1½ times heavier than water and is generally kept in steel cylinder called chlorinator. Chlorinators using liquid chlorine, take the chlorine form the container, mix it with a small amount of water to form a strong chlorine solution. The chlorine solution is fed into distribution system by means of a water injector (Fig. 12.18).

Advantages: The liquid chlorine is now-a-days invariably and universally used because of the following advantages-

(i) Can be easily stored for long periods, without any risk of deterioration.
(ii) Cheap and easily available.
(iii) Occupies less space for storage.
(iv) No sludge is formed on its application.
(v) Very powerful disinfectant and may remain in water as residual for enough time, especially if NH_3 is also present in water.

(b) Bleaching powder/Hypochlorites: Calcium and Magnesium hypochlorites may be used for chlorinating small public supplies. When hypochlorite is dissolved in water, it forms hypochlorite ions

$$\underset{\text{Calcium hypochlorite}}{Ca(OCl)_2} \xrightleftharpoons{H_2O} Ca^{2+} + 2OCl^-$$

OCl^- may combine with hydrogen ions present in water.

$$OCl^- + H^+ \underset{pH > 8}{\overset{pH > 7}{\rightleftharpoons}} \underset{\text{(Germicide)}}{HOCl.}$$

HOCl as well as OCl^- cause disinfectioning of water. Commercially available calcium hypochlorite called Hig Test Chlorite or Perchloron, contains 60% to 70% of available chlorine.

Another compound used for chlorination is bleaching powder. When freshly prepared, it contains 30% of available chlorine.

$$\underset{\text{(Bleaching powder)}}{CaOCl_2 + H_2O} \longrightarrow Ca(OH)_2 + Cl_2.$$

$$Cl_2 + H_2O \longrightarrow \underset{\text{(Germicide)}}{HOCl} + HCl$$

It is however, unstable and on exposure to air and light loses its chloride content.

But bleaching powder and hypochlorites are generally not used in modern days for treating water because of the following *disadvantages*;

(i) On account of their lime content, they increase the pH of water and make it hard.
(ii) Contain very low amounts of chlorine.
(iii) Bleaching powder deteriorates on prolonged storage.

They may, however be used for treating small supplies for swimming pools, storage reservoirs etc and also as emergency disinfectants.

(c) Use of chlorine tablets: Chlorine tablets may also sometimes be used to disinfect small quantities of water, but they are expensive. They are available under various trade names, such as halazone tablets.

These can be used to disinfect overhead storage tanks, or for treating ground water supplies.

(d) Use of Chloramines: Chloramines are quite stable. When chlorine and NH_3 are mixed in the ratio of 2:1 by volume, chloromines are formed.

$$Cl_2 + NH_3 \longrightarrow ClNH_2 + HCl.$$

Due to their stability, they remain in water as residuals for a sufficient time. (compared to chlorine which evaporates after some time). They thus provide a greater safeguard against recontamination. However, they are weaker disinfectants compared to free chlorine.

The **advantages** of using chloramines are-

(i) They do not cause bad odour and bad tastes even used in excess.

(ii) Very useful when phenols are present. Free chlorine reacts with phenol to form chlorophenols which impart unpleasant taste to water. But chloramines do not form chlorophenols and hence no such tastes are developed.

(e) Use of ClO_2 Gas: It is very effective and powerful disinfectant, about 2½ times stonger than chlorine. It is produced by passing chlorine gas through sodium chlorite.

$$2NaClO_2 + Cl_2 \longrightarrow 2NaCl + 2ClO_2.$$

It is costly and unstable and has to be used immediately after its production. However, since the action of ClO_2 remains unaffected by high pH values (6 to 10), it can be used for treating highly alkaline waters.

Dechlorination: Overchlorination i.e > .2ppm after the break point produces unpleasant odour and tastes in water. Dechlorination means removing the chlorine from water. The dechlorination process may either be carried out to such on extent that sufficient residual chlorine .1 to .2ppm do remain in water. Otherwise additional chlorine will generally be added to maintain such residues.

Dechlorination is done by using dechlorinating agents such as SO_2 gas, $Na_2S_2O_3$, Na_2SO_3 etc. The reactions are-

$$SO_2 + Cl_2 + 2H_2O \longrightarrow H_2SO_4 + 2HCl$$

$$Na_2SO_3 + Cl_2 + H_2O \longrightarrow Na_2SO_4 + 2HCl.$$

Dechlorination can also be brought about by passing the treated water through bed of molecular carbon or by adding activated carbon to water which is then removed by filteration.

(ii) **Disinfection by Ozone:** Ozone gas is an excellent disinfectant. It can be produced by passing a high tension electric current through a stream of air in a closed chamber.

$$3O_2 \longrightarrow 2O_3$$

It is unstable and readily decomposes to give nascent oxygen.

$$O_3 \longrightarrow O_2 + [O]$$

The nascent oxygen so produced is a powerful oxidizing agent and removes the organic matter and kills the bacteria.

During the treatment, ozone gas is manufactured and bubbled through water contained in a sterilizing chamber having inlets and outlets for water.

Separate inlet for ozone is also provided. The sterilized water is removed from the top.

Ozone and chlorine are quite competative disinfectants, each having its advantages and disadvantages. Chlorine is however preferred over ozone as it is cost effective. Ozone is quite expensive but still it is used in some countries.

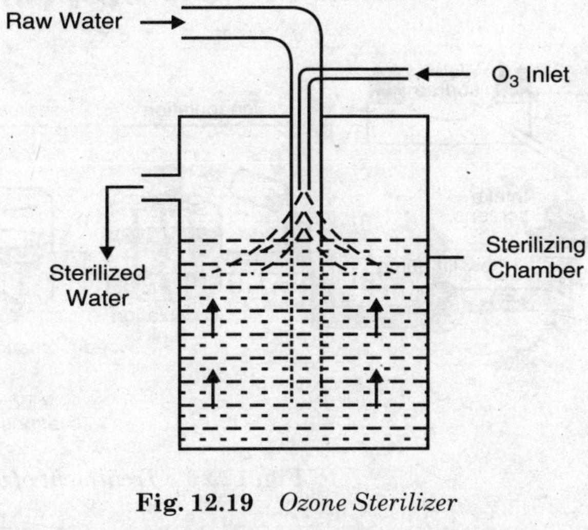

Fig. 12.19 *Ozone Sterilizer*

Advantages of using Ozone:

- Ozone removes the colour, taste and odour from water, in addition to killing bacteria.
- Ozone being unstable does not leave any residue.
- The excess of ozone is not harmful as it decomposes into oxygen.

Disadvantages

- It is very expensive and hence cannot be used for disinfecting municipal water supply.
- No residuals can be maintained and hence no safety against recontaminations.
- Cannot be supplied in cylinders but require, an apparatus ozoniser, to be installed at the treatment plants.

(iii) **Ultra-Violet Rays:** Ultraviolet rays (uv) are found in sun light but can be produced by passing electric current through mercury enclosed in quartz bulbs. Mercury vapour lamps enclosed in quartz bulbs can therefore be used as a good source of ultraviolet rays. These rays are highly effective in killing all types of bacteria. For treatment, the water is passed several times around the quartz bulbs emitting uv rays. These rays can penetrate upto 10cm of distance and hence the depth of water over the bulb should not exeed 10cm. Disinfection with ultra-violet rays does not impart any additional taste or odour to the water as no chemicals are added. But, this is very costly method and is unsuitable for treating large supplies. However, this method is used for treating small quantities of water at places where cost is a minor factor.

(iv) **Potassium Permanganate:** This is used as a disinfectant where water is contaminated with lesser amount of micro organisms. It is added in small doses such as 0.05 to 0.10ppm even to filtered and chlorinated water. It is a popular disinfectant for disinfecting well water in villages. The addition of $KMnO_4$ produces pink colour. However, if pink colour disappears, it shows that more of organic matter is present in water and more quantity of $KMnO_4$ should be added.
$KMnO_4$ is cheap and quite easy to use. But it cannot remove all the bacteria from water. This method of disinfection is also, therefore not recommended except for rural areas, where the water is contaminated with lesser amount of bacteria.

(v) **Boiling:** Water for domestic purposes on a smaller scale may be sterilized by boiling the filtered water for about 20 minutes. This method is very costly and hence can be used only in individaul cases.

Apart from the above methods iodine and bromine tablets are also added to water to kill bacteria. Basic steps, mentioned above for the treatment of municipal water are shown in Fig. (12.20).

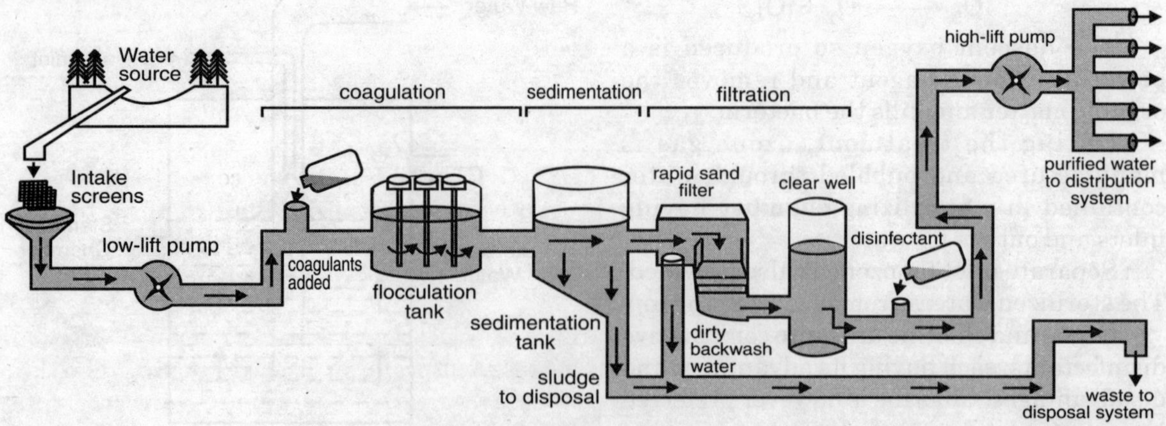

Fig. 12.20 *Treatment of water for drinking purpose*

Besides these normal treatment methods, certain other special treatments are sometimes required in order to remove the minerals, tastes odour and colours etc from water. Some of them are briefly discussed below:

(a) **Aeration:** Under the process of aeration, water is brought in intimate contact with air so as to absorb oxygen and to remove CO_2. It also helps in killing bacteria to a certain extent. The aeration of water can be carried out with the help of spray nozzles. In this method, water is sprinkled in air or atomosphere through special nozzles which breaks the water into droplets, thus permitting the escape of dissolved gases and absorption of oxygen.

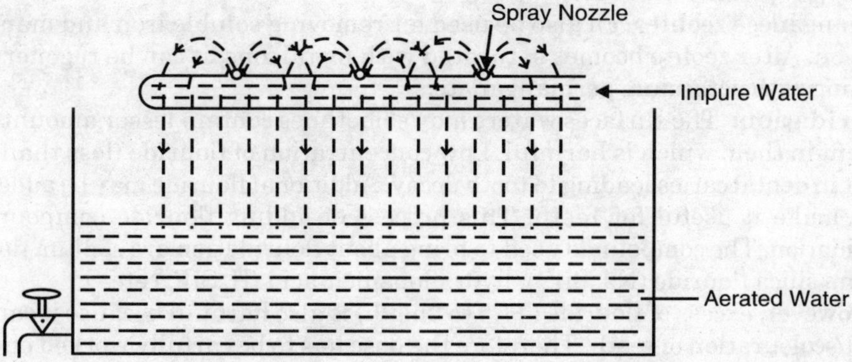

Fig. 12.21 *Aeration through Spray Nozzles*

(b) **Treatment with Activated Carbon:** Activated carbon is specially treated carbon, which possess the property of absorbing and attracting impurities, such as gases, liquids and finely divided solids. It is widely used for removing tastes and odours from public supplies.

Activated carbon is prepared by heating or charring wood or saw dust at about 500°C in a closed vessel and then slowly burning (oxidizing) it under controlled conditions at 800°C. Thus all the hydro carbons, which may interfere with the absorption of organic matter are removed and the residue is called the activated carbon. It is highly porous and absorptive. The activated carbon is mostly used in the powdered form and is added to water before the filteration.

With the passage of time, carbon looses its absorptive power and has to be rejuvenated. During regeneration, the contaminants are usually burned from the surface of the carbon granules, though in some cases a solvent is used to remove them.

The effectiveness of activated carbon is directly related to the surface area available to attract the molecules or particles of contaminants. The most commonly used is granular activated carbon, which has an enormous surface area. A single handful of granular activated carbon has an internal surface area of about one acre.

(c) **Treatment with Copper Sulphate:** $CuSO_4$. $7H_2O$ helps to remove colours, tastes and odours from water. It may be added to the treated water in the distribution pipes, but its main advantage is obtained by adding it to the open reservoirs or lakes. The function of $CuSO_4$. $7H_2O$ is to check the growth of algae, which on decomposition may impart bad odours to water.

(d) **Removal of Iron and Manganese:** Iron and manganese salts are generally found dissolved together in well water or anaerobic reservoir water in dissolved state. When the water containing iron or manganese is exposed to air, these slowly are converted into their oxides. They are objectionable because-

- They cause formation of stains on cloths washed with such waters due to the deposition of red/brown coloured oxides of iron/manganese.
- They cause incrustation of water mains due to deposition of their oxides.
- They impart metallic taste to water.

If present in small quantities, they can be removed during aeration, followed by coagulation, sedimentation and filteration. The soluble ferrous and manganous compounds present in water get oxidized to ferric and manganic compounds which are insoluble and hence can be sedimented out easily.

Softening by ion exchanges removes iron but is not recommended because if Fe is oxidized, it poisons the resins.

Manganese zeolite, can also be used for removing soluble iron and manganese from solution. After zeolite bcomes saturated with metal ions, it can be regenerated by back washing with potassium permanganate.

(e) **Flouridation:** The surfaces waters may sometimes contain lesser amounts of flourides present in them which is harmful. Low concentration of flouride (less than 1mg/L) may result in dental caries leading to tooth decay. Additional flouride may be added to water so as to make it useful for teeth. This process of adding flouride compounds is called flouridation. The compounds used to bring about flouridation are sodium flouride (NaF), sodium silica flouride (Na_2SiF_6), hydroflousilicic acid (H_2SiF_6) etc.

However, excess of flouride (>1.5 to 2mg/L) is also harmful because it causes spotting and discolouration of teeth. Therefore the dose has to be carfully worked out.

Soft water should not be supplied for drinking purposes. Softened water increases the potential for leaching metal from pipes, solder and plumbing fixtures.

$$2Pb + 2H_2O + O_2 \rightleftharpoons 2Pb(OH)_2$$

soft water attacks the lead used in plumbing.

The solubility of lead in water is diminished in the presence of HCO_3^- ions. If water used for drinking purposes contains about 85 ppm of $CaCO_3$ equivalent hardness.

12.8 DESALINATION OF BRACKISH WATER

Water containing high concentrations of dissolved solids is known as brackish water. Since brackish water have a peculiar saltly taste, it is therefore completely unfit for drinking. Depending upon the amount of dissolved solids, water is graded as

(i) Fresh water (having < 1000 ppm of dissolved solids)
(ii) Brackish water (having > 1000 ppm but < 3500ppm of dissolved solids)
(iii) Sea Water (having > 35000 ppm of dissolved solids)

The process of removing dissolved solids (NaCl) from water is known as desalination process. It is well known fact that with the development of civilization, our fresh water demand is increasing day by day, while the supply remains almost the same. Therefore, it becomes evident that the available fresh water is becoming scarce. This problem, can be solved by converting sea water into fresh water through desalination.

Desalination is a costly process. The water obtained by desalination proves to be much costlier than the naturally available treated water. Inspite of high cost, desalination has been used at various places around the world.

Methods of Desalination: The various methods generally adopted for the conversion of salt water into fresh water are-

1. Distillation
2. Freezing
3. Electro dialysis
4. Reverse osmosis.

1. **Distillation:** Brackish water is boiled in huge evaporators and the vapours are condensed to get pure water. This method of obtaining pure water is expensive for common practice.
2. **Freezing:** This method is based on the principle that when salt water freezes, the ice formed in the beginning is almost free from salt. The ice crystals are separated, washed and melted to get pure water. The quantity of water obtained is satisfactory, but the cost of production is high.
3. **Electrodialysis Method:** This is one of the important method of desalination of water. It is based on a standard technique of removal of ions from solution which may be applied at room temperatures. This is based on the fact that the ions present in saline water migrate towards oppositely charged electrodes through ion selective membranes under the influence of an applied electric field.

A schematic diagram of a selective anion exchange membrane is shown in Fig. 12.22:

The electrodialysis cell consists of alternate cation and anion-permeable membranes ie the membranes which allow to pass either positive or negative ions but not both. In an anion exchange membrane anions are exchanged for Cl^- ions in solution but the membrane is not permeable to Na^+ ions as these are repelled by the fixed positive charges inside the membrane (Fig. 12.22). In electrodialysis cell cathode is placed near cation-permeable membrane and anode is placed near anion-permeable membrane. When a strong electric field is applied to the saline water, the sodium ions (Na^+) start moving towards cathode, and chloride ions (Cl^-) start moving towards anode through the membranes. As a result, the concentration of the ions decreases in compartments 3,5 and it increases in the compartments 2,4,6 as shown in the figure 12.26. The desalinated water, free of ions is collected separately whereas the concentrated water in the compartments 1,3,5 etc is replaced from time to time by next lot of saline water.

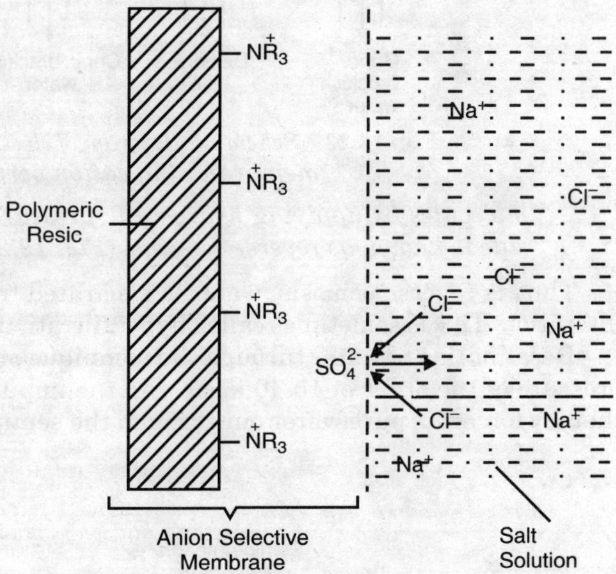

Fig. 12.22 *Schematic Diagram of Anion Exchange Membrane*

Advantages

(i) When very high purification is not required this method is best suited.

(ii) It is a compact unit which is easy to operate.

(iii) The cost of installation of plant is economical.

4. **Reverse Osmosis:** When two solutions having different concentrations are separated by a semipermeable membrane, flow of solvent molecules from the region of low concentration to high concentration takes place, untill the concentration becomes equal on the both sides. The phenomenon is called osmosis. Osmosis is thus the movement of water molecules from pure water to the solution side. The driving force in this phenomenon is called osmotic pressure. Water can also be made to flow in the reverse direction i.e from solution side to the pure water by applying a hydrostatic pressure in excess of osmotic pressure, on the concentrated or solution side. *This process of reversing the flow of solvent*

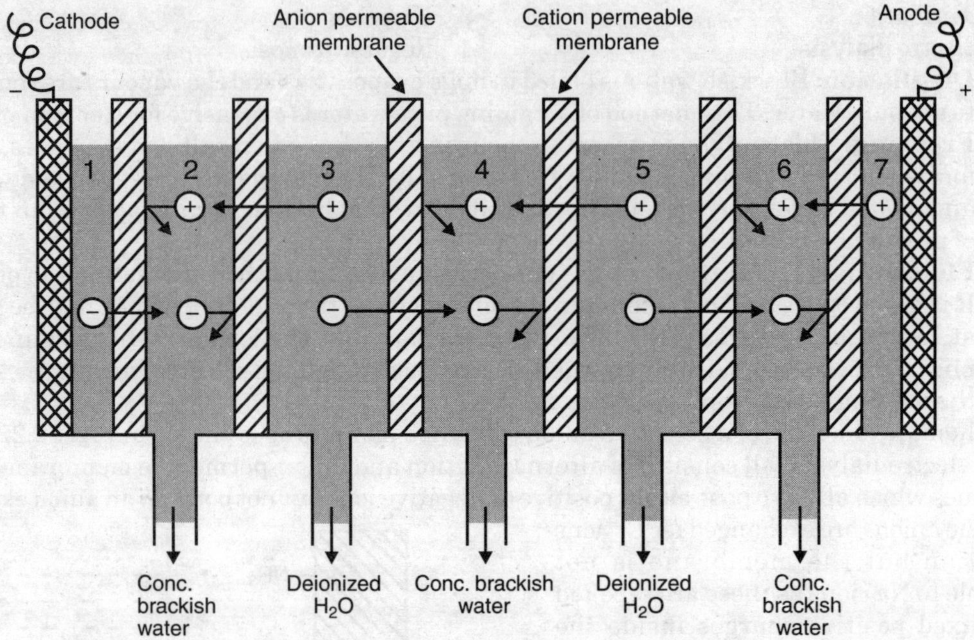

Fig. 12.23 *Schematic diagram of electrodialysis cell: anion permeable membrane and cation permeable membrane.*

molecules by applying hydrostatic pressure in excess of osmotic pressure, on solution side is known as reverse osmosis. (Fig. 12.24 b).

Thus is reverse osmosis, water is separated from the dissolved salts through membrane filteration. This is sometimes called hyper filtration or super filtration.

Desalination of water through the technique of reverse osmosis involves the application of pressure of the order of 15-40 kgcm^{-2} to the impure water (sea water/brackish water) which thereby forces its pure water out through the semipermeable membrance. A major problem in

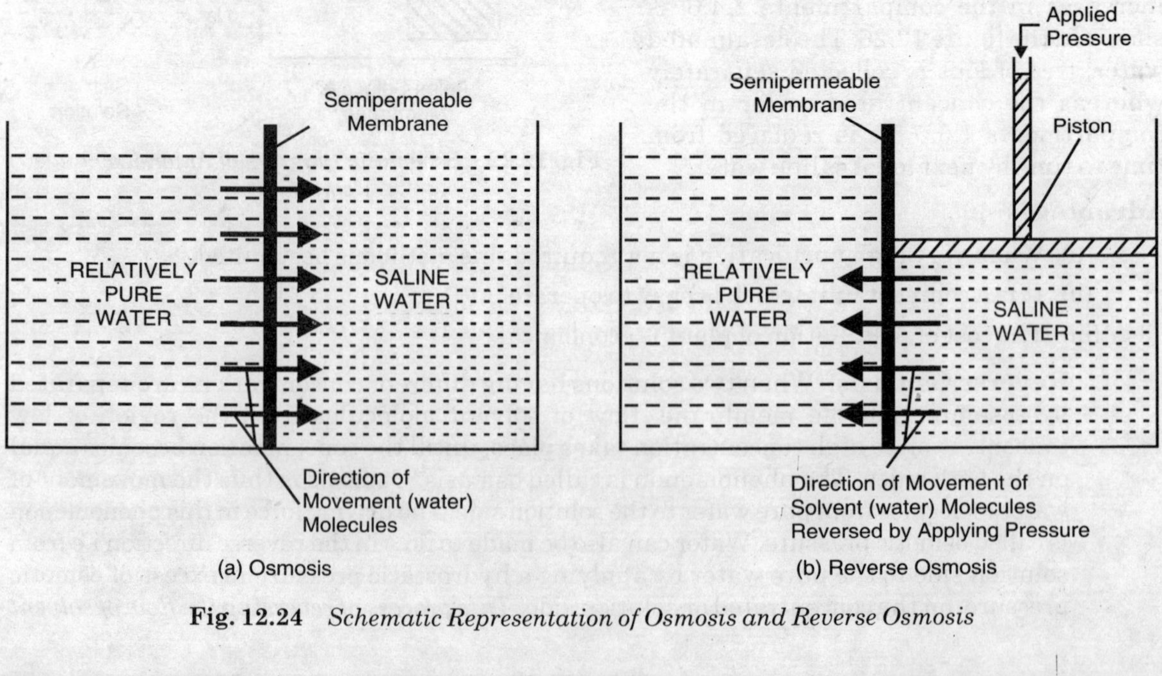

Fig. 12.24 *Schematic Representation of Osmosis and Reverse Osmosis*

reverse osmosis is to find membranes strong enough to withstand the high pressure required and at the same time impermeable to the solute particles dissolved. Synthetic membranes made up of nylon or cellulose acetate have been developed to meet these requirements. More recently the membranes made up of polymethacrylates and polyamide polymers are being used.

Advantages

(i) Purification through reverse osmosis removes ionic as well as non ionic, colloidal and high molecular weight organic matter.

(ii) The energy requirement is 30% lower than that of distillation process.

(iii) The maintenance cost includes mainly the replacement of semipermeable membrane the life time of which is quite high, extending to about 2 years.

Thus purification of water through reverse osmosis is gaining importance because of its simplicity, high reliability and low capital cost.

DOMESTIC ULTRA VIOLET PURIFICATION SYSTEM

The use of ultra-violet (uv) rays is now a days, a common method adopted for disinfection of drinking water at homes, offices, schools and various organizations. Excellent sterilization occurs when uv rays of sufficient intensity irradiate a thin film of water. UV rays sterilization unit is installed immediately before the point of use. A common uv sterilization system is shown in Fig. 12.25. Usually it contains the following units-

(i) A filter Candle.

(ii) Activated Carbon Chamber

(iii) Ultraviolet sterilization chamber

Water is first passed through the filter candle, which has multilayered filtering system to strain out physical impurities present water. Then it is passed through the activated carbon chamber which removes colour, odour, organic impurities and free gases. And finally it is subjected to u.v chamber where u.v rays kill the micro organisms. The rate of flow of water is controlled through the U-tube to ensure complete disinfection.

An electronic monitoring system monitors the purification process.

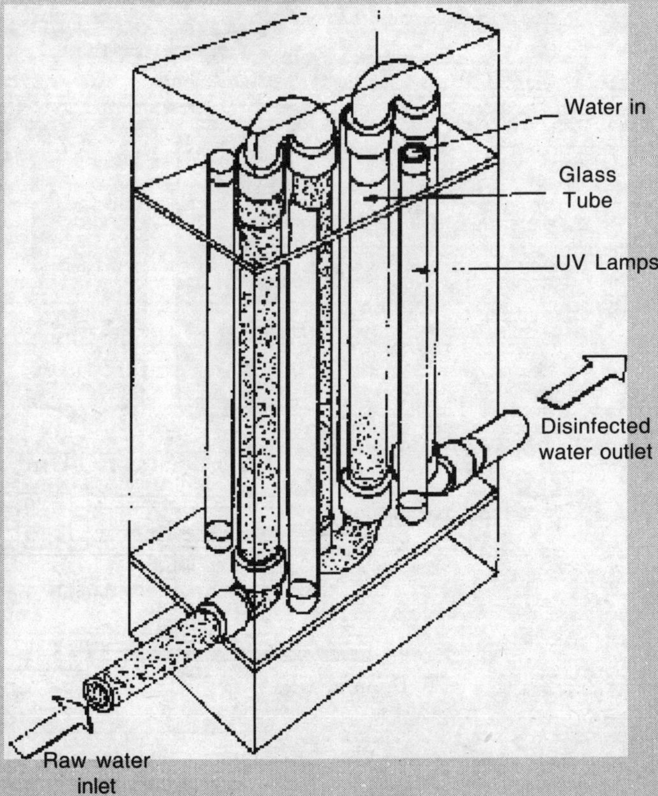

Fig. 12.25 *U.V. Sterilizer*

RO SYSTEMS

Drinking Water Purification through Reverse Osmosis Units

Another method for purification of drinking water is through Reverse Osmosis Units: The reverse Osmosis process is highly effective in removing a wide range of contaminants from water including heavy metals, chemicals, dissolved salts and micro-organisms. The water is forced through an extremely fine semi-permeable membrane rejecting the contaminated water and leaving only pure, clean and healthy water. Reverse Osmosis is the only effective way to remove fluroride from water.

Such systems typically include a number of steps:

- a sediment filter to remove particulate contaminants such as dirt, rust, slime and sand.
- an activated carbon filter to remove organic chemicals which can give your water its objectionable taste, smell and odour. It also removes chlorine which will attack and degrade reverse osmosis membranes.
- a reverse osmosis (RO) filter which is a thin film composite membrane.
- optionally an ultra-violet lamp is used for disinfection of any microbes that may escape filtering by the reverse osmosis membrane.

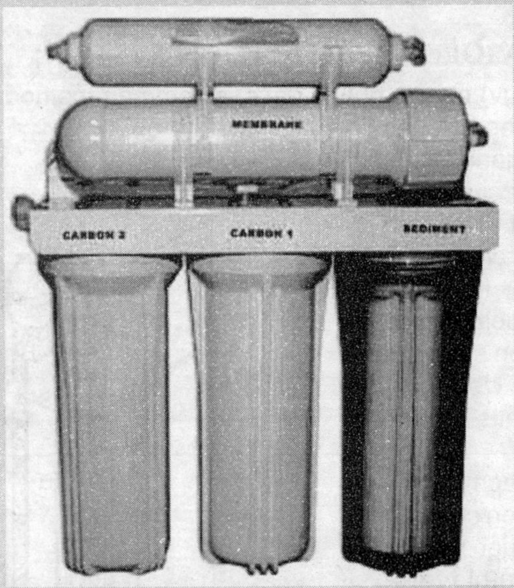

(a) RO Unit

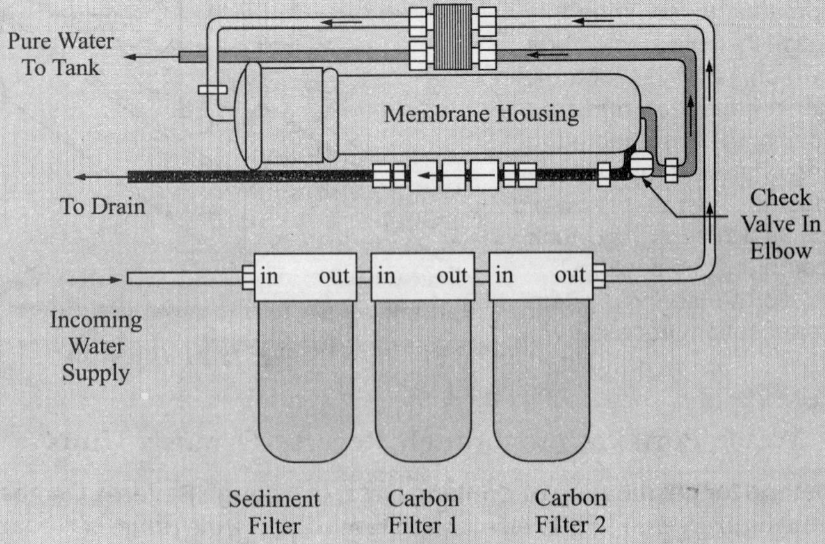

(b) Direction of Flow of Water

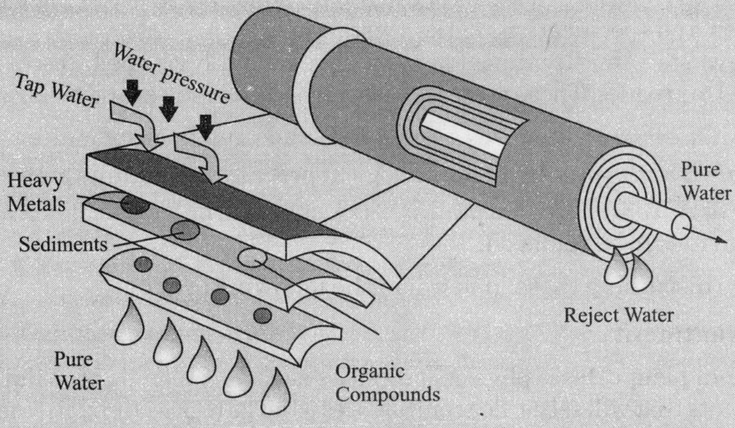

(c) The Membrane

Fig. 12.26 *(a) Reverse Osmosis Unit (b) Direction of Flow of Water in RO (c) Internal View of the Membrane*

12.9 SEWAGE TREATMENT

12.9.1 Waste Water Treatment

Until 1900's human wastes were simply dumped as raw sewage into the nearest stream or river. But later on when a number of diseases were linked to the waste waters, the treatment of waste water was taken up seriously.

The purpose of waste water treatment is to remove the contaminants from water so that the treated water can meet the acceptable quality standards. The quality standards usually depend upon whether water will be reused or discharged in to a receiving stream..

The waste water is often referred to as *sewage.* Sewage is the liquid water which includes human and domestic waste waters, industrial wastes, ground wastes and street washings. Sewage contains 99.9 per cent water and rest being the organic, inorganic compounds in dissolved colloidal or suspended states. It is this small percentage of pollutants that brings about polluted conditions.

Constituents of Sewage

There are various types of chemicals founds in waste water, depending upon the source from where it is being discharged. Depending upon the nature of the sewage it can be grouped as:

(i) *Domestic sewage:* Domestic sewage includes human excreta as well as discharges from kitchens, baths, lavatories, etc. from public and private buildings.
(ii) *Industrial Sewage:* Industrial sewage or waste water may be regarded as the liquid waste water in which industrial effluents are present as main wastes
(iii) *Storm water:* This is the rain water which flows as run-off from streets, open yards, etc.
(iv) *Combined Sewage:* Combined sewage is the combination of domestic sewage, industrial waste and storm water.

12.9.2 Waste Water Treatment Processes

Available waste water treatment processes can be broadly classified as:

- Physical processes
- Chemical processes
- Biological processes

These processes consists of a series of unit operations. These are applied in different combinations composition and specifications of the waste water.

Further, the waste waters treatment processes are generally grouped according to the water quality they are expected to produce. These processes are (physical, chemical and biological) grouped as:

(i) *Primary Treatment:* Removes suspended solids and floating matter
(ii) *Secondary Treatment:* Involves biological process and organic matter is removed.
(iii) *Tertiary or Advanced treatment:* Physical, chemical or biological processes depending upon the impurities to be removed.

We would be discussing these in detail in the following section.

1. Primary Treatment

Primary treatment plant utilises physical processes such as screening or sedimentation to remove a portion of pollutants that will settle, float or that are too large to pass through simple screening devices.

Primary treatment comprises two steps, Pretreatment and Sedimentation.

(a) *Pretreatment:* Pretreatment consists of screening and grit removal. Screening removes large floating objects such as rags, sticks, wood and other large floating and suspended solids. A typical screen consists of a parallel steel bars spaced anywhere from 2 to 7 cm apart. This may be followed by a wire mesh screen. The material collected on the screens is disposed off. One way to dispose off the material is to use a device called comminuter, which grinds the coarse material in to small pieces which flow along with water and handled in sedimentation tank. After screening, the waste water passes into a grit chamber where the velocity of water is reduced or it is detained for a few minutes. Here sand, grit and other heavy materials settle down. The grit settling chambers are periodically disconnected from the main system to remove grit manually, for possible use in land-filling, road making and on sludge drying beds. Grit also is a food manure for growing crops.

(b) *Sedimentation:* From the grit chamber, the sewage passes to a primary settling tank known as sedimentation basin. Here the velocity of the water is reduced considerably to allow most of the suspended solids to settle out by gravity. The most common equipment used include horizontal flow sedimentation tanks, similar to the one described under domestic water. The water is detained in the horizontal flow tanks for 2-3 hours resulting in removal of 50% of the suspended solid matter. An efficient sedimentation tank or clarifier removes about 80-90% of the suspended solids and 40% of organic matter. The solids that settle are called primary sludge or raw sludge.

Sedimentation aids: The removal of finely divided solids is difficult by simple sedimentation technique. In such cases, mechanical flocculation or chemical coagulation is employed.

In *mechanical flocculation*, the waste water is passed through a sedimentation tank which is fitted with rotating paddles which rotate slowly at a speed of 0.43 m/s. This slow mechanical stirring allows the finely divided solid particles to coalesce into larger particles and settle out.

In *chemical coagulation*, certain chemical substances, known as coagulants are added. Coagulation used for sewage treatment are similar to the ones used in water treatment. The coagulants react with colloidal matter in the sewage to form floc. The floc entraps the smaller particles and eventually settles down as sludge. Common coagulants used in sewage treatment are:

Alum $Al_2(SO_4)_3 \cdot 18H_2O$, Copperas $FeSO_4 \cdot 7H_2O$, hydrated lime, ferric chloride, and chlorinated copperas, $FeSO_4.Cl$ (mixture of ferric sulphate and chloride).

Apart from using coagulants, coagulant aids are also used sometimes for efficient removal of finely divided particles. The commonly used coagulant aids are, activated silica polyelectrolytes (polymers of cynamide or methacrylic acids), and hydrolysed high molecular weight polymers. The coagulant aids draw a number of suspended particles together because of their long chain molecular structure. They also neutralise the charge on the particles by them bringing them closure to form large particles.

In *industrial waste water*, sometimes the primary treatment also includes equalisation and neutralisation

Equalisation: Sometimes different types of wastes are produced by some industries. To apply uniform treatment, different effluents are held in big tanks for certain periods and are mixed thoroughly to produce homogeneous equalised effluent.

Neutralisation: Acidic wastes are neutralised with lime stone and alkaline wastes are neutralised by treatment with sulphuric acid or CO_2 or waste boiler flue gas. If both acid and alkaline wastes are produced in the nearby plants then mutual neutralisation by mixing there is them cheapest method of neutralisation.

2. Secondary Treatment

In primary treatment, only those materials are taken care of, which can be removed by simple physical or mechanical action. However, in waste water much of the organic material is dissolved or in colloidal form which is not removed by primary treatment. This organic material is oxygen demanding and must be removed further. Thus the main purpose of the secondary treatment is to remove the organic matter. This is achieved through biological processes which are quite similar in concept to the natural biodegradation of organic matter by aerobic bacteria. The various changes which are brought by secondary treatment include:

— Coagulation of the finely divided or colloidal matter.
— Oxidation of the organic matter to CO_2
— Conversion of nitrogenous organic matter to ammonia, which is eventually converted into nitrite and nitrate.
— Anoerobic digestion of the sludge so obtained.

There are three commonly used approaches, all of which take advantage of the biological processes. These processes are:

(i) Trickling filters.
(ii) Activated Sludge Process
(iii) Oxidation Ponds (Lagoons).

(i) Trickling Filters (Aerobic filteration)

A trickling filter consists of a rotating distribution arm that sprays the liquid over a circular bed of rocks or other coarse material (Fig. 12.27)

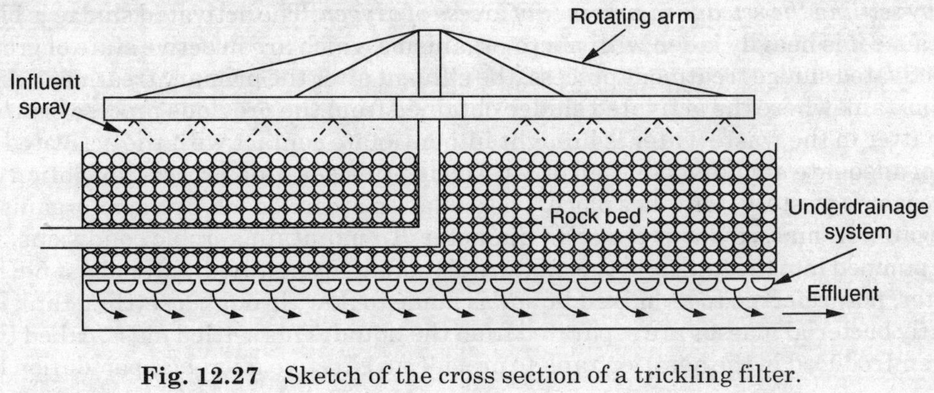

Fig. 12.27 Sketch of the cross section of a trickling filter.

The rock bed is 1 to 3 meters deep with enough open spaces between rocks to allow air to circulate easily. The individual rocks becomes coated with a layer of biological slime or a gelatinous film that absorbs and consumes the wastes trickling through the bed. The slime consists mainly of aerobic micro-organisms (bacteria, algae, protozoa, etc.) which are collectively called **zooglea** and thrive on the nutrients supplied by the waste water. The in fluent (waste water) is sprinkled over this bed packing. As the liquid trickles over the packing, oxygen and the dissolved organic matter diffuses into the slimy layer and are oxidised by the micro-organisms to form CO_2, NO_3, etc., which diffuse back out of film and appear in the filter effluent. As the micro-organisms utilise the organic matter, the thickness of the slime layer increases and it gets detached from the surface of the rock and is carried away along with the effluent, which is referred to as *Humus*. The effluent along with detached slimy layer and treated water is collected at the bottom of the filter and passed on to settling tank, where the detached slimy layer settles down and is removed. The slimy layer (sludge) is then pumped to the sludge digestion unit.

More recently, plastic filters have been used for trickling filters. The plastic filters are light and can be stacked many times higher than the conventional bed rock which crumbles under its own weight if stacked high. These filters, made with the plastic media are sometimes referred to as 'biological towers'.

Advantages of the Process

- Trickling filters are simple to operate and can produce BOD removal to the extent of 65 to 85%
- Constant monitoring is not required.
- Effluents so produced are of better quality.

Limitations

- The microbial film formed is sensitive to temperature changes in winters.
- Efficiency of the filter is dependent upon the composition of waste, pH, size uniformity of the filtering medium and the supply of air.
- Cost of construction is high
- Trickling filters are used for treating industrial waste water from dairy, brewery, food processing, pulp and paper mills, pharmaceuticals, petrochemicals, etc.

(ii) Activated Sludge Process

This is the most versatile biological oxidation method, employed for the treatment of waste water containing organic matter. It is a biological waste water treatment process in which a mixture of waste water and activated sludge is agitated and aerated. *The activated sludge is the sludge obtained by settling the sewage in presence of excess of oxygen.* The activated sludge is biologically active because it is heavily laden with micro-organisms which are in active state of growth.

The activated sludge treatment process, the effluent after the primary treatment, is passed to an aeration tank where the activated sludge obtained from the previous process is present. The organic matter in the waste water is brought into intimate contact with the activated sludge in presence of adequate amount of oxygen or air. During this aeration process, oxidation of organic matter present is waste water takes place. The oxidation is carried out by micro-organisms which are aggregated in masses about .1mm in diameter. To maintain aerobic conditions, the air or oxygen in pumped into the tank and the mixture is thoroughly agitated. After about 6-8 hours, the waste water, now referred to as 'mixed liquor' is made to flow through a settling tank where the solid, mostly bacterial masses are separated from the liquid. This settled mass, called the sludge, is partly reintroduced in the aeration tank to provide an effective microbial population for a fresh treatment cycle and is therefore termed as activated sludge. Rest of the sludge is digested in a sludge digester, along with the primary sludge obtained from primary treatment. Fig. 12.28.

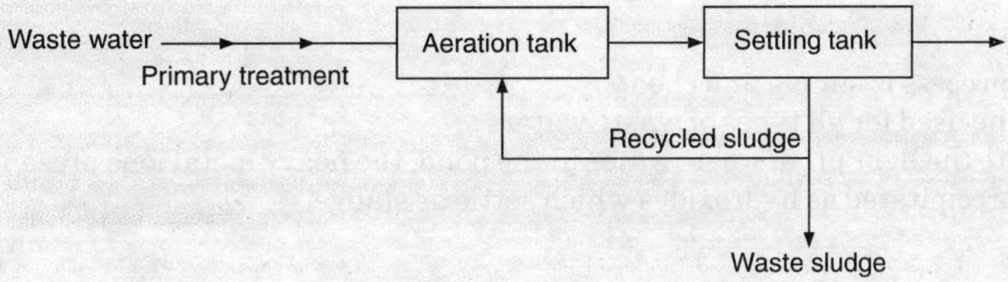

Fig. 12.28 Activated sludge process

Advantages

- The primary advantage of the activated sludge process is, good effluent quality. The effluent after going through activated sludge has little BOD (< 20mg/L)
- It takes less area as compared to trickling water filters.
- The activated sludge process equipment is less expensive.

Limitation

- For the process to be efficient, at least 0.5 ppm oxygen must be present all time.
- The optimum pH range 6.5 to 9.0 has to be maintained throughout.
- Low temperature slows down the rate of metabolism, while high temperature increases the metabolic activity to such an extent that the oxygen is consumed fast, leading to anaerobic conditions.
- The presence of detergents (which are not biodegradable) lead to the formation of foam, making the process difficult.
- The disadvantage of activated sludge process is the production of a huge amount of sludge, which should be digested and disposed off.
- Table 12.7 gives a comparison between trickling filters and activated sludge process for water treatment.

Table 12.7: Comparison of trickling filters with activated sludge systems

Trickling filters	*Activated sludge systems*
1. Bacterial growth is fixed on the media	1. Bacterial growth is suspended as a dispersed floc
2. All solids from the settler are wasted	2. Solids from the settler are partially recycled
3. Less sensitive to shock loading, more stable	3. More sensitive to shock loadings, required closer process control
4. Produce odours	4. Produce spray clouds
5. Less effective in removing disease-causing organisms	5. More effective in removing pathogens than trickling filters
6. Low operating costs	6. High operating costs

(iii) Oxidation Ponds

Oxidation ponds are large shallow ponds, typically 1-2 m deep, where the partially treated waste water is decomposed by micro-organisms. The waste water enter the pond at one and of the treated waste water is collected at the other end. The organic matter is oxidized by the micro-organisms present in the pond (such as pseudmonas, flavo bacterium, etc). The decomposition of the organic matter near the surface is aerobic, whereas near the bottom, it is anaerobic. Such ponds having a mix of aerobic and anaerobic conditions are called facultative ponds. The oxygen required for aerobic decomposition comes from algal photosynthesis. However deeper ponds known as lagoons are mechanically aerated. Fig. 12.29.

Advantages

- The process is simple and cheap.
- Can be used for all types of waste waters
- Due to the light pH of waste water in the pond, the heavy metal ions present in waste water are precipitated as hydroxides which settle as sludge.

Limitations

- The oxidation ponds require larger space.
- Anaerobic conditions may lead to release of bad odours.

The main drawback of the above secondary treatment processes is the formation of sludge. The collection, processing and disposal of sludge can be the most costly and complex aspect of waste water treatment.

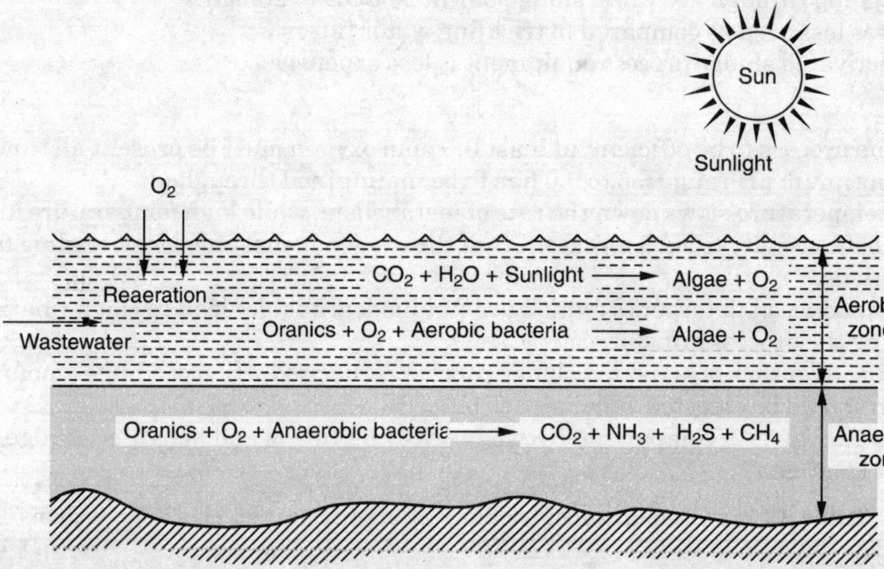

Fig. 12.29 Schematic diagram of an oxidation pond.

Sludge Treatment and Disposal

Sludge is the watery residue from the primary sedimentation tank and humus tank from secondary treatment. The quantity of sludge produced may be as high as 2 per cent of the original volume of waste water, depending upon the treatment process used. One of the primary goals of sludge treatment is to separate as much of the water from the solids as possible. In addition to reducing the water content of the sludge, it may be stabilised so that its biological activity and tendency towards putrefaction is reduced.

The traditional method of sludge digestion is **anaerobic digestion.** It involves the microorganisms that thrive in absence of oxygen. The organic material in the sludge is digested by these micro-organisms under anaerobic conditions to give carbon dioxide and methane gas. The components of the sewage which can be converted into gases are called volatile solids. Though the sludge digestion is complex, it can be summarised in two steps.

- First the complex organics such as fats, proteins, and carbohydrates are converted to organic fatty acids by acid forming bacteria.
- Then these acids are converted into methane, CO_2 etc by methane forming bacteria.

Most treatment plants use a sludge digester in which the sludge is maintained at 35° for 30 days at a pH of 7.0 to 8.0. CH_4, CO_2 and NH_3 are liberated as the end products (Fig. 12.30)

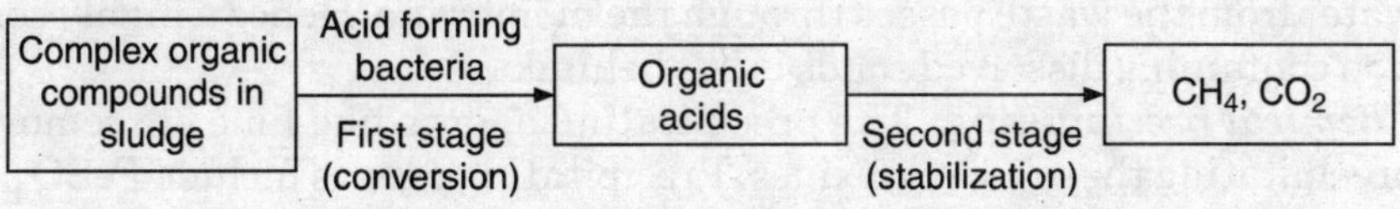

Fig. 12.30 Anaerobic digestion

Digested sludge is removed from the anaerobic digester. This sludge contains 90 to 93% water and is *dewatered*. Dewatering is accomplished by mechanical methods, the most common being centrifugation and filteration, which includes pressure filteration and vacuum filteration. Drying beds are also commonly used. The dewatered sludge is sent for ultimate disposal. Wet sludge is sprayed on to crop land where it functions as fertiliser. Dried sludge may be used as a landfill or a soil conditioner.

Tertiary Treatment

Any treatment that follows primary and secondary treatment is considered tertiary treatment. Tertiary treatment improves the quality of the effluent further.

The effluent after secondary treatment plant still contains suspended solids (20-40 mg/L) which may settle on the stream or river bed and inhibit certain forms of aquatic life. Some amount of BOD, significant amount of nutrients, dissolved solids, traces of organic chemicals and other contaminants are also present. The emphasis on recovery of valuables from industrial wastewaters have created the need for tertiary treatment.

The type of tertiary treatment depends upon the specific goal which include removal of:

- suspended solids
- bacteria
- dissolved organic solids
- toxic substances
- nutrients (phosphorus and nitrogen)

Depending upon the required quality of final effluent a number of techniques can be used. Some of which are described in brief as follows:

(i) **Removal of suspended solids (Microstraining):** This can be achieved by **microstraining.** The filter media consists of finely woven stainless steel fabric. The treated waste is allowed to pass through it. The solids which are retained on the fabric are washed into a trough, which recycles the solids to the sedimentation tank.

(ii) **Removal of dissolved solids:**

(a) *Adsorption:* Dissolved solids can be organics or inorganic. The soluble organics are removed by **adsorption** on activated carbon. Special adsorbents are commercially available for the removal of toxic heavy metals from industrial waste water.

(b) *Solvent Extraction:* It is sometimes used to recover phenolic materials from waste waters of refineries and coke plants. The waste water is intimately brought in contact with a solvent having high affinity for the solute. The layers (aqueous and the solvent layer containing solute).

(c) *Ion Exchange:* Ion exchange is extensively used to remove hardness and iron and manganese salts from drinking water as described in chapter 1. The same technique has been extended to waste water treatment for the removal and recovery of waste water treatment.

(d) *Reverse Osmosis:* When waste water containing dissolved solids is allowed to pass through a semi-permeable membrane at a pressure, which is more than osmotic pressure, the water from the waste passes through the membrane. Hence a highly concentrated solution containing dissolved solids is left behind.

(e) *Chemical precipitating:* The precipitating agents like lime etc remove heavy ions by precipitating these as hydroxides. Precipitating agents include $FeSO_4$, alum and ferric chloride.

(iii) **Removal of Nutrients**

(a) *Nitrogen Removal:* All forms of nitrogen in wastewater are harmful because plants can utilize the inorganic forms as nutrients, NH_3 can be utilized by bacteria resulting in reduced oxygen in water.

One of the method of nitrogen removal is **ammonia stripping**. Ammonia is present in natural water as ammonium in (NH_4^+). This NH_4 is changed to ammonia gas by raising the pH (the $\overline{O}H$ concentration) of the waste water by adding quick lime.

$$NH_4^+ + \overline{O}H \rightarrow NH_3\uparrow + H_2O$$

The ammonia gas is liberated

Another approach of nitrogen removal is nitrification, i.e. to convert NH_4^+ to NO_3^-, followed by anaerobic stage in which micro-organisms convert nitrates to nitrogen gas (N_2)

$$NH_4^+ + 2O_2 \xrightarrow{bacteria} NO_3^- + 2H^+ + H_2O$$

$$2NO_3^- + \text{organic matter} \rightarrow N_2\uparrow + CO_2 + H_2O$$

(b) *Phosphorus Removal (Chemical precipitation):* Phosphorus is present in the form of orthophosphates ($H_2PO_4^-$, HPO_4^{2-} and PO^{3-}) Phosphates are removed by adding coagulants usually alum [$Al_2(SO_4)_3$] or lime [$Ca(OH)_2$].

$$Al_2(SO_4) + 2PO_4^{3-} \rightarrow 2AlPO_4\downarrow + 3SO_4^{2-}$$

(iv) **Removal of bacteria:** Bacteria are removed by retaining the effluents in maturation ponds or lagoons for specified period of times.

QUESTIONS

SHORT ANSWER QUESTIONS

1. What are the major sources of water?
2. What type of dissolved impurities are present in water?
3. What are the specifications of water to be used in Alcoholic Distilleries, and confectionaries?
4. What are the specifications of water to be used in textile industries?
5. Why is rain water a purest form of natural water ?

   Ans. Rain water is obtained through a process similar to distillation hence it is the purest from of natural waters.
6. What do you understand by hardness of water ?

7. What is the cause of hardness of water ?
8. Distinguish between temporary and permanent hardness
9. Why hardness is expressed in terms of $CaCO_3$ equivalents. How is it calculated ?
10. Determine the $CaCO_3$ equiv. of 83 mg of $Mg(HCO_3)_2$ (Molecular weight = 146)

    Ans. $CaCO_3$ equivalents of 83 mg of Mg $(HCO_3)_2$

$$= 83 \times \frac{100}{146} = 56.8 \text{ mg}$$

11. What are the units of hardness?
12. What is the relation between these units?
13. What is the principle of hardness of water?
14. What is the structure of Eriochrome Black T.
15. Why $NH_3 - NH_4Cl$ buffer solution is added in determination of hardness of water by EDTA method.
16. What type of ligand is EDTA?
17. Why the complex of Ca^{2+} or Mg^{2+} with EDTA is more stable than the complex of Ca^{2+} or Mg^{2+} with EBT?

    Ans: EDTA is a hexadentate ligand. It forms 4 or 6 bonds with the divalent cations (Ca^{2+} or Mg^{2+}) thus entrapping it like a claw of a crab. Such type of compounds are known as chelate compounds which are very stable.
18. What is the cause of alkalinity in natural waters?
19. Why alkalinity of water cannot be due to the simultaneous presence of $\overline{O}H$, CO_3^{2-} and HCO_3^-.
20. What type of alkalinity is present if phenolphthalein alkalinity is more than half the methyl orange alkalinity $\left(P > \frac{1}{2}M\right)$.
21. During titration of a water sample with N/10 H_2SO_4 using phenolphthalein indicator zero ml of H_2SO_4 was used. What do you inter from this.

    Ans: Absence of CO_3^{2-} and $\overline{O}H$. Only HCO_3 may be present.
22. What are the disadvantages of using hard water in boiler operations?
23. Differentiate between scale and sludge?
24. What are the salts responsible for the formation of scales?

    Ans: $Mg(OH)_2$ and $CaSO_4$
25. Why should natural water not be fed to boiler?
26. What is conditioning or internal treatment of hard water ?
27. What is carbonate conditioning?
28. Why phosphate conditioning is better than carbonate conditioning?
29. Why is calgon conditioning is better than phosphate conditioning?

    Ans: In calgon conditioning soluble complexes of Ca^{2+} and Mg^{2+} are formed whereas in phosphate conditioning precipitates of calcium or magnesium phosphate are formed which give rise to sludge which has to be removed. No such problem of sludge removal present in calgon conditioning.
30. What are the reasons of boiler corrosion?
31. How dissolved oxygen can be removed by chemical and mechanical means?
32. What do you mean by caustic embrittlement?
33. How can you prevent caustic embrittlement?
34. How caustic embrittlement is controlled by adding sodium sulphate?
35. What is priming and foaming ?
36. How silica can be removed from water?
37. What are the various techniques employed for treating water externally ?

38. What do you mean by zeolite?
39. Why zeolite process is sometimes referred to cation-exhange process?
40. Discuss regeneration of zeolite?
41. What are the disadvantages of zeolite process.
42. What are Ion Exchange Resins.
43. What do you mean by cation and Anion Exchangers.
44. How are exhausted ion-exhange resins are regenerated?
45. Compare zeolite and ion-exchange process.
46. What are the advantages of ion-exchange process.
47. In ion exchange process, water is usually passed through cation exchanger and then through anion exchanger and not vice versa. Explain.

    Ans: Cation exchangers are attacked by alkalies. Thus if water is passed through anion exchanger first, the alkalies so produced will damage the cation exchanger.
48. What are the specifications of water for drinking purposes?
49. Why soft water cannot be used for drinking purposes?

    Ans: For drinking purposes, some amount of $CaCO_3$ (about 85 ppm) should be there otherwise the soft water reacts with the lead used in plumbing.

$$2Pb + 2H_2O + O_2 \rightarrow 2Pb(OH)_2$$

Presence of HCO_3^- or CO_3^- diminishes the dissolution of lead in drinking water.

50. What is sedimentation and coagulation?
51. Distinguish between primary and secondary standards of drinking water?
52. What are the coagulants used for treating water for drinking purposes.
53. What do you mean by the term disinfection?
54. Why is boiled water is not 100% safe drinking purposes?

    Ans: Boiling kills germs but do not provide protection against contamination during storage etc.
55. Why chloramine is preferred over bleaching powder during chlorination?
56. What is break point chlorination?
57. What is the process of dechlorination?
58. How water is disinfected using $KMnO_4$.
59. What do you mean by the term desalination?
60. What are the advantages.
61. What is reverse osmosis as applied to water purification ?
62. What is the main advantage of reverse osmosis over ion exchange process?

    Ans: Ion exchange process removes ions whereas reverse osmosis also removes colloidal and high molecular weight organic matter along with ions.

LONG ANSWER QUESTIONS

1. What do you understand by hardness of water ?

   What is its cause? Distinguish between temporary and permanent hardness.
2. What are the specifications of water for industrial use?
3. Give the chemical reactions involved in determination of hardness of water by EDTA titration.
4. What are the disadvantages of hard water ? How is the hardness of water removed by ion-exchange method ?
5. How is water softened by lime-soda process ? Describe its types and suitable chemical reactions.
6. (a) What are the external and internal treatment required for prevention of scale formation in boiler.

   (b) Comment on the statement that internal treatment is only a corrective or supplementary method of treatment for boiler feed water.

7. (a) Give the chemical reaction involved during :
   (i) softening of water by ion-exchange resins, and
   (ii) regeneration of exhausted ion-exchange resins.
   (b) Mention advantages and disadvantages of ion-exchange process of softening water over lime-soda process.
8. (a) What are the disadvantages of using hard water in boiler operation ? Disuss any four methods of disinfection of water.
   (b) Specifications of domestic and industrial water are different. Explain.
   (c) Mention the requisites of water used for paper mills, textile industry and sugar industry.
9. (a) What is meant by phosphate conditioning ?
   (b) What is calgon? What is its application in water treatment?
   (c) What is the role of $NaAlO_2$ in internal treatment of boiler water?
10. (a) What is the ideal pH range for phosphate conditioning?
   (b) Identify from the following compounds, those suitable for treating acidic, neutral and highly alkaline water: NaH_2PO_4, Na_2HPO_4 and Na_3PO_4.
11. (a) What is caustic embrittlement? What are the contributing factors for caustic embrittlement?
   (b) How can the caustic embrittlement be prevented?
12. (a) What are the different factors contributing to boiler corrosion? How can the boiler corrosion be minimised?
   (b) What is meant by deaeration? How can it be effected mechanically?
   (c) Suggest some chemical reagents for the removal of dissolved oxygen. Which one is the best? Why?
   (d) Suggest some feasible chemical reagents for the removal of dissolved CO_2 in water.
13. (a) What is priming? How is it caused? What are its disadvantages? How can it be minimised ?
   (b) What do you mean by foaming. How it can be minimised?
14. (a) What are cation and anion exchangers?
   (b) What are the merits of ion exchange process over the other water softening methods?
   (c) What is meant by exhaustion of cation and anion exchangers? How can they be regenerated?
15. Describe the principle and procedure involved in the zeolite process for the treatment of water. What are the limitations, advantages and disadvantages of the process ?
16. Why is demineralization process preferred over zeolite process for softening of water for use in boilers.
17. (a) Write, in brief, how is water purified for drinking purpose by municipalities ?
   (b) What is break-point chlorination? What are its advantages?
18. (a) Define the term disalination. With neat diagram, describe desalination by electrodialysis process.
   (b) What do you mean by reverse osmosis? How it help in desalination of water? What are its advantages?
19. Distinguish between :-
   (a) Temporary hardness and permanent hardness.
   (b) Purification and softening of water.
   (c) Boiler scales and sludges
   (d) Cold lime soda process and hot lime soda process
   (e) Softening and demineralization of water
20. (a) What are the requisites of drinking water?
   (b) How is natural water sterlized by : (i) Chlorine, (ii) bleaching powder, (iii) chloramine, and (iv) ozone?
21. (a) State two harmful effects of silica present in water? How it can be removed.
   (b) What is the basis of chlorination of water explain.
22. Explain the following steps involved in purification of water (a) Sedimentation (b) Coagulation (c) Filteration

23. Explain the terms : blow down operation, make up water, priming and foaming.
24. (a) Give chemical reactions involved during the softening of water by lime-soda process
    (b) Why is hot lime-soda process better than the cold process? Why is sodium aluminate added in cold process? Discuss its function.
25. How is water analyzed for alkalinity. How the alkalinity due to various ions can be determined?
26. What is meant by disinfection in treating domestic water supply? What is its importance? What are the chemicals which are used as disinfectants and what are their comparative merits and demerits?
27. Describe a method for the removal of suspended impurities and micro organisms from water.
28. Describe the methods of determining total solids, alkalinity and chlorides in sample of water.

PROBLEMS

On Hardness of Water

29. A water sample contains 204 mg of $CaSO_4$ per liter. Calculate the hardness in terms of $CaCO_3$ equivalents. **(150 ppm)**
30. How many grams of $FeSO_4$ dissolved per litre gives 210.5 ppm of hardness? **(0.31996g/L)**
31. A sample of water on analysis was found to contain the following impurities in mg/L :

Impurity	$(Ca(HCO_3)_2$	Mg $(HCO_3)_2$	$CaSO_4$	$MgSO_4$
Quantity	10	8.5	12	14
Mol. Wt.	162	146	136	120

Calculate the temporary, permanent & total hardness of water in mg/L?
(11.99, 20.49 and 32.48 mg/L)

32. A sample of water is found to contain 40.5 mg/L $Ca(HCO_3)_2$, 46.5mg/L $mg(HCO_3)_2$, 27.6 mg/L $MgSO_4$, 32.1 mg/L $CaSO_4$ and 22.45 mg/L $CaCl_2$ Calculate the total hardness of water? **(123.68 ppm)**
33. A sample of water on analysis has been found to contain the following in ppm :

$$Ca(HCO_3)_2 = 12.5;\ CaCl_2 = 8.2;\ MgSO_4 = 2.6.$$

Calculate the temporary and permanent hardness in degree French?
(1.504 and 1.507 degree French)

34. 100ml of water sample has a hardness equivalent to 12.5ml of 0.08N $MgSO_4$ What is the hardness in ppm? **(500ppm)**

On Determination of Hardness of Water

35. 0.28 g of $CaCO_3$ was dissolved in HCI and the solution was made to one litre with distilled water. 100 ml of the above solution required 28 ml of EDTA solution on titration. 100 ml of the hard water sample required 33 ml of the same EDTA solution on titration. After boiling 100ml of this water, cooling, filterning and then titration required 10 ml of EDTA solution. Calculate the temporary and permanent hardness of water.
(Hardness ; Total = 330, Perm. = 100, Temp. = 230ppm)
36 A standard hard water contains 15g of $CaCO_3$ per lilter. 20 mL a this required 25mL of EDTA solution. 100 ml of sample water required 18 ml EDTA solution. The same sample after boiling required 12 mL EDTA solution. Calculate the temporary hardness of the given sample of water, in terms of ppm. **(Temporary hardness = 720 ppm)**
37. 50 ml of standard hard water containing 1 mg of pure $CaCO_3$ per ml consumed 20 ml of EDTA. 50 mL of water sample consumed 25 mL of the same EDTA solution, using Eriochrome T as indicator. Calculate the total hardness of water sample in ppm. **(1250 ppm)**

38. Calculate the hardness of a water sample, whose 10 mL required 20 mL EDTA. 20 mL of $CaCl_2$ solution, whose strength is equivalent to 1.5 g of $CaCO_3$ per litre, required 30 ml of EDTA solution. **(2000 ppm)**

39. One gm of $CaCO_3$ was dissolved in dil. HCl and the solution diluted to one litre. 50 mL of this solution required 45 mL of EDTA solution, while 50 ml of the sample water required 18 mL EDTA solution. On the other hand, 50 ml of boiled sample water when titrated against EDTA consumed 9 mL of solution. Calculate each type of hardness in ppm. **(Perm. = 200 ppm; Temp.= 200 ppm)**

40. 100 mL of a sample of water required 15mL of 0.01M EDTA for titration using Erichrome T as indicator. In another experiment, 100 mL of the same sample was boiled to remove the temporary hardness. The cold sample (100mL), required 8 mL of 0.01 M EDTA solution for titration. Calculate (i) the total hardness (ii) permanent hardness **(150 ppm ; 80 ppm)**

41. 250 ml of a sample of water on EDTA titration with Erio-T as indicator consumed 13 ml of 0.022 M EDTA till end point is reached. Calculate the hardness of water (1ml of 0.01 M EDTA = 1 mg of $CaCO_3$) **(114.4 ppm)**

On Determination of Alkalinity of Water

42. 200 ml. of a sample of water required 20 ml. N/50 H_2SO_4 using methyl orange as indicator but did not give any colouration with phenolphthalien. What type of alkalinity is present ? Express the same in mg/L. **(HCO_3 = 100 ppm)**

43. 200 ml of a sample required 20 ml of N/50 HCl using methyl orange as indicator. Another 200 ml of the same sample required 8 ml of N/50 HCl using phenolphthalien as indicator. Express the alkalinity in terms of mg of $CaCO_3$ per litre. **[80ppm(CO_3^{2-}) and 20 ppm (HCO_3^-)]**

44. A sample of water was alkaline both to phenolphthalien and methyl orange 0.50 ml of this water sample required 15 mL of N/50 H_2SO_4 for phenolphthalien end point and another 10 mL for complete neutralisation. Calculate the type of alkalinity in ppm. **[400 ppm (CO_3^{2-}) and 100 ppm ($\overline{OH}$)]**

45. A water sample is not alkaline to phenolphthalien. However, 100 ml of the sample water on titration with N/50 HCl, required 16.9 ml to obtain the end point, using methyl orange as indicator. What are the types and the amounts of alkalinity present in the sample. **(HCO_3^- = 169mg/l)**

46. A water sample is alkaline to both phenolphthalien as well as methyl orange. 100 ml of water sample on titration with N/50 HCl required 4.7 ml of the acid to phenolphthalien end point. When a few drops of methyl orange are added to the same solution and the titration further continued, the yellow colour of the solution just turned red after addition of another 10.5 ml of the acid solution. Elucidate on the type and extent of alkalinity present in the water sample. **($CO3^{--}$ = 94 ppm and HCO_3^- = 58 ppm)**

47. 100 ml of a water sample, on titration with N/50 H_2SO_4 gave a titre value of 5.8 ml to phenolphthalein end point and another 100 ml sample on titration with same acid gave a titre value of 11.6 ml to methyl orange end point. Calculate the alkalinity of the water sample in terms of $CaCO_3$ and comment on the type of alkalinity present. **(CO_3^{2-} = 116 ppm)**

48. 100 ml of water sample, on titration with N/50 H_2SO_4 using phenolphthalein as indicator, gave the end point when 5 ml of acid were run down. Another lot of 100 ml of the sample also required 5 ml of the acid to obtain methyl-orange end point. What type of alkalinity is present in the sample and what is its magnitude? **(OH^- = 50 ppm)**

Based on Zeolite Process of Water Softening

49. A zeolite softner was completely exhausted and was regenerated by passing 100 litres of sodium chloride solution, containing 120 g/L of NaCl. How many litres of a sample of water of hardness 500 ppm can be softened by this softener? **(20,512 L)**

50. The total hardness of 1,000 litres of water was completely removed by a zeolite softener. The zeolite softener required 30 litres of sodium chloride solution, containing 15 g/litre of NaCl for regeneration. Calculate the hardnes of water. **(384.6 ppm)**
51. 10,000 litres of hard water was softened by zeolite process. The zeolite required a total amount of 8 litres of NaCl solution containing 150 g/L of NaCl for regeneration. Calculate the hardness of water. **(102.5 ppm)**
52. A completely exhausted zeolite softener requires 120 litres of NaCl solution having 100 g/L of NaCl. How many litres of water having hardness 500 ppm can be softened by the zeolite. **(20,518 L)**

❑❑❑

UNIT 5

Some Materials of Industrial Importance

13 POLYMERS

"Every great advance in science has issued from a new audacity of imagination".

— John Dewey

CHAPTER OUTLINE

The word 'polymer' is derived from the classical Greek words *poly* meaning 'many' and *meres* meaning 'parts'. Simply stated, a polymer is a molecular compound with high molecular mass, ranging into thousands and millions of gram. It is composed of a large number of repeating units of identical structure called *monomers*. For example, polythene is a polymer formed by linking together of a large number of ethene molecules.

$$\underset{\text{Ethene (Monomer)}}{n\,CH_2 = CH_2} \longrightarrow \underset{\text{Polyethene (Polymer)}}{-\!\!\left(CH_2 - CH_2\right)\!\!-}$$

Certain polymers, such as proteins, cellulose, and silk are found in nature and hence are called natural polymers, whereas, a number of polymers are produced by synthetic routes and are called synthetic polymers. In some cases, naturally occuring polymers can also be prepared synthetically. For e.g. rubber, which is also known as polyisoprene.

Both natural and synthetic polymers impinge on every aspect of our existence. The idea is often expressed that we live in "a plastic age". The range of polymeric materials is so wide that they have found applications in practically every branch of industry. About 80% of today's chemists, including biochemists, work with polymers.

The number of repeating units in a chain formed in a polymer, is known as the **"degree of polymerization"**. Polymers with high degree of polymerization are termed as "high polymers" and those with low degree of polymerization are called oligopolymers.

13.1 CLASSIFICATION OF POLYMERS

Synthetic polymers have been classified into various groups, in various manners.

Classification of polymers

- (i) On the basis of Chemical structure
  - Homopolymer
  - Copolymer
- (ii) On the basis of Polymeric structure
  - Linear
  - Branched
  - Crosslinked
- (iii) On the basis of arrangement of monomers
  - Block copolymers
  - Graft copolymers
- (iv) On the basis of tacticity
  - Isotactic polymer
  - Syndiotactic polymer
  - Atactic polymer
- (v) On the basis of thermal behaviour
  - Thermoplastic
  - Thermoscts
- (vi) On the basis of ultimate form
  - Plastic
  - Elastomers
  - Fibers

(i) On the basis of the chemical structure of their backbone, polymers are of two types.

Homopolymer: If a polymer consists of identical monomer, the polymer is termed as homopolymer.

Copolymer: If the polymer is a mixture of more than one type of monomer it is termed as copolymer.

$$\sim A - A - A - A - A \sim \qquad \sim A - B - A - B - A - B \sim$$

Homopolymer Copolymer

(ii) On the basis of their polymeric structures, the polymers can be classified as linear, branched or crosslinked.

$\sim$A — A — A — A — A$\sim$

Linear Homopolymer

$\sim$A — B — A — B — A — B$\sim$

Linear Copolymer

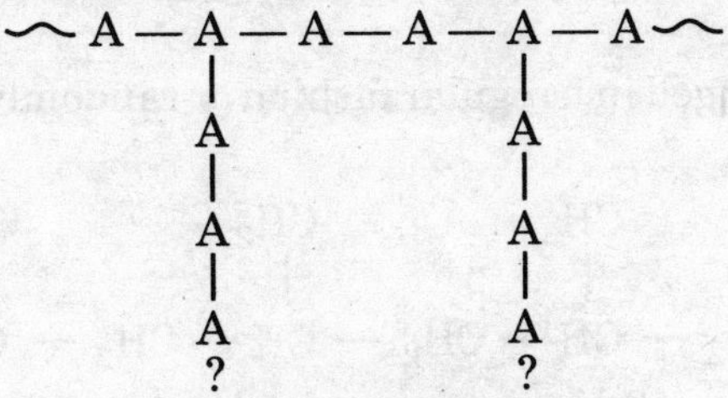

Branched Chain Homopolymer

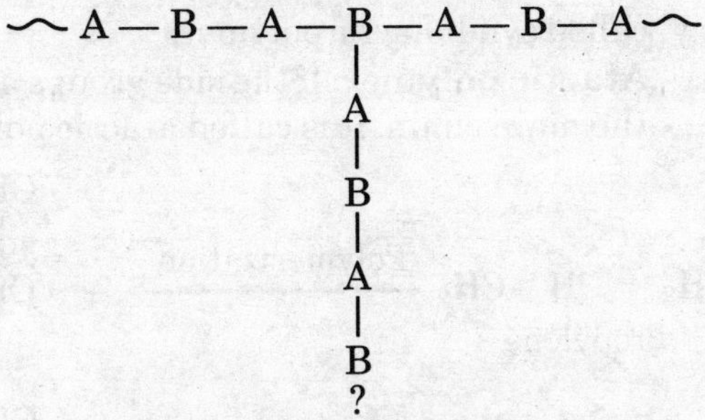

Branched Chain Copolymer

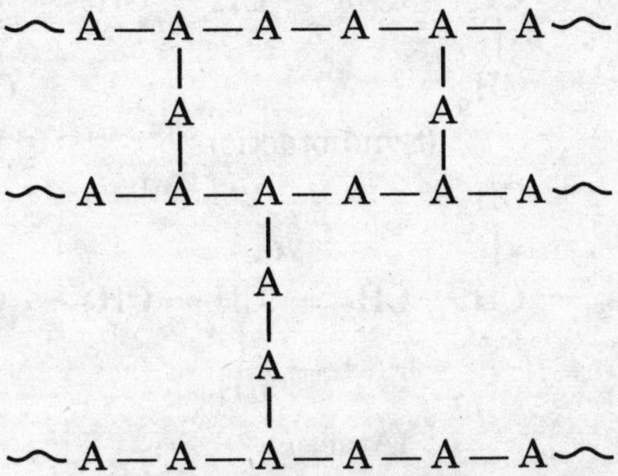

Cross Linked Homopolymer

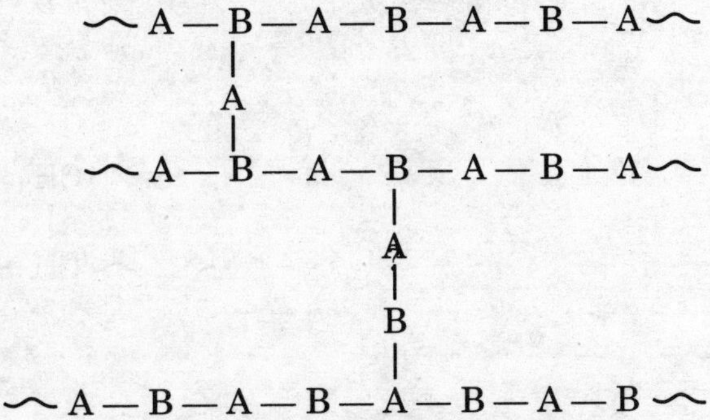

Cross Linked Copolymer

(iii) Depending upon the arrangement of the monomers in the case of copolymers, the polymers are divided into block copolymers and graft copolymers.

Block Copolymers: Linear polymers in which the identical monomeric units occur in relatively long sequences are called block co-polymers.

— A — A — A — B — B — B — A — A — A — B — B — B —

Block Copolymer

Graft Copolymers: Graft copolymers are branched copolymers in which the backbone is formed from one type of monomer and branches are formed of the other, i.e. the monomer segments on the branches and backbone are not the same.

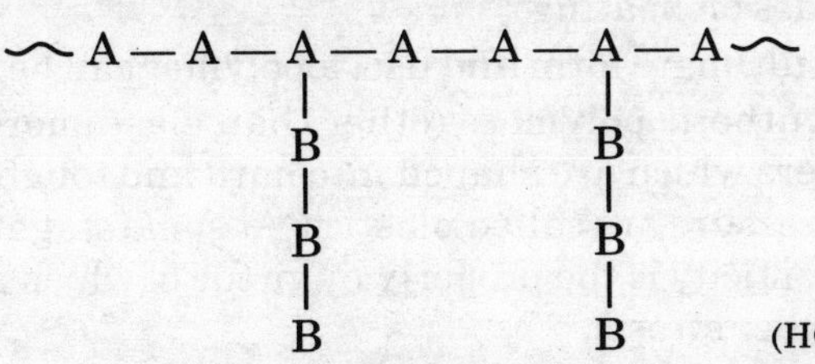

(HOMOCHAIN POLYMER)

(HETROCHAIN POLYMER)

Graft Copolymer

(iv) Depending upon the orientation of monomer units in a polymer molecule: With respect to the main chain the polymers are classified as:

Isotactic polymer: If the side groups of the monomers lie on the same side of the chain, it is called an "isotactic" polymer.

Syndiotactic polymer: If the side groups are arranged in an alternate fashion, it is called syndiotactic polymer.

Atactic polymer: If the side groups are arranged in irregular fashion or randomly around the main chain, it is called atactic polymer.

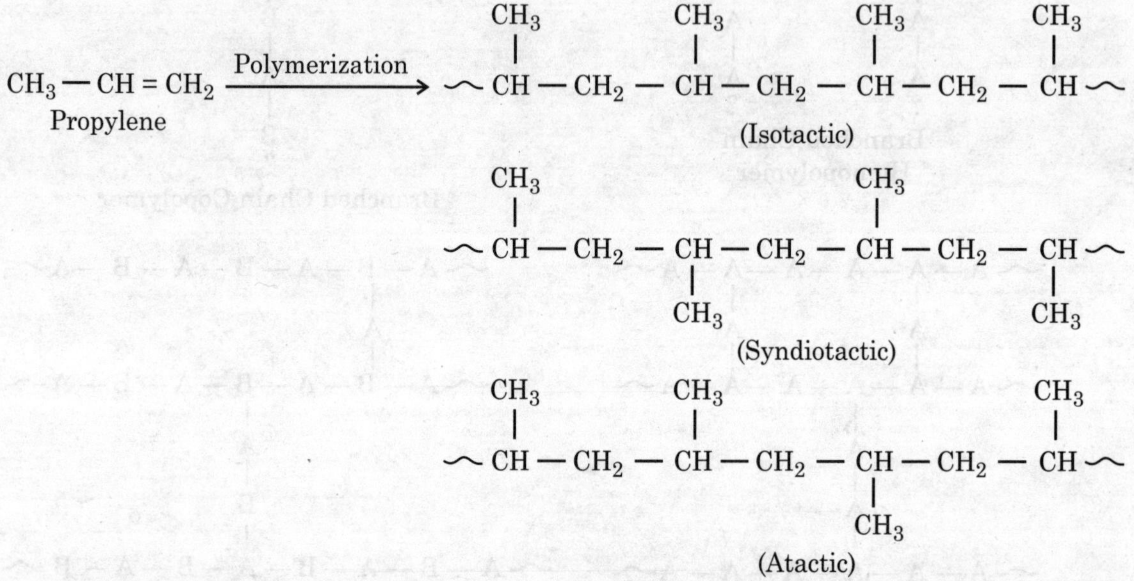

(v) If the main chain is composed of the atoms of the same specie, the polymer is called homochain and if the main chain is mads up of different atoms then they are called heterochain polymers.

$$\sim C-C-C-C-C- \sim$$

$$\sim C-C-O-C-O-C\sim$$

(vi) According to their behaviour on heating, the polymers are divided into two groups.

Thermoplastics: Those polymers which soften on heating and become plastic so that they can be converted to any shape by moulding are known as thermoplastics.

Thermosets: Those polymers which change irreversibly into hard and rigid materials on heating and cannot be reshaped, once they are set are called thermosets. They form infusible and insoluble mass on heating.

(vii) Depending on their ultimate form and use, a polymer can be classified as:

Plastics: All the synthetic polymers (other than elastomers) are usually referred to as plastics. The polymers which are shaped into hard and tough utility articles by the application of heat and pressure are called plastics. At some stage of manufacture, these are in plastic condition (Plasticity is the property by virtue of which a material undergoes permanent deformation under stress).

Elastomers: The polymers which are capable of being stretched rapidly at least 150 percent of their original length without breaking and return to their original shape on release of stress are known as elastomers. Synthetic rubbers constitute important examples of this group.

13.2 FUNCTIONALITY

The functionality of a molecule is the number of reactive sites it has. For a substance to act as a monomer, it must have two reactive sites. In otherwords, they should have a functionality of atleast two. A compound assumes functionality because of the presence of reactive functional groups like –OH, –COOH, $–NH_2$, –SH etc. The number of functional group present in a compound defines its functionality. For example-

Compound	*Chemical Formula*	*Functionality*
Acetic Acid	CH_3 COOH	1
Malonic acid	HOOC CH_2 COOH	2
Ethyl alcohol	C_2 H_5 OH	1
Ethylene glycol	$HOCH_2$ CH_2 OH	2
Lactic acid	CH_3 CH (OH) COOH	2
Tartaric acid	HOOC $(CHOH)_2$ COOH	4

Some compounds, however, do not contain any reactive functional groups but the presence of double or triple bonds in the molecules makes them bifunctional or polyfunctional. Hence these act as monomers. For example, in olefins the double bond can be considered as a site for two free valencies. When a double bond is broken, two single bonds become available for combination Thus,

$$CH_2 = CH_2 \longrightarrow —CH_2 – CH_2—$$

(i) 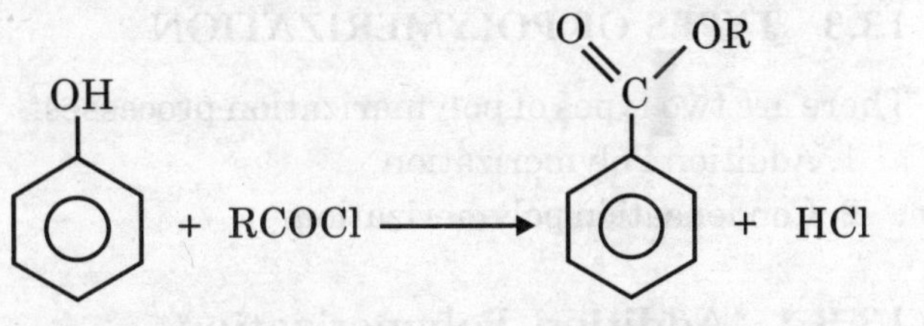

(ii) 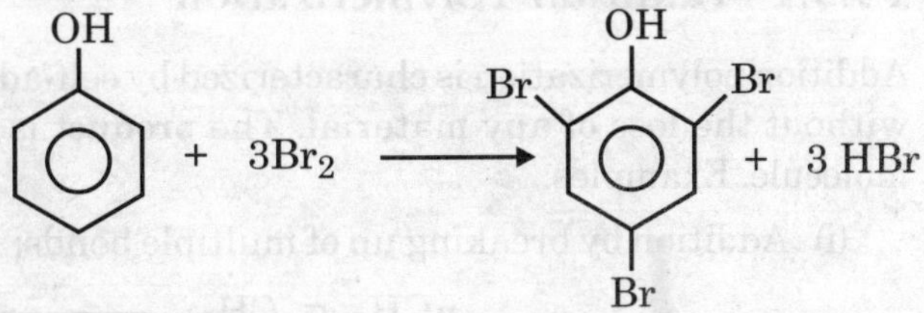

Acelytene, on the other hand, has a functionality of four.

$$CH \equiv CH \longrightarrow \text{---}CH = CH\text{-----} \longrightarrow \text{---}\ CH — CH\ \text{---}$$

There are some other compounds in which the presence of easily replaceable hydrogen atoms impart functionality. For example, phenol. It has got –OH groups as a functional group but can also undergo substitution reaction at three sites replacing its three hydrogen atoms. For example, in reaction (i), phenol exhibits monofunctionality and in reaction (ii), it exhibits a functionality of three.

Depending on the functionality of the monomers used, we get linear, branched or three dimensional cross-linked polymers.

(i) In case of bifunctional monomers, each molecule can add two more molecules and give a linear polymer as shown:

$$n \bullet A \bullet \longrightarrow \bullet A — A — A — A — A \bullet$$

Bifunctional Monomer   Linear polymer

(• represent functionality)

(ii) In case of trifunctional monomers, we get a cross-linked polymer:

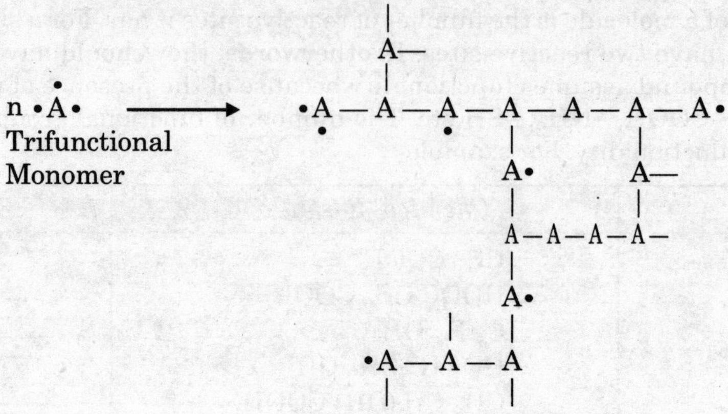

Cross-linked polymer

13.3 TYPES OF POLYMERIZATION

There are two types of polymerization processes:

1. Addition Polymerization,
2. Condensation polymerization.

13.3.1 Addition Polymerization

Addition polymerization is characterized by self-addition of the monomer molecules to each other, without the loss of any material. The product is an exact multiple of the original monomeric molecule. Examples:

(i) Addition by breaking up of multiple bonds:

$$\underset{\text{(ethene)}}{nCH_2 = CH_2} \longrightarrow \underset{\text{(Polyethene)}}{\left(CH_2 - CH_2 \right)_n}$$

(ii) Addition by opening up of ring structures:

$$CH_3OH + (n+1)\ \underset{\diagdown\ O\ \diagup}{CH_2 - CH_2} \longrightarrow CH_3O \left[CH_2 - CH_2 - O \right]_n CH_2CH_2OH$$

Vinyl compound ($CH_2 = CHX$), allyl compounds ($CH_2 = CH.CH_2X$), olefins ($CH_2 = CHR$) and dienes ($CH_2 = CR\text{–}CH = CH_2$) typically undergo addition polymerization. Since a majority of these fall under the 'vinyl' category, addition polymerization is also customarily termed as vinyl polymerization.

MECHANISM OF ADDITION POLYMERIZATION

Addition polymerization is a typical chain reaction and consists of three important steps:

(i) Initiation or the formation of active centres.

$$A \longrightarrow A^* \text{ (Excited momomer)}$$

(ii) Chain propagation

$$A^* + A \longrightarrow A - A^*$$

$$A - A^* + A \longrightarrow A - A - A^* \quad \text{(Activated growing chain)}$$

(iii) Chain termination or removal of active centre.

$$\sim A - A - A^* \longrightarrow \sim A - A - - A \text{ (Inactivated chain)}$$

The process of addition polymerization can be brought about by three methods:

1. Free radical mechanism
2. Ionic mechanism (Anionic and Cationic)
3. Coordination mechanism

The three types of mechanisms for addition polymerization, depending upon the active centres formed, are discussed below.

1. **Free Radical Addition Polymerization:** The free radical addition polymerization involves the formation of free radicals in the initiation step, which can be produced by decomposition of compounds called initiators. The free radical polymerization has three principal steps:

(i) **Initiation:** Initiation in a free radical polymerization consists of two steps. The first is the generation of the free radical species, followed by the second step which involves the addition of a single monomer molecule to the free radical.

The generation of free radical takes place as a result of homolytic dissociation of an initiator, which may take place in the presence of heat energy, light energy or catalysts. A number of low molecular weight compounds, comprising mainly of azo compounds, peroxides, hydroperoxides, peracids and peresters are useful as initiators. An important example of a free radical initiator is benzoyl peroxide which dissociates as:

$$C_6H_5\overset{\overset{O}{\|}}{C} - O - O - \overset{\overset{O}{\|}}{C} - C_6H_5 \xrightarrow{\Delta} 2C_6H_5 - \overset{\overset{O}{\|}}{C} - \dot{O}$$

In general, the dissociation of the initiator (R–R) to form free radical specie can be represented as

$$\underset{\text{(Initiator)}}{R - R} \longrightarrow \underset{\text{(Free radicals)}}{2R^*}$$

In the second step of initiation, the free radical so formed gets attached to the monomer molecule (M) to form the chain initiating species

$$\underset{\text{(free radical)}}{R^* + CH_2} = \underset{\text{Monomer}}{CH_2} \longrightarrow R - CH_2 - \dot{C}H_2$$

(ii) **Propagation:** In the next step, additional monomer units are added to the initiated monomer species, as.

$$R - CH_2 - \underset{\times}{\underset{|}{\dot{C}H}} + CH_2 = \underset{\times}{\underset{|}{CH}} \longrightarrow R - CH_2 - \underset{\times}{\underset{|}{CH}} - CH_2 - \underset{\times}{\underset{|}{\dot{C}H}}$$

$$R - CH_2 - \underset{\times}{\underset{|}{CH}} - CH_2 - \underset{\times}{\underset{|}{CH}} \quad CH_2 = \underset{\times}{\underset{|}{CH}} \longrightarrow R - CH2 - \underset{\times}{\underset{|}{CH}} - CH2 - \underset{\times}{\underset{|}{CH}} - CH2 - \underset{\times}{\underset{|}{\dot{C}H}}$$

The propagation step involves a continuing attack on fresh monomer molecules which, in turn, keep successively adding to the growing chain one after another.

(iii) **Termination:** Propagation will continue until some termination process takes place. The propagating polymer chain is terminated by any of the following reactions.

(*a*) *By combination:* Since the decomposition of the initiator produces many free radicals at the same time, each one of them can initiate and propagate the growth of a number of chains simultaneously. The two growing chains may come close and collide with each other and get deactivated.

$$R\!-\!\!\left[CH_2-\underset{\times}{CH}\right]_{n1}\!CH_2-\underset{\times}{\dot{C}H} + R\!-\!\!\left[CH_2-\underset{\times}{CH}\right]_{n2}\!CH_2-\underset{\times}{\dot{C}H} \longrightarrow$$

$$R\!-\!\!\left[CH_2-\underset{\times}{CH}\right]_{n1}\!CH_2-\underset{\times}{CH}-\underset{\times}{CH}-CH_2\!-\!\!\left[\underset{\times}{CH}-CH_2\right]_{n2}\!R$$

(*b*) *By disproportionation:* In some cases, one hydrogen atom from one growing chain is abstracted by the other growing chain and results in the formation of two polymer molecules, one saturated and other unsaturated. In this case the termination process results in the formation of two polymer molecules of shorter chain length as against a single molecule of a longer chain length obtained by first method.

$$R\!-\!\!\left[CH_2-\underset{\times}{CH}\right]_{n1}\!CH_2-\underset{\times}{\dot{C}H} + R\!-\!\!\left[CH_2-\underset{\times}{CH}\right]_{n2}\!CH_2-\underset{\times}{\dot{C}H} \longrightarrow$$

$$R\!-\!\!\left[CH_2-\underset{\times}{CH}\right]_{n1}\!CH_2-\underset{\times}{CH_2} + R\!-\!\!\left[CH_2-\underset{\times}{CH}\right]_{n2}\!CH=\underset{\times}{CH}$$

Thus the products so formed do not contain any reactive sites and may be termed as dead polymers as these cannot grow further.

(c) In addition to the above two methods of chain termination, chain transfer by hydrogen abstraction from solvent molecule, may also terminate the growing chain.

$$R\!-\!\!\left[CH_2-\underset{\times}{CH}\right]_{n1}\!CH_2-\underset{\times}{\dot{C}H_2} + \underset{\text{solvent}}{SH} \longrightarrow R\!-\!\!\left[CH_2-\underset{\times}{CH}\right]_{n1}\!CH_2-\underset{\times}{CH_3} + \dot{S}$$

In this reaction the original growing chain is terminated whereas the new chain gets initiated by the solvent free radical, hence this is called termination by chain transfer. Certain chemical compounds which are capable of inhibiting or completely stopping the chain growth process are called radical inhibitors. Phenolic compounds are widely used as inhibitors, because phenolic hydrogen can be easily abstracted by the polymer chain. The phenolic free radical so formed is highly stabilized and no further reaction is possible. Hydroquinone is often added as inhibitor.

Polythene, polypropylene, styrene etc are some of the polymers that can prepared by free radical polymerization.

Branching and crosslinking during the free-radical addition polymerization:

2. **Ionic Addition Polymerization:** Ionic polymerization follow the same basic steps as free-radical addition polymerization i.e initiation, propagation and termination. In this process, the initiation step involves the formation of carbonium ion or the carbanion as the active site and hence the polymerization can be termed as cationic or anionic polymerization respectively.

(i) **Cationic Polymerization**

Cationic polymerization takes place in the presence of catalyst that is regenerated at the end of polymerization (in contrast to free radical and anionic polymerization where the catalyst added become incorporated into the terminated polymer chain) Polymerization of vinyl monomers with an electron-donating group (such as $-OC_2H_5-OCH_3$, $-C_6H_5$ etc.) may take place by a cationic mechanism which involves the following steps.

Initiation: This step involves the formation of a carbonium ion by the attack of a proton on the monomer. Any strong Lewis acid like BF_3, $AlCl_3$ etc can be used as a catalyst. In this case a cocatlyst like water is required to provide the proton source.

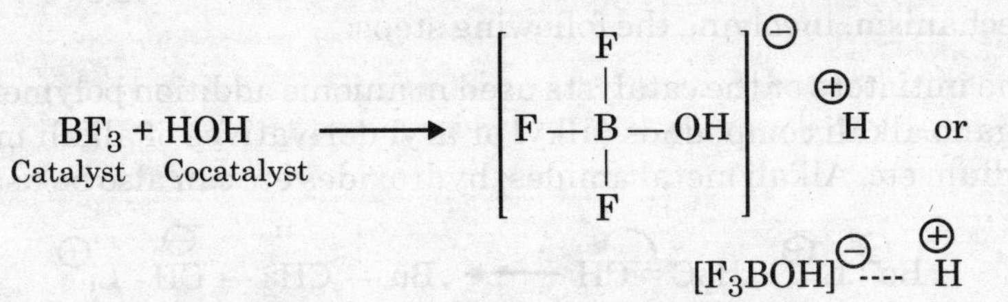

H^+ is the initiator and $[F_3\,BOH]^-$ is the counter ion.

The H^+ now attacks the π electron of the monomer and forms the carbonium ion. The counter ion $[F_3\,BOH]^-$ anion faces the growing chain end.

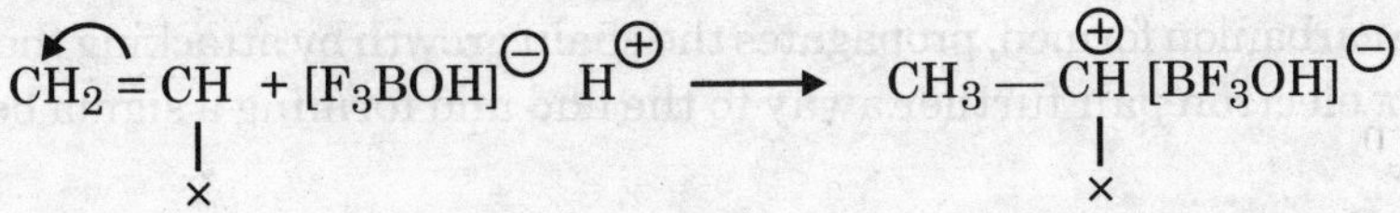

Propagation: The propagation reaction involves the addition of monomer units and simultaneous transfer of the charge to the newly added monomer unit.

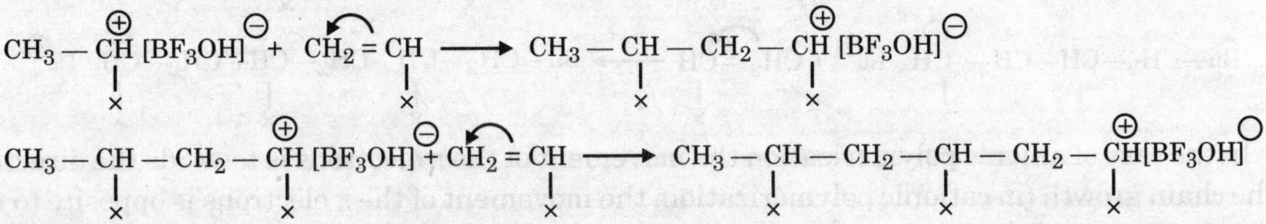

More and more units are added up and the chain keeps on growing. The electron pair of the adding monomer units are pulled in a direction opposite to the growth of the chain and the positive charge keeps on moving in the direction of the chain growth. The counter ion $[BF_3\,OH]^-$, moves along the positive charge all the time.

Termination: Termination takes place when the growing polymer chain having C^+ collides with an anion, which may be added deliberately or with the counter ion. The possible reactions may be:

(a) Donation of a proton to the counter ion by the growing polymer chain: This results in the formation of a double bond at the end of growing polymer molecule.

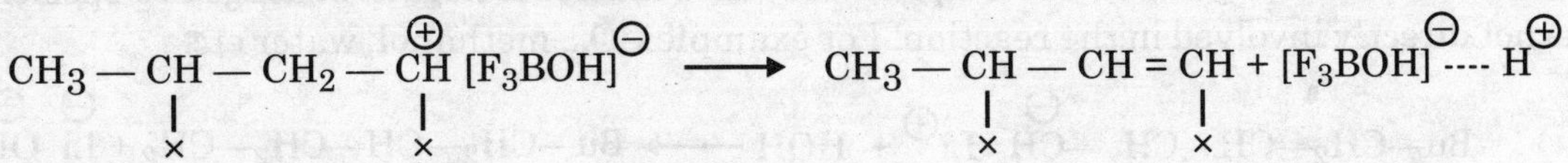

Thus the initiator or catalyst is regenerated.

(b) Formation of covalent bond between the carbonium ion and the counterion (coupling):

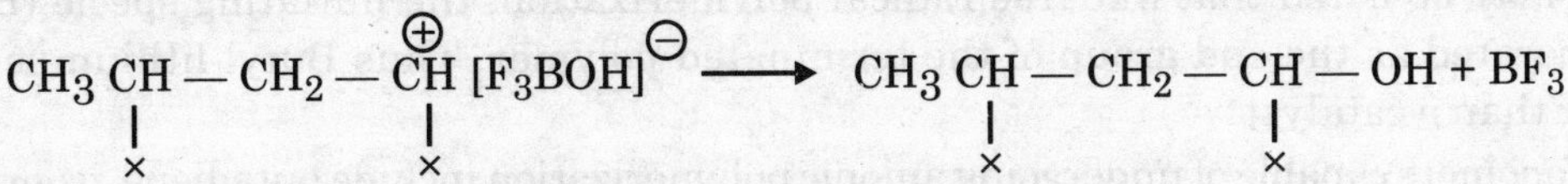

In this case also the regeneration of the initiator takes place.

Examples of the monomers that can undergo cationic polymerization are isobutylene, styrene, methyl styrene and many vinyl ethers. Cationic polymerization proceeds at a very fast rate even at low temperatures. For example, polymerization of isobutylene is completed within a few seconds at –100° C.

(ii) Anionic Polymerization

Anionic polymerization involves the formation of the carbanion in the initiation step. Polymerization of vinyl monomers with an electron withdrawing group can proceed by an anionic mechanism, involving the following steps:

Initiation: The initiators or the catalysts used in anionic addition polymerization are strong bases such as organo alkali compounds (alkyl or aryl derivatives of alkali metals) like n-butyl lithium, ethyl sodium etc. Alkali metal amides, hydroxides etc can also be used.

$$\overset{\ominus}{Bu}\ \overset{\oplus}{Li} + H_2C = \underset{\times}{\underset{|}{CH}} \longrightarrow Bu - CH_2 - \underset{\times}{\underset{|}{\overset{\ominus}{CH}}}\ \overset{\oplus}{Li}$$

The electron pair in this case is pushed to the end of the molecule forming the carbanion. Li^+ is the counter ion.

Propagation: The carbanion formed, propagates the chain growth by attacking the second monomer unit, pushing its π electron pair further away to the end and forming a sigma bond with the new monomer unit.

$$Bu - CH_2 - \underset{\times}{\underset{|}{\overset{\ominus}{CH}}}\ \overset{\oplus}{Li} + CH_2 = \underset{\times}{\underset{|}{CH}} \longrightarrow Bu - CH_2 - \underset{\times}{\underset{|}{CH}} - CH_2 - \underset{\times}{\underset{|}{\overset{\ominus}{CH}}}\ \overset{\oplus}{Li}$$

$$Bu-CH_2-\underset{\times}{\underset{|}{CH}}-CH_2-\underset{\times}{\underset{|}{\overset{\ominus}{CH}}}\ \overset{\oplus}{Li} + CH_2 = \underset{\times}{\underset{|}{CH}} \longrightarrow Bu-CH_2-\underset{\times}{\underset{|}{CH}}-CH_2-\underset{\times}{\underset{|}{CH}}-CH_2-\underset{\times}{\underset{|}{\overset{\ominus}{CH}}}\ \overset{\oplus}{Li}$$

In the case of anionic polymerization the movement of the π electrons is towards the direction of the chain growth (in cationic polymerization, the movement of the π electrons is opposite to the chain growth).

Termination: The termination in anionic polymerization is not a spontaneous process. If the starting reagents are pure and no impurities are present, propagation can proceed indefinitely or until all the monomer is consumed. Thus, anionic polymerization is also referred to as "living polymerization", because the carbanions at the chain ends remain potentially active. If a fresh quantity of monomer is added, polymerization again goes on until the added monomer is consumed.

Termination is generally accomplished by the transfer of negative charge to a species which is not directly involved in the reaction. For example CO_2, methanol, water etc.

$$Bu-CH_2-\underset{\times}{\underset{|}{CH}}-CH_2-\underset{\times}{\underset{|}{\overset{\ominus}{CH}}}\ \overset{\oplus}{Li} + HOH \longrightarrow Bu-CH_2-\underset{\times}{\underset{|}{CH}}-CH_2-\underset{\times}{\underset{|}{CH_2}} + \overset{\ominus}{Li}\ \overset{\oplus}{OH}$$

It may be noted that like free radical polymerization, the initiating specie (Bu) has been incorporated as the end group of the terminated polymer. Thus Butyl-lithium is an initiator rather than a catalyst.

Monomers capable of undergoing anionic polymerization include butadiene isoprene, styrene, acrylonitrile.

The strong electron withdrawing groups have stabilizing effect on the carbanion so formed and hence anionic polymerization takes place with the monomers having – CN, – COOR, and – CH = CH_2 groups.

Moreover, the reactivity of carbanions with oxygen, carbon dioxide is very high, hence the system should be free of these impurities.

The mechanism of anionic polymerization can be further illustrated by polymerization of acrylonitrile in the presence of butyl lithium.

Initiation:

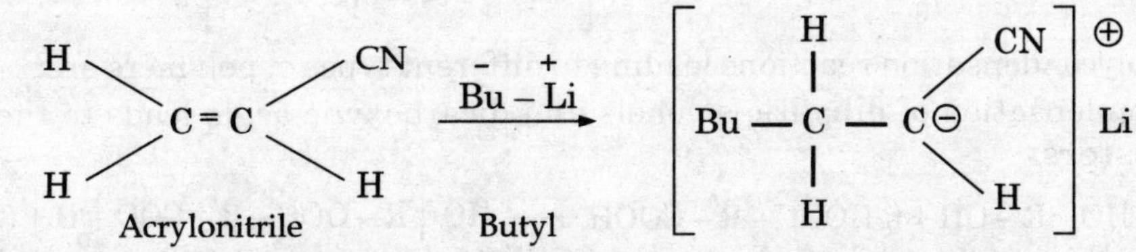

Acrylonitrile Butyl Lithium Stablized Anion

Propagation:

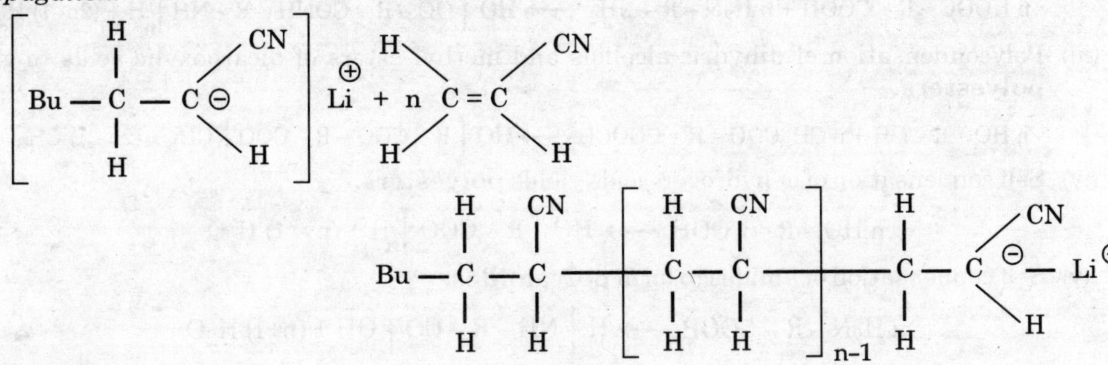

Termination:

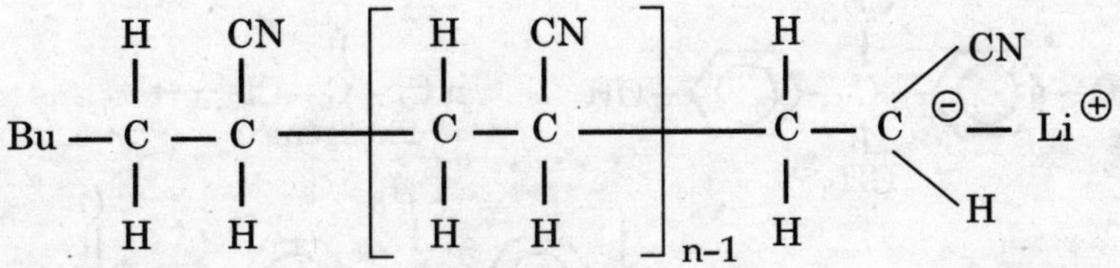

13.3.2 Condensation Polymerization

Condensation polymerization is brought about by monomers containing two or more reactive functional groups with the elimination of small molecules like H_2O, HCl etc. For example,

$$n\ HO-\overset{O}{\overset{\|}{C}}-(CH_2)_4-\overset{O}{\overset{\|}{C}}-OH + n\ H_2N-(CH_2)_6-NH_2$$

(adipic acid) (hexamethylene diamine)

$$HO\left[\overset{O}{\overset{\|}{C}}-(CH_2)_4-\overset{O}{\overset{\|}{C}}-NH-(CH_2)_6-NH\right]_n H + (2n-1)H_2O$$

nylon 6, 6

It is seen from the above reaction that the polymer formed still contains both the reactive functional groups at its chain ends and hence, is 'active' and not 'dead'.

In the above example, the different type of reactive functional groups, viz., $-NH_2$ and $-COOH$ are present on two different monomers. In some cases, both the reactive groups may be present on the same monomer and hence lead to self condensation reaction. For example, polycondensation of amino acids.

$$n\ H_2N \left(CH_2 \right)_6 COOH \longrightarrow H \left[HN - (CH_2)_6 - CO \right] OH + (n-1)\ H_2O$$

Some polycondensation reactions leading to different types of polymers are:

(i) Polycondensation of dihydric alcohols and dicarboxylic acids leads to the formation of **polyesters.**

$$n\ HO - R - OH + n\ HOOC - R' - COOH \longrightarrow HO \left[R - OOC - R' - COO \right]_n H + (2n-1)\ H_2O$$

(ii) Polycondensation of dicarboxylic acids and diamines to give **polyamides.**

$$n\ HOOC - R - COOH + n\ H_2N - R - NH_2 \longrightarrow HO \left[OC - R - CONH - R - NH \right]_n H + (2n-1)\ H_2O$$

(iii) Polycondensation of dihydric alcohols and methyl esters of dicarboxylic acids to give **polyesters.**

$$n\ HO - R - OH + n\ CH_3COO - R' - COOCH_3 \longrightarrow HO \left[R - COO - R - COO \right]_n CH_3 + (2n-1)\ CH_3OH$$

(iv) Self condensation of ω-hydrogen acids yields **polyesters.**

$$n\ HO - R - COOH \longrightarrow HO \left[R - COO \right]_n H + (n-1)\ H_2O$$

(v) Self condensation of amides to form **polyamides.**

$$n\ H_2N - R - COOH \longrightarrow H \left[NH - R - CO \right] OH + (n+1)\ H_2O$$

(vi) Polycondensation of bisphenol A and phosgene to form **polycarbonate.**

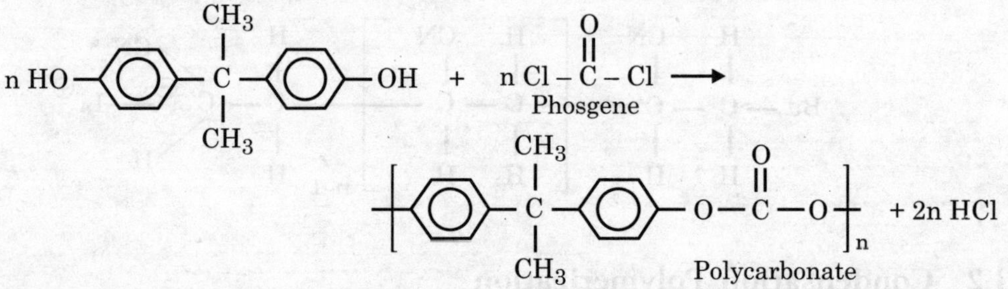

13.4 CHAIN GROWTH POLYMERIZATION AND STEP GROWTH POLYMERIZATION

Recently, another classification scheme based on polymerization kinetics has been adopted over the traditional addition and condensation categories. According to new scheme, all polymerization mechanisms are classified as either chain growth or step growth polymerizations.

Chain Growth Polymerization: During chain growth polymerization, high molecular weight polymer is formed early during polymerization and polymer yield gradually increases with time. Chain growth polymerizations require the presence of an initiating molecule that is used to activate the monomer molecule at the start of the polymerization. Most of the addition polymers are chain growth polymers and hence in most of the books the term chain growth and addition polymerization are used interchangeably.

Step Growth Polymerization: In step growth polymerization, high molecular weight polymer is formed only near the end of the polymerization. Most of the condensation polymers are step growth polymerization, but a number of important exceptions exist. In step growth polymerization, the reaction takes place in stepwise manner and polymer build-up is slow (unlike chain growth, where the polymer build-up is very rapid).

Apart from condensation reactions discussed earlier, step growth polymerization include some addition reactions also. Example, the non-condensation step growth polymerization of a polyurethane prepared by ionic addition of a diol (1, 4 butanediol) to a diisocyanate (1, 6-hexane diisocyanate.)

$$\underset{\text{1,4 butanediol}}{n\,HO-(CH_2)_4-OH} + \underset{\text{1,6 hexane diiso cyanate}}{O=C=N-(CH_2)_6-N=C=0}$$

$$\underset{\text{Polyurethane}}{\left[O-(CH_2)_4-O-\overset{\overset{O}{\|}}{C}-NH-(CH_2)_6-NH-\overset{\overset{O}{\|}}{C}\right]_n}$$

13.5 PLASTICS

As was discussed earlier, plastics are a group of artificially synthesized materials, which can be moulded into any desired form, when subjected to heat and pressure in the presence of a catalyst. At some stage in their manufacture or fabrication, these materials exist in plastic condition and hence their name.

The resins are the basic binding materials which form the major part of the plastics and determine the type of treatment needed in the moulding operations. However, there is no sharp line of demarcation between plastics and resins, and the two terms are considered synonymous. From engineering point of view there are two types of resins-

(i) Thermoplastic resins.
(ii) Thermosetting resins.

Plastic materials or the plastic articles depending on the bases of the type of resins used in their preparation, are called thermoplastics or thermosetting resins/plastics.

(i) **Thermoplastics:** Thermoplastics or resins soften on heating and become plastic so that they can be converted to any desired shape by moulding. On cooling, they become hard and rigid but their hardness is a temporary property subject to change with increase or decrease in temperature. In thermoplastic resins, the chemical structure or the molecular weight are not changed during the heating or moulding operations, because the changes involved are of physical nature. Only the secondary bonds between the individual molecular chains are broken on heating which results in their softening. On cooling, these secondary bonds are again formed and hence they become hard again.

Some common examples of thermoplastics are polyethene, polypropylene, polystyrene, polyvinyl chloride, teflon, polyacrylonitrile, etc.

Plasticizer: Sometimes, the substances called plasticizers are added to improve plasticity and flexibility of the plastics. The plasticizing properties are believed to be mainly due to the neutralization of intermolecular forces of attraction in the resin molecules. A variety of organic materials are used as plasticizers including non-drying vegetable oils, tributyl phosphate, triphenyl phosphate, triacetin, camphor, esters of oleic or stearic acid.

(ii) **Thermosetting resins:** Thermosetting resins are those which upon heating, change irreversibly into hard and rigid materials on heating. After cooling, if the set article is again heated, it will not soften again. The thermosetting resins during moulding acquire three dimensional cross linked structures with predominantly covalent bonds, which retain their strength even on heating. Thus, the three dimensional network structure joined by strong covalent bonds, make the thermosetting resins harder, stronger and more brittle than thermoplastic resins. If heating of a thermoset resin is prolonged, these undergo change in chemical composition to give a hard, infusible and insoluble mass.

Some common examples include phenol formaldehyde, malamine formaldehyde, alkyds, epoxy resins etc.

Table 13.1 Difference Between Thermoplastics and Thermosetting Resins

	Thermoplastic	*Thermosetting Resin*
1.	They are generally long chain linear polymers.	They have three dimensional crosslinked net work structure joined by strong covalent bonds.
2.	On heating, they soften readily because secondary forces between the chain can be easily broken.	They do not soften on heating because cross links and strong covalent bonds retain their strength on heating. They can, however, he charred on strong heating.
3.	They can be softened, reshaped and even reused by reheating to a suitable temperature.	They can not be reshaped and reused because they do not soften on heating and retain their shape and structure.
4.	They are generally weak, soft and less brittle.	They are usually hard, strong and more brittle.
5.	They can be reclaimed from waste.	They can not be reclaimed from waste.
6.	They can dissolve in some organic solvents.	They are insoluble almost in all organic solvents because of the presence of cross linking and strong covalent bonds.

13.6 COMMERCIALLY IMPORTANT THERMOPLASTICS

1. **Polyethene:** Ethylene, one of the most important petrochemicals, may be polymerized by a number of methods to produce thermoplastic polymer, polyethene.

$$\underset{\text{ethene}}{n\,CH_2 = CH_2} \longrightarrow \underset{\text{Polyethene}}{\left(CH_2 - CH_2 \right)_n}$$

The starting material (monomer) is obtained either by dehydration of alcohol or by the cracking of petroleum.

There are two varieties of polyethylenes, viz., low density polyethylene (LDPE) and high density polyethylene (HDPE).

(a) LDPE: Low density polyethylene is prepared by polymerizing ethylene at high pressures (1000-5000 atm) at a temperature of 250° C in the presence of oxygen as a free radical initiator. LDPE consists of highly branched structure which does not allow the polymeric chains to pack close together and hence results in low density (0.91to 0.925 g/cm^3) of the polymer.

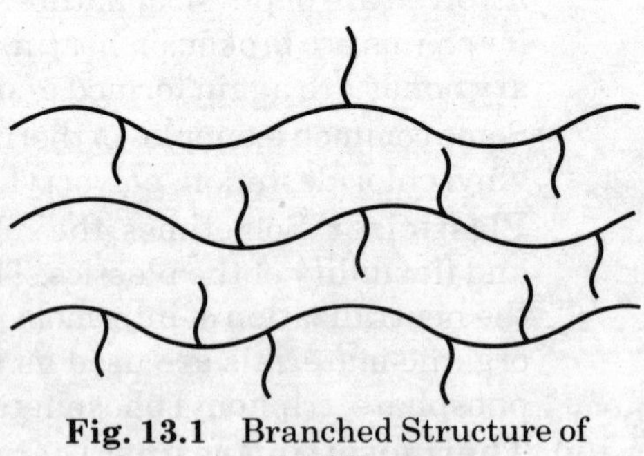

Fig. 13.1 Branched Structure of LDPE.

Properties:

(i) Low density polyethylene melts at 110–125° C and is only 40–50% crystalline.

(ii) It has low density (0.91 g/cm^3).

(iii) It is a transparent polymer having moderate tensile strength and high tonghness.

(iv) It is chemically inert, slightly flexible and poor electrical conductor.

(v) Its elasticity is preserved even at lower temperatures
(vi) It is moisture resistant.

Uses:

(i) LDPE films are mainly used for packing and wrapping frozen food, textile products etc because of their flexibility, chemical and moisture resistance.
(ii) The non-polar nature of the polymer makes it suitable for insulation for electric wires and cables.
(iii) Its inertness to chemicals and resistance to breakage makes it a useful material for the manufacture of 'squeeze bottles' and in many attractive containers.
(iv) It is also used for making pipes for agricultural, irrigation and cold water plumbing and chemical plants.

(b) **HDPE:** High density polyethylene is prepared by polymerizing ethylene at low pressure (1200-2000 atm) and at a temperature of 180–200° C in the presence of metal oxide catalysts. HDPE can also be produced by using Ziegler Natta catalyst by coordination polymerization. In this second method ethylene is heated to 60-200° C under a pressure of 6-7 atmospheres in the presence of a catalyst consisting of triethylaluminium and titanium tetrachloride (Ziegler-Natta catalyst)

$$n\,CH_2=CH_2 \xrightarrow[\substack{6\text{–}7\ \text{atm} \\ (C_2H_5)_3Al\,+\,TiCl_4}]{60-200°C,} \left(CH_2-CH_2 \right)_n$$

The polyethene so produced consists of linear chains of polymer molecules which pack close together resulting in high density polyethene (0.97 g/cm^3).

Properties:

(i) High density polyethene has melting point in the range of 145–150° C and is 90% crystalline.
(ii) It possesses density in the range of 0.941 to 0.965 g/cm^3.
(iii) It is much stiffer than LDPE and has higher tensile strength and hardness.
(iv) It is chemical resistant and has significantly lower gas permeability.
(v) Insulating and moisture resistant properties are similar to LDPE.
(vi) Due to high crystallinity, it has low gas and water vapour permeability.

Uses:

(i) HDPE is used in the manufacture of toys, crates closures, caps, buckets etc because it is more tough and stiffer than LDPE.
(ii) Its grease proof nature and chemical inertness makes it suitable for wrapping material for food products.
(iii) Low gas permeability makes it suitable for domestic gas piping.
(iv) It is also used for manufacturing bottles for milk, household chemicals and drug packaging.
(v) It is used in the manufacture of overhead tanks for water storage.

Whenever high tensile strength and stiffness are required HDPE finds better use than LDPE.

2. **Polypropylene:** Polypropylene is probably the lightest known industrial polymer and can be prepared in isotactic, syndiotactic or atactic forms. The monomer propylene is obtained as a byproduct from gasoline refineries. Polypropylene (isotactic) is produced by polymerizing propylene in the presence of Ziegler-Natta catalyst [Al (C$_2$ H$_5$)$_3$–TiCl$_4$)]

$$n\,\underset{}{\overset{\overset{\displaystyle CH_3}{|}}{CH}}=CH_2 \xrightarrow{Al(C_2H_5)_3-TiCl_4} \left(\overset{\overset{\displaystyle CH_3}{|}}{CH}-CH_2 \right)_n$$

Properties:

(i) Polypropylene is stereoregular (isotactic) and highly crystalline polymer (slightly lower than HDPE).
(ii) Isotactic polypropylene (IPP) melts at 160-170° C.
(iii) It exhibits high stiffness, hardness and tensile strength.
(iv) Its moisture resistance is comparable to that of polyethylene.
(v) Polypropylene is resistant to many chemicals such as alkalies, acids and oils. But due to presence of methyl groups attached to alternate carbon atoms, it is susceptible to oxidation.
(vi) Its electrical properties are similar to polyethylene.
(vii) Films of polypropylene are impermeable to a number of vapours and gases.

Uses:

(i) The higher melting point of polypropylene allows it to be used in products that must be steam sterilized such as baby bottles, flasks etc.
(ii) Its fibres are extremely strong by weight and hence are used in making ropes and filaments.
(iii) Polypropylene films are used for packaging due to high clarity and their impermeable nature to a number of vapours and gases. It is also used for wrapping confectionary items (bread, biscuits etc) and potato chips.
(iv) Due to high strength, rigidity and temperature and chemical resistance it is used for the production of chemical process equipments such as pipes and tanks.
(v) It is used for the preparation of textile machinery parts (bobbins, spools, spindles etc).
(vi) It has excellent resistance to continued flexing and hence is used for joints in boxes, containers etc.
(vii) When blended with suitable elastomer it is widely used for making bumpers, radiator grills etc.
(viii) It is also used as an insulating material for electrical wires.
(ix) It is also used for the manufacture of drinking glasses, dinner wares etc.

3. **Polystyrene:** Polystyrene is also known as polyvinyl benzene. The monomer styrene is produced from ethylene and benzene. When ethylene is passed into benzene in the presence of $AlCl_3$ catalyst, ethyl benzene is formed, which on passing over iron oxide catalyst at high temperatures is converted into styrene.

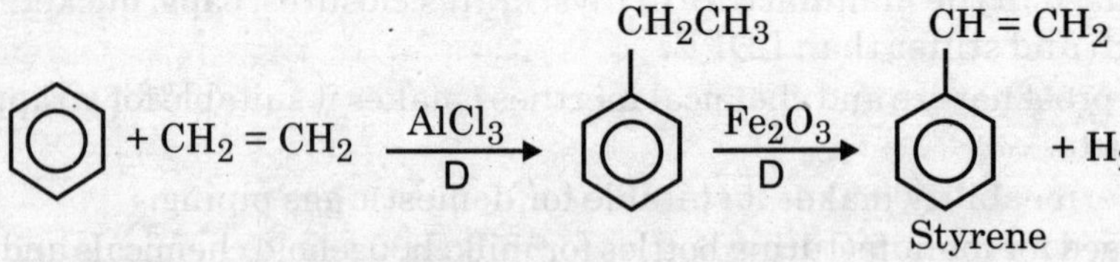

Polystyrene is prepared by polymerization of styrene in the presence of benzoyl peroxide as initiator.

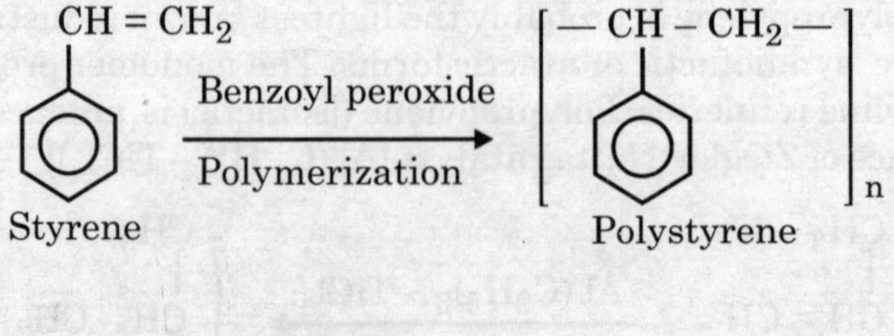

Properties:

(i) Polystyrene is transparent and has high refractive index (1.6) which gives it high brilliance.

(ii) It has got low heat distortion temperature (85° C) and hence articles made of polystyrene cannot be sterilized with steam.

(iii) It consists generally of linear chains and is chemically inert and resistant to most alkalies, oxidising and reducing agents and acids.

(iv) It is susceptible to photooxidative degradation, which results in brittleness and yellowing. Degradation is initiated by the phenyl groups, which absorb UV radiation from sunlight resulting in the formation of free radicals.

(v) It has got excellent moisture resistance.

(vi) It has good electrical insulating properties.

(vii) It can be dyed to different colours.

(viii) It has the unique property of transmitting light of all wavelengths through curved sections.

(ix) Polystyrene is attacked by a number of solvents.

Uses:

(i) Polystyrene is widely used in the manufacture of articles such as lids, jars, bottles, combs, brush handles, radio and television cabinets.

(ii) It is used for preparing synthetic ion exchange resins in combination with p-divinyl benzene. The resulting crosslinked polymer on sulphonation gives cation exchange resin.

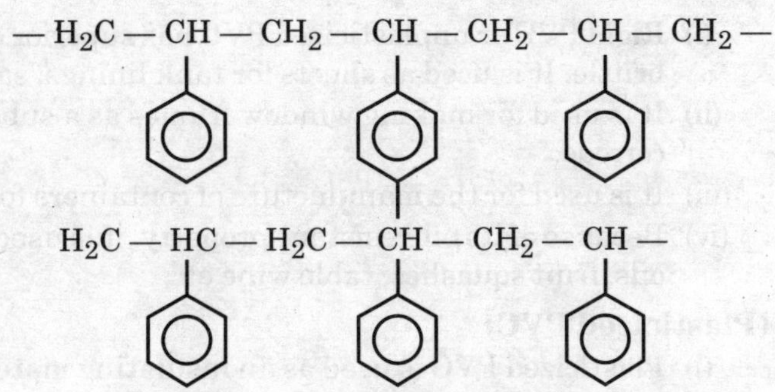

styrene-divinyl benzene copolymer

whereas, when styrene-divinyl benzene copolymer is treated with formaldehyde and HCl in presence $ZnCl_2$ as catalyst, it results in anion exchange resin.

(iii) An important use of polytyrene is the manufacture of foam and bead for insulation and packaging material. The polymer is foamed by using blowing agents, usually pentane. This foamed expandable polystyrene (EPS) is used for disposable drinking cups, cushioned packaging (for delicate equipment such as cameras and electronic equipments), and thermal insulation.

(iv) High-impact polystrene (HIPS), produced by combination of polystyrene with rubbery material such as polybutadiene (SBR), contributes to improved impact properties and is used for door liners, inner liners of refrigerators, etc.

(v) Polystyrene copolymerized with acrylonitrite and butadiene (ABS), due to its high impact resistance and high dimensional stability at elevated temperatures is used in the manufacture of door handles, radiator grills, loudspeaker housings etc.

(vi) Styreneacrylonitrile (SAN) copolymer is a transparent plastic with a very good impact strength and is used for moulding crockery items and machine components.

4. **Polyvinyl Chloride (PVC):** Polyvinyl chloride is one of the cheapest and most widely used plastic globally. The monomer vinyl chloride is a gas, boiling at –14° C. Polyvinyl chloride is prepared by the polymerization of vinyl chloride by heating its water emulsion in presence of a small amount of benzoyl peroxide.

$$\underset{\text{Vinyl chloride}}{CH_2=\underset{|}{\overset{Cl}{C}}H} \xrightarrow{\text{Polymerization}} \underset{\text{Polyvinyl chloride}}{\left[-CH_2-\overset{Cl}{\overset{|}{C}}H-\right]_n}$$

Polyvinyl chloride is hard and horny material owing to the high intermolecular attraction forces, but when plasticised with tricresyl phosphate it becomes very useful product. Thus, PVC is available in two general grades, viz, rigid and flexible.

Properties:

(i) It is colourless, odourless and non inflammable.
(ii) PVC is partially syndiotactic and have low crystallinity.
(iii) PVC is thermally not very stable and degrades above 200° C, with the evolution of HCl.
(iv) It is resistant to atmospheric oxygen, inorganic acids and alkalies.
(v) It has greater abrasion and flex resistance compared to leather.
(vi) It is water proof and washable.
(vii) It is less brittle and lighter when compared to glass.

Uses: (Rigid PVC)

(i) Rigid PVC or unplasticized PVC has superior chemical resistance and high rigidity but is brittle. It is used as sheets for tank linings, safety helmets, mudguards.
(ii) It is used for making window frames as a substitute for wood because the frames do not corrode.
(iii) It is used for the manufacture of containers for detergents, cosmetics, mineral water.
(iv) Because of its oil-resistant property, it is used for storing consumable liquids like edible oils, fruit squashes, table wine etc.

(Plasticized PVC)

(i) Plasticized PVC is used as an insulating material.
(ii) Because it is washable and has greater abrasion resistance therefore it is used for making ladies-handbags, bathroom curtains, kitchen upholstery.
(iii) Used for upholstery for seats in public transport vehicles.
(iv) In the film form it is used in membrane separation technology.

Properties of PVC can be modified through chemical modification or copolymerization. For example chlorinated PVC (CPVC) increases the heat-distortion temperature of PVC which can be used for the manufacture of hot water pipe.

5. Polyvinylacetate (PVA): Polyvinyl acetate is produced by heating vinyl acetate in the presence of small amount of benzoyl peroxide.

$$\underset{\text{Vinyl acetate}}{n\,CH_2=CH-COOCH_3} \xrightarrow{\text{Polymerization}} \underset{\text{Polyvinyl acetate (PVA)}}{\left(CH_2-\overset{COOCH_3}{\overset{|}{C}}H\right)_n}$$

Properties

(i) Polyvinyl acetate is a solid, clear resin, and has no taste or odour and is soft sticky material.
(ii) It is insoluble in water and is saponified with great difficulty and hence is not absorbed by the digestive system. Moreover if at all saponification takes place, the products polyvinyl alcohol and acetic acid are not harmful. Due to the above stated reasons, polyvinyl acetate is used as a basic material for chewing gums and surgical dressings.
(iii) It is resistant to atmospheric oxygen, water and chemicals.

Uses

(i) Polyvinyl acetate is used for the manufacture of lacquers, paints and adhesives. Polyvinyl acetate adhesives are used for bonding of textile to textile, leather to leather, paper to paper, paper to textile, leather to wood etc.

(ii) It is used as a basic material for chewing gums and surgical dressings.

(iii) It is also used for finishing of textile fabrics.

(iv) On copolymerization with polyvinyl chloride, it is converted into a polymer having varying degree of softness. For example, vinylite which is a copolymer of polyvinyl acetate and polyvinyl chloride, is especially useful as a surface coating material for metals.

(v) It is mainly used for the manufacture of polyvinyl alcohol.

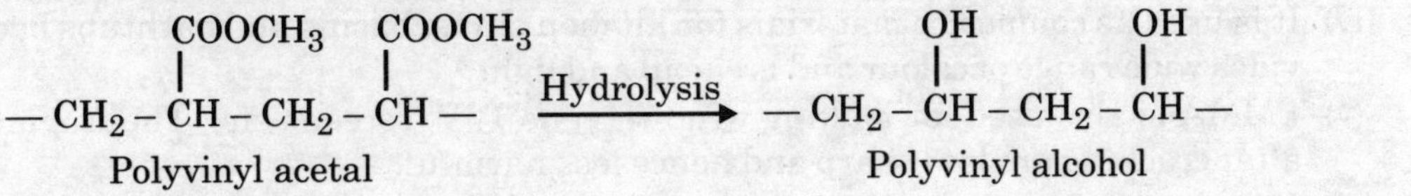

Polyvinyl alochol is used in the preparation of polyvinyl butyral on condensation of polyvinyl alcohol with butyraldehyde.

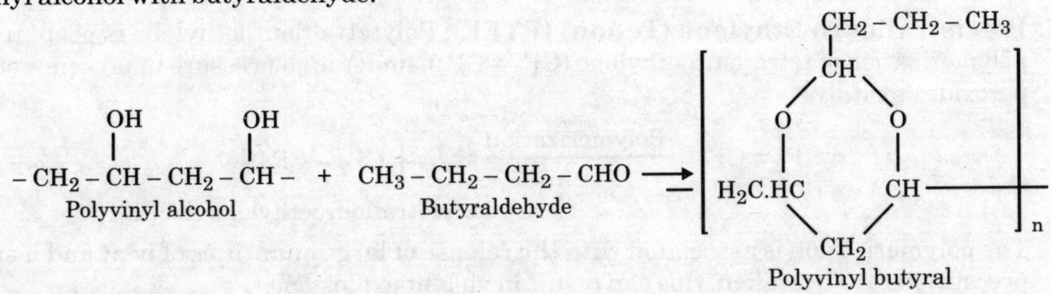

Polyvinyl butyral is tough and elastic and is used for laminating safety glass. It holds the broken pieces of glass in case of accident and thus minimizes the danger from flying glass fragments.

6. Polymethylmethacrylate (PMMA): Polymethyl methacrylate (PMMA) is one of the important thermoplastic from commercial point of view, because of its outstanding optical properties. It is prepared by the polymerization of methylmethacrylate in presence of free radical initiator such as peroxides

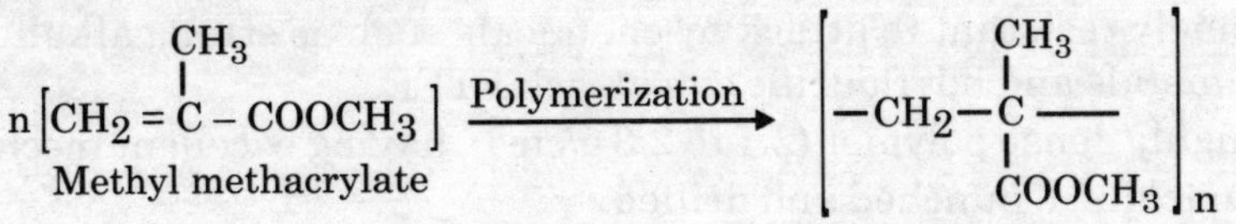

Polymethylmethacrylate

Properties

(i) It is a colourless transparent plastic with an excellent outdoor life period and good strength.

(ii) Owing to the presence of bulky side groups, it is amorphous in nature.

(iii) It can be depolymerized to yield back the monomer.

(iv) It has high softening point of about 130-140° C but it becomes rubber-like at a temperature above 65° C.

(v) The main feature of this plastic is its high optical transparency and capacity of transmitting light accurately, even in curved sections.

(vi) It has good mechanical properties but has poor scratch resistance as compared to glass.
(vii) It has low chemical resistance to hot acids and alkalies.
(viii) It has excellent weathering properties.

Uses

(i) PMMA is the nearest alternative to glass as it does not shatter and can be moulded into any shape. Infact, PMMA sheets known as plexiglass or lucite are used as an alternative to glass.
(ii) PMMA finds major use in automotive industry (e.g. housing for rear lamps, profiles, light fittings, ceiling lighting signal-light etc).
(iii) It is used for making hard contact lenses due to its high transparency.
(iv) It is used as composite materials for kitchen sinks, basins and bathtubs because it provides wide range of colour and is cheap and light.
(v) PMMA is also used for making wind screens T. V. screens etc. The fragments formed after cracking are less sharp and hence less harmful.
(vi) It is also used for decorative purposes such as in jewellery etc.
(vii) It is also used in bone splints, artificial eyes, dentures etc.

7. Polytetraflouro Ethylene (Teflon) (PTFE): Polytetra flouroethylene is prepared by the polymerization of tetraflouroethylene ($CF_2 = CF_2$), under high pressure in presence of benzoyl peroxide as catalyst.

$$n\,CF_2 = CF_2 \xrightarrow{\text{Polymerization}} \left(CF_2 - CF_2 \right)_n$$

Polytetraflouroethylene (PTFE)

The polymerization is associated with the release of large quantities of heat and if sufficient precautions are not taken, this can result in violent explosions.

Properties

(i) PTFE has a highly regular structure and is highly crystalline.
(ii) Due to the presence of highly electronegative flourine atoms and regular configuration of PTFE molecule there are very strong attractive forces between different chains, which give the material extreme toughness.
(iii) PTFE has high softening point of about 327° C. Above 400° C it decomposes without changing into vapour.
(iv) It is extremely resistant to attack by chemicals such as strong alkalies and acids. Only hot alkali metals and hot flourine can attack PTFE.
(v) PTFE is highly dense polymer (2.1 to 2.3 g/cm^3), having excellent mechanical properties.
(vi) It can be machined, punched and drilled.
(vii) It has a very low coefficient of friction.
(viii) It has extremely good electrical insulating properties.
(ix) It is slippery and waxy to touch.
(x) It does not melt like other thermoplastics and hence behaves somewhat like a thermosetting polymer.
(xi) It has non-adhesive characteristics.

Uses

(i) Due to its chemical inertness, it is used in chemical carrying pipes, tubings, stop-cock for burettes.

(ii) Its extreme inertness and stability to temperature makes it a suitable material for making gaskets and seals.
(iii) It is used as an insulating material for transformers, cables, wires, fittings etc.
(iv) Because of its low coefficient of friction, it is used in non-lubricating bearings.
(v) Its most important application is, as non-stick coating for pots and pans. PTFE is applied as a coating over the rollers and mixing equipments used for preparing sticky materials such as adhesives and jams.
(vi) Its fibre is used to form belts, filter cloth etc.

8. **Polycarbonates (PC):** The polycarbonates are amorphous polymers with interesting engineering properties. Polycarbonates are the polyesters of phenol and carbonic acid and are prepared by the condensation of diphenoxy methylene derivatives with diphenyl carbonate. Bis phenol-A [2, 2–bis (4–hydroxyphenyl) propane] on interaction with diphenyl carbonate, gives a polycarbonate as follows:

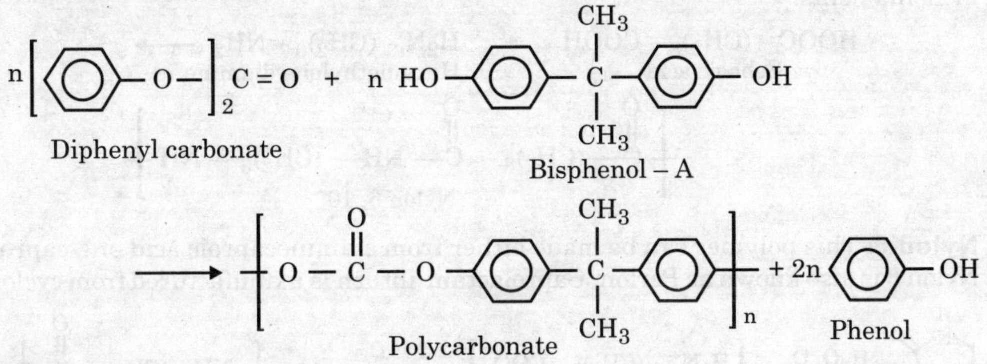

Properties

(i) Polycarbonate is an amorphous polymer.
(ii) It has very high impact strength.
(iii) It is resistant to water.
(iv) It has high light transmittance.
(v) Polycarbonate melts at around 265° C.
(vi) It has good dimensional stability.
(vii) It has good electrical insulation characteristics.

Uses

(i) Polycarbonate, due to its electrical properties and dimensional stability is used for mouldings for computers, calculating machines, for moulding compact disks etc.
(ii) It is used for the manufacture of hair drier, and electric razor housing.
(iii) As it is not stained by the usual domestic beverages and fruit juices, it is used for making food processor bowls, coffee maker and food mixer housings.
(iv) Due to its high temperature stability and toughness it is used for microwave cookware and transparent baby bottles.
(v) It is also used in cameras (lens holders, shutter assembly etc) safety shields, and for the manufacture of unbreakable caps of pens.

9. **Polyamides:** Polyamides are a group of polymers which contain the amide (–CONH–) linkage in the main polymer chain. Polymers of this type are the synthetic linear aliphatic polyamides which are capable of fibre formation.

Nylon is the generic name for synthetic polyamides capable of forming fibres. Polyamides (nylons) are prepared by the polycondensation between dicarboxylic acids and diamines. Nylons are designated by two numbers, the first representing the diamine. Polyamides of commercial importance are Nylon 6, 6; Nylon 6, 10, Nylon 6 etc.

(i) **Nylon 6,6:** Nylon 6,6 is prepared by the polycondensation of hexamethylene diamine with adipic acid.

$$\underset{\text{Adipic acid}}{HOOC-(CH_2)_4-COOH} + \underset{\text{Hexamethylene diamine}}{H_2N-(CH_2)_6-NH_2} \longrightarrow$$

$$-\overset{O}{\overset{\|}{C}}-(CH_2)_4-\overset{O}{\overset{\|}{C}}-NH-(CH_2)_6-NH-$$

(ii) **Nylon 6, 10:** Nylon 6, 10 is prepared by the condensation between hexamethylene diamine and sebacic acid. Nylon 6, 10 is generally not used as a fibre, but it is suitable for monofilaments.

$$\underset{\text{Sebacic acid}}{HOOC-(CH_2)_8-COOH} + \underset{\text{Hexamethylene diamine}}{H_2N-(CH_2)_6-NH_2} \longrightarrow$$

$$\underset{\text{Nylon 6, 10}}{\left[-\overset{O}{\overset{\|}{C}}-(CH_2)_8-\overset{O}{\overset{\|}{C}}-NH-(CH_2)_6-NH-\right]_n}$$

(iii) **Nylon 6:** This polymer can be made either from ε-aminocaproic acid or ε-caprolactam. Nylon 6 is also known as Perlon. Caprolactam inturn is manufactured from cyclohexane.

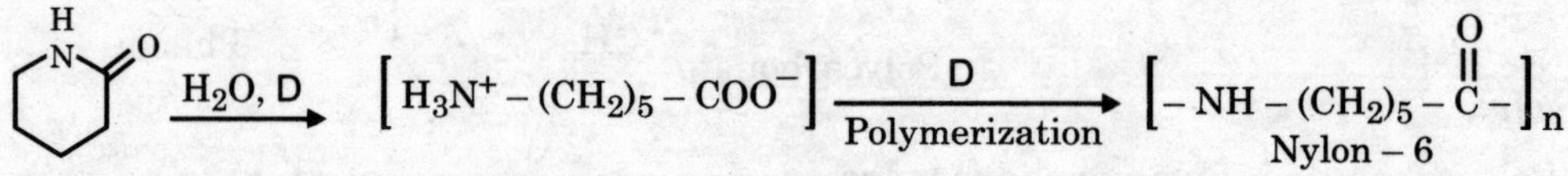

(iv) **Nylon-11:** It is prepared by self condensation of ω-amino undecanoic acid.

$$\underset{\text{– Aminoundecanoic acid}}{n[H_2N-(CH_2)_{10}-COOH]} \xrightarrow{\text{Self condensation}} \underset{\text{Nylon-11}}{\left(HN-(CH_2)_{10}-\overset{O}{\overset{\|}{C}}\right)_n}$$

Properties of Nylons

(i) Nylons are characterized by a combination of high strength, elasticity, tonghness and abrasion resistance.

(ii) They have very good moisture resistance.

(iii) They are insoluble in common solvents.

(iv) The molecular chains are held together to each other by hydrogen bonds having linear structure, permitting side-by-side alignment.

(v) They are very flexible and retain their original shape after use.

(vi) Nylons provide excellent resistance to wear and abrasion.

(vii) Due to the presence of hydrogen bonding, all nylons are water sensitive. Water acts as plasticizer, which reduces tensile strength and dimensional stability.

Uses

(i) Nylon 6, 6 is used as a plastic as well as a fibre. Its most important fibre applications include automobile tire cords, ropes, threads, cords having high tenacity and good elasticity.

(ii) Nylon 6,6 is also used to make textile fibres for use in dresses, undergarments, socks etc.

(iii) Nylon 6,6 being a tough plastic, is used as a substitute for metals in gears and bearings etc. Nylon 6,6 can be moulded in one piece whereas similar item in metal require assembling of several parts. Moreover, nylon bearings and gears perform quietly and need little or no lubrication.

(iv) It is also used in making rollers, slides and door latches.

(v) Nylon 6, 10 is not used much as synthetic fibre, but is used to manufacture articles like brushes and bristles.

(vi) Nylon 11, is used as textile fibre.

(vii) Due to its good barrier properties, nylon-6 film is used for packaging some oxygen sensitive food.

(viii) Due to its high tenacity, nylon is suitably used for parachute fabric.

(ix) Nylons are also used in making transmission belting, ribbons for electric-typewriters, spindle bands on textile spinning machines etc because of its high durability.

(x) Nylon is also used for making insect screens because of their biological resistance.

Aromatic polyamides, often called aramids have been produced to improve the heat and flammability resistance of nylons. These polymers have very high melting points (greater than 500° C). One such nylon is Kevlar. It is prepared by polycondensation between aromatic dichloride and aromatic diamines.

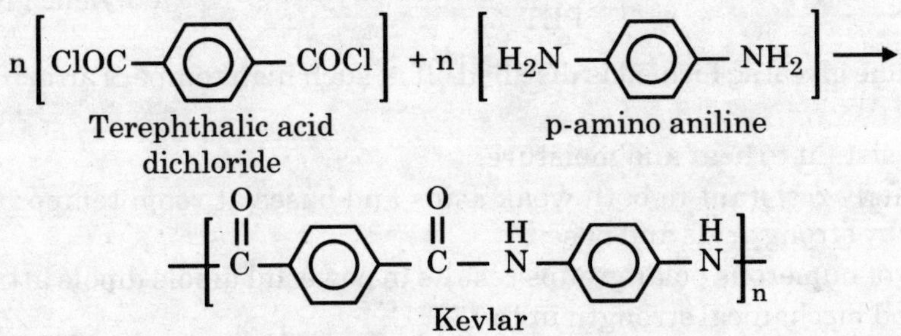

Kevlar is exceptionally strong and decomposes only above 500° C. It is 5 times stronger than steel and it is used as a substitute for steel in belted radial tires as well as in bullet resistant vests. It is also used in making car parts such as tyres, brackes, clutch linings etc.

13.7 COMMERCIALLY IMPORTANT THERMOSETTING RESINS

1. **Polyesters:** Polyester resins are the condensation products of dicarboxylic acid with dihydroxy alcohols. The industrially most important polyster is poly (ethylene terephthalate) (PET), with the trade name Terylene or Dacron. PET can be prepared from terephthalic acid and ethylene glycol, either by direct esterification or by catalyzed ester inter change.

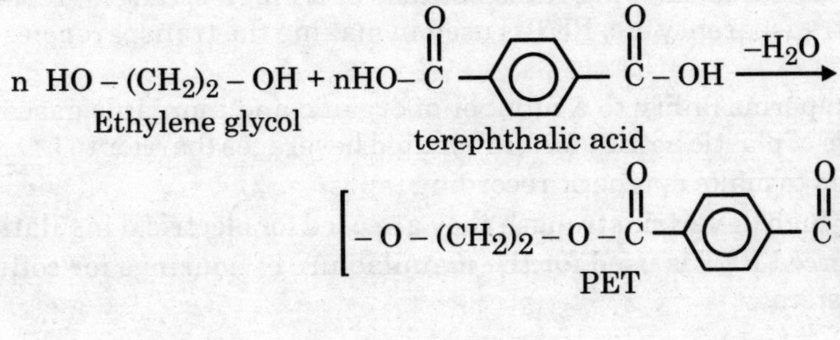

Now a days, dimethyl terephthalate is used instead of terephthalic acid. This is because of the fact that terephthalic acid is nonvolatile and is very slightly soluble in most of the solvents. The reaction takes place in two successive stages:

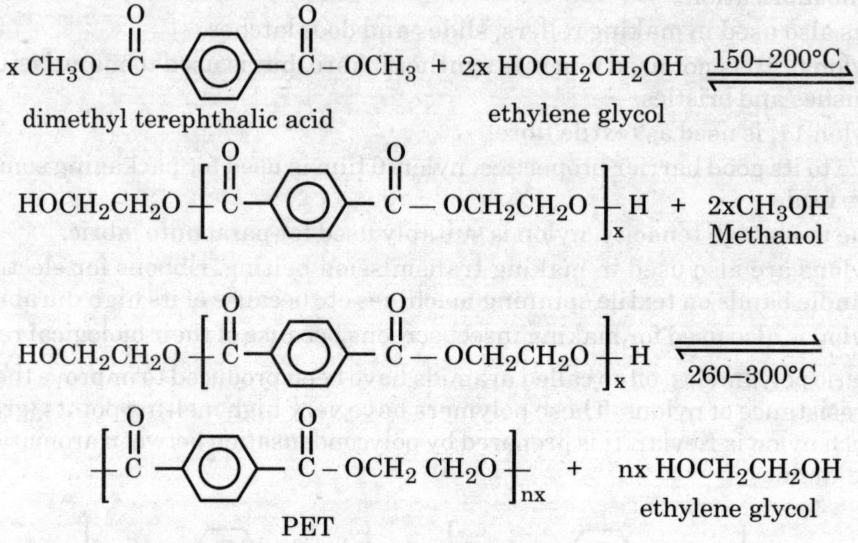

The enthylene glycol so formed is distilled off at such high temperature.

Properties

(i) PET is resistant to heat and moisture.

(ii) It is strongly resistant to both weak acids and bases at room temperature but readily attacked by strong acids and bases.

(iii) Presence of numerous polar groups results in powerful dipole dipole attractions in it.

(iv) It has good mechanical strength upto 175° C.

(v) It is resistant to almost all kinds of insect attacks.

(vi) It can be dyed at 100° C or in presence of a carrier.

(vii) The fibre has good wrinkle resistance due to good elasticity.

(viii) PET fibres are abrasion and oxidation resistant.

Uses:

(i) PET is mostly used for making synthetic fibres like Terylene, Dacron etc. Garments made from its fibres resist the formation of wrinkles and have many desirable properties such as warmth to touch, low moisture absorption (and hence dry rapidly), crease resistance, shape retention, and mildew proof. Moreover, these may be blended with other fibres, such as wool and cotton.

(ii) PET is also useful for making films because of its high strength, resistance to tearing durability, transparency etc. PET is used in making the transparencies of overhead projector.

(iii) Due to its impermeability to a number of organic and inorganic gases it is used in the manufacture of plastic bottles for carbonated beverages (barrier to CO_2 and water)

(iv) It is also used to make magnetic recording tapes.

(v) Due to their high dielectric strength they are used for electrical insulation.

(vi) Glass reinforced PET is used for the manufacture of housings for coffee machines, car heaters, toasters.

2. Phenolic Resins: Phenol formaldehyde resins are the most important and extensively used synthetic plastics and perhaps the oldest synthetic plastics. In these resins one of the monomers is always a phenol molecule. One of the most important member of this class is *phenol formaldehyde resin.*

The phenol formaldehyde reaction is catalyzed by alkalies and acids. The nature of the product depends on several factors, the major one being the nature of the catalyst and the proportion of the reactants (phenol formaldehyde ratio). The formation of phenol formaldehyde resin comprises of the following steps:

(i) **Methylolation:** When phenol and formaldehyde react together, the first step is the entry of methylol (CH_2 OH) groups in ortho and para positions to the hydroxy group. The reaction takes place in the presence of acid or alkali. Depending upon phenol formaldehyde ratio, various phenol-alcohols may be formed.

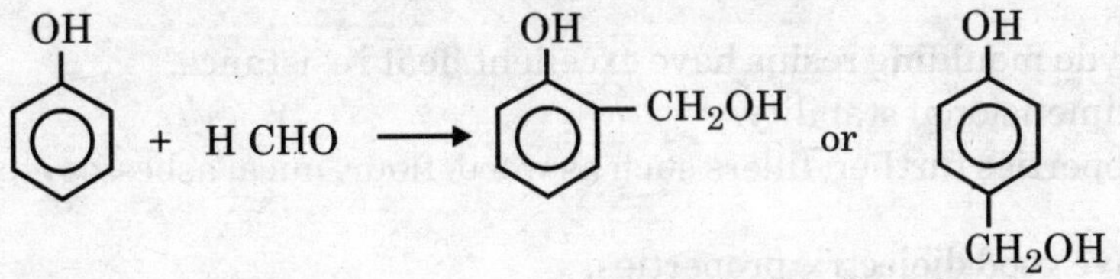

Monomethylol phenols

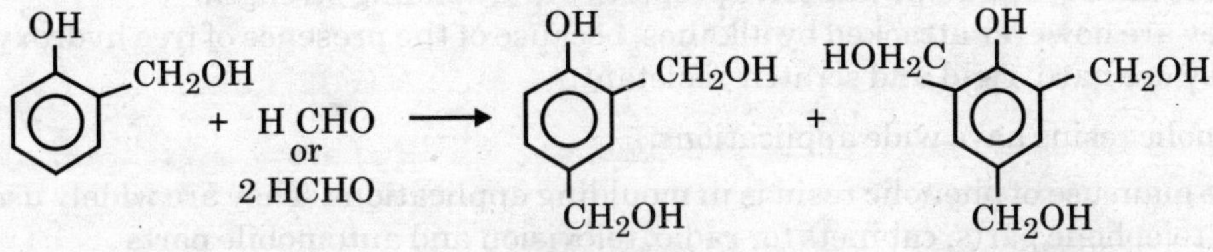

Dimetheylol phenols

Trimethylol phenols

(ii) **Novolac formation:** Depending upon the ratio of phenal formaldehyde (P/F) different resins, namely novolac and resol resins are obtained. In the presence of acid catalyst, when the P/F ratio is greater than unity, the methylol derivatives condense with phenol to form a linear polymer with little methylol groups. The product is thermoplastic in nature and is known as novolac.

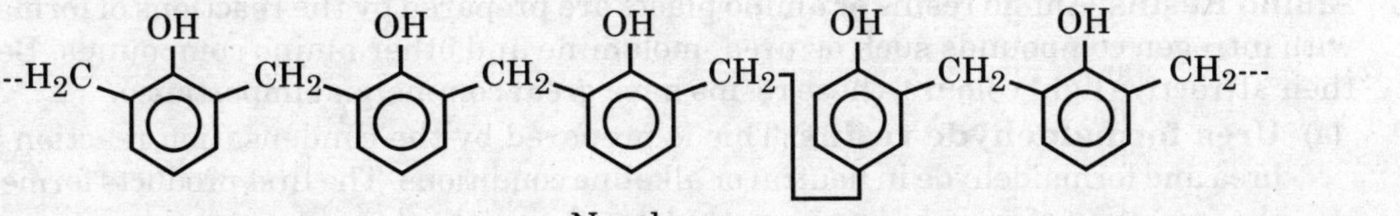

Novolac

(iii) **Resole formation:** In the presence of alkaline catalysts with P/F ratio less than 1, the methylol phenols condense to form linear structure called resoles. The resoles are soluble in the reaction mixture and have excess of methylol groups capable of further reaction during continued heating. Resols, therefore have limited shelf-life. However, the reaction may be stopped at the desired step by cooling.

(iv) **Resite formation:** On further heating, resole is transformed into a three dimensional polymer, called resite, due to the presence of active groups and mobile hydrogens Resite is an infusible and insoluble material.

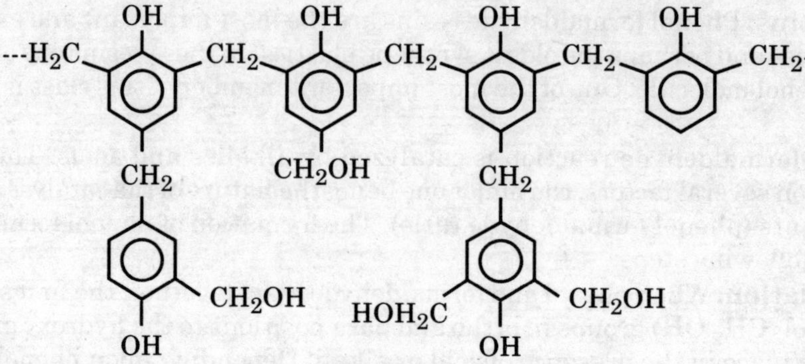

Resite

Properties

(i) Phenol formaldehyde moulding resins have excellent heat resistance.
(ii) These have high dimensional stability.
(iii) To improve the properties further, fillers such as wood, flour, mica, asbestos etc are added to them.
(iv) Phenolic resins have good dielectric properties.
(v) These have remarkable adhesive properties and bonding strength.
(vi) They are however attacked by alkalies, because of the presence of free hydroxy groups.
(vii) They are hard, rigid and scratch-resistant.

Uses Phenolic resins have wide applications.

(i) The main use of phenolic resin is in moulding applications. They are widely used in making telephone parts, cabinets for radio, television and automobile parts
(ii) Due to their adhesive properties and bonding strength, they are used for producing brake linings, abrasive wheels and sand paper.
(iii) They are used for making electric insulator parts like switches, plugs, switch boards, heater-handles etc.
(iv) Phenolic resins are used for impregnating paper, wood and other fillers.
(v) These are also used in varnishes, paints and protective coatings.
(vi) These are used in the production of ion exchange resin for water softening.

3. **Amino Resins:** Amino resins or amino plasts are prepared by the reactions of formaldehyde with nitrogen compounds such as urea, melamine and other amino compounds. Because of their attractive light colours, these resins have great commercial importance.

(a) **Urea formaldehyde resins:** This is prepared by the condensation reaction between urea and formaldehyde in neutral or alkaline conditions. The first products formed during the formation of resin are monomethylol and dimethyol ureas.

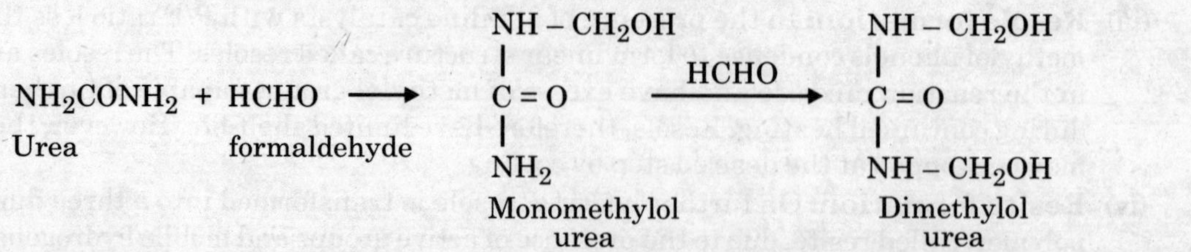

(i) Polymerization can take place from mono or dimethylol urea, or possibly through both, with the formation of long chains.

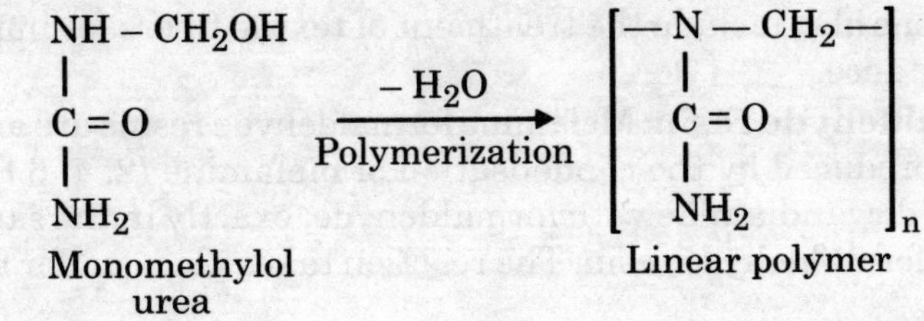

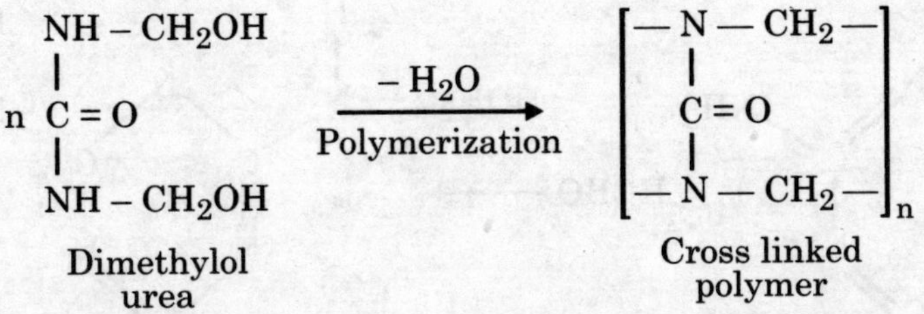

(ii) Thermosetting of the resin takes place during the moulding process. For moulding, the methylol derivatives are compounded with fillers, plasticizers, pigments etc and then are subjected to heat and pressure, resulting into a hard, infusible product. A fully cross linked urea-formaldehyde resin can be represented as

as

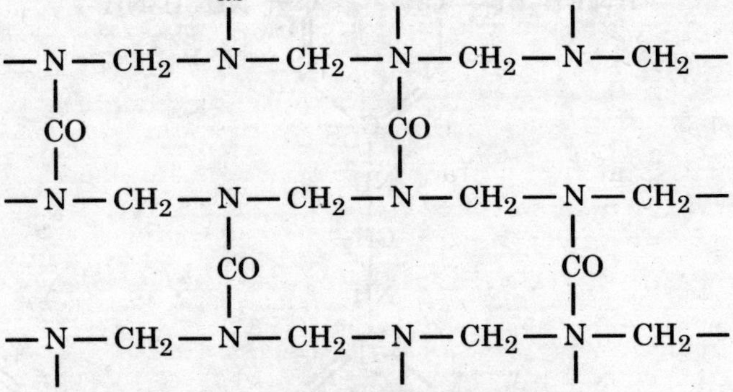

Urea-formaldehyde resin
(Cross linked polymer)

Properties

(i) The urea formaldehyde resins are light in colour and enable the production of light or pastel coloured objects.
(ii) Their heat and moisture resistance are lower than those of phenolic resins.
(iii) They have good electrical insulating properties.
(iv) They are resistant to oxidation to oil, grease and weak acids.
(v) They are hard, resist abrasion and scratching.
(vi) They have good adhesive properties.

Uses

(i) Urea-formaldehyde resins are widely used to make moulded articles like vacuum flask cups, jugs buttons, bottle caps etc where the darker colour of phenolics may be objectionable.
(ii) They are also used for adhesive applications for the production of plywood and laminating.
(iii) They are used for the manufacture of coated abrasive paper and binder for foundry cores.
(iv) They are used for the manufacture of cation exchange resins.
(v) These also find use in the manufacture of electrical switches, plugs and insulating foams.

(vi) Their applications also include the treatment of textile fibers for improving their shrink and crease resistance.

(b) **Melamine-formaldehyde resin:** Melamine formaldehyde resins are another class of amino resins. They are produced by the condensation of melamine (2, 4, 6 triamino - 1, 3, 5 triazine), a trimer of dicyandiamide with formaldehyde, exactly in the same manner as in the formation of urea formaldehyde resin. The reaction takes place under mild alkaline or neutral conditions.

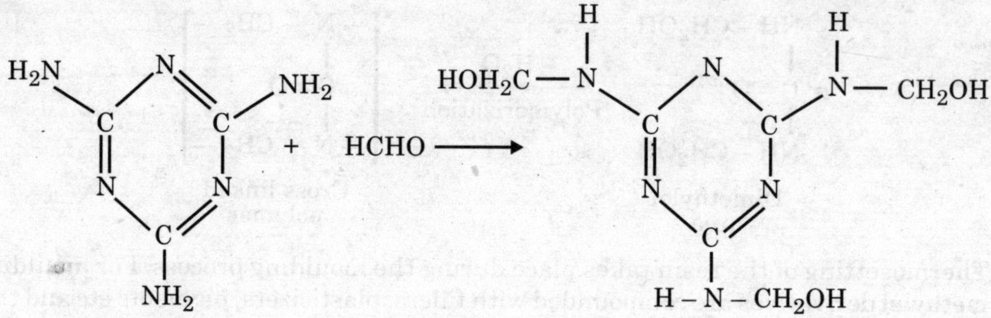

Melamine Trimethylol melamine

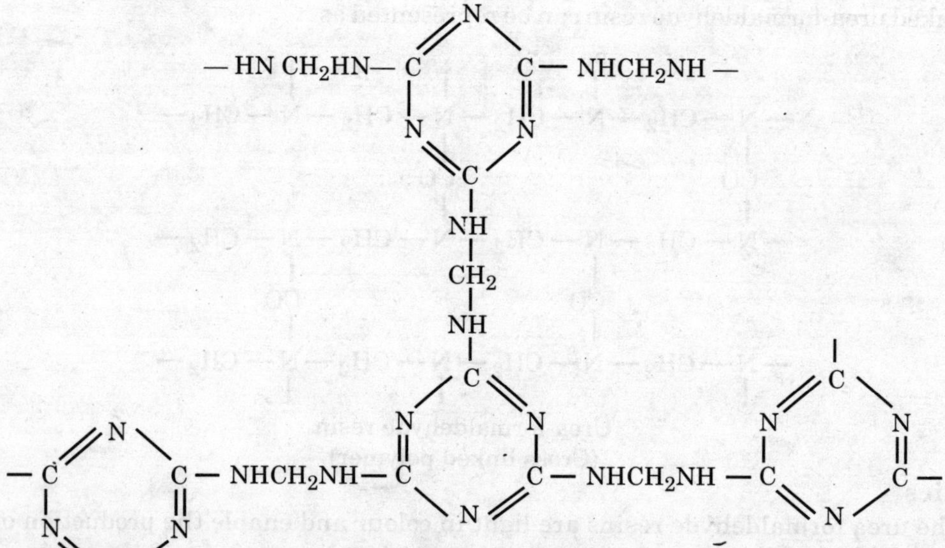

Cross linked melamine

Properties:

(i) The melamine resins have better hardness, heat resistance, moisture resistance than the urea-formaldehyde resins. However these are expensive. Rest of the properties are similar to urea-formaldehyde resins.

Uses

(i) Typical applications of urea formaldehyde resins include the production of decorative plastic ware, laminated worktops.

(ii) They are used in electrical fittings.

(iii) They are used for resin bonded grinding wheels.

(iv) They are used for textile finishing to provide crease resistance, stiffness, water repellency etc.

(v) They are also used in automotive finishes.

(vi) Melamine resin is also used as a tanning material for leather. The presence of small amount of this resin in the tanning material improves the strength of the leather.

(vii) They are also used for the preparation of lacquers.

4. Epoxy Resins: Epoxy resins are so called because their starting material is epichlorohydrin and the polymer before crosslinking contains epoxide groups.

The epoxy resins are formed by the reaction of epichlorohydrin and bisphenol A.

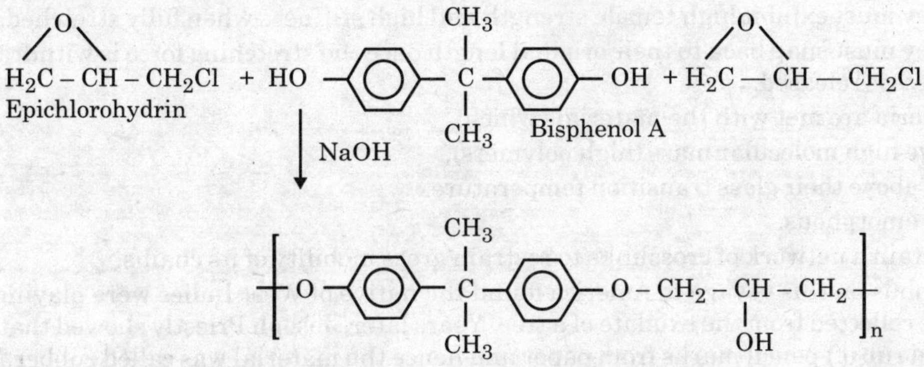

Epoxy resin

The linear epoxy resins are converted into thermosets by reacting with many types of materials including polyamines, polyamides, poly-sufides etc. to give three dimensional crosslinked structures. This is called curing of epoxy resins.

Properties

(i) Epoxy resins have remarkable chemical resistance due to the presence of very stable ether linkages.

(ii) They are tough and heat resistant.

(iii) The polar nature of the molecules produces excellent adhesion quality in them.

(iv) They are highly flexible because the reactive groups (epoxide and hydroxyl groups) are comparatively widely-spaced.

(v) They have good electrical properties and dimensional stability.

Uses

(i) Epoxy resins are excellent structural adhesives and hence are used as adhesives (araldite) for glass, metals etc.

(ii) They are used as surface coatings because of their outstanding toughness, flexibility, adhesion and chemical resistance.

(iii) Their electronic applications include their use as encapsulating material for electronic components.

(iv) Because of their toughness, epoxy resins are used to provide skid-resistant surfaces for highways.

(v) Epoxy resins are applied over cotton, rayon and bleached fabrics to impart crease resistance and shrinkage controls.

(vi) The structural application of epoxy resin is the production of glass fibre reinforced articles with better mechanical strength.

13.8 RUBBERS (ELASTOMERS), NATURAL AND SYNTHETIC

It was discussed in the beginning of the chapter that depending on the ultimate form and use, a polymer can be classified into two broad categories: Plastics and Elastomers (rubbers).

Elastomers are the polymers which are capable of being stretched rapidly at least 150 precent of their original length without breaking and return to their original shape on release of stress. For example, rubber band is a typical elastomeric material. When stretched, it is elongated to many times its length. But when the force is released it gets back to its original length.

The properties of typical elastomers are defined by the following:

- They must stretch rapidly and considerably when stretched, reaching high elongations (500-1000%).
- They must exhibit high tensile strength and high stiffness when fully stretched.
- They must snap back to their original length once the stretching force is withdrawn or the stress is released.

These criteria are met with the materials which:

- have high molecular mass (high polymers).
- are above their glass transition temperature.
- are amorphous.
- contain a network of crosslinks to restrain gross mobility of its chains.

Columbus on this voyage to America found the native of West Indies were playing with the solid mass collected from the exudate of a tree. Years later, Joseph Priestly showed that this solid mass could rub off pencil marks from paper and hence the material was called rubber. The main source of natural rubber is the species of tree known as 'Hevea Basiliensis' even today, which account for multi billion dollar rubber industry.

Rubber or elastomers can be obtained naturally or synthetically. Consequently we have natural rubber and synthetic rubber.

13.8.1 Natural Rubber

Natural rubber of nearly same characteristics occur in the inner bark and to a lesser extent in the leaves and roots of a number of tropical plants. But the latex from the tree 'Hevea Basiliensis' is proved to be the main source of rubber. Latex is an emulsion of poly hydrocarbon droplets in an aqueous solution and looks like milk. The negatively charged particles of polyhydrocarbons, a colloidal dispersion, stabilizes the emulsion.

Processing

The latex is collected by making small incisions in the trees in such a manner so as to allow the latex to accumulate in small cups. The percentage of rubber in it is 25-35%. A fully grown tree gives about 5 gms of rubber in a day.

The average composition of the latex is;

Water = 60%; Rubber = 35%; Proteins, enzymes and nucleic acid = 3%; Fatty acids and esters = 1% and inorganic salts = 0.5%.

In these trees, the rubber is formed from polymerization of isoprene i.e (2-methyl-1,3-butadiene) to produce poly-(cis)isoprene by a biochemical reaction in which a particular type of enzyme acts as a catalyst.

$$CH_2 = \underset{\displaystyle CH_3}{\underset{|}{C}} - CH = CH_2$$

Isoprene (2 – methyl – 1, 3 – butadiene)

The latex is first sieved to remove the impurities such as leaves, bark etc. From 25-35% dry rubber content the latex is diluted to contain about 15-20% of rubber content. The rubber is then separated from the latex by the process of coagulation. The acetic acid or formic acid is added as coagulants and the rubber is coagulated to soft white mass. About 1 g of acetic acid/formic acid is added per 200 kg of rubber material. This coagulated mass is the crude rubber and consists of 90-95% rubber, 2-4% proteins and 1-2% resins. The crude rubber is wached and treated for producing the following types of rubber depending upon the process.

1. **Crepe Rubber:** Crepe rubber is prepared by adding a small amount of sodium bisulphite ($NaHSO_3$) to the coagulated mass to bleach the rubber. Then the rubber is passed through a creping machine. The creping machine is rolling machine which consists of two rollers having longitudinal grooves. Thus, after passing through it, the spongy rubber is converted into a sheet having a rough surface which resembles the crepe paper and hence the name.
2. **Smoked Rubber:** The smoked rubber is produced by carrying out the coagulation of latex in the tanks provided with vertical grooves on sides and fitted with the metal plates which run across the width of the tank. First the diluted latex is poured into the tanks, with the metal plates removed. Then the coagulant is added (acetic or formic acid) and the plates are inserted. The tanks are kept undisturbed for 15-16 hours. The tough slabs are formed. These slabs are passed through a series of rolling machines and finally rubber sheets having ribbed pattern are obtained. The ribbed surface prevents the sticking of the rubber with each other. These sheets are then dried in smoke houses at about 50°C in the smoke obtained from burning wood or coconut shells. The smoked rubber so obtained is transluscent and ambre coloured.

GUTTA PERCHA

Gutta percha is another variety of natural rubber. It is obtained from the mature leaves of dichopsis gutta and palagum gutta trees. The hevea rubber and gutta percha, both are composed of isoprene units with structures shown in the figure? But **natural rubber** is *cis-isomer* where a bending back of successive isoprene units take place giving the molecule a coiled structure. The **gutta percha**, on the other hand is a *trans-isomer* which consist of straight chains, giving a rod like structure to the molecule. The rod-like structure results into close packing of adjacent chains. Thus gutta percha is more crystalline.

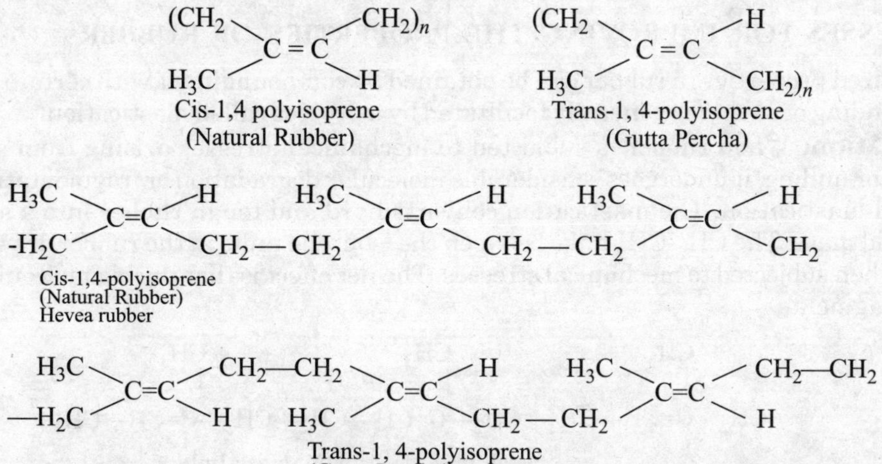

Fig. 13.2 Configuration of isoprene units in natural rubber and gutta-percha.

Processing of Gutta Percha

The mature leaves obtained from the trees Dichopsis gutta or palagum gutta are carefully ground in mills and are then treated with water maintained at 70ºC for about half an hour when the gutta percha is extracted in water. The solution is dropped into cold water and gutta percha floats on the surface. Alternatively, gutta percha can be recovered by solvent extraction.

Properties of natural (Hevea) rubber and gutta percha:

	Natural Rubber (Hevea rubber)	Gutta Percha
1.	Natural rubber is highly soft and elastic material	Gutta percha is a hard thermoplastic material at room temperature but it softens and becomes tacky at about 1000ºC
2.	It is soluble in carbon disuphide and petrol.	It is soluble in petrol.
3.	It is cis polyisioprene	It is transpolyisoprene.

Drawbacks of Raw Rubber: Raw rubber has a number of drawbacks when put to use. Some of the drawbacks are:

(i) Crude rubber is soft at high temperatures but becomes brittle at low temperatures. So it can be used in a limited temperature range.

(ii) It is too weak with a tensile strength of 200 kg/cm^2.

(iii) It is not resistant to mineral oils, organic solvents.

(iv) It has large water absorption capacity.

(v) Due to oxidation in air, it deteriorates and hence is durability its decreased.

(vi) It is also attacked by the oxidizing agents like nitric acid, conc. sulphuric acid, chromic acid etc.

(vii) It undergoes permanent deformation when stretched. This may be due to stretching the individual chains slip past each other and get separated resulting in the breakage of the material.

However the above drawbacks can be overcome by improving the properties by the addition of suitable materials and further heat treatments.

Some of the processes for improving the properties of rubber are discussed below:

PROCESSES FOR IMPROVING THE PROPERTIES OF RUBBER

The desired properties in rubber can be obtained by compounding it with certain chemicals. The compounding of rubber is generally facilitated by a process called mastication.

Mastication: When rubber is subjected to mechanical stresses arising from say, high-speed stirring or milling, it undergoes considerable molecular degradation or fragmentation. The process is called mastication. The mastication converts hard and tough rubber into a soft, supple and semisolid mass. The CH_2-CH_2 links between the isoprene units in the rubber are weak and hence break when subjected to mechanical stresses. The net effect is that the big molecule is broken into small fragments.

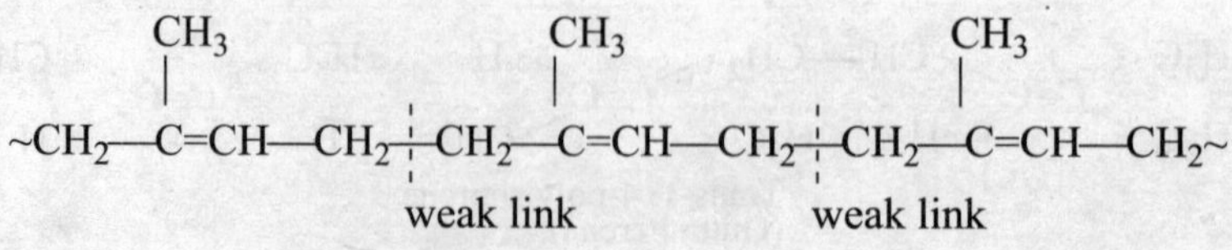

Polyisoprene

After mastication is complete the compounding agents are added to the rubber.

Compounding: The process involving the incorporation of chemicals such as plasticizers, vulcanizing agents, stabilizers, fillers, etc., into the rubber is known as compounding of the rubber. Out of above *vulcanization* is very significant and requires a detailed study.

Vulcanization

The process of vulcanization which revolutionized the rubber industry was discovered by Charles Goodyear in 1839. The process by which a network of crosslinks is introduced into an elastomer is called vulcanization. Charles Goodyear used sulphur to crosslink polyisoprene molecules in natural rubber. The added sulphur combines chemically at the double bonds bringing about changes in its properties profoundly. Vulcanization transforms an elastomer which is a weak thermoplastic mass of no use into a strong, elastic and tough rubber. The tensile strength, stiffness, durability, resistance to changes in temperature, elasticity and chemical resistance of a vulcanized rubber are increased manifolds as compared to unvulcanized rubber.

The introduction of sulphur links in between the polymer chains restrict the intermolecular movement which brings about stiffening of rubber.

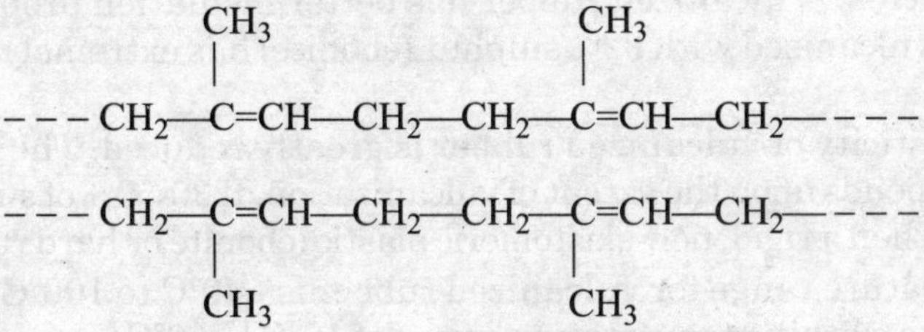

Unvulcanized Rubber

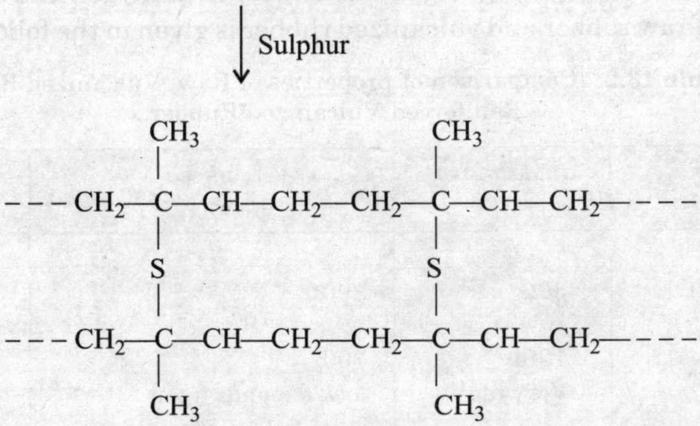

Vulcanized Rubber

The extent of stiffness of vulcanized rubber depends upon the amount of sulphur added. For example, a tyre rubber may contain 3 to 5% sulphur whereas Vulcanite or ebonite (hard rubber) contains about 32% sulphur. The extensive crosslinking in hard rubber makes it so stiff that there is practically no elasticity in it. Fig. shows the effect of vulcanization on rubber. Ordinary rubber molecules are bent and convoluted (a) When stretched the molecules can slip past one another. But vulcanization introduces crosslinks between these and hence on stretching the molecular chains cannot slip past each other.

(a) Unvulcanized Rubber (b) Vulcanized Rubber (Unstretched) (c) Vulcanized Rubber (Stretched)

Advantages of Vulcanization

Vulcanized rubber has got ideal elastomeric properties which increase its utility manifolds. The main advantages of vulcanized rubber are:

1. **Tensile strength:** Tensile strength is the measure of the ability of a polymer to withstand pulling stresses. Vulcanized rubber has got good tensile strength and extensibility. The tensile strength of the raw rubber is 20 kg/cm^2 which increases to 2000 kg/cm^2 on vulcanization.
2. **Resilience:** It is the property of a elastomer to return to its original shape on the release of stress. Resilience of vulcanized rubber improves manifolds as compared to raw rubber.
3. **Water Absorption Capacity:** The water absorption capacity of vulcanized rubber is greatly reduced.
4. Vulcanized rubber is resistant to organic solvents such as benzene CCl_4, fats and oils. Though it swells in these solvents but to a lesser extent than raw rubber.
5. **Insulation Properties:** Vulcanized rubber has better insulation properties than raw rubber. Raw rubber vulcanized with 32% sulphur (ebonite) has extremely good insulating properties.
6. **Elasticity:** The elasticity of vulcanized rubber is greatly reduced. The elasticity of the vulcanized rubber depends upon the extent of vulcanization. If 30-50% of sulphur is present in vulcanized rubber then a rigid, non-elastomeric plastic (ebonite or hard rubber) is formed.
7. The working temperature range for vulcanized rubber is –40°C to 100°C as compared to raw rubber which has working temperature range of 10°C to 60°C.
8. It is resistant to abrasion, ageing and reactivity with oxygen and ozone.

A comparison of raw rubber and vulcanized rubber is given in the following Table 13.2:

Table 13.2 Comparison of properties of Raw, Vulcanized Rubber, Reinforced Vulcanized Rubber.

	Property	*Raw Rubber*	*Vercanized Rubber*	*Reinforced Valcanized Rubber*
1.	Tensile strength (psi)	300	3000	4500
2.	Elongationat break (%)	1200	800	600
3.	Elasticity	Very high	low, depends upon extent of vulconization	Very low, depends upon extent of vulcanizatin and reinforcement
4.	Water absorption	large	small	small
5.	Solvent resistence (hydrocarbons)	sluble	swells only	swells only
6.	Chemical resistance	Poor	Much better	Much better

13.8.2 Synthetic Rubbers

The polymers possessing similar physical properties as that of natural rubber are called synthetic rubbers. As natural rubber suffers from many disadvantages, attempts to find out a substitute for natural rubber began very early.

Greville William (1860) was first to isolate isoprene as a decomposition product by heating rubber to a high temperature. This gave a clue that rubber can be synthesized by the polymerization of monomers like isoprene. By the end of 19th century the fact was confirmed when polymerization of isoprene yielded rubber like product.

Since then attempts were on to prepare synthetic rubber by other monomers as well. Synthetic rubbers have been produced by the polymerization of a large number of conjugated dienes resembling isoprene. The foremost of all is, Buna Rubber (polymer of butadiene) which was prepared by the action of sodium on butadiene. That is why synthetic rubber is known as Buna rubber—'Bu' for Butadiene and 'Na' for sodium. However, the product prepared had poor tensile strength, and ageing characteristics. Improvements in properties were made by developing *copolymers* of butadiene. Several types of synthetic rubber are available today and some of them are described below:

1. SBR or Buna-s Rubber

Styrene-butadiene rubber (SBR) is the first commercially made synthetic rubber. It is a copolymer obtained by the free-radical polymerization of two monomers, styrene and butadiene. The two components are allowed to react in a mixing vessel containing an aqueous solution of an emulsifying agent in presence of free radical initiator such as, cumene hydroperoxide or p-menthane hydroperoxide.

According to the temperature used, two processes have been developed, the hot process (at 50°C) and the cold process. In the cold process the temperature is kept in the range –18°C to 5°C. The rubber is known as cold rubber. Cold process is more efficient than the hot process because chain length can be controlled more efficiently. The SBR copolymer contains about 75% butadiene and 25% styrene and is a random copolymer. Both type of rubbers 'cold' as well as 'warm rubbers' can be vulcanized by conventional methods using sulphur. A reinforcing filler is added to achieve good physical properties.

Properties:

(i) SBR possesses high abrasion loses resistance and high load bearing capacity.
(ii) At elevated temperatures SBR almost 2/3 of their tensile strength whereas natural rubber only loses 1/4th.
(iii) Its resilience is poorer than that of natural rubber which restricts its use in tire industry.
(iv) The weatherability of SBR is better than that of natural rubber.
(v) 'Cold' SBR is superior to the standard product because it contains less chain branching and crosslinking.
(vi) SBR gets easily oxidized in presence of traces of ozone present in atmosphere.
(vii) It swells in oils and solvents.

Uses:

Of all the synthetic elastomers, SBR enjoys the major share of the market. SBR has the following applications.

(i) The principal use of SBR is in the production of tires and tire products which are typically blends of both natural and synthetic rubbers. SBR is used for lighter duty tyres but is inferior to the natural rubber in its use for truck tyres.
(ii) For many mechanical goods, it is superior to natural rubber and is used because of its easier processing.

(iii) It is used for making hose, shoe soles, gaskets, floor tiles, carpet backing, tank linings etc.

(iv) It is widely used for electrical insulation, although its properties are not good as those of butyl rubber.

2. Buna-N or Nitrile Rubber (NBR) or GR-A

The nitrile rubbers are polymers of butadiene and acrylonitrile in the ratio of 75% and 25% respectively. It is prepared by emulsion polymerization using cumene hydroperoxide as initiator. Hydrogen peroxide and ferros sulphate can also be used as initiators. Vulcanization is done by conventional means by adding sulphur.

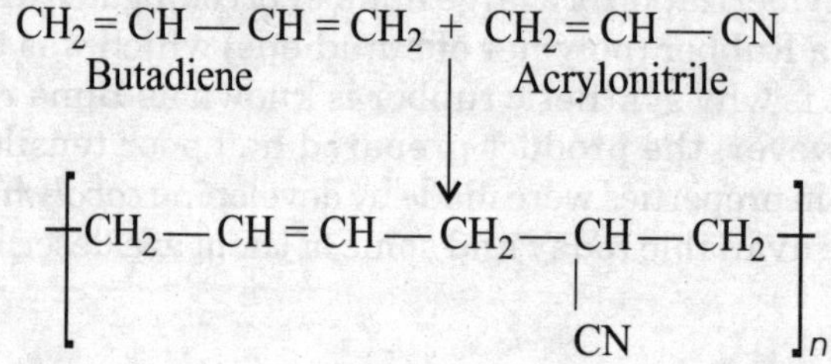

Properties:

(i) Nitrile rubbers are noted for their oil resistance. However the oil resistance varies with composition. If the acrylonitrile content is between 18% then the rubber has fair oil resistance and if it is more than 40% then it is extremely oil resistant.

(ii) It possesses excellent resistance to heat, sunlight, acids and salts.

(iii) Because of the presence of cyano groups (-CN), it is less resistant to alkalies than natural rubber.

(iv) It has good tensile strength and abrasion resistance.

(v) The rubbers are less resilient than natural rubbers.

(vi) It has good heat resistance and if properly protected by antioxidants, it shows resistance to oxidative degradation as well.

Uses:

(i) Nitrile rubber is used primarily for its oil resistance. These are extensively used for fuel tanks, gasoline hoses, gaskets, printing rollers and high attitude aircraft components.

(ii) They can be used in contact with water and antifreeze solutions.

(iii) They are also used in the form of latex for impregnating paper, textile and leather.

(iv) It can also be used as blend component in tyre manufacture.

3. Neoprene or GR-M Rubber

This is also known as polychloroprene or polychlorobutadiene or duprene. Neoprene was the first synthetic rubbers developed in the United States.

Preparation: Neoprene or polychloroprene is prepared by polymerization of chloroprene, which is 2-chloro-1,3 butadiene.

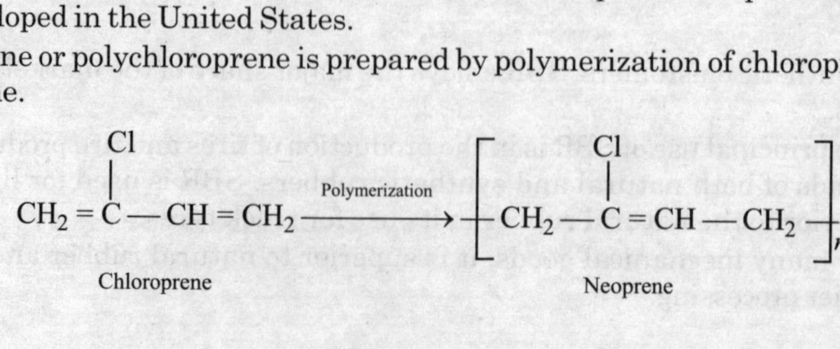

The preferred configuration produced during the polymerization is trans 1-4 polychloroprene. The polymerization of chloroprene takes place through free radical emulsion polymerization.

The starting material, chloroprene is manufactured by the addition of hydrogen chloride to vinyl acetylene, which inturn is prepared by dimerization of acetylene.

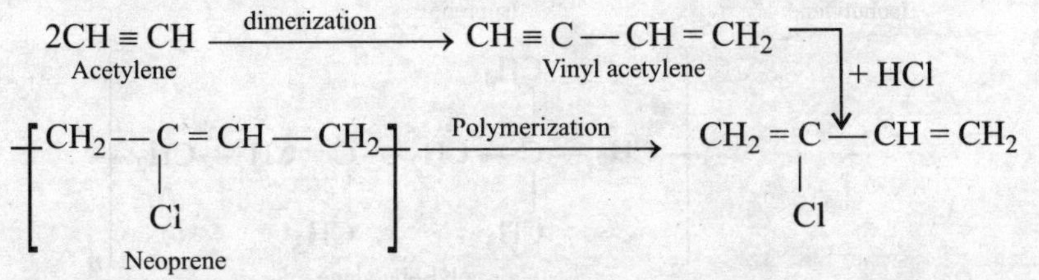

The vulcanization of neoprene is different from that of other elastomers in that it can be vulcanized by heat alone. Although some types are polymerized in presence of sulphur which introduces crosslinks in the polymer. Neoprene can also be vulcanized using zinc oxide or magnesium oxide.

Properties: Neoprene is closely related to the natural rubber and can replace rubber because similarity in the properties of the two. Some of the properties of Neoprene are:

(i) As H atom in the natural rubber is replaced by Cl atom in Neoprene, Neoprene has superior resistance to vegetable and mineral oils.
(ii) It is superior to natural rubber as far as ageing and effects of high temperatures are concerned.
(iii) Neoprene is soluble in polar solvents due to the presence of chloro-group.
(iv) Vulcanized neoprenes have high tensile strength (3500-4000 psi) in the absence of carbon black.
(v) Vulcanized neoprene is extremely resistant to oxidative degradation if suitably protected.
(vi) It has good weathering and ozone resistance.
(vii) It has got excellent oil resistance, though inferior to nitrile rubber but better than natural rubber or SBR.

Uses

Neoprene is used for:

(i) It is used in oil resistant wire and cable coatings, industrial hoses, shoe holes.
(ii) Its latex is used for the manufacture of gloves and coated fabrics.
(iii) It is used for making tubings for carrying corrosive gases and oils.
(iv) It is used for the manufacture of sponges, conveyer belts, linings of reaction vessels etc.

4. Butyl Rubber (GR-I)

Butyl rubber belongs to the class of elastomers which do not have unsaturated chain structure. However, a small amount of diene monomer that provides a site for vulcanization is included by means of copolymerization.

Preparation

Butyl rubber or polyisobutylene are prepared by the copolymerization of isobutylene with a small amount (1.5 to 4.5%) of isoprene. Isoprene is added to render them vulcanizable.

It is prepared by low temperature cationic polymerization using Friedal-crafts catalyst at –80°C.

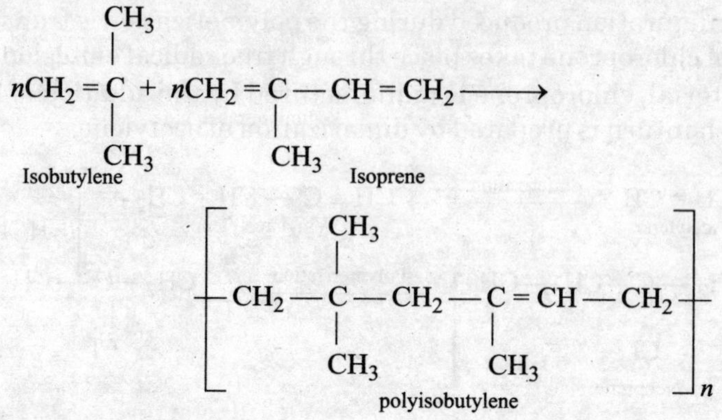

Properties

The absence or low concentrations of double bonds in the main chain of butyl rubber results in high resistance to attack by a number of chemicals. It has the following properties.

(i) Polyisobutylene or butyl rubber is amorphous under normal conditions but crystallize on stretching.

(ii) It exhibits outstanding resistance to attack by oxygen and ozone due to low concentration of double bonds. It is less sensitive to oxidative ageing as compared to other elastomers except silicons.

(iii) Butyl rubber becomes soft on oxidative degradation.

(iv) It exhibits low gas permeability due to closepacked linear paraffinic chains.

(v) It has high ozone and acid resistance.

(vi) Due to very low unsaturation, it cannot be hardened much on vulcanization.

(vii) Unstabilized polyisobutylenes are degraded by light or heat to sticky low-molecular weight products. Therefore usual rubber antioxidants should be added to stabilize the polymer.

(viii) It has excellent insulating properties as predicted from its non-polar, saturated nature.

(ix) It is soluble in hydrocarbons such as benzene.

(x) The tear resistance of butyl rubber is quite good and is retained well at high temperature and for long times, in contrast to natural rubber.

Uses

(i) Butyl rubber is used for inner tubes for tires because of its impermeability to gases.

(ii) It is used for wire and cable insulation.

(iii) It is used for linings of tanks, hoses etc.

5. Polysulphide Rubber (Thiocol or GR-P)

Polysulphide Rubbers are the condensation products of ethylene dihalides and alkali sulphides. These are linear condensation polymers.

Preparation

Polysulphide rubber is obtained by the reaction between sodium polysulphide and ethylene dichloride

Thus sulphur forms the part of the polymer chain. During the reaction NaCl gets eliminated. Vulcanization cannot be achieved by sulphur because polymer chain is not unsaturated. Hence it is not a hard rubber. However vulcanization can be achieved by reacting with ZnO which reacts with sulphur links forcing crosslinks.

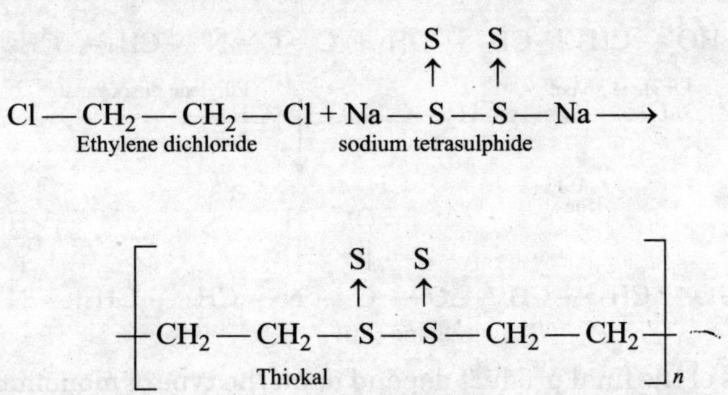

Properties

The properties of the materials depend on the length of the aliphatic groups and the number of sulphur atoms present in it. With four sulphur per monomer, the product is rubbery, whereas if only two sulphur atoms are present per monomer, the polymer do not behave as elastomer. In that case at least four methylene groups are needed in the dihalide to induce elastomeric properties to the products. Some of the important properties are:

(i) Thiokol is resistant to swelling.
(ii) It is resistant to oils and solvents. Fuel oils, lubricating oils, gasoline and kerosene have no effect on thiokol.
(iii) Thiokol films are impermeable to gases to a large extent.
(iv) It cannot be vulcanized and hence does not form hard rubber.

Thiokol rubbers from many disadvantages:

(i) It has disagreeable odour (mercaptan odour).
(ii) It has poor heat resistance, poor abrasion resistance and low tensile strength.
(iii) It tends to lose its shape under continuous pressure.

However, recent developments of Thiokol have overcome these disadvantages.

Uses

(i) Polysulphides or Thiokol is used to make sealants, gaskets, balloons, fabric coating.
(ii) Due to low permeability it is used to make gasoline hoses.
(iii) Liquid Thiokol can be cured at room temperature to tough solvent-resistant rubbers. These liquid compounds are used as gasoline tank sealants and liners for aircraft and in variety of other sealing applications.
(iv) The liquid polymers, when combined with oxidizers burn with great intensity and generate large volume of gas. Thus it can be used in solid-fuel rocket propellents.

6. Polyurethane (PU)

Polyurethanes are polymers containing the group (carbonate group)

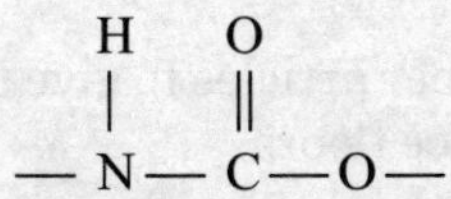

These are known as urethane linkages.

Preparation: The polyurethanes are formed through the reaction of diisocyanate with the compound containing atleast two active hydrogen atoms such as diols or diamines. The reaction of diisocyanate with glycol is represented as:

$$HO-CH_2-CH_2-OH + O=C=N-CH_2-CH_2-N=C=O$$

Ethylene glycol Ethylene diisocyanate

$\downarrow$ Polymerization

$$\left[O-CH_2-CH_2-O-\overset{\overset{\displaystyle O}{\|}}{C}-\overset{\overset{\displaystyle H}{|}}{N}-CH_2-CH_2-\overset{\overset{\displaystyle H}{|}}{N}-\overset{\overset{\displaystyle O}{\|}}{C} \right]_n$$

Polyurethane (Pu)

The properties of the final product depend upon the type of monomers and their chain length. A wide variety of polymers can be prepared ranging from rigid to elastic polymers. Crosslinking takes place through isocyanate groups which are capable of reacting with hydrogen atoms attached to the polymers leading to vulcanization. The polyurethanes can be produced in the form of fibers, elastomers or foams.

The foamed plastic can be obtained by blowing air/gas through molten polyurethane. This foamed plastic is also known as *thermocol*.

Note: Thermocol is also prepared by blowing air/gas through polystyrene.

Alternatively, for the production of foamed polyurethane little water is mixed with the reaction mixture, which reacts with isocyante group to evolve CO_2. The CO_2 produced foams the polymer as it sets.

$$O=C=N-R-N=C=O + 2H_2O \longrightarrow H_2N-R-NH_2 + 2CO_2$$

Properties

(i) Polyurethanes have high strength, good resistance to gas and aromatic hydrocarbons.

(ii) They have excellent resistance to ozone and oxygen but are susceptible to microbial attack.

(iii) The polyurethanes can be prepared in different forms varying from hard coatings to flexible rubbers and rigid foams, each having different properties:

Coatings: Coatings are hard, glassy and solvent resistant. They have good flexibility and impact resistance.

Elastomers: Elastomers have excellent abrasion resistance and are resistant to oils, greases and solvents.

Fibers: Fibers are obtained by reacting di-isocyanate with long chain glycols which are usually polyesters or polyethers.

Foams: Foams can be flexible or tough depending upon the polymer and the type of crosslinking produced. They are light, tough and resistant to heat.

Uses

(i) The polyurethane foams have found the widest use. The flexible foams are used as upholstery material for furniture, insulation and crash pads.

(ii) The rigid polyurethane foams are used as core material as insulation cover in refrigerated appliances and vehicles.

(iii) Flexible polyurethanes can also be used for coating rubber articles to give them additional resistance to abrasion, such as in gymnasium and dance floors.

(iv) Polyurethane resin is used to insulate the wires polyurethane coated wires can be soldered directly.

(v) Polyurethane fibers with unusually high elasticity are used in foundation garments and swim wear. They have replaced rubber latex thread in this use.

(vi) Flexible urethane foams are used as cushions for furniture and automobilies and has replaced rubber foams in these applications because of improved strength, lower density and easier fabrication.

(vii) Polyurethanes are also used in making matteresses and pillows.

(viii) As they are radiation resistant, they are also used in encapsulation for integrated-circuit devices.

(ix) Its applications in medical field include its use in dental materials, artificial heart and skin.

7. Chlorosulphonated Polyethylene Rubber (Hypalon)

Preparation: When polyethylene is treated with a mixture of chlorine and sulphur dioxide, chlorosulphonated polyethylene is formed.

$$\left(CH_2 - CH_2 \right) + Cl_2 + SO_2 \xrightarrow{-HCl}$$

$$\left[CH_2 - \underset{\underset{Cl}{|}}{CH} - \left(CH_2 - CH_2 \right)_n CH_2 - \underset{\underset{SO_2Cl}{|}}{CH} \right]$$

The chlorosulphonation can be carried out either on the solid material or in solutions.

Properties: The chlorosulphonylation of ethene provides two modifications:

(a) The chlorine atoms breaks the regularity of polyethene chain structure changing the crystalline polyethylene to a elastomer.

(b) The sulphonyl groups which are introduced, provide the sites for crosslinking.

The chlorosulphonyl ethane has the following properties:

(i) Its resistance to ozone is better than neoprene and butyl rubber.

(ii) It has good heat resistance.

(iii) It has good chemical resistance and flame resistance.

(iv) It is resistant to solvents.

Uses

It is used in chemical industry.

13.9 CONDUCTING POLYMERS

In general, polymers are found to be insulators. Infact electricity wires are coated with polymers because of the insulating properties of the latter.

In 1800s, it was observed that the conductivity of natural rubbers-which is normally an excellent insulator-could be significantly increased by adding carbon black which has natural conductivity. In 1930s, natural rubber filled with acetylene black was used as an antistatic device in hospitals where it was necessary to control the danger of sparks due to the building of static electricity.

Earlier (before 1973) the only polymer capable of showing appreciable electrical conductivity (~10^3 S cm^{-1}) was polysulphurnitride $(SN)_x$. However this polymer could not generate much interest among scientists because of the explosive nature of its precursor monomer, S_2N_2.

During the past 15 years, however, a new class of organic polymers has been devised with the remarkable ability to conduct electrical current. Some of these conductive polymers are already under development for practical applications, which include rechargeable batteries, electrolytic capacitors and 'smart windows' that absorb sunlight in summer. Thus conductive polymers have tremendous potential for future scientific and technological development.

Now, the major question is that, how the electric current passes through the conducting polymers. A common feature of all conducting polymer is a conjungated π-electron system (consisting of alternating single and double bonds) along the polymer chain backbone or ring structure. Among many polymers known to be conductive, **polyacetylene, polyanaline, polypyrrole, poly (phenylenevinylene)** have been studied the most. However, the conductive polymer that actually launched this new field was polyacetylene.

Prof. Alan J. Heeger, Prof Alam. G. MacDiarmid, and Prof Hideki Shirakawa in 1977 discovered that partial oxidation of polyacetylene with iodine, chlorine or bromine vapours made polyacetylene 10^9 times more conductive. The process of transforming a polymer to its conductive form via chemical oxidation or reduction is called **doping** by analogy to somewhat related doping of semiconductors. Thus, the conductive form of polyacetylene is called "doped polyacetylene." The conductivity of doped polyacetylene was found to be as high as 10^3 S cm^{-1}, which was higher than that of any known polymer. By comparison, silver has conductivity of 10^6 S cm^{-1}. The conductivity of doped polyacetylene is comparable to that of copper on an equal weight basis. The three scientists received the Nobel prize-2000 in chemistry, for their discovery and development of electrically conductive polymers.

In general, conducting polymers can be classified into the following categories:

1. **Conjngated π-electrons conducting polymers:** As is discussed earlier, a key property of a conductive polymer is the presence of conjugated double bonds along the backbone of the polymer. The orbitals of conjngated π-electrons, overlap over the entire backbone of the polymer, resulting into the formation of valence bands as well as conduction bands, extending over the entire polymer molecule.

   The energy spacing between the highest occupied and lowest unoccupied band is called the band gap. The highest occupied band is called the **valence band** and the lowest unoccupied band is the **conduction band**. If the band gap is narrow, the electrons from the valence band are excited to conduction band giving rise to conductivity. If the band gap is too wide, the substance is an insulator.

   In an electric field, conjugated π electrons get excited and electrons have sufficient energy to jump the gap and reach conduction band.

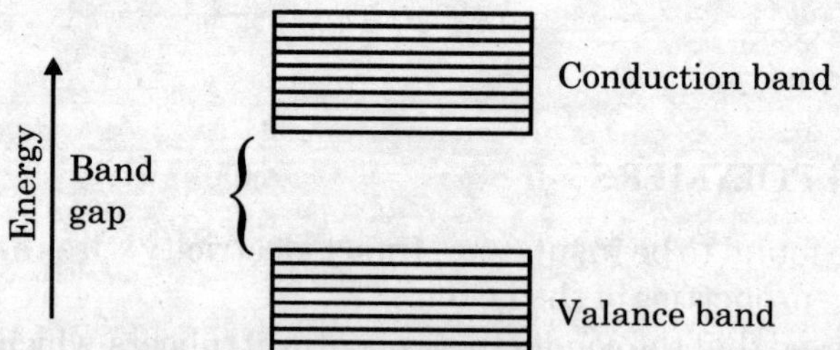

In otherwords electrons thereby can be transported through the solid polymeric material and hence polymer becomes conducting. The most common examples are polyacetylene and polyaniline. However a number of heterocyclic aromatic conducting polymers are also being studied.

Examples:

1.

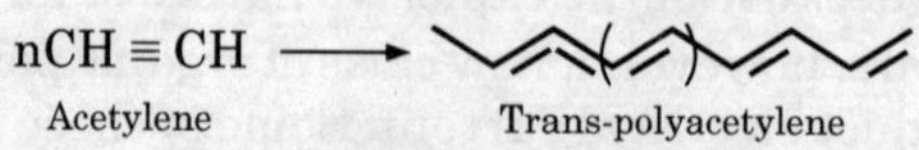

2.

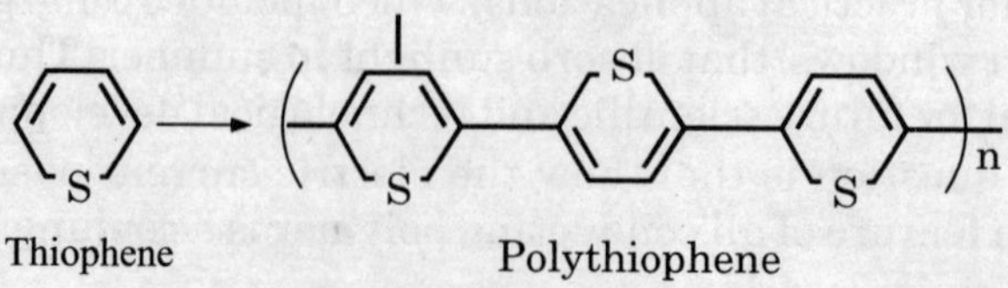

3. 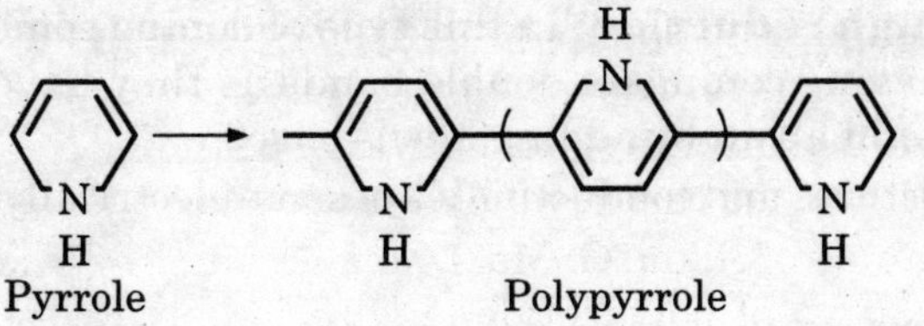

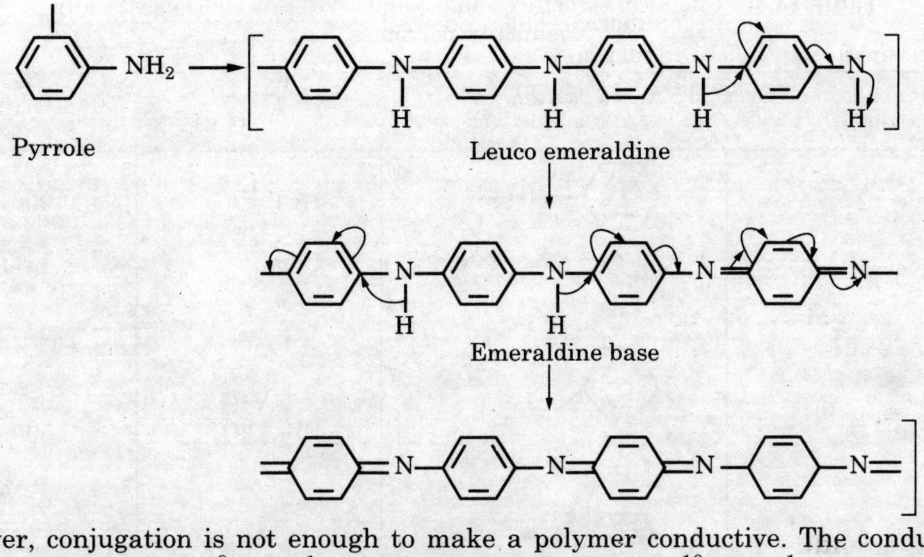

However, conjugation is not enough to make a polymer conductive. The conductivity of polyacetylene is 1.7×10^{-9} S cm^{-1} and that of polyaniline is 10^{-10} S cm^{-1}. The conductivity is greatly enhanced through doping.

2. **Doped conducting polymers:** It was shown that electrical conductivity of polyacetylene could be increased by a factor of 10^{12} S cm^{-1} when it was doped with an electron donor, such as alkali metal ion or an electron acceptor, such as arsenic pentaflouride (AsF_5) or iodine.

The process of doping can be carried out by two methods:

(i) **Oxidation**: In this process, doping typically removes some electrons from π bond of the conjugated double bonds as they have low ionization potentials. The holes so created can move along the molecule i.e. the polymer becomes electrically conductive. Thus, when an electron is removed from the top of the valence bond of a conjugated polymer, such as polyacetylene or polypyrrole, a valency (hole or radical cation) is created, causing partial delocalization, extending over several monomeric units. The radical cation is called 'polaron'. It stabilizes itself by polarizing the medium around it and hence the name. The polorons are mobile and can move along the polymer chain by rearrangement of double and single bonds and hence the polymer become conducting. The oxidation process is generally brought about by Lewis acids such as $FeCl_3$ etc.

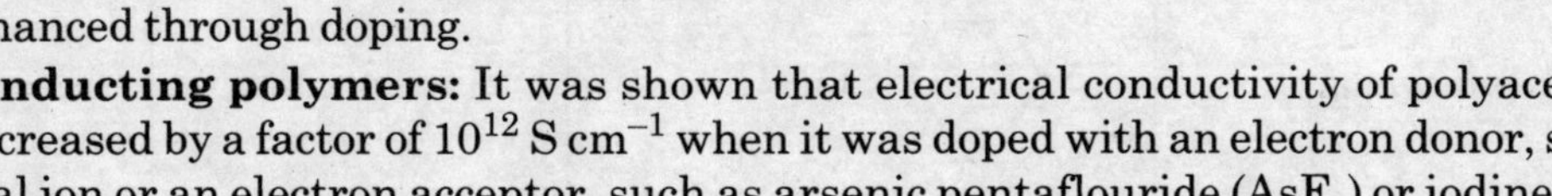

Propagation of a polaron through a conjugated polymer chain by shifting of double bonds.

(ii) **Doping through reduction**: In this type of doping some electrons are introduced into the polymer having conjugate double bonds as they have high electron affinities. This process is brought about by adding Lewis bases.

The chemical structures and conductivities of some electrically conducting polymers are given in (Table 13.3).

Table 13.3: Chemical structures and conductivities of some electrically conductive polymers

Conductive polymer	*Repeating unit*	*Dopants*	*Conductivity ($S\ cm^{-1}$)*
Polyaniline	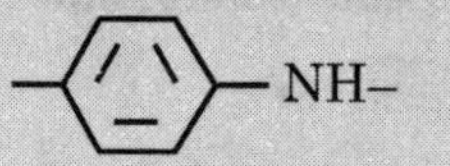	HCl	200
trans-polyacetylene	 Li, AsF_5	I_2, Br_2,	10^4
Polypyrrole	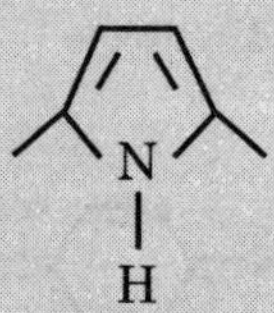	BF_4, ClO_4	500-7500
Poly (p-phenylene)		AsF_5, Li, K	10^3
Poly (p-phenylene vinylene)	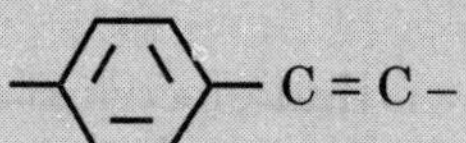	AsF_5	10^4

It should be noted that polyaniline exists in four oxidation states of which, only the emeraldine salt is conducting which require protonic doping of the imine nitrogen.

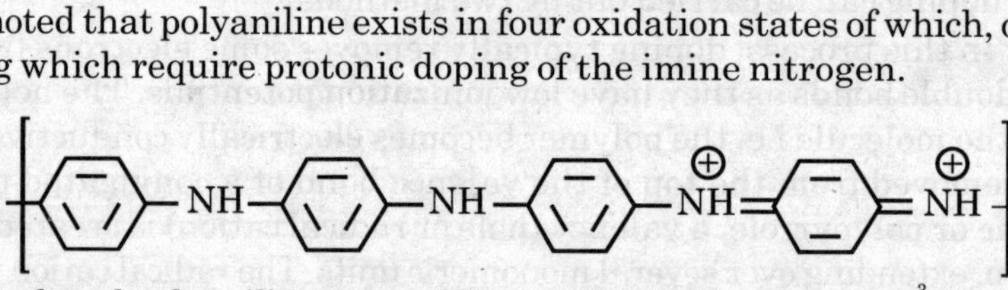

doped polyaniline (emeraldine salt) (Conductivity ~10^3 Scm^{-1})

Because of its conductivity, it is also known as *synthetic metal*.

3. **Conductive element-filled polymers:** These are the polymers which are filled with conducting elements, such as carbon black, metallic fibres, metal oxides etc. The polymer acts as a binder to hold the conducting elements together. These polymers possess reasonable conductivity. Moreover, these are low in cost, light in weight and mechanically durable and strong. Carbon black which is used as a filler is special conducting grade carbon. This carbon has very high surface area and is highly porous.
   However, the addition of the filler affect the properties of the polymer such as tensile strength, and impact strength etc. This problem can be overcome by the use of blended conducting polymers.
4. **Blended conducting polymer:** Conducting polymers can be blended with the conventional polymers to produce blends. These blends so formed possess better physical, chemical and mechanical properties.

ENGINEERING APPLICATIONS OF CONDUCTING POLYMERS

Potential applications of conductive polymers are quite significant. This is due to the fact that these are generally lighter, more flexible and easier to fabricate than many of the materials, which are conventionally used. Some of the promising applications include:

(i) **Rechargeable batteries:** The use of conducting polymers as cathodes and solid electrolytes in batteries for automotive and other applications is highly advantageous as an alternative to lead acid batteries. The polymeric batteries have high reliability, light weight, nonleakage of electrolyte solution and are small in size.

(ii) Some conducting polymers like polyaniline show different colours in different oxidation forms. This electrochromic property can be used to produce "**smart windows**," and electrochromic displays. (Smart windows are the windows which change colour in response to sunlight or temperature changes).

(iii) These are also used in **electroluminance displays** like in mobile phones etc.

(iv) Other possible uses of conductive polymers include **analytical sensors**, (for pH, SO_2, glucose etc) conductive paints, etc.

(v) These can also be used in **Light Emitting Diodes** (LEDs), electromagnetic shielding and non-linear optical material (Non-linear optical property means that the optical properties varies with the intensity of light).

(vi) Can also be used in photovoltaic devices e.g in Al/polymer/Au photovoltaic cells.

(vii) Their wider commercial utilization awaits further increase in their conductivities and improvements in the mechanical properties and processability. One of the promising near-future applications of conductive polymers would be in the integrated circuits. The integrated circuits have attained the minimum pattern dimension of 200 nm. But the dream is to have the pattern on a single molecule, which can only be realised with the conductive polymers. This will increase the speed and dynamic memory of computers by a factor of 10^8.

13.10 MOLECULAR WEIGHT OF POLYMERS

The molecular weight of polymers is related to the chain length and the extent of crosslinking between different chains. The extent of cross-linking depends on the concentration of the monomer during polymerization. The length of the polymer chain depends on the random encounter between the monomer and the reactive site of the chain. Due to the randomness, some polymeric chains grow longer than the others. Thus, a polymer sample is a mixture of different molecules having different chain lengths. Hence molecular weight of a polymer is always expressed as some sort of average of molecular weight. The molecular weights can be expressed as–

(i) Number Average Molecular Weight M_n:

$$\bar{M}_n = \frac{\text{Total Mass of the polymer sample}}{\text{Number of molecules present in the sample}}.$$

or

$$\bar{M}_n = \frac{n_1 M_1 + n_2 M_2 + n_3 M_3 + \ldots}{n_1 + n_2 + n_3 + \ldots}$$

$$= \frac{\Sigma n_i M_i}{\Sigma n_i}$$

where n_1, n_2, n_3 etc. are the number of molecular species having molecular mass M_1, M_2, M_3 etc ... respectively. Measurement of an appropriate colligative property affords the Number Average Molecular Weight.

(ii) Weight Average Molecular weight Mw:

$$\bar{M}_W = \frac{W_1M_1 + W_2M_2 + W_3\ M_3 + \ldots}{W_1 + W_2 + W_3 + \ldots} = \frac{W_1\ M_1 + W_2M_2 + W_3M_3 + \ldots}{W}$$

where W_1 = Total mass of the species having molecular weight M_1.
$W_1 = n_1M_1$
W = Total weight of the polymer sample

$$\therefore \quad \bar{M}_W = \frac{\sum n_i\ M_i^2}{W}$$

(iii) Viscosity Average Molecular Weight, $\bar{M}_v$:

$$\bar{M}_v = \left[\frac{\sum Ni\ Mi^{(1+a)}}{\sum Ni\,Mi}\right]^{1/a}$$

where Ni is the number of molecules of molecular weight Mi a is the exponent in Mark-Kuhn-Houwink equation i.e. $[\eta] = KM^a$, where η is the intrinsic viscosity and K is a constant.

(iv) Degree of polymerization (*DP*): It is the average number of monomer units in a chain. It can be number average degree of polymerization (DP_n) or weight average degree of polymerization ($\overline{DP}_W$).

$$\overline{DP}_n = \frac{\bar{M}_n}{M_o} \text{ and } \overline{DP}_W = \frac{\bar{M}_W}{M_o}$$

where $\bar{M}_n$, $\bar{M}_W$ and $\bar{M}_o$ are number average, weight average and molecular weight of the monomer respectively.

NUMERICAL PROBLEMS

Example 1. Calculate the number average molecular weight, if two polymers, having mass equal to 100 to 10,000 are mixed.

Solution: $\bar{M}_n = \dfrac{100 + 10,000}{2} = 5050.$

Example 2. Calculate the number and weight average molecular wights of a polymer sample containing 20% of polymer *A* and 80% of polymer *B*. The molecular weights of *A* and *B* are 3000 and 30,000 respectively.

Solution: $\bar{M}_w = \dfrac{0.2 \times 3000 + 0.8 \times 30,000}{0.2 + 0.8} = 24600.$

For finding out $\bar{M}_n$ we have to calculate the number of molecules of *A* and *B*.

Number of molecules of *A* having

$$\frac{\text{molecular mass 30,000}}{\text{Number of molecules of } B \text{ having molecular mass 3000}} = \frac{80 \times 3000}{20 \times 30,000} = \frac{2}{5}$$

Thus for every 2 macromolecules of mass 3000, there are 5 molecules of mass 30,000. Hence,

$$\bar{M}_n = \frac{5 \times 3000 + 2 \times 30{,}000}{(5+2)} = 10714.$$

Example 3. Equal weights of polymer molecules with molecular weights 20,000 g/mol and 200,000 g/mol are mixed. Calculate $\bar{M}_n$ and $\bar{M}_w$.

Solution: Let the weight of one of the polyemer = W_1

Weight of the other polymer = W_2

$$W_1 = W_2$$

Let it $W_1 = W_2 = 400{,}000$ g

Number of molecules of first monomer

$$n_1 = \frac{400{,}000}{20{,}000} = 20$$

Number of molecules of second polymer

$$n_2 = \frac{400{,}000}{202{,}000} = 2$$

$$\bar{M}_n = \frac{n_1M_1 + n_2M_2}{n_1 + n_2} = \frac{20 \times 20{,}000 + 2 \times 200{,}000}{20 + 2}$$

$$\bar{M}_w = \frac{n_1M_1 + n_2M_2}{n_1M_1 + n_2M_2} = \frac{20 \times (20{,}000) + 2 \times (200{,}000)^2}{20 \times 20{,}000 + 2 \times 200{,}000}$$

$$= \frac{20 \times 2 \times 10^8 + 2 \times 2 \times 10^{10}}{8 \times 10^5} = \frac{4 \times 10^9 + 4 \times 10^{10}}{8 \times 10^5} = \frac{4 \times 10^9\ (1+10)}{8 \times 10^5}$$

$= 110{,}000$ g/mol.

Example 4. If 1000 g of a polymer of molecular weight 1000 g/mole is mixed with 1000 g of another polymer of molecular weight 10^6 g/mole, what is the ratio of $\overline{M}_w / \overline{M}_n$.

Solution:

Polymer	*Weight*	*Mol.wt.*	*No. of moles in*
1	1000 g	1000 g/mol	1 (n_1)
2	1000 g	10^6 g/mol	$10^{-3}(n_2)$

$$\bar{M}_n = \frac{n_1M_1 + n_2M_2}{n_1 + n_2} = \frac{(1 \times 1000) + (10^{-3} \times 10^6)}{1 + 10^{-3}} = \frac{2000}{1.001} \approx 2000$$

$$\bar{M}_w = \frac{n_1M_1^2 + n_2M_2^2}{n_1M_1 + n_2M_2} = \frac{1 \times (1000)^2 + 10^{-3} \times (10^6)^2}{(1000 + (10^6 \times 10^{-3}))}$$

$$= \frac{10^6 + 10^9}{1000 + 1000} = \frac{10^6(1+10)}{2000} = 5.5 \times 10^5$$

$$\frac{\bar{M}_w}{\bar{M}_n} = \frac{5.5 \times 10^5}{2000} = 2.75 \times 10^2$$

Problem: A polydisperse sample of polystyrene is prepared by mixing three monodisperse samples in the following proportions:

1 g 10,000 molecular weight, 2 g 50,000 molecular weight, 2 g 100,000 molecular weight. Using this information, determine the following:

(a) Number-average
(b) Weight-average.

13.11 POLYMER BLENDS

Polymer blends are mixtures of at least two macromolecular species, polymers and/or copolymers. For practical reasons, the name *blend* is given to a system only when the minor component content exceeds 2 wt%.Polymer blends constitute ~36 wt% of the total polymer consumption.

Why Blends?

- In the economically advanced countries, plastics have displaced conventional materials for most applications.
- Today the market pressure forces the resin manufacturers to provide better,more economic materials with superior combinations of properties, not as a replacement for wood or steel, but rather to replace the more traditional polymers.
- This has resulted in:
  - Increased scale of production.
  - Use of multicomponent and multiphase materials.
  - New processing methods.

Polymer melt blends may be miscible or immiscible.

Miscible blends (homogeneous) form solutions and there is no distinct phase morphology to be of concern.

Immiscible blends (heterogeneous) are characterized by two or more phases that are separated by interfaces.

In homogeneous blends, both blend components lose part of their identity and the final properties usually are the average of both blend components. In heterogeneous blends, the properties of all blend components are present. Weaknesses of one polymer can to a certain extent be overcome by strengths of the other. In exceptional cases, some properties of the either homogeneous or heterogeneous blend can be better than those of the individual components.

Miscibility of Polymers

Homogeneous miscibility in polymer blends requires a negative free energy of mixing:

$$\Delta G_{mix} = \Delta H_{mix} - T.\Delta S_{mix}$$

However, if two high molecular weight polymers are blended, the gain in entropy, ΔS_{mix} , is negligible, and the free energy of mixing can only be negative if the heat of mixing, ΔH_{mix} is negative. In other words, the mixing must be exothermic, which requires specific interactions between the blend components. These interactions may range from strongly ionic to weak and nonbonding interactions, such as hydrogen bonding, ion-dipole, dipole-dipole and donoracceptor interactions. Usually, only Van der Waals interactions occur.

Types of blends

Basically, three different types of blends can be distinguished.

- *Completely miscible blends,* for which $\Delta H_{mix} < 0$ due to specific interactions. A well known example of a blend which is miscible over a very wide temperature range and in all compositions is PS/PPO, that combines the heat resistance, the inflammability and the toughness of PPO with the good processability and the low cost of PS.
- In *partially miscible blends* a (small) part of one blend component is dissolved in the other. This type of blend, which exhibits a fine phase morphology and satisfactory properties, is referred to as *compatible.* An example is the PC/ABS blends, which combine the heat resistance and toughness of PC with the low temperature impact, processability, stress cracking resistance and low cost of ABS.
- By far most blends are fully *immiscible.* They have a coarse phase morphology, the interface is sharp, and the adhesion between both blend phases is poor, so that these blends are useless without being *compatibilised.*

Most polymer blend systems are immiscible because of the low entropies of mixing associated with mixing chain-like molecules to produce homogeneous solution. It is possible to modify the interfaces between liquids with specific additives. Such additives are generally molecules with hydrophobic and hydrophilic sections that align along interfaces between the two liquid phases. They reduce interfacial tension and allow the polymers to mix. These polymeric interfacial agents, are known as *compatibilizing* agents.

Advantages of Blending

1. Providing materials with full set of desired properties at the lowest price.
2. Improving specific properties, viz. impact strength or solvent resistance.
3. Offering the means for industrial and/or municipal plastics waste recycling.
4. Blending also benefits the manufacturer by offering: Improved processability, product uniformity, and scrap reduction

13.12 DENDRIMERS

Dendrimers are a new class of polymeric materials. They are highly branched, monodisperse macromolecules. The structure of these materials has a great impact on their physical and chemical properties. As a result of their unique behaviour dendrimers are suitable for a wide range of biomedical and industrial applications.

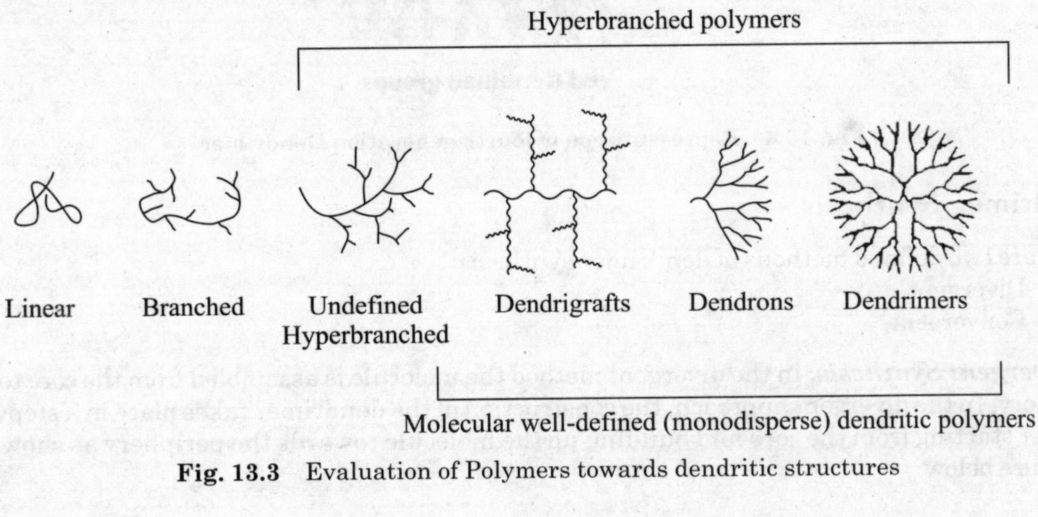

Fig. 13.3 Evaluation of Polymers towards dendritic structures

A dendrimer is generally described as a macromolecule, which is characterized by its highly branched 3D structure that provides a high degree of surface functionality and versatility. The term originates from 'dendron' meaning a tree in Greek They are highly branched, monodisperse macromolecules. Dendrimers having a well-defined finite structure belongs to a special case of hyperbranched polymers (see Figure 13.3).

Terms Used

1. Dendifrafts: Dendrigrafts are centred around a linear polymer chain, to which branches consisting of copolymer chains are attached.
2. Dendrons: *Dendrons* is the term used about a dendritic wedge without a core, the dendrimer can be prepared from assembling two or more dendrons.(Figure)
3. Generation: The number of focal points when going from the core towards the dendrimer surface is the generation number i.e a dendrimer having five focal points when going from the centre to the periphery is denoted as the 5th generation dendrimer. The core part of the dendrimer is sometimes denoted generation "zero", or in the terminology presented here "G0". (Figure 13.3)

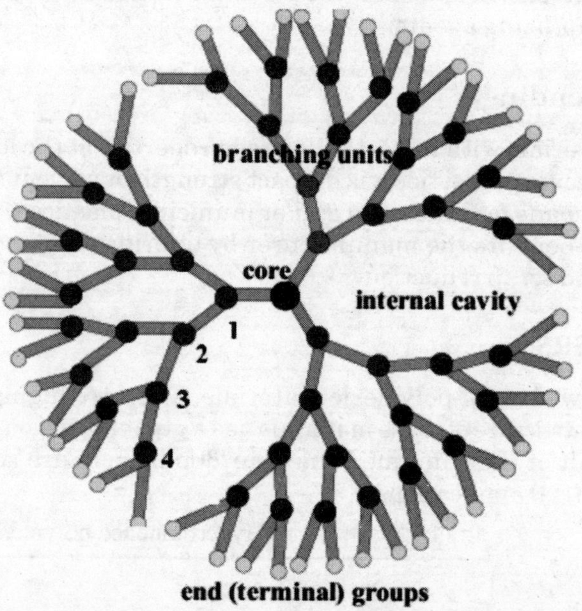

Fig. 13.4 Representation of fourth generation Dendrimer

Dendrimer Synthesis

There are two defined methods of dendrimer synthesis:

1. Divergent
2. Convergent.

1. Divergent Synthesis: In the divergent method the molecule is assembled from the core to the periphery; In the divergent approach, the construction of the dendrimer takes place in a stepwise manner starting from the core and building up the molecule towards the periphery as shown in the figure below:

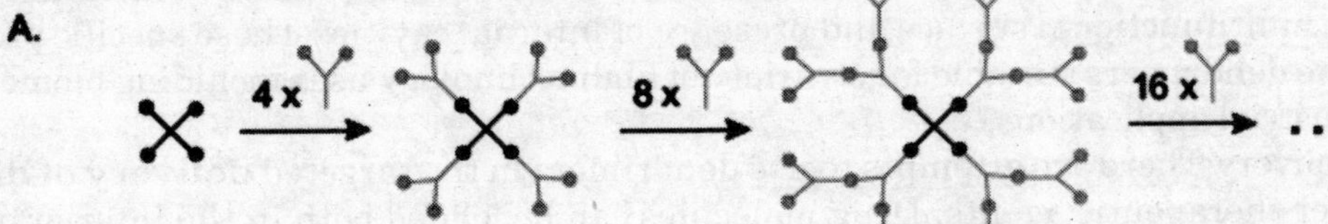

2. Convergent Synthesis: In the convergent approach, the dendrimer is constructed stepwise, starting from the end groups and progressing inwards. When the growing branched polymeric arms, called dendrons, are large enough, they are attached to a multifunctional core molecule

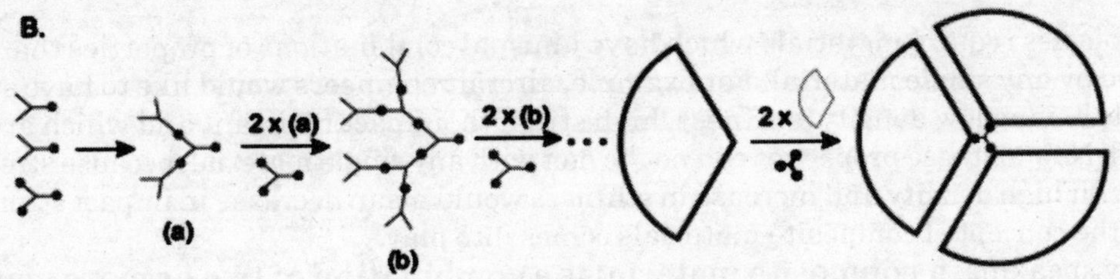

Comparison

In contrast to the divergent method, the convergent method construct a dendrimer from the surface and inwards towards the core.The final part of the convergent synthesis ends up at the core, where two or more dendritic segments (dendrons) are joined together, creating the dendrimer. Whereas in the divergent method, the dendrimer is synthesized beginning from the core.

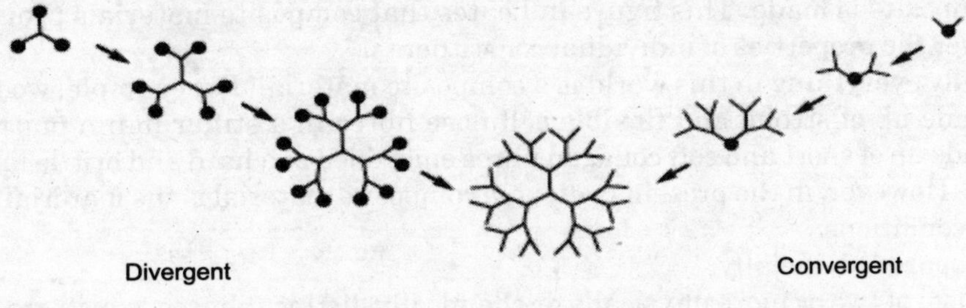

Applications

The structure of these materials has a great impact on their physical and chemical properties. As a result of their unique behaviour dendrimers are suitable for a wide range of biomedical and industrial applications.

- Dendrimers, like macromolecules found in biology, respond to the surrounding chemical environment showing altered conformational behaviour upon changes in, *e.g.* pH, solvent polarity and ionic strength.
- Dendrimers have some unique properties because of their globular shape and the presence of internal cavities. The most important one is the possibility to encapsulate guest molecules in the macromolecule interior.
- Light Harvesting Dendrimer :It is possible to create dendrimers which can act as extremely efficient light-harvesting antennae Absorbing dyes are placed at the periphery of the dendrimer and transfer the energy of light to another chromophore located in the core.

- Many potential applications of dendrimers are based on their unparalleled molecular uniformity, multifunctional surface and presence of internal cavities.These specific properties make dendrimers suitable for a variety of high technology uses including biomedical and industrial applications.
- Drug Delivery:There are attempts to use dendrimers in the targeted **delivery of drugs** and other therapeutic agents. Drug molecules can be loaded both in the interior of the dendrimers as well as attached to the surface groups. Dendrimers can be used as coating agents to protect or deliver drugs to specific sites in the body or as time-release vehicles for biologically active agents.

13.13 COMPOSITES

Modern technologies require materials which have unusual combinations of properties that cannot be provided by any single material. For example, aircraft engineers would like to have structural materials having low density, stiffness, high strength, impact resistant and which are not easily corroded. Now all these properties can not be met with any single material, because strength is associated with high density and increase in stiffness would mean decrease in impact strength. This is where the concept of composite materials comes into play.

Generally speaking, **a composite material is a combination of two or more materials that exhibit a significant proportion of the properties of all the constituent materials**. A combination of two or more distinct materials can provide better combination of properties such as strength, heat resistance or stiffness etc. Such composite material systems result in a performance which cannot be achieved by any of the individual constituents. Thus, the composite materials offer the great advantage of combining different materials to tailer make the desired properties as per the required specification.

As an illustration, in Fig. 13.5 a comparison of different properties of individual constituents and their composite is made. This figure indicates that composite materials provide great improvement over the properties of individual constituents.

Practically everything in this world is a composite material. For example, wood is a fibrous composite made up of strong and flexible cellulose fibres in a stiffer lignin matrix. Bone is a composite made up of short and soft collagen fibres embedded in a hard and brittle mineral matrix called apatite. However, in the present context, a composite material is the material which meets the following conditions:

(i) It is prepared artificially.

(ii) It consists of two or more physically or chemically distinct phases, which are separated by a distinct interface.

(iii) It has properties which are not depicted by any of its individual components.

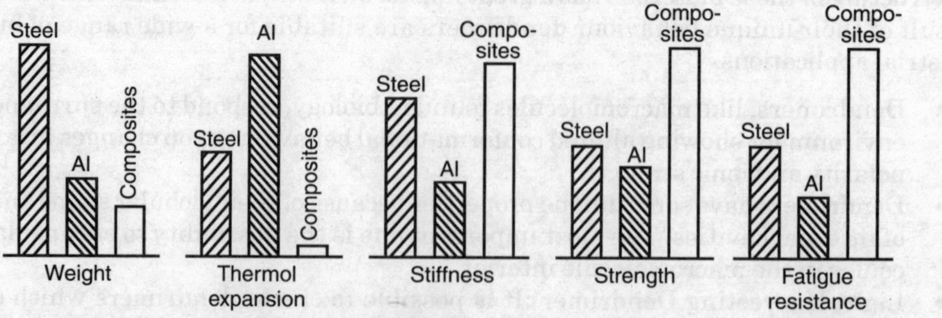

Fig. 13.5 Comparison between conventional materials and composite materials.

13.13.1 Constituents of Composites

Composite materials generally comprise of just two phases

(i) The constituent forming the body of the composite, called the **'matrix'** and

(ii) The structural constituent called **'dispersed phase'** or the reinforcing material.

The properties of the composites are determined by the properties of the constituent phases, their relative amounts and their distribution and orientation.

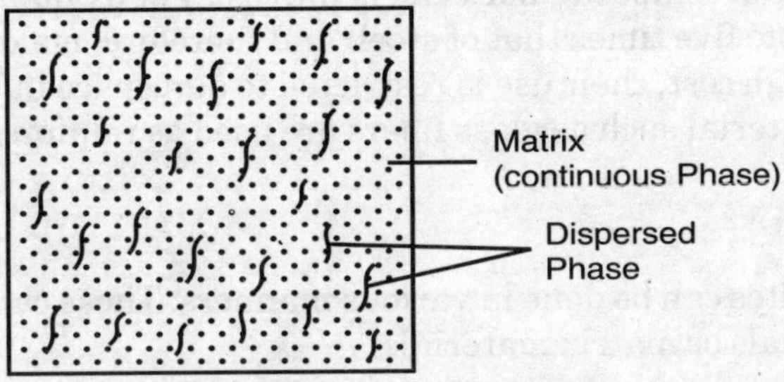

Fig. 13.6 Representation of a composite material

The most common example of a composite material is the **fiber glass reinforced plastic**. The plastic acts as the matrix and fibre glass is the dispersed phase. The plastic alone is relatively weak and has a low elastic modulus, bends and stretches easily. The glass fibers provide the strength and stiffness and their modulus of elasticity may be 50 times greater than that of plastic. Glass fibers take most of the load when the composite is stressed.

(i) **Matrix Phase**

The matrix phase of composites has several important functions like-

- It binds the reinforcing material (dispersed phase) together,
- protects the reinforcing material from surface damage due to mechanical abrasion or chemical reaction.
- It separates the individual particles of the reinforcing materials.
- It acts as a medium by which external load is transmitted and distributed to the dispersed phase.
- It prevents the propagation of brittle cracks because of its plasticity.

The matrix material should meet the following criteria.

- Its elastic modulus should be lower than that of the dispersed phase :
- It should be ductile.
- The adhesive bonding forces between the fiber and the matrix must be high. The ultimate strength of the composite depends on the magnitude of these forces.

Examples of some materials used as matrix include-polymers, metals and ceramics. Thus we have—

- Polymer matrix composites (PMC)
- Metal matrix composites (MMC)
- Ceramic matrix composites (CMC)

(ii) **The Dispersed Phase (reinforcement)**

To improve the strength of a matrix, the reinforcement must be stronger and stiffer than the matrix. These should possess high tensile strength.

Properties of the composites, to a large extent, depend upon the properties of the reinforcing materials, their geometry, which is described by shape, size and size distribution and orientation.

They are chosen for their high strength and modulus. The physical form and shape of the reinforcements vary greatly depending on many factors. The most effective reinforcements are *long fibres*, and the composites so formed are called *fibre reinforced composites*. Some of the fibers used as reinforcement are glass fibres, carbon fibers, boron fibers and ceramic fibers. Other reinforcing materials include whiskers, metallic wires, particles, flakes etc. But still, fibers form the most commonly used reinforcing materials because the composites so formed are stiffer and stronger than the composites formed by any other reinforcing material. Another factor which is responsible for the choice of composite material is the cost. For example, carbon fibers have very high modulus values (upto five times that of steel) and therefore, make excellent reinforcement. However, due to their high cost, their use is restricted to aerospace industry. For non-aerospace applications cheaper material such as glass fibers are used as reinforcing materials.

13.13.2 Classification

Classification of composites can be done in various manners. These can be classified on the basis of the reinforcing materials or matrix materials.

(A) On the basis of reinforcing materials the composites can be classified as:

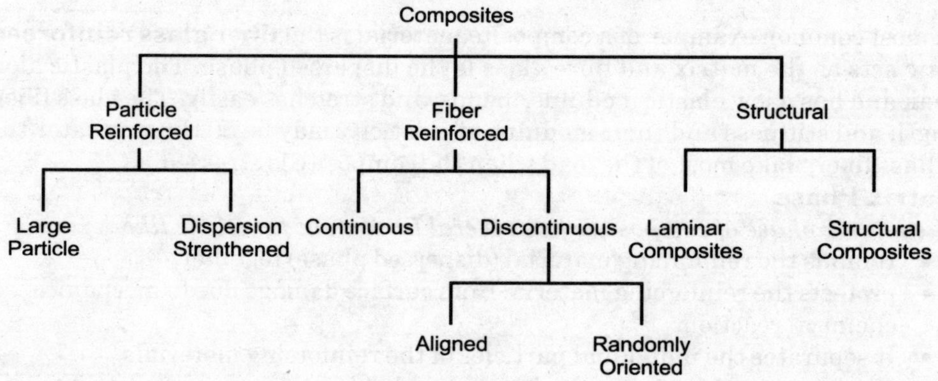

Fig. 13.7 Classification scheme on the basis of reinforcing materials

As in evident from the Fig. 13.7 that reinforcements can be in the form of particles, fibers or structural. Following is the brief description of these materials:

1. **Particle Reinforced Composites**

   When the reinforcing material is in the form of large or small particles suspended in a matrix, the composites are called particle reinforced composites. The particles can be either metallic or non metallic. These are of two types :

   (a) **Large-particle Composites:** The term "Large"is used to indicate that particle-matrix interactions cannot be treated on the atomic and molecular level. For most of these composites the particulate phase is harder and stiffer than the matrix. The reinforcing particles can have quite a variety of geometries, but they should be of approximately the same dimensions in all directions (equiaxed). For effective reinforcement, the particles should be small and evenly distributed throughout the matrix. Common examples of large-particle composites include:

   automobile tyres in which polymer matrix contains spherical particles of carbon with 20-50 mm diameters. The carbon increases the tensile strength, toughness and abrasion resistance.

   Cermet is another important particle reinforced composite which is composed of extremely hard particles of a refractory carbide, such as tungsten carbide or titanium carbide, embedded in a matrix of metal such as cobalt or nickel. These composites are utilized extensively as cutting tools for hardened steels. No single material could possibly provide the combinaiton of properties possessed by a cermet. In these composite materials (cermets) the hard carbide particles provide the cutting surface, but, being extremely brittle, are not themselves capable of withstanding the cutting stresses. The metal matrix thus enhances the toughness, which also isolates the carbide particles from one another and prevents particle to particle crack propagation.

Another example of large-particle composites is *concrete*, which is composed of cement matrix and particulates of sand and gravel.

(b) **Dispersion strengthened composites:** In these type of composites, very small particles of the range 10 to 100 nm are used as strengthening material. The particle-matrix interactions occur on the atomic or molecular level. The dispersed phase i.e. the reinforcing particles may be metallic or non-metallic. The strengthening mechanism involves interactions between the particles and dislocations within the matrix. The strengthening is retained at elevated temperatures and for longer time periods because the material for dispersed phase is so chosen, which is unreactive with the matrix phase. While the matrix bears the major part of the applied load, the dispersed particles hinder the motion of dislocations. This helps in restricting the plastic deformation and improving the tensile strength and hardness.

Fine metallic or non-metallic particles are used as dispersed particles for strengthening metals and metal alloys. For example, the high temperature strength of nickel alloy is improved by adding thoria (ThO_3) in the form of finely dispersed particles in 3 volume%.

2. **Fiber-Reinforced Composites :** Fibers constitute an important class of reinforcing materials. These composites provide high strength and stiffness on a weight basis. The use of fiber as reinforcing materials in due to

- The small diameters of fiber with respect to other microstructural units.
- The aspect ratio (length/diameter) of fibers is very high and hence, a very large fraction of the applied load is transferred via the matrix to the fibers which are stiff and strong.
- A very high degree of flexibility permits the use of a variety of techniques for the formation of composites, using the fibers.

Two most important characteristics of fiber reinforced composites are Fiber length and Fiber orientation.

(i) **Fiber Length:** The applied load on the composite is transmitted to the fibers by the matrix phase and hence, some critical fiber length is necessary for effective strengthening and stiffening of composites. The critical length l_c is dependent on the fiber *diameter d*, its *tensile strength* σ_f and on the *fiber matrix bond strength* (shear yield strength) Tc as per the following equation:

$$l_c = \frac{\sigma_f d}{T_c}$$

For a number of fibers this critical length is of the order of 1 nm which is, in most of the cases, ranges between 20 to 150 times of the fiber diameter. If the fiber length is greater than the critical length, the fibers are termed as **continuous** and if it is consideraly less than the critical length, then the fibers are termed as **discontinuous** or short fibers.

(ii) **Fiber Orientation :** The strength and other properties of the fiber reinforced composites are significantly influenced by the orientation and arrangement of fibers relative to one another. There are two extreme possibilites with respect to the orientation of fibers:

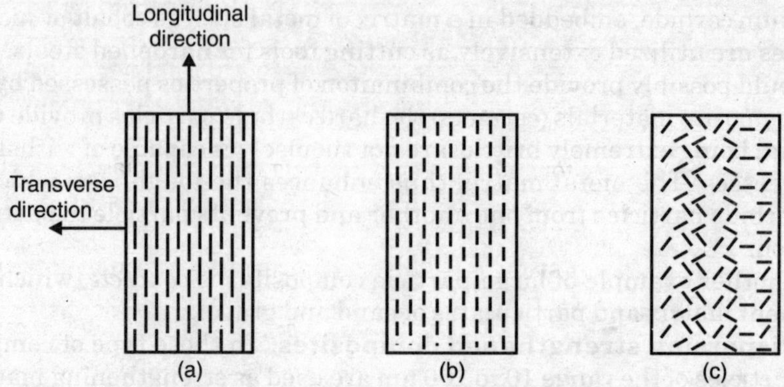

Fig. 13.8 Schematic representations of (a) continuous and aligned (b) discontinous and aligned, and (c) discontinuous and randomly oriented fiber reinforced composites

- A parallel alignment of the longitudinal axis of the fibers in a single direction.
- A totally random alignment.

Continuous fibers are normally aligned whereas discontinuous fibers may be aligned or randomly oriented. Fig. 13.8 represents the three types of fiber composites schematically.

(a) **Continuous and aligned Fiber Composites:** A uniform fiber distribution results in better overall composite properties. The properties of continuous and aligned composites are highly anisotropic, that is, they depend on the direction in which they are measured. In the longitudinal direction, i.e. along the alignment, maximum strength and reinforcement is achieved. In the transverse direction, as the fiber reinforcement is virtually non-existent, hence strengthening is least.

(b) **Discontinuous Fiber Composites :** The reinforcing property of discontinuous fiber composites (aligned as well as random) is lower than that of continuous fiber composites, but still these are used extensively because of very low cost as compared to continuous fibers. Further, in case of discontinuous (short) fiber composites, intricate shapes can be formed which are rather impossible with continuous fiber reinforcement. Different type of fibers used for reinforcing materials are discussed in the next section (13.3)

3. **Structural reinforced composites :** A structural composite consists of both, the homogeneous material and composite material. These two different materials are bonded together in layers. The properties of these composites are not only governed by the properties of the constituent materials but also on their geometrical arrangements. Broadly structural composites are classified as :

(i) Laminar composites and
(ii) Sandwitch composites.

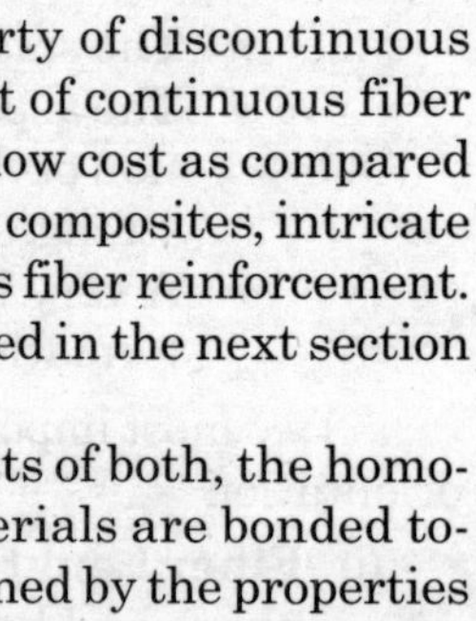

Fig. 13.9 Stacking of successive oriented fiber-reinforced layers for producing a laminar composites

(i) **Laminar Composites:** In laminar composites, two dimensional layers of two different materials are stacked together to form the composite structure. Each single layer is known as lamina. A single lamina is unsuitable for any useful purpose. The laminae are stacked and glued together such that the orientation of the high strength direction varies with each successive layer as shown in Fig. 13.9. The individual unidirectional lamina piles are oriented in such a way that the resulting structural component has the desired mechanical or physical characateristic in different direction.

Examples of laminar composites include plywood, cladded metals, plastic based laminates. Adjacent sheets in plywood are aligned with the grain direction at right angles to each other. The plastic based laminates are very useful in high temperature applications. These are formed by impregnating silicons in layers of glass or asbestos.

(ii) **Sandwitch Panels:** These composites consists of two strong outer sheets or faces, separated by a layer of less dense material called core. The core material is of lower strength and lower stiffness. The face materials are generally fiber reinforced plastics, plywood, titanium, steel and aluminium alloys. For core material, synthetic rubbers, foamed polymers etc are used. The core can also be composed of honey comb structure made up of thin foils that have been formed into interlocking hexagonal cells with their axes oriented in a perpendicular direction to the face planes as shown in Fig. 13.10.

The core serve to separate the "faces" and hence resist any deformations perpendicular to the face plane.

Sandwitch panels are used extensively in aircraft for wings. They are also used in roofs, floor and walls of buildings. Use of lighter composites in aircraft structures results in energy savings. For a given aerodynamic configuration of an aircraft, there is a direct correlation between aeroplane weight and fuel consumption. Example, in 1986 aircraft voyager completed non stop flight around the world without refueling because of the use of sandwitch composite in its manufacture, resulting in lower weight and hence, fuel saving.

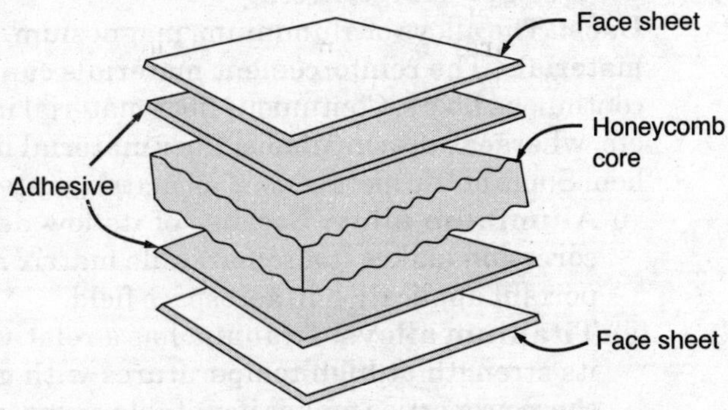

Fig. 13.10 Schematic diagram showing the construction of a honeycomb core sandwitch panel

Sandwitch composites also find uses in safety glasses. Safety glass is a layer of PVB sandwitched between two layers of glass. The glass in the composite protects the plastic from scratching and gives it stiffness. The plastic provides the toughness to the composites.

(B) On the basis of matrix material, composites can be classified as:

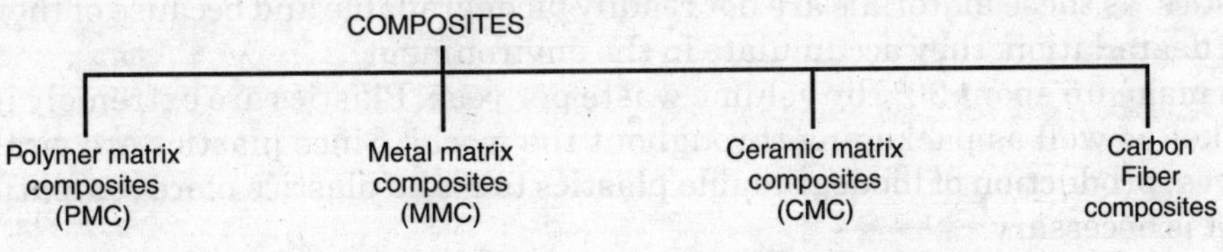

Fig. 13.11 Classification scheme on the basis of matrix material.

1. **Polymer-matrix composites (PMC):** Polymer matrix composites (PMCs) are extensively used as engineering structural materials. The polymers which are used as the matrix material can be thermoplastic or thermoset. The most widely utilized polymer resins are the polyesters and vinyl esters. Epoxies have also been used as matrix resins for advanced composite materials because of ease in processing and excellent mechanical properties. Polyether ketones, polyphenylene sulphide and polyetherimide thermoplastics are used as matrix materials for high temperature applications.

A variety of reinforcements can be used with both thermoset and thermoplastic PMCs. High performance fibers such as carbon, boron, aramid, glass fiber etc are used as reinforcing materials. Nevertheless, glass fiber reinforced polymers represent the largest class of PMCs.

Uses: Glass fiber reinforced polymer matrix are used in a wide variety of industries. Tanks and vessels in the chemical process industry, as well as process and effluent pipelines are made of glass fiber reinforced polymer matrix composites. Use of PMCs in military and commercial helicopters has been most significant. The main driving force in such applications is that of weight reduction. Boron/epoxy composites are used in sporting goods, e.g., golf clubs and tennis rackets.

2. **Metal Matrix Composites (MMC) :** Metal matrix composites consist of metal or metal alloy as matrix material. MMCs are of interest these days because they offer the opportunity to combine the desirable properties of two materials which are not available with the single material. The use of metals as matrix materials is due to their high operating temperature, nonflammability, and greater resistance to degradation by organic fluids. Metal matrix composites are costlier than the polymer matrix composites and hence, their use is somewhat restricted.

   Uses: The alloys of aluminium, magnesium, titanium and copper are employed as matrix materials. The reinforcement materials can be particulates, short fibers or whiskers or continuous fibers. Continuous fiber material include carbon, silicon carbide, boron, alumina etc, wheraeas discontinuous fiber material include chopped fibers and alumina and carbon. Some of the metals used as matrix materials are.

   (i) **Aluminum alloy:** Because of its low density, excellent strength and resistance to corrosion makes it a remarkable matrix material. The composites so formed find important application in aerospace field.

   (ii) **Titanium alloys:** Titanium has a relatively high melting point (1672^0C) and retains its strength at high temperatures with good oxidation and corrosion resistance. All these properties makes it suitable as matrix material for aerospace applications. Titanium alloys are used in jet engines, fuselage parts etc. However, it is an expensive material.

13.14 BIO-DEGRADABLE POLYMERS

Plastics are being used all over the world. The same durability properties which make plastics ideal for many applications such as in packaging, building materials and commodities, as well as in hygiene products, can lead to waste disposal problems in the case of traditional petroleum-derived plastics, as these materials are not readily biodegradable and because of their resistance to microbial degradation, they accumulate in the environment.

Plastics make up about 20% by volume waste per year. Plastics are extremely important to the job market as well as packaging throughout the world. Since plastics are vital to people's everyday lives, production of biodegradable plastics to make plastics more compatible with the environment is necessary.

The term biodegradation is often used to denote degradation occurring in a biological environment. Biodegradation (i.e. biotic degradation) is a chemical degradation of materials (i.e. polymers) provoked by the action of microorganisms such as bacteria, fungi and algae. The most common definition of a biodegradable polymer is "a degradable polymer wherein the primary degradation mechanism is through the action of metabolism by microorganisms."

Generally, biodegradable polymers are those that can be broken down by nature either by hydrolytic processes or enzymatic processes producing nontoxic by-products.In recent years there has been an increase in interest in biodegradable polymers.

Two classes of biodegradable polymers can be distinguished: synthetic or natural polymers. Natural polymers are available in large quantities from renewable sources, while synthetic polymers are produced from non renewable petroleum resources.In general natural polymers offer fewer advantages than synthetic polymers.

Biodegradation takes place through the action of enzymes and/or chemical deterioration associated with living organisms. This event occurs in two steps.

- The first one is the fragmentation of the polymers into lower molecular mass species by means of either abiotic reactions, i.e. oxidation, photodegradation or hydrolysis, or biotic reactions, i.e. degradations by microorganisms.
- This is followed by bioassimilation of the polymer fragments by microorganisms and their mineralisation.

Biodegradability depends not only on the origin of the polymer but also on its chemical structure and the environmental degrading conditions.

Biodegradable Polymers Derived from natural and synthetic polymers

Biodegradable plastics include natural or synthetic polymers (i.e., starch-based materials, cellulose- derived polymers, bacterial polyesters and a range of synthetic polymers).

*1. **Natural polymers or agro-polymers*** Natural polymers are formed in nature during the growth cycles of all organisms. Natural biodegradable polymers are called biopolymers. Polysaccharides, as starch and cellulose, represent the most characteristic family of these natural polymers. Other natural polymers as proteins can be used to produce biodegradable materials. These are the two main renewable sources of biopolymers. Another resource is lipids. These natural polymers include polysaccharides, such as cellulose, chitin, starch, alginate, galactan, hyaluronic acid, dextran, and gellan, obtained from plants, animals, and microbial sources and polyesters, such as poly(-hydroxybutyrate) and poly(b -hydroxybutyrate-cob -hydroxyvalerate), produced by numerous bacteria.

*2. **Synthetic BiodegradablePolymers:*** Developments in science and technology, especially over the last two decades, have increased the amount of synthetic polymers produced worldwide each year. Each year approximately 140 million tonnes of synthetic polymers are produced The presence of these substances in the environment brings about important problems, including a challenge to wastewater treatment plants and pollution of groundwater and surface water. Solid waste management has resulted in a growing commercial interest in the development of biodegradable materials for consumer products. Among these aliphatic polyesters are one important class of biodegradable polymers which are commercially potential biomaterials. The common examples of biodegradable polymers are polyhydroxy butyrate (PHB), polyhydroxy butyrate –co-b-hydroxy valerate (PHBV), polyglycolic acid (PGA), polylactic acid (PLA), poly (Î-caprolactone) (PCL), etc.

Mechanism of Degradation

The most important organisms in biodegradation are fungi, bacteria and algae. Natural polymers (i.e., proteins, polysaccharides, nucleic acids) are degraded in biological systems by oxidation and hydrolysis. Biodegradable materials degrade into biomass, carbon dioxide and methane. In the case of synthetic polymers, microbial utilization of its carbon backbone as a carbon source is required .

The biodegradation process can be divided into (1) aerobic and (2) anaerobic degradation (Fig. 13.12).

Aerobic biodegradation:

Polymer + $O_2 \rightarrow CO_2 + H_2O$ + biomass + residue(s)

Anaerobic biodegradation:

Polymer $\rightarrow CO_2 + CH_4 + H_2O$ + biomass+ residue(s)

Bacteria important in the biodegradation process include, *inter alia, Bacillus* (capable of producing thick-walled endospores that are resistant to heat, radiation and chemical disinfection), *Pseudomonas, Klebsiella, Actinomycetes, Nocardia, Streptomyces, Thermoactinomycetes, Micromonospora, Mycobacterium, Rhodococcus, Flavobacterium, Comamonas, Escherichia, Azotobacter* and *Alcaligenes*.

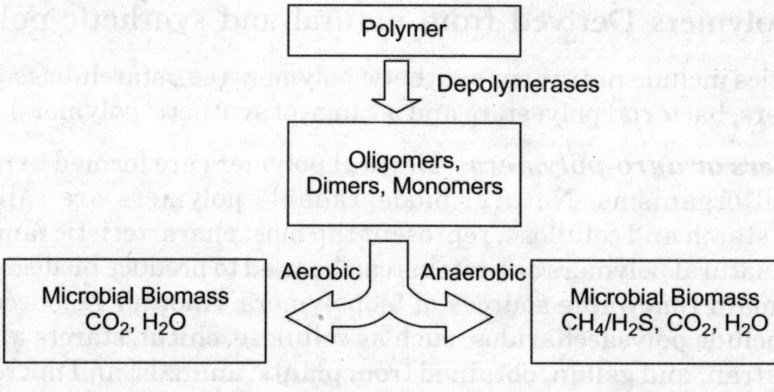

Fig. 13.12 Polymer degradation in aerobic and anaerobic conditions

Fungi active in the biodegradation process are*Sporotrichum, Talaromyces, Phanerochaete, Ganoderma, Thermoascus, Thielavia, Paecilomyces, Thermomyces, Geotrichum, Cladosporium, Phlebia, Trametes, Candida, Penicillium, Chaetomium,* and *Aerobasidium*

Temperature is one of the most important factors affecting microorganism growth. Also important are sources of carbon and nitrogen, and pH.

Uses

Biodegradable polymers are used mainly for medical goods such as surgical sutures, tissue in growth materials or for controlled drug release devices, plasma substitutes etc. The decomposition reactions usually involve hydrolysis (either enzymatically induced or by non-enzymatic mechanisms) to non-toxic small molecules which can be metabolized by or excreted from the body. These are also finding use in agriculture materials (such as films, seed coatings), fast food wrappers, personal hygiene products, etc.

13.15 ADHESIVES

An adhesive may be defined as a material which when applied to surfaces of materials can join them together and resist separation.The surfaces joined by the adhesive are known as *adherends.* Adhesive Bonding is the surface-to-surface joining of similar or dissimilar materials using a substance which usually is of a different type, and which adheres to the surfaces of the two adherents to be joined, transferring the forces from one adherent to the other.

The possibilities, advantages and disadvantages of adhesive bonding compared to other joining techniques are summarized in Table 13.4

One of the many advantages of bonding is that little or no heat is needed to create the joint. As a result, the material structure of the adherents to be joined is not macroscopically affected, and deformations or internal stress – which generally are related to the application of heat – rarely occur. Even those materials with finished surfaces or coated materials can easily be bonded without any heat supply. From this point of view, there are no limits with regard to the combinations of materials that can be joined.

Table 13.4 *Characteristic features of adhesive joints*

Advantages	*Disadvantages*
• The adherents are not affected by heat • Uniform stress distribution • Possibility to join large surfaces • Possibility to join different materials • Possibility to join very thin adherents • Gas-proof and liquid-tight • Now crevice corrosion • No contact corrosion • No precise fits of the adherent surfaces are necessary • Good damping properties • High dynamic strength	• Limited Stability to heat • Long-term use may alter the properties of the bond-line • Cleaning and surface preparation of the adherents in necessary in many cases • Specific production requirements to be met • Specific clamping devices are often required to fix the joint • Nondestructive quality testing in only possible to a certain extent

13.15.1 Theories of Adhesion

Four mechanisms or theories of Adhesion are prevalent:

1. Mechanical Interlocking
2. Diffusion Theory
3. Chemical Adhesion
4. Electrostatic

1. Mechanical Interlocking: This theory postulates that the adhesion is achieved as a consequence of flowing of an adhesive into a rough surface and the resulting 'interlocking'. Adhesive materials fill the voids or pores of the surfaces and hold surfaces together by interlocking. Thus this mechanical anchoring between the adhesive and the adherend prevents the removal of adhesive from the substrate.

- Applicable in situations where the adherends are impermeable to the adhesive.
- Surface of adherend is usually rough.
- Interphase may exhibit a "lock and key" effect
- Rubber tire repair kit is a good example.
- Rough surface increases area of physical contact.

2. Diffusion Theory: Some materials may merge at the joint by diffusion. This may occur when the molecules of both materials are mobile and soluble in each other. This would be particularly effective with polymer chains where one end of the molecule diffuses into the other material. One instructive example is that of polymer-on-polymer surfaces. Diffusive bonding in polymer-on-polymer surfaces is the result of sections of polymer chains from one surface interdigitating with those of an adjacent surface.

3. Chemical adhesion: This is widely credited and much investigated theory .According to this theory, in the event of intimate contact between the adhesive and the adherend, the adhesive strength arises as a result of interatomic and intermolecular forces at the interface. The forces between the adhesive and the adherend are usually grouped into two categories (1) primary forces and (2) secondary forces.

- Primary forces include ionic (600-1100 kJ/mole), covalent (60-700 kJ/mole) and metallic bonds (110-350 kJ/mole).
- Secondary bonds may include Van der Waals forces (0.02-40 kJ/mole), hydrogen bonds (10-40 kJ/mole), Lewis acid-base interactions (up to 80 kJ/mole), dipole-dipole and dipole-induced dipole (up to 20 kJ/mole) interactions.

Chemical adhesion occurs when the surface atoms of two separate surfaces form ionic, covalent, or hydrogen bonds. The engineering principle behind chemical adhesion in is that if surface molecules can bond, then the surfaces will be bonded together by a network of these bonds.These attractive ionic and covalent forces are effective over only very small distances – less than a nanometer.

4. Electronic theory: This theory is based on the argument that the electrostatic forces arising from the junction potentials between the contacting adhesive and substrate will contribute significantly towards the forces required to rupture the bonds

Adhesive Action

The principle component of an adhesive is an organic polymer, or one or more (usually two) compounds which can chemically react to produce a polymer. At the time of application the adhesive or sealant must be a liquid, as this enables it to make intimate molecular contact with the adherends; that is it must wet the surfaces. It must then harden (cure) to a cohesive solid.

Curing (hardening) takes place within the bulk of the adhesive and adhesion occurs at the interface. Van der Waals forces will always contribute to adhesion as these are the normal attractions between atoms and molecules; this provides the physical adsorption theory of adhesion.

Classification

Adhesives and sealants can be classified by the manner in which they harden.

- Some harden through solvent evaporation (Drying Adhesives)
- Some harden through application of pressure.
- Some change chemically .
- Some harden as they cool.

Adhesives can be classified based on the manner of hardening or on the chemical nature of the adhesives. Broadly, the adhesives can be classified as:

1. Drying adhesives: There are two types of adhesives that harden by drying: *solvent based adhesives* and *polymer dispersion adhesives*, also known as *emulsion adhesives*. Solvent based adhesives are a mixture of ingredients (typically polymers) dissolved in a solvent. White glue, contact adhesives and rubber cements are members of the *drying adhesive* family. As the solvent evaporates, the adhesive hardens. Depending on the chemical composition of the adhesive, they will adhere to different materials to greater or lesser degrees.

Polymer dispersion adhesives are milky-white dispersions often based on polyvinyl acetate (PVAc). They are used extensively in the woodworking and packaging industries; also used with fabrics and fabric-based components, and in engineered products such as loudspeaker cones.

2. *Pressure sensitive adhesives:* *Pressure sensitive adhesives* (PSA) form a bond by the application of light pressure to marry the adhesive with the adherend. They are designed with a balance between flow and resistance to flow. The bond forms because the adhesive is soft enough to flow (i.e. "wet") to the adherend. The bond has strength because the adhesive is hard enough to resist flow when stress is applied to the bond. Once the adhesive and the adherend are in close proximity, molecular interactions, such as van der Waals forces, become involved in the bond, contributing significantly to its ultimate strength.Examples of permanent applications include safety labels for power equipment, automotive interior trim assembly, and sound/vibration damping films.

Pressure sensitive adhesives can also be designed for either permanent or removable applications. Removable adhesives are designed to form a temporary bond, and ideally can be removed after months or years without leaving residue on the adherend. Removable adhesives are used in applications such as surface protection films, masking tapes, bookmark and note papers, price marking labels, promotional graphics materials, and for skin contact (wound care dressings).

3. *Contact adhesives:* Contact adhesives must be applied to both surfaces and allowed some time to dry before the two surfaces are pushed together. Some contact adhesives require as long as 24 hours to dry before the surfaces are to be held together. Contact adhesives are also called cements because they are usually applied to both surfaces to be bonded. Following evaporation of the solvent, the two surfaces may be joined to form a strong bond with high resistance to shearing forces. Contact cements are used extensively in the assembly of automotive parts, furniture, leather goods, and decorative laminates. They are effective in the bonding of plastics.Natural rubber and polychloroprene (Neoprene) are commonly used contact adhesives

4. *Hot adhesives:* *Hot adhesives*, also known as *hot melt adhesives*, are thermoplastics applied in molten form (in the 65-180 °C range) which solidify on cooling to form strong bonds between a wide range of materials. Based on thermoplastic resins, which melt at elevated temperatures without degrading, these adhesives are applied as hot liquids to the adherend. Commonly used polymers include polyamides, polyesters, ethylene-vinyl acetate, polyurethanes, and a variety of block copolymers and elastomers such as butyl rubber, ethylene-propylene copolymer, and styrene-butadiene rubber.

Hot-melts find wide application in the automotive and home-appliance

5. *Chemically reactive adhesives:* *Multi-component adhesives* harden by mixing two or more components which chemically react. This reaction causes polymers to cross-link into acrylics, urethanes, and epoxies. There are several commercial combinations of multi-component adhesives in use in industry. Some of these combinations are:

- Polyester resin – polyurethane resin
- Polyols – polyurethane resin
- Acrylic polymers – polyurethane resins

The individual components of a multi-component adhesive are not adhesive by nature. The individual components react with each other after being mixed and show full adhesion only on curing. The multi-component resins can be either solvent-based or solvent-less.

Natural and Synthetic Adhesives

Natural

Natural adhesives are made from organic sources such as vegetable matter, starch (dextrin), natural resins or from animals e.g. casein or animal glue. They are often referred to as bioadhesives.

(a) *Starch and Dextrin:* Starch and dextrin are extracted from corn, wheat, potatoes, or rice. They constitute the principal types of vegetable adhesives, which are soluble or dispersible in water and are obtained from plant sources throughout the world. Starch and dextrin glues are used in corrugated board and packaging and as a wallpaper adhesive.

(b) *Natural Gums:* Substances known as natural gums, which are extracted from their natural sources, also are used as adhesives. Guar gum is one of the few products having very good export potential. Guar is a leguminous plant being grown in semi aried regions of India, Pakistan, USA. etc. Agar, a marine-plant colloid (suspension of extremely minute particles), is extracted by hot water and subsequently frozen for purification. Algin is obtained by digesting seaweed in alkali and precipitating either the calcium salt or alginic acid. Gum arabic is harvested from acacia trees that are artificially wounded to cause the gum to exude. Another exudate is natural rubber latex, which is harvested from *Hevea* trees. Most gums are used chiefly in water-remoistenable products.

(c) *Casien:* Casein based glues are made by precipitating casein from milk protein using the acetic acid from vinegar. This forms curds, which are neutralized with a base, such as sodium bicarbonate (baking soda), to cause them to unclump and become a thicker plastic-like substanceCasein is mainly used to adhere glass bottle labels.

Synthetic:

Although natural adhesives are less expensive to produce, most important adhesives are synthetic. Synthetics can be modified in many ways and are often combined to obtain the best characteristics for a particular application.

The polymers used in synthetic adhesives fall into two general categories—*thermoplastics and thermosets.*

- Thermoplastics provide strong, durable adhesion at normal temperatures, and they can be softened for application by heating without undergoing degradation. Thermoplastic resins employed in adhesives include nitrocellulose, polyvinyl acetate, vinyl acetate-ethylene copolymer, polyethylene, polypropylene, polyamides, polyesters, acrylics, and cyanoacrylics.
- Thermosetting systems, unlike thermoplastics, form permanent, heat-resistant, insoluble bonds that cannot be modified without degradation. Adhesives based on thermosetting polymers are widely used in the aerospace industry.

  Thermosets include phenol formaldehyde, urea formaldehyde, unsaturated polyesters, epoxies, and polyurethanes.
- Elastomer-based adhesives can function as either thermoplastic or thermosetting types, depending on whether cross-linking is necessary for the adhesive to perform its function. The characteristics of elastomeric adhesives include quick assembly, flexibility, variety of type, economy, high peel strength, ease of modification, and versatility. The major elastomers employed as adhesives are natural rubber, butyl rubber, butadiene rubber, styrene-butadiene rubber, nitrile rubber, silicone, and neoprene.

The adhesives can also be grouped into six structural-adhesive families:

1. Epoxies, which exhibit high strength and good temperature and solvent resistance, these are made up of a base resin and a hardener. Most epoxies will set fully at room temperature but some require heat to do so.
2. Polyurethanes, which are flexible, have good peeling characteristics, and are resistant to shock and fatigue, — these are available as a single pack or two pack adhesive. These adhesives are strong but are more flexible than epoxies.
3. Acrylics, a versatile adhesive family that bonds to oily parts, cures quickly, and has good overall properties, there are two types of acrylic adhesives – two part acrylic adhesives or acrylic adhesives that require exposure to UV light in order to cure fully. Acrylic adhesives are good for creating bonds that have very good peel strengths as well as good shear and impact strengths.

4. Anaerobics, or surface-activated acrylics, which are good for bonding threaded metal parts and cylindrical shapes, these are designed to be used in situations where no oxygen is present. Because they set when no oxygen is present they come in containers that are designed to "breathe" so that the adhesive cannot set before it is used
5. Cyanoacrylates, which bond quickly to plastic and rubber but have limited temperature and moisture resistance, commonly called superglue, set extremely quickly. These only require a very small amount to be used in order to achieve strong bonds.
6. Silicones, which are flexible, weather well out-of-doors, and provide good sealing properties. Each of these families can be modified to provide adhesives that have a range of physical and mechanical properties, cure systems, and application techniques. — these can be single pack or two pack. Single pack silicone adhesives are called room temperature vulcanising silicones and they cure by reacting to the moisture in the air. Two pack silicone adhesives are higher performing and are good for bonding metal, glass, and ceramics.

Table 13.5 lists some common types of adhesives and their uses.

Table 13.5 *Common adhesives and their uses.*

TYPES OF ADHESIVES		
Adhesive Type	*Sources/Properties*	*Common Uses*
Animal glue	Obtained from animal byproducts such as bones, blood, and hooves	Binding of abrasives in sandpaper and other grinding materials
Casein	Main protein in milk	Labels on beer bottles that do not come off in ice water, yet are recyclable
Starch	From corn and maize	Corrugated cardboard bonding
Natural rubber	Not "sticky enough" by itself but is used as an additive in other adhesives	Self-adhesive envelopes and other pressure-sensitive adhesives; adhesives that bond to substrates on contact (like tapes)
Butyl rubber/	It is elastomeric—it stretches isobutylene	Additive for hot-melt adhesives, window sealants, and pressure-sensitive adhesives
Amino resins	Water-soluble adhesives	Bonding of layers in plywood and the bonding of particles in particle board
Polyurethane	A flexible adhesive	Bonding soles to the bodies of shoes; also used in food packaging
Polyvinyl acetate	Common "white" glue	Book bindings and labels
Polyolefin/ethylene copolymer	No solvents involved	Hot melts
Acrylates or anaerobic adhesives	Cure when air is removed	Adhesive used to keep nuts tight on bolts, such as those within ATMs and heavy machinery
Silicone	Both an adhesive and a sealant and only common adhesive that is based on silicon rather than carbon	Bathtub and shower sealants; also many car applications, such as oil pans and head gaskets

Fevicol:

Fevicol type adhesives are synthetic resins and latex adhesives made from PVA. PVA (Polyvinyl Acetate) is used in adhesives for bookbinding, paper bags, milk cartons, drinking straws, envelopes, gummed types, folding boxes, labels and common products like foils, films, paper board etc. There are relatively of low cost, readily available, have wide compatibility and excellent adhesive characteristics. Synthetic adhesives produced in India are mostly based on vinyl acetate monomers. They account for 80 % of total production of adhesives.

Araldite

This is an adhesive containing the epoxy group. Epoxy resins are obtained by condensation of excess epichlorohydrin and bisphenol A, in the presence of alkali catalyst.

The major use of epoxy resins are as adhesives for surfaces such as glass and metals and surface coatings. They can be used both in moulding and laminating techniques to make glass fibre reinforced articles with better mechanical strength, chemical resistance and electrical insulating properties. Epoxy resins are also used in industrial flooring, highway surfacing and patching materials

QUESTIONS

1. Define the terms: Polymer; Monomer; Degree of polymerization; copolymerization.
2. Define functionality of a molecule.
3. How are polymerization reactions classified on the basis of mechanism?
4. What are addition and condensation polymerization processes? Explain with example.
5. Differentiate between thermoplastics resins and thermosetting resins. Classify the following as thermoplastic or thermosetting:
   polyethylene, nylon, polyester, urea-formaldehyde resin, teflon, polystyrene, bakelite.
6. What do you mean by tacticity? Explain an isotactic, a syndiotactic and an atactic polymer.
7. Explain (a) cationic polymerization, (b) anionic polymerization, (c) coordination polymerization.
8. Explain step growth and chain growth polymerization with examples.
9. What are the chemical and structural differences between thermosetting and thermoplastic polymers. Name three important examples of each class.
10. What are the monomers used for each of the following polymers. Give commercial application of each:
    (a) Bakelite (b) Nylon-6
    (c) Nylon 6-6 (d) Terylene
    (e) Teflon (f) PVC
11. (a) List out the difference between LDPE and HDPE.
    (b) What are the materials from which an epoxy adhesive is prepared.
    (c) Name any flourine containing polymer. State whether it is thermoplastic or thermosetting.
12. Give properties and uses of the following polymers:
    (a) Nylon 6-6 (b) PMMA
    (c) PVC (d) Teflon
    (e) Polyester
13. Give main properties and uses of:
    (a) HDPE (b) Plexiglass
    (c) Epoxy resins (d) Polypropylene.

14. Distinguish between nylon-6 and nylon 6,6.
15. What are the commercial uses of plastics.
16. Write a short note on coordination polymerization.
17. Mention the uses of urea-formaldehyde resins and polyurethane.
18. Differentiate between homopolymer and copolymer. Illustrate your point with examples.
19. Explain:
    (a) Why can not thermosetting resins be reused and reshaped?
    (b) Why is bakelite used in electrical appliances.
    (c) Why is teflon highly chemical resistant.
    (d) Why is PVC generally plasticized?
20. Distinguish between the following:
    (a) Addition and condensation polymerization
    (b) Anionic and cationic polymerization.
    (c) LDPE and HDPE
    (d) Thermoplastic and thermosetting resins.
21. Give properties and uses of the following:
    (a) PVC (b) PVA
    (c) Polystyrene (d) PMMA
22. Outline briefly on the applications of plastics in engineering and industry.
23. What is polyamide? Discuss various types of nylons.

Composites

1. What are composite materials? How are they classified on the basis of reinforcing material?
2. Discuss the classification of composite on the basis of matrix material?
3. Explain laminated composites giving their properties.
4. What are fiber reinforced composites?
5. What do you mean by metal matrix composites.
6. Write short notes on :
    (a) Carbon fiber as reinforcing materials.
    (b) Sandwitch Panels
    (c) Polymer matrix composites
    (d) Metal matrix composites
7. What are the uses of Aramid Fiber composites.
8. What do you mean by particle reinforced composites? What are dispersion strengthened composites?
9. Name a few alloys used as matrix materials for composite. Also give reasons for their use as matrix materials.
10. What are the advantages of using glass as reinforcing materials?
11. Why is surface protection of glass important?
12. Briefly discuss structural composites.
13. What are composite materials? Discuss some important types of fibre-reinforced composites.
14. (a) How is a composite defined? Give a broad classification of composite material
    (b) What are the factors on which the properties of a composite material depend?

□□□

14

SILICONES

"Everyone takes the limits of his own vision for the limits of the world."

Asthur Schopenhauer

CHAPTER OUTLINE

14.1 SILICONES-POLYSILOXANES

An outstanding feature of organic polymer chemistry is the ability of carbon atoms to link together to form chains and rings. This property of self linkage is known as catenation which is responsible for one of the rapidly growing branch of chemistry i.e polymer chemistry.

Similarly catenation is an important feature of some of the elements other than carbon, belonging to group 14, 15 and 16, though the extent of catenation is much less as compared to carbon. Therefore, very long chains of homoatomic molecules are seldom formed, but polymers having heteroatomic chains nevertheless constitute by and large, the chemistry of inorganic polymers.

Introduction: The chemistry of silicones is based on the classical researches of Frederick Kipping (1901-44) Silicon tetrachloride, $SiCl_4$ when fully hydrolysed gives silica, SiO_2–

$$SiCl_4 + 4H_2O \longrightarrow Si(OH)_4 \xrightarrow{-2H_2O} SiO_2$$

But the hydrolysis of alkyl substituted chlorosilanes do not lead to the expected alkyl silicon compound analogous to a ketone, but to long chain polymers called silicones.

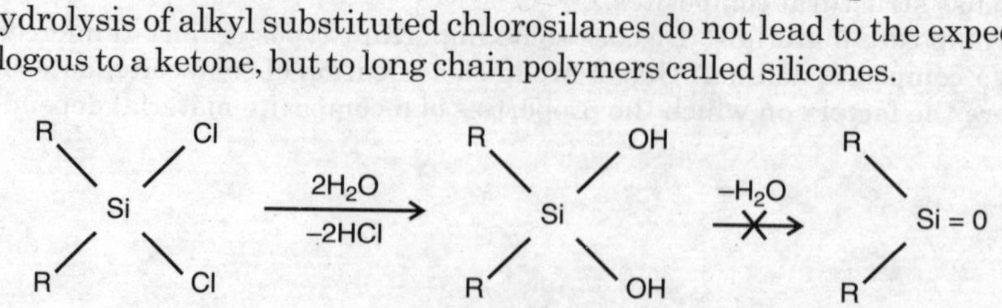

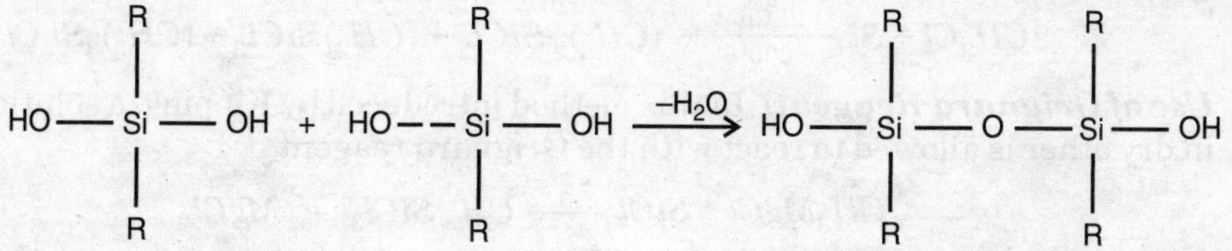

Since an active OH group is left at each end of the chain, polymerization reaction continues and a long chain product is formed.

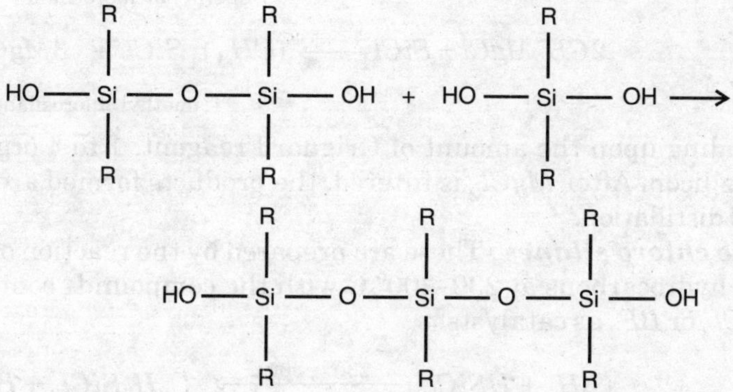

These products are known as silicones or poly siloxanes. Thus silicones are a group of neutral organosilicon polymers, based on network of alternating silicon and oxygen atoms, having the general formula :

$[R_x\,SiO_{(4-x)/2}]n$ where n is very large while x varies from 0 to 3.

Silicones can be categorized both as organic as well as inorganic polymers. These polymers resemble inorganic polymers in having high percentage of ionic character of Si–0 bonds and they resemble organic polymers in having organic groups on silicon atoms. The silicon products possess certain unusual, but very useful properties such as remarkable stability over wide range of temperature (from –70° C to 250° C), good water repellency, chemical and physiological inertness, good resistance to the effect of weathering, low vapour pressure and desirable low temperature characteristics. The broad spectrum of properties combined with the many forms in which silicones can be formulated (eg. liquids, gums, resins and rubbers) lead to their wide spread application in a broad range of processes and industries.

Preparation: Silicones are generally prepared through the hydrolysis of alkyl or aryl substituted chloro silanes which in turn are obtained through various methods. Thus the preparation of silicones requires two steps–

(a) Preparation of intermediates.

(b) Hydrolysis/polymerization of intermediates.

(a) ***Preparation of Intermediates:*** There are three commercial methods for the manufacture of organochlorosilanes.

  (i) ***Direct Method:*** Used for the preparation of methyl chlorosilanes. In this method methylchloride is reacted with silicon in the presence of copper as a catalyst at a temperature of 300° C. This is a highly exothermic reaction and provisions must be made for the removal of heat.

$$2CH_3Cl + Si \xrightarrow[300°C]{Cu} (CH_3)_2SiCl_2 + (CH_3)SiCl_3 + (CH_3)_3Si\,Cl$$

(ii) ***Use of Grignard Reagent:*** In this method introduced by Kipping, a solution of $SiCl_4$ in dry ether is allowed to react with the Grignard reagent

$$CH_3MgCl + SiCl_4 \longrightarrow \underset{\text{Methyltrichlorosilane}}{CH_3\,SiCl_3} + MgCl_2$$

$$2CH_3MgCl + SiCl_4 \longrightarrow \underset{\text{Dimethyl dichlorosilane}}{(CH_3)_2\,SiCl_2} + 2Mg\,Cl_2$$

$$2CH_3MgCl + SiCl_4 \longrightarrow \underset{\text{Trimethyl chlorosilane}}{(CH_3)_3\,SiCl} + 3MgCl_2$$

Depending upon the amount of Grignard reagent, 1 to 4 organic groups can be added to silicon. After $MgCl_2$ is filtered, the products formed are seperated through fractional distillation.

(iii) ***Aromatic chloro silanes :*** These are prepared by the reaction of benzene and other aromatic hydrocarbons at 230–300° C with the compounds containing *Si–H* bonds, using $AlCl_3$ or BF_3 as catalysts.

$$C_6H_6 + HSiCl_3 \xrightarrow{230° - 300\,C} C_6H_5SiCl_3 + H_2$$

(b) ***Hydrolysis of Intermediates :*** Hydrolysis of different intermediates under controlled conditions yield different products. Since these intermediates are very reactive and inflammable, they are preserved under dry inert gases such as nitrogen.

(i) ***Dimethyldichlorosilane $(CH_3)_2SiCl_2$:*** This is mainly a chain building unit. On controlled hydrolysis $(CH_3)_2SiCl_2$ gives rise to straight chain polymers. As an active *OH* group is left at each end of the chain, polymerization continues and finally a linear thermoplastic polymer is formed.

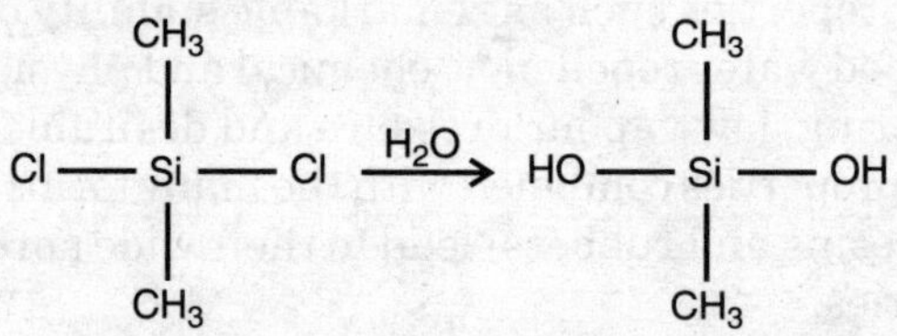

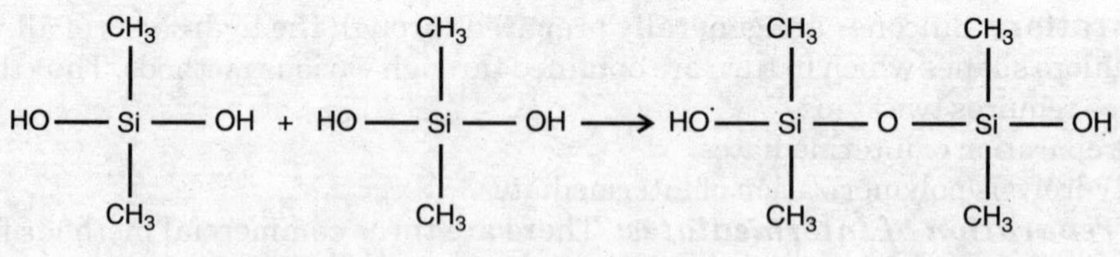

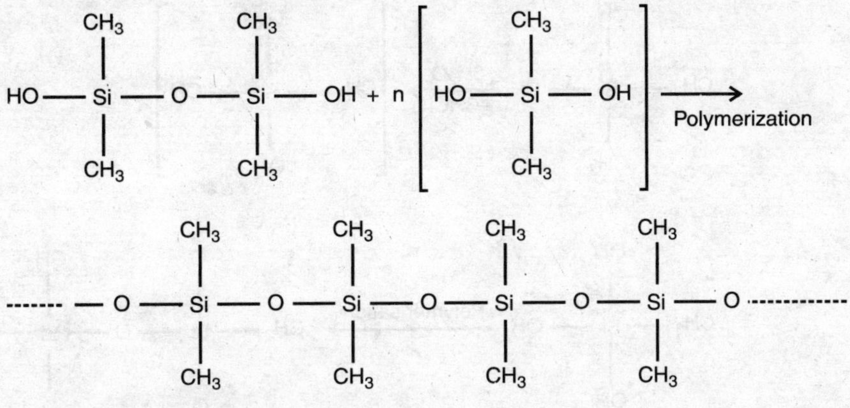

Linear Thermoplastic Polymer

Hydrolysis under carefully controlled conditions can produce cyclic structures, with rings containing three to six silicon atoms.

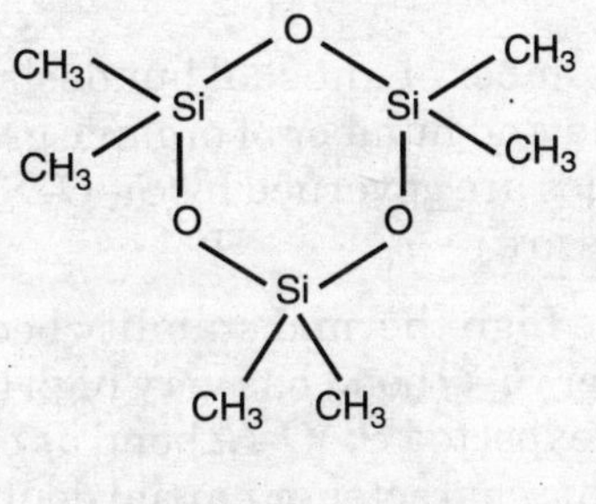

Triscyclo dimethyl Siloxane

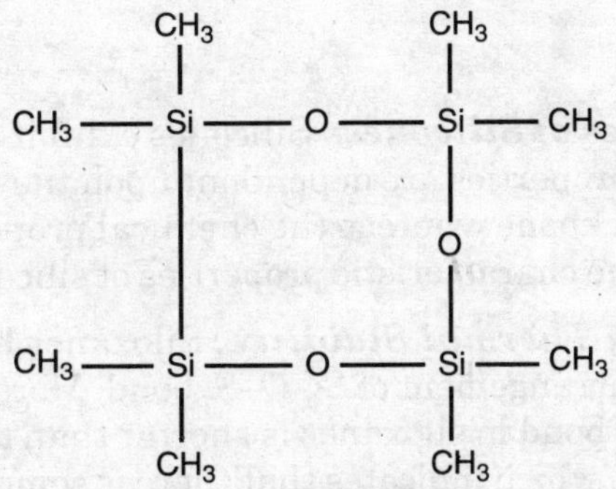

Tetrakis cyclo dimethyl Siloxane

(ii) ***Trimethylmonochlorosilane*** $\boldsymbol{(CH_3)_3SiCl}$**:** Hydrolysis of this compound yields trimethylsilanol as a volatile liquid which condenses to give hexamethyldisiloxane. Since the resulting compound does not have any *OH* group, it cannot polymerize further. Therefore it is also used as a "chain stopper".

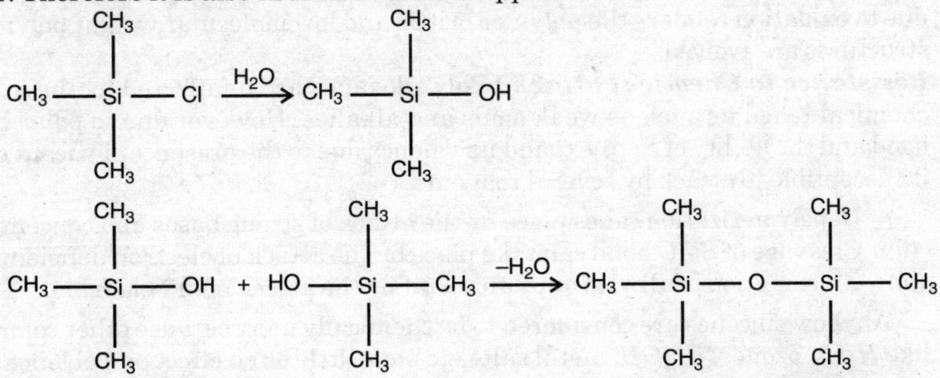

Hexamethyl disiloxane

(iii) ***Methyltrichlorosilane*** $\boldsymbol{CH_3\,SiCl_3}$**:** Hydrolysis of this compound results in the formation of crosslinked polymer due to the presence of three functional groups per molecule. Its amount used in polymerization, determines the amount of crosslinking that can be obtained.

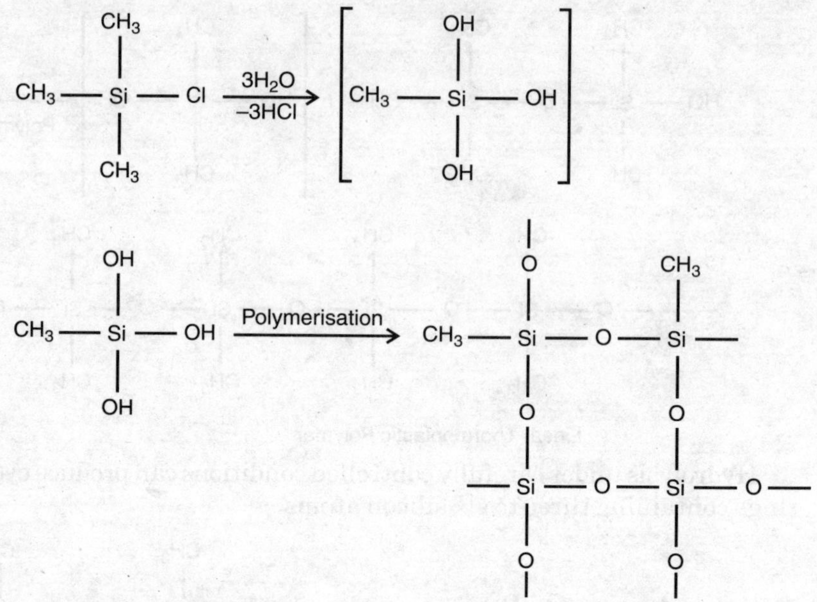

Properties of Silicones: Silicones exhibit a number of unusual but desirable properties. The physical properties are dependent upon the kind and number of organic groups attached to the silicon backbone whereas the chemical properties are governed by *Si–O–Si* bonds in siloxanes. Some of the characteristic properties of silicones are :

(1) ***High Thermal Stability :*** Siloxanes have high thermal stability because of stable silica like arrangement of *Si–O–Si* bond. Moreover *Si–O* bond has very high bond energy. The *Si–O–Si* bond in siloxanes is shorter than the expected *Si–O–Si* bond as calculated from their radii, which indicates that there is some ionic character or partial double bond character in the bond which confers stability to the bond. In addition, they remain flexible in the temperature range of –90° C to 250° C which is a much wider range than for natural rubber.

(2) ***Resistance to Oxidation :*** Siloxanes are quite resistant to attack by oxygen. On heating in air to 250° C oxidation of organic groups takes place which leads to cross linking. However the presence of aromatic groups on silicon increases the stability towards oxygen. Crosslinking due to oxidation renders the polymer brittle and low molecular weight polymers and cyclic structures are evolved.

(3) ***Resistance to Chemical attack :*** Poly siloxanes are unaffected by the majority of mild chemical reagents such as weak acids and alkalies. However due to polar nature of *Si–O* bond and the ability of *Si* to expand its valency due to the presence of vacant orbitals makes it susceptible to attack by several reagents.

(i) Depolymerization takes place by the attack of strong bases and concentrated acids.
(ii) Cleavage of *Si–O* bond can take place by the attack of electron deficient compounds.
(iii) Chlorination of alkyl group can result by the attack of dry chlorine.

Anyhow silicones are considered to be chemically inert because other common chemicals like H_2O_2, ozone, NH_4OH, metal salts etc have little or no effect on polysiloxanes.

(4) ***Chemical Properties :*** As is discussed above, silicones are generally chemically inert. However, reagents like Grignard reagent, lithium aluminium hydride can cleave the *Si–O* bond. For example–

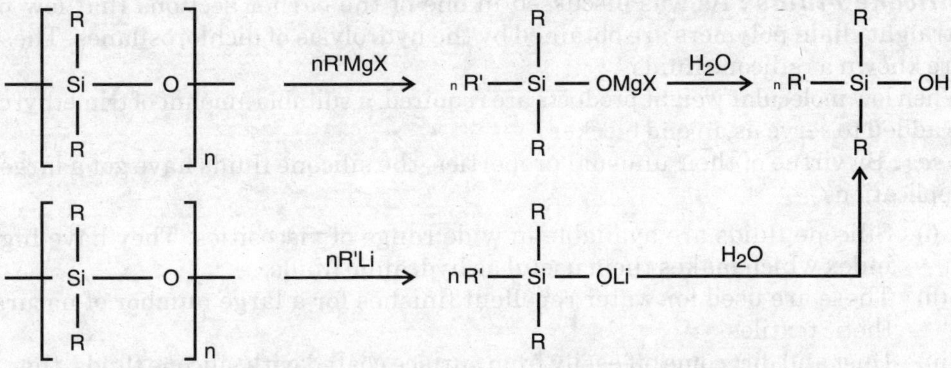

(5) ***Electrical Insulating Properties :*** Polysiloxanes are good electrical insulators. They are more stable to heat than organic polymers and hence are used for high temperature insulation. They do not break down at high temperatures ordinarily but if they do, they do not produce conducting materials as carbon does.

(6) ***Variation in viscosity with temperature :*** Silicon fluids exhibit very small variations in viscosity with temperature. Thus, these can function over a considerably wide range of operating temperatures. This property is attributed to the molecular structure of silicones. Silicones have regularly coiled structure with weak Vander Waals' forces in between the linear chains. At low temperatures the molecules have kinked and coiled structure, which opens up at high temperatures. The opening up causes greater interaction and entanglement of polymer molecules, thus increasing the viscosity. This increase partly compensates the normal decrease in viscosity caused by increase in temperature. Thus, the net result is that the viscosity of silicones does not change much on increasing the temperature i.e silicones show very small variations in viscosity with temperature.

Large size of silicon atom facilitates the coiling which is not possible in its linear organic analogs.

(7) ***Water repellency :*** Silicon resins are highly water repellent. Silicone chain is surrounded by organic groups and thus high content of hydrocarbon is responsible for this property. The high degree of water repellency coupled with the spreading efficiency due to low surface tension allows the deposition of silicon fluid films over plastics, metals, glass etc. When glass is coated with polysiloxanes, the oxygen atom of *Si–O–Si* chain adheres to the glass surface while the organic groups are directed outwards.

(8) ***Specific Gravity :*** The specific gravity for silicone greases and fluid resins lies between 1.03 and 1.09, whereas for silicon rubbers its value lies between 1.7 to 2.1

Types of Silicones And Their Uses

The silicones are generally classified on the basis of their physical state and appearance at room temperature. Depending upon the length of the chain, amount of crosslinking, nature of organic groups attached to the silicon atom, they can be obtained in the following forms–

(1) Liquid silicones or silicone fluids.

(2) Silicone Greases.

(3) Silicone Rubber or Elastomers.

(4) Silicone Resins

(1) ***Silicone Fluids :*** As was discussed in one of the earlier sections that low molar mass straight chain polymers are obtained by the hydrolysis of dichlorosilanes. These polymers are known as silicone fluids.

When low molecular weight products are required, a suitable amount of trimethyl chlorosilane is added to serve as an end blocker.

Uses : By virtue of their unusual properties, the silicone fluids have got a large number of applications.

(i) Silicone fluids are available in wide range of viscosities. They have high viscosity index which makes them useful as hydraulic fluids.
(ii) These are used for water repellent finishes for a large number of natural and synthetic textiles.
(iii) Dust and dirt come off easily from surface coated with silicone fluids, they are used in polishes for cars and furniture.
(iv) They have excellent dielectric properties and are thermally stable and hence are used in capacitors and small transformers.
(v) They are used in high temperature water baths as they can withstand high temperature.
(vi) They are also used as high temperature lubricants. Silicones with phenyl groups are better lubricants.
(vii) Silicone fluids have antifoam properties (e.g. dimethylsilicone fluids) and are used to eliminate foam in petroleum oils.
(viii) Silicon fluids are nontoxic.

2. ***Silicone Greases :*** These are modified silicone fluids and are formed by adding fillers such as silica, carbon black, lithium soap etc to silicone oils. Silicone greases vary in properties depending on the type of fluid and thickner used.

Uses :

(i) Greases have a wide range of service temperature range from –100 to 450° C and hence can be very useful where effective lubrication is required at very high and very low temperatures.
(ii) Greases with lithium soap filler can function from –75° C to 150° C and are used for ball bearings, operating and high speed and high temperatures.
(iii) Greases with carbon black filler are heat resistant and are used in high temperature conveyers and oven doors etc.
(iv) The greases in which silica is used as filler have been used as electrical greases in car ignition systems.

3. ***Silicone Rubbers or Elastomers :*** Silicon rubbers are the polymers containing high molecular weight dimethyl silicon polymer, mixed with a filler, usually finely divided SiO_2 and a peroxide as curing agent. Peroxide helps to form few bridges between the linear chains, SiO_2 is added to reinforce the polysiloxane structure. The mechanism of formation of silicon rubbers from straight chain compound can be depicted as–

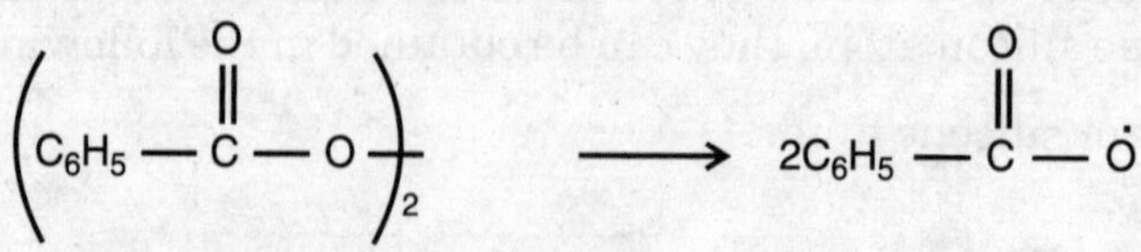

Benzoyl peroxide

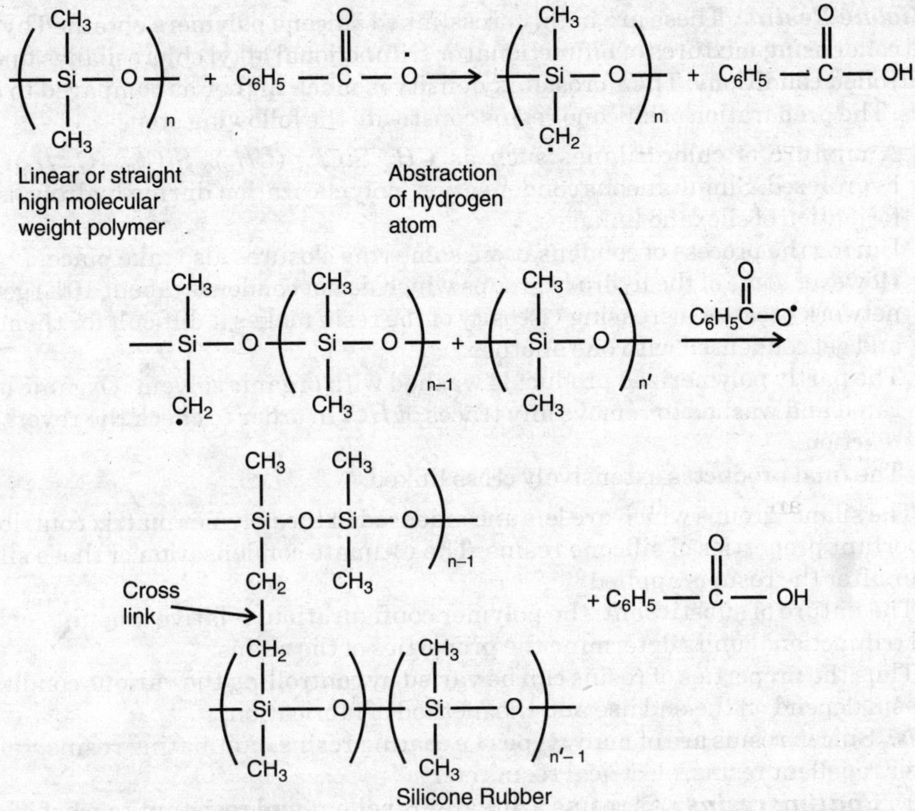

Uses: Silicone rubbers are exceptionally resistant to prolonged exposure to weathering, boiling water, dilute acids and alkalies. They have got remarkable stability at higher temperatures. Some of the uses of silicone rubbers are–

(i) Because of their property to withstand heat, they are used in sealing joints and seams of jet aircrafts. They are also useful for insulating electrical parts.

(ii) Silicone rubbers are very resistant to many chemicals. That is why they are used especially when chemical inertness is required.

(iii) They remain flexible in the temperature range of 90–250° C and hence used in making tyres for fighter aircrafts, because they prevent damage on landing whereas ordinary rubber tyre becomes brittle and undergoes disintegration.

(iv) Silicon rubber is available in variety of forms like thin sheets, mouldable stocks, smooth or heavy pastes. These may be dispersed in organic solvents to give free flowing dispersion which can be applied to surfaces by brushing, spraying or dipping. The pastes are used for coating organic and inorganic fabrics. Coated fabrics are used to make hoses for carrying hot gas or liquid.

(v) Pastes can provide an electrically insulating coating as it can adhere well to ceramic, glass and metalware.

(vi) Their heat resistance, resistance to mineral oils makes them exceptionally useful as O–rings, pressure cooker seals etc.

(vii) Silicon rubbers are also used for making artificial heat valves, transfusion tubings and in implant surgery.

However, the tensile strength of silicone rubbers is inferior to organic rubber at normal temperature.

4. ***Silicone Resins:*** These are highly crosslinked silicone polymers obtained by hydrolysing and condensing mixtures of bifunctional or trifunctional alkyl chlorosilanes under carefully controlled conditions. Their crosslink density is much higher as compared to silicone rubbers. The preparation of silicone resins constitute the following steps –
   (a) A mixture of chlorosilanes such as $CH_3\ SiCl_3$, $(CH_2)_2\ SiCl_2$, $(C_6\ H_5)CH_3\ SiCl_2$ is hydrolyzed. Simultaneous condensation, polymerization during hydrolysis leads to the formation of siloxane linkages.
   (b) During the process of condensation, some ring closures also take place.
   (c) However some of the hydroxy groups which do not condense (about 10%) get fixed in the network because increasing viscosity of the resin makes it difficult for them to approach and get condensed with one another.
   (d) The partly polymerized product is washed with organic solvent. Organic layer is separated and washed to remove any traces of HCl in order to check the reversibility of the reaction.
   (e) The final product is extensively cross linked.

The silanol groups which are left uncondensed in the polymer matrix contribute towards important properties of silicone resins. The ultimate condensation of these silanol groups occur after the resin is applied.

The nature of substituent, the polymer configuration, relative amounts of bifunctional and trifunctional units determine the properties of the resins.

Thus the properties of resins can be varied by controlling the various conditions. Choice of resin depend on the end use and the method of fabrication.

Uses: Silicon resins are of many types i.e coating resins, laminating resins, release resins, water repellent resins, electrical resins etc.

(i) ***Coating resins :*** As resins are water repellent and resistant to most of the aqueous chemicals they are used as films on exhausts, stoves, boilers, steam pipes etc where conditions are too severe for the use of usual type of protective coatings. As they can withstand the temperature of the order of 500° C they can be applied on high temperature processing equipments.

(ii) ***Laminating resins :*** Resin laminated structures are used in electrical industry in the form of insulated boards and rods. Methyl phenyl silicone resins are used in the manufacture of heat resistant glass cloth laminates, especially for electrical applications.

(iii) ***Release resins :*** These are used in bread pans, candy pans etc.

(iv) ***Water repellent resins :*** These resins are applied as dilute solutions to coat the pores in masonry and concrete. Due to their water repellency they keep the masonry dry.

(v) ***Electrical resins*** : For water proof and high temperature insulation, silicon resins are used. Impregnation of silicone resins are also used in the treatment of capacitor insulation and motor windings.

QUESTIONS

1. What are inorganic polymers? Discuss their general properties. How do they differ from organic polymers.
2. Discuss in detail linear and cyclophosphazenes. Give an account of the bonding present in phosphazenes.
3. Discuss the structures, properties and uses of phosphazenes.
4. What are phosphonitrilic chlorides? Draw the structures of trimer and tetramer.
5. What are silicones and how are they classified? Which fundamental property of silicon makes the formation of silicones possible.

6. Give a brief account of the applications of silicones in technology.
7. Write short notes on :
   (a) Silicone fluids (b) Silicone resins
8. How are silicones prepared? Mention their important properties.
9. Discuss the applications of:
   (a) Silicon greases (b) Silicon fluids
   (c) Silicon resins
10. What are chalcogenide glasses? Give an account of the properties of network polymers of sulphur.
11. Wrtie short notes on :
   (a) Polymeric sulphur. (b) Polyhosphonitrilic chlorides.
   (c) Silicone rubbers.

❑❑❑

15 METALLIC-CORROSION AND ITS CONTROL

"What lies behind us and what lies before us are small matters compared to what lies within us."

Ralph Waldo Emcrson

CHAPTER OUTLINE

15.1 CORROSION

Corrosion can be defined as degradation or deterioration of a metal by chemical or electrochemical reaction with its environment (gaseous or liquid medium).

Degradation or deterioration means reduction in the useful properties of the material which include:

- Weakening of the material due to loss of crossectional area.
- Loss of properties such as malleability, ductility.
- Decaying of surfaces of metals.
- Cracking of a polymer due to sunlight.

The materials affected by corrosion are metals, polymers (plastics or rubbers etc.), ceramics (concrete) etc. But metals, the most useful class of materials corrode most frequently and hence this discussion is mainly devoted to corrosion in metals.

Some of the common examples of corrosion in metals are:

Rusting of Iron: When exposed to the atmospheric conditions, a layer of reddish scale and powder of Fe_3O_4 is formed on the surface.

Formation of green film on the surface of copper: A green layer of basic carbonate consisting of $CuCO_3 + Cu(OH)_2$ is formed on the surface of copper when exposed to moist air

Tarnishing of silver: When silver is exposed to the atmosphere, a black coating is formed.

15.2 WHY METALS CORRODE

All metals except gold, platinum and palladium corrode spontaneously when exposed to atmosphere, which contains 21% of oxygen. Metals, with the exception of gold, platinum and palladium are found in nature in combined forms as their oxides, carbonates, sulphides, chlorides or silicates. The metals in the combined form are known as minerals. The natural combined form of the metal i.e., mineral is more stable than the metal. During the extraction of the metal from the mineral, a considerable amount of energy is required. As energy is consumed in the extraction process, it implies that metal has a higher energy state than the mineral or compound of the metal. Thus, the metal which is in higher energy state (unstable) tends to revert back to its stable natural combined form which is of low energy, when it comes in contact with the elements in the environment. *In other words, corrosion process is the return of the refined material to its natural state.*

Greater is the amount of energy required to separate the metal from its mineral or ore, greater will be the tendency of the metal to revert back to its natural state resulting in degradation of the metal.

Metallic Ore (Mineral or metal in combined from) [or Stable State] $\xrightarrow{\text{Extration of Metal}}$ Pure Metal [Unstable State]

Pure Metal [Unstable State] $\xrightarrow{\text{Corrosion (Reaction with atmosphere)}}$ Metallic Ore (Mineral or metal in combined from) [or Stable State]

15.3 CONSEQUENCES OF CORROSION

Corrosion has a number of economic and social consequences which include:

(i) Enormous waste of machineries and different types of metallic materials.

(ii) It leads to unpredictable machinery failure which may lead into dangerous situations such as loss of life.

(iii) It leads to decrease in efficiency of machine and frequent replacement of corroded equipment, which is an expensive proposition.

(iv) It may cause leakage of inflammable gas from the corroded pipe line resulting into fire hazards and leakage of toxic product.

(v) It causes contamination of potable water.

(vi) It has been estimated that 25% of annual world production of iron is wasted due to corrosion.

From the above discussion it becomes clear that study of corrosion and its prevention is essential to both the engineers and metallurgists. For this, an understanding of the mechanism of corrosion is imperative.

15.4 THEORIES OF CORROSION

A number of different views have been put forward to explain as to how corrosion takes place. Some say, corrosion is oxidation or electrochemical phenomena while others say, it is a chemical attack. Still others have opinion that it is electrical phenomena. Each is partially true. The different theories of corrosions are discussed below.

15.4.1 Chemical or Dry Corrosion

Chemical or dry corrosion is the simplest case of corrosion, where corrosion takes place by direct chemical attack. This type of corrosion occurs mainly when metal surface come in direct contact with the atmospheric gases such as oxygen, halogens, hydrogen sulphide, nitrogen etc. or with anhydrous inorganic liquid. The corrosion product formed can be insoluble, soluble or liquid product.

Insoluble corrosion product implies that a solid film of the corrosion product is formed on the surface of the metal, which goes on thickening as corrosion proceeds. After sometime this layer protects the metal from further corrosion because rate of penetration of the corroding environment decreases and fresh metal is not exposed to it.

When the corrosion product is soluble or liquid, it is removed from the surface as soon as it is formed, exposing fresh metal to environment. This leads to excessive corrosion.

Dry or chemical corrosion is generally of three main types:

Chemical corrosion
- Oxidation corrosion
- Corrosin by other gases
- Liquid metal corrosion

15.4.1.1 Oxidation Corrosion

Oxidative corrosion takes place by direct action of oxygen on metals, at low or high temperature, forming oxides of the metal. It generally takes place in absence of moisture. The reaction of a metal with air or pure oxygen to form the corresponding oxide can take place spontaneously in case of alkali and alkaline earth metals even at ordinary temperatures. However, most of the metals react with oxygen only at high temperatures. Almost all the metals except Ag, Au, and Pt, get oxidized at higher temperatures.

Mechanisms of Dry Corrosion

The growth of the oxide film on metal takes place at the surface as:

- The oxygen is adsorbed at the surface of the metal by physical adsorption.
- When the temperature rises the physical adsorption turns into chemisorption, wherein metal gets oxidized to metal ions ($M \rightarrow M^{n+} + ne$)
- The electrons are taken up by oxygen which gets reduced to oxide ion ($nO_2 + 2ne^- \rightarrow 2nO^2$)
- The M^{n+} O^{2-} combine to form metal oxide (M_2O_n)
- The reactions can be written as

$$2M \longrightarrow 2M^{n+} + 2\,ne^- \quad \text{(Oxidation of metal)}$$

$$n\,O_2 + 2ne^- \longrightarrow 2n\,O_2 \quad \text{(Reduction of oxygen)}$$

$$\underset{\text{Metal}}{2M} + \underset{\text{Oxygen}}{O_2} \longrightarrow \underset{\text{Metal ion}}{2M^{n+}} + \underset{\text{Oxide ion}}{nO^{2-}}$$

$$\text{Metal Oxide } (M_2O_n)$$

- The metal oxide scale so formed forms a barrier between the metal and the oxygen and restricts further oxidation or corrosion of the metal. For oxidation to continue

— either metal must diffuse outwards through the scale to the surface.

— or oxygen must diffuse through the scale to the underlying metal.

- As the metal cation is much smaller in size, the outward diffusion of metal ion is much more rapid than inward diffusion of oxide ion. Hence corrosion continues by rapid diffusion of highly mobile cations. (Fig. 15.1)

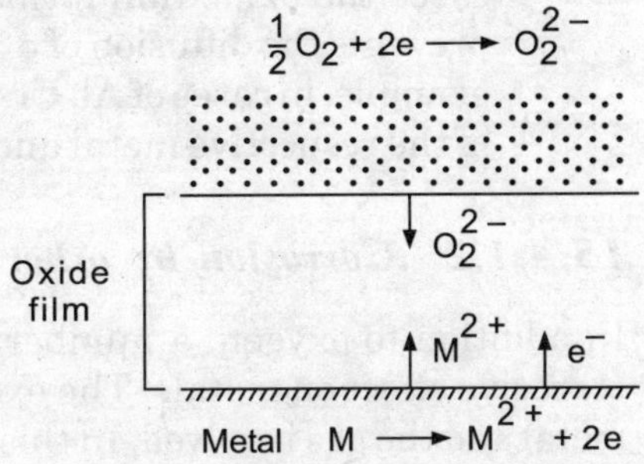

Fig.15.1 Mechanism of Growth of Oxide film

Nature of the oxide film formed plays an important part in corrosion process. If the thickness of the oxide layer is less than 300 Å then such layer is called a **film** and if the thickness of the film exceeds 300 Å than it is known as **scale**. The nature of the film decides the continuation of corrosion process as discussed below:

(i) **Stable film:** The stable oxide film is fine grained in structure and adheres tightly to the metal surface. This can be impervious in nature, thereby preventing the penetration of oxygen to the underlying metal. In otherwords, such a film forms a protective coating over the metal and prevents further oxidation or corrosion. The oxide films on Al, Cr, Cu, Pb, Sn etc. are stable and adhere to the surface. Therefore these metals are employed for alloy formation with other reactive metal to protect them from corrosion.

(ii) *Unstable film:* The oxide film formed on the surface of metal decomposes back into metal and oxygen and hence the film is called unstable film.

$$\text{Metal Oxide} \rightleftharpoons \text{Metal} + \text{Oxygen}$$

Hence in such cases corrosion is not possible. Unstable film is formed by the oxides of Ag, Au and Pt and hence these do not undergo corrosion.

(iii) *Volatile film:* The metal-oxide film volatizes from the surface as soon as it is formed. Hence the underlying metal is exposed to the further attack by oxygen leading to rapid and continuous corrosion. For example, molybdenum oxide (MoO_3) is volatile and hence molybdenum undergoes excessive corrosion.

(iv) *Porous film:* Sometimes the oxide film so formed is porous or have cracks. Hence the diffusion of oxygen takes place through these pores leading to further oxidation or corrosion. The metal oxides of metals like alkali and alkaline earth metals fall under this category.

Pilling-Bedworth Rule

The oxide layer can be protective or non-protective depending upon the ratio of volume of the metal oxide to the volume of the metal consumed. This is known as Pilling-Bedworth ratio.

$$\text{Pilling} - \text{Bedworth ratio} = \frac{\text{Volume of the metal oxide formed}}{\text{Volume of the metal consumed}}$$

According to Pilling-Bedworth's rule,

- If the volume of the oxide film formed is less than the volume of the respective metal from which it is formed, the ratio is less than unity then the oxide film will be non-protective in nature. In such cases the oxide film formed over the surfaces of the metals is porous in nature and oxygen can diffuse through these pores to bring about further attack on the metal. For example, in case of alkali and alkaline earth metals such as Na, Ca, Mg (except Al) oxide film produced has smaller volume than that of the metal from which it is formed. Consequently, the oxide film develops cracks and pores in its structure.

- If the metal oxide film formed is equal or greater than the respective metal from which it is formed the ratio is more than unity, the metal oxide film is protective in nature. In such cases the oxide film formed is continuous, adhering and non-porous. This type of film prevents the diffusion of oxygen and hence the rate of corrosion by oxidation decreases. For example, in cases of Al, Cr, and W etc., the volume of oxide layer is greater than the volume of the respective metal and hence is protective in nature.

15.4.1.2 Corrosion by other gases

In addition to oxygen, a number of other gases such as CO_2, SO_2, Cl_2, H_2S, F_2 etc. also have corrosive effect on metals. The extent of corrosion depends on the chemical affinity between the metal and the gas involved. In this case also the film so formed may be protective or non-protective in nature.

- If the film formed is protective in nature, the extent of attack or corrosion decreases. For example, chlorine gas attacks silver metal surface to form non-porous protective film which prevents further corrosion.
- If the film formed is non-protective in nature then it destroys the whole metal eventually. For example, chlorine attacks Sn to form $SnCl_4$ film which is volatile and hence leaves the metal surface exposing the fresh metal for further attack.
- H_2S attacks steel forming FeS layer which is porous in nature.

15.4.1.3 Liquid-metal corrosion

Sometimes the corrosion also results due to the action of flowing liquid metal at high temperatures on solid metals or alloys. Liquid sodium is used as heat transfer media in nuclear reactor. It is shown to cause corrosion in carrier tubes. The corrosion reaction involves either dissolution of a solid metal by a liquid metal or by penetration of the liquid metal into the solid metal causing corrosion in solid metal.

15.4.2 Electrochemical Corrosion or Wet Corrosion

The corrosion of metals in aqueous medium seems to be the electrochemical in nature. This type of corrosion is also known as immersed corrosion and generally takes place when:

(i) a metal is in contact with the moist air or any liquid medium or
(ii) when two different metals or an alloy is in contact with a solution.

The most familiar example of electrochemical corrosion is the *formation of rust on iron*. We all know that iron pipes corrode in wet soil or moist atmosphere. In this type of corrosion the metal is oxidized to a soluble form or ionic state.

Before the corrosive process takes place, the following conditions must be satisfied for the formation of electrochemical cell:

(i) There must be a cathode and an anode.
(ii) There must be an electrical potential between the anode and cathode.
(iii) There must be a metallic path electrically connecting the anode and cathode.
(iv) The anode and cathode must be immersed in an electrically conductive medium.

Once these conditions are satisfied a "corrosion cell" is formed. The corrosion takes place as discussed below:

(a) **Anodic reactions:** At anodic area, oxidation reaction takes place where metal atoms lose their electrons to the environment and pass into the solution.

$$\underset{\text{(Metal)}}{M} \longrightarrow \underset{\text{(Metal ion)}}{M^{n+}} + ne^-$$

Thus metal is destroyed by dissolving or changing into metallic ions. *Hence, corrosion takes place at anodic areas or metal, which forms the anode, is corroded.*

(b) **Cathodic reactions:** The electrons released in the anodic reactions are consumed in the reactions at the cathode or cathodic reactions. Depending upon the nature of the corrosive environment, cathodic reactions may result in:

(i) *Evolution of hydrogen:* In absence of oxygen and in acidic environment evolution of hydrogen takes place. The H^+ ions of acidic solution take up electrons released at anode and are eliminated as hydrogen gas

$$2H^+ + 2e^- \longrightarrow H_2 \uparrow \quad \text{(in acidic medium)}$$

In alkaline or neutral solution in absence of oxygen, the reaction will be:

$$2H_2O + 2e^- \longrightarrow H_2 + 2OH^- \quad \text{(in neutral or alkaline medium)}$$

Example: Rusting of iron in acidic medium in absence of oxygen takes place as follows,

$$Fe \longrightarrow Fe^{2+} + 2e^- \quad \text{(Anode reaction)}$$

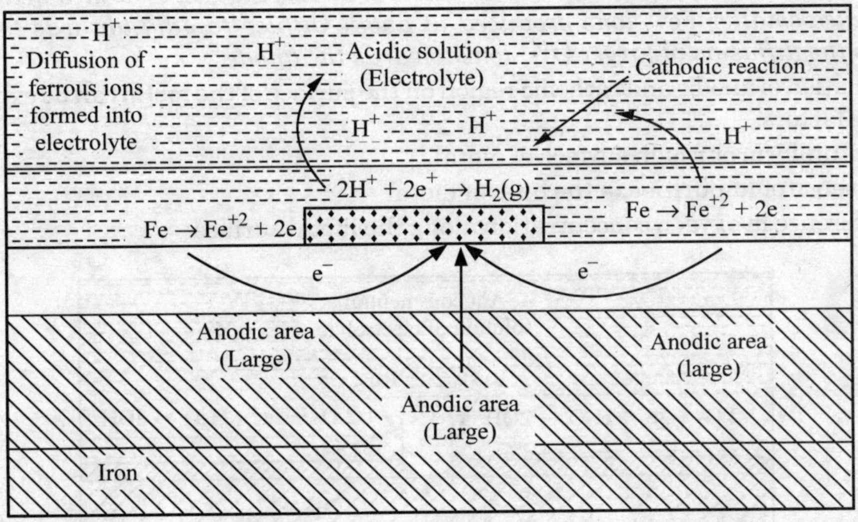

Fig. 15.2 Mechanism of wet Corrosion by Hydrogen Evolution

The electrons that are released, flow through the metal surface (electronic conductor) from anode to cathode. The H^+ ions of **acidic** solution take up these electrons and are eliminated as hydrogen gas.

$$2H^+ + 2e^- \longrightarrow H_2 \uparrow \quad \text{(Cathodic reaction)}$$

The overall reaction is: $\quad Fe + 2H^+ \longrightarrow Fe^{2+} + H_2 \uparrow$

Hence, in this type of corrosion hydrogen ions are displaced from the acidic solution by metal ions.

(ii) *Absorption of oxygen:* This type of corrosion reaction takes place in the presence of atmospheric oxygen. In neutral aqueous solution or weakly alkaline solution, containing oxygen the reaction, $2H^+ + 2e^- \longrightarrow H_2$, does not take place and oxygen gets reduced as, $2H_2O + O_2 + 4e^- \longrightarrow 4\bar{O}H$

Example: Rusting of iron in neutral aqueous solution in presence of oxygen can be explained, with the following reactions.

$$Fe \longrightarrow Fe^{2+} + 2e^- \quad \text{(Anode reaction)}$$

The electrons travel to cathode through the metal surface and react with oxygen,

$$\frac{1}{2}O_2 + H_2O + 2e^- \longrightarrow 2OH^- \quad \text{(Cathodic reaction)}$$

The Fe^{2+} produced at anode travel to the cathodic region through the moisture and react with OH^- ions produced at cathode as follows:

$$Fe^{2+} + 2OH^- \longrightarrow Fe(OH)_2\downarrow$$

The $Fe(OH)_2$ produced is oxidized as,

$$Fe(OH)_2 + O_2 + 2H_2O \longrightarrow 4Fe(OH)_3 \text{ or } 2Fe_2O_3 \cdot 3H_2O \text{ (Rust)}$$

$Fe(OH)_3$ corresponds to rust, which is hydrated iron (III) oxide of variable composition, Fe_2O_3, nH_2O. The degree of hydration of the iron (III) oxide affects the colour of the rust which may vary from black (Fe_2O_4) to yellow or reddish brown ($Fe_2O_3 \cdot H_2O$).

It may be noted that although corrosion occurs at anode, the rust is formed near the cathode. This is because smaller Fe^{2+} *ions produced at anode diffuse rapidly towards the cathode as compared to the diffusion of larger* OH^- *ions towards the anode.*

The formation of anodic and cathodic region on the surface of the metal can be explained due to the following factors:

(i) Non-uniform or rough surface of metal.

(ii) Presence of impurities in the bulk of metal

(iii) Difference in oxygen concentration along the metal surface.

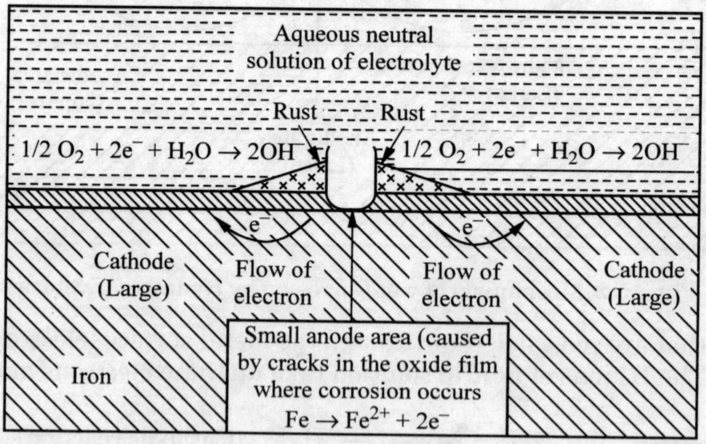

Fig. 15.3 Mechanism of wet corrosion by absorption of oxygen

Difference between Chemical Corrosion and Electrochemical Corrosion

Chemical Corrosion	*Electrochemical Corrosion*
1. It takes place in dry conditions	It takes place in presence of wet conditions such as presence of aqueous solution or electrolytes.
2. It takes place by direct chemical attack of environment on the metal	It takes place through the formation of a series of galvanic cells.
3. It can takes place on homogeneous and heterogeneous surface	It takes place on heterogeneous metal surfaces only.
4. Corrosion product accumulates at the same spot where corrosion has occured	Corrosion product accumulates at the cathode whereas corrosion occurs at the anode.
5. Uniform corrosion takes place	Non uniform corrosion takes place.

15.5 PASSIVITY

In an electrochemical series, metals are arranged from top to bottom in increasing order of their reduction electrode potentials. A metal high in the series is more anodic and undergoes corrosion faster than the metal below it. For example, Zn corrodes faster than Fe which inturn corrodes faster than Sn. This is because in electrochemical series, Zn is placed high above followed by Fe, which inturn is followed by Sn.

However, there are number of *exceptions*. For example, Ti is placed above Ag in electrochemical series, and is expected to corrode faster than Ag. But it is just the reverse i.e., Ti is corroded to a lesser extent than Ag. This can only be explained by passivity of the metals. *A metal is said to be passive in a certain environment if it exhibits much lower corrosion rate than what is expected from its position in the electrochemical series.*

However, on changing the environmental a passive metal can be rendered active.

Passivity of the metal is due to the formation of a highly protective, but very thin film on the surface of the metal under particular environmental conditions. The film so formed at the surface makes the metal or the alloy, act like a noble metal.

For example, metals like chromium, aluminium, nickel, titanium and a wide variety of stainless steel alloys containing chromium, exhibit outstanding corrosion resistance in *oxidizing* environments. These metals and alloys form insoluble oxide films on their surface in presence of oxygen and hence are rendered passive. (Some other metals such as zinc, cadmium, tin, uranium and thorium have also been observed to exhibit passivity effects under limited conditions.)

On changing the environment i.e., *in reducing environment*, all these metals become *chemically active* and undergo rapid corrosion. Fig. 15.4 illustrates the effect of passivity on corrosion of a metal. Fig. 15.4(a) shows the behaviour of the normal metal i.e. metal which do not show possivity effect. For example, metal immersed in an air-free acid solution.

Fig. 15.4 (b) illustrates the typical behaviour of a metal which exhibits passive effect. Initially, the metals shows corrosion rate similar to normal metal. But once the protective oxide film is formed it is rendered passive and corrosion rate shows a sudden decrease. This is the beginning of the passive region. Further increase in oxidizing power produce little, if any, change in corrosion rate of the material. However, at very high concentrations of the oxidizers, the corrosion rate again increases. This region is termed as transpassive region.

Significance: Usually 10^3 to 10^6 times reduction in corrosion rate is observed on passivity. From an engineering standpoint, the metals most susceptible to this kind of behaviour are the common engineering and structural materials, including iron, nickel, silicon, chromium and al-

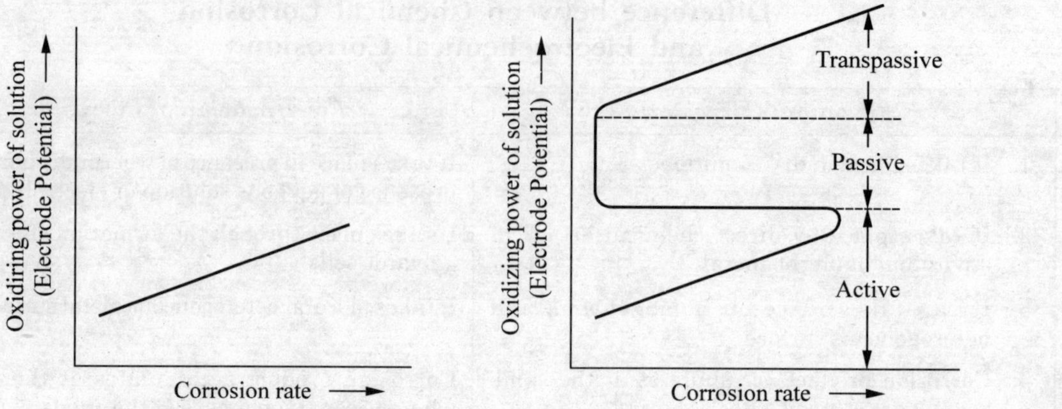

(*a*) Corrosion rate of the normal metal (*b*) Corrosion rate of an active-passive metal

Fig. 15.4

loys of these metals. Passivity has been used successfully to develop new methods of preventing corrosion and to predict corrosion resistance. Passive oxide films are artificially produced on certain metals and alloys by treatment in acidic or alkaline medium in presence of strong oxidizing agents. For example, the action of conc. HNO_3 on Al produce a thin protective oxide film, thereby making it passive. Metals which are susceptible to corrosion can also be made passive by alloying with one or more metals which are passive or resist corrosion. For example, iron is rendered passive when alloyed with any of the transition metals such as Cr, Ni and Mo.

Iron is heavily attacked by dilute HNO_3 but in conc. HNO_3, it forms a very thin protective passive film and behaves as a noble metal.

15.6 GALVANIC SERIES

As is discussed in the previous section that electrochemical series can provide the basis for predicting the corrosion of the metals. The metal lying at the top of the series is more anodic and undergoes corrosion faster than the metal below it in electrochemical series. However, due to the formation of protective oxide film at the surface of some metals (passivity), these metals behave exactly opposite to that predicted by the electrochemical series. Hence a more reliable series for predicting corrosion behaviour in different environments is the '**Galvanic Series**'. This series gives the reliable information about the corrosion behaviour of metals and alloys in different environments. The galvanic series for sea water is given in table:

Table 15.1 Galvanic Series in Sea Water

ANODIC/MOST ACTIVE	Magnesium
	Zinc
	Aluminium
	Cadmium
	Duralumin
	Steel
	Lead Tin Solder
	Lead
	Tin
	Copper and copper alloys

	Cupro-nickel
	Bronze
	Nickel (passive)
	Passive stainless steel
	Silver
	Titanium
	Graphite
	Gold
CATHODE OR NOBLE METALS	Platinum

15.7 FACTORS INFLUENCING CORROSION

Corrosion depends upon number of factors which include (i) nature of the metal and (ii) environmental factors

15.7.1 Nature of the Metal

A number of properties of metal affect corrosion rate.

(a) **Position in Galvanic Series:** Extent of corrosion depends upon the position of the metal in galvanic series. When two metals are in electrical contact in presence of an electrolyte, the metal higher up in galvanic series becomes anodic and suffers corrosion. Moreover, greater the difference in their positions in the electrochemical series, faster is the corrosion of the anodic metal.

(b) **Purity of the metal:** The rate of corrosion increases with the extent of impurities. Impurities accelerate the corrosion due to the formation of minute/tiny electrochemical cells because of 'heterogeneity' caused by them. The anodic part gets corroded.

For example, very pure and smooth zinc will not corrode in very pure hydrochloric acid. But zinc metal containing impurities of Pb or Fe corrodes rapidly. The figures below show the effect of impurities on corrosion of aluminum by hydrochloric acid.

% Aluminium	99.998	99.97	99.2
Relative Corrosion rate	1	1000	30,000

However pure metals are expensive and they are usually weak. Therefore, pure metals cannot be used in engineering applications where strength and cost are main considerations.

(c) **Physical state of the metal:** The physical state of the metal such as orientation of grains, grain structures, localized stresses, scratches etc., highly influence the rate of corrosion. For example grain boundaries are attacked more rapidly than the grain faces when exposed to a corrosive. Similarly, smaller the grain-size of the metal or alloy, greater will be its solubility as compared to the macroscopic crystal and hence, greater will be its corrosion.

Further the areas under stress, even in case of pure metals tend to be anodic and undergo corrosion.

(d) **Nature of the oxide film:** It was discussed earlier that in aerated atmosphere, almost all the metals get covered with an oxide film.

- If the specific volume of the metal oxide film formed is lesser than the volume of the metal atom, then the film will be porous. This porosity will easily allow oxygen to diffuse into the metal and cause further corrosion. For example alkaline earth metals such as Be, Mg, Ca etc., form oxides which have smaller volumes than the metals from which they are formed and hence are easily corroded even at low temperatures.

- On the other hand if the specific volume of the metal oxide film formed is less than that of the metal from which it is formed, it will not allow the oxygen to diffuse into it. Thus, this type of oxide film protect the metal from further oxidation or corrosion. For example, Al and heavy metals form the films whose specific volume is more than the metal itself and hence this film protects the metal from further corrosion.

The ratio of the volume of the metal oxide to the metal is known as specific volume ratio. If,

Specific volume ratio > 1, the metal will be protected from corrosion
Specific volume ratio < 1, the metal will corrode further.

(e) **Area Effect:** Another important factor in corrosion is the area effect i.e., relative areas of anode and cathode. If the area of cathode is large as compared to anode then the corrosion of the anode is rapid and severe. This is because, when the cathodic area is large, it will demand more electrons, which can only be met by brisk and rapid corrosion of anodic area, may be 100 or 1000 times greater than if the anodic and cathodic areas were equal in size. For example, in copper plates (cathode) with steel (anode) rivets, the steel rivets are completely corroded because the smaller steel rivets are coupled to large copper cathodic area.

(f) **Solubility of the products of corrosion:** If the corrosion product is soluble in the corroding medium then corrosion proceeds at a faster rate. Whereas insoluble corrosion product suppresses the corrosion as it functions as physical barrier. Further, if the product formed interacts with the medium to form insoluble product then also corrosion is suppressed. For example the corrosion of Pb in H_2SO_4 proceeds at power much lower rate because of the formation of insoluble $PbSO_4$.

(g) **Volatility of corrosion product:** The volatility of the corrosion product has a marked significance on corrosion rate. It the corrosion product is volatile, it leaves the surface of the metal as soon as it is formed. Hence it leaves the underlying fresh metal surface exposed to further attack. This results into excessive corrosion. One of the volatile corrosion product MoO_3 is formed by the oxidation corrosion of molybdenum.

15.7.2 Environmental Factors

The nature and aggressiveness of the environment determine to a large extent the degree of corrosion. Some of the common environment variables are:

(a) **Temperature:** Temperature increases the rate of almost all chemical reactions. Therefore rate of corrosion also increases with increase in temperature.

At higher temperatures even the passive metal can become active and get corroded. In high pressure boilers intergranular corrosion takes place only at high temperatures.

(b) **Humidity:** One of the important environment variable in corrosion is relative humidity. Corrosion increases with the increase in humidity of the atmosphere. The effects of humidity on corrosion of enameled steel is apparent in the following table.

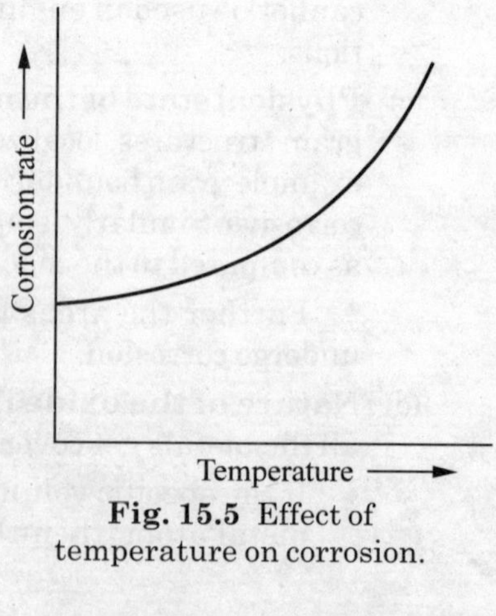

Fig. 15.5 Effect of temperature on corrosion.

Relative Humidity	0 – 65	65 – 80	80 – 90	100
Corrosion (Appearance)	No corrosion	Thin corrosion filaments	Wide corrosion filaments	Blisters due to corrosion

Though presence of moisture increases the corrosion rate, but it should be emphasized that metal does not rust if covered with ice. This is due to the fact that electrochemical attack (corrosion) requires water in the form of liquid.

The relative humidity above which the atmospheric corrosion rate of metal increases sharply is known as **critical humidity.** The increase in corrosion rate due to increase in relative humidity is due to the following reasons:

- Moisture acts as a solvent for oxygen CO_2 etc., present in the atmosphere and helps in setting up the electrochemical corrosion cell.
- Further, the metal oxide film possesses the property of absorbing moisture which can enhance the corrosion. Moreover, rain water may also wash away oxide film from the surface partially, leading to enhanced corrosion.
- Solids like NaCl etc dissolve in water and increase its conductivity and hence increase the rate of cossosion.

(c) **Presence of impurities in atmosphere:** Atmospheric corrosion increases if air is polluted by smoke or gases like H_2S, SO_2, sodium and other chlorides in marine atmosphere.

- NaCl and $(NH_4)SO_4$ if present absorb moisture and act as strong electrolytes.
- Some dust particles containing charcoal are capable of adsorbing SO_2 and so enhance the corrosion process.
- The presence of H_2S, H_2SO_4, HCl near the industrial areas increase the acidity of the liquid adjacent to the metal surface and hence increase its conductivity. This results into increase in corrosion rate.

(d) **Effect of pH:** Lower is the pH, greater is the corrosion. This implies that acidic media having pH less than 7 are more corrosive than alkaline or neutral media. Zinc is easily corroded even in weakly acidic solutions such as carbonic acid.

However, amphoteric metals like Al, Pb etc., go into the solution as complex ions at higher alkalinities. For example, zinc suffers corrosion in acidic medium and corrosion rate can be reduced by increasing the pH, but only up to a certain value. It suffers minimum corrosion at pH 11. On increasing the alkalinity further, it goes into the solution as complex ions.

(e) **Nature of the ions present:** The nature of the ions of the electrolytes greatly affect the corrosion.

- Chloride ions present in the medium destroys the passive film and expose the metal to corrosion.
- The anions such as silicates in the medium lead to the formation of insoluble product such as silicon gel which inhibits the corrosion
- Many metals and alloys can undergo corrosion in the presence of cyanide ions. Stress corrosion cracking of steel is frequently reported in cyanide containing solutions.
- Iron undergoes rapid corrosion in the presence of NH_4^+ salts.

(f) **Conductance of the corroding medium:** Conductance of the corroding medium plays an important role in corrosion. Dry sandy soils have lower conductance as compared to clay or mineralized soils. Hence the metallic structures buried under mineralized or clay soils undergo faster corrosion compared to dry sandy soils.

(g) **Effect of Oxygen and Oxidizers:** The effect of oxygen and oxidizers have been discussed in some details under the heading 'passivity'. Fig. 15.6 illustrates the effect of oxidizers an corrosion rate.

The graph can be divided into three sections. Section 1 is the behavior exhibited by normal metals when they exist in active state. When oxidizer is added, some metals achieve passivity due to the formation of oxide film which is protective in nature, represented by section 2. Section 3 is characteristic of the metals such as copper in acid solution which do not passivate.

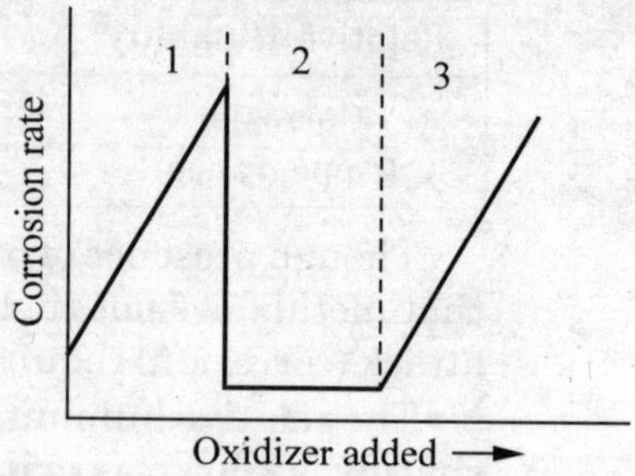

Fig. 15.6 Effect of oxidizers or oxygen on corrosion rate

(h) **Effect of Velocity:** Velocity of the environment plays an important role in erosion corrosion. Increase in velocity generally result in increased corrosion attack. For example, hardened straight chromium stainless steel valves and plugs give excellent service in most steam application. But, if the steam velocity is increased, corrosion increases leading to 'wire drawing.' High velocity may also reduce surface protection thereby increasing the corrosion.

However, velocity can decrease the corrosion attack in some cases by preventing the deposition of dirt which would cause corrosion. Increasing velocity may also sweep away the corrosive environment thus reducing corrosion. For example, under stagnant conditions, steel is attacked by nitric acid autocatalytically, because of the formation of nitrous acid as a cathode reaction product. Increasing velocity sweeps away the nitrous acid and thus removes one of the corrosive agents.

(i) **Formation of Oxygen Concentration Cell:** The rate of corrosion increases due to differential aeration by the formation of concentration cells. The corrosion by differential aeration has already been discussed. The corrosion due to differential aeration or at the surfaces where oxygen cannot diffuse readily is predominant in:

- metal washers where oxygen cannot diffuse
- buried pipe-lines passing from one type of soil to another
- crevice corrosion and waterline corrosion.

15.8 PROTECTION AGAINST CORROSION (CORROSION CONTROL)

The process of corrosion is slow but the losses incurred are enormous. In fact our economy would be drastically changed if there were no corrosion. But while the corrosion is inevitable, its cost can be considerably reduced. The main responsibility of a corrosion engineer is to control corrosion. Corrosion engineering mainly deals in applying scientific principles to prevent or control corrosion damage.

Methods for Controlling Corrosion

The various methods by which corrosion can be controlled and prevented are:

1. Protection by proper designing.
2. Material selection.
3. Modifying the environment.
4. Modification of the properties of the metal.
5. Use of inhibitors.
6. Use of protective coatings.
7. Cathodic protection or Electrochemical protection.

These methods are discussed below in detail.

1. **Protection by Proper Designing**

For controlling the corrosion, proper designing of the equipment plays a significant role. Some of the factors which should be considered for proper designing are:

(i) The contact between different metals and alloys should not be made, especially if they are far apart in the galvanic series. Where it cannot be avoided delete an efficient electrical insulator should be used. If this is not followed then the more active metal will get corroded causing local corrosion.

Example: Plastic washers are preferred over metal washers because plastic washers act as insulators and prevent corrosion.

(ii) The rate of corrosion also depends upon the relative sizes of the anode and the cathode. As far as possible, if two dissimilar metals are in contact, *the combination of small anode and large cathode should be avoided.* (refer to section 15.9) Screws, bolts, nuts and welds should therefore be made from a more noble material than the materials which are joined through them. Thus the screws nuts and bolts will act as cathodes and because of their small areas (cathodes), a limited cathodic reaction occurs leading to decreased rate of corrosion.

Example: Copper rivets are used to fasten steel sheets and not vice versa because copper is nobler than steel and hence acts as cathode. Due to the small area of the copper rivets, the cathodic reaction is limited and anodic reaction of steel proceeds slowly. Whereas if steel rivets are used for joining copper sheet, the small areas of the rivet anode will keep on furnishing electrons to the large cathodic copper sheet thereby corroding steel rivets completely.

(iii) The presence of crevices between adjacent parts of structure, even in the case of the same metal, should be avoided. The welded butt joints of the same metal/alloy should be used instead of riveted or bolted joints otherwise crevice corrosion may result. *In other words welding is better joining technique than mechanical fastners.*

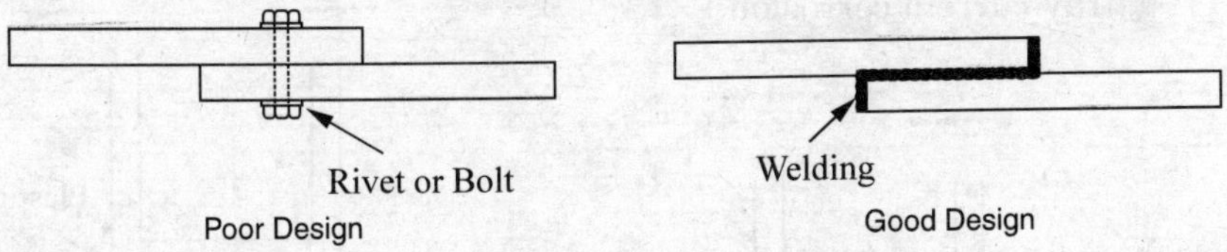

Fig. 15.7 Different techniques of joining the metals/alloys

(iv) As moisture is the main factor governing corrosion, the design should be such that retention of moisture should be as low as possible. Properties should be so designed that water is drained completely. For that, sharp corners should be avoided.

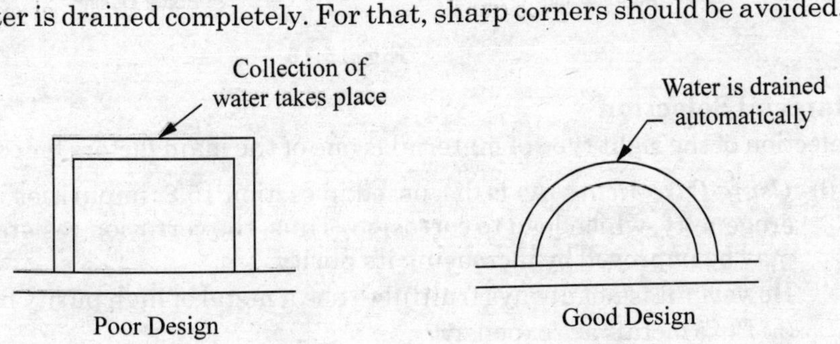

Fig. 15.8 Water stagnation

Storage containers should be designed in such a way that they can be completely drained and cleaned.

(v) Sharp corners and recesses should be avoided because dust or solid particles may accumulate leading to the corrosion by differential aeration.

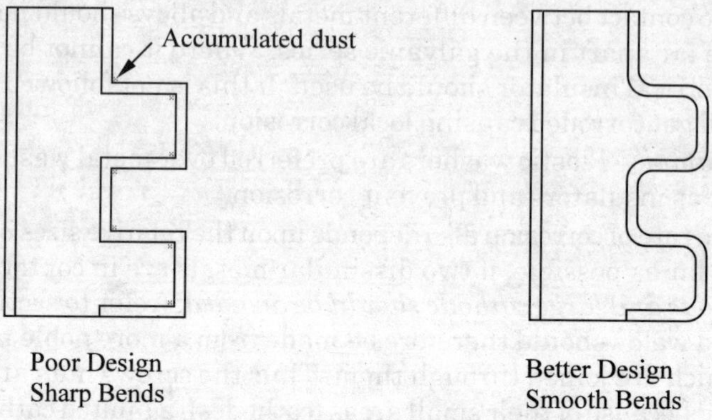

Fig. 15.9

(vi) The anodic parts should be readily replaceable or should be made thicker for longer life.

(vii) Increasing pipe diameters helps from mechanical standpoint, by decreasing velocity of the flowing fluids and ensuring laminar flow (refer to section 15.9 for the effects of velocity) Pipelines should be free from obstruction or crevices, in which water may stay giving rise to corrosion by differential aeration.

(viii) The equipment should be supported on stands to allow circulation of air.

(ix) Insulation of the materials should be proper otherwise leakage of current takes place. The site from which the leakage takes place becomes anodic and leads to corrosion (stray-current corrosion.)

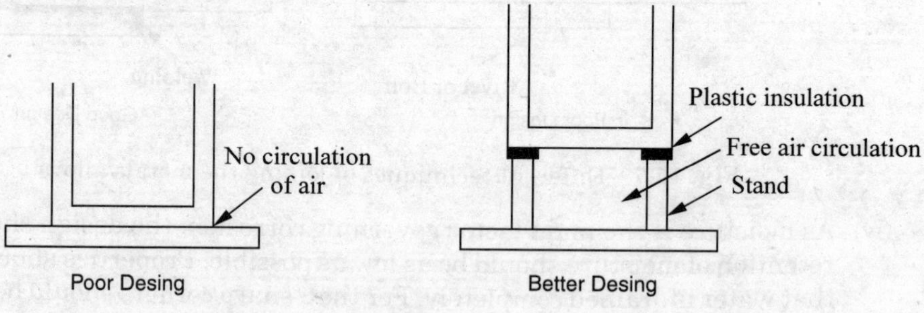

Fig. 15.10

2. Material Selection

Selection of the right type of material is one of the main factors for corrosion control.

(i) *Using Pure Metal:* As is discussed in section 15.8, impurities in a metal cause heterogeneity, which lead to corrosion. Thus, the corrosion resistance of a given metal may be improved by increasing its purity.

However it is not always fruitful to use a metal of high purity because:

(a) Pure metals are expensive

(b) They are usually weak and do not posses the required strength. For example, one would not build a bridge of pure iron.

(ii) *Using metal alloys:* Corrosion resistance of most commercial metals can be increased by alloying them with suitable alloying elements. (Alloys are mixtures of two or more metals or elements.) There are two kinds of alloys:

(a) *Homogeneous alloys are* solid solutions in which the components are completely soluble in one another e.g., stainless steel.

(b) *Heterogenous alloys* are the mixtures of two or more separate phases. The components of such alloys are not completely soluble and exist as separate phases.

The choice between the two types depends on the mechanical properties desired. Solid-solution alloys are more corrosion resistant than alloys with two or more phases because no galvanic coupling can take place in former case.

Several corrosion-resistant alloys have been developed depending upon the desired properties.

(a) Chromium is used as alloying metal for iron and steel as it produces an exceptionally coherent oxide film which protects the steel from further attack.

(b) Silicon iron alloy called (15% Si and 85% Fe) is highly resistant to the acids because of the formation of silicon oxide film on its surface.

(c) Cupro-nickel alloys (70% Cu + 30% Ni) containing 0.2% Fe are used for condenser tubes and for bubble trays used in fractionating columns.

3. Use of Inhibitors

Corrosion inhibitors are the substances which when added in small quantities to the corrosive environments decrease the corrosion rate. Inhibitors are organic as well as inorganic substances that dissolve in the corroding medium.

Inhibitors are generally classified on the basis of electrode at which their action is predominant. Thus we have two types of inhibitors, Cathodic inhibitors and Anodic inhibitors.

(i) *Anodic Inhibitors:* The anodic reaction takes place when the ions from the anodic metal pass into the solution. (For example, $Fe \rightarrow Fe^{2+} + 2e^-$). The anodic inhibitors are the compounds which react with these metal ions to produce insoluble compounds. This insoluble compound gets adsorbed on the metal surface forming a film or a protective barrier thereby reducing the corrosion rate.
The examples of anodic inhibitors include molybdates, phosphates and chromates of alkali and transition elements. Substances like Na_2CrO_4, $NaNO_2$ act as anodic inhibitors either by forming less soluble compounds with the anodic metal ions (as discussed above) or they may also retard the corrosion by oxidative repair of the oxide film.

The organic compounds can also acts as anodic inhibitors and example include sodium benzoate.

Limitations of anodic inhibitors: Anodic inhibitors are not considered satisfactory due to the following reasons:

(a) The inhibitors can prevent corrosion only above a certain critical concentration of solution.

(b) If insufficient concentration is used then the film so formed may not cover all the anodic area and corrosion of small areas left exposed suffer intense corrosion and cause severe pitting. (due to small anodic area and large cathodic area)

(ii) *Cathodic Inhibitors:* The catholic reaction in acidic medium is represented by:

$$2H^+ + 2e^- \longrightarrow H_2$$

Thus corrosion can be controlled by slowing down the diffusion of H^+ ions to the cathode. The diffusion of H^+ ions is considerably reduced by organic inhibitors such as amines, mercaptans, heavy metal soaps, substituted ureas and thioureas. These substances are capable of being adsorbed at the metal surfaces and thus restrict the diffusion of H^+ ions. Antimony or arsenic oxides are used as cathodic inhibitors because of the formation of adherent film on the surface of the metal.

In neutral solutions, the cathodic reaction is

$$H_2O + \frac{1}{2}O_2 + 2e^- \rightleftharpoons 2\overline{O}H$$

In this environment, the corrosion can be controlled by either removing the oxygen from the corroding medium or by not conducting the electrons.

The removal of oxygen is achieved by adding reducing agents such as sodium sulphite or by de-aeration.

$ZnSO_4$ and $MgSO_4$ can also be added which react with $\overline{O}H$ ions and the respective hydroxides precipitated on the cathode surface. The important cathodic inhibitor is $Ca(HCO_3)_2$ which is found in water.

$$HCO_3^- + \bar{O}H \longrightarrow H_2O + CO_3^{2-}$$

Compared to anodic inhibitors cathodic inhibitors are quite safe. This is because, even if they are used in concentrations less than required for complete protection, the partially covered cathodic sites decrease the surface of the cathode whereas anodic area remains practically the same. Thus corrosion decreases due to area effect.

(iii) *Vapour Phase Inhibitors:* These are the inhibitors which readily vapourize and form a protective layer of inhibitor on the metal surface. Vapour phase inhibitors (VPI) are used to prevent corrosion in enclosed spaces and also during storage, packing, shipping etc. Small metals are protected from corrosion by keeping them in the envelopes made from paper impregnated with a suitable vapour phase inhibitor.

Dicyclohexylammonium nitrile and cyclohexyl amine carbonate are amongst the most widely used VPIs.

4. Cathodic Protection or Electrochemical Protection

For the structures immersed in soils or electrically-conducting liquids, the cathodic protection is the best way to control corrosion. Cathodic protection is to force the metal to be protected to behave like a cathode as a result of which it is not corroded.

Cathodic protection can be achieved in two ways:

(i) By using galvanic or sacrificial anode (Galvanic Protection).

(ii) By using impressed current.

(i) *Galvanic Protection:* In galvanic protection or sacrificial anodic protection, the metal structure to be protected is made the cathode by connecting it by a wire to a more anodic material. As a result, the corrosion takes place at anodic material and the metal structure is saved from corrosion. The more active metal used for the purpose is known as *sacrificial anode*. The sacrificial anode is replaced by a fresh one when it gets corroded or consumed from time to time. Most commonly metals used for sacrificial anodic protection are Mg, Zn, Al and their alloys.

The application of cathodic protection include:

(a) Protection of buried pipelines, underground cables.

(b) Protection of marine structures, ship hulls etc.

(c) Magnesium rods are inserted into domestic water boilers or tanks to prevent rusting.

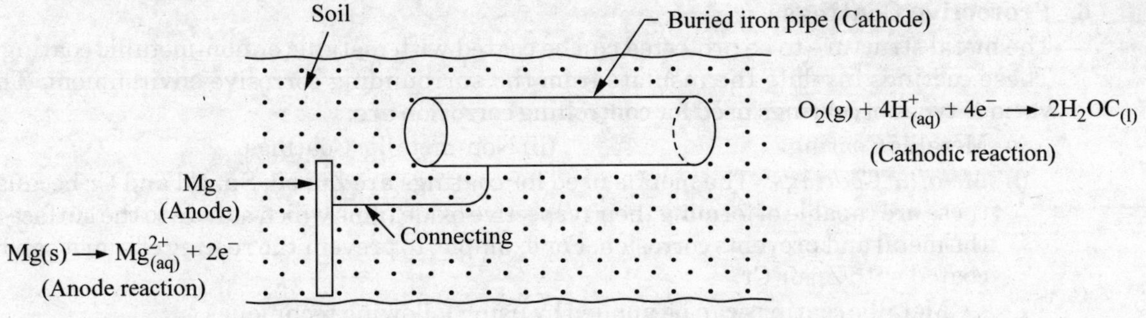

Fig. 15.11 Galvanic Protection using Sacrificial Anode

(ii) *Impressed Current Cathodic Protection:* This method, known as electrolytic method of cathodic protection, the object to be protected is made cathode by connecting it to negative terminal of DC source. The positive terminal is connected to a insoluble anode like graphite, scrap iron or platinum. The objective is to apply the current from an external source (insoluble anode) in the opposite direction to that of the corrosion current. The insoluble anode is buried in the soil or in the conducting medium adjacent to the metal which is to be protected (Fig. 15.12) These are placed in such a manner that a uniform potential is obtained over the entire metal surface. The electrons flow to the metal structure and as a result it act as cathode and is protected. The anode has to be replaced periodically.

The impressed current cathodic protection has been applied to protect.

(a) Buried pipelines, transmission line towers.

(b) Water pipes, water tanks, marine piers etc.

This kind of technique is particularly useful for large structures which require long term applications.

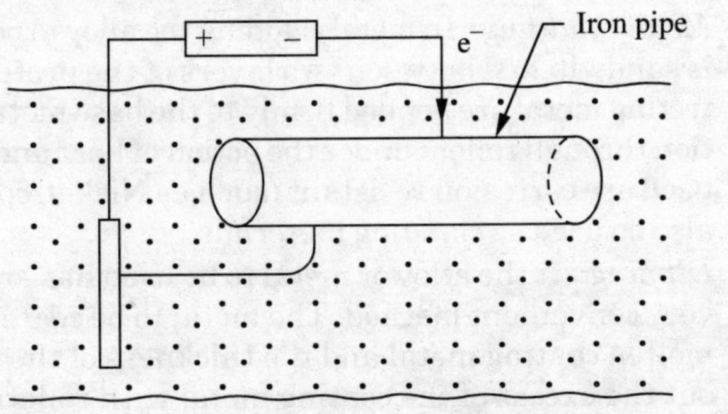

Fig. 15.12 Impressed Current Cathodic Protection

5. Anodic Protection

Anodic protection is applicable to the metals which can be passivated by the deposition of oxide films. In anodic protection, a protective oxide film is grown on the surface of the metal structure to be protected. This is done by electrochemical means i.e., by applying current in a direction that makes the metal structure more anodic. Under these conditions the metal structure is passivated when the required potential is maintained and hence rate of corrosion decreases.

Anodic protection has been applied in case of steel, stainless steel aluminium and chromium. It must be remembered that potential has to be accurately maintained so that it just induces passivity.

6. Protective Coatings

The metal structure to be protected can be coated with metallic or non-metallic coatings. These coatings insolate the insulate from the surrounding corrosive environment. The various types of coatings used for controlling corrosion are:

(i) Metallic Coatings (ii) Non-metallic Coatings

(i) *Metallic Coatings:* The metals used for coatings are Zn, Ni, Sn, Cd and Cr because these are capable of forming their respective oxide films which adhere to the surface of the metal and prevents corrosion. For example, to prevent the rusting of iron it can be coated with Zn or Cr.

Metallic coatings can be applied by using following techniques

(a) *Electroplating:* This is the most widely used method of coating metals to protect them from corrosion. The metal to be plated is made the cathode in a plating bath

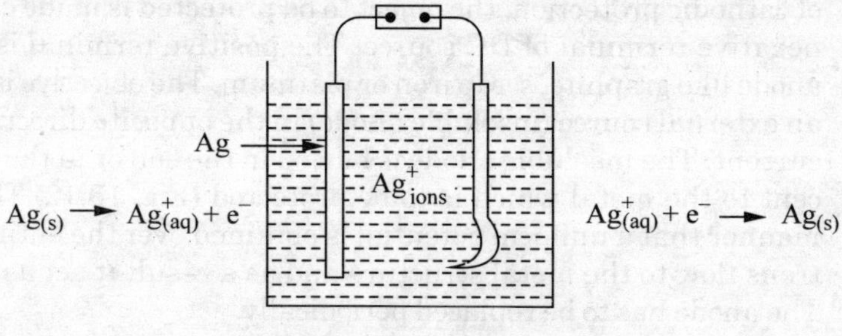

Fig. 15.13 Electroplating

containing the ions of the metal to be deposited. The anode is either an inert material or coating metal to be deposited. When current is passed, the coating metal ions migrate to the cathode and get deposited in the form of thin layer. Fig. 15.16 shows the silver plating of a spoon

(b) *Metal Cladding:* In metal cladding the alloy to be protected, (known as base metal) is sandwitched between two layers of the protecting metal. These layers of protecting metal are bonded firmly to the base metal by pressing the whole combination through rollers under the action of heat and pressure. The cladding material used are corrosion resistant (such as Nickel, copper, lead, silver etc.) Alloys can also be used as cladding materials.

(c) *Dipping:* It the alloy or metal to be used has low melting point then dipping is a very convenient method. The metal to be coated is passed through a bath of the molten coating metal and the thickness of the coating is adjusted by squeezing out the excess of the coating metal with rollers. Coating of zinc on iron by this process is called galvanizing.

(d) *Spraying:* A stream of 'atomized' droplets is blown on to the surface of the metal to be protected. This type of coating can be applied to the finished structures such as tanks, ships and towers etc.

(ii) *Non-Metallic Coatings:* The non-metallic coatings may be inorganic or organic coatings and fall under the following categories.

(a) *Organic Coatings:* Organic coatings are inert organic barriers which provide corrosion resistance as well as decorative value to the structure which is coated. The organic coatings include; paints, varnishes, enamels, and lacquers.

(b) *Chemical Conversion Coatings:* These are inorganic surface-barriers produced by chemical or electrochemical reactions at the surface of the metal. These coatings provide the protection to the base metal from corrosion and also provide a surface ideal for painting or other decorative measures. Examples of chemical coatings include, phosphate coating, chromate coating etc. The phosphate coating is produced by the chemical reaction of base metal with aqueous solution of phosphoric acid and zinc or manganese phosphate. Such coatings are applied to iron, steel and zinc. The chemical reaction between the phosphating solution and the base metal results in the formation of a surface film consisting of crystalline zinc-iron or manganese-iron phosphates.

ASHOKA PILLAR

This pillar was built by samrat Ashoka who ruled around 280 B.C, and is known by his name Ashoka Pillar. It is situated near Qutab Minar in Delhi. This pillar is the testimony of the advanced metallurgical technology ancient India possesed.

The pillar is solid shaft of iron, sixteen inches in diameter and 23 feet high. What is most astounding about it is that it has never rusted even though it has been exposed to wind and electrolysis! The technique used to cast such a gigantic, solid pillar is also a mystery, as it would be difficult to construct another of this size even today. With today's advances only four foundries in the world could make this piece and none were able to keep it rust free.

The pillar stands as mute testimony to the highly advanced scientific knowledge that was known in ancient India. The pillar is said to be 1600 years old.

A protective layer of 'misawite'— a compound made up of iron, oxygen and hydrogen on the steel pillar, which is said to contain phosphorus- is claimed as the reason for the non-corrosive existence.

QUESTIONS

SHORT ANSWERS QUESTIONS

1. Why some metal have tendency to get corroded ?
2. What are the consequences of corrosion ?
3. The oxide film formed by alkali and alkaline earth metals such as Na, Ca, Mg etc. develops cracks and pores leading to excessive corrosion whereas the oxide film formed by Cr, W etc. is protective in nature explain.
4. Discuss mechanism of rusting of iron
5. List differences between chemical corrosion and electrochemical corrosion.
6. What is pilling-bedworth rule ?
7. What are the various factors that contribute to the rusting of iron ?
8. Iron corrodes faster than aluminium, even though iron is placed below aluminium in the electrochemical series.

   Ans: The film formed by aluminium on oxidation is non porous, very thin, tightly adhering which protects the metal from further corrosion, whereas iron forms non-adhering film thus exposing the metal surface to atmospheric attack.
9. Which part of an electrochemical cell formed by two metals, suffer corrosion cathode or anode.

   Ans. Anode
10. What do you understand by the term-passivity ?
11. Presence of moisture increases the corrosion rate but metals do not rust if covered with ice
12. Steel connected to copper plumbing corrodes faster.
13. With increase in temperature, the rate of metallic corrosion increases.

    Ans: Because with increase in rate of reaction, the rate of reaction as well as diffusion increases which increase the rate of corrosion.
14. Impure metals corrode faster than pure metal under identical conditions. Give reason.

    Ans: Impurities in a metal cause heterogenity and form minute electrochemical cells at the exposed parts causing anodic part to get corroded.

15. Rusting of iron in saline water is quicker than ordinary water.
    Ans: Presence of NaCl (an electrolyte) increases the conductivity of water and hence speed up rusting.
16. Copper equipment should not possess steel bolt. Give reason.
    Ans: Iron acts as anode and is corroded. Moreover anodic area (steel bolt) is very small and hence corrosion is more rapid because demand for electrons be the cathodic area can only be fulfilled by the smaller anodic area by undergoing corrosion more rapidly.
17. Why plastic washers are preferred over metal washers.
18. Welding is better joining technique than mechanical fasteners.
    Ans: Welding by the same metal prevents crevice corrosion, which may occur if mechanical fasters are used.
19. Why chromium is used as an alloying metal for steel and iron
20. What is impressed current cathodic protection ?
21. Can we use aluminium in place of zinc for cathodic protection of iron from rusting ?
    Ans: In galvanic series the electrode potential of aluminium is more than that of zinc and hence we can use aluminium.

LONG ANSWER QUESTIONS

1. Discuss the various factors which influence corrosion.
2. What do you mean by 'dry corrosion' and 'wet corrosion' ?
3. What is the difference between rusting and corrosion?
4. The nature of the oxide film decides the rate of corrosion. Explain.
5. What is pilling-bedworth rule ?
6. What are the reasons for the formation of anodic and cathodic region on the surface of the metal ?
7. What is the difference between chemical and electrochemical corrosion ?
8. Explain the mechanism of hydrogen evolution and oxygen absorption in electrochemical corrosion ?
9. What is meant by the term passivity? What is its significance ?
10. Galvanic series is a preferred way of predicting the corrosion of the metals as compared to electrochemical series. Comment.
11. Impurities increase the rate of corrosion. Explain.
12. If the area of the cathode is larger than the anode, the corrosion is severe and very rapid. Explain.
13. How does temperature and humidity affect the corrosion rate ?
14. What are the various factors influencing corrosion ?
15. The welded joints of the same metal/alloy to be joined are preferred over bolted joints-why?
16. What are corrosion inhibitors ? Discuss cathodic and anodic inhibitors ?
17. What do you mean by vapour phase inhibitors ?
18. Discuss the importance of design and material selection in controlling corrosion.
19. A pure metal rod half immersed vertically in water starts corroding at the bottom. Give reason.

16

ELECTROCHEMICAL CELL/ELECTROLYSIS

To accomplish great things, we must not only act, but also dream; not only plan, but also believe.

— ***Anatole France***

CHAPTER OUTLINE

16.1 INTRODUCTION

Electrochemistry is a branch of physical chemistry that studies the relationship between chemical and electrical phenomena and the laws of their interaction.

We know that on passing electric current through an electrolyte solution a chemical reaction takes place. In other words, an electric current can cause a chemical reaction to take place under suitable conditions. The phenomena is called *electrolysis*. The reverse is also true i.e. we can generate electric current by the use of a chemical reaction. Hence, depending upon these two types of electrochemical processes, there are two types of cells such as:

(*i*) Electrolytic cells
(*ii*) Electrochemical cells.

(*i*) **Electrolytic Cells:** A device in which electrical energy is converted into chemical energy by passing an electric current to produce the desired chemical change is called an electrolytic-cell and the process is called electrolysis.

The refining of metals, electroplating with silver, gold and chromium or production of many chemical substances such as chlorine, sodium hydroxide, hydrogen peroxide etc., are carried out in electrolytic cells. The charging of storage batteries is also based on electrolytic cells.

(*ii*) **Electrochemical Cells:** An electrochemical cell is a device which converts chemical energy into electrical energy. It is also called *galvanic cell*. Daniell cell is a well known example of galvanic cell.

16.2 REDOX REACTION

Various chemical reactions occur through redistribution of electrons among the reacting substances. Any substance that loses electrons is said to be *oxidized* and the one which gains electrons is said to be *reduced*.

However, in a chemical reaction, a substance can lose electrons only if there is present another substance, which can gain electrons. This implies that oxidation can take place only if reduction also occurs at the same time or vice-versa. These reactions in which reduction and oxidation take place are called redox reactions.

$$Zn_{(s)} + Cu^{2+}_{(aq)} \longrightarrow Zn^{2+}_{aq} + Cu_{(s)}$$

In this $Zn_{(s)}$ loses two electrons whereas Cu^{2+} gains two electrons, and hence this is a redox reaction. The redox reactions are of two types:

(*a*) **Direct Redox Reactions:** The redox reactions in which oxidation and reduction take place in the same vessel are called direct redox reactions. Example, displacement of copper from $CuSO_4$ solution when a zinc rod is dipped in it.

$$Zn + CuSO_4 \longrightarrow ZnSO_4 + Cu_{(s)} \quad \text{or} \quad Zn_{(s)} + Cu^{2+} \longrightarrow Zn^{2+} + Cu_{(s)}$$

The above reaction can be split into two halves,

$$Zn \longrightarrow Zn^{2+} + 2e^- \text{ (oxidation half reaction)}$$
$$Cu^{2+} + 2e^- \longrightarrow Cu \text{ (reduction half reaction)}$$

(*b*) **Indirect Redox Reaction:** If redox reaction is allowed to take place in such a way that oxidation reaction takes place in one beaker and the reduction half reaction in another beaker, the electrons given out by former will be taken by latter and a current will flow. These indirect redox reactions form the basis of electro-chemical cells.

16.3 ELECTRODE POTENTIAL

A metal rod or plate in contact with its own ions is called electrode. When a metal is in contact with the solution of its own ions, it can undergo oxidation or reduction. *The tendency of an electrode to lose or gain electrons is called electrode potential.* Since each electrode represents a half cell, therefore, electrode potential is also called potential for half cell.

Consider a copper rod dipped in the solution of its own ions, i.e. Cu^{2+} ions. The copper rod has a tendency to lose electrons to form Cu^{2+} ions which pass into the solution. Thus, copper rod acquires a net negative charge as electrons, which are released, get accumulated around the Cu rod. These electrons will attract Cu^{2+} ions from the solution. This indicates that copper rod will not keep on losing electrons and a equilibrium will soon be established.

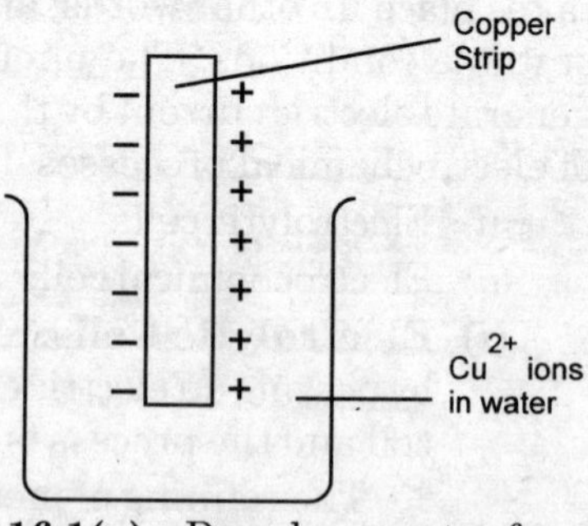

Fig. 16.1(a) Development of a pote-ntial difference between metal and its ions.

$$Cu_{(s)} \rightleftharpoons Cu^{2+}_{(aq)} + 2e^-$$

When such an equilibrium is reached, a separation of positive and negative charges would occur. This separation creates a potential difference between metal rod and the solution. This potential difference is called the *electrode potential*.

Thus, electrode potential of a metal is the measure of tendency of a metallic electrode to lose or gain electrons, when it is in contact with its own salt. It magnitude depends upon:

(*a*) The nature of metal and ions
(*b*) Concentration of ions
(*c*) Temperature.

Electrode potential measured under standard conditions i.e. 1 molar concentration of metal ions and a temperature of 298 K is called *standard electrode potential* and is denoted by E_o.

Moreover, the tendency of an electrode to *lose* electrons is a direct measure of its tendency to get oxidised and this tendency is called *oxidation potential*. Similarly, the tendency of an electrode to *gain* electrons is the measure of its tendency to get reduced, and this tendency is known as *reduction potential*. The value of reduction potential is negative of its oxidation potential.

i.e. $$E_{oxdn} = - E_{red}$$

Generally, the term standard electrode potential refers to reduction potential.

16.4 ELECTROCHEMICAL CELL

Electrochemical cell or the galvanic cell is a device which converts free energy of a physical or chemical process into electrical energy. Electrochemical cell usually consists of two electrodes immersed in one or more suitable electrolytes. When the electrodes are connected externally, a chemical reaction occurs in the cell resulting into oxidation at one electrode and reduction at the other electrode. The electrode at which oxidation occurs is called *cathode* and the electrode at which reduction occurs is called anode.

A typical example of galvanic cell is that of Daniell cell (Fig. 16.1(b)). It consists of a zinc rod dipping in $ZnSO_4$ solution, where oxidation takes place and a copper rod dipped in $CuSO_4$ solution, where reduction takes place. In other words, each electrode may be regarded as a half cell.

The metallic rods in the two beakers are connected to an ammeter by means of an insulated wire. The solutions in two beakers are connected by an inverted U-tube known as salt bridge. The tube is filled with a solution of electrolyte such as KNO_3, KCl, NH_4NO_3 or solidified solution of an electrolyte agar-agar and gelatine. The ends of the U-tube are plugged with glass wool or cotton wool. The salt in the bridge is such that it does not chemically react with either of the two solutions connected by it.

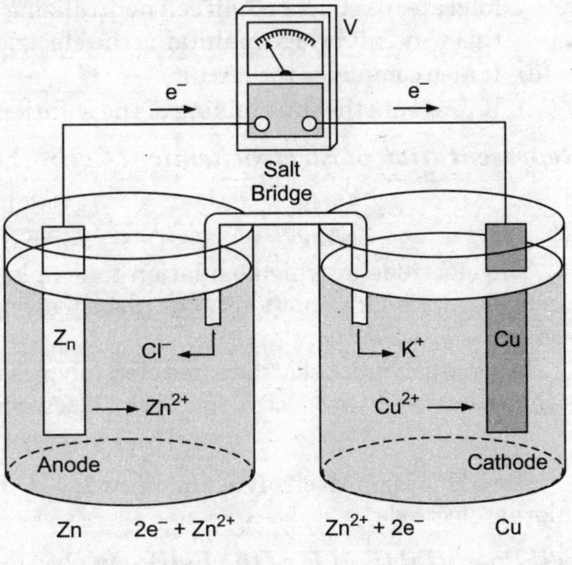

Fig. 16.1(b) Daniell cell

As soon as the circuit is closed, i.e. the two electrodes are connected, a current starts flowing which is shown by the deflection in the ammeter. The electric current flows due to the passage of electrons from zinc to copper rod. The electrode reactions involved as:

At anode	$Zn \longrightarrow Zn^{2+} + 2e^-$	(Oxidation)
At cathode	$Cu^{2+} + 2e^- \longrightarrow Cu$	(reduction)
	$Zn + Cu^{2+} \longrightarrow Cu + Zn^{2+}$	

Since electrons, i.e., negative electricity flows from zinc to the copper rod, the conventional current, i.e., positive electricity flows from copper to zinc rod. Thus zinc rod serves as negative electrode and copper rod serves as positive electrode.

As shown in the figure, in an electrochemical cell the two half reactions take place in two different compartments. The electrode where oxidation takes place (the anode) is assigned the *negative charge* because oxidation is accompanied by liberation of negatively charged electrons. Each time an ion leaves the electrode surface, two electrons are left on the electrode surface making it negatively charged. Similarly, the electrode where reduction takes place (the cathode) is assigned the *positive charge* because the metal cations from the solution tend to withdraw electrons from electrode surface and get deposited as metal atoms on electrode. This process decreases the availability of electrons on the electrode surface and thus leaves the electrode surface positively charged.

Note: However, in an *electrolytic cell*, the situation will be just opposite. The anode would acquire positive charge and cathode would acquire negative charge.

Function of Salt Bridge

(*i*) It maintains the electrical neutrality of the two electrolyte solutions. In the Daniell cell, which consists of $Zn/ZnSO_4$ and $Cu/CuSO_4$ electrodes, the Zn metal undergoes oxidation to release electrons to the external circuit and Zn^{2+} ions are added to $ZnSO_4$ solution. To maintain the electrical neutrality the negative ions (Cl^- or SO_4^{2-}) from the salt bridge migrate to $ZnSO_4$ solution and neutralize the excess positive charge in the oxidation half cell. Similarly in reduction half cell, Cu^{2+} ions from $CuSO_4$ solution are reduced. This leads to an excess of SO_4^{2-} ions in $CuSO_4$. The positive ions (K^+ or NH_4^+) from the salt bridge migrate to reduction half cell neutralizing the excess negative charge in $CuSO_4$ solution. In this way salt bridge maintains the electrical neutrality of the electrolytes.

(*ii*) It also completes the circuit.

(*iii*) It prevents the intermixing of the solutions of both the half cells.

Representation of Electrochemical Cell: The electrochemical cell is represented as

Anode	Salt bridge	Cathode
Zn/Zn^{2+}	KNO_3	Cu^{2+}/Cu

The electrode at which oxidation takes place (anode) is written on the left hand side. The electrode at which reduction takes place (cathode) is written on the right hand side of the salt bridge.

In some galvanic cells, the same electrolyte is used in two half cells. In such cases, no separation is shown between two electrolytes of two electrodes for example

$$Pt, H_2 \text{ (1 atm)}; HCl \text{ (0.1 M)}; Cl_2, Pt$$

Here HCl is an electrolyte and provides H^+ ions to the hydrogen electrode and Cl^- ion to the chlorine electrode.

Cell Potential (E.M.F. of the Cell): An electrochemical cell is made up of two electrodes, one of these electrodes must have a higher electrode potential (higher tendency to lose electrons) than the other electrode. As a result of this potential difference, the electrons flow from the electrode at higher potential to the electrode at lower potential. A cell that has not reached equilibrium, can do the electrical work till a transfer of electron takes place. *The electron transfer depends upon the potential difference between two electrodes. This potential difference is called the cell potential and* measured in *volts*. This potential is also known as *electromotive force of the cell*. When the cell potential is large, a given number of electrons travelling between the electrodes can do a large amount of electrical work. When the cell potential is small, the work done by the same number of electrons will be small. When the cell potential is zero, the reaction is at equilibrium and no work is done. The total cell potential can be written as the sum of the two electrode potentials. Thus

$$E_{cell} = E_{cathode} + E_{anode}$$

The total cell potential is a measure of the driving force of a chemical reaction whereas $E_{cathode}$ and E_{anode} are representatives of oxidising and reducing capabilities of concerned reactants, respectively. Now, we will explain how to measure the electrode potential of a single electrode.

The maximum electrical work done by a system is given by value of ΔG

$$W_{max} = \Delta G \text{ at the constant temperature and pressure.}$$

Therefore, the e.m.f. of a cell or cell potential is a measure of overall tendency which determines the spontaneity of a reaction to take place.

16.5 ELECTROCHEMICAL SERIES

The measurement of reduction potential can be used to calculate the thermodynamic functions and standard potential of the cell.

Now, we know that the standard potential of the electrode can be measured with reference to standard hydrogen electrode, which has been assigned the value zero. IUPAC recommends the adoption of the reduction potential as electrode potentials of the corresponding electrodes. When electrodes are arranged in increasing order of their standard (reduction) electrode potential in a series, the series is called *electrochemical series*. The electrochemical series is also called *activity series* (Table 16.1).

Table 16.1 Standard Electrode Potentials (Reduction) at 25°C.

Metalion		***Potential in volts***	
$Li^+ + e^- \rightarrow Li$	(base)	–3.05	(Anode)
$K^+ + e^- \rightarrow K$		–2.93	
$Ca^{2+} + 2e^- \rightarrow Ca$		–2.90	
$Na^+ + e^- \rightarrow Na$		–2.71	
$Mg^{2+}\ 2e^- \rightarrow Mg$		–2.37	
$Al^{3+} + 3e^- \rightarrow Al$		–1.66	
$Zn^{2+}\ 2e^- \rightarrow Zn$		–0.76	
$Cr^{3+} + 3e^- \rightarrow Cr$		–0.74	
$Fe^{2+} + 2e^- \rightarrow Fe$		–0.44	
$Ni^{2+} + 2e^- \rightarrow Ni$		–0.23	
$Sn^{2+} + 2e^- \rightarrow Sn$		–0.14	
$Pb^{2} + 2e^- \rightarrow Pb$		–0.13	
$F^{3+} + 3e^- \rightarrow Fe$		–0.04	
$H^+ + e^- \rightarrow 1/2H$		0.00	
$Cu^{2+} + 2e^- \rightarrow Cu$		+0.34	
$Ag^+ + e^- \rightarrow Ag$		+0.80	
$Pt^{4+} + 4e^- \rightarrow Pt$		+0.86	
$Au^+ + e^- \rightarrow Au$		+1.69	
$1/2F_2 + e^- \rightarrow F^-$	(Noble)	+2.87	(Cathodic)

16.6 COMMERCIAL CELLS/BATTERIES

As in evident from the previous section, there are many kinds of electrodes. All these electrodes may combine in different ways to give a large number of galvanic cells. Combination of two or more galvanic cells electrically connected to work together, to produce electric energy is called a *battery*. Commercially, batteries are used as a source of electrical energy for various purposes. Various combinations of electrolytes and electrodes are employed to make batteries for specific purposes. Thus commercial batteries can be made from different types of electrochemical cells.

The term battery should not be used while referring to a single cell. The batteries act as a portable source of electrical energy.

There are many different types of batteries ranging from the relatively large 'flash light' batteries to the miniature versions used for wrist watches or calculators. Batteries consist of either of two types of cells:

(*i*) Primary cells
(*ii*) Secondary cells.

16.6.1 Primary Cells

In primary cells, there is fixed amount of energy stored and once that energy has been used up, the cell is considered dead. In other words, the chemical reaction occurs only once and the cell then becomes dead and cannot be used again. Some of the important primary cells are described below:

(a) Dry Cell (Leclanche Cell): This is the most familiar type of dry cell. It consists of a zinc anode which is draped as a container. The cathode is comprised of a graphite rod surrounded by MnO_2 a moist electrolyte paste of NH_4Cl and $ZnCl_2$. Starch is added to the electrolyte (NH_4Cl, $ZnCl_2$ and MnO_2) to make it a thick paste like. To reduce leakage the cell is enclosed in polypropylene cylinder.

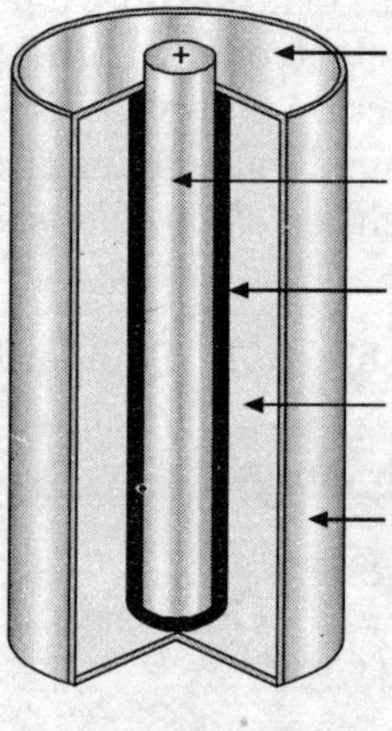

Fig. 16.2 Dry cell.

The cell reactions are:

$$Zn_{(s)} \longrightarrow Zn^{2+}_{(aq)} + 2e^- \qquad \text{(At Aanode)}$$

$$2MnO_2 + 2NH_4^+ + 2e^- \longrightarrow Mn_2O_3 + 2NH_3 + H_2O \qquad \text{(At Cathode)}$$

Thus, the reduction reaction occurs within the moist paste comprised of NH_4Cl and MnO_2. The NH_3 which is liberated reacts with Zn^{2+} from $ZnCl_2$ to form the complex $[Zn(NH_3)_2]\ Cl_{2(s)}$. The cell can be represented as:

$$Zn/Zn^{2+},\ NH_4^+/MnO_2/C$$

The dry cell 'couple' produces about 1.5 volts.

Dry cells do not have indefinite life. The acidic NH_4Cl corrodes the zinc container even when not in use.

In *alkaline version*, a modified form of dry cell, the NH_4Cl is replaced by KOH or NaOH. Thus corrosion of Zn container is prevented because zinc does not dissolve readily in basic medium.

The half cell reactions of alkaline cell are:

$$Zn + 2\bar{O}H \longrightarrow ZnO + H_2O + 2e^- \qquad \text{(Anode)}$$

$$2MnO_2 + 2e^- + H_2O \longrightarrow Mn_2O_3 + 2\bar{O}H \qquad \text{(Cathode)}$$

The alkaline dry cell lasts much longer compared to Leclanche cell.

(b) Mercury Cell: The type of cell commonly used for calculations, hearing aids, digital watches and cameras is the mercury cell. The mercury cell, also known as Ruben Cell was developed in 1947. The anode consists of zinc-mercury amalgam and cathode is comprised of a paste of HgO and carbon. A paste of KOH and ZnO forms the electrolyte.

The cell reaction is:

$$Zn_{(amalgam)} + 2\bar{O}H \longrightarrow ZnO_{(s)} + H_2O + 2e^- \qquad \text{(At Anode)}$$

$$HgO_{(s)} + H_2O + 2e^- \longrightarrow Hg(l) + 2\bar{O}H \qquad \text{(At Anode)}$$

$$Zn_{(amalgam)} + HgO_{(s)} \longrightarrow ZnO_{(s)} + Hg_{(l)}$$

Mercury cells are much more expensive to manufacture than are dry cells or alkaline cells. This is due to high cost of the mercury (II) oxide they contain. At 25°C the cell voltage is 1.344 V. Since the overall cell reaction does not involve any ion in solution whose concentration can change, the cell shows constant potential throughout its life. This makes it useful in sensitive instruments.

Similarly is *silver button cell* the anodic reaction is the same as mercury cell whereas anodic reaction involves $Ag_2O_{(s)}$.

$$Zn_{(s)} + 2\bar{O}H_{(aq)} \longrightarrow ZnO(s) + H_2O_{(l)} + 2e^-$$

$$Ag_2O_{(s)} + H_2O_{(l)} + 2e^- \longrightarrow 2Ag_{(s)} + 2O\bar{H}_{(aq)}$$

$$E_{Cell} = 1.3 \text{ V (Mercury)}, 1.6 \text{ V (Silver)}$$

Other types of primary cell include lithium cell which has been developed for special uses where long life, small size and very low power are required. Cameras, heart pacemakers and electronic wrist watches are powered by these cells.

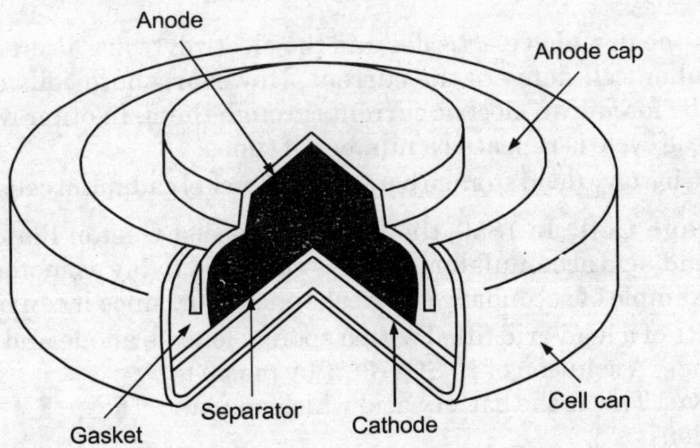

Fig. 16.3 Mercury cell.

A brief account of primary cells is given in Table 16.2

Tabel 16.2: Primary Cells

Type	*Anode (–)*	*Cathode (+)*	*Electrolyte*	*Voltage*	*Uses*
(1) *Leclanche or dry cell* Zn | $(ZnCl_2)$, NH_4Cl | MnO_2,C	Zinc case $Zn(s) \rightarrow Zn^{2+}(aq) + 2e^-$	Carbon rod in contact with carbon & MnO_2 $NH_4^+(aq) + MnO_2(s) + e^- \rightarrow NH_3(aq) + MnO(OH)(s)$	Paste of NH_4Cl, $ZnCl_2$ (acid electrolyte)	1.5 volts	Low drain appliances such as radios, torches, liquid crystal calculators, toys
(2) *Alkaline cell* Zn, ZnO | KOH (paste) | MnO_2,C	Brass rod in contact with powdered zinc $Zn(s) + 2OH^-(aq) \rightarrow ZnO(s) + H_2O(l) + 2e^-$	Steel case in contact with carbon & MnO_2 $MnO_2(s) + H_2O(l) + e^- \rightarrow MnO(OH)(s) + OH^-(aq)$	Paste of 7 M KOH (alkaline electrolyte)	1 volt	High current or high total capacity appliances such as electronic phot-graphic flash units, tape recorders, radios, shavers. Can deliver higher current then Leclanche cell without severe volt-age drop.
(3) *Mercury button Cell*	Steel cap in contact with powdered zinc	Steel case in contact with powdered HgO $HgO(s) + H_2O(l) + 2e^- \rightarrow Hg(l) + 2OH^-(aq)$	Paste of KOH (alkaline electrolyte)	1.3 volt	Watches, pacemakers, hearing aids, micro-phones.
(4) *Silver button cell* Zn, ZnO | KOH(paste) | Ag_2O, Ag	Steel cap in contact with powdered zinc $Zn(s) + 2OH^-(aq) \rightarrow ZnO(s) + H_2O(l) + 2e^-$	Steel case in contact with powdered Ag_2O $Ag_2O(s) + H_2O(l) + 2e^- \rightarrow 2Ag(s) + 2OH^-(aq)$	Paste of KOH (alkaline electrolyte)	1.3 volts	Watches, pacemakers, hearing aids. As cell discharges, concentrations do not Change as reactants and products are solids. $[OH^-]$ in electrolyte does not change. So, the voltage of this cell remains constant throughout its lifetime.

16.6.2 Secondary Cells

A secondary cell is one in which electrodes and the electrolyte are altered by the chemical action that takes place when cell delivers the current. However, these cells can be restored to their original condition by forcing an electric current through them. In other words, such batteries can be recharged and the cycle is repeated a number of times.

The automobile battery (lead storage battery) and nickel cadmium cells fall under this category.

***(i)* Lead Storage Cell:** In 1859, the French physicist Gaston Plante (1834-89) invented a device called the lead-acid accumulator, which is still used today as motor car battery with minor changes. It is an example of secondary electrochemical cell, since it can be recharged.

The cell consists of a lead grid filled with a spongy lead as anode and a lead grid packed with lead oxide as cathode. A solution of H_2SO_4 (38% by mass) is used as an electrolyte. This is the battery acid which one can buy at service stations.

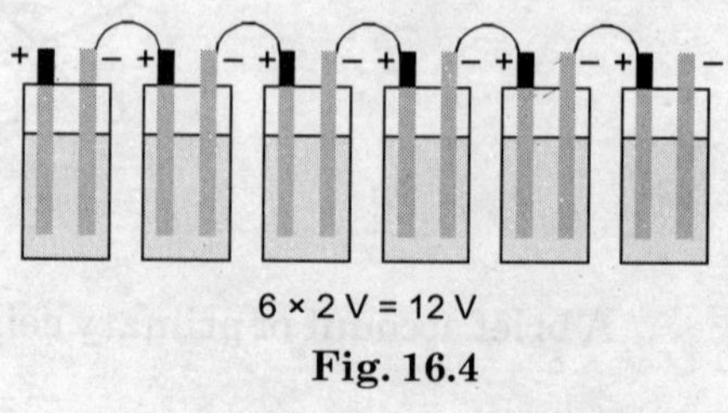

6 × 2 V = 12 V

Fig. 16.4

The battery consists of 6 such cells connected in series, each cell having an emf of about 2 V, giving 12 V as the overall emf of the battery.

When the battery is discharging, i.e., when it is supplying a current, the reactions are

At the anode:

$$Pb(s) + SO_4^{2-}(aq) \longrightarrow PbSO_4(s) + 2e^-$$

At the cathode:

$$PbO_2(s) + 4H^+(aq) + SO_4^{2-}(aq) + 2e^- \longrightarrow PbSO_4(s) + 2H_2O\ (l)$$

The overall reaction (during discharge) is, therefore,

$$Pb(s) + PbO_2(s) + 4H^+\ (aq) + 2\,SO_4^{2-}\ (aq) \longrightarrow PbSO_4(s) + 2H_2O\ (l)$$

Note that the electrolyte is sulphuric acid (H_2SO_4), which is gradually used up. Both electrodes become coated with an insoluble layer of lead sulphate, which, being an insulator, would eventually ruin the battery.

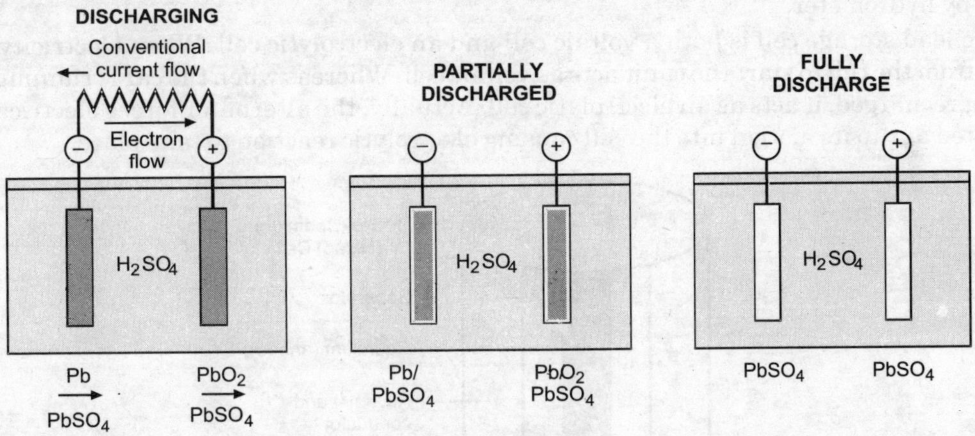

Fig. 16.5

As the battery provides energy, it is discharged and this lead to the gradual formation of lead sulphate at the electrodes, and a steady decrease in the concentration of the sulphuric acid:

If the battery is totally discharged, both electrodes will have been converted to lead sulphate. It will no longer be able to provide a current, and cannot be recharged. It is only good for scrap!

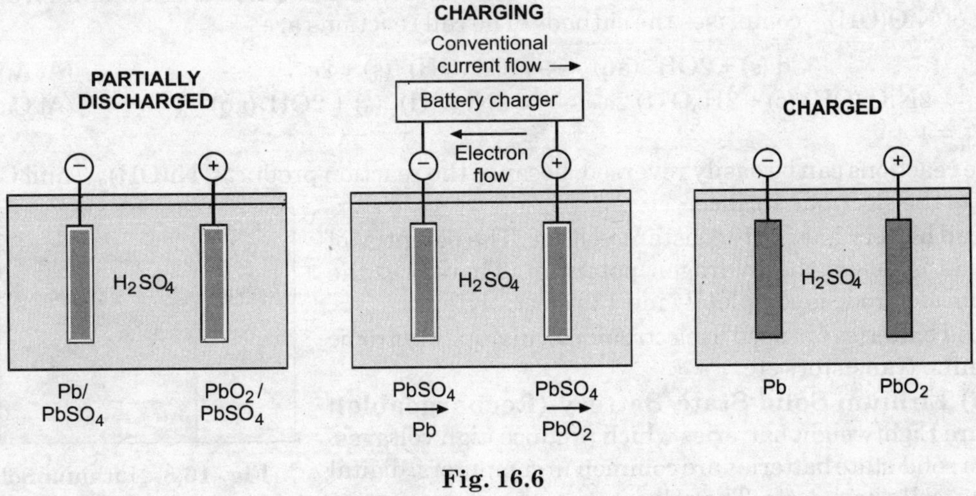

Fig. 16.6

When the battery is being charged, (using a battery charger or the motor car's alternator), electrons are supplied to the anode of the battery and this converts the lead sulphate to lead. At the same time, the lead sulphate at the anode is converted to lead dioxide. That is the cell reactions are reversed and the cell operates like an electrolyte cell.

The reactions are:

$$PbSO_4(s) + 2e^- \longrightarrow Pb(s) + SO_4^{2-}(aq) \qquad \text{(at cathode)}$$

$$PbSO_4(s) + 2H_2O(O) \longrightarrow PbO_2(s) + 4H^+(aq) + SO_4^{2-}(aq) + 2e^- \qquad \text{(at anode)}$$

$$2PbSO_4 + 2H_2O \longrightarrow Pb(s) + PbO_2(s) + 4H^+(aq) + SO_4^{2-}(aq)$$

The concentration of H_2SO_4 decreases during discharging whereas it increases during charging. Thus the extent to which the battery is discharged can be checked by measuring the density of H_2SO_4 by hydrometer.

The lead storage cell is both a voltaic cell and an electrolytic cell. When electricity is being drawn from the cell to start the car it acts as voltaic cell. Whereas when the car is running, the cell is being recharged, it acts as an electrolytic cell. Actually, the alternator forces electrical energy generated due to its motion into the cell causing electrolytic reaction to take place.

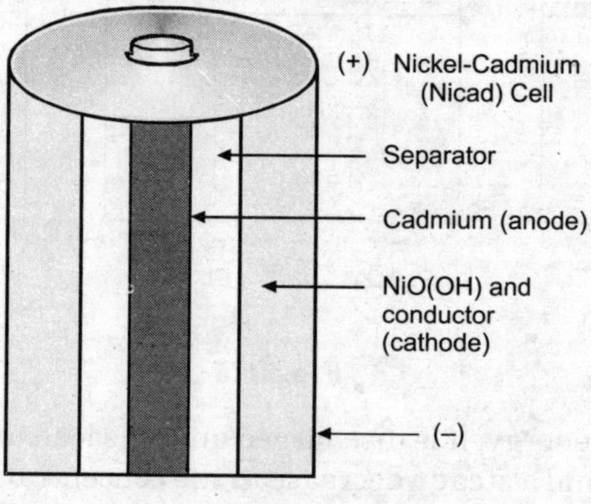

Fig. 16.7 Nicad cell.

***(ii)* Nickel-Cadmium (Nicad) Cell:** Nickel-cadmium cell consists of a cadmium anode and a paste of $NiO(OH)_{(s)}$ comprises the cathode. The cell reactions are:

$$Cd\ (s) + 2OH^-\ (aq) \longrightarrow Cd\ (OH)_2\ (s) + 2e^- \qquad \text{(At Anode)}$$

$$2NiO\ (OH)\ (s) + 2H_2O\ (I)\ 2e^- \longrightarrow 2Ni\ (OH)_2\ (s) + 2OH^-(aq) \qquad \text{(At Cathode)}$$

$E_{cell} = 1.4$ V

The reactions can be easily reversed, because the reaction products, $Ni(OH)_{2(s)}$ and $Cd(OH)(s)$ adhere to the electrode surface.

Nicad battery has fairly constant voltage. The potential of Cd anode is below the hydrogen potential. Therefore, the cadmium electrode is completely inert to electrolyte.

Nicad batteries are used in electronic calculators, electronic flash units, transistors etc.

***(iii)* Lithium Solid-State Battery (Rechargeable):** These are Lightweight batteries which produce high voltages. Lithium solid-state batteries are common in computers, digital cameras, cell phones, etc. The cell reactions are:

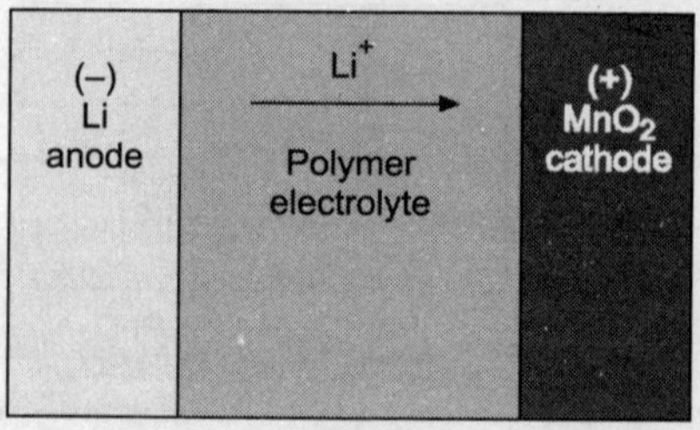

Fig. 16.8 Lithium Solid-state Battery

$$Li(s) \longrightarrow Li^+ \text{ (in solid electrolyte)} + e^- \quad \text{(Anode)}$$
$$MnO_2(s) \longrightarrow Li^+ + e^- \longrightarrow LiMnO_2(s) \quad \text{(Cathode)}$$

A brief account of secondary batteries is given in Table 16.3.

Table 16.3

Type	*Anode (–)*	*Cathode (+)*	*Electrolyte*	*Voltage*	*Uses*
(1) *Lead-acid battery (or accumulator)* Pb, $PbSO_4$ | $H_2SO_4(aq)$ | PbO_2, Pb	Lead plates $Pb(s) + HSO_4^-(aq) \rightarrow PbSO_4(s) + H^+(aq) + 2e^-$	Lead plates covered with PbO_2 $PbO_2(s) + 3H^+(aq) + HSO_4^-(aq) + 2e^- \rightarrow PbSO_4(s) + 2H_2O(l)$	4M H_2SO_4	2.1 volts per cell (6 cells in series makes up a 12 volt car battery)	Motor vehicle batteries. As cell delivers current, lead sulfate deposits on electrodes and $[H_2SO_4]$ falls. Six 2 V cells connected in series make up a 12 V battery. To obtain high current output from the battery, the surface area of the electrodes in contact with the electrolyte must be large and the plates close together.
(2) *Nickel-cadmium cell (NiCad)* Cd, $Cd(OH)_2$ | KOH | NiO_2, $Ni(OH)_2$, Ni	Grid covered with cadmium $Cd(s) + 2OH^-(aq) \rightarrow Cd(OH)_2(s) + 2e^-$	Grid covered with nickel (III) hydroxide $NiO(OH)(s) + H_2O(l) + e^- \rightarrow Ni(OH)_2(s) + OH^-(aq)$	KOH	1.25 volts	Video cameras, phones, cordless drills, laptop computers

16.7 FUEL CELLS

These are the electrical cells which convert the energy from the combustion of fuels into the electrical energy. It differs from an ordinary battery that the reactants are not contained within the cell, but are continually supplied from outside i.e. the fuel and the oxidant are supplied from outside the cell to the electrodes at which they undergo reactions.

Fuel cells is not a primary cell, since neither the reactants nor products are stored. It is not a secondary cell which can be recharged. Rather, the reactants have to be fed continuously and the product have to be removed. So it is reasonable to keep fuel cell in a separate category.

The fuel cell converts the chemical energy of the fuel directly to electricity. The essential process in a fuel cell is:

Fuel + Oxygen ⟶ Oxidation product + Electricity

One type of fuel cell, used in space crafts and advanced non-linear submarines, is hydrogen-oxygen fuel cell.

Hydrogen-Oxygen Fuel Cell: It consists of two inert porous electrodes which are made up of either graphite impregnated with finely divided Pt or a 75/25 alloy of Pd with Ag or Ni. The electrolyte solution is 25% KOH. Hydrogen and oxygen are bubbled through the anode and cathode compartment respectively. The cell reactions are:

$$H_2(g) \longrightarrow 2H^+_{(aq)} + 2e^- \quad \text{(Anode)}$$
$$O_2(g) + 4e^- + 4H^+_{(aq)} \longrightarrow 2H_2O_{(l)} \quad \text{(Cathode)}$$
$$2H_{2(g)} + O_{2(g)} \longrightarrow 2H_2O_{(l)} \quad \text{(Overall reaction)}$$

The theoretical e.m.f. of the cell is 1.23 V

$$E^o = E^o_{oxd.} + E^o_{red} = 0.83\ V + 0.40\ V = 1.23\ V$$

However, in actual practice, the e.m.f. of the cell is about 0.8 V to 1.0 V.

The only by-product formed in fuel cell is water which is used by the spacecraft or submarine crews.

Fig. 16.10 shows a basic design of a fuel cell. However, is practice, it is very complex piece of equipment.

Let us look in more detail at the key cell components:

(a) The Fuel: Oxygen is usually taken from the air (filtered and cleaned before introduction to the cell) and hydrogen is generally made from the decomposition of a hydrocarbon source (such as methane). Any hydrogen rich material can serve as possible fuel source such as methanol, ethanol, propane etc.

(b) The Electrodes: The function of the electrodes is to bring about reaction between the reactant (fuel or oxygen) and the electrolyte, without itself being consumed or corroded. They must also, by definition, be an electronic conductor and bring the three phases (gaseous fuel, liquid electrolyte and electrode itself) to contact. They must be good electron sources or sinks.

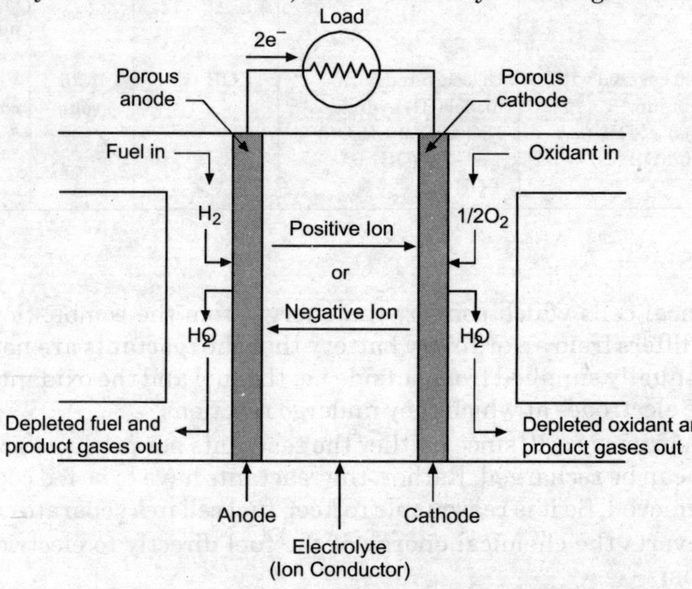

Fig. 16.9 Schematic of an individual fuel cell.

There are several methods for gas-liquid interface stabilisation which rely on capillary action. A biporous electrode allows liquid to penetrate the small pores and a gas pressure is applied that drives the liquid out of the large pores. All this is set-up within the structure of the electrode. Then the merging of the two phases may be achieved in several ways. The electrolyte will tend to form a thin wetting film over part of the internal surface of the electrode. The reactant gas, sparingly soluble in the electrolyte, can diffuse through this film and reach the electrode surface, where a solid-liquid reaction can occur. The electrode structure have to be designed to maximise the area of the wetted film (Fig. 16.10).

Fig. 16.10

(c) ***The Electrolyte:*** A large range of fuel cells have been developed using a variety of different electrolytes. These developments have been driven by the desire to improve the economics of the cells and the potential applications at a range of different temperatures. In fact fuel cells are usually classified in terms of their electrolyte contents. Electrolytes most often used are aqueous KOH or H_2SO_4 or ion exchange resin saturated with water.

Depending upon the electrolyte used the fuel cell can be of different types. Table 16.4 gives an account on the different types of fuel cells.

Table 16.4 Fuel Cells

Type	*Anode (–)*	*Cathode (+)*	*Electrolyte*	*Voltage*	*Uses*
Alkaline fuel cell	Porous carbon anode with catalyst in contact with $H_2(g)$ $H_2(g) + 2OH^-(aq) \rightarrow 2H_2O(l) + 2e^-$	Porous carbon cathode with catalyst in contact with $O_2(g)$ $O_2(g) + 2H_2O(l) + 4e^- \rightarrow 4OH^-(aq)$	Hot KOH (*aq*)	~ 1 volt	• Promising future • More efficient than coal or gas-fired power stations. • Non-polluting • Porous electrodes lined with catalysts, eg, Ni or Pt, making it expensive
Acid fuel cell	Porous carbon anode with catalyst in contact with $H_2(g)$ $H_2(g) \rightarrow 2H^+(aq) + 2e^-$	Porous carbon cathode with catalyst in contact with $O_2(g)$ $O_2(g) + 4H^+(aq) + 4e^- \rightarrow 2H_2O(l)$	hot H_3PO_4 (*aq*)	~ 1 volt	• Systems required to continuously supply reactants and remove products • CH_4, CH_3OH, C_3H_8 can be used instead of $H_2(g)$

16.8 ELECTROLYSIS

Electrolysis refers to the decomposition of a substance by an electric current. In other words , the process of decomposing a compound using electricity is called **electrolysis**.

$$2H_2O_{(l)} \rightarrow 2H_{2(g)} + O_{2(g)}$$

Electrolysis is a most important industrial process with a wide application. However, the largest application is that of the manufacture of chlorine and sodium hydroxide from concentrated aqueous sodium chloride in the Chlor-alkali industry.

16.8.1 Electrolytic Cells

The apparatus used to bring about electrolysis is called *electrolytic cell* (Fig. 16.18). It consist of a glass vessel in which the electrolyte in the form of an aqueous solution or in the molten state is taken. Two metal rods are dipped into it and are connected to a source of electricity, i.e., battery. These rods are called electrodes; the one connected to the negative pole of the battery is called cathode and the other connected to the positive pole of the battery is called anode.

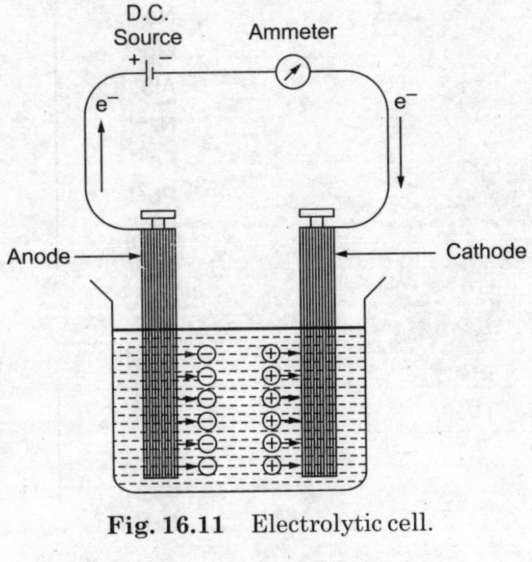

Fig. 16.11 Electrolytic cell.

16.8.2 Mechanism of Electrolysis

Whenever an electrolyte (acid, base or salt) is dissolved in water or is taken in the molten state, the electrolyte dissociates to produce positively and negatively charged ions. On passing electric current, the positively charged ions move towards the cathode an hence are called **cations** whereas negatively charged ions move towards the anode and hence are called **anions.** On reaching their respective electrodes, ions lose their charge and become neutral particles/species. The cations accept electrons form the cathode to become neutral species whereas the anions give electrons to the anode to become neutral species. Thus, *oxidation occurs at the anode while reduction takes place at the cathode. The conversion of ions into neutral species at their respective electrodes is called the* **primary change.** The product formed as a result of primary change may be collected as such or it may undergo a **secondary change** to form the final products.

Predicting Products in Electrolysis

In the example above, of molten sodium chloride, there is one cation Na^+ and one anion Cl^-. Therefore, the products can easily be calculated.

However, if a compound is in an **aqueous** solution, then we have two more ions, OH^- and H^+. *So which ion will appear at the electrode will be decided by the electrochemical series.*

In the case of the cathode, the cations which are reduced most easily back to their atoms will constitute the cathodic reaction. Out of OH^- and H^+ ,hydrogen ions are most easily reduced so the following cathode reaction will occur:

$$2H^+_{(aq)} + 2e^- \rightarrow H_{2(g)}$$

In the case of the anode, the anions which are most easily oxidised back to atoms will constitute the anode reaction. Chloride ions are most easily oxidised, out of Cl^- and OH^- ions, hence the following anode reaction will take place:

$$2Cl^-_{(aq)} \rightarrow Cl_2 + 2e^-$$

This leaves $Na^+_{(aq)}$ and $OH^-_{(aq)}$ in solution i.e. sodium hydroxide.

Electrochemical series or reactivity series

K^+
Na^+
Mg^{2+}
$Al3^+$
Nz^{2+}
Fe^{2+}
Pb^{2+}
$H^+_{(aq)}$
Cu^{2+}
Ag^+

$M^{X\neq} + X^{e-} = M$

This becomes easier as we descend the table

SO_4^{2-}
NO_3^-
Cl^-
Br^-
I^-
OH^-

$A^{X\neq} = A + X^{e-}$

This becomes easier as we descend the group

Electrolysis of Some of the Salts

(i) ***Electrolysis of molten lead bromide:*** In the molten state, $PbBr_2$ exist as Pb^{2+} and Br^- ions. On passing electricity, Pb^{2+} ions move towards cathode and Br^- ions move toward anode. On reaching cathode, each Pb^{2+} ion picks up two electrons from the cathode, and becomes neutral lead atom (Primary change) Thus, lead is liberated at the cathode.

On reaching anode, each Br^- ion gives an electron to the anode and becomes a neutral bromine atom (**Primary change).** Since bromine atoms are unstable, therefore, two bromine atoms combine to form a bromine molecule **(secondary change).** Thus, *bromine gas is evolved at the anode.*

The reactions occurring at the two electrodes may be shown as follows:

At cathode: $\underset{\text{(reduction, primary change)}}{Pb^{2+}(l) + 2e^- \longrightarrow Pb(l)}$

At anode: $\underset{\text{(oxidation, primary change)}}{Br^-(l) - e^- \longrightarrow Br}$

$$Br + Br \longrightarrow \underset{\text{(secondary change)}}{Br_2(g)}$$

Overall reaction: $Pb^{2+}(l) + 2Br^-(l) \xrightarrow{\text{Electrolysis}} Pb(l) + Br_2(g)$

or $PbBr_2(l) \xrightarrow{\text{Electrolysis}} \underset{\text{At cathode}}{Pb(l)} + \underset{\text{At anode}}{Br_2(g)}$

The lead obtained at the cathode is in the molten state.

(ii) ***Electrolysis of molten sodium chloride:*** Molten sodium chloride contains Na^+ and Cl^- ions.

$$NaCl(l) \rightleftharpoons Na^+(l) + Cl^-(l)$$

On passing electricity, Na^+ ions move towards cathode while Cl^- ions move towards anode. On reaching the cathode, each Na+ ion picks up one electron from the cathode and becomes a neutral atom. Thus, *sodium metal is liberated at the cathode.*

On reaching the anode, each Cl^- ions gives an electron to the anode and becomes a neutral chlorine atom. Since chlorine atoms are unstable, therefore, two atoms of chlorine combine to form a chlorine molecule. Thus, Cl_2 is *evolved at the anode.*

The reaction occurring at the two electrodes may be show as follows:

At cathode: $Na^+ + e^- \longrightarrow \underset{\text{(reduction, primary change)}}{Na}$

At anode: $Cl^- - e^- \longrightarrow \underset{\text{(oxidation, primary change)}}{Cl}$

$$Cl + Cl \longrightarrow \underset{\text{(Secondary change)}}{Cl_2(g)}$$

Overall reaction: $2Na^+(l) + 2Cl^-(l) \xrightarrow{\text{Electrolysis}} 2Na(l) + Cl(g)$

or $2NaCl(l) \xrightarrow{\text{Electrolysis}} \underset{\text{At cathode}}{2Na(l)} + \underset{\text{At anode}}{Cl_2(g)}$

The metal obtained at the cathode is in the molten state

(iii) **Electrolysis of acidulated water.** Water is only weakly ionized and hence is poor conductor of electricity but the presence of an acid makes it a better conductor. This is because water acidified with sulphuric acid contains three types of ions, i.e., H^+, OH^- and SO_4^{2-} produced as follows:

$$H_2O(l) \rightleftharpoons H^+(aq) + OH^-(aq)$$
$$H_2SO_4(aq) \longrightarrow 2H^+(aq) + SO_4^{2-}(aq)$$

When electric current is passed through acidulated water, H^+ ions move toward cathode, OH^- and SO_4^{2-} ions move towards anode. Now, since the *discharge potential of* OH^- *ions is much lower than that of* SO_4^{2-} ions, therefore, OH^- *ions are discharged* at anode while SO_4^{2-} ions remain in the solution. Similarly, H^+ ions are discharged the cathode. The reaction, occurring at the two electrodes may be written as:

At cathode: $H^+(aq) + e^- \longrightarrow H$
(reduction, primary change)

$H + H \longrightarrow H_2(g)$
(secondary change)

At anode: $OH^-(aq) - e^- \longrightarrow OH$
(oxidation, primary change)

$4OH \longrightarrow 2H_2O + O_2$
(secondary change)

Overall reaction: $4H^+(aq) + 4OH^-(aq) \xrightarrow{\text{Electrolysis}} 2H_2(aq) + 2H_2O(l) + O_2(g)$

or $2H_2O(l) \xrightarrow{\text{Electrolysis}} \underset{\text{At cathode}}{2H_2(g)} + \underset{\text{(At anode)}}{O_2(g)}$

(iv) **Electrolysis of an aqueous solution of sodium chloride.** Sodium chloride and water ionize as under:

$$NaCl(aq) \longrightarrow Na^+(aq) + Cl^-(aq)$$
(almost completely ionized)

$$H_2O(l) \rightleftharpoons H^+(aq) + OH^-(aq)$$
(only slightly ionized)

On passing electricity, $Na^+(aq)$ and $H^+(aq)$ move towards the cathode while $Cl^-(aq)$ and $OH^-(aq)$ ions move towards the anode.

At Cathode. Both $Na^+(aq)$ and $H^+(aq)$ are present near the cathode. Since the discharge potential of H^+ ions is lower than that of Na^+ ions, therefore, H^+ ions are discharged in preference to Na^+ ions.

$$H^+ + e^- \longrightarrow H$$
(reduction, primary change)

$$H + H \longrightarrow H_2(g)$$
(secondary change)

Thus, H_2 *gas is evolved at the cathode while* Na^+ *ions remain in the solution.*

At anode. Both Cl^- and OH^- ions are present near the anode. Since the discharge potential of Cl^- ions is lower than that of OH^- ions, therefore, Cl^- ions are discharged in preference to OH^- ions.

$$Cl^- - e^- \longrightarrow Cl$$
(oxidation, primary change)

$$Cl + Cl \longrightarrow Cl_2(g)$$
(secondary change)

Thus, Cl_2 *gas is evolved at the anode while* OH^- *ions remain in the solution.*

In nutshell, during electrolysis of aqueous sodium chloride, Cl_2 is liberate at the anode, H_2 is liberated at the cathode while sodium hydroxide remains in the solution.

(v) **Electrolysis aqueous copper sulphate solution using inert electrode, e.g., platinum electrodes.** Copper sulphate and water ionize as under:

$$CuSO_4(aq) \longrightarrow Cu^{2+}(aq) + SO_4^{2-}(aq)$$
(almost completely ionized)

$$H_2O(l) \rightleftharpoons H^+(aq) + OH^-(aq)$$
(only slightly ionized)

On passing electricity, Cu^{2+}(aq) and H^+(aq) move towards the cathode while SO_4^{2-} ions and OH^- ions move towards the anode.

At cathode. Both Cu^{2+} and H^+ ions are present near the cathode. Since the discharge potential of Cu^{2+} ions is lower than that of H^+ ions, therefore, Cu^{2+} ions are discharged in preference to H^+ ions,

$$Cu^{2+} + 2e^- \longrightarrow Cu \text{ (primary change)}$$

Thus, copper is deposited on the platinum electrode.

At anode. Both SO_4^{2-} and OH^- ions are present near anode. Since the discharge potential of OH^- ions is lower than that of SO_4^{2-} ions, therefore, OH^- ions are discharged in preference to SO_4^{2-} ions.

$$4OH^- + 4e^- \longrightarrow 4OH$$
(oxidation, primary change)

$$4OH \longrightarrow 2H_2O(l) + O_2(g)$$
(secondary change)

Thus, during electrolysis of aqueous $CuSO_4$ solution, Cu is deposited at the cathode, O_2 is liberated at the anode while the solution contains sulphuric acid.

It may be noted here that in the example (iii) to (v) discussed above, the electrodes used are those which are not attacked by ions, i.e., graphite and platinum. However, if the electrodes are attacked by ions, then the reactions taking palace during electrolysis are different. For example, let us consider the electrolysis of an *aqueous solution of copper sulphate using copper electrodes.*

(iv) **Electrolysis of an aqueous solution of copper sulphate using copper electrodes.** Copper sulphate and water ionize as under:

$$CuSO_4(aq) \longrightarrow Cu^{2+}(aq) + SO_4^{2+}(aq)$$
(almost completely ionized)

$$H_2O(l) \rightleftharpoons H^+(aq) + OH^-(aq)$$
(only weakly ionized)

On passing electricity, Cu^{2+} and H^+ move towards cathode while OH^- and SO_4^{2-} ions move towards anode.

At cathode. Both Cu^{2+} and H^+ ions are present near the cathode. But the discharge potential of Cu^{2+} ions is lower than that of H^+ ions, therefore, Cu^{2+} ions are discharged in preference to H^+ ions.

$$Cu^{2+}(aq) + 2e^- \longrightarrow \underset{\text{(reduction, primary change)}}{Cu(s)}$$

Thus, copper is deposited on the cathode

At anode. Unlike electrolysis of $CuSO_4$ using platinum electrodes, no ions are liberated here, Instead, anode itself undergoes oxidation (i.e., loses electrons) to form Cu^{2+} ions which go into the solution. This is due to the reason that Cu is more easily oxidised that both OH^- and SO_4^{2-} ions.

$$Cu(s) \longrightarrow \underset{\text{(Oxidation, primary change)}}{Cu^{2+}(aq) + 2e^-}$$

Thus, the net result of electrolysis is that copper is deposited at the cathode from the solution and an equivalent amount of copper from the anode dissolves into the solution to form Cu^{2+} ions. This principle of electolysis widely used in electroplating and purification of less reactive metal such as Pb, Cu, Ag, etc. The impure metal is made the anode that dissolves on passing current and pure metal is deposited at the cathode.

Examples of Industrial Electrolytic Processes

Today, applied electrochemistry is of great value for the economy. Electrolytic processes are used to manufacture a number of valuable materials. Electrochemical power sources (batteries) are used widely in many technical fields. Some of these processes are:

(1) Caustic Soda Manufacture:

The electrolysis of brine is carried out on a huge scale for the industrial production of chlorine and caustic soda (sodium hydroxide).

In aqueous sodium chloride solution, four ions are present Na^+, Cl^-, OH^-, H^+.

Reaction at the cathode:

$$2H^+_{(aq)} + 2e^- \longrightarrow H_{2(g)}$$

Reaction at the anode:

$$2Cl^-_{(aq)} \longrightarrow Cl_{2(g)} + 2e^-$$

The ions remaining in solution are OH^- and Na^+ which combine to form NaOH

$$\underset{\text{salt}}{2NaCl} + \underset{\text{water}}{2H_2O} \xrightarrow{\text{DC Power}} \underset{\substack{\text{sodium}\\ \text{hydroxide}}}{2NaOH} + \underset{\text{chlorine}}{Cl_2} + \underset{\text{hydrogen}}{H_2}$$

To produce NaOH it is necessary to prevent reaction of the NaOH with the chlorine.

The electrolysis cell used for this reaction is called the **diaphragm cell.** In the diaphragm cell process the anode area is separated from the cathode area by a permeable diaphragm to prevent the sodium hydroxide making contact with the chlorine. The brine is introduced into the anode compartment and flows through the diaphragm into the cathode compartment. A diluted caustic soda leaves the cell. One of the main disadvantages of the diaphragm cell is that it produces low quality sodium hydroxide (containing sodium chloride) at only 12% concentration, and so additional energy is needed to purify and concentrate it before it can be sold. The sodium hydroxide must usually be concentrated to 50% and the salt removed. The salt separated from the caustic brine can be used to saturate diluted brine.

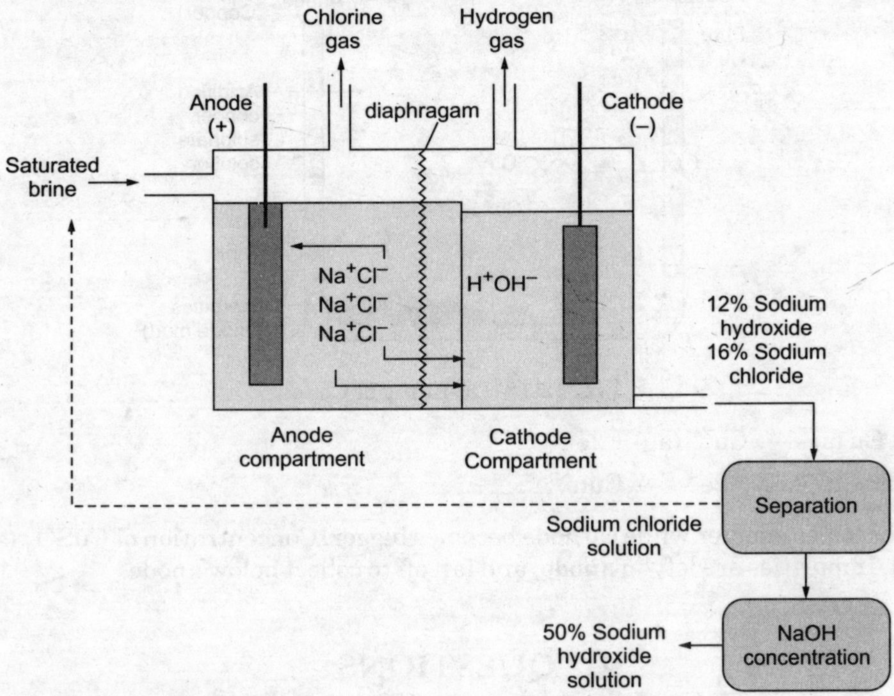

Fig. 16.12 Diaphragm Cell for the production of NaOH

Current annual world production of chlorine by electrolysis is over 30 that of alkali is 35 megatons, and it increases by 2 to 3% per year. This industry consumes about 100 billion kilowatt-hours of electrical energy per year.

(2) Refining of Metals:

Many of the metals such as copper, silver, gold, aluminium, lead, etc., are purified by electrolytic method. This is perhaps the most important method.

The impure metal is made anode while a thin sheet of pure metal acts as a cathode. The electrolytic solution consists of generally an aqueous solution of a salt or a complex of the metal. On passing the current, the pure metal is deposited on the cathode and equivalent amount of the metal gets dissolved from the anode. Thus, the metal is transferred from anode to cathode through solution. The soluble impurities pass into the solution while the insoluble one, especially less electropositive impurities collect below the anode as **anodic mud or anode sludge**.

Example: Refining of Blister copper (98% pure)Impure copper is made the anode; a piece of pure copper is used as the cathode. Electrolyte is $CuSO_4$ (aq) or a solution of a copper salt. The apparatus is set up as shown in Fig. 16.13. On passing the current through the electrolyte, the pure metal from the anode dissolves into the electrolyte. An equivalent amount of pure metal from the electrolyte is deposited on the cathode. The soluble impurities go into the solution, whereas, the insoluble impurities settle down at the bottom of the anode and are known as anode mud.

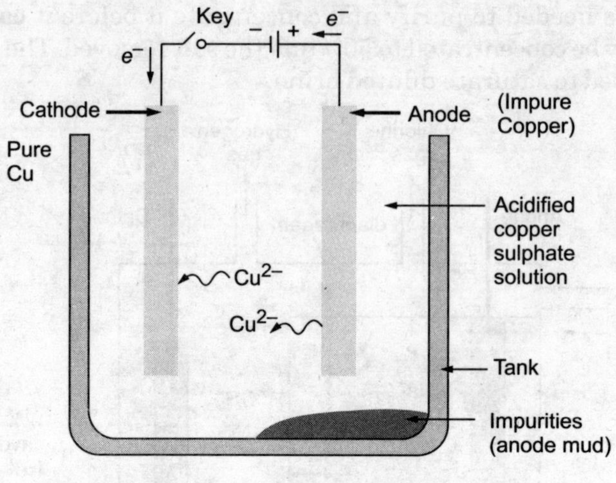

Fig. 16.13 Refining of Copper

Anode: $Cu\ (s) \longrightarrow Cu^{2+}\ (aq) + 2e^-$

Cathode: $Cu^{2+}(aq) + 2e^- \longrightarrow Cu(s)$

Anode becomes smaller while cathode becomes bigger. Concentration of $CuSO_4$ (aq) remains unchanged. Impurities are left on anode, and fall off to collect below anode.

QUESTIONS

1. What is a galvanic cell? Describe a Daniell cell.
2. Differentiate between an electro chemical cell and an electrolytic cell
   Give two important differences between e.m.f. and potential difference
3. What is an elecrochemical series ? Discuss its three important applications.
4. (a) What is alkaline battery? Describe the construction and working of Ni-Cd battery with relevant reactions taking place during discharge. Mention its advantages and applications.
   (Calicut, June 01; VT April 2L; Aug. 01)
   (b) Explain the construction of Ni-Cd cell. **(Kerala, April 97)**
5. Discuss the electrolysis of molten and aqueous sodium chloride. What are the products formed.

17 OILS, FATS AND LUBRICANTS

"What we observe is not nature itself, but nature exposed to our method of questioning."

Werner Heisenberg

CHAPTER OUTLINE

17.1 FRICTIONAL RESISTANCE

When the surface of a metal is allowed to move tangentially with respect to another metal, with which it is in contact, it experiences a resistance to its motion. This resistance is known as frictional resistance.

Actually, all the material surfaces, no matter how smooth they appear, show many irregularities in the form of peaks and valleys (of the order of 1000 A°). When the two metal surfaces come in contact with each other, the area of contact is only a small fraction of the apparent contact area, because the contact is through the peaks as shown in Fig. 17.1.

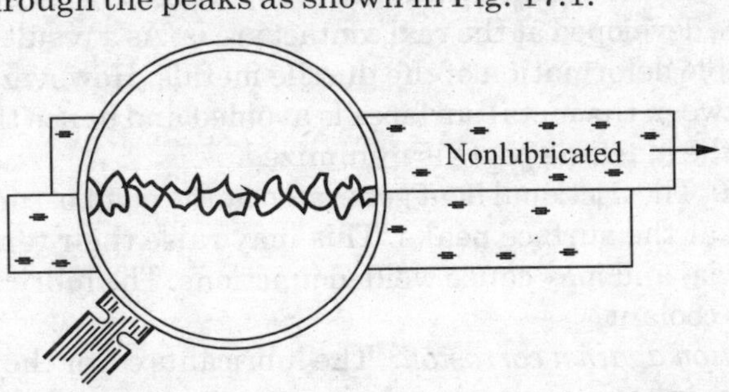

Fig. 17.1 Contact between peaks

These real areas of contact between the two surfaces are called *asperites*. When two metal surfaces slide over one another, the interlocking of the aspirites or peaks cause the resistance to the motion and give rise to frictional resistance. Moreover, due to sliding of one metal over the other, the peaks also get broken, resulting in wear and tear of the surfaces.

Further, the local pressure at the asperites may be sufficiently high to cause fusion of the peaks which leads to the formation of *welded junctions* between asperites. Additional force is then required to break these junctions. This further adds to frictional resistance.

Frictional resistance results in and loss of energy, a lot of wear and tear of surfaces of moving parts thereby causing loss in efficiency of machine. Sometimes, the moving parts get heated up and may also cause welding of two surfaces due to heat.

17.2 LUBRICANTS

In order to reduce frictional resistance and wear and tear of machine parts, a layer of fluid is introduced in between the moving parts. This layer of fluid is known as *lubricant*. The lubricant prevents direct contact of the rubbing surfaces by forming a film in between. Lubricant also acts as a coolant by removing the frictional heat generated due to rubbing of the surfaces.

Thus a lubricant can be defined as any substance introduced between two moving or sliding surfaces which minimizes the frictional resistance. The process of introducing the lubricant is known as lubrication (Fig. 17.2). The study of the phenomena and mechanism of friction, wear and lubrication of surfaces in relative motion is called *Tribology*, the main objective of which is to minimize the wear and tear of moving machine parts by reducing friction.

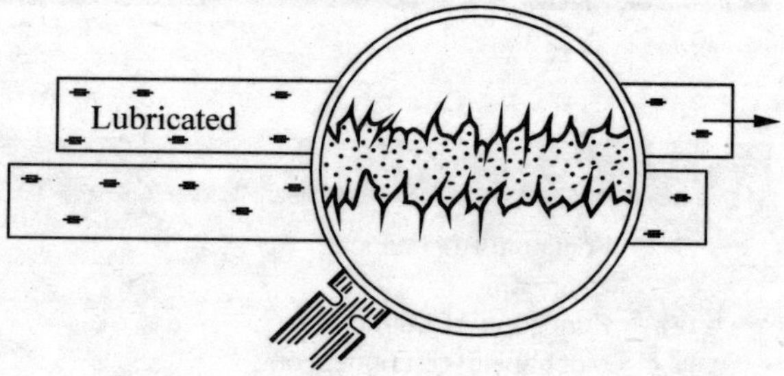

Fig. 17.2 Effect of lubrication

17.3 FUNCTIONS OF LUBRICANTS

(i) *Lubricants reduce the frictional resistance:* To reduce the frictional resistance, is one of the main functions of lubricants. Lubricants prevent the direct contact between the sliding metal surfaces so that the small peaks of the two metals do not interlock and hence frictional resistance is reduced.

(ii) *It reduces the surface deformation, wear and tear:* When two solid surfaces are in contact the real area of contact is very small as compared to the apparent area of contact, because the real contact is only at a limited number of asperites (Fig. 17.1). Even under light loads, high pressures are developed at the real contact areas. As a result the asperites are crushed causing appreciable deformation of the ductile metals. However, through lubrication the direct contact between the metal surfaces is avoided and hence the deformation as well as wear and tear of the machine parts is minimized.

(iii) *It acts as a coolant:* The frictional heat generated at the point of contact between the rubbing parts is localized at the surface peaks. This may raise their temperature to the melting point of the material and may cause welded junctions. The lubricating oil absorbs the heat and hence acts as coolant.

(iv) *It provides protection against corrosion:* The lubricants cover the metal surface and hence the attack of atmospheric corrosive environment on metal surface is prevented. Thus lubricants provide protection against corrosion.

(v) *It acts as a seal in some cases:* The lubricants also act as a seal preventing the leakage of gases. For example, in internal combustion engines the lubricant used between piston and cylinder not only reduces the friction but prevents the leakage of gases from the cylinder.

(vi) *It improves the efficiency of the machine:* Since the use of lubricant minimizes the liberation of frictional heat, it reduces expansion of metal. Moreover it avoids welded joints and hence reduces the maintenance and running cost of the machine.

17.4 MECHANISM OF LUBRICATION

Efficient lubrication plays a vital role in engineering. Depending upon the operating conditions and the lubricant characteristics, lubrication are of the following types:

1. Hydrodynamic or fluid film lubrication.
2. Thin film or boundary lubrication.
3. Extreme pressure lubrication.

1. Hydrodynamic or Fluid Film Lubrication

In hydrodynamic lubrication the two sliding metal surfaces are separated from each other by a thick film of fluid (~ 1000 A° thick). This film of lubricant in adsorbed on the metal surface and held by Vander Waals forces. For example, if hydrocarbon oils are used as lubricant, the long hydrocarbon chains are oriented outwards.

Multilayer adsorption over this monolayer also occurs and the molecules are randomly arranged with increasing distance from the metal surface.

Thus the two metal surfaces are separated by the oil film and direct surface to surface contact between the two metals is prevented and hence the peaks or the asperities do not interlock. This consequently reduces the friction. Whatever friction is there after lubrication, is only due to the internal resistance between the particles of lubricant. In the above example we may say that the friction is only along the tails of hydrocarbon chain. The coefficient of friction in such cases is as low as 0.001 to 0.03 as compared to 0.5 to 1.5 for sliding metal to metal surfaces. The longer the hydrocarbon chain, greater will be the separation between the surfaces and a consequent decreased coefficient of friction.

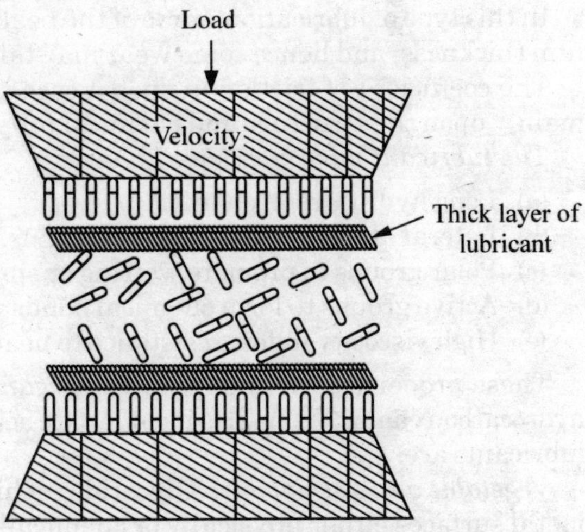

Fig. 17.3 Fluid-film lubrication

The lubricants used for fluid film lubrication are hydrocarbon oils. However the viscosity of the hydrocarbon oil decreases with rise in temperature, therefore these oils are blended with selected long chain polymers to maintain the viscosity of the oil in all seasons of the year. Moreover, the hydrocarbon oils which are the fractions from petroleum refining contain small quantities of unsaturated hydrocarbons and hence get oxidised forming gummy products. Hence antioxidants are also blended with the hydrocarbons used as lubricants.

Fluid film lubrication is provided in delicate instruments such as watches, clocks, light machines (sewing machines, scientific instruments) etc.

2. Thin Film or Boundary Lubrication

In some cases, under the conditions of high load and low speed or low viscosity of the lubricant, the lubricant tends to be squeezed out of position or a continuous film of lubricant cannot persist between the two surfaces. In such cases, lubrication is maintained by a very thin film of the lubricant adsorbed on the surface *by physical or chemical forces or both.* These adsorbed layers avoid direct metal to metal contact. Though, the moving surfaces are not completely separated but then points of contact are considerably decreased resulting in lowering of frictional resistance.

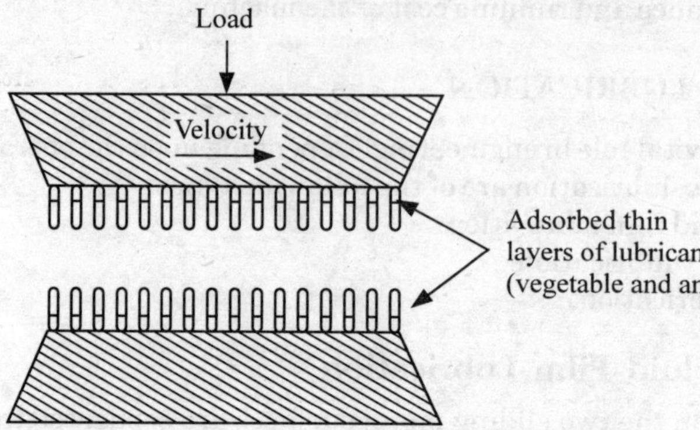

Fig. 17.4 Boundary Lubrication

The thickness of the boundary film may not exceed one or two molecular layers Fig. 17.4.

In this type of lubrication, some of the peaks in the surface of the metal may be higher than the film thickness, and hence some wear may take place.

The coefficient of friction in such a case is generally 0.05 to 0.15. This friction is dependent mainly upon the structure and chemical properties of the oil to be used for boundary lubrication.

The lubricant molecules should have:

(a) Long hydrocarbon chains.
(b) Lateral attraction between the chains.
(c) Polar groups to promote wetting or spreading over the surface.
(d) Active groups to form chemical bonds with the metal surface.
(e) High viscosity index, resistance to heat and oxidation, good oiliness and low pour-point.

These properties are met with the *organic acids* which have both, polar groups as well as hydrocarbon chains. Examples include oleic acid and stearic acid. Other lubricants used in boundary lubricants are:

Vegetable and animal oils: Glycerides of higher fatty acids and their soaps get adsorbed on the metal surfaces either physically or chemically forming a thin film, which acts as lubricant.

Mineral oils: Mineral oils are also used for boundary lubrication. Although fatty acids have more oiliness than mineral oil, fatty acids tend to break down at higher temperatures. Hence these cannot be used in internal combustion engines. Mineral oils can be used, as they are thermally more stable. The oiliness of mineral oils can be increased by adding a small amount of fatty acids.

Graphite and MoS_2 are also used for boundary lubrication. These can bear compression as well as high temperatures. These are solid lubricants and are either used alone or as stable suspension in oil.

3. Extreme Pressure Lubrication

If the sliding surfaces are subjected to very high pressure and speed, excessive frictional heat is generated. Under such conditions, the liquid lubricants fail to stick and may decompose and even vaporize. In order to provide effective lubrication under these conditions, special additives called "extreme pressure additives" are used along with the lubricants. The active chemicals used are compounds of chlorine such as chlorinated esters, sulphur such as sulphurized fats and oils, and phosphorus such as tricresol phosphate. At prevailing high temperatures, these additives form solid surface films of chlorides, sulphides or phosphides as the case may be. These metallic compounds possess high melting points and serve as good lubricants under the conditions of extreme pressure and temperature.

These lubricants have an additional advantage that if the low shear strength films formed are broken by rubbing action, they are immediately replenished.

Extreme pressure additives are also needed in cutting fluids used as lubricants in machining of tough metals. These are also used in wire drawing. For example, graphite is used for drawing wires made up of mild steel.

17.5 CLASSIFICATION OF LUBRICANTS

A wide variety of lubricants have been used for lubrication purposes. On the basis of their physical state lubricants can be classified as:

1. Liquid lubricants or lubricating oils.
2. Semi-solid lubricants or greases
3. Solid lubricants.

17.5.1 Lubricating Oils

The lubricating oils or liquid lubricants are used to reduce the friction and wear between two sliding metallic surfaces. Apart from reducing the friction they also act as cooling medium. Lubricating oils can further be divided into three categories viz;

(i) Vegetable and animal oils.
(ii) Mineral oils or Petroleum oils.
(iii) Blended oils.

A brief discussion of all the three types is given below:

(i) **Animal and Vegetable Oils:** These oils contain glycerides of higher fatty acids. These oils were used as lubricants before the development of petroleum industry, after which they were replaced by mineral oils. Animal and vegetable oils decompose on heating but do not distil and hence are also known as fixed oils. These have the following limitations when used as lubricants:

- They are expensive
- They undergo oxidation and form gummy and acidic products, which are disastrous to the metal parts
- They get thickened on coming in contact with air.
- They hydrolyze in presence of water.

Because of these limitations they are not widely used now a days. However both animal and vegetable oils have a property called "oiliness" by virtue of which the oil sticks to the surface of the metal even under high temperature and high load. Hence they are blended with the mineral oils to improve the oiliness of the latter.

Vegetable oils are obtained by crushing the seeds and washing the crushed seeds with a suitable solvent (solvent extraction). Therefore the washings are distilled to separate oil and the solvent.

Animal oil is obtained by a process called rendering in which the enclosing tissue is broken by a combined action of steam and water.

Both the types of oils, vegetable as well as animal oils require further treatments such as cooling, filteration, neutralization and coagulation.

Some of the important vegetable and animal oils along with their uses are given in Table 17.1.

Table. 17.1 Animal and Vegetable Oils

Vegetable Oils	*Uses*	*Remark*
1. Olive Oil	Good lubricant for machine parts working under low pressure and high speed.	Expensive and tends to decompose on storage
2. Castor Oil	Used as a lubricant for bearing and machine parts subjected to low pressure and high speeds.	Possesses excellent oiliness and is soluble in alcohol and not in petroleum. Oxidizes into gummy products.
3. Palm Oil	Lubricant for delicate article such as watches, clocks, etc.	Pleasant smelling oil.
4. Rape seed oil	Lubricant for steam cylinders and delicate apparatus.	Most extensively used fatty oil as lubricant.

(i) **Mineral Oils:** These oils are obtained by fractional distillation of crude petroleum. These are obtained as heavy residual fraction. This heavy residual fraction is subjected to vacuum distillation and the lubricating oil is extracted. The length of the hydrocarbon chain varies

Animal Oils	*Uses*	*Remarks.*
1. Tallow oil	Used as lubricant as such or can be blended with mineral oils.	Obtained by compression of cattle or sheep fat under high pressure.
2. Lard oil	Used as a cutting oil and for lubricating ordinary machine parts.	Obtained from fats, kidneys and intestines of pigs.
3. Neats foot oil	Used for lubricating instruments such as clocks and sewing machines.	Obtained from the hoofs of slaughtered cattle and sheep by extraction process.
4. Whale Oil	Used for lubrication of light machinery	Obtained by different parts of sperm whale flesh. It is relatively a cheap oil.
5. Sperm Oil	Good spindle oil and is suitable for light machinery.	Lightest of fatty oils.

from 12 to 50 carbon atoms. Mineral oils are used most widely as lubricants because they are cheaper and are available in abundance. However they have poor oiliness which can be increased by blending them with higher fatty acids.

The lubricating oils obtained from mineral residues have a large number of impurities such as paraffin waxes, resins, asphalts etc. These impurities are removed by different processes:

(a) *Dewaxing.* Waxes if present in the lubricating oil increase the setting point and also increase the pour point and hence lubricant cannot be used in low temperatures. The wax can be removed by dewaxing, in which oil is mixed with a suitable solvent (propane, trichloroethylene) and is then refrigerated. The wax crystallizes and is removed by filteration. The solvent present in oil is then recovered by distillation.

(b) *Acid Refining.* The asphaltic constituents in mineral oils undergo decomposition at higher temperatures causing carbon deposition and sludge formation. These are removed by agitating the oil with the required amount of H_2SO_4. Some of the unwanted impurities dissolve in acid while others react chemically to form tarry sludge. The oil is separated through filteration and is neutralized with a calculated quantity of ammonia or caustic soda. The oil is decolourized by passing through fuller's earth at 108-230°C.

(c) *Solvent Refining.* The other undesirable impurities such as naphthenic, asphaltic and resinous constituents can be removed by solvent refining. In this refining the mineral oil is mixed with a suitable solvent in which it is immiscible but the impurities are highly soluble. The mixture of solvent and the oil is left undisturbed for some time when two layers are separated-the oil layer free from impurities and the solvent layer containing the dissolved impurities. The oil layer is recovered and distilled to remove the small amount of solvent which may be present. The solvent is recovered from the solvent layer through distillation and the impurities left behind as residue can be used as a fuel oil.

Solvent refining has number of advantages over acid refining as it is cheaper and gives higher yield of refined oil. However solvent refined oils are generally have less oxidation resistance due to removal of natural oxidation inhibitors present in it. Moreover these oils possess lower oiliness. Thus it becomes imperative to add some additives to improve their desirable properties.

(ii) **Blended Oils:** Small quantities of certain substances are added to the lubricating oils to improve their desirable properties. These substances are called additives. The oils to which additives are added are known as blended oils or compounded oils. These blended oils are capable of giving desired lubricating properties required for a particular machinery. Some of the additives along with their different functions are tabulated in table 17.2.

Table 17.2: Additives and their functions

Additive Type	*Chemical Used*	*Functions*	*Application*
1. Antioxidant or inhibitor	Aromatic phenolic or amino compounds.	Retard the oxidation of lubricating oils by getting themselves preferentially oxidised.	Added in lubricants used in internal combustion engines.
2. Oiliness-carrier	Vegetable oils like castor oil and fatty oils like stearic acid, palmitic acid	Increase the oiliness of the lubricating oil	Added in lubricants used in boundary lubrication.
3. Viscosity-index improver	Polymers and copolymers of olefins.	Increases the viscosity index	Added in the machines working under broad range of temperature change.
4. Pour-point depressant	Alkylated naphthalene and phenolic polymers, polymethacrylates	Lowers the pour point of the oil	Reduce interlocking
5. Detergents and deflocculants	Basic calcium and barium salts of phosphates and sulphonates	Reduce or prevent deposits	Used in engines operated at high temperatures.
6. Dispersants	High molecular weight amines and amides, polymers	Prevent or retard sludge formation	In lubricating oils used in internal combustion engine.
7. Metal deactivators	Amines, sulphides or phosphides	Stop the catalytic effect of metals on oxidantion rate	Protect the metal surfaces by forming a protective layer over the metal surface

8. Antifoaming agents	Glycol, glycerol, silicon polymers	Prevent the formation of stabel foams.	
9. Corrosion inhibitors	Organo metallic compounds, zinc dithiophosphates,	Protect the metal in contact with the lubricant from corrosion.	In lubricating oils used for bearing etc.
10. Rust inhibitors	Amine phosphates, Alkyd succinic acid, fatty acids.	Protect ferrous metals from rusting.	
11. Extreme Pressure additives	Sulphurized fats, organic phosphorus compounds etc.	React chemically with the metal, forming a surface layer which prevent seizure and welding of the metals.	
12. Thickness	Polystyrene, Polyesters, soaps of lithium, sodium, calcium and barium	Given high viscosity to the lubricating oil.	Used for forming greases.
13. Abrasion inhibitors	Tricresyl phosphate	They provide abrasion resistance.	
14. Emulsifiers	Sodium salts of sulphonic acids and sodium salts of organic acids	Help the formation of emulsions fo lubricating oils with water.	

17.5.2 Semi Solid Lubricants or Greases

In textile, paper and food mills, if liquid lubricants are applied to the metal surfaces of the machinery, there is always a possibility of spilling or spurting of lubricating oil. This is detrimental to the product being manufactured. In these cases the lubricating oil is thickened with the thickening agent and the semi-solid so obtained is known as 'grease'. Greases are ideal in such cases as these do not spill or splash. Greases are also employed in the situation where the compounds have to be sealed against the entry of dirt, water etc.

Greases can be defined as a semisolid combination of a lubricating oil and a soap (of lithium, sodium, calcium or barium etc.) as a thickening agent. Other substances used as thickener include carbon black, asphaltenes, siloxanes and flouro carbons. These substances act as structural elements which trap the oil and give stiffness to the oil so that it clings to the moving parts. The thickening agents are also known as gelling agents. Thus soaps which are used as thickeners or gelling agents give a interconnected structure held together by intermolecular forces. At high temperature, the interconnected forces cease to exist because the soap dissolves in the oil and hence greases get liquified as the temperature rises.

Preparation of Greases

Greases are generally prepared by saponification of fat (such as fallow oil or fatty acids) with alkali (NaOH, $Ca(OH)_2$) followed by addition of hot lubricating oils with constant mixing. Sufficient oil is added to obtain the desired consistency. The nature of the soap decides the temperature upto which grease can be used and its consistency.

Some of the important aspects of greases are as follows:

(i) Consistency of the greases can vary from a heavy viscous liquid to that of a solid mass. The consistency changes suddenly with change in temperature because the interaction of gelling agents and the viscosity of the oil changes with increase in temperature.

(ii) Greases have a tendency to separate into oils and soaps on storage.
(iii) Greases are less stable to oxidation than the component lubricating oil present. Thus antioxidants or inhibitors are added to grease.
(iv) Most of the thickening agents are hydrophillic in nature and hence greases absorb water. Thus the greases are poor water resistant.
(v) Greases have higher shear or frictional resistance than oils and therefore can support much heavier loads at lower speeds.
(vi) Greases do not require as much attention as oils and hence are convenient to use.
(vii) Greases cannot dissipate heat from the bearing effectively as compared to oils.

Applications of Greases

Greases are mainly used in the following:

(i) When a machine is worked at slow speeds and high pressures the lubricating oil cannot remain in place and in these situations greases are used for lubrication.
(ii) Greases are used in bearings and gears that work at high temperatures.
(iii) They are used in machines used in the manufacture of paper, textiles or edible articles because unlike oil, there is no danger of spilling of grease.
(iv) They are used in bearing which required to be sealed against entry of dust, dirts, grit or moisture.

Classification of Greases

The consistency, temperature upto which grease can be used are all dependent upon the nature of soap and hence greases are classified on the basis of soap used in their manufacture. Important classes of greases are:

(i) Calcium-based or cup greases.
(ii) Soda-based greases.
(iii) Lithium based greases.
(iv) Axle greases.

The soaps used in their preparation and their uses are given in Table 17.3.

Table 17.3: Types of greases and the thickening agents used

Type of Greasem	*Oil and thickening agent used*	*Remarks*
1. Calcium based greases	Emulsions of petroleum oils with calcium soaps	Cheapest, water resistant and can be used upto 80°C, suitable for lubricating water pumps, tractors etc.
2. Soda based greases	Petroleum oils thickened by mixing sodium soaps.	Not resistant to water and hence cannot be used in the machine parts exposed to wet conditions. Due to high melting points can be used upto 175°C. Suitable for use in ball bearings.

3. Lithium based greases	Petroleum oils thickened by mixing lithium soaps	Have good water resistance and good thermal stability. They are excellent multipurpose greases for a wide range of applications. However, they are expensive.
4. Axle greases	Resin mixed with lime.	Cheapest of all greases, water resistant and find applications in equipment working under high loads and low speeds.
5. Non-soap greases	Oil mixed with carbon black, silica gel etc.	Suitable for high temperature applications. Talc, mica etc. are added as filters.

17.5.3 Solid Lubricants

There are number of situations where lubricating oils and greases cannot be used satisfactorily. For example:

(i) Where the operating temperatures and pressures are too high, the liquid lubricants cannot be used because these are easily combustible.
(ii) Where a liquid or semi solid lubricant cannot be maintained in position.
(iii) Where the contamination of lubricating oil by the entry of dust particles is not acceptable such as in aerospace devices.

Under these conditions, to minimize the friction, *solid lubricants* are used. The solid lubricants are used either in dry powder form or mixed with water or oil.

The most important solid lubricants that have widely been used are:

(i) Graphite
(ii) Molybdenum disulphide

Their lubrication properties are due to their layered structure or laminar structure.

(i) *Graphite:* Graphite consists of a network of hexagonal carbon rings separated from the adjacent layer by a distance of about 3.4 A° (Fig. 17.5). The adjacent layers are held together by weak Vander Waals' forces. Consequently the parallel layers slide over one another easily. Graphite is soapy to touch, non-inflammable and not oxidized in air below 375° C. When graphite is used as a lubricant between uneven surfaces, it fills into the valleys on the surface making the surface smooth. Moreover as there is no strong bonding between the adjacent layers of graphite, thus these layers slide past each making the surface lubricated. Graphite is used either as powder, as suspension, or as grease. Suspension of graphite is prepared in oil or water using *tannin* as emulsifying agent. Its suspension in water is called 'aquadag' and in oil is called 'oildag'. (dag-*deflocculated adhesion graphite*). Oil dag is used in internal combustion engines because it provides a tight fit contact by forming a film between the piston and rings. Aquadag is used where lubricant free from oil is needed. Graphite greases are used for lubrication at very high temperatures.

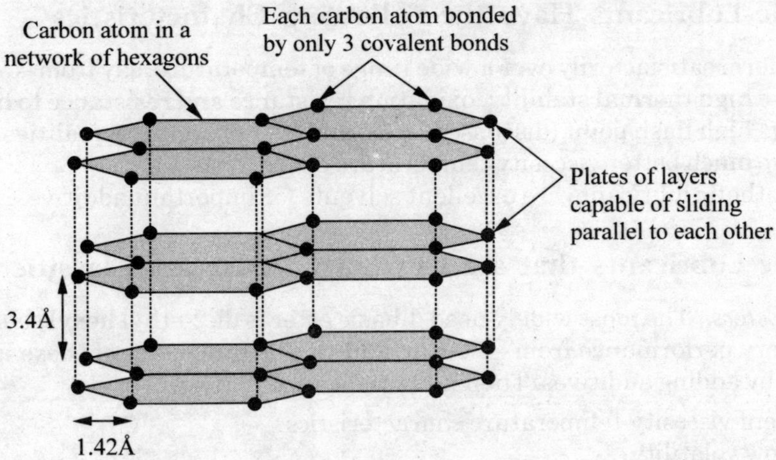

Fig. 17.5 Multiple layered structured of graphite

(ii) *Molybdenum Disulphide:* MoS_2 has a sandwitch like structure in which a layer of Mo atoms lies between two layers of sulphur atoms (Fig. 17.6). It has low coefficient of friction and is thermally stable in air upto 400° C. It is also used as a powder or mixed with solvents or greases. It is used as dispersion in petroleum oils or water. Above 800°C, MoS_2 gets oxidized to MoO_3 which is slightly abrasive. Therefore, for higher temperatures, graphite is preferred.

MoS_2 is used in airframe lubrication and wire-drawing dies. A solid-film lubricating surface useful for space vehicles is made from 70% MoS_2 + 7% graphite bonded with 23% silicates which can with stand extreme temperatures, low pressures and nuclear radiations

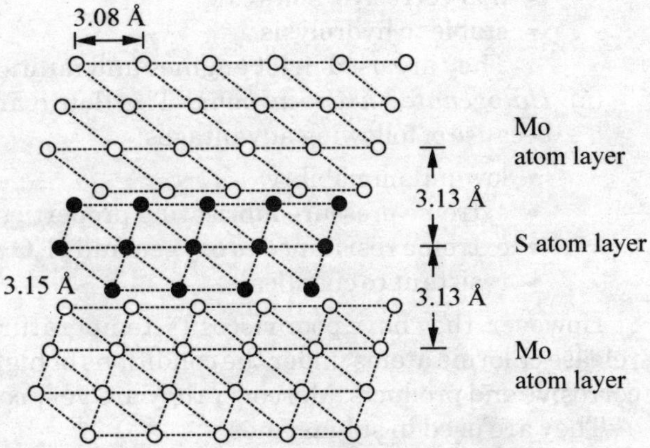

Fig. 17.6 Sandwitch-like structure of molybdenum disulphide.

Other solid lurbicants beside graphite and molybdenum disulphide are soapstone, talc, mica etc. Teflon, copper, phthatocyanine are examples of solid organic lubricants.

17.6 SYNTHETIC LUBRICANTS

Mineral lubricants are used extensively and satisfactorily in engineering applications because of their ready availability, desirable viscosity temperature characteristics and lower costs. However, under severe operating conditions, the mineral lubricants, even with suitable additives, have been found to be unsatisfactory. Such operating conditions include extreme temperatures, wide temperature fluctuations in machinery. For example in air craft turbines, which require lubricants to have low viscosities and high working temperature range (– 55° C to + 260° C). These requirements have led to the development of synthetic lubricants.

Synthetic lubricants are the lubricants prepared to meet the special requirements *i.e.* which can work under severe or extreme conditions satisfactorily. However synthetic lubricants are used only for special applications because of their high cost and some of them are toxic and corrosive.

The Synthetic Lubricants Have the Following Characteristics

(i) They perform satisfactorily over a wide range of temperature, say from – 50° C upto 260° C.
(ii) They have high thermal stability, oxidation resistance and resistance to hydrolysis.
(iii) They have high flash point (discussed later) and rust preventive qualities.
(iv) They have much better viscosity temperature characteristics.
(v) Many synthetic lubricants are excellent solvents for important additives.

The Synthetic Lubricants that are Currently in use are Classified as

(i) *Dibasic Esters:* The most widely used dibasic ester is di-2-ethyl hexyl sebacate, which has satisfactory performance from – 50°C to 230°C. The properties of these lubricants can be modified by adding additives. Their advantages are:

- excellent viscosity-temperature characteristics
- very low volatility
- high temperature stability
- non-corrosive, non-toxic
- stable to hydrolysis.

They are used in jet engines and military equipment.

(ii) *Halogenated hydrocarbons:* The flouro and chloro hydrocarbons are used as lubricants because of following advantages:

- low-inflammability
- extreme pressure lubricating properties
- extreme resistance to oxygen and H_2O_2
- resistant to chemicals.

However, they have poor viscosity-temperature characteristics and chloro-hydrocarbons may release chlorine atoms under the conditions of high loading and high temperatures and result into corrosive end products. Moreover they are very costly.

They are used in submarines.

(iii) *Organo-phosphate esters:* Phosphatic esters are used because of their fire-resistant characteristics. They can be used as extreme pressure additives and are also used in air craft hydraulic oils.

(iv) *Polyglycols and their derivatives:* These can be water soluble or water insoluble depending upon the carbon chain length. Higher polyalkylene oxides are water insoluble but can absorb lot of water whereas polythylene glycol, poly-propylene glycol are water soluble and can be used in water diluted lubricants.

The advantages of these lubricants are:

- low pour points
- no effect on rubber and other gasket materials
- comparable to petroleum oils in thermal stability, fire resistance and lubricity
- cheaper.

However, they decompose at high temperatures into volatile parts which undergo oxidation subsequently. Hence they are not useful as high temperature lubricants. They are used for combustion engines, gears, compressors and pumps. They are also used as aircraft turbine lubricants because of their thermal stability.

(v) *Silicones:* Silicones are used as lubricants because of their excellent properties which are as follows: They

- are oxidation resistant upto 200° C.
- possess high viscosity index

- are chemically more inert than other synthetics
- hamve smallest change in viscosity with temperature.

Silicon fluids are used for rubber and plastic surfaces such as slide rules, bushings, bearings etc. They are also used as moisture repellent lubricants for clocks, timers and other electronic devices.

(vi) *Amines, imines and amides:* They have got high viscosity index and low pour point. They can be used in the temperature range – 50°C to 250°C.

17.7 LUBRICATING EMULSIONS

Emulsion is a heterogeneous system consisting of disperse phase and dispersion medium. It consists of two immiscible liquids, the one being dispersed as fine droplets in the other. It is stabilized by adding a third component called emulsifying agent. The liquid which forms the droplets is called *dispersed phase* and the surrounding liquid is known as the continuous or *dispersing phase.* The size of the dispersed droplets can vary from 1 to 6 μ.

There are two types of emulsions which are used as lubricants :

1. **Oil-in water emulsions or cutting emulsions:** These are obtained by adding an oil containing about 3-20 % water soluble emulsifying agent (*e.g.*, water soluble soap) to a suitable quantity of water.
   In various types of machining operation, (milling, boring, turning etc.), the tool employed gets heated to a very high temperature. This may cause oxidation and subsequent corrosion of metal. In order to prevent overheating, efficient cooling and lubrication is provided. This is usually done by employing emulsions of oil-in water which are called *cutting fluids* or cutting emulsions. Oil lubricates whereas water act as a coolant. Fatty oils such as lard oil and sperm oil are very good for cutting oils.
2. **Water-in-oil emulsions or cooling liquids:** These are prepared by mixing together water and oil containing 1 to 10% of water insoluble emulsifiers (*e.g.*, alkaline earth metal soaps). There are very useful as compressors and pneumatic tool lubricants as they give internal cooling effect due to the evaporation of water.

17.8 SOAPS AND DETERGENTS

Soaps

Soap, by chemical classification, is a surfactant. A surfactant is a material that can greatly reduce the surface tension of water. Any substance which have cleaning properties in water i.e. which can remove dirt or grease from a body is called detergent. Soap is also detergent and oil based soapy detergent are known as Soaps.

Chemically ,soaps are water-soluble sodium or potassium salts of fatty acids. Soaps are made of fats and oils, or their fatty acids, by treating them chemically with a strong alkali.

Synthesis

Saponification of fats and oils is the most widely used soap making process. This method involves heating fats and oils and reacting them with a liquid alkali to produce soap and water plus glycerine.

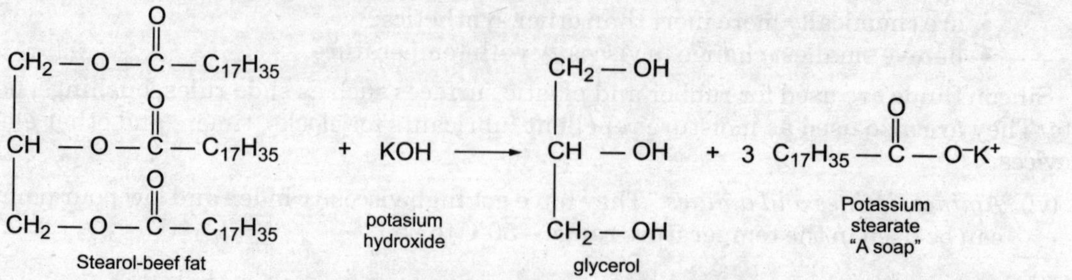

A by-product of this reaction is glycerol which is useful as in foods, as an antifreeze, as a tobacco-moistening agent, and in the manufacture of nitroglycerine and dynamite.

The triglycerides (or triesters of fatty acids) are the raw material for the production of soap. Tallow and coconut oil are the principal fatty materials in making soap.

Hard and Soft Soaps

When the alkali is sodium hydroxide, a sodium soap is formed. Sodium soaps are "hard soaps". When the alkali is potassium hydroxide, a potassium soap is formed. Potassium soaps are softer and are found in some liquid hand soaps and shaving creams.

The carboxylate end of the soap molecule is attracted to water. It is called the hydrophilic (water-loving) end. The hydrocarbon chain is attracted to oil and grease and repelled by water. It is known as the hydrophobic (water-hating) end.

Cleaning Action of Soap

Soap is effective as a cleaning agent because it is amphiphilic; it is partly polar and partly nonpolar. Soap molecules contain an intensely polar "head" (the ionic part) and a non-polar "tail" (the long hydrocarbon chain, usually 10-18 carbons, depending on which fatty acid is used). The polar part is often referred to as being hydrophilic (water-loving) whereas the tail is called hydrophobic (water-hating) because of its non-polar character. The hydrophobic tails (hydrocarbon chain) attach themselves to any substance which is non-polar (grease). The tails surround the grease with the polar heads in the aqueous phase, forming the structure shown in the figure . With agitation the grease is pulled from the clothing allowing the grease to be completely surrounded and moved into the aqueous phase. This is then washed away leaving only clean clothes behind.

The structure formed by the detergent in water is called a micelle.

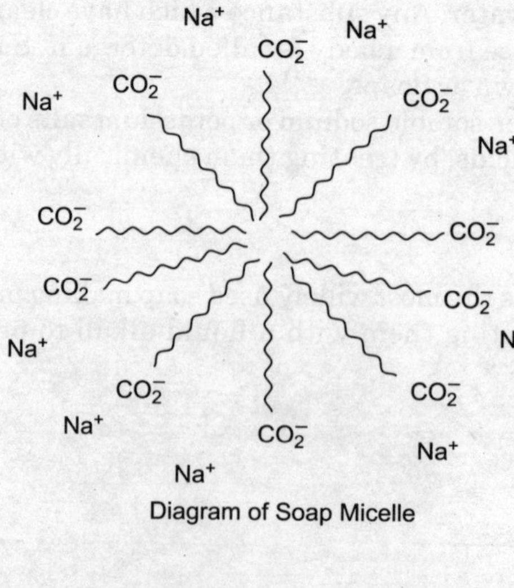

Diagram of Soap Micelle

Limitation of Soaps

Hard water" contains ions such as calcium, magnesium, and iron and they render soap ineffective. The sodium ion of the soap is replaced by one of these new ions, and the resulting salt is no longer soluble. It then drops out of solution in the form of insoluble calcium soap and can no longer clean anything.

Detergents

There are other materials that also have cleaning capacity like soaps. A detergent is a cleaning agent similar to a soap, but with a general structure R-SO_4^-, Na^+, where R is a long-chain alkyl group. The molecules of detergents also have polar and nonpolar ends. They clean like soaps except that their calcium and magnesium salts are generally more soluble in water than their soap counterparts. In recent years many different detergents have been introduced for use in cleaning.

Soaps and detergents are very similar in their chemical properties. However, there is a significant difference between them, which soaps are produced from natural products, and detergents are synthetic, or man-made. Synthetic detergents usually have a long hydrocarbon chain with between 12 and 20 carbon atoms. At the head of this chain is a polar group that is a strong hydrogen bonder to water but does not form strong ionic bonds with the ions of hard water. Detergents do not precipitate out of solution so the micelles are present and cleaning power remains.

Detergents can be anionic, cationic and non ionic as shown in Fig. 17.7.

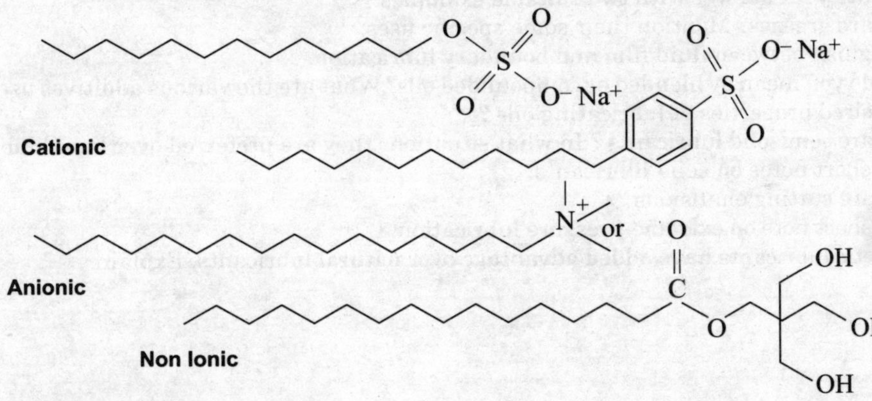

Fig. 17.7 Cationic, Anionic and non ionic detergents

Soaps	*Detergents*
They contain sodium carboxylate (COONa) group.	The contain sodium sulphonate (SO_3Na) group.
They are not suitable for washing with hard water.	They are suitable for both hard and soft water.
They have relatively weak cleansing action.	They have strong cleansing action.
They are biodegradable.	Most of them are non-biodegradable.

QUESTIONS

1. Explain thin layer mechanism of Lubrication.
2. Explain mechanism of fluid and boundary lubrication.
3. Write short notes on :
   (a) Flash and fire point of lubricant
   (b) Cloud and pour point of lubricant
4. Discuss the principle underlying lubricants. Describe method of determining the viscosity index of lubricants.
5. Write notes on : solid lubricants
   Write short notes on flash and fire point.
6. What are lubricants? How are they classified? Explain the following properties of lubricants and give their significance:
   (1) Viscosity (2) Viscosity Index;
   (3) Pour point
7. Write short notes on :
   Cloud and pour point
8. Write short notes on :
   Viscosity and viscosity index
9. Write short notes on Flash and fire point and their determination
10. Define viscosity and viscosity index. How are these properties of lubricating oil improved?
11. What are lubricants and how are these classified? Discuss the function of lubricants.
12. What are solid lubricants? How the chemical structure of lubricants contribute to its functions? Illustrate your answer with two suitable examples.
13. What are greases? Mention their some specific uses.
14. Distinguish between fluid film and boundary lubrication.
15. What do you mean by blended or compounded oils? What are the various additives used to improve the desired properties of lubricating oils ?
16. What are semisolid lubricants ? In what situations they are preferred over liquid lubricants ?
17. Write short notes on solid lubricants.
18. What are cutting emulsions ?
19. Write short note on extreme-pressure lubrication.
20. Synthetic lubricants have added advantage over natural lubricants. Explain.

❑❑❑

PRACTICALS

INTRODUCTION TO SAFETY AND WORKING IN THE CHEMICAL LAB

The chemistry laboratory can be a place of discovery and learning. However, by the very nature of laboratory work, it can be a place of danger if proper common-sense precautions aren't taken.

I. LABORATORY CLOTHING

1. Wear shoes that fully cover the feet. Sandals and clogs are not adequate. Shoes provide a great deal of initial protection in the case of dropped containers, spilled chemicals, and unseen hazards on the floor. Use old clothes, which are not too loose, especially at the sleeves. Laboratory coats or aprons must be worn over clothes. Snaps or fasteners are preferable to buttons for quicker removal in case of an emergency. Tie back long hair so that it will not fall into flames or chemicals
2. Safety goggles **must be worn at all times** while in the laboratory. This rule must be followed whether you are actually working on an experiment or simply writing in your lab notebook. You must wear safety goggles provided by the chemistry department.
3. Contact lenses are not allowed. Even when worn under safety goggles, various fumes may accumulate under the lens and cause serious injuries or blindness.

II. CONDUCT

1. Eating, drinking, and smoking are strictly prohibited in the laboratory.
2. **Unauthorized Experiments:**Do only the experiment which has been assigned by the laboratory instructor. Never do any unauthorized experiment in place of the one assigned by the instructor. Do not change the designated procedure without the advice of the instructor.
3. Never taste anything. Never directly smell the source of any vapor or gas; instead by means of your cupped hand, waft a small sample to your nose. Do not inhale these vapors but take in only enough to detect an odor if one exists.
4. Coats, backpacks, etc., should not be left on the lab benches and stools. There is a hook rack along the back wall at either end of the lab. There are coat racks just inside the each entrance to the balance room at the back of the lab. Beware that lab chemicals can destroy personal possessions.
5. Always wash your hands before leaving lab.
6. Learn where the safety and first-aid equipment is located. This includes fire extinguishers, fire blankets, and eye-wash stations.
7. Notify the instructor immediately in case of an accident.

III. PROPER HANDLING OF CHEMICALS AND EQUIPMENT

1. Handling of Liquid Chemicals:

Take an appropriate container to the reagent shelf. Avoid measuring volumes of strong acids and alkaline solutions with your graduated cylinder held at eye level. Support your graduated cylinder on your bench. Add hazardous liquids a little at a time, inspecting after each addition.

Reagent in dropper bottle

If the general supply bottle is equipped with a dropper, use it, but be sure that the dropper never touches your container or the contents in it. Never put it down on the bench top, but return it immediately the right reagent bottle.

Reagent in a stopper bottle

If the general supply bottle is equipped with a stopper, the stopper should either be held during the transfer or placed on its flat top. Do not lay the stopper on its side on the bench top. Pour chemicals from the general supply bottle into your container. Be sure that the proper stopper is returned to the supply bottle; do not interchange stoppers.

Mixing

If liquid chemicals are to be mixed with water, always add the concentrated chemical to water rather than water to chemical. This keeps the new solution dilute at all times and avoids many accidents. Usually addition should be done slowly, using small quantities. It is especially important to add acid to water because of the heat generated.

Pipetting

Liquids are drawn into the pipet by applying a slight vacuum at the top, using a small rubber suction bulb but NEVER THE MOUTH.Use pipette fillers.

Heating

Liquids in beakers and flasks can be heated by placing them on a ring or tripod stand on wire gauze with the container preferably supported by a clamp. Liquid should never be heated in a graduated cylinder or in other columetric glassware.

Disposal

Check with your laboratory instructor before disposing of any chemicals down the drain. If the liquid chemical can be disposed of in the sink, dispose of it by rinsing it down the sink with large quantities of water. Avoid unnecessary splashing during this process by pouring the chemical directly down the drain while the water is running vigorously.

2 Handling of Solid Chemicals

Take an appropriate container to the reagent shelf where the general supply is kept. Solids are somewhat more difficult to transfer than are liquids, so a wide-mouthed container such as a beaker is preferable.

During the transfer, hold the stopper or lay it on the bench without contaminating the stopper. Solid chemicals are most easily poured by tipping the general supply bottle and slowly rotating it back and forth. Mere tipping of the bottle alone often causes large chunks to come out very suddenly which leads to spills. If you use your own spatula, be sure that it is absolutely clean. Return the proper stopper to the general supply bottle; do not interchange stoppers.

Mixing

If the solid is to be mixed with a liquid, add the solid to the liquid. Additions should be made in small quantities except in special circumstances.

Disposal

If the laboratory instructor directs you to dispose of any solid chemicals in the sink, flush it down the drain with copious amounts of running water. All other solids should be disposed of in special containers provided for this purpose.

3. Handling Glassware

Accidents involving glassware are a leading cause of laboratory injuries. These can be avoided by following a few simple procedures. In general, be certain that you have received proper instructions before you use glass equipment designed for specialised tasks that involve unusual risks or potential injury.

- Handle and store glassware carefully so as not to damage it or yourself.
- When inserting glass tubing into rubber stoppers, corks or when placing rubber tubing on glass hose connections:
- Protect hands with a heavy glove or towel.
- Lubricate tubing or stopper with water or soap solution and be sure that the ends of the glass tubing are fire-polished.
- Hold hands close together to limit movement of glass should fracture occur.
- Substitute plastic connections for glass whenever possible to decrease the risk of injury.
- When dealing with broken glass wear hand protection when picking up the pieces.

4. Handling Heating Devices

Electrical devices that supply heat in laboratories include:

Hotplates
Heating Mantles
Hot-Air Guns
Oil Baths

Improper use of any one of these could result in fire or burns to the user.

Before using any heating device:

- Check to see if the unit has an automatic safety shutoff in case of overheating.
- Note the condition of electrical cords and have them replaced as required.
- Make sure the apparatus has been maintained as required by the manufacturer.
- Check to see that all heating units in use without automatic shut-off have been turned off before leaving an area for any extended period of time.

IV. CHEMICAL SPILLS IN THE LABORATORY

In all cases, immediately alert your neighbours and laboratory instructor of the spill. Locate spill cleanup materials. Laboratories should be equipped with spill cleanup kits. Wear the appropriate personal protective equipment (e.g., gloves, goggles) when cleaning up spills.

Non volatile and non flammable materials

If the material is not particularly volatile, nor toxic, and poses no fire hazard the liquid can be cleaned up by using an absorbent material which neutralizes them, for example, sodium bicarbonate solution or powder for acids, or sodium thiosulfate solution for bromine. Rubber or plastic gloves should be worn while using absorbent materials. A dustpan and brush should be used to remove the absorbent material. Then, the contaminated area should be cleaned with detergent and water and the area mopped dry.

Volatile, flammable and toxic spill materials

Alert everyone in the laboratory to extinguish flames, disconnect spark-producing equipment, shut down all experiments, and evacuate the laboratory. The laboratory instructor and safety personnel will handle the clean up.

Acid Spills

Apply neutralizer (or sodium bicarbonate) to perimeter of spill. Mix thoroughly until fizzing and evolution of gas ceases. NOTE: It may be necessary to add water to the mixture to complete the reaction. Neutralizer has a tendency to absorb acid before fully neutralizing it. Check mixture with pH indicator paper to assure that the acid has been neutralized.

Transfer the mixture to a plastic bag, tie shut, fill out a waste label, and place in the fume hood. Notify supervisor.

Caustic Spills

Apply neutralizer to perimeter of spill. Mix thoroughly until fizzing and evolution of gas ceases. Check mixture with pH indicator paper to assure that the material has been completely neutralized.

Transfer the mixture to a plastic bag, tie shut, fill out a waste label, and place in the fume hood. Notify supervisor.

Solvent Spills

Apply activated charcoal to the perimeter of the spill. Mix thoroughly until material is dry and no evidence of liquid solvent remains.

Transfer absorbed solvent to a plastic bag (if compatible), tie shut, fill out and attach a waste label, and place in the fume hood. Notify supervisor.

V. CHEMICAL SPILLS ON A PERSON:

Over the body

Within seconds, quickly remove all contaminated clothing while person is under safety shower. Flood the affected body area with cold water for at least fifteen minutes. If pain continues or resumes, flood with more water. Wash off chemicals with a mild detergent solution. Do not apply any materials such as neutralizing agents or salves, to the area. Obtain medial assistance immediately.

On small area of body

Immediately flush area thoroughly with cold water. Wash with a mild detergent solution. If there is no visible burn, wash out the area with warm water and soap.

In the eyes

You will need to assist the person who has chemicals spattered in the eyes. Immediately drench the eyes at the nearest emergency eyewash station. Force the eye or eyes open to get water into them. The speed of your response to this emergency is extremely important. Notify the laboratory instructor of the accident immediately.

VI. SWALLOWING CHEMICALS:

The laboratory instructor should determine what specific substance is ingested.

The individual should be forced to drink copious amounts of water while en route to medical assistance. The Health Center or Hospital should be notified while the individual is in transit as to what chemicals are involved.

VII. BURNS

For burns by hot objects, flames or chemical, flush the affected area with cold water for several minutes. Notify the laboratory instructor of the burn and he will arrange transportation to the infirmary if necessary.

VIII. FIRE

Give assistance to people first. If the person clothes are on fire, guide him/her without running to the fire blanked station or to the safety shower and drench him. Do not hesitate because of such insignificant things as shrinking sweater, ruined hairstyles, or soggy discomfort. While the victims are being cared for, other available people should try to shut off or reduce the fuel supply to the fire. Get a fire extinguisher and direct its spray toward the base of the fire. If the fire is too big to extinguish, have the laboratory instructor call the fire department and sound the fire alarm.When the fire is out, be sure all extinguishers used are tagged as empty and are replaced.

IX. INJURY OR ILLNESS

Render assistance if necessary. For minor cuts, wash them thoroughly, apply a good antiseptic, and a band-aid. For major cuts, severe bleeding or serious illness, send someone for help and administer first aid. Only a physician is trained to treat serious injury or illness. Notify the instructor immediately.

X. CLEANLINESS

1. Keep the lab bench clean at all times. If a solution, a solid, or liquid chemical is spilled on the bench or on the laboratory floor, clean up the spill immediately. Any chemical spilled on your skin or your clothing, should be washed immediately and thoroughly. Notify the laboratory instructor of the spill.
2. When leaving the laboratory, wipe the bench top thoroughly. Make sure that your work area is clean and free of spilled chemicals or scraps of paper. Wash your hands with detergent or soap and water

EXPERIMENT NO. 1

AIM Determination of hardness of water by titration with standard EDTA solution.

CHEMICALS 0.1 M EDTA, Eriochrome Black T, water sample, buffer solution.

THEORY The hardness of water can be determined by *complexometric titration. EDTA is used as complexing reagent. The Ca^{2+} and Mg^{2+} present in water are titrated with EDTA using Eriochrome Black T as indicator.

Estimation of hardness by EDTA method is based on the following principle

(i) First, the indicator Eriochrome Black-T, which is a blue coloured dye, forms an unstable complex with calcium or magnesium ions in hard water at a pH of 9 to 10. The complex is wine red in colour.

Mg^{2+}/Ca^{2+} + Eriochrome Black T → [Mg/Ca – Eriochrome Black T]
Unstable Complex (Wine red)

(ii) As this solution is titrated against EDTA, the free Ca^{2+} and Mg^{2+} ions in water form stable metal-ion EDTA, complex

[Mg^{2+}/Ca^{2+} + EDTA → [Mg/Ca – EDTA]
Stable Complex
(Colourless)

Once the free metal ions are complexed, the EDTA replaces Ca and Mg ions from the unstable indicator complex also, to form a stable complex, with the result, the indicator is set free. Since the free indicator is blue in colour at the above mentioned pH, the endpoint is the appearance of blue colour.

[Mg/Ca-Eriochrome Black T] + EDTA
→ [Mg/Ca – EDTA] + Eriochrome Black T
(Blue)

Thus the amount of EDTA used corresponds to the hardness of water.

The temporary hardness is removed by boiling and after the removal of precipitate by filteration, the permanent hardness in the filterate is determined by titration with EDTA as above.

Temporary hardness will be given by the difference of total hardness and permanent hardness. Hardness in water is generally expressed in $CaCO_3$ equivalent.

PROCEDURE

1. Rinse and fill the burette with EDTA solution
2. Pipette out 25 ml of hard water into volumetric flask and add 20 ml of distilled water and 10 ml of buffer solution.

3. Add 2-3 drops of indicator Eriochrome Black T.
4. Titrate against 0.1 M EDTA solution till the colour of the solution changes from wine red to blue.
5. Note the final reading and repeat to get three concordant readings.

OBSERVATIONS

Volume of hard water taken for each titration = 25 ml

Volume of 0.01 M EDTA used (from titration) = V_1 ml

Observation table

S. No.	*Volume of the solution taken in the titration flask (ml)*	*Burette Readings*		*Volume of the titrant used (Final-Initial reading) (ml)*
		Initial Reading	*Final Reading*	
1.				
2.				
3.				

CALCULATIONS

EDTA forms 1 : 1 complex with Ca^{2+} and Mg^{2+},

$\therefore$ 1000 ml of 1M EDTA = 1 Mole of $CaCO_3$ = 100 gm of $CaCO_3$

1 ml of 0.01 M EDTA = $100 \times \frac{1}{1000} \times \frac{1}{100}$ gm of $CaCO_3$ = 1mg of Ca CO_3

V ml of 0.01 M EDTA = 1.00 × Vmg of $CaCO_3$ = x mg of $CaCO_3$.

25ml of hard water contains = x mg of $CaCO_3$

1000 ml of hard water contains = $\frac{x}{25} \times 1000$ mg of $CaCO_3$ = y mg/L

RESULT The total hardness of water = y mg/L

PRECAUTIONS

Same as in experiment no. 1.12 in complexometric titrations, chapter-1.

Instructions for the Lab Assistant

Provide the following

1. **0.1 M EDTA solution.** Dissolve 37.225 *g* of EDTA (Disodium salt) in distilled water and make up the volume to 1L.
2. **Indicator solution:** Dissolve 0.5g of the dye in 100 ml of methanol.
3. **Buffer solution:** Dissolve 70g of NH_4Cl in 570 ml of conc. NH_3 and dilute the solution to one litre with distilled water.

EXERCISE

1. Define hardness of water?
2. What do you mean by temporary and permanent hardness?
3. What is the principle of EDTA titration?
4. Why is it necessary to add buffer solution to the water sample in determination of hardness by EDTA method.
5. What is the structure of Eriochrome black T?

Ans.

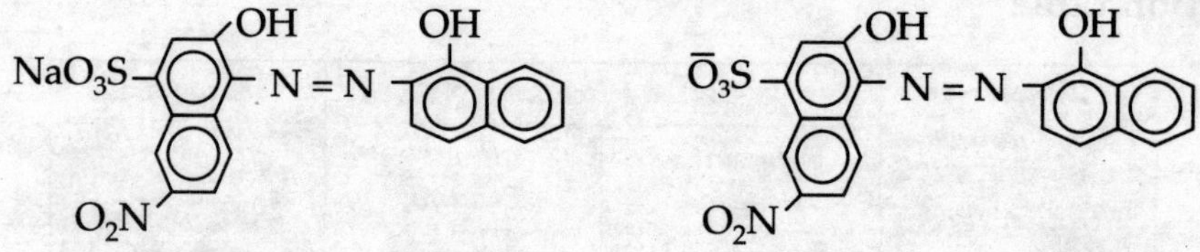

(a) Black (unionized form)

(b) Wine red (ionized form in aqueous medium)

Eriochrome Black T (sodium 1– (1–hydroxy–2–naphthylazo) 6 - nitro - 2 - naphthol - 4 -sulphonate

6. How would you find out permanent hardness of water?
7. What are the disadvantages of hard water.
8. If hard water sample does not contain Mg^{2+}, what will be the effect on the end point with EBT as indicator.

Ans. *The colour change from wine red to blue is not sharp if only Ca^{2+} ions are present. Therefore, Mg^{2+} ions have to be present to produce a sharp end point. For this a small amount of $MgCl_2$ is added to the EDTA solution before it is standardized.*

EXPERIMENT 2

AIM **Determination of iodide content in iodized salt through iodometric titration with sodium thiosulphate**

APPARATUS Burette, pipette, conical flask, beakers etc.

CHEMICALS N/10 $Na_2\ S_2\ O_3$, KI, dil acetic acid, starch solution, bleaching powder solution.

THEORY To determine the iodide content, iodine is liberated by adding sulphuric acid to a solution of iodized salt. Potassium iodide solution is added to keep the iodine in the dissolved state. Iodine liberated is titrated with sodium thiosulphate solution to form sodium iodide and sodium tetrathionate. Starch is used as an external indicator.

$$KIO_3 + 5KI + 3H_2SO_4 \longrightarrow 3K_2SO_4 + 3I_2 + 3H_2O$$

$$KI + I_2 \longrightarrow KI_3$$

$$2Na_2S2O_3 + I_2 \longrightarrow 2NaI + Na_2S4O_6$$

PROCEDURE

1. Rinse and fill the burette with N/10 $Na_2S_2O_3$ solution.
2. Pipette out 25 ml of bleaching powder solution in a conical flask.
3. Add 25 ml of distilled water, followed by 1 – 2g of solid KI and 10 ml of glacial acetic acid.
4. Keep the solution in dark for 3-4 minutes.
5. Titrate this solution with N/10 $Na_2S_2O_3$ till the brown colour of the solution changes to pale yellow.
6. Add 1-2ml of starch. The solution will turn blue.
7. Continue titration, till the blue colour disappears.
8. Repeat to get three concordant readings.

OBSERVATIONS

Normality of $Na_2S_2O_3$ solution $N_1 = 0.1N$

Observation Table.

S. No.	*Volume of the solution taken in the titration flask (ml)*	*Burette Readings*		*Volume of the titrant used (Final-Initial reading) ml*
		Initial Reading	*Final Reading*	
1.				
2.				
3.				

Volume of $Na_2S_2O_3$ used $V_2 = . . .$ ml.

CALCULATIONS

$$N_1V_1 = N_2V_2$$

(Bleaching powder) $(Na_2S_2O_3)$

$$N_1 \times 50 = \frac{1}{10} \times V_2$$

or

$$N_1 = \frac{1}{10} \times \frac{V_2}{50}$$

Strength of Iodine $Cl_2 = N_1 \times$ Eq.wt

$= Ni \times 127$ g/L

Precentage of iodine = (Strength of iodine/w) × 100

where w is the weight of bleaching powder (given) dissolved in 250 ml of distilled water for preparing the solution.

RESULT The available chlorine in the given sample is = ...%

Instructions For The Lab Assistant

Provide the following:

1. **Solution of bleaching powder:** Weigh accurately 5g of bleaching powders into a watch glass. Add a little distilled water to form a paste. Transfer it in 250 ml flask with the help of water. Make up the volume to 250 ml with distilled water. A turbid solution will be obtained.

2. N/10 $Na_2S_2O_3$ solution
3. Starch solution
4. KI (Solid)
5. Glacial acetic acid.

EXERCISE

1. *What are iodine titrations?*

Ans. The titrations involving the use of iodine are termed as iodine titrations. Iodine is a mild oxidising agent and is used in volumetric estimations.

2. *What type of substances are estimated using iodine titrations?*

Ans. Iodine is a much weaker oxidizing agent than potassium permanganate and dichromate. It is used to estimate the amount of various reducing agents like sodium thiosulphate, sodium sulphite, sodium arsenite and antimony salts in their solutions.

Iodine can also be employed to estimate the strength of oxidizing agents as well.

3. *What is the difference between iodimetery and iodometery?*

Ans. Iodine Titrations are of two types:

(a) Iodimetry. All such titrations which involve the direct titration of iodine with a reducing agent are grouped under iodimetry. Iodimetry is employed to determine the strength of reducing agents such as sodium thiosulphate, sulphite, arsenite, stannous and ferrous ions in neutral or moderately acidic solutions.

(b) Iodometry: In these titrations, iodine is liberated as a result of chemical reaction involving oxidation of an iodide ion by a strong oxidizing agent in neutral or acidic medium. The liberated iodine is then titrated with a reducing agent. Iodometry is used to estimate the strength of oxidizing agents. For example, the estimation of Cu^{2+} with thiosulphate–

$$2Cu^{2+} + 4I^- \rightarrow Cu_2I_2 + 2I_2$$

$$I_2 + 2S_2O_3^{2-} \rightarrow S_4O_6^{2-} + 2I^-$$

4. *What type of indicator is used in these type of titrations?*

Ans. Indicator: The indicator used in these types of titrations is starch solution which is used as an internal indicator. Starch forms a dark blue complex with triiodate ions. The blue colour disappears when there is no more of free iodine.

5. *Why iodine titrations are performed in cold?*

Ans. Since iodine is volatile, titrations should be performed in cold. Moreover, the indicator starch also loses its sensitivity at high temperatures.

6. *Iodometers should not be performed in strongly alkaline solutions?*

Ans. Iodometry titrations cannot be performed in strongly alkaline solutions. This is because iodine reacts with alkalies to give hypoiodite which is a strong oxidizing agent and may oxidize thiosulphate ions.

$$I_2 + 2NaOH \rightarrow NaIO + NaI + H_2O$$

or $$I_2 + 2\bar{O}H \rightarrow I\bar{O} + I^- + H_2O$$

$$S_2O_3^{2-} + 4I\bar{O} + 2\bar{O}H \rightarrow 4I^- + 2SO_4^{2-} + H_2O$$

7. *Why always excess of KI is used?*

Ans. This is because the solubility of I_2 in water is very low. But in the presence of KI it forms an unstable complex salt, KI_3. This soluble complex will release iodine slowly during titration due to reversible reaction,

$$KI + I_2 \rightleftharpoons KI_3$$

$$I^- + I_2 \rightleftharpoons I_3^-$$

When free iodine is used up in the reaction with thiosulphate, the equilibrium shifts to the left releasing more I_2.

8. *Why the solution is allowed to stand for 2 - 3 minutes before the titration?*

Ans. Since the rate of reaction between I^- ion and the oxidizing agent is low, therefore the solution is allowed to stand for 2 – 3 minutes before titrating.

9. *Why the reaction mixture is kept in dark?*

Ans. The reaction mixture is kept in the dark before titrating. This is due to the reason that light accelerates the oxidation of iodide ions to iodine by atmospheric oxygen

$$4I^- + 4H^+ + O_2 \rightarrow 2I_2 + 2H_2O$$

10. *Why iodine is not used as self indicator, even though it is strongly coloured?*

Ans. It is possiblke to use iodine solution as indicator but its brown colour becomes very faint towards end point and hence is difficult to detect. It is better to use a little starch, which gives intense bluecolour with iodine. The colour disappears when the last traces of iodine has been reduced.

11. *What is the advantage of using starch as indicator in iodine titrations?*

Ans. Starch reacts with iodine in the presence of iodide ion to form an intensely colored blue complex, which is visible at very low concentrations of iodine, making it a very good indicator in both direct and indirect iodometric titrations.

12. *The iodine titrations should be performed as pH below 8?*

Ans. The blue color produced is also destroyed by alkali and heat, so titrations involving Starch indicator should be conducted at a pH below 8 and at temperatures near room temperature (this will also minimize loss of volatile iodine from the titration vessel).

13. *Why starch should also not be used in a strongly acid medium.*

Ans. Starch should also not be used in a strongly acid medium because hydrolysis of the starch occurs.

14. *Why starch gives bluecolour with iodine in presence of iodide?*

Ans. This is due to the formation of Starch iodine complex.

Starch can be separated into two fractions - amylose and amylopectin.Amylose in starch is responsible for the formation of a deep blue color in the presence of iodine. The iodine molecule slips inside of the amylose coil to form a blue coloured complex.

Only iodine element in the presence of iodide ion will give the characteristic blue black colour. Neither iodine element alone nor iodide ions alone will give the colour result.

15. *Why starch is added near the endpoint of an iodine titration and not in the beginning?*

Ans. This is because the concentration of iodine is very high in the beginning and it is strongly adsorbed by colloidal starch particles. It is not released even at the end point thus making the detection difficult. The additioin of the starch solution is thus delayed till the reaction is nearly complete (near the end point) as shown by a colour change of solution from deep red-brown to pale yellow. Moreover, starch tends to decompose at high concentrations of iodine.

16. *Why only freshly prepared starch solution should be used?*

Ans. Aqueous starch suspensions decompose within a few days, primarily because of bacterial action. The decomposition product is glucose, a reducing agent. This may interfere with a redox titration.

17. *What is the substitute for starch as an indicator in iodometric titrations.*

Ans. Sodium starch glycollate.

18. *Why sodium thiosulphate solution is not considered a primary standard. What is its equivalent weight in iodometric titration?*

Ans. Sodium thiosulphate crystals are efflorescent, i.e. lose their water of crystallisation on keeping in air and decompose due to bacterial action in solution. This it is not a primary standard. The equivalent weight of $Na_2S_2O_3$ in iodometry will be $\frac{2 \times \text{Molecular weight}}{2}$ since the reaction involves 2 electrons per two moles of $Na_2S_2O_3$.

Information about chemicals used

Sodium Thiosulphate

Sodium thiosulphate ($Na_2S_2O_3$) is a colour less *crystalline* compound that is more familiar as the pentahydrate, $Na_2S_2O_3 . 5H_2O$, an *efflorescent, monoclinic* crystalline substance, also called sodium hyposulphite or "hypo".

Sodium thiosulphate has a tendency to cake on storage due to the presence of surface moisture.

Synthesis

Sodium thiosulphate is produced chiefly from liquid waste products of *sodium sulphide* or suphur dye manufacture. It is also produced from *sodium carbonate* and *sulphur dioxide*.

Uses

1. Because sodium thiosulphate dissolves *silver* salts, its major use is in *photography* for developing *film*. (As a fixer)
2. It is also used in the *tanning* of *leather*, in chemical manufacture in *bacteriological* water assessment, and in treating *cyanide* poisoning.
3. As an aid in the bending of tubes.

Safety Concern

1. This material may be harmful if swallowed. It is moderately toxic when ingested.
2. It may also irritate the eyes or lungs. Flush the eye with plenty of water.

EXPERIMENT NO. 3

AIM **Determination of Alkalinity in water sample.**

APPARATUS Burette, pipette, conical flask, beakers.

CHEMICALS N/10 HCl, phenolphthalein and methyl orange.

THEORY Determination of alkalinity due to different ions is based on the titration of the water sample against a standard acid making selective use of indicators. The indicators used are– phenolphthalein and methyl orange*. The reactions taking place may be represented by the following equations:

$$\left.\begin{array}{l}\left.\begin{array}{l}\text{(a) } \overline{O}H + H^+ \rightarrow H_2O \\ \text{(b) } CO_2^{2-} + H^+ \rightarrow HCO_3^-\end{array}\right] P \\ \text{(c) } HCO_3^- + H^+ \rightarrow H_2CO_3\end{array}\right] M$$

- The volume of the acid used upto phenolphthalein end point corresponds to the reaction (a) and (b) i.e. complete neutralization of $\overline{O}H$ ions and neutralization of CO_3^{2-} ions upto HCO_3^- stage (half neutralization of CO_3^{2-} ions).

***Note.** (For the theory of acid base indicators refer to Acidimetry- Alkalimetry, chapter 1, page 9

- The volume of the acid used upto methyl orange end point corresponds to the reaction (a), (b), (c) i.e. complete neutralization of $\bar{O}H$, CO_3^{2-} and HCO_3^- ions.

Thus from the respective volume of the acid used, the respective strengths of various ions can be determined.

A known volume of the sample is titrated against a standard acid using phenolphthalein as indicator and alkalinity is found out in terms of $CaCO_3$ equivalents by using normality equation. This alkalinity is called *phenolphthalein alkalinity (P)*. Similarly the sample is titrated against a standard acid using methyl orange as indicator and alkalinity is calculated in terms of $CaCO_3$ equivalents. This alkalinity is called *methyl orange alkalinity (M)* or total alkalinity of water sample.

From the measurement of phenolphthalein alkalinity and methyl orange alkalinity, it is possible to calculate the magnitude of various forms of alkalinity present in water sample,

(i) Alkalinity due to HCO^- only

(ii) Alkalinity due to CO_3^{2-} only

(iii) Alkalinity due to CO_3^{2-} and HCO_3^-

(iv) Alkalinity due to CO_3^{2-} and $\overline{O}H$

(v) Alkalinity due to $\overline{O}H$ only.

Discussion.

Case (i): When phenolphthalein alkalinity = 0, this implies that volume of the acid used till phenolphthalein end point is 0. This can only happen if CO_3^{2-} and $\overline{O}H$ both are absent. Whatever alkalinity is present is due to HCO_3^- ions and can be detected using methyl orange as indicator.

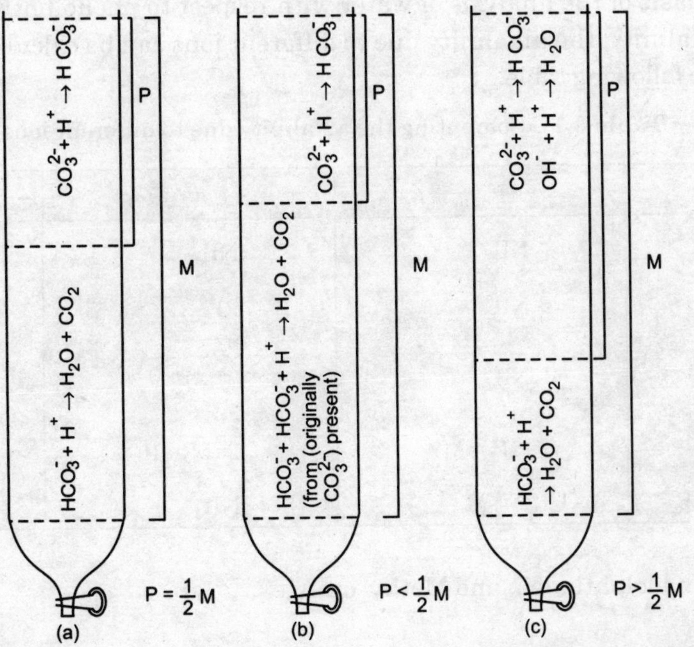

Fig. 3.1 Relative amounts of acid used, using phenolphthalein and methyl orange as indicator

Case (ii): When $P = \frac{1}{2} M$, this implies that only CO_3^{2-} ions are present.

Neutralization using phenolphthalein as indicator reaches upto HCO_3^- stage. Same amount of acid will be further used when methyl orange is used as indicator to neutralize HCO_3^- to H_2O and CO_2 as represented in fig: 3-1 (a) The strength of CO_3^{2-} will be given by M or 2P

Case (iii): When $P < \frac{1}{2} M$, this implies that besides CO_3^{2-} ions, HCO_3^- ions are also present. The acid used uptil phenolphthalein end point corresponds to half neutralization of CO_3^{2-} ions. Further titration using methyl orange as indicator corresponds to neutralization of HCO_3^- obtained from CO_3^{2-} ions and HCO_3^- originally present in the reaction mixture as shown in Fig. 3.1(b)

Alkalinity due to $CO_3^{2-} = 2P$

Alkalinity due to $HCO_3^- = (M - 2P)$

Case (iv) When $P > \frac{1}{2} M$. This implies that besides CO_3^{2-} ions, $\bar{O}H$ ions are also present. $\bar{O}H$ will be completely neutralized by the acid whereas CO_3^{2-} will be neutralized upto the HCO_3^- stage using phenolphthalein as indicator. On further titration with the acid using methyl orange as indicator, the neutralization of HCO_3^- takes place as shown in Fig. 3.1(c)

Alkalinity due to $CO_3^{2-} = 2\,(M - P)$.

Alkalinity due to $\bar{O}H = M - 2\,(M - P) = (2P - M)$

Case (v): When P = M implies that only $\bar{O}H$ ions are present. Alkalinity due to $\bar{O}H = P = M$.

Thus, on the basis of the analysis of water with respect to phenolphthalein alkalinity and methyl orange alkalinity, the alkalinity due to different ions can be calculated. The results are summarized in the following table.

Table 3.1 Computing the alkalinity due to different ions

Alkalinity	$\bar{O}H$ *(ppm)*	CO_3^{2-} *(ppm)*	HCO_3^- *(ppm)*
$P = 0$	NIL	NIL	M
$P = \frac{1}{2} M$	NIL	2P	NIL
$P < \frac{1}{2} M$	NIL	2P	(M – 2P)
$P > \frac{1}{2} M$	(2P – M)	2 (M – P)	NIL
P = M	P = M	NIL	NIL

INDICATORS Phenolphthalein and Methyl orange

END POINT Pink to colourless (Phenolphthalein), Yellow to red (Methyl orange)

PROCEDURE

1. Pipette out 20 ml of water sample into a conical flask. Add 1 – 2 drops of phenolphthalein.
2. Rinse and fill the burette with N/10 HCl.
3. Titrate the water sample in conical flask with N/10 HCl till the pink colour just disappears.
4. Note down the reading and repeat to get three concordant readings.
5. Again take 20ml of the water sample in conical flask and add methyl orange indicator to it.
6. Titrate it using N/10 HCl till a red colour is obtained.
7. Record the observation and repeat to get three concordant readings.

Observations

(a) Using Phenolphthalein

Normality of the acid used = N/10

S. No.	*Volume of the solution taken in the titration flask (ml)*	*Burette Readings*		*Volume of the titrant used (Final-Initial reading) (ml)*
		Initial Reading	*Final Reading*	
1. 2. 3.				

Volume of the acid used = Vml

(b) Using Methyl Orange

Normality of the acid used = N/10

S. No.	*Volume of the solution taken in the titration flask (ml)*	*Burette Readings*		*Volume of the titrant used (Final-Initial reading) ml*
		Initial Reading	*Final Reading*	
1. 2. 3.				

Notes 1. If no colour develops on addition of phenolphthalein to the water sample, it means that phenolphthalein alkalinity is zero and hence do not titrate the sample for phenolphthalein alkalinity.

2. Methyl orange can also be added to the solution in which phenolphthalein was added, after the phenolphthalein end point is reached. In that case the total volume of the acid used to methyl orange end point would be, the volume of acid used upto phenolphthalein end point plus the volume of acid used upto methyl orange end point.

CALCULATIONS

Phenolphthalein Alkalinity in terms of $CaCO_3$ equivalents:

$$\begin{aligned} &\text{(Acid)} \quad \text{(Water sample)} \\ N_1V_1 &= N_2V_2 \\ \frac{1}{10} \times V_1 &= N_2 \times 20 \\ N_2 &= \frac{1}{10} \times \frac{V_1}{20} = \frac{V_1}{200} \end{aligned}$$

Strength in terms of $CaCO_3$ equiv. $= N_2 \times$ Eq.wt of $CaCO_3$

$$= \left(\frac{V_1}{200}\right) \times 50 \text{ g/L} = x \text{ g/L}$$

$\therefore$ Phenolphthalein alkalinity (P) $= x \times 1000$ mg/L $= x \times 1000$ ppm.

(II) Methyl orange Alkalinity in terms of $CaCO_3$ equivalents:

$$\begin{aligned} &\text{Acid} \quad \text{water sample} \\ N'_1V'_1 &= N'_2V'_2 \\ \frac{1}{10} \times V'_1 &= N'_2 \times 20 \\ N'_2 &= \frac{1}{200} \times V'_1 \end{aligned}$$

Strength in terms of $CaCO_3$ equiv. $= N'_2 \times 50$ g/L

$$= \left(\frac{V'_1}{200} \times 50\right) \text{ g/L} = y \text{ g/L}$$

Methyl orange alkalinity or total alkalinity of water sample $= y \times 1000$ mg/L

$= y \times 1000$ ppm

RESULT

Phenolphthalein alkalinity = ... ppm of $CaCO_3$

Mehtyl orange alkalinity = ... ppm of $CaCO_3$.

To find out the alkalinity in terms of individual ions, we find out to which case the values of P and M falls and from table 2.1, we calculate the amount of individual ions.

Examples

1. If P = 0, the alkalinity is due to bicarbonate ions.

   $\therefore$ Bicarbonate Alkalinity in terms of $CaCO_3$ eq. = M.

   Bicarbonate ion concentration $= \dfrac{M \times 61}{50}$ mg/L.

   (The equiv. weight of HCO_3^- is 61 and of $CaCO_3$ is 50)

2. If $P = \frac{1}{2}M$, then alkalinity is only due to CO_3^{2-}. Carbonate alkalinity in terms of $CaCO_3$ equiv = 2P or M

Carbonate ion concentration $= 2P \times \frac{30}{50}$ mg/L.

(The equiv. weight of $CO_3^{2-} = 30$)

3. If $P < \frac{1}{2}M$, then the alkalinity is due to CO_3^{2-} and HCO_3^-

Carbonate alkalinity in terms of $CaCO_3$ equiv = 2P

Carbonate ion concentration $= 2P \times \frac{30}{50}$ mg/L

Bicarbonate alkalinity in terms of $CaCO_3$ equiv = (M – 2P)

Bicarbonate concentration $= (M - 2P) \times \frac{61}{50}$ mg/L

4. $P > \frac{1}{2}M$, then the alkalinity is due to $\bar{O}H$ and CO_3^{2-}

Hydroxide alkalinity in terms of $CaCO_3$ equiv = (2P – M)

Hydroxide concentration $= (2P - M) \times \frac{17}{50}$ mg/L

(Equiv. weight of $\bar{O}H = 17$)

CO_3^{2-} alkalinity in terms of $CaCO_3$ equiv. = 2 (M – P)

CO_3^{2-} concentration $= 2(M - P) \times \frac{30}{50}$ mg/L

5. If P = M, the alkalinity is due to hydroxide ions

∴ Hydroxide alkalinity in terms of $CaCO_3$ eq. = P or M

Hydroxide ion concentration $= P \times \frac{17}{50}$ mg/L

EXERCISE

1. Alkalinity of water cannot be due to the simultaneous presence of $\bar{O}H$, CO_3^{2-} and HCO_3^-. Explain.
2. Write the structural formula of phenolphthalein. In what forms does it exist in acidic and alkaline medium.
3. What is the source of alkalinity in water.
4. A 200 ml of a water sample required 20 ml. of N/50 H_2SO_4 using methyl orange as indicator but did not give any colouration with phenolphthalein . What type of alkalinity is present. Express in mg/L.

Note. P and M values in the above calculations are taken in mg/L.

ACIDITY OF WATER

Acidity in water is caused by the presence of mineral acids, free carbon dioxide, sulphates of iron and aluminium etc in waters. Some salts like $MgCl_2$ etc also liberate HCl due to hydrolysis.

$$MgCl_2 + 2H_2 \rightarrow Mg(OH)_2 + 2HCl$$

Acidic water reacts with iron in the pipes or in the boilers and hence cause corrosion. Moreover, in the lime soda process of water softening, the amounts of lime and soda to be added depend on the amount of mineral acids also, apart from other factors. Thus determination of acidity is important.

EXPERIMENT NO. 4

AIM **Determination of rate constant for acid catalysed hydrolysis of ethyl acetate through titration.**

THEORY The hydrolysis of ethyl acetate catalyzed by dil. HCl can be represented as:

$$CH_3COOC_2H_{5(aq.)} + H_2O_{(l)} \rightarrow CH_3COOH_{(aq.)} + C_2H_5OH_{(aq.)}$$

In the above reaction the concentration of water being in excess, is very high and practically remains constant during the reaction. Concentration of H^+ which catalyze the reaction also remains constant. Thus the rate of reaction only depends upon the concentration of the ester i.e. it is a *first order reaction.*

$$\text{rate} = k\,[CH_3COOC_2H_5]$$

Reactions of the above type in which one of the reactant is present in large excess, are though bimolecular but follow the kinetics of first order are called *pseudounimolecular reactions.*

The value of rate constant is calculated from the equation of first order reaction.

$$k = \frac{2.303}{t} \log \frac{C_o}{C_o - x} \qquad \text{...(1)}$$

where C_o = Conc. of the ester

$C_o - x$ = Conc. of the ester at a time t.

Since acetic acid is formed as a result of hydrolysis, the kinetics of the reaction can be followed by determining the amount of acetic acid produced at definite intervals of time. For this, known amounts of the ester and the mineral acid are mixed and equal volumes of the reaction mixture are withdrawn at definite intervals. This reaction mixture so withdrawn is titrated against standardized alkali (NaOH) using phenolphthalein. The volume of the alkali used for each titration corresponds to the amount of acid used as catalyst and the amount of acetic acid produced at that time. The final reading is taken after keeping the remaining reaction mixture in water bath for 24hrs. or keeping it in hot water at 50–60° C for about 30 minutes.

Let V_o = Volume of NaOH used at time $t = 0$

V_t = Volume of NaOH used at time t

V_∞ = Volume of NaOH used at time $t = \infty$

Now, $V_o \alpha$ The amount of HCl present in the reaction mixture.

$V_t \alpha$ The amount of HCl present in the reaction mixture + acetic acid produced at time t

$V_\infty \alpha$ The amount of HCl present in the reaction mixture + acetic acid produced after the completion of reaction.

Therefore,

Amount of acetic acid produced when the hydrolysis is complete

$$= V_\infty - V_o \qquad ...(2)$$

Amount of acetic acid produced at time $t = V_t - V_o$ ...(3)

Further,

$V_\infty - V_o \alpha$ The initial conc. of the ester, C_0. ...(4)

$(V_\infty - V_o) - (V_t - V_o) = (V_\infty - V_t) \alpha$ The conc. of ester at time t, $C_0 - x$ ...(5)

Using equations (4) and (5) in equation (1) we get,

$$k = \frac{2.303}{t} \log \frac{V_\infty - V_o}{V_\infty - V_t} \qquad ...(6)$$

The value of k is determined using equation (6) for the different sets of values corresponding to different intervals of time.

From equation (6)

$$\log (V_\infty - V_t) = \log (V_\infty - V_o) - \frac{kt}{2.303} \qquad ...(7)$$

Graphically, k can be calculated by plotting $\log (V_\infty - V_t)$ against t. As per equation (7) this should give a straight line where

$$\text{slope} = \frac{-k}{2.303}$$

or
$$k = -2.303 \times \text{slope}.$$

APPARATUS 50 cm^3 burettes, 100cm^3 conical flasks, 250cm^3 conical flask, graduated pipette, stopwatch, thermometer, water bath or thermostat.

CHEMICALS Ethyl acetate, hydrochloric acid (0.5N), Phenolphthalein indicator, sodium hydroxide (0.1 N), ice cubes (prepared from CO_2 free water.)

PROCEDURE

1. Take 50 cm^3 of 0.5 N HCl in 250 cm^3 conical flask and 10ml of ethyl acetate in a test tube.
2. Cork both of them and place them in thermostat maintained at room temperature. Let the two solutions attain the temperature of the bath.
3. In the mean time, rinse and fill a burette with 0.1 N NaOH solution. Fix it on a burette stand. Also arrange five conical flasks of 50 cm^3 capacity and put 25ml of ice cold water in one of them.

4. Now pipette out 5ml of the ester kept in the thermostat and transfer it to the conical flask containing hydrochloric acid. Start the stop watch when nearly half the volume of ester has been added.
5. Mix the contents quickly and at once pipette out 5ml of the reaction mixture into the conical flask containing 25 cm^3 of ice cold water. This freezes or arrests the reaction (due to lowering of temperature)
6. Now titrate it quickly against 0.1 N NaOH using phenolphthalein as indicator. Appearance of light pink colour, marks the end point. The titre value represents V_o.
   (The volume of sodium hydroxide used in the above titration nearly corresponds to the concentration of hydrochloric acid in the reaction mixture before the hydrolysis starts.)
7. Similarly, after a time interval of 10 minutes, again pipette out 5ml of the reaction mixture into a conical flask in which 25 cm^3 of ice cold water has already been added. Titrate it against 0.1 N NaOH as before using phenolphthalein as indicator.
8. Repeat the process of pipetting 5cm^3 of reaction mixture into a conical flask containing phenolphthalein and 25 ml of ice cold water and titrating against 0.1 N NaOH after every 10 minutes for 1 hour. Each of the titre values from second onwards constitute V_t
   (The titre value required for each titration after *t* minutes corresponds to the volume of alkali used to neutralize hydrochloric acid and acetic acid formed after time *t*).
9. To obtain the volume of NaOH required when the hydrolysis is complete, the corked flask containing the reaction mixture is kept in a water bath at a temperature of 50–60° C for about 30 minutes to 1 hour. [To save time, this can be done just after first reading by withdrawing 20 ml of the reaction mixture in a separate flask and keeping it in water bath at 50°– 60°C for 30 minutes to 1 hour] Cool the conical flask to room temperature and titrate 5 cm^3 of this mixture against 0.1 N NaOH as above. In this titration, there is no need to add ice cold water. The last titre value represents the infinite reading V_∞.

OBSERVATION

Room temperature = ... °C

Observation Table

Time (Minutes)	*Burette Reading*		*Titre Value*	*log ($V_\infty - V_t$)*
	Initial	*Final*		

CALCULATIONS

The value of k is now calculated according to the first rate expression,

$$k = \frac{2.303}{t} \log \frac{V_\infty - V_o}{V_\infty - V_t}$$

1. The values of k are calculated at different intervals of time t

   (i) $k =$ ...sec^{-1}

   (ii) $k =$ ...sec^{-1}

   (iii) $k =$ ...sec^{-1}

2. A graph between t versus log $(V_\infty - V_t)$ is plotted. A straight line is obtained. Value of k is also calculated from the graph

   $k = -2.303 \times \text{slope} =$ ... sec^{-1}

RESULT k (by calculation) = ... sec^{-1}

k (by graphical method) = ...sec^{-1}

Instructions for the Lab. Assistant

1. **Preparation of 0.5 N HCl solution.** Transfer 45cm^3 of conc. HCl (11N) in a measuring flask containing approx. 500 cm^3 of distilled water. Make up the volume to 1dm^3 (1 litre).
2. **0.1 N NaOH.** Dissolve 8g of NaOH pellets in 2 litres of distilled water.

   Standardize it with N/10 oxalic acid solution
3. **Phenolphthalein indicator.** Dissolve 1g of phenolphthalein in 100 ml of ethanol. Add 100 ml of water stirring it continuously.

EXERCISE

1. *What do you mean by rate of reaction?*

Ans. The rate of reaction is defined as the change in concentration of reactant or product per unit time.

2. *What are the factors which affect rate of reaction?*

Ans. Rate of reaction is influenced by:
(i) Nature of reactants (ii) Concentration of reactants (iii) Temperature (iv) Presence of catalyst.

3. *What do you mean by rate constant?*

Ans. It is defined as the rate of reaction when the concentration of each reactant is taken as unity. Rate constant is sometimes also referred to as specific reaction rate.

4. *Define order of reaction.*

Ans. The sum of the concentration terms on which the rate of a reaction actually depends as observed experimentally is called the order of the reaction.

5. *Distinguish between order and molecularity of a reaction.*

Ans. The number of atoms, ions or molecules that must collide with one another simultaneously so as to result into chemical reaction is called molecularity of the reaction.

Order of reaction is experimentally determined quantity whereas molecularity can be calculated by simply adding the molecules of the rate determining step (slowest step)

Order of reaction need not be a whole number whereas molecularity is always a whole number.

6. *What are the units of rate constant of (i) zero order (ii) first order (iii) second order and third order*

Ans.(i) L mol^{-1} s^{-1} (ii) s^{-1} (iii) mol L^{-1} s^{-1} (iv) mol^2 L^{-2} s^{-1}

7. *What are pseudo-unimolecular reactions?*

Ans. The reactions which are not truly first order but under certain conditions become reactions of the first order are called pseudounimolecular reactions. These are bimolecular reactions which follow the kinetics of the first order reaction.

8. *Give examples of pseudounimolecular reactions.*

Ans. (i) Hydrolysis of ethyl acetate

(ii) Inversion of cane sugar

9. *Why the reaction is speeded up with increase in temperature?*

Ans. At temperature increase causes greater fraction of the molecules to have energy equal to or greater than activation energy.

10. *Which of the following will react: (i) fastest, (ii) at highest rate?*

(a) 1 mol A + 1 mol B in a 1L vessel, (b) 2 mol A + 2 mol B in 2L vessel, (c) 0.2 mol A + 0.2 mol B in a 0.1 L vessel.

Ans. (*i*) (*b*), since there is more reactant present so more products are produced per unit time.

(ii) (c), since the vessel with greatest concentration experiences greatest change in concentration per unit time.

11. *How does half-life period of a first order reaction vary with temperature? Explain.*

Ans. Half-life *decreases* with *rise* of temperature, because $t_{1/2} \propto 1/k$ and k increases with rise of temperature.

12. *How is rate constant related to concentration of reactants?*

Ans. $k = \dfrac{1}{C^{n-1}}$, where n = order of reaction.

13. *For the reaction in acid solution:* $HCOOCH_3 + H_2O \rightarrow HCOOH + CH_3OH$, *the rate* = $k\,[HCOOCH_3][H^+]$. *Why does* $[H^+]$ *appear in the rate law, when it does not appear in the balance equation?*

Ans. H^+ (a *catalyst* for the reaction) actually acts as a reactant in an early intermediate stage of the reaction and is then released back into the solution at a later stage.

14. *Why is reaction speeded up in presence of a catalyst?*

Ans. A catalyst *lowers* the activation energy, therefore more molecules in the reactants possess sufficient energy to react.

EXPERIMENT 5

AIM Synthesis of benzimidazole and record of its UV spectrum

BENZIMIDAZOLE is a *heterocyclic aromatic organic compound.* This bicyclic compound consists of the fusion of *benzene* and *imidazole*

THEORY The usual synthesis involves condensation of *o-phenylenediamine* with *formic acid*

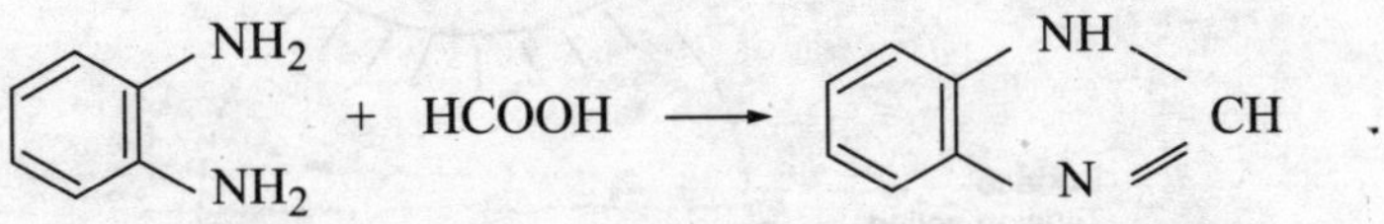

With acetic acid, 2-methylbenzimidazole, m.p. 173-174°, is formed .

PROCEDURE

1. Place 27 g. of o-phenylenediamine in a 250 ml. round- bottomed flask and add 17-5 g. (6 ml.) of 90 per cent, formic acid A).
2. Heat the mixture on a water bath at 100° for 2 hours.
3. Cool, add 10 per cent sodium hydroxide solution slowly, with constant rotation of the flask, until the mixture is just alkaline to litmus.
4. Filter off the crude benzimidazole , wash with ice-cold water, drain well and wash again with 25 ml. of cold water.
5. Dissolve the crude product in 400 ml. of boiling water, add 2 g. of decolourising carbon, and digest for
6. 16 minutes. Filter rapidly at the pump through a preheated Buchner funnel and flask.
7. Cool the filtrate to about 10°, filter off the benzimidazole, wash with 25 ml. of cold water, and dry at 100°.
8. The yield of pure benzimidazole, m.p. 171-172°, is 25 g.

EXPERIMENT NO. 6.1A

AIM Comparison of viscosity and surface tension of two liquids and their variation on addition of surfactants and soap of known concentration.

THEORY When a liquid is allowed to flow through a capillary tube, a drop is formed at its lower end. It increases to a certain size and falls off. The size of the drop depends on the radius of the capillary and the surface tension of the liquid. The surface tension acting along the circumference of the capillary tube supports the drop in the upward direction. (Fig. 6.1)

The measurement of surface tension of liquid is based on the fact that the drop of the liquid at the lower end of capillary falls down when weight of drop becomes just equal to the surface tension.

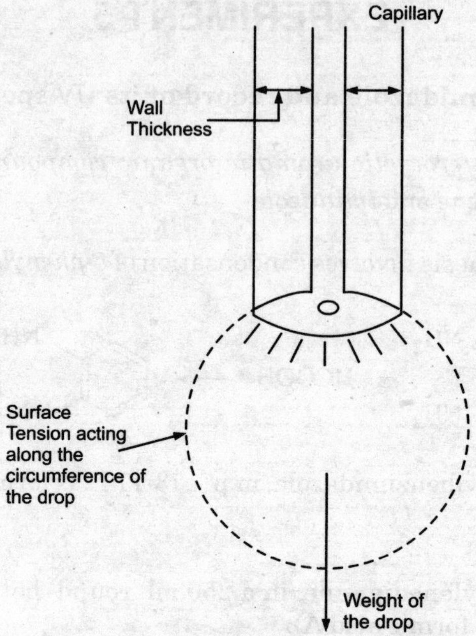

Fig. 6.1 Forces acting on the formed in stalagmometer

1. Now the force of gravity exerted on the drop, is given by v.d.g.

   Where v is the volume of the drop, d its density and g its gravity. The force of gravity will pull down the drop .

2. The force tending to uphold the drop is given by $2\pi r\gamma$.

   where $2\pi r$ is circumference of the capillary of radius r.

   When the two forces are balanced.

$$2\pi r\gamma = \text{v.d.g} \quad \text{...(1)}$$

If n is the number of drops in volume V of the liquid then volume of each drop, will be

$$v = \frac{V}{n},$$

From eq. (1) we have

$$\therefore \quad 2\pi r\gamma = \frac{V}{n}\,\text{dg.} \quad \text{...(2)}$$

Consider two liquids of densities d_1 and d_2 having the surface tension γ_1 and γ_2. Let the number of drops counted for the same volume V of the two liquids be n_1 and n_2 respectively. Then,

$$\therefore \quad 2\pi r\gamma_1 = \frac{V}{n_1}\,d_1 g \text{ for 1}^{\text{st}} \text{ liquid} \quad \text{...(3)}$$

$$\therefore \quad 2\pi r\gamma_2 = \frac{V}{n_2}\,d_2 g \text{ for 2}^{\text{nd}} \text{ liquid} \quad \text{...(4)}$$

Dividing equation (3) by (4), we have

$$\frac{\gamma_1}{\gamma_2} = \frac{n_2}{n_1} \times \frac{d_1}{d_2}$$

Where γ_1 and γ_2 are the surface tensions of two individual liquids and d_1 and d_2 are their densities, respectively. Thus, for the determination of surface tension of any liquid, the number of drops produced from equal volume of two liquids and their densities must be known. Knowing the surface tension of the reference liquid (e.g water) the surface tension of the given liquid can be determined

APPARATUS Stalagmometer, beakers, rubber tubing, pinch cock, relative density bottle, thermometer.

Stalagmometer. Stalagmometer is essentially a pipette with a capillary at the lower end and a bulb in the middle. The end is flattened and grounded carefully so that there is a large dropping surface. There are two marks A and B above and below the bulb respectively (Fig. 6.2). These marks define the volume to be delivered for counting the drops. In case the rate of flow of the liquid is rapid leading to rapid drop formation, a piece of rubber tubing with a pinch cock is attached to the top of the stalagmometer and the rate of flow is adjusted by altering the pressure of the screw so that the drops are formed at a rate not greater than 15 drops per minute.

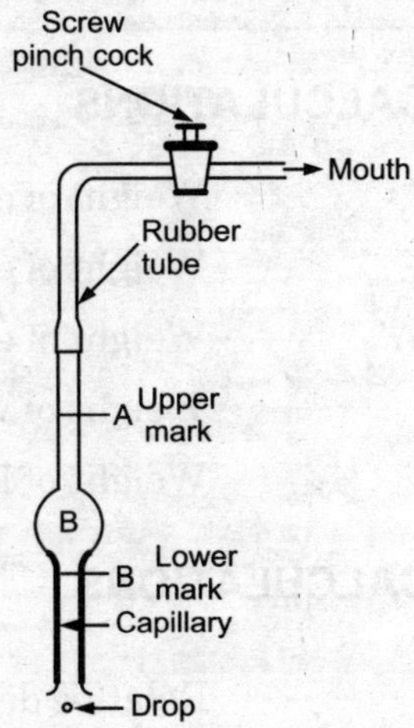

Fig. 6.2 Stalagmometer

(ii) *Relative Density Bottle.* The density of liquid is conveniently measured by means of a relative density bottle or specific gravity bottle. It is a round bottomed glass vessel, fitted with a glass cork containing a fine capillary

PROCEDURE

1. Clean the stalagmometer first with NaOH, then with chromic acid solution to remove oil, grease etc. sticking on the inner capillary surface. Wash it with distilled water and finally with acetone and dry by passing air. (In case water is used as reference liquid there is no need to dry the apparatus)
2. Immerse the lower end of the stalagmometer in a beaker containing distilled water. *Suck the water until it rises above the mark A (Control the rate of flow of water with the help of screw pinch cock so that the number of drops per minute is about 15-20)*
3. Now let the water flow out and start counting the number of drops when the water meniscus just reaches the mark A, and counting is stopped when the water meniscus passes the lower mark B. Let the number of drops be n_w.
4. Repeat to get three readings.
5. Clean the stalagmometer and dry it. Fill it with liquid until it rises above the mark A and count the number of drops when a fixed volume of liquid flows between the mark A and B, as before. Let the number of drops be n_l.

6. Weigh the relative density bottle and note down its weight.
7. Fill it with water and weigh it again.
8. Remove water and dry the relative density bottle. Now fill it with the given liquid and weigh.

Observation Table

S.No.	*Water*		*Given Liquid (C_2H_2OH)*	
	No. of Drops	*Mean*	*No. of Drops*	*Mean*
1.				
2.				
3.	n_w	n_l		

CALCULATIONS

Weight of empty specific gravity bottle = ω_1g

Weight of empty specific gravity bottle with water = ω_2g

Weight of empty specific gravity bottle with liquid = ω_3g

$\therefore$ Weight of water = $(\omega_2 - \omega_1)$g

Weight of liquid = $(\omega_3 - \omega_1)$g

CALCULATIONS

$$\text{Relative density of the liquid} = \frac{d_l}{d_\omega} = \frac{\omega_3 - \omega_1}{\omega_2 - \omega_1}$$

$$\frac{\gamma_l}{\gamma_w} = \frac{d_l}{d_w} \times \frac{n_w}{n_l}$$

where γ_l, d_l, n_l are the surface tension, density and number of drops of liquids.

γ_w, d_w, n_w are the surface tension, density and number of drops of water.

By putting the respective values, the surface tension of the liquid relative to water is found out.

Absolute surface tension of the liquid.

$$\gamma_1 = \frac{n_\omega}{n_l} \times \frac{d_1}{d_\omega} \times \gamma_\omega = \ldots\text{Nm}^{-1}.$$

The value of γ_w is obtained from the table. 6.1

Table 6.1 Surface tension of water at different temperatures.

Temp	5	10	15	20	25	30	35	40
Surface tension (dynes/cm)	75.09	74.36	73.62	72.88	72.14	71.70	70.66	69.92

RESULT The surface tension of the given liquid is. . . dynes cm^{-1}

PRECAUTIONS

1. Surface tension is highly sensitive to contamination of even minute traces of grease and impurities. Stalagmometer therefore should be thoroughly cleaned with chromic acid.
2. Stalagmometer should be in a vertical position throughout the counting process.
3. Drop formation should be adjusted at a slower rate and should not exceed fifteen drops per minute.

EXERCISE

1. *Define surface tension.*

Ans. The surface tension γ is the magnitude *F* of the force exerted parallel to the surface of a liquid divided by the length *L* of the line over which the force acts:

$$\gamma = \frac{F}{L}$$

2. *What are the units of surface tension?*

Ans. **The common units for surface tension are dynes/cm or N/m. These units are equivalent.** Units of surface tension (N-m^{-1}) are equivalent to joules per square metre (J-m^{-2}). This means that surface tension can also be considered as **surface energy**. If a surface with surface tension γ is expanded by a unit area, then the increase in the surface's stored energy is also equal to γ.

3. *What are the forces responsible for surface tension?*

Ans. The **cohesive** forces between liquid molecules are responsible for the phenomenon known as surface tension.

4. *What is the surface tension of water?*

Ans. The **surface tension** of water is 72 dynes/cm at 25°C. It would take force of 72 dynes to break a surface film of water 1 cm long

5. *Why is Surface Tension important?*

Ans. If any formulation changes at the molecular level then the surface tension will change. If the formulation changes due to the addition of another chemical, the addition of a surfactant, or if anything contaminates the fluid in question, then the surface tension will change. Measuring surface tension is a direct indicator of the quality of any chemical and any formulation.

6. *Why hot water is considered better cleaning agent?*

Ans. Hot water is a better cleaning agent because the lower surface tension makes it a better "wetting agent" to get into pores and fissures rather than bridging them with surface tension. Soaps and detergents further lower the surface tension.

7. *What is the difference between adhesive and cohesive forces? On the basis of these forces discuss capillary action.*

Ans. Molecules in liquid state experience strong intermolecular attractive forces. When those forces are between like molecules, they are referred to as *cohesive forces*. For example, the molecules of a water droplet are held together by cohesive forces, and the especially strong cohesive forces at the surface constitute **surface tension**.

When the attractive forces are between unlike molecules, they are said to the *adhesive forces*. The adhesive forces between water molecules and the walls of a glass tube are stronger then the cohesive forces, then this leads to an upward turning meniscus at the wall of the vessel and contribute to capillary action.

8. *How small insects can walk on water?*

Ans. Small insects such as the water strider can walk on water because their weight is not enough to penetrate the surface.

9. *Why it is advised, not to touch the tent during rainy season?*

Ans. Common tent materials are somewhat rainproof because the surface tension of water will bridge the pores in the finely woven material. But if we touch the tent material with our finger, we break the surface tension and the rain will drip through.

10. *How the property of surface tension can be used as a test for jaundice?*

Ans. Normal urine has a surface tension of about 66 dynes/cm but if bile is present (a test for jaundice), it drops to about 55. In the Hay test, powdered sulphur is sprinkled on the urine surface. It will float on normal urine, but sink if the surface tension is lowered by the bile.

11. *Discuss capillary action.*

Ans. Capillary action or capillarity is the ability of a narrow tube to draw a liquid upwards against the force of **gravity**.

Capillary action is the result of **adhesion** and **surface tension**. Adhesion of water to the walls of a vessel will cause an upward force on the liquid at the edges and result in a meniscus which turns upward. The surface tension acts to hold the surface intact, so instead of just the edges moving upward, the whole liquid surface is dragged upward.

12. *Which phenomena is responsible for the shape of liquid droplets?*

Ans. **Surface tension** is responsible for the shape of liquid droplets. Although easily deformed, droplets of water tend to be pulled into a spherical shape by the **cohesive** forces of the surface layer.

13. *Why the surface of water has a concave shape when it is confined in a tube?*

Ans. When liquid water is confined in a tube, its surface (meniscus) has a concave shape because water wets the surface and creeps up the side.

14. *Why mercury does not wet the glass?*

Ans. Mercury does not wet glass because the cohesive forces within the drops are stronger than the adhesive forces between the drops and glass. When liquid mercury is confined in a tube, its surface (meniscus) has a convex shape because the cohesive forces in liquid mercury tend to draw it into a drop.

15. *What are surfactants?*

Ans. Surfactants, also known as wetting agents, lower the **surface tension** of a liquid, allowing easier spreading, and the **interfacial tension** between two liquids. The term surfactant is a contraction of "Surface active agent". Surfactants are usually **organic** compounds that are **amphipathic,** meaning they contain both **hydrophobic** groups (their "tails") and **hydrophilic** groups (their "heads"). Therefore, they are typically sparingly soluble in both organic solvents and water. Surfactants reduce the surface tension of water by **adsorbing** at the air-water interface.

16. *Why does the surface tension of water decrease on adding detergents? What is the phenomenon behind decrease in surface tension of water?*

Ans. In the solid and liquid phase, water molecules are bonded to each other by virtue "hydrogen bonding." The result is a kind of "skin" on liquid water's surface. Soaps and detergents have a polar head to which water is attracted and a non-polar tail that is hydrophobic — water repelling. When these substances are present, they weaken the strength of the skin by interfering with hydrogen bonding between water molecules because the polar end of the soap or detergent is also attracted to the water.

17. *What is the effect of temperature on surface tension?*

Ans. Surface tension arises due to cohesive forces within the molecules. With increase in temperature, kinetic energy of the molecules of liquid increases and hence force of attraction between them decreases. Hence surface tension of a liquid decreases with rise in temperature.

EXPERIMENT NO. 6B

AIM **To determine the composition of a liquid mixture A and B (acetic acid and water) by surface tension method.**

APPARATUS Stalagmometer, thermometer, beaker etc.

CHEMICALS Pure A (acetic acid) pure B (water) and the given solution.

THEORY When the surface tension of two liquids constituting the mixture do not differ greatly, the surface tension of the mixture varies with composition.

A number of known solutions of A (acetic acid) and B (water) are prepared and the number of drops for each one of these is calculated as in the previous experiment. The number of drops of unknown solution is also determined. Let it be x. A graph is plotted between the number of drops against the percent composition. From the graph the concentration of the unknown solution can be determined.

PROCEDURE

1. Prepare a number of solutions of A and B containing 90%, 80%, 70%, 60%, 50%, 40% etc. of the component A.
2. Determine the number of drops of each of the solution by stalagmometer as described in the previous experiment. Also determine the number of drops of unknown mixture.

OBSERVATIONS

Composition of the mixture		*Number of drops*
A	*B*	
90%	10%	
80%	20%	
70%	30%	
60%	40%	
50%	50%	
unknown	—	

CALCULATION Plot a graph between the number of drops against the percentage composition. The point corresponding to the number of drops of the unknown mixture is found. A perpendicular is drawn on the composition axis from that point and the value gives the composition of the unknown mixture.

RESULT The composition of the given mixture is .....

PRECAUTIONS Same as in experiment 6.1

Note The composition can also be found out by determining the surface tension of each of the solution and then plotting a graph between surface tension versus composition. (The density of each solution has to be measured in this case).

EXPERIMENT NO. 6C

AIM **Determination of coefficient of viscosity of the given liquid by Ostwald's Viscometer method.**

THEORY The Ostwald's Viscometer method is based on Poiseuille's equation. This relates the rate of flow of a liquid through a capillary tube with the coefficient of viscosity and is expressed by the equation,

$$\eta = \frac{\pi r^4 \, tP}{8Vl} \quad ...(7)$$

where V = Volume of the liquid of viscosity η flowing in time t, through a capillary tube of radius r and length l

P = Hydrostatic pressure of the liquid.

Thus the determination of the absolute viscosity by means by Poiseuille's expression involves the determination of V, r, t, l and P.

The method is however tedious and laborius one. Hence, a simpler method is used wherein we compare the viscosities of the two liquids. If the coefficient of viscosity of one liquid is known, then that of the other can be calculated.

If t_1 and t_2 are the flow times required to flow for equal volumes of two liquids through the same length of a capillary tube then from equation (7) we have

$$\eta_1 = \pi P_1 r^4 t_1/8Vl$$

$$\eta_2 = \pi P_2 r^4 t_2/8Vl$$

or

$$\frac{\eta_1}{\eta_2} = \frac{P_1 t_1}{P_2 t_2} \quad ...(8)$$

Now

$$P = hdg \quad ...(9)$$

where h is the height of the liquid column (It is constant for all the liquids if taken for identical points and liquids are taken in equal volumes for a particular set of observations.)

g = acceleration due to gravity

d = density of the liquid

Since in this case for two liquids h and g are same hence,

$$\frac{\eta_1}{\eta_2} = \frac{d_1}{d_2}\frac{t_1}{t_2}$$

This will give us the relative viscosity of the given liquid. Relative viscosity has no units. The absolute viscosity $\eta = \frac{d_1}{d_2} \cdot \frac{t_1}{t_2} \cdot \eta_2$

APPARATUS Ostwald's viscometer, stop watch, specific gravity bottle, 10cm^3 pipette, distilled water, rubber tubing.

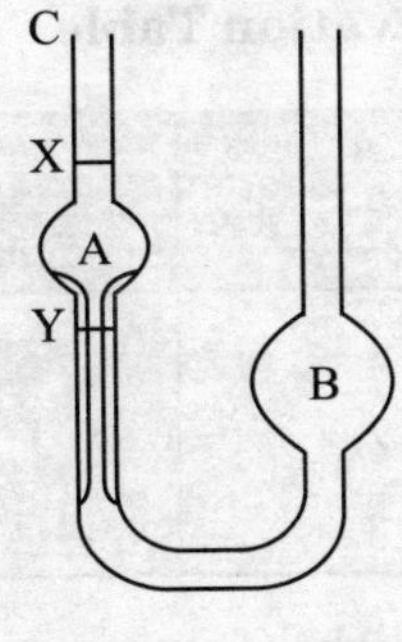

Fig. 6.3 Viscometer

Description. (i) *Ostwald's Viscometer:* It consists of a capillary tube connected at its upper end with a bulb A and its lower end with a U-tube provided with bulb B. Bulb B is of larger size than A (Fig. 6.3) Marks X and Y are etched on the capillary tube above and below the bulb A. The bulbs are necessary to maintain the hydrostatic pressure during flow of liquid. Through the capillary, the liquid flows with a measurable speed.

(ii) *Relative Density Bottle.* The density of liquid is conveniently measured by means of a relative density bottle or specific gravity bottle. It is a round bottomed glass vessel, fitted with a glass cork containing a fine capillary Fig. 6.4

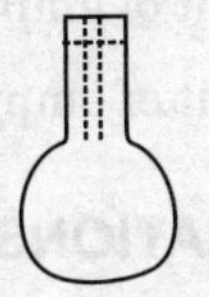

Fig. 6.4 Relative Density Bottle

PROCEDURE

1. Clean the viscometer with chromic acid ($K_2Cr_2O_7$ + Conc. H_2SO_4) and then wash it several times with distilled water. It is finally washed with alcohol and ether and then dried.
2. Attach a piece of clean rubber tube to the end C and clamp the viscometer vertically in air.
3. Now introduce a sufficient volume of the given liquid with the help of a pipette in bulb B so that the bend portion of the U-tube and more than half of bulb B are filled up.
4. Through the rubber tube, suck up the liquid until it rises above the mark X. Make sure that there is no air bubble inside the liquid.
5. Now, allow the liquid to fall freely through the capillary upto the mark X. Start the stop watch and note the time t_1 for the flow of the liquid from mark X to mark Y.
6. Repeat the experiment thrice. The values should be concordant.
7. Remove the liquid and clean and dry the viscometer again.
8. Repeat the experiment by taking the same volume of the distilled water and note down the time taken t_2 for flow of water from mark X to Y. Repeat thrice.
9. Weigh the relative density bottle and note down its weight.
10. Fill it with the given liquid and weigh it again.
11. Remove the liquid. Wash it with chromic acid and then distilled water. Dry in the oven. Now fill it with the distilled water and weigh it

OBSERVATIONS

Room temperature = ... °C.

Observation Table

S	Given liquid		Water	
No.	*Time of Flow*	*mean (t_1)*	*Time of flow*	*Mean (t_2)*
1.				
2.				
3.				

Weight of empty specific gravity bottle = ... ω_1g

Weight of empty specific gravity bottle + given liquid = ... ω_2g

Weight of empty specific gravity bottle + distilled water = ... ω_3g

CALCULATIONS

$$\frac{\text{Density of solution}}{\text{Density of water}} = \frac{d_1}{d_2} = \frac{\omega_2 - \omega_1}{\omega_3 - \omega_1}$$

$$\text{Relative viscosity of the given liquid} = \frac{\eta_1}{\eta_2} = \frac{d_1}{d_2} \times \frac{t_1}{t_2}$$

$$\text{Absolute Viscosity} = \frac{d_1}{d_2} \times \frac{t_1}{t_2} \times \eta_2$$

RESULT The relative viscosity of the given solution is ... at ... °C.

The absolute viscosity of the given solution is ... Pas at ... °C.

EXERCISE

1. *Define Viscosity*

Ans. Viscosity is the resistance that one part of a liquid flowing with one velocity offers to another part of liquid flowing with different velocity.

2. *What are the units of viscosity?*

Ans. Units of Viscosity:

(i) In the C.G.S system the unit of viscosity is dyne cm^{-2} s and is known as poise.

$$\text{poise} = \text{dyne cm}^{-2}\text{ s}$$
$$= \text{gcm s}^{-2}.\text{ cm}^{-2}\text{ s} = \text{g cm}^{-1}\text{ s}^{-1}$$

In other word 1 poise = 1g cm^{-1} s^{-1}

(ii) In S.I system the unit of viscosity is Pascal second (Pas).

Note.
$$1\text{ poise} = 1\text{g cm}^{-1}\text{ s}^{-1} = 1\ (10^{-3}\text{ kg})\ (10^{-2}\text{ m})^{-1}\text{ s}^{-1}$$
$$= 10^{-1}\text{ kg m}^{-1}\text{ s}^{-1}$$
$$= 10^{-1}\text{ Pas}$$

Hence 1 poise = 10^{-1} Pas.

3. *What do you mean by fluidity?*

Ans. The reciprocal of viscosity is called fluidity and is represented by ϕ so that,

$$\phi = \frac{1}{\eta}$$

Fluidity is the measurement of the ease with which the liquid flows. Its units are *Poise*$^{-1}$.

4. *How viscosity of a liquid varies with temperature?*

Ans. The viscosity of a liquid decreases with increase of temperature. As the temperature rises, the kinetic energy of the molecules increases and intermolecular forces of attraction become weak, resulting in subsequent decrease in the viscosity.

The variation of viscosity with temperature is best expressed by empirical relation given by Arrhenius

$$\eta \propto e^{Ea/RT}$$

$$\eta = Ae^{Ea/RT} \quad \text{....(6)}$$

where A is a constant of proportionality, T is the temperature and R is gas constant.

In contrast to liquids, η increases with increasing temperature in the case of gases.

5. *What is the effect of pressure on viscosity of a liquid?*

Ans. The viscosity of the liquids increases with increase in pressure. As the pressure is increased the molecules of the liquid come closer and it becomes difficult for liquid molecules to move around and thus it becomes more difficult for them to flow.

6. *How viscosity is related to chemical constitution?*

Ans. Though there is no significant relationship between viscosity and chemical constitution, however some general results have been observed.

(i) The viscosity increases as the size of the molecules increases.

(ii) In the compounds having chain isomerism, the n-isomers have generally greater viscosity than the branched chain.

However, moiecules with flexible chains have very high viscosity due to entanglement of side chains.

(iii) Non-polar molecules have lower viscosities than polar molecules.

(iv) Presence of intermolecular hydrogen bonding increases the viscosity of the liquid.

(v) Presence of impurities increases the viscosity. Viscosity is greatly enhanced due to the presence of lyophilic colloids.

7. *What are the applications of viscosity measurements?*

Ans.

- Viscosity Measurements are useful in predicting the behaviour of lubricants and oils in hot and cold conditions.
- Viscosity is also used for determination of molar masses of polymers.
- For distinguishing between branched and chain isomers. Chain isomers generally have high viscosities.

DETERMINATION OF NEUTRALIZATION OF LUBRICANT OF OR IODINE NUMBER OF AN UNSATURATED OIL

NEUTRALIZATION NUMBER

The acidity or alkalinity of lubricating oil is determined in terms of neutralization number. Determination of acidity is expressed as Acid Number or Acid Value and determination of basicity is expressed as Base Number or Base Value.

Acid Number or Acid Value. Acid value of a lubricating oil is the number of milligrams of potassium hydroxide required to neutralize all the free acid present in 1 g of the oil.

Base Number or Base Value. Base value of a lubricating oil is the number of milligrams of a titrating acid required to neutralize all the basic constituent present in 1 g of the oil.

Out of the two, determination of acidity is more common. Fatty oils consist mostly of glycerol or other esters of higher fatty acids and in some cases, considerable amounts of free acids are present in the lubricating oil if these are not refined carefully i.e. acids and bases used in refining process have not been completely removed.

Significance

- The small intrinsic acidity may not be harmful in itself but the degree to which it increases in used oils is usually taken as a measure of deterioration of the oil due to oxidation or contamination.
- The acid content of the fatty oils increases with time due to hydrolysis with moisture and is therefore a rough indicator of age of the oil.
- The deterioration in the flavour of edible oils with time is due to the increase in free acid concentration formed by hydrolysis or oxidation.
- Periodic determination of Acid number is thus helpful in deciding the state, where the lubricant needs replacement.

However the presence of free acid in lubricating oils is very rare and indicates contamination. A lubricating oil must possess acid value **less than 0.1**.

EXPERIMENT NO. 7A

AIM **Determination of Acid Value (Acidity) of an oil.**

APPARATUS Burette, pipette, conical flask, beakers water bath.

CHEMICALS N/10 KOH, Neutral ethyl alcohol, phenolphthalein indicator.

THEORY The acid value of a lubricating oil is defined as number of milligrams of KOH required to neutralize the free acid present in 1 g of the oil. For the determination of Acidity, a known weight of lubricating oil is dissolved in a suitable solvent and is titrated with a standard solution of KOH using phenolphthalein as indicator.

$$RCOOH + KOH \rightarrow RCOOK + H_2O$$
$$H^+ + {}^-OH \rightarrow H_2O.$$

PROCEDURE

1. Weigh the empty beaker in which the oil is to be taken.
2. Transfer about 150 drops of the lubricating oil (about 5 g) into the beaker and weigh again. Find out the exact weight of the oil taken by subtracting the weight of empty beaker from it.
3. Add 50 ml of the solvent (neutral alcohol) to the oil and shake vigorously to dissolve the oil. Transfer the contents into a 250 ml conical flask.
4. Heat the flask over a water bath for 30 minutes. Cool the flask to room temperature.
5. Add 10-12 drops of phenolphthalein indicator.
6. Titrate with the standard N/10 KOH solution until a faint permanent pink colour appears at end point. Note down the volume of alkali used. Let it be y ml.
7. Run a blank titration on the same volume of hot solvent. Let the volume of KOH used be z ml.

OBSERVATIONS AND CALCULATIONS

Initial weight of the beaker = x_1 g

Final weight of the beaker and oil = x_2 g.

Weight of the oil = $x_2 - x_1$ g

Volume of N/10 KOH used in the test oil = y ml

Volume of N/10 KOH used in the blank titration = z ml

Therefore volume of N/10 KOH used for titrating the acid in the oil = $(y - z)$ ml.

$$\text{Acid value} = \frac{\text{ml of KOH used} \times \text{Normality} \times 56}{\text{weight of oil taken}}$$

$$= \frac{(y-z)}{\text{weight of oil}} \times \frac{1}{10} \times 56 = \frac{(y-z) \times 5.6}{\text{weight of oil}}$$

where 5.6 represents the amount of KOH in mg present per ml of N/10 KOH solution. (1000 ml of 1 N KOH ≡ 56 g of KOH)

RESULT The acid value of the given oil is ....mg.

PRECAUTIONS

1. KOH should be standardized just before the titration.
2. If the oil is coloured (red or black) phenolphthalein is not a satisfactory indicator. In that case use p-naphthol-benzoin as indicator where the end point will be from orange to green.

3. If the pink colour fades repeatedly after subsequent additions of alkali, the titration should be completed rapidly and the first appearance of pink colour should be taken as the end point.

EXERCISE

1. Define Acid number. How is it determined?
2. When phenolphthalein is used as an indicator, the pink colour at the end point sometimes fades away after a few seconds. Explain

Ans. *Sometimes the oil contain the materials which are attacked by the alkali. If this extra drop is consumed by the reactable materials, pink colour disappears.*

3. What do you mean by base value?
4. What is the significance of determination of acid value?

SAPONIFICATION VALUE/NUMBER

Saponification value or number of an oil is the number of milligrams of potassium hydroxide required to saponify 1 g of oil. (Saponification is the hydrolysis of an ester with KOH to give alcohol and sodium or potassium salt of the acid).

Mineral oils are the mixture of hydrocarbons and hence do not react with KOH and are not saponifiable. On the other hand vegetable and animal oils are the mixtures of glyceryl esters of fatty acids and require large amount of alkali to get hydrolyzed. Their saponification values are very high.

Significance

- The determination of saponification value helps us to ascertain whether the oil under reference is mineral or vegetable oil or a compounded oil containing vegetable and mineral oil.
- Each of the oil has its own specific saponification number. Any deviation from this value in a given sample indicates the probability and extent of adulteration.

EXPERIMENT NO. 7B

AIM **Determination of saponification value of an oil.**

APPARATUS Condensor, water bath, round bottomed flask, burette, pipette, conical flask.

CHEMICALS Standard HCl solution (N/2), standard KOH solution (N/2) and phenolphthalein.

THEORY Saponification number is the number of milligrams of potassium hydroxide required to saponify 1 g of oil. The saponification value of an oil is determined by refluxing a known amount of sample with known excess of standard KOH solution. The alkali consumed is then determined by titrating the unconsumed KOH against standard HCl using phenolphthalein as indicator.

$$RCOOH + KOH \rightarrow RCOOK + H_2O$$

$$H^+ + \bar{O}H \rightarrow H_2O.$$

PROCEDURE

1. Transfer about 5 g of accurately weighed oil sample in a conical flask.
2. Add 50 ml N/2 alcoholic KOH and 50 ml of alcohol into the flask through a pipette.
3. Add the same amount of alcohol and N/2 KOH in another flask for blank determination.
4. Fit the two flasks with water condensers and reflux the contents on a water bath for a minimum period of 1 hour.
5. Cool the contents slightly and disconnect the condenser. Rinse the condenser with a small amount of distilled water into the condenser.
6. Add 10-12 drops of phenolphthalein indicator and titrate the contents of the two flasks with N/2 HCl until the pink colour just disappears.

OBSERVATIONS AND CALCULATIONS

Weight of the empty beaker = w_1 g

Weight of the beaker and oil = w_2 g

Weight of the sample taken = $w_2 - w_1$ g.

Volume of alcoholic KOH added to both the flasks = 50 ml.

Volume of N/2 HCl used in sample determination = V_1 ml

Volume of N/2 HCl used in blank determination = V_2 ml

Volume of N/2 HCl equivalent to KOH used in saponifying $(w_2 - w_1)$ g of oil

$$= (V_2 - V_1) \text{ ml.}$$

$$\text{mg of KOH present in 1000 ml of N/2 KOH} = \frac{1}{2} \times 56 \text{ g} = 28 \text{ g} = 28 \times 1000 \text{ mg}$$

$$\text{mg of KOH present in 1 ml of N/2 KOH} = \frac{28}{1000} \times 1000 = 28 \text{ mg}$$

$$\text{mg of KOH present in } (V_2 - V_1) \text{ ml of N/2 KOH} = (V_2 - V_1) \times 28 \text{ mg}$$

$$\text{Saponification value of the sample} = \frac{(V_2 - V_1) \times 28}{\text{Amount of oil}}.$$

RESULT The saponification value of the given oil is ....mg.

PRECAUTIONS

1. Boiling should be done at a slow rate.

2. To hasten the process of saponification, the sample flask should be occasionally shaken during refluxing.
3. The alcoholic solution of KOH should be freshly prepared.

EXERCISE

1. What is saponification value?
2. Why mineral oils are not saponifiable?
3. What is the significance of determination of saponification number?

IODINE VALUE

Iodine value or iodine number is the number of grams of iodine equivalent to the amount of Iodine Monochloride (ICl) absorbed by 100 g of the oil.

As is well known, oils and fats are mixtures of mixed glycerides and can be represented as:

CH_2OOCR

$CHOOCR'$

CH_2OOCR''

These are formed by glycerol and monobasic fatty acids. The acids present in combination with glycerol have all even number of carbon atoms (C_4 to C_{22}) and may be unsaturated or saturated. Example, $C_{15}H_{31}COOH$ (palmitic acid) $C_{17}H_{35}COOH$ (stearic acid), $CH_3(CH_2)_4CH{=}CHCH_2CH{=}CH(CH_2)_7.COOH$ (linoleic acid) etc.

The oils having predominance of short chain saturated fatty acids or long chain fatty acids with a considerable degree of unsaturation are liquids at room temperature and are called oils. Others are solid at room temperatures and are called fats.

The degree of unsaturation is reported in terms of Iodine Number.

Significance

- As each oil has a specific iodine value so the determination of iodine value can be utilized to detect the extent of contamination in any specific oil.
- It can be used to classify oils based on iodine value because unsaturation confers drying characteristics on oils.

Iodine Value	Type of oil	Examples
> 150	Drying oil	Linseed oil, tung oil
100 to 150	Semidrying oil	Castor oil (dehydrated), soyabean oil
< 100	Non Drying oil	Coconut oil, olive oil

- When a thin film of an oil with a high degree of unsaturation (high Iodine Value) is spread on a surface, it takes up oxygen at the double bond sites and gets oxidized and polymerized into a hard, tough, infusible mass. Thus these oils having high I.V. are used in varnishes and paints.
- To use oil as lubricant its Iodine Value should be low so that it does not deteriorate.

EXPERIMENT NO. 7C

AIM **Determination of Iodine Value of an oil.**

APPARATUS Burette, pipette, conical flask, beakers. etc.

CHEMICALS N/10 $Na_2S_2O_3$ solution, Wij's solution, (solution of ICl in glacial acetic acid) Potassium iodide solution (10%), chloroform or carbon tetrachloride, starch solution (freshly prepared).

THEORY Iodine Value is the number of grams of iodine equivalent to the amount of iodine monochloride (ICl) absorbed by 100 g of the oil.

Determination of iodine value is done through iodometric titration*. A known weight of the oil is dissolved in $CHCl_3$ or CCl_4 and treated with a known excess of Wij's solution. One molecule of ICl adds to the double bond.

After the completion of the reaction, KI is added and unreacted ICl liberates I_2 from it

$$ICl + KI \rightarrow I_2 + KCl.$$

The liberated iodine is titrated against $Na_2S_2O_3$ using starch as indicator.

PROCEDURE

1. Remove the moisture present in the oil sample by straining it through a filter paper.
2. Weigh about 5 g of the sample oil and transfer it to a conical flask. Add 25 ml of CCl_4 and 25 ml of Wij's solution to it. In another flask take the same volumes of CCl_4 and Wij's solution for blank determination.
3. Close the flask and mix the contents by swirling. Keep in dark for about one hour at a temperature below 30°C. The glass stopper may be moistened with KI solution.
4. Now add 20 ml of KI solution to both the flasks and 100 ml of distilled water.
5. Titrate the liberated iodine with N/10 $Na_2S_2O_3$ solution till the colour of the solution becomes pale yellow.
6. Now add the indicator, i.e. starch solution, the solution will turn blue.
7. Continue the titration till the blue colour disappears.
8. Note down the volume of $Na_2S_2O_3$ used, in both the cases, i.e. test solution and blank.

OBSERVATIONS AND CALCULATIONS

Weight of the weighing tube = w_1 g

Weight of the weighing tube and oil = w_2g

Weight of the sample = $(w_2 - w_1)$g

Volume of $Na_2S_2O_3$ used to titrate excess iodine in sample determination = x ml.

Volume of $Na_2S_2O_3$ used to titrate total iodine in blank titration = y ml.

Volume of N/10 $Na_2S_2O_3$ equivalent to iodine (or ICl) consumed by $(w_2 - w_1)$g of oil = $(y - x)$ ml.

Iodine present in (y – x) ml of N/10 iodine solution $= (y - x) \times 127 \times \frac{1}{10} \times \frac{1}{10^3}$ g

(Equivalent weight of iodine = 127)

$$\therefore \text{Iodine Value} = \frac{(y-x) \times 127}{(w_2 - w_1)} \times \frac{1}{10} \times \frac{1}{10^3} \times 100 = \frac{(y-x)}{(w_2 - w_1)} \times 1.27$$

RESULT The iodine value of the given oil is............

PRECAUTIONS

1. Wij's solution should never be sucked through mouth. It should be added with the help of burette.
2. Wij's solution should be added in about 100% excess than the expected requirement.
3. Wij's solution should be added in the two flasks simultaneously.

EXERCISE

1. What do you mean by iodine value?
2. What is Wij's solution?
3. Why is glass stopper moistened with KI solution?

Ans. *To prevent the leakage of I_2 vapour.*

4. What is the significance of determination of Iodine Value?

Note. (For details on iodometric titration, please refer to chapter-1, iodometric titrations.)

Note to the Lab Assistant. For the preparation of various solutions ($Na_2S_2O_3$ starch solution etc.) refer to the chapter on iodometry. (students are also advised to go through the chapter on iodometry in volumetric titrations for complete understanding of such type of titrations).

Wij's solution. Wij's solution is available commercially or it can be prepared by dissolving 1.6 g of commercially available iodine monochloride (ICl) in 100 ml of A.R. glacial acetic acid. Wij's solution is stored in amber coloured stoppered bottles in a dark place below 30°C

EXPERIMENT 8

AIM Detection of functional groups in an organic compound by wet tests

QUALITATIVE ANALYSIS OF AN ORGANIC COMPOUND

Qualitative analysis of an organic compound involves the detection of various elements present in it. The elements commonly present in organic compounds are carbon, hydrogen, oxygen, nitrogen, halogens, sulphur and sometimes phosphorous. The carbon and hydrogen are present in almost all the organic compounds. Generally their presence is assumed in all the organic compounds, and hence detection of these elements is not necessary.

(A) Detection of Carbon and Hydrogen

A small quantity of compound is mixed with about ten times its weight of copper oxide (CuO). The mixture is taken in a hard glass test tube fitted with a delivery tube having a small bulb. The other end of the tube is immersed in freshly prepared lime water. If the lime water turns milky, the organic compound contains carbon.

Further, if small drops of water collects in the cooler part of the test tube, it indicates the presence of hydrogen also. The water can be tested by dusting a small amount of anhydrous copper sulphate over the upper parts of the test tube. Even if a small amount of water is present, the anhydrous copper sulphate will turn blue.

Reactions involved:

$$\underset{\text{in the compound}}{C + 2CuO} \xrightarrow[\Delta]{} CO_2 + 2Cu$$

$$\underset{\text{in the compound}}{2H + CuO} \xrightarrow[\Delta]{} H_2O + Cu$$

$$CO_2 + \underset{\text{lime water}}{Ca(OH)_2} \longrightarrow \underset{\text{milky}}{\downarrow CaCO_3} + H_2O$$

$$5H_2O + \underset{\text{(anhydrous)}}{CuSO_4} \longrightarrow \underset{\text{Blue}}{CuSO_4 \cdot 5H_2O}$$

(B) Detection of Nitrogen, Sulphur and Halogens

All these can be detected by **Lassaigne's Test.**

Preparation of Lassaigne's Extract or Sodium Extract:

(i) Take a freshly cut piece of dry sodium metal of the size of a small pea and dry it by placing it between the folds of filter paper.

(ii) Hold vertically a fusion tube (or a small hard glass test tube 3" × $\frac{3}{8}$" — ignition tube) in a pair of iron tongs. Drop into it small dried piece of sodium metal.

(iii) Heat the fusion tube (or ignition tube) gently from below in a Bunsen flame until sodium melts to a shining globule.

(iv) Drop a small quantity of the powdered organic compound (5–10 mg or 1–2 drops) into the fusion tube. Heat the tube first gently and then strongly until the lower end of the fusion tube becomes dull red hot.

(v) Plunge the hot end of the fusion tube into a porcelain dish containing about 10 ml. of *distilled* water.

(vi) When the bottom of the fusion tube has cracked and the reaction with left-over sodium, if any has subsided, heat the contents of the dish to boiling (if necessary, break the tube with the help of glass rod)

(vii) Filter and keep the filtrate as reserve solution and call it **Sodium Extract** (SE).

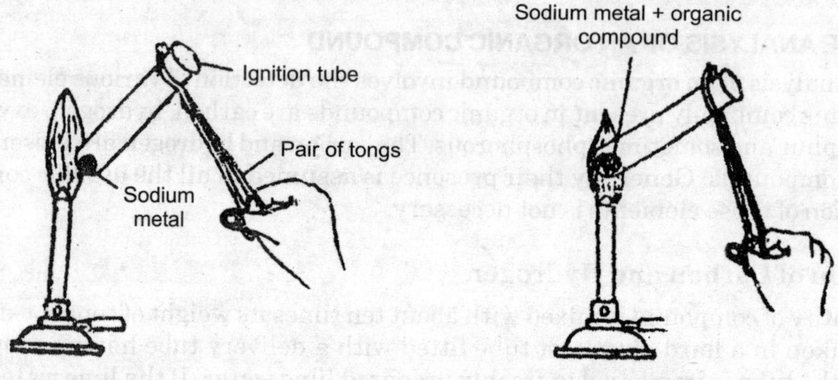

Fig. 2 Fusing the organic substance and a piece of sodium metal in a fusion tube.

The Lassaigne's or sodium extract is, therefore, tested for these element, i.e., nitrogen, sulphur and halogens as detailed below in the Table 8.1

Table 8.1 Detection of nitrogen, sulphur and halogen from Lassaigne's extract

Experiment	Observation	Inference
Test for nitrogen To a portion of the sodium extract in a test tube, add 2 ml of of freshly prepared concentrated solution of *ferrous sulphate.* In case no green precipitate is formed, add 2 drops of *sodium hydroxide.* Boil for a minute and immediately add dil. H_2SO_4 and shake.	(i) A deep blue or greenish-blue colouration is obtained. (ii) A blood-red colouration is obtained.	(i) Presence of nitrogen confirmed. (ii) Presence of both nitrogen and sulphur confirmed.
Tests for sulphur (i) *Lead acetate test:* Acidify another portion of sodium extract with acetic acid and to the solution add lead acetate solution.	A black ppt. is formed	Presence of sulphur confirmed.
(ii) *Sodium nitroprusside test:* To another portion of the sodium extract add a few drops of freshly prepared dilute solution of ammoniocal sodium nitroprusside solution.	A deep violet colouration is produced.	Presence of sulphur confirmed.
Tests for halogens (i) *Silver nitrate test:* Acidify another portion of the sodium extract with conc. HNO_3 (slight excess) and	(i) A curdy white ppt. soluble in excess of NH_4OH. (ii) Presence of bromine confirmed.	(i) Presence of chlorine confirmed, (ii) A pale yellow ppt. sparingly soluble in NH_4OH.

boil to expel HCN and H_2S (if present due to nitrogen and sulphur). Cool the solution and add silver nitrate solution to it	(iii) A yellow ppt. insoluble in NH_4OH	(iii) Presence of iodine confirmed.
(ii) *Carbon disulphide *test.* Acidify another portion of the sodium extract with dil. H_2SO_4 and add about 1. ml. of CS_2 and 2 drops of fresh chlorine water into it and shake.	(i) Violet colour is imparted to CS_2 layer. (ii) Orange colour is imparted to CS_3 layer.	(i) Presence of iodine confirmed. (ii) Presence of bromine confirmed.

Safety Concern

1. Take only a small piece of sodium metal for fusion otherwise unreacted metal will spurt when the red hot tube is plunged into distilled water in a dish.
2. While heating the tube, keep the mouth of the tube away from your face or your neighbour.
3. While plunging the red hot tube into the dish containing distilled water, be careful and keep your face away from the dish since the unreacted hot sodium metal may cause explosion on coming in contact with water.
4. In the detection of nitrogen, always use freshly prepared ferrous sulphate solution.
5. Sometimes the coloured precipitate is not clearly visible. In that case filter the solution. If coloured (blue or green) particles of the ppt. are visible on the filter paper, presence of nitrogen is confirmed.
6. **Beilstein's Test for Halogens.** For halogens, the Beilstein's test is sometimes performed although it is not a sure test. This test, of course, is very sensitive for detecting even the traces of halogens.

Push one end of a fine copper wire into a cork which can serve as a holder. Oxidize the free end of the copper wire, by heating, in the flame of a burner, until it cases to impart any green colour to the flame. When cold, dip the wire in the substance, to be tested. Heat in the outer edge of the flame. The compound first burns with luminous flame and then, if it contains halogen, gives a green colour owing to the vaporization of cuprous halides or cyanides formed.

A few substances such as urea, thiourea, naphthalene, some pyridine derivatives, etc., which do not contain halogen, also give a green colour.

If the compound does not give any green colour, absence of halogen is confirmed.

Explanation of The Above Tests

(a) Lassaigne's test

The following reactions occur when sodium is fused with an organic compound.

Carbon and nitrogen from the organic compound react with sodium to form sodium cyanide.

$$Na + C + N \rightarrow NaCN$$

Halogen and sulphur present in the organic compound react with sodium to form sodium halide and sodium sulphide respectively.

$$Na + X \rightarrow NaX \quad (X = Cl, Br, I)$$

$$2Na + S \rightarrow Na_2S$$

When nitrogen and sulphur both are present, sodium thiocyanate is formed.

$$Na + C + N + S \rightarrow NaCNS$$

These substances, being soluble, are extracted with water and identified.

(1) For nitrogen

The following reactions take place when the filtrate is treated first with ferrous sulphate solution, then with ferric chloride and finally with excess of dilute sulphuric acid.

$$2Na + 2H_2O \rightarrow 2NaOH + H_2$$

$$FeSO_4 + 2NaOH \rightarrow Fe(OH)_2 + Na_2SO_4$$

$$Fe(OH)_2 + 2NaCN \rightarrow Fe(CN)_2 + 2NaOH$$

$$Fe(CN)_2 + 4NaCN \rightarrow \underset{\text{sodium ferrocyanide}}{Na_4[Fe(CN)_6]}$$

$$3Na_4[Fe(CN)_6] + 4FeCl_3 \rightarrow \underset{\substack{\text{ferricferrocyanide} \\ \text{(prussian blue)}}}{Fe_4[Fe(CN)_6]_3} + 12NaCl$$

The alkaline solution should not be acidified with HCl because prussian blue colour changes to greenish.

(2) For sulphur

(i) Lead Acetate test

$$NaS + (CH_3COO)_2Pb \rightarrow \underset{\substack{\text{lead sulphide} \\ \text{(black)}}}{PbS \downarrow} + 2CH_3COONa$$

(ii) Nitroprusside test

$$Na_2S + Na_2[Fe(CN)_5NO] \rightarrow \underset{\substack{\text{sodium sulpho nitroprusside} \\ \text{(violet colour)}}}{Na_4[Fe(CN)_5NOS]}$$

(3) For halogens

Before adding $AgNO_3$ solution, the filtrate is boiled with conc. HNO_3 to decompose NaCN, if present. Otherwise, a white precipitate will be formed even in absence of halogen.

$$NaCN + HNO_3 \rightarrow NaNO_3 + HCN$$

(a)

$$NaCl + AgNO_3 \rightarrow \underset{\text{curdy white}}{AgCl \downarrow} + NaNO_3$$

$$NaBr + AgNO_3 \rightarrow \underset{\text{pale yellow}}{AgBr \downarrow} + NaNO_3$$

$$NaI + AgNO_3 \rightarrow \underset{\text{yellow}}{AgI \downarrow} + NaNO_3$$

(b)

$$2NaI + Cl_2 \rightarrow 2NaCl + \underset{\text{violet}}{I_2}$$

$$2NaBr + Cl_2 \rightarrow 2NaCl + \underset{\text{brown}}{Br_2}$$

Chlorine is not detected by the above test.

(4) For nitrogen and sulphur when both are present

$$C + N + S \xrightarrow{Na} NaCNS$$

(a)

$$3NaCNS + FeCl_3 \rightarrow \underset{\substack{\text{ferric thiocyanate} \\ \text{(red colour)}}}{Fe(CNS)_3} + 3NaCl$$

(b)

$$NaCNS + HCl \rightarrow \underset{\text{thiocyanic acid}}{HCNS} + NaCl$$

$$4HCNS + Co(NO_3)_2 \rightarrow \underset{\substack{\text{cobalti-thiocyanic acid} \\ \text{(blue)}}}{H_2[Co(SCN)_4]} + 2HNO_3$$

EXERCISE

1. What is the purpose of fusing the organic compound with sodium metal for preparing this solution?

Ans: By fusing the organic compound with sodium metal, the elements in it are changed into the sodium salts which are soluble and can thus be easily extracted.

2. What is Lassaigne's extract?

Ans: The fused mass obtained on igniting an organic compound with sodium is extracted with water. This extract is called Lassaigne extract or sodium extract.

3. What is the colour of the Lassaigne's extract?

Ans: It is colourless since it contains colourless sodium salts.

4. Can potassium be used for fusing of an organic compound in place sodium?

Ans: Potassium being more reactive than sodium in not used since it creates difficulties during fusion.

5. Can calcium be used in place of sodium for fusing an organic compound?

Ans: Calcium being less reactive them sodium is not used as the fusion will not be complete.

6. Why only distilled water is used for preparing Lassaigne's extract?

Ans: Tap water contain chlorides and hence the extract will be contaminated with chloride impurity Therefore only distilled water should be used in this extract.

7. Why is fresh solution of ferrous sulphate used in the test for nitrogen?

Ans: Because ferrous sulphate solution, on keeping, gets oxidised to ferric sulphate by atmosphoric oxygen.

8. What are the elements you detect with Lassaigne tests and what are sodium salts formed?

Ans: The elements detected are, nitrogen, sulphur, chlorine, bromine and iodine. Then sodium salts formed are NaCN, Na_2S, NaCl, NaBr, NaI respectively.

If nitrogen and sulphur both are present. Na CNS is formed.

9. Why is the sodium metal placed in kerosene oil?

Ans: Sodium metal is very reactive, towards moisture and air. On exposure to them, it forms sodium hydroxide. In kerosene oil these reactions are avoided.

10. Why is the sodium extract alkaline?

Ans: The organic compound is fused with metallic sodium. It is then extracted with distilled water. The unreacted sodium, reacts with water and forms alkaline solution.

11. What element is indicated when Lassaigne's extract gives deep blue colour with ferrous sulphate solution.

Ans: Nitrogen.

12. The Lassaigne's extract of an organic compound gives a blood red colour with ferrous sulphate. What do you infer?

Ans: Nitrogen and sulphur are present in the compound.

EXPERIMENT 9

AIM Identification of an organic compound

DETECTION OF FUNCTIONAL GROUPS

Detection of functional groups is an important step in the analysis of organic compounds. It largely depends on the correct determination of the constituent elements. On the basis of elements present, the organic compounds are classified into following groups:

A. Compounds containing C, H, and [O]
B. Compounds containing C, H, N and [O]
C. Compounds containing C, H, halogens and [O]
D. Compounds containing C, H, N, S and [O]

[A] When only Carbon, Hydrogen and/or Oxygen are present

B.NO.	*Class*	*Functional group*	*Tests*
1.	Carboxylic acid	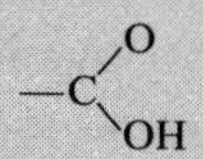	(i) **Litmus Test.** Shake a small amont of the given substance with water and dip a blue litmus paper in it. If the litmus paper turns red, it shows presence of carboxylic group (Phenols being acidic in nature also give this test). (ii) **Sodium bicarbonate test.** Add a pinch of the given substance if solid (or a few drops if liquid), to 5 ml. of cold 50% solution (aqueous) of $NaHCO_3$ and shake gently. Effervescence indicates presence of carboxylic group.
2.	Phenols	Phenolic—OH	(i) blue litmus changes to red but no effervescence with $NaHCO_3$. (ii) **$FeCl_3$ Test.** To 1 ml. of aqueous or alcoholic solution of compound add 2-3 drops of neutral or very dilute solution (aqueous) of ferric chloride. A blue, green, red or violet colour indicates presence of phenolic group (—OH). (iii) **Libermann's Test.** Take 0.2 gm. of compound and a few crystals of $NaNO_2$ in a dry test tube and heat gently for a minute. Cool the mixture, add 1 ml. of conc. H_2SO_4 and shake. A deep green or blue colour develops, upon dilution with water the solution turns red. To it add excess of dil. NaOH solution, the red solution again becomes deep green or blue. It indicates the presence of phenolic group. (iv) **Phthalein Test.** Take 0.2 gm. each of compound and phthalic anhydride in a dry *t.t.*, and 0.5 ml. of conc. H_2SO_4 and heat for a minute. Cool. Make it alkaline with dil. NaOH soln. pour a few drops of this soln. in 20 ml. water taken in a breaker.

			Characteristic colour shows the presence of phenolic group, *e.g.* Pink colour—Phenol. o-Cresol Blue colour—Catechol m-Cresol Fluorescent green—Resorcinol No colouration—*P*-Cresol.
3.	Carbohydrate	....	(i) **Blackening with H_2SO_4.** Heat 0.2 gm. of compound with 1 ml. conc. H_2SO_4. An immediate charring and blackening of the solution shows presence of carbohydrate. (ii) **Mosich's Test.** To 2 ml. aq. solution of compound add 1 ml. of Molisch's reagent and shake well. Now add carefully 1-2 ml. of conc. H2SO4 (taken in a separate *t.t.*) from the side of the *t.t.* Allow to stand. A reddish violet ring at the junction of two liquids shows the presence of carbohydrate. (iii) To 1 ml. aqueous soln. of compound add 1-2 drops of an-throne solution, shake and warm the mixture gently. A green or bluish green colour shows the presence of carbohydrate.
4.	Aldehydes	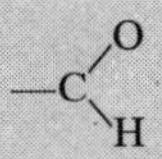	(i) **Schiff Test.** To 1 ml. aqueous or alcoholic solution of compound add 2 ml. of Schiff reagent and shake. A deep red or violet colour shows the presence of aldehydic group (Never heat during this test). (ii) **Tollen's Test.** To about 2 ml. of Tollen's reagent add about 0.5 cm. of compound and heat the mixture on water bath. Formation of silver mirror or blackish ppt. indicates the presence of aldehydic group. (This test is also given by reducing sugars). (iii) **Fehling Solution Test.** Add Fehling solution B to A until a blue ppt. first formed is redissolved to give a deep blue solution. To it add a little of and heat for 2 mts. Reddish brown ppt. shows the presence of aldehydic group. (This test is also given by reducing sugars.)
5.	Ketones	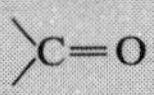	(i) No colour with Schiff reagent. (ii) **2, 4 Dinitrophenyl hydrazine Test.** Add 1-2 drops (0.5 gm.) of O.S. to 2 ml. of 2, 4 dinitrophenyl hydrazine solution (Brady's reagent). Shake vigorously. Heat and cool. A red, yellow or orange coloured ppt. Indicates the presence of ketonic group.
6.	Esters	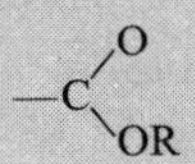	(i) **Hydroxamic acid Test (Feigl Test).** Take a pinch or two drops of the compound in a test tube, add 1 ml. of 5% methanolic solution of hydroxylamine hydrochloride. Now add a few drops of 2N methanolic KOH till the solution is just alkaline and boil. Cool and acidify with dil.

B.NO.	Class	Functional group	Tests
			HCl. Now add 2-3 drops of $FeCl_3$ solution. Deep red or violet colour is produced. It indicates the presence of ester group. (ii) **Hydrolysis Test.** Dissolve 0.5 gm. of O.S. in 2 ml. alcohol. To it add 4.5 drops of dil. NaOH soln. and 2.3 drops of phenolphthalein. A pink colour is produced. Now put the *t.t.* in boiling water bath for 5 mts., the pink colour either fades or disappears. This shows presence of ester group.
7.	Alcohols	—OH	(i) In a dry *t.t.* take 2 ml. of O.S., add 1 gm. of anhydrous sodium sulphate (to absorb water if already present) and filter. To the filtrate add a small piece of sodium. Effervescence due to evolution of H_2 shows the presence of an alcohol.
8.	Hydrocarbons	....	No specific group test.

[B] When Carbon, Hydrogen, Nitrogen and Oxygen are present

B.NO.	Class	Functional group	Tests
1.	Amides	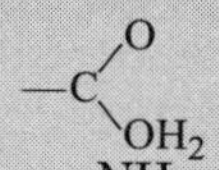	To 0.2 gm. or 0.5 ml. of O.S., add 1 ml. of aq. NaOH and heat. Smell of ammonia shows presence of amide.
2.	Primary	$—NH_2$	(i) **Carbylamine Test.** Take a small amount of substance, 4 ml. alcoholic soln. of KOH and 2 drops of chloroform in a *t.t.*, shake and heat gently. An intolerable offensive odour of carbylamine shows the presence of primary amine. (ii) **Nitrous Acid Test.** Take 0.2 gm. of O.S. in 2 ml. of dil. HCl and cool. Now add 10% aq. $NaNO_2$ soln. Formation of brisk effervescence shows presence of aliphatic primary amine. (iii) **Diazotisation Test.** Dissolve 0.2 gm. or 0.5 ml. of OS: in 2.3 ml. of dil. HCl. Cool under tap water. Now add 2 ml. of 2.5% $NaNO_2$ soln. (aqueous), cool again and add 0.5 ml. of alkaline b-naphthol solution. An orange-red or a red dye shows the presence of **aromatic** primary amine.
3.	Secondary amines	> NH	(i) No carbylamine test. (ii) **Nitrous Acid Test.** Take 0.2 gm. or 0.5 ml. of O.S. add 2 ml. dil. HCl and cool under tap water. Add 2 ml. of 2.5% $NaNO_2$ soln. yellow only drops are obtained. This shows presence of secondary amine. (This test is given by both aliphatic & aromatic secondary amines.) (iii) **Libermann's Nitroso.** 0.5 ml. of O.S. add 2 ml. dil. HCl and cool under tap water. Add 2.5% aqueous solution of $NaNO_2$ gradually with shaking until yellow oil separates at the bottom. Decant off the aqueous layer. Now take 2 drops

			of this only nitroso compound in a dry *t.t.*, add 0.5 gm. of phenol and warm gently for a few seconds. Cool and add 1 ml. of conc. H_2SO_4. A greenish blue colour is obtained which changes upon dilution with water. On adding excess of NaOH soln., the greenish blue colour is restored. This shows the presence of secondary amine.
4.	Tertiary amines	$>\!\!-N$	(i) Carbylamine test negative. (ii) Dissolve 0.2 gm. or 0.5 ml. of O.S. in 2.5 ml. of dil. HCl and cool under tap water. Add 2.5 ml. of 2.5% aqueous solution of $NaNO_2$ gradually with shaking and observe;— (a) No reaction indicates presence of aliphatic tertiary amine. (b) Production of green or brown coloured salt indicates presence of aromatic tertiary amine.
5.	Anilide	Ar NHCOR	(i) **Carbylamine Test.** Heat a mixture of 0.5 gm. of O.S. 4 ml. of alcoholic KOH soln. and 1 ml. chloroform. Bad smell of carbylamine comes out. (ii) **Tafel's Test.** Take 0.2 gm. of O.S. add 5 ml. of conc. H_2SO_4 and shake. Add 0.2 gm of powdered potassium dichromate. A red or violet green on standing shows the presence of an anilide. (iii) **Hydrolysis Test.** Take 0.2 gm. of O.S. add 5 ml. of dil. HCl, boil and cool. Add 2 ml. of 2.5% $NaNO_2$ solution, cool again and add 0.5 ml. of alkaline β-naphthol soln. An orange-red dye shows the presence of an anilide. (Anilides are hydrolysed on boiling with HCl and thus liberates primary amino group which gives the dye test.)
6.	Nitro compounds	$—NO_2$	(i) To R.G. of O.S. add 1 ml. aq. solution of NaOH and warm. An yellow colour shows presence of nitro group. (ii) **Mulliken Test.** Take 0.2 gm. or 2.3 drops of O.S. in a *t.t.*, and 4.5 ml. of alcohol, a pinch of zinc dust and 10 drops of 10% $CaCl_2$ soln. (NH_4Cl soln.) Boil and cool. Filter it directly into 2 ml. of Tollen's reagent taken in another *t.t.* On warming this content a grey or black ppt. or silver mirror is produced which confirms presence of nitro group. (iii) **Azo Dye Formation.** In a boiling tube take 0.2 gm. or 3 drops of O.S. 3 ml. conc. HCl, 2 ml. water, and 1 gm. solid stannous chloride (or tin granules). Heat the contents on water bath for 14 mts. Filter and cool. To the filtrate and 2.5% aq. $NaNO_2$ solution drop by drop to of alkaline β-naphthol soln. An orange-red or red dye shows the presence of nitro group. (The dye is formed due to liberation of primary amino group by reduction of nitro group.)

[C] When Nitrogen and Sulphur are present along with C, H, and [0]

1.	Thioureas	—HN—C(=S)—NH	(i) Simple thiourea upon boiling with NaOH solution gives off ammonia. (ii) Boil a little of the substance with dil. NaOH soln., cool and lead acetate. A brown or black colour or ppt. indicates the compound may be a thiourea.
2.	Amino Sulphonic acids	$—NH_2$ and $—SO_2$ groups	Dissolve a little of the O.S. in dil. HCl. Cool and add $NaNO_2$ solution so as to complete the diazotisation. Now add alkaline solution of β-naphthol. Red or orange dye is formed.

[D] When Halogen is present along with C, H_2 and [0]

1.	Halogenated hydrocarbons (*a*) Aliphatic halogen compounds RX	....	Boil 0.5 ml. of O.S. with 5 ml. of alcoholic NaOH soln. for about 10 mts. Cool, dilute the soln. with water, add excess of dil. HNO_3 and then $AgNO_3$ soln. A ppt. of silver halide is obtained.
	(*b*) Aromatic halogen compounds	....	Depends on the type of the additional functional group present.

EXPERIMENT 10

AIM **Synthesis of p-nitroacetanilide from acetanilide**

THEORY Nitration of acetanilide leads to the formation of p-nitroacetanilide.

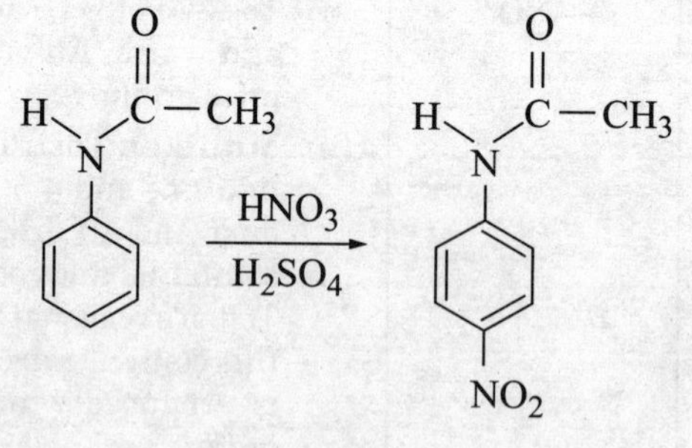

MECHANISM In this electrophilic reaction , the nitronium ion is formed *in situ* by dehydration of nitric acid.Sulfuric acid is the dehydrating agent. The nitronium ion is a very powerful electrophile and will react with the À electrons of the aromatic ring of aniline. These reactions are shown below:

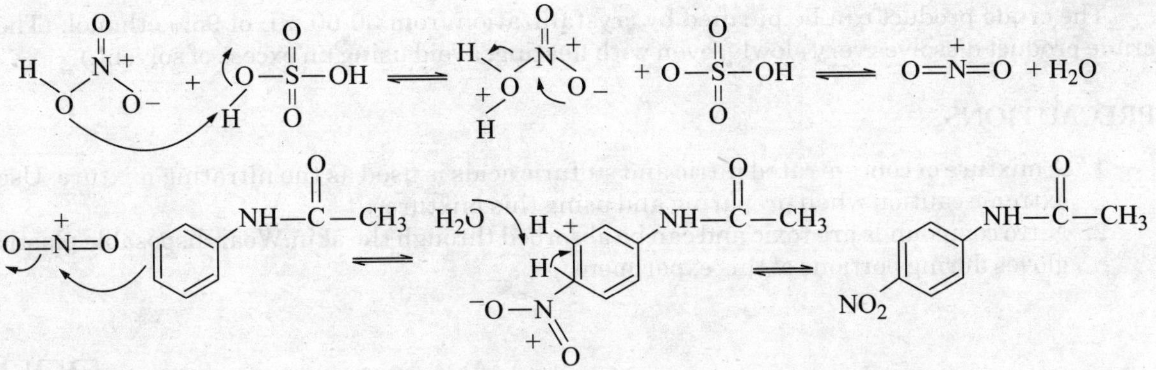

APPARATUS 250-mL beaker,dropper or disposable pipet,two 50-mL and one 125-mL Erlenmeyer flasks,10-mL graduated cylinder,thermometer,vacuum filtration assembly,watch glass

CHEMICALS acetanilide,disodium hydrogen phosphate, 95% ethanol, glacial acetic acid, conc. nitric acid, conc. sulfuric acid,

PROCEDURE

1. Place 6.5 g of acetanilide in a 125-mL Erlenmeyer flask,
2. Add 10 mL of glacial acetic acid
3. Warm the flask on a hot plate in a fume hood until the acetanilide dissolves.
4. Cool the flask in an ice bath to about 20°C; then add 10 mL of cold, conc.sulfuric acid. The temperature of the mixture will rise to about 60°C.
5. Chill the solution to about 10°C in an ice bath. (The solution will become very viscous.)
6. Mix 3.5 mL of conc. nitric acid and 5 mL of conc. sulfuric acid in a 50-mL flask, and chill the flask in an ice bath.
7. When both solutions are cold, slowly add the HNO_3, H_2SO_4 solution, 1 mL at a time, to the acetanilide solution.
8. Keep the reaction flask in an ice bath so that the temperature of the reaction mixture is maintained between 10-20°C . Stir the reaction mixture carefully after each addition. The entire addition requires about 15 minutes.

Allow the reaction flask to stand at room temperature for 30 minutes. Monitor the temperature; if it rises above 25°C, chill the flask in an ice bath. Should the rechilling be necessary, allow the flask to stand for 30 minutes or more at room temperature after the rechilling.

Pour the reaction mixture into a 250-mL beaker containing 100 mL of water and 25 g of cracked

Ice whereby the crude nitroacetanilide is at once precipitated

Using a large Buchner funnel, filter the heavy lemon-yellow precipitate with vacuum.

Press out as much aqueous acid from the filter cake as possible with a spatula or clean cork while suction is being applied .

Transfer the filter cake to a clean 250-mL beaker, and add 100 mL of 15% aqueous disodium hydrogen phosphate.

Stir the mixture to a paste-like consistency and refilter using vacuum. Wash the beaker with two 30-mL portions of cold water.

Finally, wash the filter cake with an additional 50 mL of cold water then dry the solid overnight on a watch glass. Determine the yield and melting point.

The crude product can be purified by crystallization from 30-60 mL of 95% ethanol. (The crude product dissolves very slowly, even with heating; avoid using an excess of solvent.)

PRECAUTIONS

1. A mixture of concentrated nitric and sulfuric acids is used as the nitrating mixture. Use extreme caution when preparing and using this mixture.
2. Nitro compounds are toxic and can be absorbed through the skin. Wear disposable plastic gloves during portions of this experiment.

❑❑❑

INDEX

❑❑❑

Win Prizes!

ATTENTION READERS

We request you for your frank assessment regarding some of the aspects of the book given as under:

ENGINEERING CHEMISTRY (MTU)

By

Dr. Sunita Rattan

Please fill up the given space in neat capital letters.

(i) What topics of your syllabus that are important from you examination point of view are not covered in the book ?

(ii) What are the chapters wherein the treatment of the subject matter is not systematic or organised or updated ?

(iii) Have you come across misprints/mistakes/factual inaccuracies in the book? Please specify the chapters and the page numbers.

(iv) What other books on the same subject have you found/heard better that the present book ? Please specify in terms of price and quality.

(v) Further suggestions and comments for the improvement of the book:

(PS: If need be please attach a separate sheet to write your views.)

Other Details :

(i) Who recommended you the book ? (Please tick in the box near the option relevant to you.)

☐ Teachers ☐ Friends ☐ Bookseller

(ii) Name of the recommending teacher, his designation and address. ....
....
....
....
....

(iii) Name and address of the bookseller you purchased the book form :
....
....
....
....

(iv) Name and address of your institution (Please mention the University or Board, as the case may be)
....
....
....
....

(v) Your name and complete postal address :
....
....
....
....

(vi) Course you are enrolled in :
....
....
....
....

The best assessment will be awarded each month. The award will be in the form of our publications as decided by the Editorial Board.

Please mail the filled up coupon at your earliest to:

S.K. Kataria & Sons®

4760-61/23, Ansari Road, Daryaganj, New Delhi-110002
Phone : +91-11-23243489, 23269324; Telefax : +91-11-23243489
e-mail: katariabooks@yahoo.com; katariabook@gmail.com
Website: www.skkatariaandsons.com